Advances in Behavioral Biology

Volume 58

For other titles published in this series, go to
www.springer.com/series/5558

Henk J. Groenewegen • Henk W. Berendse
Alexander R. Cools • Pieter Voorn
Antonius B. Mulder

Editors

The Basal Ganglia IX

Springer

Editors
Hendrik Jan Groenewegen
Department of Anatomy & Neurosciences
Neuroscience Campus Amsterdam
VU University Medical Center Amsterdam
PO Box 7057, 1007 MB
Amsterdam, Netherlands
hj.groenewegen@vumc.nl

Pieter Voorn
Department of Anatomy & Neurosciences
Neuroscience Campus Amsterdam
VU University Medical Center Amsterdam
PO Box 7057, 1007 MB
Amsterdam, Netherlands
p.voorn@vumc.nl

Henk W. Berendse
Department of Neurology
Neuroscience Campus Amsterdam
Vrije Universiteit Medical Center
PO Box 7057, 1007 MB
Amsterdam, Netherlands
hw.berendse@vumc.nl

Antonius B. Mulder
Department of Anatomy & Neurosciences
Neuroscience Campus Amsterdam
VU University Medical Center Amsterdam
PO Box 7057, 1007 MB
Amsterdam, Netherlands
ab.mulder@vumc.nl

Alexander R. Cools
Department of Cognitive Neuroscience
Radboud University Nijmegen
Medical Centre
Geert Grooteplein 21a, 6525 EZ
Nijmegen, Netherlands
a.cools@cns.umcn.nl

ISBN 978-1-4419-0339-6 e-ISBN 978-1-4419-0340-2
DOI 10.1007/978-1-4419-0340-2
Springer Dordrecht Heidelberg London New York

Library of Congress Control Number: 2009926039

Cover illustration: Medium-sized spiny neuron in the shell of the nucleus accumbens of the rat. The neuron was recorded *in vivo* upon its response to hippocampal stimulation and subsequently filled by juxtacellular application of neurobiotin. Courtesy Jean-Michel Deniau, Anne-Marie Thierry and Yvette van Dongen.

Printed on acid-free paper

Springer is part of Springer Science+Business Media (www.springer.com)

Preface

This volume 'Basal Ganglia IX' is derived from the proceedings of the Ninth Triennial Meeting of the International Basal Ganglia Society (IBAGS). The meeting was held from 2 to 6 September 2007 at Hotel Zuiderduin, Egmond aan Zee, the Netherlands. Basal ganglia researchers meet each other during the years on several occasions and in various combinations. Yet, the triennial IBAGS meetings have become a strong tradition and an attractive occasion to specifically consider the progress in our understanding of the basal ganglia in health and disease, and to meet colleagues and friends with a common interest. For the 9th IBAGS meeting more than 240 delegates were registered, many of them young scientists, with an interest in the basal ganglia, and with many different expertises and backgrounds. Based on the input from the Scientific Advisory Committee and the Council, the Local Organizing Committee was able to put together a diverse and attractive scientific programme with more than 160 posters and 47 oral presentations. Many different aspects and a wide range of questions regarding functions and dysfunctions of the basal ganglia were dealt with in either oral presentations or posters. The attractive format of former IBAGS meetings was maintained in having no parallel oral sessions and ample time for poster viewing. Contributions to the present Volume were open to all presenters during the meeting. The book, consisting of 45 chapters, is a good reflection of the wide range of subjects related to the basal ganglia dealt with during the meeting.

It was originally planned that Dr. Lennart Heimer would open the conference with a plenary lecture on the 'Changing Faces of the Basal Ganglia'. Unfortunately, Lennart passed away on March 12, 2007. Here we want to memorize the major contributions of Lennart Heimer to the concepts of the parallel organization of cortical–basal ganglia circuits and limbic–motor interactions, and the functional–anatomical characterization of the major neuronal components of the basal forebrain, including the ventral striato-pallidal system, the extended amygdala and the magnocellular basal forebrain cholinergic system. His ultimate goal was to further our understanding of the organization of the brain in the context of debilitating neurological and psychiatric disorders such as Parkinson's disease, schizophrenia, and drug abuse. The importance of these concepts is reflected in their influence on the presently widely utilized surgical and pharmacotherapeutic approaches of these neurological and psychiatric diseases. Therefore, we dedicate this volume to Lennart Heimer for his great contribution to the functional anatomy of the basal ganglia.

Scientific interactions prosper in an open and pleasant social atmosphere. Many of the participants, senior or junior, took part in the social activities. The dancing to Caribbean music was very pleasant and animating. We hope that on the basis of the scientific and social interactions new ideas and collaborations have evolved.

Finally, we want to thank all contributors to this volume for their pleasant cooperation and the publishing editor Ann Avouris for her assistance in finishing this job. We further thank all sponsors for their contributions which in particular enabled us to promote participation of junior researchers to participate in the meeting. We hope that the contents of this volume will further inspire and stimulate discussions and new interdisciplinary research on the basal ganglia.

The next meeting will take place in the East coast of the USA in 2010 and will be organized by Jim Tepper and Elizabeth Abercrombie. We are all looking forward to another exciting IBAGS meeting.

Amsterdam, The Netherlands

Henk Groenewegen
Henk Berendse
Alexander Cools
Pieter Voorn
Antonius Mulder

IBAGS IX

The 9th Triennial Meeting of the International Basal Ganglia Society was held at Hotel Zuiderduin, Egmond aan Zee, the Netherlands
2nd-6th September 2007

President Henk J. Groenewegen
Secretary Henk W. Berendse

Local Organizing Committee
Henk J. Groenewegen
Henk W. Berendse
Alexander R. Cools
Pieter Voorn
Antonius B. Mulder

Scientific Organizing Committee

Peter Heutink, Amsterdam
Erik Ch. Wolters, Amsterdam
Adriaan A. Lammertsma, Amsterdam
Carol A. Seger, Fort Collins, CO
Patricio O'Donnell, Baltimore, MD
Barbara J. Knowlton, Los Angeles, CA
Etienne C. Hirsch, Paris
Jochen Roeper, Frankfurt
J.Paul Bolam, Oxford
Barry J. Everitt, Cambridge
Louk J.M.J. Vanderschuren, Utrecht
Jose A. Obeso, Pamplona
Jose L. Lanciego, Pamplona
Andre Parent, Quebec City
Jean-Michel Deniau, Paris
Kevin N. Gurney, Sheffield

The following organizations are gratefully acknowledged for their financial assistance

Cambridge Electronic Design Limited, International Brain Research Organization (IBRO), Leica Microsystems, MBF Bioscience, Medtronic, Noldus Information Technology, Parkinson Patienten Vereniging, Royal Dutch Academy of Sciences (KNAW), Solvay Pharma.

Contents

Part III Pharmacological and Receptor Studies in the Basal Ganglia

Contributors

Avital Adler Department of Physiology, The Hebrew University-Hadassah Medical School, Jerusalem, 91120, Israel, and The Interdisciplinary Center for Neural Computation, The Hebrew University, Jerusalem 91904, Israel

Fernando Alonso-Frech Section of Neurology, Hospital Universitario de Fuenlabrada, Madrid, Spain

Lazaro Alvarez Movement Disorders, Neurophysiology and Neurosurgery Units, Centro Internacional de Restauracion Neurologica (CIREN), La Habana, Cuba

Dorian Aur Department of Clinical Neurological Sciences, University of Western Ontario, 339 Windermere Road, A10-026, London, ON, N6A 5A5 Canada

Kristin Baer Department of Anatomy with Radiology, Faculty of Medical and Health Sciences, University of Auckland, Private Bag 92019, Auckland, New Zealand

Izhar Bar-Gad Gonda Brain Research Center & Goodman Faculty of Life Sciences, Bar-Ilan University, Ramat-Gan, 52900, Israel, bargadi@mail.biu.ac.il

Arcangelo Benigno Dipartimento di Medicina Sperimentale, Sezione di Fisiologia Umana, "G. Pagano", Università degli Studi di Palermo, Corso Tuköry 129, 90134 Palermo, Italy

Hagai Bergman Department of Physiology, The Hebrew University-Hadassah Medical School, Jerusalem, 91120, Israel, and The Interdisciplinary Center for Neural Computation, The Hebrew University, Jerusalem 91904, Israel, and Eric Roland Center for Neurodegenerative Diseases, The Hebrew University, Jerusalem 91904, Israel

Debra A. Bergstrom National Institutes of Health, National Institute of Neurological Disorders and Stroke, Neurophysiological Pharmacology Section, Bethesda, MD, USA

J. Paul Bolam MRC Anatomical Neuropharmacology Unit, University of Oxford, Mansfield Road, Oxford OX1 3TH, UK

Livia Brusa Clinica Neurologica, Department of Neuroscience, University Tor Vergata, Rome, Italy

Ashley I. Bush The University of Melbourne, Melbourne, VIC 3010, Australia, and Mental Health Research Institute of Victoria, 155 Oak Street, Parkville, VIC 3052, Australia

Daniela Cano-Sotomayor Instituto de Neurobiología, Universidad Nacional Autónoma de México, Juriquilla, Qro. 76230, México

Anna R. Carta Department of Toxicology and Center of Excellence for Neurobiology of Addiction, University of Cagliari, 09124 Cagliari, Italy acarta@unica.it

Maurizio Casarrubea Dipartimento di Medicina Sperimentale, Sezione di Fisiologia Umana, "G. Pagano", Università degli Studi di Palermo, Corso Tuköry 129, 90134 Palermo, Italy

Robert A. Cherny The University of Melbourne, Melbourne, VIC 3010, Australia, and Mental Health Research Institute of Victoria, 155 Oak Street, Parkville, VIC 3052, Australia

Marie-Francoise Chesselet Departments of Neurology and Neurobiology, The David Geffen School of Medicine at UCLA, Los Angeles, CA 90095-1769, USA, mchesselet@mednet.ucla.edu

Michael A. Clements Department of Physiology, Anatomy and Genetics, University of Oxford, Oxford OX1 3PT, UK

Stephanie J. Cragg Department of Physiology, Anatomy and Genetics, University of Oxford, Oxford OX1 3PT, UK

Giuseppe Crescimanno Dipartimento di Medicina Sperimentale, Sezione di Fisiologia Umana, "G. Pagano", Università degli Studi di Palermo, Corso Tuköry 129, 90134 Palermo, Italy

Marieke de Boer Cognitive Neurophysiology-CNCR, Department of Anatomy and Neurosciences, VU University Medical Center, PO Box 7057, 1007 MB Amsterdam, The Netherlands

Jean-Michel Deniau Institut National de la Santé et de la Recherche Médicale, U114, Chaire de Neuropharmacologie, Collège de France, Paris, France jean-michel.deniau@college-de-france.fr

Giuseppe Di Giovanni Dipartimento di Medicina Sperimentale, Sezione di Fisiologia Umana, "G. Pagano", Università degli Studi di Palermo, Corso Tuköry 129, 90134 Palermo, Italy, g.digiovanni@unipa.it

Vincenzo Di Matteo Istituto di Ricerche Farmacologiche "Mario Negri", Consorzio Mario Negri Sud, Santa Maria Imbaro (CH), Italy

Jonathan O. Dostrovsky Department of Physiology, University of Toronto, Toronto, ON, Canada, and Toronto Western Research Institute, Toronto, ON, Canada, j.dostrovsky@utoronto.ca

Benjamin Drukarch Department of Anatomy and Neurosciences, Neuroscience Campus Amsterdam, VU University Medical Center, PO Box 7057, 1007MB Amsterdam, The Netherlands

Christian Duval Département de kinanthropologie, Université du Québec à Montréal, and Centre de Recherche Institut Universitaire de Gériatrie de Montréal, Montréal, QC, Canada, duval.christian@uqam.ca

Ennio Esposito Istituto di Ricerche Farmacologiche "Mario Negri", Consorzio Mario Negri Sud, Santa Maria Imbaro (CH), Italy

Richard Exley Department of Physiology, Anatomy and Genetics, University of Oxford, Oxford OX1 3PT, UK, richard.exley@dpag.ox.ac.uk

Richard L.M. Faull Department of Anatomy with Radiology, Faculty of Medical and Health Sciences, University of Auckland, Private Bag 92019, Auckland, New Zealand

Alison Fenney Neuroscience, McMaster Institute for Neuroscience Discovery and Study, Hamilton, ON, Canada

Leif H. Finkel Department of Bioengineering, University of Pennsylvania, Philadelphia, PA, USA

David I. Finkelstein The University of Melbourne, Melbourne, VIC 3010, Australia, and Mental Health Research Institute of Victoria, 155 Oak Street, Parkville, VIC 3052, Australia, dfinkelstein@mhri.edu.au

Inna Finkes Department of Physiology, The Hebrew University-Hadassah Medical School, Jerusalem 91120, Israel

Sheila M. Fleming Departments of Neurology and Neurobiology, The David Geffen School of Medicine at UCLA, Los Angeles, CA 90095-1769, USA

Lucia Frau Department of Toxicology and Center of Excellence for Neurobiology of Addiction, University of Cagliari, 09124 Cagliari, Italy

Sarah Jane French Department of Pharmacology, University of Oxford, Mansfield Road, Oxford OX1 3QT, UK, sarah.french@pharm.ox.ac.uk

Weiping Gai Department of Anatomy with Radiology, Faculty of Medical and Health Sciences, University of Auckland, Private Bag 92019, Auckland, New Zealand

Salvatore Galati Clinica Neurologica, Department of Neuroscience, University Tor Vergata, Rome, Italy

Adriana Galvan Yerkes National Primate Research Center and Department of Neurology, Emory University, Atlanta, GA 30322, USA

Ray T. Gilbert Department of Anatomy with Radiology, Faculty of Medical and Health Sciences, University of Auckland, Private Bag 92019, Auckland, New Zealand

Magda Giordano Instituto de Neurobiología, Universidad Nacional Autónoma de México, Juriquilla, Qro. 76230, México, giordano@servidor.unam.mx

Henk J. Groenewegen Department of Anatomy and Neurosciences, VU University Medical Center, Neuroscience Campus Amsterdam, The Netherlands

Lifeng Gu Department of Neurology and Neurosurgery, Montreal Neurological Institute, McGill University, Montreal, QC, Canada H3A 2B4

Jorge Guridi Department of Neurology and Neurosurgery, Clinica Universitaria and Medical School, Neuroscience Centre, CIMA, University of Navarra, Pamplona, Spain

Glen R. Hanson National Institute on Drug Abuse, National Institutes of Health, Bethesda, MD 20892, USA

Tessa Hartog Department of Anatomy and Neurosciences, Neuroscience Campus Amsterdam, VU University Medical Center, 1007 MB Amsterdam, The Netherlands

Henrike Hartung Department of Pharmacology, University of Oxford, Mansfield Road, Oxford OX1 3QT, UK

Nobuhiko Hatanaka Division of System Neurophysiology, National Institute for Physiological Sciences, and Department of Physiological Sciences, The Graduate University for Advanced Studies, Okazaki, Aichi 444-8585, Japan

Mojgan Hodaie Toronto Western Research Institute and Toronto Western Hospital, Division of Neurosurgery, University of Toronto, Toronto, ON, Canada

Xing Hu Yerkes National Primate Research Center and Department of Neurology, Emory University, Atlanta 30322, USA

William D. Hutchison Department of Physiology, University of Toronto, and Toronto Western Research Institute, Toronto, ON, Canada

Ken-ichi Inoue Tokyo Metropolitan Institute for Neuroscience, Tokyo Metropolitan Organization for Medical Research, Fuchu, Tokyo 183-8526, Japan

Zvi Israel Department of Neurosurgery, Hadassah University Hospital, Jerusalem, Israel

Hirokazu Iwamuro Division of System Neurophysiology, National Institute for Physiological Sciences, Okazaki, Japan, and Department of Neurosurgery, Graduate School of Medicine, The University of Tokyo, Tokyo, Japan

Xiao-Tao Jin Division of Neuroscience, Yerkes National Primate Research Center and Department of Neurology, Emory University, 954, Gatewood Road NE, Atlanta, GA 30322, USA, Xjin@rmy.emory.edu

Mandar Jog Department of Clinical Neurological Sciences, University of Western Ontario, 339 Windermere Road, A10-026, London, ON, Canada N6A 5A5, mandar.jog@lhsc.on.ca

Allert J. Jonker Department of Anatomy and Neurosciences, Neuroscience Campus Amsterdam, VU University Medical Center, PO Box 7057, 1007MB Amsterdam, The Netherlands

Mati Joshua Department of Physiology, The Hebrew University-Hadassah Medical School, Jerusalem, 91120, Israel, and The Interdisciplinary Center for Neural Computation, The Hebrew University, Jerusalem 91904, Israel mati@alice.nc.huji.ac.il

Kristen A. Keefe Department of Pharmacology and Toxicology, University of Utah, Salt Lake City, UT 84112, USA

Raymond P. Kesner Department of Psychology, University of Utah, Salt Lake City, UT 84112, USA

Zafar U. Khan Department of Cell Biology, Faculty Sciences, University of Malaga, Malaga, Spain

Andrea C. Kil Department of Anatomy and Neurosciences, Neuroscience Campus Amsterdam, VU University Medical Center, PO Box 7057, 1007MB Amsterdam, The Netherlands

Michele A. Kliem Yerkes National Primate Research Center, Emory University, Atlanta, GA 30322, USA, mkliem@emory.edu

Bogdan P. Kolomiets Institut National de la Santé et de la Recherche Médicale, U114, Chaire de Neuropharmacologie, Collège de France, Paris, France

Myriam Lafreniere-Roula Department of Physiology, University of Toronto, Toronto, ON, Canada

Heyne Lee Yerkes National Primate Research Center, Emory University, Atlanta, GA 30322, USA

Lee Wei Lim Departments of Neurosurgery and Neuroscience, Maastricht University, Maastricht, The Netherlands

Gerardo Lopez Movement Disorders, Neurophysiology and Neurosurgery Units, Centro Internacional de Restauracion Neurologica (CIREN). La Habana, Cuba

Andres M. Lozano Toronto Western Research Institute and Toronto Western Hospital, Division of Neurosurgery, University of Toronto, Toronto, ON, Canada

Kelvin C. Luk Department of Neurology and Neurosurgery, Montreal Neurological Institute, McGill University, Montreal, QC, Canada H3A 2B4 Canada

Raul Macias Movement Disorders, Neurophysiology and Neurosurgery Units, Centro Internacional de Restauracion Neurologica (CIREN). La Habana, Cuba

Duncan A.A. MacLaren School of Psychology, University of St Andrews, St Mary's Quad, South Street, St Andrews, Fife, KY16 9JP, UK

Francesco Marzetti Clinica Neurologica, Department of Neuroscience, University Tor Vergata, Rome, Italy

Colin L. Masters The University of Melbourne, Melbourne, VIC 3010, Australia, and Mental Health Research Institute of VIC, 155 Oak Street, Parkville, Vic. 3052, Australia

John S. McKenzie Department of Physiology, The University of Melbourne, Melbrune, Vic. 3010, Australia, jsmcke@unimelb.edu.au

Juan Mena-Segovia MRC Anatomical Neuropharmacology Unit, University of Oxford, Mansfield Road, Oxford OX1 3TH, UK juan.mena-segovia@pharm.ox.ac.uk

Rea Mitelman Department of Physiology, Hadassah Medical School and the Interdisciplinary Center for Neural Computation, the Hebrew University, Jerusalem, Israel, ream@alice.nc.huji.ac.il

Shigehiro Miyachi Primate Research Institute, Kyoto University, Inuyama, Aichi 484-8506, Japan

Micaela Morelli Department of Toxicology and Center of Excellence for Neurobiology of Addiction, University of Cagliari, and CNR Institute of Neuroscience Cagliari, Via Ospedale 72, 09124 Cagliari, Italy

Vincenzo Moschella Clinica Neurologica, Department of Neuroscience, University Tor Vergata, Rome, Italy

Jason T. Moyer Department of Bioengineering, University of Pennsylvania, Philadelphia, PA, USA

Antonius B. Mulder Cognitive Neurophysiology-CNCR, Department of Anatomy and Neurosciences, VU University Medical Center, PO Box 7057, 1007 MB Amsterdam, The Netherlands, ab.mulder@vumc.nl

Atsushi Nambu Division of System Neurophysiology, National Institute for Physiological Sciences, and Department of Physiological Sciences, The Graduate University for Advanced Studies, Okazaki, Aichi 444-8585, Japan nambu@nips.ac.jp

Bijli Nanda Hindustan Institute of Medical Sciences and Research, Knowledge Park-III, Greater Noida, Up-201306

Jose A. Obeso Department of Neurology and Neurosurgery, Clinica Universitaria and Medical School, Neuroscience Centre, CIMA, University of Navarra, Pamplona, Spain, jobeso@unav.es

Chihiro Ohye Hidaka Hospital, Functional and Gamma Knife Surgery Center, Takasaki, Gunma, Japan, hidaka.sps@tiara.ocn.ne.jp

Haruo Okado Tokyo Metropolitan Institute for Neuroscience, Tokyo Metropolitan Organization for Medical Research, Fuchu, Tokyo 183-8526, Japan

Ryuichi Okiyama Department of Neurology and Neurosurgery, Tokyo Metropolitan Neurological Hospital, Fuchu, Tokyo 183-0042, Japan

Jean-Francois Paré Division of Neuroscience, Yerkes National Primate Research Center and Department of Neurology, Emory University, 954, Gatewood Road NE, Atlanta, GA 30322, USA

Antonella Peppe IRCCS Foundation S. Lucia, Rome, Italy

Mariangela Pierantozzi Clinica Neurologica, Dept Neuroscience, University Tor Vergata, Rome, Italy

Massimo Pierucci Istituto di Ricerche Farmacologiche "Mario Negri", Consorzio Mario Negri Sud, Santa Maria Imbaro (CH), Italy

Annalisa Pinna CNR Institute of Neuroscience Cagliari, Via Ospedale 72, 09124 Cagliari, Italy

Dinesh Raju Yerkes National Primate Research Center, Emory University, Atlanta, GA 30322, USA

Mark I. Rees Department of Anatomy with Radiology, Faculty of Medical and Health Sciences, University of Auckland, Private Bag 92019, Auckland, New Zealand

Anton Reiner Department of Anatomy & Neurobiology, College of Medicine, The University of Tennessee Health Science Center, Memphis, TN 38163, USA areiner@utmem.edu

Matthew D. Riedy Program in Neuroscience, University of Utah, Salt Lake City, UT 84112, USA

Maria Cruz Rodriguez-Oroz Department of Neurology and Neurosurgery, Clinica Universitaria and Medical School, Neuroscience Centre, CIMA, University of Navarra, Pamplona, Spain

Vladimir V. Rymar Department of Neurology and Neurosurgery, Montreal Neurological Institute, McGill University, Montreal, QC, Canada H3A 2B4

Abbas F. Sadikot Department of Neurology and Neurosurgery, Montreal Neurological Institute, McGill University, Montreal, QC, Canada H3A 2B4

Nobuhito Saito Department of Neurosurgery, Graduate School of Medicine, The University of Tokyo, Tokyo, Japan

Sumito Sato Department of Neurosurgery, Kitazato University School of Medicine, Kanagawa 228-8555, Japan

Carol A. Seger Department of Psychology, Colorado State University, Fort Collins, CO 80523, USA, seger@lamar.colostate.edu

Thibault Sesia Departments of Neurosurgery and Neuroscience, Maastricht University, Maastricht, The Netherlands

Tohru Shibazaki Hidaka Hospital Functional and Gamma Knife Surgery Center, Takasaki, Gunma, Japan

Osamu Shouno Honda Research Institute Japan Co., Ltd., 8-1 Honcho, Wako-shi, Saitama 351-0188, Japan, shouno@jp.honda-ri.com

Yoland Smith Yerkes National Primate Research Center, and Department of Neurology, School of Medicine, Emory University, Atlanta, GA, USA

Paolo Stanzione Clinica Neurologica, Department of Neuroscience, University Tor Vergata, Rome, Italy, and IRCCS Foundation S. Lucia, Rome, Italy

Alessandro Stefani Clinica Neurologica, Department of Neuroscience, University Tor Vergata, and IRCCS Foundation S. Lucia, Rome, Italy Stefani@uniroma2.it

Mathijs Stegeman Cognitive Neurophysiology-CNCR, Department of Anatomy and Neurosciences, VU University Medical Center, PO Box 7057, 1007 MB Amsterdam, The Netherlands

Heinz Steiner Department of Cellular and Molecular Pharmacology, Rosalind Franklin University of Medicine and Science, The Chicago Medical School, North Chicago, IL 60064, USA, Heinz.Steiner@rosalindfranklin.edu

Yoshihisa Tachibana Division of System Neurophysiology, National Institute for Physiological Sciences, and Department of Physiological Sciences, The Graduate University for Advanced Studies, Okazaki, Aichi 444-8585, Japan

Masahiko Takada Systems Neuroscience Section, Primate Research Institute, Kyota University, Inuyame, Aichi 484–8506, Japan takada-ms@igakuken.or.jp

Hiroshi Takahashi Department of Neurology and Neurosurgery, Tokyo Metropolitan Neurological Hospital, Fuchu, Tokyo 183-0042, Japan

Sayuki Takara Division of System Neurophysiology, National Institute for Physiological Sciences, and Department of Physiological Sciences, The Graduate University for Advanced Studies, Okazaki, Aichi 444-8585, Japan

Sonny Tan Departments of Neurosurgery and Neuroscience, Maastricht University, Maastricht, The Netherlands

Makoto Taniguchi Department of Neurology and Neurosurgery, Tokyo Metropolitan Neurological Hospital, Fuchu, Tokyo 183-0042, Japan

Yasin Temel Departments of Neurosurgery and Neuroscience, Maastricht University, Maastricht, The Netherlands

Toru Terao Department of Neurology and Neurosurgery, Tokyo Metropolitan Neurological Hospital, Fuchu, Tokyo 183-0042, Japan

Anne-Marie Thierry Institut National de la Santé et de la Recherche Médicale, U114, Chaire de Neuropharmacologie, Collège de France, Paris, France

Patrick L. Tierney National Institutes of Health, National Institute of Neurological Disorders and Stroke, Neurophysiological Pharmacology Section, Bethesda, MD, USA

Robert S. Turner
Department of Neurobiology & Center for the Neural Basis of Cognition, University of Pittsburgh, Pittsburgh, PA 15261, USA, rturner@pitt.edu

Pepijn van den Munckhof Department of Anatomy and Neurosciences, Neuroscience Campus Amsterdam, VU University Medical Center, PO Box 7057, 1007 MB Amsterdam, The Netherlands, and Department of Neurosurgery, Academic Medical Center, Amsterdam, 1105 AZ, The Netherlands
p.vandenmunckhof@amc.nl

Marcel van der Roest Cognitive Neurophysiology-CNCR, Department of Anatomy and Neurosciences, VU University Medical Center, PO Box 7057, 1007 MB Amsterdam, The Netherlands

Louk J.M.J. Vanderschuren Rudolf Magnus Institute of Neuroscience, Department of Neuroscience and Pharmacology, University Medical Center Utrecht, Utrecht, The Netherlands

Yvette C. van Dongen Department of Anatomy and Neurosciences, VU University Medical Center, Neuroscience Campus Amsterdam, The Netherlands

Rosa M. Villalba Yerkes National Primate Research Center, Emory University, Atlanta, GA 30322, USA, rvillal@emory.edu

Veerle Visser-Vandewalle Departments of Neurosurgery and Neuroscience, Maastricht University, Maastricht, The Netherlands

Rinske Vlamings Departments of Neurosurgery and Neuroscience, Maastricht University, Maastricht, The Netherlands

Pieter Voorn Department of Anatomy and Neurosciences, Neuroscience Campus Amsterdam, VU University Medical Center, PO Box 7057, 1007 MB Amsterdam, The Netherlands

Erno Vreugdenhil Division of Medical Pharmacology, Leiden/Amsterdam Centre for Drug Research, Leiden University, Leiden, The Netherlands

Henry J. Waldvogel Department of Anatomy with Radiology, Faculty of Medical and Health Sciences, University of Auckland, Private Bag 92019, Auckland, New Zealand, h.waldvogel@auckland.ac.nz

Judith R. Walters National Institutes of Health, National Institute of Neurological Disorders and Stroke, Neurophysiological Pharmacology Section, Bethesda, MD, USA, waltersj@ninds.nih.gov

Nienke S. Weiss Department of Anatomy and Neurosciences, Neuroscience Campus Amsterdam, VU University Medical Center, PO Box 7057, 1007 MB Amsterdam, The Netherlands

Thomas Wichmann Yerkes National Primate Research Center, Emory University, Atlanta, GA 30322, USA, and Department of Neurology, School of Medicine, Emory University, Atlanta, GA

Simon Wilkins The University of Melbourne, Melbourne, VIC 3010, Australia, and Mental Health Research Institute of Victoria, 155 Oak Street, Parkville, VIC 3052, Australia

Ingo Willuhn Department of Cellular and Molecular Pharmacology, Rosalind Franklin University of Medicine and Science, The Chicago Medical School, North Chicago, IL 60064, USA

Charles J. Wilson Department of Biology, University of Texas at San Antonio, One UTSA Circle, San Antonio, TX 78249, USA, Charles.Wilson@utsa.edu

David I.G. Wilson School of Psychology, University of St. Andrews, St. Mary's Quad, South Street, St Andrews, Fife, KY16 9JP, UK

Philip Winn School of Psychology, University of St. Andrews, St. Mary's Quad, South Street, St. Andrews, Fife KY16 9JP, UK, pw@st-andrews.ac.uk

John A. Wolf Department of Psychiatry, University of Pennsylvania, Philadelphia, PA, USA, wolfjo@upenn.edu

Fusako Yokochi Department of Neurology and Neurosurgery, Tokyo Metropolitan Neurological Hospital, Fuchu, Tokyo 183-0042, Japan fyokochi-tmnh@umin.ac.jp

Adam Zaidel Interdisciplinary Center for Neural Computation, and Department of Physiology and Eric Roland Center for Neurodegenerative Diseases, The Hebrew University, Hadassah Medical School, Jerusalem, Israel adam@alice.nc.huji.ac.il

Part I
Functional Organization of the Basal Ganglia

You Cannot Have a Vertebrate Brain Without a Basal Ganglia

Anton Reiner

Abstract Early twentieth century theories of telencephalic evolution maintained that the parts of the telencephalon appeared in serial order in vertebrate phylogeny – the globus pallidus in fish, the striatum in amphibians, and the cerebral cortex in reptiles. The notion of serial appearance of the parts of the basal ganglia is reflected in the frequent use of the terms paleostriatum and neostriatum as synonyms for globus pallidus and caudate–putamen, respectively. Moreover, in older theories of basal ganglia evolution, the cerebral cortex was thought to supplant the basal ganglia in motor control in the mammalian lineage, resulting in the replacement of a supposed reptilian stereotyped motor repertoire with the adaptable mammalian motor capacity. Modern evidence, in fact, shows that both striatum and pallidum have been the basal ganglia constituents since the earliest jawed fish. Moreover, the striatal part of the basal ganglia appears to have been organized into separate direct and indirect striatal output pathways by this time as well. Modern findings also show that during most of its evolutionary history (i.e., prior to mammals), the basal ganglia mediated its role in motor control via output pathways to motor layers of the midbrain roof, via midbrain targets of the striatum such as the substantia nigra pars reticulata. Output pathways to the pallium emerged in the amniote lineage, but remained of lesser significance in living members of the sauropsid lineage (birds and reptiles) than the circuitry to the tectum. With the appearance of neocortex in mammals, however, basal ganglia outputs to motor cortex via thalamus became of greater significance, especially in primates, in which a parallel expansion of cerebral cortex and basal ganglia occurred.

A. Reiner (✉)
Department of Anatomy and Neurobiology, The University of Tennessee
Health Science Center, 855 Monroe Avenue, Memphis, TN 38163, USA
e-mail: areiner@utmem.edu

H.J. Groenewegen et al. (eds.), *The Basal Ganglia IX*, Advances in Behavioral Biology 58,
DOI 10.1007/978-1-4419-0340-2_1,

1 Introduction

Early twentieth century comparative neuroanatomists formulated a theory of telencephalic evolution (Fig. 1) based strictly on the size, position, and cytological appearance of the major telencephalic regions (Edinger et al. 1903; Ariëns-Kappers et al. 1936). This theory resulted in terminologies for both mammalian and nonmammalian telencephala and notions about their functional organization that proved enduring. In this older terminology, the globus pallidus in mammals is alternatively referred to as the paleostriatum, while the caudate-putamen is called the neostriatum. These terms stemmed from the notion that the parts of the telencephalon appeared in serial order during vertebrate evolution, globus pallidus

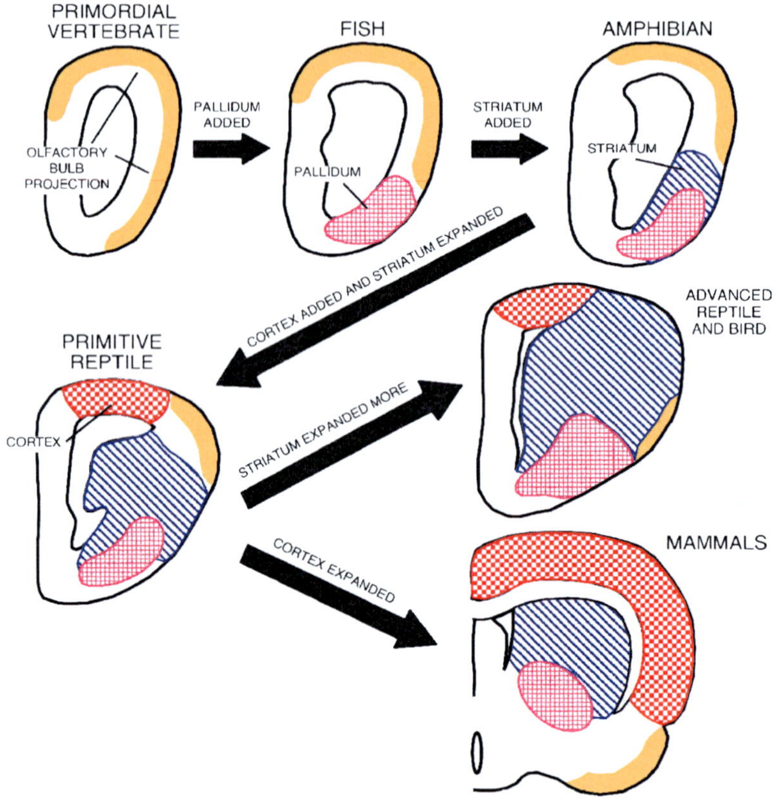

Fig. 1 Schematic illustrating telencephalic evolution as viewed by early twentieth century neuroanatomists, with individual brain regions color coded. This older and now outdated theory proposed that the major parts of the telencephalon evolved in serial order, with the ancestral telencephalon being devoted entirely to olfaction. The globus pallidus was said to be added in fish, the striatum in amphibians, and a rudimentary cerebral cortex in reptiles. Reptiles and birds were presumed to have expanded the basal ganglia (more so birds), while mammals were thought to have expanded the cerebral cortex into neocortex

in jawed fish, striatum in amphibians, and cerebral cortex in reptiles (Fig. 1) (Ariëns-Kappers et al. 1936). Mammals were thought to have elaborated cerebral cortex into neocortex, while birds were thought to have elaborated the basal ganglia by addition of a new territory known as the hyperstriatum. In mammals, the cerebral cortex was thought to take over behaviors that had been carried out by the basal ganglia in nonmammals (Ariëns-Kappers et al. 1936; Herrick 1948).

Comparison of Nissl-stained sections through pigeon and rat telencephalon reveal why the early twentieth century neuroanatomists concluded that virtually the entire avian telencephalon was part of the basal ganglia. In mammals, the neocortex is organized into distinct layers of differing neuronal size, shape, and packing density, while neurons in the striatal part of the basal ganglia are more uniform in size, shape, and packing density (Fig. 2). As a result, much of the avian telencephalon was assumed to be part of the striatum and named accordingly (Fig. 2). With the advent of silver staining of degenerating fibers in the 1950s, it was recognized that the role of the basal ganglia in mammals in motor control involved cortical input and in large part was mediated by a return projection to motor cortex (Carman et al. 1963; Nauta and Mehler 1966). Neurochemical methods developed in the 1960s revealed a major nigral input to striatum that used dopamine as a neurotransmitter, and led to the discovery that loss of this input was the basis of Parkinson's disease (Carlsson 1959). Understanding of the cellular make up of the basal ganglia, the neurotransmitters used by its neurons, and the delineation of its circuitry at a cellular level accelerated greatly with the application of immunohistochemistry in the 1970s, and in situ hybridization histochemistry in the 1980s to the study of the basal ganglia (Parent 1986; Reiner and Anderson 1990; Gerfen 1992; Graybiel 1990; Swanson and Petrovich 1998). Recent advances have also been made in identifying the genes controlling the regional identity and development of the striatum, globus pallidus, and cerebral cortex. These genes include *Dlx1* and *Dlx2*, which specify the striatum and pallidum, and *Nkx2.1*, which specifies globus pallidus (Rubenstein et al. 1994). Other genes (such as *Emx1*, *Emx2*, and *Tbr1*) control pallial identity, and are expressed at high levels in developing cerebral cortex.

The neuroanatomical tools that have been used to clarify the cellular neurochemistry and functional circuitry of the basal ganglia of mammals and determine the genetic control of regional identity in the developing mammalian telencephalon have been used to study telencephalic organization and development in members of other vertebrate groups as well. These studies have dramatically revised our understanding of basal ganglia evolution, and are reviewed in the following sections.

2 Where is the Basal Ganglia in Nonmammals?

Whereas it was once thought that most or all of the lateral telencephalic walls in birds were basic ganglia, it is now clear that only the ventrolateral sector of the avian telencephalon contains both a striatum and a globus pallidus, as defined by

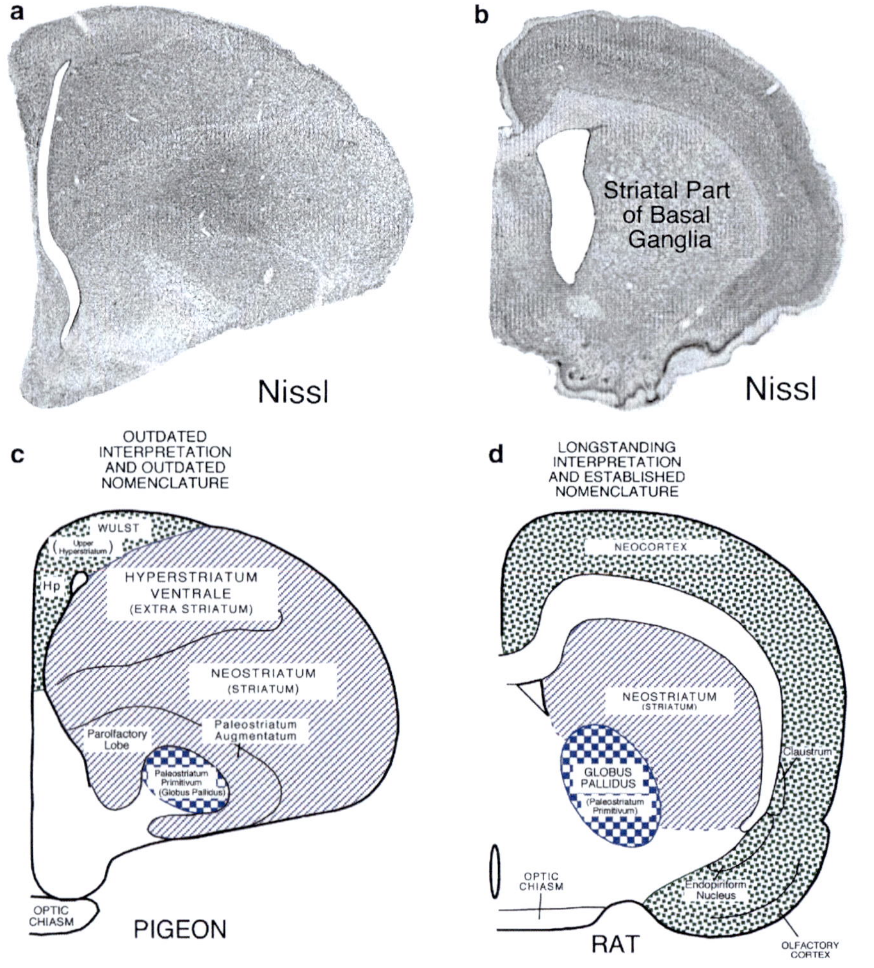

Fig. 2 A pair of low power images of transverse Nissl-stained sections through the adult pigeon telencephalon (**a**), and adult rat telencephalon (**b**), and a pair of schematic line drawings of midtelencephalic transverse brain sections of pigeon and rat illustrating: (**c**) the outdated interpretation of the organization of the telencephalon in birds (Ariëns-Kappers et al. 1936) and the outdated nomenclature that view engendered for the telencephalon of birds, (**d**) the longstanding interpretation of mammalian telencephalic organization and the established nomenclature consistent with that view. Comparison of (**a**) and (**b**) shows that bird telencephalon conspicuously lacks a laminated structure resembling mammalian cerebral cortex. Rather, much of the avian telencephalon resembles the striatal part of mammalian basal ganglia in its histological appearance. These features of the telencephalon in birds led comparative neuroanatomists in the early part of the twentieth century to conclude that avian telencephalon largely consisted of hypertrophied basal ganglia (Ariëns-Kappers et al. 1936), and name its subdivisions accordingly. As indicated by a comparison of (**c**) and (**d**), the terms devised for specific telencephalic regions in birds were seemingly appropriate at the time that most of the telencephalon of birds was thought to represent hypertrophied basal ganglia

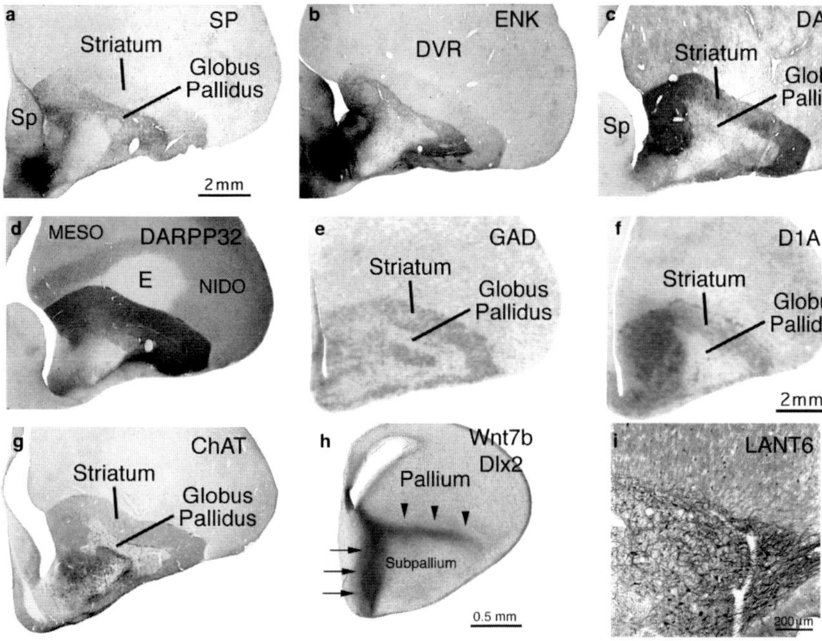

Fig. 3 Images of frontal sections through the right hemisphere of avian telencephalon illustrating the enrichment of the striatum and/or pallidum in SP (**a**), ENK (**b**), dopaminergic fibers and terminals (**c**), DARPP32 (**d**), GAD (**e**), the D1A dopamine-receptor (**f**), choline acetyltransferase (**g**), Dlx expression (**h**), and LANT6+ neurons (**i**). Images (**a**) and (**b**) show sections from pigeon that had been immunostained for SP (**a**) or enkephalin (ENK) (**b**) (adapted from Fig. 2c, d from Reiner et al. 1998b). Note the intense SP+ and ENK+ immunoreactivity in the ventrolateral telencephalon that defines the region of the striatum and globus pallidus, and distinguishes the subpallium from the dorsal ventricular ridge (DVR) of the overlying pallium. Image (**c**) shows a section from pigeon that had been immunostained for dopamine (adapted from Fig. 2b from Reiner et al. 1998b), revealing conspicuous enrichment of the striatum in dopaminergic fibers and terminals. Image (**d**) shows a section from pigeon that had been immunostained for DARPP32 (adapted from Fig. 2a from Reiner et al. 1998b). Striatal projection neurons are rich in DARPP32 and the striatum and its neuropil is accordingly rich in DARPP32. Image (**e**) shows a section from chick that had been labeled by in situ hybridization for GAD (adapted from Fig. 2d from Sun et al. 2005). The densely packed striatal projection neurons express GAD, and the striatum is accordingly rich in GAD. The more widely spaced globus pallidus neurons are also rich in GAD. Image (**f**) shows a section from chick that had been labeled by in situ hybridization for the D1A dopamine receptor (adapted from Fig. 2a from Sun and Reiner 2000). The densely packed projection neurons express D1A, and the striatum is accordingly rich in D1A. Image (**g**) shows a section from pigeon that had been immunostained for choline acetyltransferase (ChAT) (adapted from Fig. 2e from Reiner et al. 2004). A subset of striatal interneurons is cholinergic and therefore rich in ChAT. The striatum is accordingly rich in ChAT. Image (**h**) shows a section from an E10 chick embryo (adapted from Fig. 2f from Reiner et al. 2004). The section had been double-labeled by in situ hybridization histochemistry for Dlx2 and Wnt7b. Dlx2 is expressed along the subventricular zone of the subpallium (*arrows*), while Wnt7b is expressed by cells at the pallial-subpallial boundary (*arrowheads*). Image (**i**) shows LANT6-immunostained neurons in pigeon globus pallidus (adapted from Fig. 5e from Reiner et al. 2004). The scale bar in (**a**) provides the magnification for (**a**)–(**d**) and (**g**), while the scale bar in (**f**) provides the magnification for (**e**) and (**f**). Abbreviations: *DVR* dorsal ventricular ridge, *Meso* mesopallium, *Nido* nidopallium, and *Sp* septum

neurochemical, hodological, and developmental molecular criteria (Fig. 3) (Reiner et al. 1998a). The striatal sector contains SP+ and ENK+ neurons (Reiner et al. 1998a, b), with the identity of the striatum further confirmed by the widespread neuronal expression of GAD (Veenman and Reiner 1994; Sun et al. 2005), and avian *Dlx1/2* (Smith-Fernandez et al. 1998; Puelles et al. 2000). Moreover, the SP+ and ENK+ striatal neurons give rise to projections to a ventrocaudal telencephalic cell group that is comparable to globus pallidus, because of this striatal input, because it contains large GABAergic neurons that also contain the neurotensin-related hexapeptide LANT6 (as do pallidal neurons in mammals), and because its neurons express *Nkx2.1* (Karten and Dubbeldam 1973; Reiner and Carraway 1987; Veenman and Reiner 1994; Reiner et al. 1998a; Puelles et al. 2000). The avian striatum receives a dopaminergic input from midbrain dopaminergic neurons, which receive a return projection from SP+ striatal neurons (Reiner et al. 1994, 1998b). As in mammals, the dopamine effects on striatum are mediated by D1 and D2 dopamine receptors, which are expressed at high levels in the avian striatum (Richfield et al. 1987; Sun and Reiner 2000). The pallium is large in birds and is divided into functionally distinct territories, some of which are sensory in nature and receive input from ascending primary sensory pathways (notably visual and auditory) via the thalamus, some of which are higher-order information processing centers, and some of which give rise to descending projections to the brainstem (Karten 1969; Reiner 2000). Higher-order and motor output parts of the pallium give rise to a massive excitatory input to the avian striatum (Veenman et al. 1995b; Veenman and Reiner 1996), and the intralaminar thalamus also provides excitatory input to the striatum in birds as well (Wild 1987; Veenman et al. 1995a, 1997; Reiner et al. 1998a). The communication between thalamus and pallium on one hand and striatum on the other appears to be mediated in birds by the same cell-type-specific glutamate receptors in striatum as in mammals (Laverghetta et al. 2006).

Birds evolved from archosaurian reptiles, of which crocodilians are the only other living representatives (Chiappe 1995). Not surprisingly, therefore, the reptilian basal ganglia closely resemble that of birds, especially those in crocodilians. For example, as in birds, the ventrolateral reptilian telencephalon contains both a striatum and a globus pallidus, by neurochemical, hodological, and developmental molecular criteria (Reiner et al. 1998a; Smith-Fernandez et al. 1998). As in birds, the striatal sector contains SP+ and ENK+ neurons (Reiner et al. 1998a), with the identity of the striatum further confirmed by the widespread neuronal expression of GAD (Bennis et al. 1991), and a homolog of *Dlx1/2* (Smith-Fernandez et al. 1998). Moreover, the SP+ and ENK+ striatal neurons give rise to projections to a ventrocaudal cell group that is comparable to globus pallidus, both because of this striatal input and because it contains large GABAergic neurons that contain LANT6 (Bennis et al. 1991; Reiner and Carraway 1987; Reiner et al. 1998a). The SP+ and ENK+ inputs overlap in globus pallidus, as they do in birds, implying that GPi-type and GPe-type neurons were intermingled in the stem amniotes. The reptilian striatum receives a substantial dopaminergic input from the midbrain tegmentum and these dopaminergic neurons in turn receive a substantial return projection from SP+

striatal neurons (Reiner et al. 1998a). As in birds and mammals, dopaminergic effects on striatum are mediated by D1 and D2 dopamine receptors, and the striatum in reptiles is enriched in D1 and D2 receptors (Richfield et al. 1987; Reiner et al. 1998a). The pallium occupies the dorsolateral sector of the telencephalon, but is less extensive in reptiles than in birds. Nonetheless, the pallium in reptiles is the source of a major excitatory input to the striatum, with the thalamus also providing excitatory input (Reiner et al. 1998a). Moreover, the glutamate receptors employed by specific types of striatal neurons in their responsiveness to this excitatory input are very similar in reptiles to those in birds and mammals (Fowler et al. 1999).

Use of neurochemical labeling methods has also revealed the location and extent of the striatum and pallidum in anamniotes. The telencephalon in the vertebrate groups immediately ancestral to amniotes (i.e., amphibians and lobe-finned fish) is, however, much smaller in size and much simpler in architecture, consisting of paired evaginated tubular structures with the neurons confined to the periventricular zone. Nonetheless, basal ganglia are present in the telencephalon of amphibians and lobe-finned fish. For example, the ventrolateral telencephalon in amphibians contains both a striatum and a globus pallidus, by neurochemical, hodological, and molecular developmental criteria (Fig. 4). The striatal sector contains SP+ and ENK+ neurons (Marin et al. 1997, 1998a, b; Reiner et al. 1998a), with the identity of the striatum further confirmed by the expression of GAD and an amphibian *Dlx1/2* (Brox et al. 2003). Moreover, the SP+ and ENK+ striatal neurons give rise to projections to a ventrocaudal cell group that appears comparable to globus pallidus, by this striatal input, because it contains large GABAergic neurons and because it expresses a homolog of *Nkx2.1* (Marin et al. 1998a, b; Gonzalez et al. 2002; Brox et al. 2003). As true in other anamniote groups, the SP+ and ENK+ inputs overlap in this pallidal field, implying that GPi-type and GPe-type neurons are intermingled. The amphibian striatum receives a dopaminergic input from midbrain dopaminergic neurons, which receive a return projection from SP+ striatal neurons (Gonzalez and Smeets 1994; Marin et al. 1998a, b; Reiner et al. 1998a). Both striatum and pallidum are, nonetheless, much more cell poor, and the dopaminergic input more modest, in amphibians than in amniotes. Moreover, excitatory input to the striatum in amphibians arises mainly from the thalamus, and very little appears to arise from the pallium. It is not surprising that pallial input to the striatum is meager in amphibians, since the dorsal pallium (which in amniotes is the territory that receives sensory thalamic input and projects to the striatum) in amphibians is meager and not a target of ascending sensory input from the thalamus (Northcutt and Kicliter 1980). These features of living amphibians, as compared to the features discussed above for reptiles, indicate that major elaborations of thalamopallial and corticostriatal circuitry, and of the basal ganglia in general, occurred by the evolutionary appearance of stem amniotes (Fig. 5).

Bony fish diverged from a common ancestor into two groups early in their evolution, the ray-finned fish (actinopterygians) and the lobe-finned fish (sarcopterygians) (Northcutt 1981). The lobe-finned fish possess the paired tubular evaginated telencephala that are likely to be characteristic of the common ancestor of the two types of bony fish, and an extinct member of the lobe-finned fish (the rhipidistians)

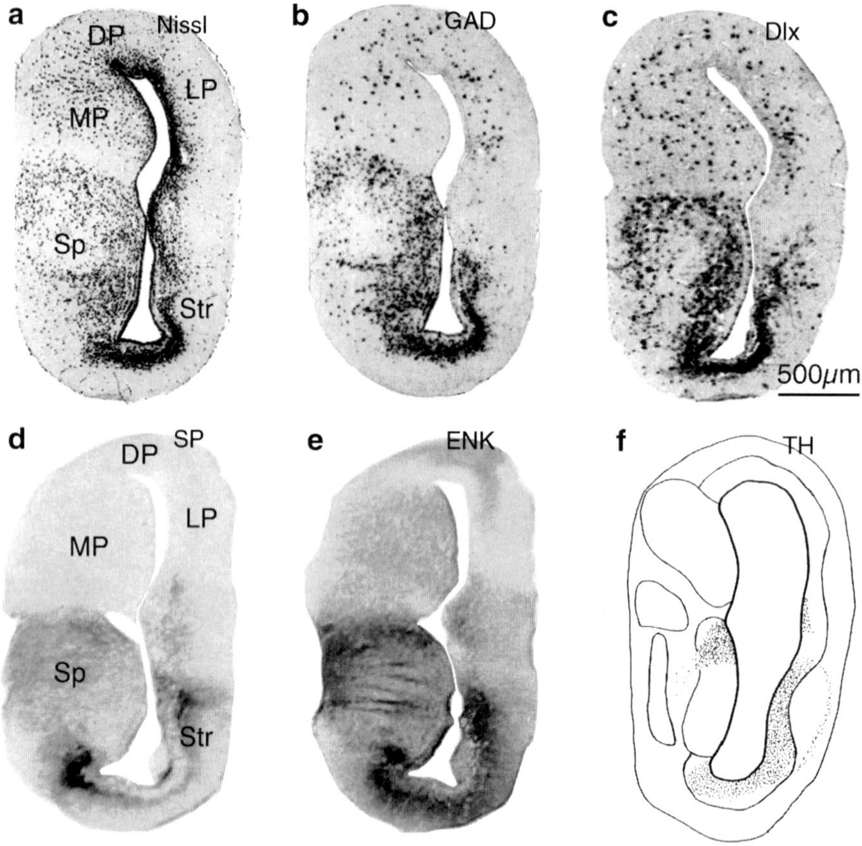

Fig. 4 Images of frog telencephalon illustrating the cytoarchitectonic organization of the telencephalon as seen in a Nissl-stained section (**a**) (adapted from Fig. 2g from Brox et al. 2003), the enrichment of the basal ganglia in GAD-expressing neurons as seen by in situ hybridization histochemistry (**b**) (adapted from Fig. 2i from Brox et al. 2003), the enrichment of the basal ganglia in Dlx-expressing neurons as seen by in situ hybridization histochemistry (**c**) (adapted from Fig. 8b from Brox et al. 2003), the enrichment of the basal ganglia in SP as seen by immunolabeling (**d**), the enrichment of the basal ganglia in ENK as seen by immunolabeling (**e**), and the enrichment of the striatum in dopaminergic fibers and terminals as seen by immunolabeling for tyrosine hydroxylase (TH) (**f**) (adapted from Fig. 2c from González and Smeets 1991). All images are of the right telencephalic hemisphere at a level containing the striatum. Image (**a**) reveals the thin lateral wall of the telencephalon and the periventricular position of its neurons. Images (**b**) and (**c**) show that the subpallium, including the septum (Sp) and striatum (Str), are defined by their conspicuous enrichment in GAD-expressing and Dlx-expressing neurons. The GAD-expressing and Dlx-expressing neurons in the pallial territories represent interneurons that have migrated in from the subpallium. Images (**d**) and (**e**) show the neuropil of the striatum is enriched in SP and ENK, reflecting the presence of SP+ and ENK+, dendrites and their local axon collaterals. Image (**f**) shows a line drawing reconstruction of a TH-immunostained section, showing the relative enrichment of the striatum in dopaminergic fibers and terminals. The *scale bar* in (**c**) provides magnification for (**a**)–(**f**). Abbreviations: *DP* dorsal pallium, *LP* lateral pallium, *MP* medial pallium, *Sp* septum, and *Str* striatum

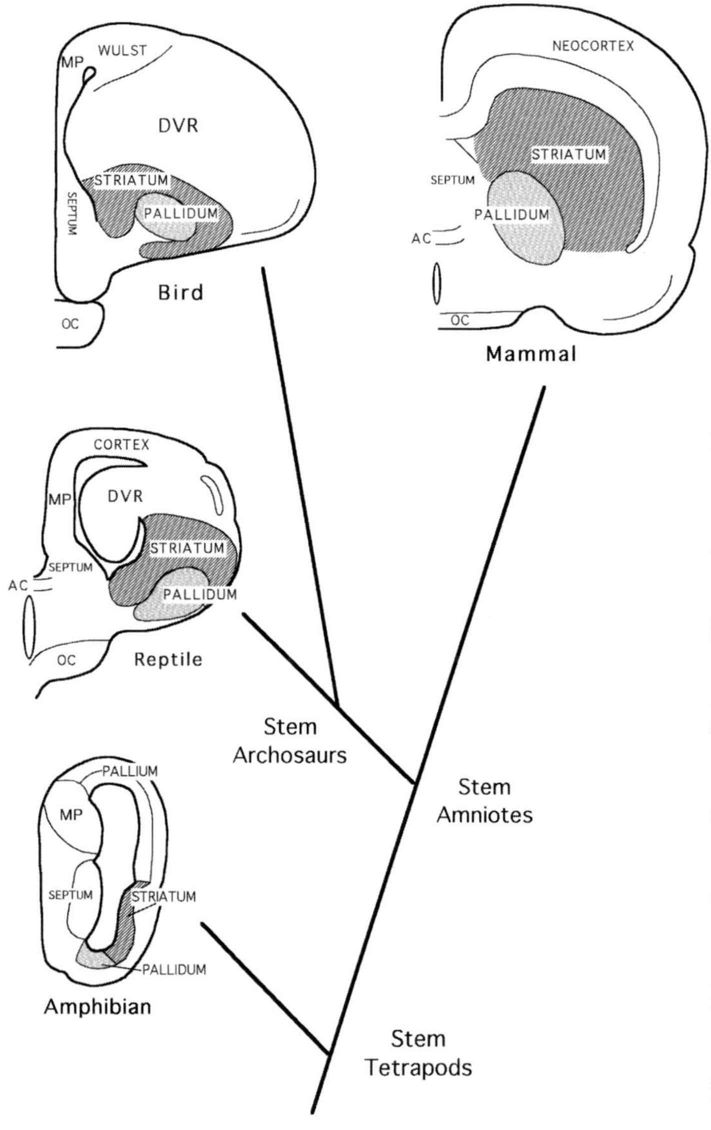

Fig. 5 Schematics of frontal sections through the basal ganglia of the right telencephalic hemisphere in representative species from four tetrapod groups, an amphibian (frog), a reptile (turtle), a bird (pigeon), and a mammal (rat), arranged according to their evolutionary divergences. The basal ganglia in all four groups consist of a striatum and a pallidum located in the basal telencephalon, beneath the pallium. The pallidum, however, tends to be more laterally located in reptiles and birds, than in amphibians and mammals. The phylogenetic distribution of pallidal laterality suggests this trait arose in the reptilian lineage and was retained in birds. Medial is to the *left* and dorsal to the *top* in all sections. Abbreviations: *AC* anterior commissure, *MP* medial pallium, and *OC* optic chiasm

was the ancestor of tetrapods (Hedges et al. 1993). The ray-finned fish, by contrast, evolved a telencephalon with an everted pallial. In both bony fish groups, nonetheless, basal ganglia have been demonstrated by modern hodological, neurochemical, and developmental methods (Figs. 6 and 7) (Reiner et al. 1998a). For lobe-finned fish, the topography and cytology of the basal ganglia resemble that in amphibians (Reiner and Northcutt 1987), while the peculiarities of the ray-finned fish telencephalon have rendered overall telencephalic topography and cytoarchitecture (including that of the basal ganglia) less reminiscent of that in other jawed vertebrates (Reiner and Northcutt 1992). For example, in lungfish, the only lobe-finned fish whose basal ganglia have been studied neurochemically (Reiner and Northcutt 1987), the ventrolateral telencephalon contains both a striatum and a globus pallidus. As in amphibians, the striatal sector contains SP+ and ENK+ neurons, and like in amphibians these are located in a neuron-rich periventricular zone (Fig. 6). As in mammals, the SP+ and ENK+ striatal neurons are likely to be enriched in GABA (Trabucchi et al. 2008). The SP+ and ENK+ striatal neurons give rise to projections to a ventrocaudal neuronal cell group that is comparable to globus pallidus, because of this striatal input and because its neurons contain LANT6. In this pallidal field, the SP+ and ENK+ inputs overlap, implying that GPi-type and GPe-type neurons are intermingled. Moreover, the striatum receives a dopaminergic input from the midbrain and these dopaminergic neurons receive a return projection from SP+ striatal neurons (Reiner and Northcutt 1987). The pallium occupies the dorsal

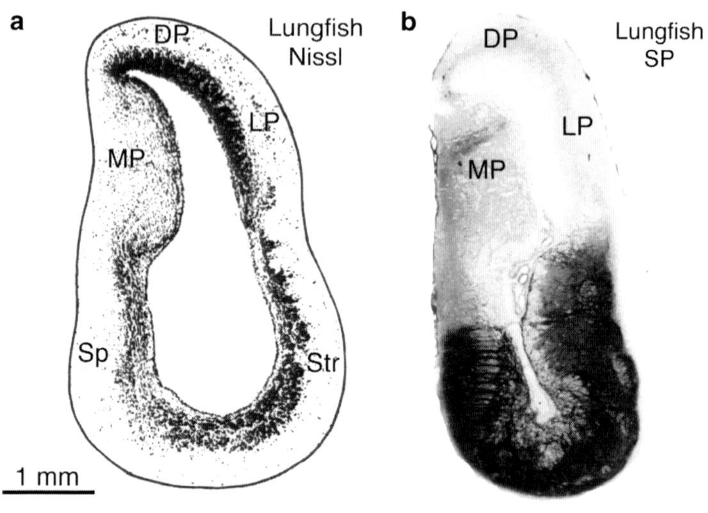

Fig. 6 Images illustrating telencephalic organization in the African lungfish *Protopterus annectens* (**a**, **b**) (adapted from Figs. 2d and 4c from Reiner and Northcutt 1987). Images (**a**) and (**b**) show coronal views of sections through the right telencephalon of the African lungfish stained for Nissl substance (**a**) and immunolabeled for SP (**b**). Note that the subpallium is defined by intense SP immunolabeling. The *scale bar* in (**a**) provides the magnification for (**a**) and (**b**). Abbreviations: *DP* dorsal pallium, *LP* lateral pallium, *MP* medial pallium, *Sp* septum, and *Str* striatum

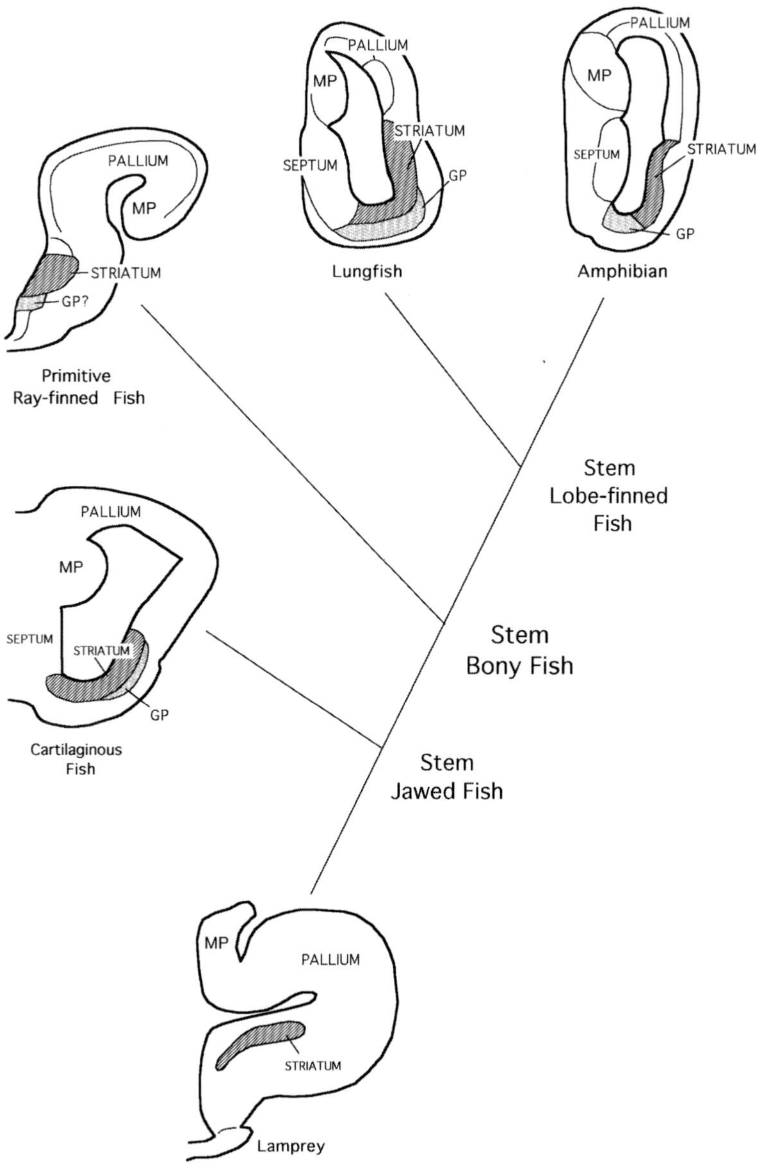

Fig. 7 Schematics of frontal sections through the basal ganglia of the right telencephalic hemisphere in representative species from five anamniote groups, a lamprey (jawless fish), a shark (cartilaginous fish), a polypterid (a primitive ray-finned bony fish), a lungfish (a lobe-finned bony fish), and a frog (amphibia), arranged according to their evolutionary divergences. The basal ganglia in all groups with an evaginated telencephalon other than lamprey (i.e., shark, lungfish, and frog) has been shown to consist of a striatum and a globus (GP) located in the basal telencephalon. Note that the pallium contains a medial (hippocampal) pallium (MP) in all amniote groups, with the hippocampal pallium located laterally in ray-finned fish due to their telencephalic eversion. While a striatum is evident in the ventral uneverted part of the telencephalon in ray-finned fish, a pallidum is not well-defined, but may reside below the striatum. Medial is to the *left* and dorsal to the *top* in all schematized sections

sector of the telencephalon in lobe-finned fish, and it is unlikely that it has a substantial projection to the striatum, since (like in amphibians) it largely lacks a dorsal pallium. In ray-finned fish, a striatum containing SP+ and ENK+ neurons likely to be GABAergic (Martinoli et al. 1990; Reiner and Northcutt 1992; Reiner et al. 1998a), having reciprocal connections with midbrain dopaminergic neurons (Reiner and Northcutt 1992; Reiner et al. 1998a; Rink and Wullimann 2001), and expressing a *Dlx1/2* homolog has been identified (Wullimann and Mueller 2004). Consistent with a dopaminergic input, the striatum in ray-finned fish is enriched in D1 and D2 dopamine receptors (Kapsimali et al. 2000; Vacher et al. 2003). A pallidal field may be present, since a part of the subpallium ventral to the striatal territory expresses an *Nkx2.1* homolog (Wullimann and Mueller 2004). The pallium in those advanced ray-finned fish with a large pallium (some teleosts) projects to striatum, while thalamic projections to the striatum appear to be present in all ray-finned fish (Northcutt 1981; Rink and Wullimann 2004).

Bone structure in living and extinct jawed fish and molecular data on living fish support a common origin of the two groups of living jawed fish, the cartilaginous fish (chondrichthyes) and bony fish (osteichthyes), from jawed fish with a bony skeleton (Hedges 2001). Cartilaginous fish are the most ancient living representatives of the jawed vertebrates, having appeared very early in the paleontological record, from which they are primarily known by their teeth (cartilage fossilizing poorly) (Hedges 2001). Cartilaginous fish are divided into two subclasses, the deep sea dwelling, primitive holocephalians (ratfish and chimeras), and the elasmobranchs (the sharks, the skates, and the rays) (Northcutt 1977). The elasmobranchs are divided into the sharks and batoids, with sharks consisting of the more primitive squalomorphs and the typically more advanced galeomorphs. Elasmobranchs are characterized by higher brain–body weight ratios than are the holocephalians, while batoids and galeomorph sharks commonly have higher brain–body weight ratios than do squalomorph sharks (Northcutt 1977). Like lobe-finned fish, cartilaginous fish possess paired tubular telencephalic hemispheres, and by neurochemical and hodological criteria the ventrolateral telencephalon contains both a striatum and a globus pallidus (Fig. 7) (Reiner and Carraway 1985; Northcutt et al. 1988; Reiner et al. 1998a). The striatal sector is located nearest the ventricle, and is cell sparse, but contains SP+ and ENK+ neurons that give rise to projections to a cell plate lying external to the striatal field (Fig. 7). At least part of this cell plate appears comparable to globus pallidus, because of this striatal input and because its neurons contain LANT6 (Northcutt et al. 1988; Reiner and Carraway 1985, 1987; Rodriguez-Moldes et al. 1993). In this pallidal field, the SP+ and ENK+ inputs overlap, implying that GPi-type and GPe-type neurons are intermingled. Moreover, the striatum receives a dopaminergic input from the midbrain and these dopaminergic neurons receive a return projection from SP+ striatal neurons (Meredith and Smeets 1987; Northcutt et al. 1988; Stuesse et al. 1994). The pallium occupies the dorsolateral telencephalon in cartilaginous fish, and its development is controlled by homologs of genes controlling pallial development in mammals (Derobert et al. 2002). The pallium, however, varies in size among cartilaginous fish species, and it is larger and more complex in more recently evolved cartilaginous fish (Northcutt

1981; Northcutt et al. 1988). Pathway tracing data are not available, but studies of axon tracts suggest that the dorsal pallium projects to the striatum in at least more advanced cartilaginous fish. In more advanced cartilaginous fish, the dorsal pallium also has been shown to receive sensory thalamic input (Northcutt 1981).

There are two extant groups of jawless fish, hagfish and lamprey. Modern taxonomic studies suggest lamprey and hagfish to be only distantly related, with lamprey being a sister group of jawed vertebrates (Forey and Janvier 1994). The differing telencephalic organization in these two agnathans groups reflects their taxonomic distance (Northcutt 1981; Wicht and Northcutt 1992, 1994). The telencephalon in lamprey is partly evaginated and possesses well-developed lateral ventricles, consistent with its taxonomic status as a sister group of jawed vertebrates. By contrast, the telencephalic hemispheres in hagfish are largely devoid of lateral ventricles, are highly laminated along their outer surfaces, and mainly receive olfactory bulb input. Neurochemical studies clearly identify a ventral telencephalic region rich in SP+ and GABAergic perikarya in lamprey (Fig. 7) (Nozaki and Gorbman 1986; Pombal et al. 1997b; Melendez-Ferro et al. 2002; Auclair et al. 2004). This region receives a dopaminergic input from the midbrain and has a return projection to these dopaminergic neurons (Pierre et al. 1994; Pombal et al. 1997a). This ventral telencephalic region also expresses lamprey *Dlx1/2*, while the region dorsal to it expresses a lamprey *Pax6* (Murakami et al. 2001; Neidert et al. 2001). Thus, this region appears to be the striatum. Although a globus pallidus in lamprey has not been demonstrated unequivocally (Nieuwenhuys and Nicholson 1998; Murakami et al. 2001), SP+ woolly fibers ventrolateral to the striatum in a field of GABAergic neurons within a region that has been called the ventral pallium delineate a field that may be pallidal (Pombal et al. 1997b). Alternatively, SP+ woolly fibers ventromedial to the striatum define a field that may be pallidal (Nozaki and Gorbman 1986). By contrast, SP and ENK immunolabeling fail to unequivocally identify a striatum or pallidum in hagfish (Wicht and Northcutt 1994). Although lamprey clearly possesses a striatum, the region is small, neuron sparse, and the midbrain dopaminergic input meager. Little is known regarding thalamopallial or palliostriatal pathways in lamprey.

3 By What Pathways Does the Basal Ganglia Control Movement in Nonmammals

Anatomical, physiological, and pharmacological data have led to circuit level models of mammalian basal ganglia function that relate the major aspects of basal ganglia cellular circuitry to normal and pathological basal ganglia function (Albin et al. 1989; Crossman 1990; DeLong 1990). In such models, striatopallidal output circuitry is recognized as being organized into two channels. One arises from substance P (SP)-containing GABAergic striatal neurons that project to the large GABAergic neurons of the internal segment of globus pallidus (GPi). The second output arises from enkephalinergic (ENK+) GABAergic striatal neurons that project to the large

GABAergic neurons of the external segment of globus pallidus (GPe). The GPe neurons have indirect outputs to GPi neurons via the subthalamic nucleus (STN), and for this reason the SP+ striatal projection to the GPi is called the direct pathway and the ENK+ striatal projection to GPi via the GPe-STN connection is called the indirect pathway. The GPi projects to motor thalamus and motor thalamus to the motor cortices. Activation of the direct pathway striatal neurons inhibits GPi neurons, thereby disinhibiting motor thalamus and thereby yielding movement. It is presumed that this circuit underlies the role of the mammalian basal ganglia in promoting desired movement. A second striatal output pathway promoting desired movement also arises from SP+ striatal neurons, which in this case project to the GABAergic pallidal-like neurons of the substantia nigra pars reticulata (SNr). These SNr neurons influence movement by projections to medial motor thalamus and by projections to deep superior colliculus neurons with descending projections to the hindbrain. The striato-SNr circuit is thus also regarded as part of the direct pathway. The striato-GPi circuit appears to promote limb and trunk movements and the striato-SNr circuit appears to influence head and eye movements. The indirect pathway in mammals exerts an excitatory influence on the GPi and the SNr (whose neurons are inhibitory), which serves to enhance their suppression of unwanted movements.

The avian basal ganglia have its major output to motor areas via a projection to a pretectal cell group called the lateral spiriform nucleus and the SNr, both of which have input to tectal neurons with descending premotor and motor projections (Figs. 8 and 9) (Reiner et al. 1998a; Jiao et al. 2000). These circuits are likely to mainly affect head and eye movements. Additionally, birds possess a correspondent of the mammalian striato–GPi–motor thalamus circuit to motor cortex (Medina et al. 1997; Medina and Reiner 2000). While motor cortex and thalamus are small in pigeons, this circuit may be better developed in avian groups with a larger telencephalon and motor cortex (e.g., owls). The striatal output to the pretectum, the SNr, and the GPi are all "direct pathway" type outputs that appear to facilitate behavior (Reiner et al. 1998a). Birds have also been shown to possess a STN and "indirect pathway" circuitry, as well (Figs. 8 and 9) (Jiao et al. 2000). The avian basal ganglia are thus organized into direct and indirect striatal output circuits resembling those in mammals. Of note is the observation that while birds

Fig. 8 (continued) immuno labeled for TH, showing a dopaminergic terminal contacting the neck of spine (S) that also receives an asymmetric synaptic contact at its tip from an excitatory type terminal containing round vesicles. Image (**e**) shows a line drawing of a coronal section through pigeon diencephalon illustrating the location of the avian subthalamus (STN) and the avian motor thalamus, called the ventrointermediate area (VIA). Image (f) shows a transverse section through the pigeon pretectum of the right side of the brain immunolabeled for ENK. Note that the neurons of SpL are rich in ENK. *Scale bar* in (**a**) provides magnification for (**a**), (**b**), and (**e**). Abbreviations: *AL* ansa lenticularis, *CO* optic chiasm, *DTZ* dorsal thalamic zone, *FPL* lateral forebrain bundle, *GLd* dorsal lateral geniculate nucleus, *GLv* ventral lateral geniculate nucleus, *Hy* hypothalamus, *QF* quintofrontal tract, *Rt* nucleus rotundus, *SMe* stria medularis, *SNc* substantia nigra pars compacta, *SNr* substantia nigra pars reticulata, *SpL* lateral spiriform nucleus, *STN* subthalamic nucleus, *TeO* optic tectum, *TO* optic tract, *TRN* thalamic reticular nucleus, and *VIA* ventrointermediate area of thalamus

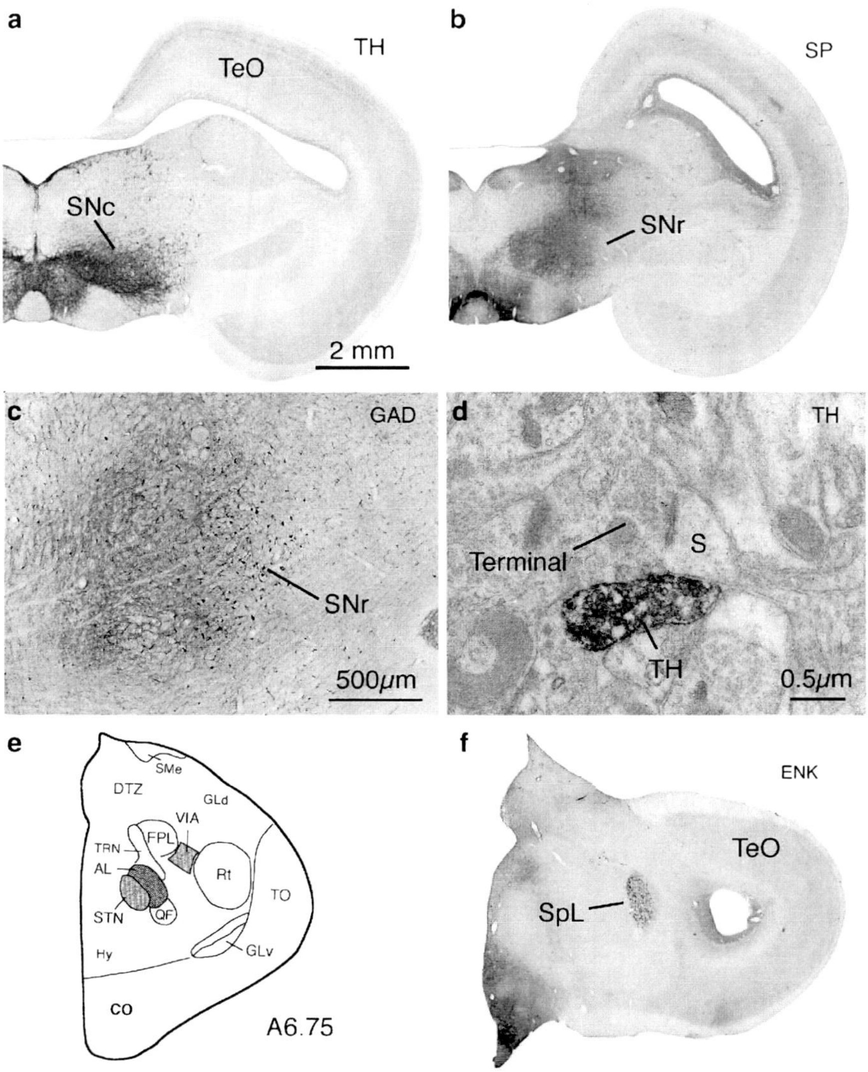

Fig. 8 Images of pigeon brain illustrating tyrosine hydroxylase-containing neurons (**a**) in substantia nigra pars compacta (adapted from Fig. 4f from Reiner et al. 2004), SP+ fibers (**b**) in the substantia nigra, GAD+ neurons (**c**) in the substantia nigra pars reticulata (adapted from Fig. 19e from Veenman and Reiner 1994), a dopaminergic terminal (**d**) in the striatum, the location (**e**) of the subthalamic nucleus and motor thalamus, and ENK+ immunolabeling (**f**) of the lateral spiriform nucleus. Image (**a**) shows a coronal section through right midbrain of pigeon immunostained for tyrosine hydroxylase (TH), illustrating the location of the dopaminergic neurons of the substantia nigra pars compacta (SNc). Image (**b**) shows a coronal section through right midbrain of pigeon immunostained for SP, illustrating the location of the terminals of the SP+ striatal projection to the substantia nigra. Image (**c**) shows a high power view of a transverse section through the right pigeon midbrain that had been immunolabeled for GAD. Note the GABAergic neurons of the substantia nigra pars reticulata (SNr), and the GAD+ striatal terminals in the substantia nigra medial to the SNr. Image (**d**) presents an EM view of a striatal section

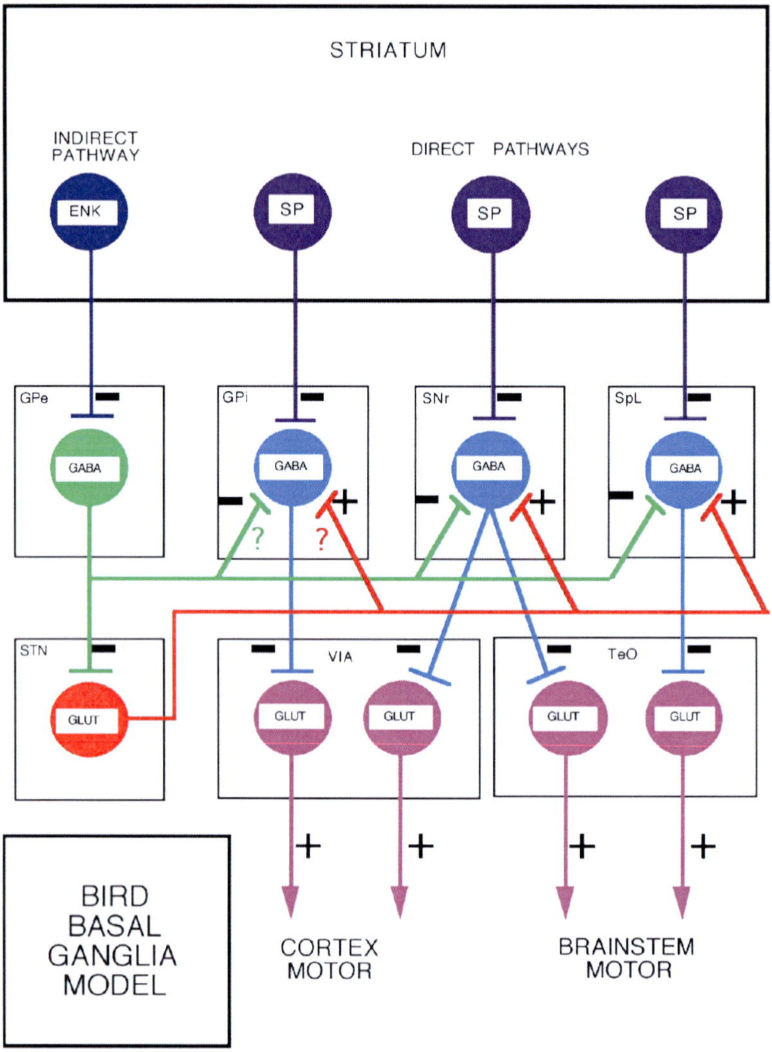

Fig. 9 Circuit diagram illustrating the functional organization of the basal ganglia in birds. The pluses and minuses indicate whether specific projections use an excitatory (*plus*) or inhibitory (*minus*) neurotransmitter. Striatal projection neuron types, which all use GABA as their primary neurotransmitter, are distinguished by their characteristic neuropeptide. As in mammals, the striatal and pallidal output circuitry in birds is organized into direct SP+ striatal outputs to pallidal neurons promoting movement and ENK+ striatal outputs to pallidal neurons inhibiting unwanted movement. The pallidal neurons of the indirect pathway project directly to the targets of the SP+ striatal neurons (i.e., GPi, SNr, and SpL) as well indirectly to these same targets via the subthalamic nucleus. In mammals, SP+ neurons target two sets of pallidal-type neurons (GPi and SNr), while in birds three sets are targeted (GPi, SNr and SpL). It is not yet certain, however, if GPe-like neurons in the avian globus pallidus (where they are intermingled with GPi-type neurons) have a projection to GPi-type neurons of globus pallidus. Such a projection is present in mammals. The striatum in birds receives a massive dopaminergic input from the midbrain, an excitatory glutamatergic input from thalamus, and an excitatory glutamatergic input from the pallium. Abbreviations: *ENK* enkephalin, *GLUT* glutamate, *GPe* external segment of globus pallidus, *GPi* internal segment of globus pallidus, *SNr* substantia nigra pars reticulata, *SP* substance P; *SpL* nucleus spiriformis lateralis, and *STN* subthalamic nucleus

are the most advanced nonmammalian group in terms of the size of their telencephalic pallium (some birds even surpassing some mammals), avian basal ganglia nonetheless have their major influence on movement via outputs to the midbrain tectum rather than telencephalic motor cortex.

The basal ganglia in reptiles also have its major output to motor areas via a projection to the pretectum and to the SNr, which are likely to affect head and eye movements by an input to tectal neurons with descending projections to brainstem premotor cell groups (Reiner et al. 1980, 1998a; Medina and Smeets 1991). The existence in both birds and mammals of a striato–GPi–motor thalamus–motor cortex loop and the demonstrated existence in reptiles of SP+ striatal neurons, a GPi, and a telencephalic motor cortex suggest that the striato–GPi–motor thalamus–motor cortex circuit was present in stem amniotes. Moreover, the presence of SP+ striatal outputs to the GPi and SNr and ENK+ striatal outputs to the GPe in reptiles and the existence of an apparent STN suggest that the overall direct-indirect pathway plan of basal ganglia functional organization was present already in stem amniotes. Again, as in birds, the major route for motor facilitation in reptiles appears to be via the midbrain tectum rather than motor cortex, and it can be inferred that this must also have been so in stem amniotes. Note that mammalian striatum lacks an evident output to the pretectum, which appears to have been lost in the lineage from stem amniotes to mammals.

In amphibians, as well, the basal ganglia appear to have their major output to motor areas via a projection to the pretectum and to the SNr, both of which affect head and eye movements by an input to tectal neurons with descending projections (Wilczynski and Northcutt 1983; Marin et al. 1998b; Reiner et al. 1998a). This is consistent with the poor development of thalamopallial and palliostriatal pathways in amphibians. Although pathway tracing data are meager, the high similarity of the morphology of the lungfish telencephalon to that in amphibians makes it likely that in lungfish as well the basal ganglia have their main impact on motor control by means of the midbrain tectum, via a projection of the striatum to the pretectum and/or SNr. Since the morphology and neurochemistry of the telencephalon in basal cartilaginous fish resembles that in amphibians and lungfish, it is likely that this is true in cartilaginous fish as well, and has been true since the appearance of jawed vertebrates. Further studies are needed to ascertain that if this is so for lamprey as well. In any event, the existing data make it clear that the basal ganglia have been a core constituent of the telencephalon since the earliest jawed vertebrates and the output of the basal ganglia to the SNr represents the ancient route by which the basal ganglia affect movements. An additional ancient circuit to the pretectum appears to have been lost in the mammalian lineage with the de-emphasis of the pretectum in mammals.

4 Major Evolutionary Steps in Basal Ganglia Organization

Basal ganglia evolution seems characterized by an increase in neuron number as the telencephalon expanded during the anamniote–amniote transition, resulting in an elaboration of cortical glutamatergic inputs and midbrain dopaminergic inputs, and an increased role for telencephalic circuitry in motor control in stem amniotes (Figs. 5 and 7). Nonetheless, the basic direct-indirect pathway circuit plan by which the basal ganglia regulate movement is likely to have already been in place in early anamniotes. A few major changes in telencephalic morphology appear to have occurred in the evolution of the mammalian basal ganglia from the inferred stem amniote condition. First, from a simple, poorly laminated state, the pallium evolved into the multilayered neocortex that came to wrap around the basal ganglia (Karten 1969; Reiner 2000). As a consequence, the basal ganglia occupy a more central position in the telencephalon than they do in nonmammals (Fig. 2). The development of the basal ganglia vs. the neocortex, however, reflects its ancestral basal position. The globus pallidus develops from an *Nkx2.1*- and *Dlx1/2*-expressing bulge [called the medial ganglionic eminence (MGE)] at the lower aspect of the lateral telencephalic wall, while the striatum develops from *Dlx1/2*-expressing bulge that does not express *Nkx2.1* just above the MGE [called the lateral ganglionic eminence (LGE)]. Unlike in nonmammals, in which GPi and GPe neurons are intermingled, in mammals these pallidal populations occupy separate sectors. There appear to be no noteworthy differences between mammals and sauropsids in the major types of neurons making up the striatum and the pallidum. In all groups, the striatum consists of two neurochemically distinct types of neurons, the medium-sized spiny SP+ and ENK+ GABAergic neurons, while the globus pallidus consists of large GABAergic neurons (Graybiel 1990; Gerfen 1992; Reiner et al. 1998a). In all amniotes, massive glutamatergic inputs to striatum arise from cortex and thalamus, and a massive dopaminergic input to striatum arises from the substantia nigra pars compacta (Gerfen 1992; Reiner et al. 1998a), and the general role of the basal ganglia in motor learning and control seems similar. Mammals do differ from birds and reptiles in that they lack an obvious correspondent of the basal ganglia output to midbrain via pretectum (Reiner et al. 1998a), and their output to motor cortices via the thalamus seems instead more prominently developed, especially in primates (Reiner et al. 1998a).

Acknowledgments I thank my various collaborators over the years who have made the parts of our research presented here possible. I also particularly thank Drs. Harvey Karten, Glenn Northcutt, Steve Brauth, Loreta Medina, and Ann Butler for their many stimulating conversations and ideas about basal ganglia evolution. Any of our work summarized here was supported by NS-19620 (A.R.), NS-28721 (A.R.), EY-05298 (A.R.), the Neuroscience Center for Excellence of the University of Tennessee, Memphis (L.M.), and the Spanish Ministry of Education and Science (L.M.), to whom we are grateful.

References

Albin RL, Young AB and Penney JB (1989) The functional anatomy of basal ganglia disorders. Trends Neurosci 12: 366–375.

Ariëns-Kappers CU, Huber GC and Crosby E (1936) The Comparative Anatomy of the Nervous System of Vertebrates, Including Man. Hafner Press, New York, NY.

Auclair F, Lund JP and Dubuc R (2004) Immunohistochemical distribution of tachykinins in the CNS of the lamprey *Petromyzon marinus*. J Comp Neurol 479: 328–346.

Bennis M, Calas A, Geffard M and Gamrani H (1991) Distribution of GABA immunoreactive systems in the forebrain and midbrain of the chameleon. Brain Res Bull 26: 891–898.

Brox A, Puelles L, Ferreiro B and Medina L (2003) Expression of the genes *GAD67* and *Distal-less-4* in the forebrain of *Xenopus laevis* confirms a common pattern in tetrapods. J Comp Neurol 461: 370–393.

Carlsson A (1959) Occurrence, distribution and physiological role of catecholamines in the nervous system. Pharmacol Rev 11: 490–493.

Carman JB, Cowan WM and Powell TPS (1963) The organization of the corticostriate connexions in the rabbit. Brain 86: 525–562.

Chiappe LM (1995) The first 85 million years of avian evolution. Nature 378: 349–355.

Crossman AR (1990) A hypothesis on the pathophysiological mechanisms that underlie levodopa- or dopamine agonist-induced dyskinesia in Parkinson's disease: Implications for future strategies in treatment. Mov Disord 5: 100–108.

DeLong MR (1990) Primate models of movement disorders of basal ganglia origin. Trends Neurosci 13: 281–285.

Derobert Y, Plouhinec JL, Sauka-Spengler T, Le Mentec C, Baratte B, Jaillard D and Mazan S (2002) Structure and expression of three *Emx* genes in the Dogfish *Scyliorhynus canicula*: Functional and evolutionary implications. Dev Biol 247: 390–404.

Edinger L, Wallenberg A and Holmes GM (1903) Untersuchungen über die vergleichende Anatomie des Gehirns. 3. Das Vorderhirn der Vögel. Abhand l.d. Senekenberate Gesellschaft, Frankfurt Am Main 20: 343–426.

Forey P and Janvier P (1994) Evolution of the early vertebrates. Sci Am 82: 554–565.

Fowler M, Medina L and Reiner A (1999) Immunohistochemical localization of NMDA and AMPA type glutamate receptor subunits in the basal ganglia of red-eared turtles. Brain Behav Evol 54: 276–289.

Gerfen CR (1992). The neostriatal mosaic: Multiple levels of compartmental organization in the basal ganglia. Ann Rev Neurosci 15: 285–320.

González, A and Smeets WJAJ (1991) Comparative analysis of dopamine and tyrosine hydroxylase immunoreactivities in the brain of two amphibians, the anuran *Rana ridibunda* and the urodele *Pleurodeles waltlii*. J Comp Neurol 303: 457–477.

Gonzalez A and Smeets WJ (1994) Catecholamine systems in the CNS of amphibians. In: WJ Smeets and A Reiner (eds) Phylogeny and Development of Catecholamine Systems in the CNS of Vertebrates. Cambridge University Press, Cambridge, pp. 77–102.

Gonzalez A, Lopez JM, Sanchez-Camacho C and Marin O (2002) Regional expression of the homeobox gene *Knx2.1* defines pallidal and interneuronal populations in the basal ganglia of amphibians. Neuroscience 114: 567–575.

Graybiel AM (1990) Neurotransmitters and neuromodulators in the basal ganglia. Trends Neurosci 13: 244–254.

Hedges SB (2001) Molecular evidence for the early history of living vertebrates. In: PE Ahlberg (ed.) Major Events in Early Vertebrate Evolution: Palaeontology, Phylogeny, Genetics and Development. Taylor & Francis, London, pp. 119–134.

Hedges SB, Hass CA and Maxson LR (1993) Relations of fish and tetrapods. Nature 363: 501–502.

Herrick CJ (1948) The Brain of the Tiger Salamander. The University of Chicago Press, Chicago, IL.

Jiao Y, Medina L, Veenman CL, Toledo C, Puelles L and Reiner A (2000) Identification of the anterior nucleus of the ansa lenticularis in birds as the homolog of the mammalian subthalamic nucleus. J Neurosci 20: 6998–7010.

Kapsimali M, Vidal B, Gonzalez A, Dufour S and Vernier P (2000) Distribution of the mRNA encoding the four dopamine D1 receptor subtypes in the brain of the European eel (*Anguilla anguilla*): Comparative approach to the function of D1 receptors in vertebrates. J Comp Neurol 419: 320–343.

Karten HJ (1969) The organization of the avian telencephalon and some speculations on the phylogeny of the amniote telencephalon. In: J Petras (ed) Comparative and Evolutionary Aspects of the Vertebrate Central Nervous System. Ann NY Acad Sci 167: 146–179.

Karten HJ and Dubbeldam JL (1973) The organization and projections of the paleostriatal complex in the pigeon (*Columba livia*). J Comp Neurol 148: 61–90.

Laverghetta AV, Toledo C, Veenman CL and Reiner A (2006) Cellular localization of AMPA-type glutamate receptor subunits in the basal ganglia of pigeons (*Columba livia*). Brain Behav Evol 67: 10–38.

Marin O, Smeets WJAJ and Gonzalez A (1997) Distribution of choline acetyltransferase immunoreactivity in the brain of Anuran (*Rana perezi*, *Xenopus laevis*) and Urodele (*Pleurodeles waltl*) amphibians. J Comp Neurol 382: 499–534.

Marin O, Smeets WJAJ and Gonzalez A (1998a) Basal ganglia organization in amphibians: Chemoarchitecture. J Comp Neurol 392: 285–312.

Marin O, Smeets WJ and Gonzalez A (1998b) Evolution of the basal ganglia in tetrapods: A new perspective based on recent studies in amphibians. Trends Neurosci 21: 487–494.

Martinoli MG, Dubourg P, Gefard M, Calas A and Kah O (1990) Distribution of GABA-immunoreactive neurons in the forebrain of the goldfish, *Carassius auratus*. Cell Tissue Res 260: 77–84.

Medina L and Reiner A (2000) Do birds possess homologues of mammalian primary visual, somatosensory and motor cortices? Trends Neurosci 23: 1–12.

Medina L and Smeets WJAJ (1991) Comparative aspects of the basal ganglia–tectal pathways in reptiles. J Comp Neurol 308: 614–629.

Medina L, Veenman CL and Reiner A (1997) Evidence for a possible avian dorsal thalamic region comparable to the mammalian ventral anterior, ventral lateral, and oral ventroposterolateral nuclei. J Comp Neurol 384: 86–108.

Melendez-Ferro M, Perez-Costas E, Villar-Cheda B, Abalo XM, Rodriguez-Munoz R, Rodicio MC and Anadon R (2002) Ontogeny of gamma-aminobutyric acid-immunoreactive neuronal populations in the forebrain and midbrain of the sea lamprey. J Comp Neurol 446: 360–376.

Meredith GE and Smeets WJAJ (1987) Immunocytochemical analysis of the dopamine system in the forebrain and midbrain of *Raja radiata*: Evidence for a substantia nigra and ventral tegmental area in cartilaginous fish. J Comp Neurol 265: 530–548.

Murakami Y, Ogasawara M, Sugahara F, Hirano S, Satoh and Kuratani S (2001) Identification and expression of the lamprey Pax6 gene: Evolutionary origin of the segmented brain of vertebrates. Development 128: 3521–3531.

Nauta WJH and Mehler WR (1966) Projections of the lentiform nucleus in the monkey. Brain Res 1: 3–42.

Neidert AH, Virupannavar V, Hooker GW and Langeland JA (2001) Lamprey Dlx genes and early vertebrate evolution. Proc Nat Acad Sci USA 98: 1665–1670.

Nieuwenhuys R and Nicholson C (1998) Lampreys, Petromyzontoidea. In: R Nieuwenhuys, HJ ten Donkelaar and C Nicholson (eds) The Central Nervous System of Vertebrates. Springer, Berlin, pp 397–495.

Northcutt RG (1977) Elasmobranch central nervous system organization and its possible evolutionary significance. Am Zool 17: 411–429.

Northcutt RG (1981) Evolution of the telencephalon in nonmammals. Ann Rev Neurosci 4: 301–350.

Northcutt RG and Kicliter E (1980) Organization of the amphibian telencephalon. In: SOE Ebbesson (ed) Comparative Neurology of the Telencephalon. Plenum, New York, NY, pp. 203–255.

Northcutt RG, Reiner A and Karten HJ (1988) Immunohistochemical study of the telencephalon of the spiny dogfish, *Squalus acanthias*. J Comp Neurol 277: 250–267.

Nozaki M and Gorbman A (1986) Occurrence and distribution of substance P-related immunoreactivity in the brain of adult lampreys, *Petromyzon marinus* and *Entosphenus tridentatus*. Gen Comp Endocrinol 62: 217–229.

Parent A (1986) Comparative Neurobiology of the Basal Ganglia. Wiley, New York, NY.

Pierre J, Rio JP, Mahouche M and Repérant J (1994) Catecholamine systems in the brain of cyclostomes, the lamprey, *Lampreta fluviatilis*. In: WJ Smeets and A Reiner (eds) Phylogeny and Development of Catecholamine Systems in the CNS of Vertebrates. Cambridge University Press, Cambridge, pp. 7–19.

Pombal MA, El Manira A and Grillner S (1997a). Afferents of the lamprey striatum with special reference to the dopaminergic system: A combined tracing and immunohistochemical study. J Comp Neurol 386: 71–91.

Pombal MA, El Manira A and Grillner S (1997b) Organization of the lamprey striatum – Transmitters and projections. Brain Res 766: 249–254.

Puelles L, Kuwana E, Puelles E, Bulfone A, Shimamura K, Keleher J, Smiga S and Rubenstein JL (2000) Pallial and subpallial derivatives in the embryonic chick and mouse telencephalon, traced by the expression of the genes *Dlx-2, Emx-1, Nkx-2.1, Pax-6,* and *Tbr-1*. J Comp Neurol 424: 409–438.

Reiner A (2000) A hypothesis as to the organization of cerebral cortex in the common reptile ancestor of modern reptiles and mammals. In: GA Bock and G Cardew (eds) Evolutionary Developmental Biology of the Cerebral Cortex. Novartis, London, pp. 83–108.

Reiner A and Anderson KD (1990) The patterns of neurotransmitter and neuropeptide co-occurrence among striatal projection neurons: Conclusions based on recent findings. Brain Res Rev 15: 251–265.

Reiner A and Carraway RE (1985) Phylogenetic conservatism in the presence of a neurotensin-related hexapeptide in neurons of globus pallidus. Brain Res 341: 365–371.

Reiner A and Carraway RE (1987) Immunohistochemical and biochemical studies on Lys^{8}-Asn^{9}-neurotensin$^{8-13}$ (LANT6)-related peptides in the basal ganglia of pigeons, turtles, and hamsters. J Comp Neurol 257: 453–476.

Reiner A and Northcutt RG (1987) An immunohistochemical study of the telencephalon of the African lungfish, *Protopterus annectens*. J Comp Neurol 256: 463–481.

Reiner A and Northcutt RG (1992) An immunohistochemical study of the telencephalon of the Senegal bichir (*Polypterus senegalus*). J Comp Neurol 319: 359–386.

Reiner A, Brauth SE, Kitt CA and Karten HJ (1980) Basal ganglionic pathways to the tectum: Studies in reptiles. J Comp Neurol 193: 565–589.

Reiner A, Karle EJ, Anderson KD and Medina L (1994) Catecholaminergic perikarya and fivers in the avian nervous system. In: WJ Smeets and A Reiner (eds) Phylogeny and Development of Catecholamine Systems in the CNS of Vertebrates. Cambridge University Press, Cambridge, pp. 135–181.

Reiner A, Medina L and Veenman CL (1998a) Structural and functional evolution of the basal ganglia in vertebrates. Brain Res Rev 28: 235–285.

Reiner A, Perera M, Paullus R and Medina L (1998b) Immunohistochemical localization of DARPP-32 in striatal projection neurons and striatal interneurons in pigeons. J Chem Neuroanat 16: 17–33.

Reiner A, Perkel DJ, Bruce LL, Butler A, Csillag A, Kuenzel W, Medina L, Paxinos G, Shimizu T, Striedter G, Wild M, Ball GF, Durand S, Güntürkün O, Lee D, Mello CV, Powers A, White SA, Hough G, Kubikova L, Smulders TV, Wada K, Dugas-Ford J, Husband S, Yamamoto K, Yu J, Siang C and Jarvis ED (2004) Revised nomenclature for avian telencephalon and some related brainstem nuclei. J Comp Neurol 473: 377–414.

Richfield EK, Young AB and Penney JB (1987) Comparative distribution of D-1 and D-2 receptors in the basal ganglia of turtles, pigeons, rats, cats, and monkeys. J Comp Neurol 262: 446–463.

Rink E and Wullimann RF (2001) The teleostean (zebrafish) dopaminergic system ascending to the subpallium (striatum) is located in the basal diencephalon (posterior tubercle). Brain Res 889: 316–330.

Rink E and Wullimann RF (2004) Connections of the ventral telencephalon (subpallium) in the zebrafish (*Danio rerio*). Brain Res 1011: 206–220.

Rodriguez-Moldes I, Manso MJ, Becerra M, Molist P and Anadon R (1993) Distribution of substance P-like immunoreactivity in the brain of the elasmobranch *Scyliorhinus canicula*. J Comp Neurol 335: 228–244.

Rubenstein JLR, Martinez S, Shimamura K and Puelles L (1994) The embryonic vertebrate forebrain: The prosomeric model. Science 266: 578–580.

Smith-Fernandez A, Pieau C, Reperant J, Boncinelli E and Wassef M (1998) Expression of the Emx-1 and Dlx-1 homeobox genes define three molecularly distinct domains in the telencephalon of mouse, chick, turtle and frog embryos: Implications for the evolution of telencephalic subdivisions in amniotes. Development 125: 2099–2111.

Stuesse SL, Cruce WLR and Northcutt RG (1994) Localization of catecholamines in the brains of Chondrichthyes (cartilaginous fishes). In: WJ Smeets and A Reiner (eds) Phylogeny and Development of Catecholamine Systems in the CNS of Vertebrates. Cambridge University Press, Cambridge, UK, pp. 21–47.

Sun Z and Reiner A (2000) Localization of dopamine D1A and D1B receptor mRNAs in the forebrain and midbrain of the domestic chick. J Chem Neuroanat 19: 211–224.

Sun Z, Wang HB, Laverghetta AV, Yamamoto K and Reiner A (2005) The distribution and cellular localization of glutamic acid decarboxylase-65 (GAD65) mRNA in the forebrain and midbrain of domestic chick. J Chem Neuroanat 29: 265–281.

Swanson LW and Petrovich GD (1998). What is the amygdala? Trends Neurosci 21: 323–331.

Trabucchi M, Trudeau VL, Drouin G, Tostivint H, Ihrmann I, Vallarino M and Vaudry H (2008) Molecular characterization and comparative localization of the mRNAs encoding two glutamic acid decarboxylases (GAD65 and GAD67) in the brain of the African lungfish, *Protopterus annectens*. J Comp Neurol 506: 979–988.

Vacher C, Pellegrini E, Anglade I, Ferriere F, Saligaut C and Kah O (2003) Distribution of dopamine D2 receptor mRNAs in the brain and the pituitary of female rainbow trout: An *in situ* hybridization study. J Comp Neurol 458: 32–45.

Veenman CL and Reiner A (1994) The distribution of GABA-containing perikarya, fibers, and terminals in the forebrain and midbrain of pigeons, with particular reference to the basal ganglia and its projection targets. J Comp Neurol 339: 209–250.

Veenman CL and Reiner A (1996) Ultrastructural morphology of synapses formed by corticostriatal terminals in the avian striatum. Brain Res 707: 1–12.

Veenman CL, Karle EJ, Anderson KD and Reiner A (1995a) Thalamostriatal projections neurons in birds utilize LANT6 and neurotensin: A light and electron microscopic double-labeling study. J Chem Neuroanat 9: 1–16.

Veenman CL, Wild JM and Reiner A (1995b) Organization of the avian "Corticostriatal" projection system: A retrograde and anterograde pathway tracing study in pigeons. J Comp Neurol 354: 87–126.

Veenman CL, Medina L and Reiner A (1997) The avian homologues of the mammalian intralaminar, mediodorsal and midline thalamic nuclei: Immunohistochemical and hodological evidence. Brain Behav Evol 49: 78–98.

Wicht H and Northcutt RG (1992) The forebrain of hagfish: A cladistic reconstruction of the ancestral craniate forebrain. Brain Behav Evol 40: 25–64.

Wicht H and Northcutt RG (1994) An immunohistochemical study of the telencephalon and the diencephalon in a myxinoid jawless fish, *Eptatretus stouti*. Brain Behav Evol 43: 140–161.

Wilczynski W and Northcutt RG (1983) Connections of the bullfrog striatum: Afferent organization. J Comp Neurol 214: 321–332.

Wild JM (1987) Thalamic projections to the paleostriatum and neostriatum in the pigeon (*Columba livia*). Neuroscience 20: 305–327.

Wullimann MF and Mueller T (2004) Teleostean and mammalian forebrains contrasted: Evidence from genes to behavior. J Comp Neurol 475: 143–162.

The Involvement of Corticostriatal Loops in Learning Across Tasks, Species, and Methodologies

Carol A. Seger

Abstract The basal ganglia contribute to a variety of forms of learning. The first goal of this chapter is to review the different tasks (instrumental conditioning, visual discrimination, arbitrary visuomotor learning, rule learning, categorization, and decision making) that have been used to study basal-ganglia-dependent learning in rodents, monkeys, and humans. These tasks have several features in common: in each, the subject is first presented with a stimulus within a behavioral context, is then required to respond with an appropriate behavior, and finally receives a reward or positive feedback for correct behavior. The second goal of this chapter is to examine how these different features (stimulus, response, and reward) involve the independent corticostriatal loops that connect the basal ganglia with cerebral cortex. The visual corticostriatal loop is involved in aspects of visual stimulus processing; the motor corticostriatal loop is involved in response selection; and the executive and motivation corticostriatal loops are involved in processing feedback and reward. The chapter concludes with a discussion of how the corticostriatal loops interact during learning.

1 Introduction

Recent years have seen an increased appreciation of the contributions made by the basal ganglia to cognitive domains other than motor processing. In particular, the basal ganglia are involved in a wide variety of learning tasks which have in common the requirement for the subject to learn via trial and error to associate a given stimulus or experimental context with a given response or behavior. In this chapter, I will first describe these tasks individually, emphasizing the features they have in common. I will then discuss the different corticostriatal loops and how each loop may subserve some of the features common to basal-ganglia-dependent learning tasks.

C.A. Seger (✉)
Department of Psychology, Colorado State University, Fort Collins, CO 80523, USA
e-mail: seger@lamar.colostate.edu

H.J. Groenewegen et al. (eds.), *The Basal Ganglia IX*, Advances in Behavioral Biology 58,
DOI 10.1007/978-1-4419-0340-2_2, © Springer Science+Business Media, LLC 2009

2 Basal-Ganglia-Dependent Learning Tasks

In this section, I will describe several of the commonly used tasks developed to study basal-ganglia-dependent learning in humans, monkeys, and/or rodents. Although these tasks are basal ganglia dependent, task performance does not rely on only the basal ganglia; instead, the basal ganglia interact with cortex and other neural systems in performing the behavior. Cortical contributions to performance will be mentioned here when they are particularly prominent, but space considerations preclude a full discussion of cortical contributions to each task.

The tasks discussed here are used along with multiple experimental methodologies. In nonhuman animal models the most prominent methodologies include experimental lesion, chemical manipulation, and electrophysiological recording of individual neurons. In human models, the most commonly used methodologies are studies of patients with neuropsychological disorders of the basal ganglia (e.g., Parkinson's disease), and functional brain imaging using functional magnetic resonance imaging (fMRI) or positron emission tomography (PET). Due to the depth of the basal ganglia within the head and the cellular anatomy of the basal ganglia, there are many research methods that cannot currently be used to study the basal ganglia, including electroencephalography, magnetoencephalography, optical imaging, and transcranial magnetic stimulation.

Across species and methodologies, learning tasks that recruit the basal ganglia typically involve three common features: presentation of a stimulus within a behavioral context, an overt motor response, and reward or feedback. First, the subject is placed into a particular behavioral context (e.g., a rat may enter a Skinner box; a human may enter the laboratory and be seated in front of a computer monitor), and is faced with particular stimuli (e.g., a lever; or a visual image on the computer screen). Second, the subject must decide what the correct behavior is and then execute the behavior (e.g., by pressing a lever; or a key on a computer keyboard). Third, the subject is given feedback or a reward indicating whether he or she responded correctly (e.g., a drop of fruit juice; or the word "Correct!" on the computer screen). Tasks differ in terms of which of these features are manipulated by the researcher and are thus considered to be of primary importance. For example, in visual discrimination learning the visual stimuli are manipulated; in arbitrary visuomotor learning, the particular motor response performed by the subject is manipulated. Nevertheless, both these tasks include all three features: the subject is presented with a stimulus, performs a response, and receives feedback.

2.1 *Tasks from the Rodent Literature*

2.1.1 Instrumental Conditioning

In instrumental conditioning, subjects learn to perform desired behaviors in response to cues in order to receive rewards. These tasks are most commonly used with rodents. A common experimental context is provided by a maze (such as a T maze, plus maze, or the Morris water maze); when placed in the maze, the subject must learn to perform the same behavior on each trial (e.g., turn right, or go to a visually cued location). The striatum has been shown to be involved in these tasks in experiments involving induced lesions of the basal ganglia (Packard et al. 1989; Packard and McGaugh 1992), chemical manipulation of neural activity and neural plasticity (Atallah et al. 2007; Brightwell et al. 2008), and electrophysiological recording of striatal neurons (Jog et al. 1999; Barnes et al. 2005).

Modern learning theory has identified two forms of instrumental conditioning: goal-directed learning, sometimes called Action–Outcome–Reward (A–O–R) learning, and habit learning, sometimes called S–R learning (Yin and Knowlton 2006; Balleine et al. 2007). Goal-directed learning tends to predominate early in training. It is reliant on the dorsomedial striatum in rodents, which is roughly homologous to the head of the caudate nucleus/anterior putamen in primates. As training progresses, goal-directed behavior shifts to habit learning. The main distinction between habit learning and goal-directed learning is that habit is resistant to incentive devaluation. For example, an animal might be trained to run a maze in order to receive water. Water can be devalued as an incentive by giving the animal ad libitum water before the session. Goal-directed responding will be impaired by incentive devaluation, whereas habitual responding is not. Habit learning is reliant on the dorsolateral striatum in rodents, which is roughly equivalent to the posterior putamen in primates. It should be noted that the use of the term "habit learning" is inconsistent across researchers and tasks: some reserve the term habit for only S–R learning, whereas others use it for any learning of repeated behavior in response to a situation.

2.2 *Tasks from the Monkey Literature*

2.2.1 Visual Discrimination

In this task the subject views two visual stimuli or objects, and has to choose one of them, typically by touching it on a computer screen (Gaffan and Eacott 1995), grasping the object, or uncovering an adjacent food well (Teng et al. 2000; Fernandez-Ruiz et al. 2001). The subject learns across trials to choose the item that is rewarded. This task is typically used with macaque monkey, which has a better developed visual system than rodents. This task was developed primarily to answer questions about visual memory and visual form processing, but has also been considered to be a monkey test of habit learning. Visual discrimination learning is often studied in conjunction with visual cortical regions. Monkeys with lesions to either the striatum or the visual cortex are impaired in visual discrimination learning (Gaffan and Eacott 1995); in addition, striatal neurons are active when performing this task (Brown et al. 1995).

2.2.2 Arbitrary Visuomotor Learning/Conditional Response

In this task the subject views a single stimulus which serves as a cue to perform a motor action; correct performance is followed by a reward. The relationship between the visual cue and the response is arbitrary (e.g., the visual stimulus does not include any information about the desired direction of movement). The emphasis in this task is on the motor action, and on the association between cue and action. Arbitrary visuomotor learning studies in monkeys have shown that both acquisition (Wise and Murray 2000) and performance of previously learned relationships (Nixon et al. 2004) are impaired by striatal lesions. Electrophysiological studies find that striatal cell activity increases during learning (Hadj-Bouziane and Boussaoud 2003; Brasted and Wise 2004). In humans, functional imaging studies of the arbitrary visuomotor association task have also found striatal activation (Toni et al. 2002; Grol et al. 2006). Arbitrary visuomotor learning often is accompanied by activation of premotor and motor cortex.

2.2.3 Rule Learning

The term rule learning is used for at least two types of experiment. One type stems from the monkey learning literature and is very similar to arbitrary visuomotor learning. In both tasks, the subject is trained to implement a particular relationship across the length of the study (Murray et al. 2000). However, in rule learning the relationship between stimulus and response is often more complex than in arbitrary visuomotor learning. For example, the rule "same" requires the subject to learn to choose whichever stimulus is the same as a previously presented stimulus. Both lesion and electrophysiological studies indicate that the striatum is important for learning and performing abstract rules (Murray et al. 2000; Muhammad et al. 2006).

A second type of rule learning experiment uses tasks that emphasize switching between rules, or reversing a previously learned association. In both humans and monkeys, the striatum is active while switching (Monchi et al. 2001; Cools et al. 2004; Seger and Cincotta 2006), and when relearning after a rule reversal (Cools et al. 2002; Pasupathy and Miller 2005). Both types of rule learning tasks often have heavy reliance on prefrontal cortical regions in conjunction with the striatum (Bunge et al. 2005).

2.3 *Tasks from the Human Literature*

2.3.1 Categorization and Classification

In categorization and classification tasks, subject view visual stimuli, and decide for each which category or group it belongs to. Typically subjects indicate the category via a button press on a computer keyboard, then receive feedback (usually verbal, either indicating whether the response was correct or incorrect, or indicating the actual category membership of the stimulus). These tasks were developed primarily to investigate the effects of stimulus similarity on learning, and the ability of subjects to generalize to novel stimuli. However, the basal ganglia are commonly involved regardless of categorical structure (Ashby and Waldron 1999; Seger 2008). Some commonly used tasks that recruit the basal ganglia are trial and error prototype learning in which stimuli are formed as distortions of a prototypical stimulus (Vogels et al. 2002), information integration learning in which stimuli are grouped on the basis of a decision bound in abstract feature space (Seger and Cincotta 2002; Ashby and Maddox 2005; Cincotta and Seger 2007; Nomura et al. 2007), probabilistic classification learning in which multiple-independent features are correlated with category membership (Knowlton et al. 1996a, b; Poldrack et al. 1999, 2001), and arbitrary categorization tasks in which stimuli in each group share no identifiable common characteristics and the category membership of each must be learned independently (Seger and Cincotta 2005).

2.3.2 Decision Making

In decision making and gambling tasks, subjects view a set of options (often pictured as cards or slot machines), choose one of them, and receive reward (often points, sometimes actual money). Unlike categorization tasks, the stimuli remain the same across the task and the emphasis is primarily on learning about the reward or feedback. Some studies manipulate the presence and valence (loss vs. gain, or reward vs. punishment) of the reward (Delgado et al. 2000; Knutson and Cooper 2005; Yacubian et al. 2007). Other studies manipulate the relationship between stimulus, decision, and reward to examine how subjects make decisions faced with ambiguity (unknown contingency) or risk (known probabilistic contingency; Delgado et al. 2005; Kuhnen and Knutson 2005; Huettel et al. 2006; Preuschoff et al. 2006). Sometimes these reward features are varied across the time course of learning to examine how subjects adjust to changing contingencies (Daw et al. 2006).

3 Roles of Corticostriatal Loops in Learning

Given the complexity of basal-ganglia-dependent learning tasks, and the many features that these tasks have in common, it is unclear whether the basal ganglia are important for all aspects of task performance, or just a subset of the task features. Furthermore, the basal ganglia are not a single isolated brain structure; instead, the basal ganglia are closely connected with cerebral cortex and other subcortical regions, and participate in multiple-independent neural networks known as corticostriatal loops. In my laboratory, we try to take fine-grained view of both learning task demands and basal ganglia anatomy in order to identify specific linkages between particular learning task features and particular corticostriatal loops. Our main methodology is fMRI, which has excellent spatial resolution that allows us to localize specific regions within the human basal ganglia.

Within the scope of this chapter, I will illustrate the patterns we have found primarily in reference to one study, which used a simple categorization task

(Seger and Cincotta 2005). In this task, the subject views a single visual stimulus, such as an abstract pattern of lines, a house, or a face. While the stimulus is visible, the subject must decide which of the two arbitrary categories it belongs to (e.g., group 1, or group 2), and press the button corresponding to the category. The subject then sees a feedback screen consisting of either the word "Right" or "Wrong." The relationship between each individual stimulus and the categories can be deterministic, random, or probabilistic. A deterministic relationship means the stimulus is always in one of the categories, and subjects receive consistent feedback to that effect. When the relationship is random the stimulus is in each category half of the time; in this situation, the subject cannot learn any regular stimulus-category pairings. In a probabilistic relationship, the stimulus might be in category 1 80% of the time and category 2 the other 20%. Subjects can learn probabilistic stimulus–category relationships, but on some trials they will receive feedback that is counter to the predominate membership. In this task, we operationally defined learning-related trials as deterministic and probabilistic trials in which the subject responded correctly and received positive feedback. In addition we examined feedback processing by comparing the response to positive ("Right") and negative ("Wrong") feedback on random trials; this manipulation serves to isolate feedback processing in the absence of learning. Figure 1 shows the patterns of recruitment of different striatal regions during successful categorization learning and during feedback valence processing in this simple categorization task.

3.1 Corticostriatal Loops: Anatomy

All lobes of the cortex project to the striatum (consisting of the caudate nucleus, putamen, and ventral striatum including the nucleus accumbens) and from there to the output structures of the basal ganglia and back to the cortex thus forming "loops." Different cortical areas project to different striatal areas (Alexander et al. 1986; Lawrence et al. 1998). Cortical regions project to the striatum in rough correspondence with the proximity of the cortical region to the particular striatal region. Although there are no strict borders between various corticostriatal loops, on the basis of the predominant cortical inputs to particular striatal regions it is possible to divide the corticostriatal systems into at least three types (Parent and Hazrati 1995): limbic (referred to here as motivational), associative, and sensorimotor (referred to here as motor). The largest of these, the associative loop, is sometimes subdivided further. In this chapter, we follow the lead of Lawrence et al. (1998) who split the associative loop into two loops; these are described in more detail below and in Table 1.

Overall, projections follow a gradient from the most anterior–medial–ventral portions on up to the most posterior–lateral–dorsal regions. In primates this corresponds to a gradient running from ventral striatum (motivational loop), to the head of the caudate/anterior putamen (executive loop), then to the body/tail of the caudate (visual loop), and finally to the posterior putamen (motor loop). In addition, there are various neurochemical processes that follow the same gradient. The most prominent

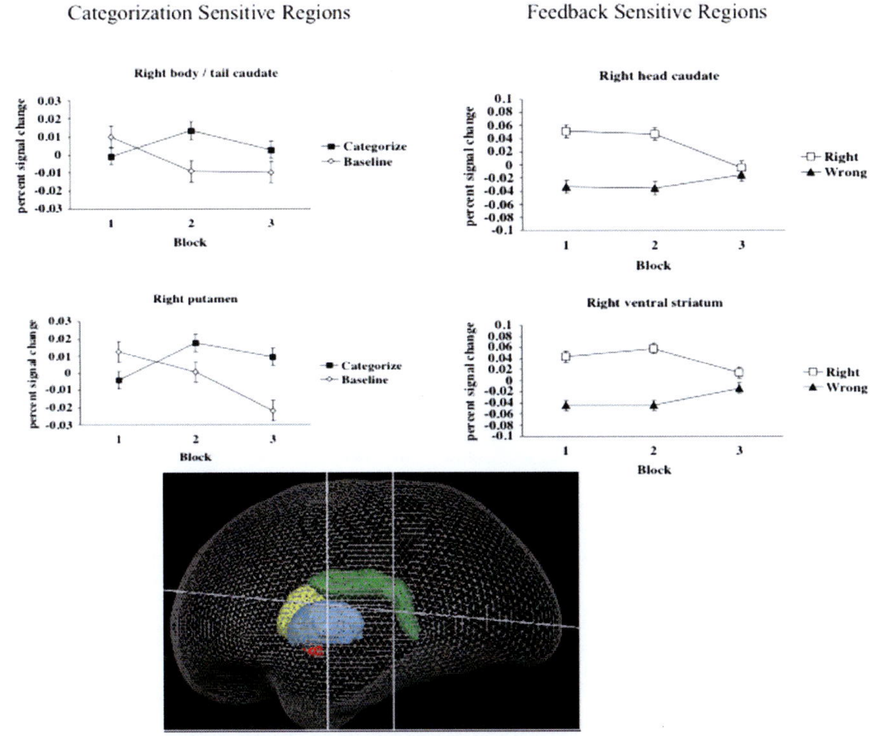

Fig. 1 *Left*: Activity in the body/tail of the caudate and putamen regions of interest during correct categorization in comparison with baseline across blocks of training. *Right*: Activity in the ventral striatum and head of the caudate regions of interest in response to positive ("Right") and negative ("Wrong") feedback on random trials across blocks of training. In all graphs, activity is measured as percent blood oxygen level-dependent (BOLD) signal change from the beginning of each trial, and is averaged for all voxels within the particular region of interest. *Bottom*: Regions of interest within the striatum, lateral view; *Red*: ventral striatum; *Yellow*: Head of the caudate; *Green*: Body and Tail of the caudate; *Blue*: Putamen. For simplicity only right-sided activity is shown; left-sided basal ganglia activity did not differ significantly

of these is the density of dopaminergic projections from the VTA/SNc (Voorn et al. 2004; Haber et al. 2006). Calbindin immunoreactivity (a measure of the presence of the calcium-binding protein, calbindin) also follows this gradient, with greatest binding in the anterior–medial–ventral portions (Karachi et al. 2002).

3.2 *Corticostriatal Loops: Function*

On a cellular level, all corticostriatal loops have a similar pattern of neural connectivity. This implies that there should be a similarity of function across loops as well. One general function served by the basal ganglia is the selection or gating of representations

Table 1 Cortical, striatal, and dopaminergic systems interacting within each corticostriatal loop

Loop name	Cortical projections Primate	Striatal target: Primate	Striatal target: Rodent homolog	Dopaminergic innervation
Motivational	Orbitofrontal	Ventral striatum	Nucleus accumbens	High
	Anterior cingulate	Nucleus accumbens	Shell and core	
	Amygdala	Ventral putamen		
	Hippocampus	Ventral caudate		
Executive	Dorsolateral prefrontal	Head of the caudate	Dorsomedial	Intermediate
	Posterior parietal	Anterior putamen		
Visual	Inferior temporal	Body of the caudate	Dorsomedial	Intermediate
	Middle temporal	Tail of the caudate	Bordering	
	Superior temporal		Dorsolateral	
	Ventrolateral prefrontal			
	Frontal eye fields			
Motor	SMA	Posterior putamen	Dorsolateral	Low
	Lateral premotor cortex			
	Primary motor			
	Primary somatosensory			

Note. *SMA* supplementary motor area

in cortex (Redgrave et al. 1999; Gurney et al. 2004). Cortex is characterized by the simultaneous activation of many different possible representations (e.g., possible movements in motor cortex, possible strategies in the frontal lobes, possible object categories in visual cortex). The basal ganglia modulate these cortical representations in two ways. First, they exert a broad tonic inhibition on these representations. Second, they can effectively select a particular representation by phasically reducing the inhibition for it. Researchers currently have identified three pathways (direct, indirect, and hyperdirect) between the striatum/STN and BG output nuclei. These allow for inhibition and selection of cortical representations across different timescales (see Frank 2006, for a review).

3.3 Visual Corticostriatal Loop

The visual corticostriatal loop links temporal and occipital extrastriate visual cortex with the body and tail of the caudate (Webster et al. 1993; Middleton and Strick 1996). In addition, visually related frontal lobe regions such as the ventrolateral prefrontal cortex (involved in visual working memory; Inoue and Mikami 2007) and the frontal eye fields (involved in eye movement planning; Lau and Glimcher 2007) also project to the body and tail of the caudate. The visual corticostriatal loop may contribute to selection within the cortex in two ways. One is by modulating representations within visual cortex, such as selection of a particular interpretation of an ambiguous visual scene or updating visual working memory. In addition, there are output projections from visual loop to motor cortex that may enable selection of appropriate motor plans on the basis of current visual processing (Ashby et al. 1998, 2007).

In our laboratory we have found that the body and tail of the caudate participate in categorization learning. As shown in Fig. 1, the body and tail of the caudate is more active overall during correct categorization than on baseline trials. Furthermore, activity in the body and tail increases across the time course of learning, and more successful learners (those who achieved greater accuracy) were more likely to recruit this area than less successful learners. We also found activity associated with learning in the body and tail of the caudate when using information integration categorization tasks (Cincotta and Seger 2007), and in simple rule learning tasks (Seger and Cincotta 2006). Current work in progress in our lab is investigating how the body and tail of the caudate interact with visual cortical regions during learning.

The visual loop has also been implicated in learning in studies involving other methodologies, species, and tasks. In humans, neuropsychological research has found that categorization is impaired in cases of basal ganglia damage (Knowlton et al. 1996a, b; Shohamy et al. 2004; Ashby and Maddox 2005), but typically these studies cannot isolate specific corticostriatal loops. Research using the concurrent visual discrimination task finds that the tail of the caudate (Brown et al. 1995; Teng et al. 2000; Fernandez-Ruiz et al. 2001) and temporal lobe visual processing area TE (Buffalo et al. 1998) both must be intact for monkeys to learn concurrent visual discrimination tasks. Conversely, visual discrimination learning is preserved when all other connections from visual areas of the inferior temporal lobe other than the connection to the striatum are severed (Gaffan and Eacott 1995).

3.4 *Motor Corticostriatal Loop*

The motor loop connects the motor and somatosensory cortexes of the frontal and parietal lobes with the putamen (Lawrence et al. 1998). Motor cortex is organized hierarchically, from primary motor cortex (involved in representing particular directions of motor effectors) to supplementary motor area (SMA) and premotor cortex (involved in motor programming) to preSMA and other prefrontal structures (involved in higher order programming and goal setting). Within the motor loop, the basal ganglia selection function is expressed in the realm of overt motor behavior (Redgrave et al. 1999; Gurney et al. 2004). Basal ganglia disorders typically result in oversuppression of movement (hypokinesia) or undersuppression of movement (hyperkinesia).

Categorization tasks require that the subject select and perform an appropriate motor responses, even in tasks in which motor performance is not the main focus of attention. In our laboratory, we have examined recruitment of the putamen during categorization learning. As shown in Fig. 1, the pattern of activation in the putamen is similar to the pattern found in the body and tail of the caudate.

This study, however, was not designed to tease apart the different roles that putamen and body and tail of the caudate may play in learning; other studies have indicated that overall the caudate nucleus is more involved during acquisition of stimulus–response relationships, whereas the putamen is more involved in skilled, habitual, or automatic performance (Haruno and Kawato 2006; Williams and Eskandar 2006; Balleine et al. 2007). Tasks in which motor demands are prominent (e.g., arbitrary visuomotor learning) are especially likely to report putamen involvement in learning (Brasted and Wise 2004). In addition, electrophysiological studies that have isolated basal ganglia responses to stimulus, response, and feedback have found that the putamen is more likely than other striatal regions to be active during the response phase (Yamada et al. 2007).

3.5 Executive Corticostriatal Loop

The executive corticostriatal loop links the head of the caudate and anterior regions of the putamen to the dorsolateral prefrontal cortex and interconnected regions of the posterior parietal cortex (Lawrence et al. 1998). Executive function involves monitoring performance, correcting for errors, and initiating, stopping, and switching between complex cognitive procedures. Within the realm of the basal-ganglia-dependent learning tasks discussed in this chapter, executive functioning is especially required for interpreting feedback or reward and using this information as a cue to initiate a strategic plan, update working memory, or switch between associations.

In our laboratory, we have investigated the role of the head of the caudate and interacting cortical regions across a series of studies. One approach we have taken is to examine how feedback valence affects caudate activity. As shown in Fig. 1, in the simple categorization task (Seger and Cincotta 2005), we found greater activity in the head of the caudate for positive feedback than for negative feedback. This difference was modulated by experience: as training continued (and presumably subjects learned that feedback would not be helpful in learning the category membership of the random stimuli), the differential sensitivity of the head of the caudate was eliminated. We found a similar pattern in a rule learning task that involved hypothesis testing: greater activity in the head of the caudate in response to positive feedback than negative (Seger and Cincotta 2006). Other studies have found a similar pattern of feedback valence sensitivity. In the domain of rule learning, Monchi et al. (2001) found greater activity for negative feedback than positive in the Wisconsin Card Sorting task, whereas tasks in which feedback valence is randomly determined (e.g., gambling tasks) generally show greater activity for positive than negative feedback (Delgado et al. 2000; Tricomi et al. 2004; Seger and Cincotta 2005).

Another approach we have taken is to compare learning tasks that involve trial and error feedback with those that are observational (that is, in which subjects are told the category membership on each trial, then tested on their learning later). We found that activity in the head of the caudate was modulated by presence of feedback during learning (Cincotta and Seger 2007). Neuropsychological studies have found that people with basal ganglia disorders are impaired in learning via feedback but learn normally on observational tasks (Shohamy et al. 2004; Smith and McDowall 2006). Electrophysiological studies have shown that cells in the head of the caudate are more likely to be active after receiving feedback than during stimulus processing or response execution (Yamada et al. 2007).

3.6 *Motivational Corticostriatal Loop*

The motivational loop connects the ventral striatum with predominantly limbic regions. In primates, the ventral striatum consists of the nucleus accumbens and the most ventral and anterior portions of the caudate and putamen. The motivational loop receives projections from ventral and medial frontal regions (including orbitofrontal cortex and the anterior cingulate), the hippocampus and the amygdala (Lawrence et al. 1998). The projections to striatum from midbrain dopaminergic regions such as the ventral tegmental area and substantia nigra pars compacta are particularly pronounced in the ventral striatum.

How selection is exhibited in the motivational loop is controversial. Some studies indicate that rather than being involved directly in response selection, the ventral striatum may maintain information from past experience, such as the results of the previous trial (Kim et al. 2007) or the stimulus and behavioral context of several seconds in the past (Nicola 2007). Research on reward processing finds that the ventral striatum is sensitive to reward valence, whereas more dorsal areas of the striatum such as the head of the caudate are sensitive to reward only when subjects must make actions contingent on the rewards (O'Doherty et al. 2004).

In our categorization study (see Fig. 1), we found that the ventral striatum was sensitive to feedback valence and showed a similar pattern as the head of the caudate across the time course of learning. In other classification learning tasks, ventral striatal activity has been found to be sensitive to the degree of prediction error for negative feedback (Rodriguez et al. 2006) and to the degree of categorization uncertainty (Grinband et al. 2006).

4 Interactions between Corticostriatal Loops

So far this chapter has focused on the role that each individual corticostriatal loop plays in learning. However, the corticostriatal loops do not act in isolation: there is interaction between the loops, and different loops take on different roles across the time course of learning. There are at least three anatomical bases for interactions between corticostriatal loops. Many of the interactions between loops follow the gradient from motivational to motor loop, as described above and in Table 1. First, the output projections from the thalamus can target a different cortical area than the one providing the input to the loop. For example, the visual loop has output projections that target premotor areas in the frontal lobe (Passingham 1993; Ashby et al. 1998). Second, there is interaction between loops at the level of projection from the striatum to basal ganglia output nuclei (Joel and Wiener 1994). Finally, there are reciprocal projections between the striatum and the SNc/VTA: the return projections from striatum to SNc/VTA terminate at cells that in turn project back to striatum, but to relatively more dorsal, lateral, and posterior regions (Haber et al. 2000).

During learning, there are shifts between corticostriatal loops. Many of these shifts appear to follow the ventromedial-anterior to dorsolateral-posterior gradient, such that activity in the executive loop precedes activity in the motor loop. In instrumental conditioning (Yin and Knowlton 2006; Balleine et al. 2007), early learning has been characterized as goal directed and shown to be reliant on the head of the caudate (dorsomedial striatum in rodents); later phases of learning have been characterized as habit learning, are relatively more automatic in the sense that it is resistant to incentive devaluation, and have been shown to be reliant on the putamen (dorsolateral striatum in rodents). Researchers using computational modeling techniques have found that the caudate nucleus is most active when learning rate is the highest (Williams and Eskandar 2006) and when subjects are making the most use of feedback to change their mental representations (Haruno and Kawato 2006). In contrast, activity in the putamen generally follows behavioral performance, increasing as accuracy increases even to the point of automaticity (Haruno and Kawato 2006; Williams and Eskandar 2006). Foerde et al. (2006) manipulated the degree of automaticity of performance by requiring subjects to perform a categorization task under dual task conditions, which limited the amount of attention they could pay to the task. In dual task conditions, in which attention was limited, learning tended to rely on the putamen, whereas in single task conditions where the subject could apply full attention to learning the caudate was recruited.

References

Alexander GE, DeLong MR and Strick PL (1986) Parallel organization of functionally segregated circuits linking basal ganglia and cortex. Ann Rev Neurosci 9: 357–381

Ashby FG and Maddox WT (2005) Human category learning. Ann Rev Psychol 56: 149–78

Ashby FG and Waldron EM (1999) On the nature of implicit categorization. Psychon Bull Rev 6: 363–378

Ashby FG, Alfonso-Reese LA, Turken AU and Waldron EM (1998) A neuropsychological theory of multiple systems in category learning. Psychol Rev 105: 442–481

Ashby FG, Ennis JM and Spiering BJ (2007) A neurobiological theory of automaticity in perceptual categorization. Psychol Rev 114: 632–656

Atallah HE, Lopez-Paniagua D, Rudy JW and O'Reilly RC (2007) Separate neural substrates for skill learning and performance in the ventral and dorsal striatum. Nat Neurosci 10: 126–131

Balleine BW, Delgado MR and Hikosaka O (2007) The role of the dorsal striatum in reward and decision-making. J Neurosci 27: 8161–8165

Barnes TD, Kubota Y, Hu D, Jin DZ and Graybiel M (2005) Activity of striatal neurons reflects dynamic encoding and recoding of procedural memories. Nature 437: 1158–1161

Brasted PJ and Wise SP (2004) Comparison of learning-related neuronal activity in the dorsal premotor cortex and striatum. Eur J Neurosci 19: 721–740
Brightwell JJ, Smith CA, Neve L and Colombo PJ (2008) Transfection of mutant CREB in the striatum, but not the hippocampus, impairs long-term memory for response learning. Neurobiol Learn Mem 89: 27–35
Brown VJ, Desimone R and Mishkin M (1995) Responses of cell in the tail of the caudate nucleus during visual discrimination learning. J Neurophysiol 74: 1083–1094
Buffalo A, Stefanacci L, Squire LR and Zola SM (1998) A reexamination of the concurrent discrimination learning task: The importance of anterior inferotemporal cortex, area TE. Behav Neurosci 112: 3–14
Bunge SA, Wallis JD, Parker A, Brass M, Crone EA, Hoshi E and Sakai K (2005) Neural circuitry underlying rule use in humans and nonhuman primates. J Neurosci 25: 10347–10350
Cincotta CM and Seger CA (2007) Dissociation between striatal regions while learning to categorize via observation and via feedback. J Cogn Neurosci 19: 249–265
Cools R, Clark L, Owen AM and Robbins TW (2002) Defining the neural mechanisms of probabilistic reversal learning using event-related functional magnetic resonance imaging. J Neurosci 22: 4563–4567
Cools R, Clark L and Robbins TW (2004) Differential responses in human striatum and prefrontal cortex to changes in object and rule relevance. J Neurosci 24: 1129–1135
Daw ND, O'Doherty P, Daya P, Seymour B and Dolan RJ (2006) Cortical substrates for exploratory decisions in humans. Nature 441: 876–879
Delgado MR, Nystrom LE, Fissell C, Noll D C and Fiez JA (2000) Tracking the hemodynamic responses to reward and punishment in the striatum. J Neurophysiol 84: 3072–3077
Delgado MR, Miller MM, Inati S and Phelps EA (2005) An fMRI study of reward-related probability learning. NeuroImage 24: 862–873
Fernandez-Ruiz J, Wang J, Aigner TG and Mishkin M (2001) Visual habit formation in monkeys with neurotoxic lesions of the ventrocaudal neostriatum. Proc Natl Acad Sci USA 98: 4196–4201
Foerde K, Knowlton BJ and Poldrack RA (2006) Modulation of competing memory systems by distraction. Proc Natl Acad Sci USA 103: 11778–11783
Frank M J (2006) Hold your horses: A dynamic computational role for the subthalamic nucleus in decision making. Neural Netw 19: 1120–1136
Gaffan D and Eacott MJ (1995) Visual learning for an auditory secondary reinforcer by macaques is intact after uncinate fascicle section: Indirect evidence for the involvement of the corpus striatum. Eur J Neurosci 7: 1866–1871
Grinband J, Hirsch J and Ferrera VP (2006) A neural representation of categorization uncertainty in the human brain. Neuron 49: 757–763
Grol MJ, deLange FP, Verstraten FAJ, Passingham RE and Toni I (2006) Cerebral changes during performance of overlearned arbitrary visuomotor associations. J Neurosci 26: 117–125
Gurney K, Prescott TJ, Wickens JR and Redgrave P (2004) Computational models of the basal ganglia: From robots to membranes. Trends Neurosci 27: 453–459
Haber SN, Fudge JL and McFarland NR (2000) Striatonigrostriatal pathways in primates form an ascending spiral from the shell to the dorsolateral striatum. J Neurosci 20: 2369–2382
Haber SN, Kim K-S, Mailly P and Calzavara R (2006) Reward related cortical inputs define a large striatal region in primates that interface with associative cortical connections, providing a substrate for incentive-based learning. J Neurosci 26: 8368–8376
Hadj-Bouziane F and Boussaoud D (2003) Neuronal activity in the monkey striatum during conditional visuomotor learning. Exp Brain Res 133: 190–196
Haruno M and Kawato M (2006) Different neural correlates of reward expectation and reward expectation error in the putamen and caudate nucleus during stimulus–action–reward association learning. J Neurophysiol 95: 948–959
Huettel SA, Stowe CJ, Gordon EM, Warner BT and Platt ML (2006) Neural signatures of economic preferences for risk and ambiguity. Neuron 49: 765–775
Inoue M and Mikami A (2007) Top-down signal of retrieved information from prefrontal to inferior temporal cortices. J Neurophysiol 98: 1965–1974

Joel D and Weiner I (1994) The organization of the basal ganglia–thalamacortical circuits: Open interconnected rather than closed segregated. Neuroscience 63: 363–379
Jog M, Kubota Y, Connolly CI, Hillegaart V and Graybiel AM (1999) Building neural representations of habits. Science 286: 1745–1749
Karachi C, Francois C, Parain K, Bardinet E, Tande D, Hirsch E and Yelnik J (2002) Three dimensional cartography of functional territories in the human striatopallidal complex by using calbindin immunoreactivity. J Comp Neurol 450: 122–134
Kim YB, Huh N, Lee H, Baeg EH, Lee D, Jung MW (2007) Encoding action history in the rat ventral striatum. J Neurophysiol 98: 3548–3556
Knowlton BK, Mangels JA and Squire LR (1996a) A neostriatal habit learning system in humans. Science 273: 1399–1402
Knowlton BK, Squire LR, Paulsen JS, Swerdlow NR, Swenson M and Butters N (1996b) Dissociations within nondeclarative memory in Huntington's disease. Neuropsychology 10: 538–548
Kuhnen CM and Knutson B (2005) The neural basis of financial risk taking. Neuron 47: 763–770
Lau B and Glimcher PW (2007) Action and outcome encoding in the primate caudate nucleus. J Neurosci 27: 14502–14514
Lawrence AD, Sahakian BJ and Robbins TW (1998) Cognitive functions and corticostriatal circuits: Insights from Huntington's disease. Trends Cogn Sci 2: 379–388
Middleton FA and Strick PL (1996) The temporal lobe is a target of output from the basal ganglia. Proc Natl Acad Sci USA 93: 8683–8687
Monchi O, Petrides M, Petre V, Worsley K and Dagher A (2001) Wisconsin card sorting revisited: Distinct neural circuits participating in different stages of the task identified by event-related functional magnetic resonance imaging. J Neurosci 21: 7733–7741
Muhammad R, Wallis JD and Miller EK (2006) A comparison of abstract rules in the prefrontal cortex, premotor cortex, inferior temporal cortex, and striatum. J Cogn Neurosci 18: 974–989
Murray EA, Bussey TJ and Wise SP (2000) Role of the prefrontal cortex in a network for arbitrary visuomotor mapping. Exp Brain Res 133: 114–129
Nicola SM (2007) The nucleus accumbens as part of a basal ganglia action selection circuit. Psychopharmacology 191: 521–550
Nixon PD, McDonald KR, Gough PM, Alexander IH and Passingham RE (2004) Cortico-basal ganglia pathways are essential for the recall of well-established visuomotor association. Eur J Neurosci 20: 3165–3178
Nomura EM, Maddox WT, Filoteo JV, Ing AD, Gitelman DR, Parrish TB, Mesulam M-M and Reber PJ (2007) Neural correlates of rule-based and information-integration visual category learning. Cereb Cortex 17: 37–43
O'Doherty J, Dayan P, Schultz J, Deichmann R, Friston K and Dolan RJ (2004) Dissociable roles of the ventral and dorsal striatum in instrumental conditioning. Science 304: 452–454
Packard MG and McGaugh JL (1992) Double dissociation of fornix and caudate nucleus lesions on acquisition of two water maze tasks: Further evidence for multiple memory systems. Behav Neurosci 106: 439–446
Packard MG, Hirsch R and White NM (1989) Differential effects of fornix and caudate nucleus lesions on two radial maze tasks: Evidence for multiple memory systems. J Neurosci 9: 1465–1472
Parent A and Hazrati L (1995) Functional anatomy of the basal ganglia. I. The cortico-basal ganglia-thalamo-cortical loop. Brain Res Rev 20: 91–127
Passingham RE (1993) The Frontal Lobes and Voluntary Action. New York, NY: The Clarendon Press
Pasupathy A and Miller EK (2005) Different time courses of learning-related activity in the prefrontal cortex and striatum. Nature 433: 873–876
Poldrack RA, Prabhakaran V, Seger CA and Gabrieli JDE (1999) Striatal activation during cognitive skill learning. Neuropsychology 13: 564–574

Poldrack RA, Clark J, Pare-Blagoev EJ, Shohamy D, Creso Moyano J, Myers C and Gluck MA (2001) Interactive memory systems in the human brain. Nature 414: 546–550
Preuschoff K, Bossaerts P and Quartz SP (2006) Neural differentiation of expected reward and risk in human subcortical structures. Neuron 51: 381–390
Redgrave P, Prescott TJ and Gurney K (1999) The basal ganglia: A vertebrate solution to the selection problem? Neuroscience 89: 1009–1023
Rodriguez PF, Aron AR and Poldrack RA (2006) Ventral-striatal/nucleus-accumbens sensitivity to prediction errors during classification learning. Hum Brain Mapp 27: 306–313
Seger CA (2008) How do the basal ganglia contribute to categorization? Their roles in generalization, response selection, and learning via feedback. Neurosci Biobehav Rev 32: 265–278
Seger CA and Cincotta M (2002) Striatal activation in concept learning. Cogn Affect Behav Neurosci 2: 149–161
Seger CA and Cincotta CM (2005) The roles of the caudate nucleus in human classification learning. J Neurosci 25: 2941–2951
Seger CA and Cincotta CM (2006) Dynamics of frontal, striatal, and hippcampal systems during rule learning. Cereb Cortex 16: 1546–1555
Shohamy D, Myers CE, Grossman S, Sage J, Gluck MA and Poldrack RA (2004) Cortico-striatal contributions to feedback-based learning: Converging data from neuroimaging and neuropsychology. Brain 127: 851–859
Smith JG and McDowall J (2006) When artificial grammar acquisition in Parkinson's disease is impaired: The case of learning via trial-by-trial feedback. Brain Res 1067: 216–228
Teng E, Stefanacci L, Squire LR and Zola S M (2000) Contrasting effects on discrimination learning after hippocampal lesions and conjoint hippocampal-caudate lesions in monkeys. J Neurosci 20: 3853–3863
Toni I, Rowe J, Stephan KE and Passingham RE (2002) Changes of cortico-striatal effective connectivity during visuomotor learning. Cereb Cortex 12: 1040–1047
Tricomi EM, Delgado MR and Fiez JA (2004) Modulation of caudate activity by action contingency. Neuron 41: 281–292
Vogels R, Sary G, Dupont P and Orban GA (2002). Human brain regions involved in visual categorization. NeuroImage 16: 401–414
Voorn P, Vanderschuren L, Groenewegen HJ, Robbins TW and Pennartz CMA (2004) Putting a spin on the dorsal–ventral divide of the striatum. Trends Neurosci 27: 468–474
Webster MJ, Bachevalier J and Ungerleider LG (1993) Subcortical connections of inferior temporal areas TE and TEO in macaque monkeys. J Comp Neurol 335: 73–91
Williams ZM and Eskandar EN (2006) Selective enhancement of associative learning by microstimulation of the anterior caudate. Nature Neurosci 9: 562–568
Wise SP and Murray EA (2000) Arbitrary associations between antecedents and actions. Trends Neurosci 23: 271–276
Yacubian J, Sommer T, Schroeder K, Gläscher J, Braus DF, and Büchel C (2007) Subregions of the ventral striatum show preferential coding of reward magnitude and probability. Neuroimage 38: 557–563
Yamada H, Matsumoto N and Kimura M (2007) History- and current instruction-based coding of forthcoming behavioral outcomes in the striatum. J Neurophysiol 98: 3557–3567
Yin HH and Knowlton BJ (2006) The role of the basal ganglia in habit formation. Nat Rev Neurosci 7: 464–476

Information Processing in the Striatum of Behaving Monkeys

Atsushi Nambu, Nobuhiko Hatanaka, Sayuki Takara, Yoshihisa Tachibana, and Masahiko Takada

Abstract Neuronal signals in the cerebral cortex are sent to the striatum, which is a major input structure of the basal ganglia. The striatum contains GABAergic spiny projection neurons and interneurons. Projection neurons, which receive excitatory glutamatergic inputs from the cortex and thalamus through both NMDA and AMPA/kainate receptors, have extensive local axon collaterals that form synapses with other neighboring projection neurons. GABAergic interneurons receive cortical inputs and innervate projection neurons. Thus, the activity of projection neurons is thought to be controlled by such GABAergic networks in the striatum. To investigate the functions of GABAergic networks, we recorded the activity of striatal projection neurons in behaving monkeys during performance of a reaching task with delay under the blockade of GABAergic and glutamatergic neurotransmission. Injection of a $GABA_A$ receptor blocker, gabazine, in the vicinity of the recorded neurons enhanced cortically evoked responses and task-related activity. Additional injection of glutamatergic blockers (a mixture of an NMDA receptor blocker, CPP and an AMPA/kainate receptor blocker, NBQX) diminished both cortically evoked responses and task-related activity to almost zero. These results suggest that the activity of striatal projection neurons is (1) mainly governed by the cortical inputs and (2) modified or adjusted by GABAergic networks in the striatum.

1 Introduction

Neuronal signals in the cerebral cortex are sent to the striatum, a major input structure of the basal ganglia. The striatum is composed of major projection neurons (80–95%) and minor interneurons (Kawaguchi et al. 1995; Wilson 2004). Projection

A. Nambu (✉), N. Hatanaka, S. Takara, and Y. Tachibana
Division of System Neurophysiology, National Institute for Physiological Sciences, Okazaki, Aichi 444-8585, Japan
e-mail: nambu@nips.ac.jp

M. Takada
Systems Neuroscience Section, Primate Research Institute, Kyoto University, Inuyama, Aichi 484-8506, Japan

H.J. Groenewegen et al. (eds.), *The Basal Ganglia IX*, Advances in Behavioral Biology 58,
DOI 10.1007/978-1-4419-0340-2_3,

neurons are inhibitory GABAergic medium spiny neurons that receive excitatory glutamatergic inputs from the cortex and thalamus through both NMDA and AMPA/kainate receptors (Bennett and Wilson 2000), and send their axons to the external and internal segments of the globus pallidus and substantia nigra pars reticulata. Spiny projection neurons have extensive local axon collaterals that form synapses with other neighboring projection neurons. On the other hand, interneurons are classified into at least four groups: cholinergic large aspiny neurons, parvalbumin (PV)-containing GABAergic aspiny neurons, somatostatin-containing GABAergic aspiny neurons, and calretinin-containing GABAergic aspiny neurons (Kawaguchi et al. 1995; Wilson 2004). Interneurons receive inputs from the cerebral cortex and innervate projection neurons. Thus, GABAergic networks in the striatum, including *feedback* inhibition through the axon collaterals of projection neurons and *feedforward* inhibition through the GABAergic interneurons, are thought to control the activity of projection neurons. However, little is known how such GABAergic networks control the activity of projection neurons in physiological states. To investigate the functions of striatal GABAergic networks, we recorded the activity of striatal projection neurons during performance of a motor task under the blockade of GABAergic and glutamatergic neurotransmission in the present study.

2 Methods

Two Japanese monkeys (*Macaca fuscata*) were used in this experiment. The experimental protocols were approved by the Animal Care and Use Committee of the Okazaki Organization of National Institutes, and all experiments were conducted according to the guidelines of the National Institutes of Health, *Guide for the Care and Use of Laboratory Animals*.

Each animal was trained to sit in a primate chair and perform the reaching task with delay. Three slots (left, center, and right) were aligned horizontally in a panel that was placed in front of the animal. A two-color (red and green) light-emitting diode (LED) was installed in the bottom of each slot. Each trial was initiated after the animal placed its hand at resting position that was located below the panel (control period). One of the three LEDs was lit with red color for 150 ms as an instruction stimulus (S1). A delay period of 700–1,950 ms followed S1. During the S1 and delay period, the monkey was required to keep its hand on the resting position. After a delay period, all the three LEDs were lit with green color for 1,200 ms as a triggering stimulus (S2). Upon the presentation of S2, the monkey was required to reach out its forelimb and touch the LED inside the slot that had been instructed previously by S1 using its index finger. The timing of hand release from the resting position (HR) and the timing of finger in the slot (FI) were detected by the infrared photoelectric sensors installed in the resting position and in the slots. If the monkey touched the correct LED within 1,200 ms, it was rewarded with a drop of juice.

After learning the behavioral task, the monkey received surgical operations to fix its head painlessly in a stereotaxic frame attached to a monkey chair, and pairs of bipolar stimulating electrodes were implanted chronically into the distal (MId) and proximal (MIp) forelimb regions of the primary motor cortex (MI) and the forelimb region of the supplementary motor area (SMA) contralateral to the dominant hand based on the physiological mapping (Nambu et al. 2000, 2002).

Single-unit recordings of putamen (Put) neurons and local applications of drugs were performed with an electrode assembly consisting of a glass-coated Elgiloy microelectrode for unit recording and two silica tubes (OD, 150 μm; ID, 75 μm) for drug delivery (Kita et al. 2004; Tachibana et al. 2008). The silica tubes were connected to two 25-μL Hamilton microsyringes that contained separately the following two drugs dissolved in saline: (1) a $GABA_A$ receptor antagonist, gabazine (0.5–1.0 mM), and (2) a mixture of an NMDA receptor antagonist, ±3-(2-carboxypiperazin-4-yl)-propyl-1-phosphonic acid (CPP, 0.5–1.0 mM) and an AMPA/kainate receptor antagonist, 1,2,3,4-tetrahydro-6-nitro-2, 3-dioxo-benzo[f]quinoxaline-7-sulfonamide disodium (NBQX, 0.5–1.0 mM). The electrode assembly was penetrated obliquely (45° from vertical in the frontal plane) into the forelimb region of the contralateral Put to the hand using a hydraulic microdrive. After the activity of a single Put neuron was isolated, the responses to cortical electrical stimulation (300-μs duration single pulse, strength of less than 0.6 mA, sometimes up to 0.7 mA, at 0.7 Hz) were observed by constructing peri-stimulus time histograms (PSTHs; bin width, 1 ms; summed for 100 times). Then, the neuronal activity when performing the reaching task with delay was recorded. The timing of neuronal firings and task events was stored in a computer at a time resolution of 1 ms. After recording neuronal activity in a control condition, a total volume of 0.2–0.5 μL of each drug was injected at a rate of 0.06–0.09 μL/min by using microinjectors (UltraMicroPump II, WPI).

3 Results

Figure 1 shows a typical example of Put neuronal activity. This neuron was silent at rest (spontaneous activity, 0.1 Hz), responded to the MIp stimulation (Fig. 1A1), and increased its activity transiently in relation to movements (Fig. 1A2). Thus, this neuron was classified as a phasically active neuron (PAN) and was considered to be a striatal projection neuron. Cortical stimulation induced excitation at a latency of 9 ms (Fig. 1A1). In the task-related activity, this neuron was silent during control and delay periods and began to increase its activity just before the hand release from the resting position (Fig. 1A2). The activity was kept high during hand movements and decreased to the control level after finger-in events. The increase in activity in left- (peak amplitude, 43 Hz) and center- (40 Hz) target trials was larger than that in right-target trials (22 Hz), and the peak amplitude ratio of the preferred target to the nonpreferred target was 1.95, indicating that the activity showed directional selectivity.

Gabazine injection in the vicinity of the recorded neuron enhanced cortically evoked excitatory responses (Fig. 1B1). The amplitude (1.0 spike/stimulus before gabazine injection and 3.3 spikes/stimulus after injection) and duration (10 ms before injection and 24 ms after injection) were both increased. However, gabazine injection had little effects on the spontaneous activity (0.1 Hz after injection)

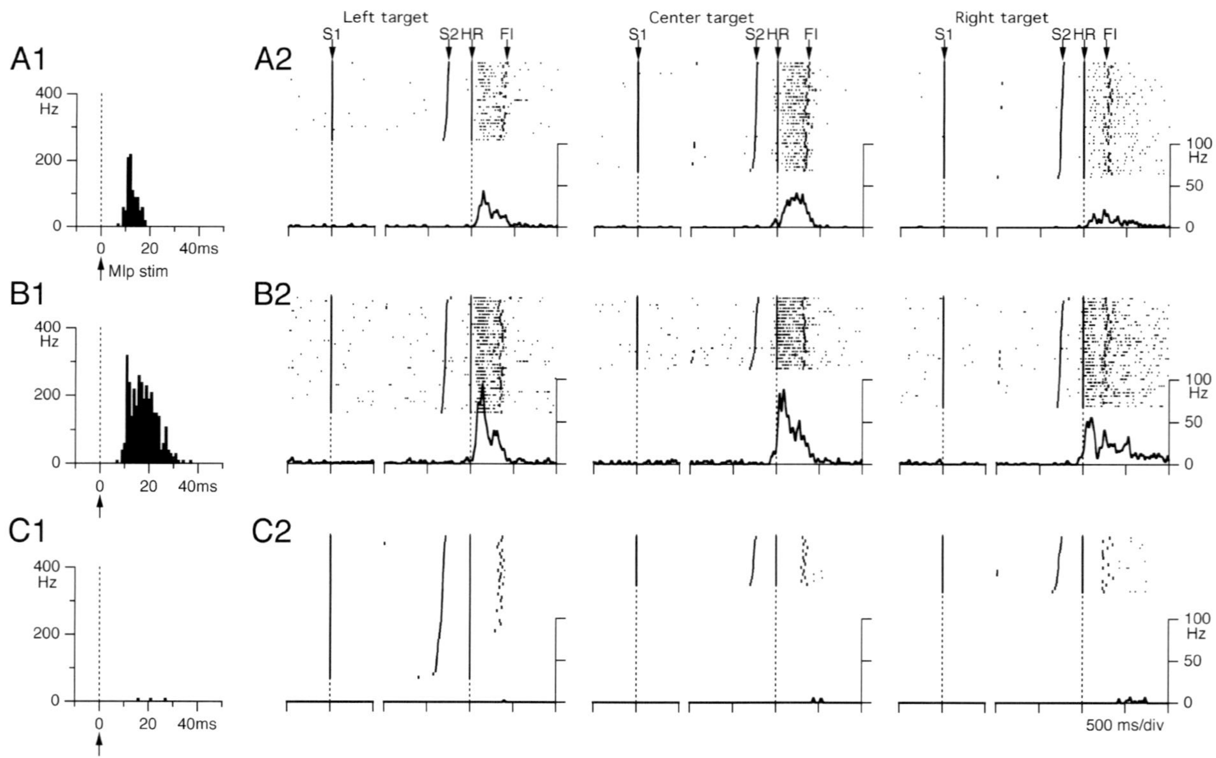

Fig. 1 (**A1**) Peristimulus time histograms (PSTHs; bin width, 1 ms; 100 times) showing responses to the stimulation of the proximal forelimb region of the primary motor cortex (MIp) at 0.6 mA. (**A2**) Raster displays showing the firing (*dots*) during task performance. Each row plots data in left-, center-, and right-target trials. In each panel, data are aligned with the S1 onset and hand release from the resting position (HR) and sorted according to the reaction time. *Vertical lines* indicate the timings of the S2 and finger-in (FI). *Continuous traces* indicate spike densities ($\sigma = 13$ ms) for the associated rasters. (**B1**, **B2**) PSTHs and raster displays after injection of gabazine (0.4 μL) in the vicinity of the recorded neuron. (**C1**, **C2**) PSTHs and raster displays after additional injection of CPP and NBQX (0.5 μL)

and the latency of cortically evoked responses (9 ms after injection). It also enhanced the amplitude of task-related activity, especially in its early phase (Fig. 1B2). Peak amplitude in left-, center-, and right-target trials was 93, 88, and 55 Hz, respectively. The peak amplitude ratio of the preferred target to the nonpreferred target decreased to 1.69, indicating that the specificity of activity to movement direction was decreased. However, gabazine injection did not change the fundamental features of the activity during task performance, such as low activity during control and delay periods, and transiently increased activity during reaching movements. After recording under gabazine injection, a mixture of CPP and NBQX was injected in the vicinity of the recoded neuron (Fig. 1C). Cortically evoked responses and movement-related activity were both diminished almost to zero.

On the basis of these observations, it was found that (1) task-related activity in the striatal projection neurons is mainly derived from excitatory inputs through ionotropic glutamatergic receptors, and (2) GABAergic networks in the striatum modify or adjust the activity of striatal projection neurons.

4 Discussion

Information processing through GABAergic networks in the striatum includes (1) feedforward inhibition mediated by striatal GABAergic interneurons and (2) collateral inhibition between striatal spiny projection neurons (Fig. 2). Gabazine injection in the present study is considered to block both GABAergic networks.

Earlier experiments demonstrated that feedforward inhibition mediated by striatal GABAergic interneurons is the most potent source of intrastriatal inhibition (Kawaguchi et al. 1995; Koós and Tepper 1999). PV-containing GABAergic interneurons receive a powerful excitatory input from the cortex and have been electrophysiologically characterized as fast-spiking interneurons that exhibited very narrow action potentials and fired a burst of action potentials when receiving excitatory inputs (Mallet et al. 2005). They primarily target the cell bodies and proximal dendrites of medium spiny projection neurons, and individual axons form a pericellular basket around spiny neuronal somata (Bennett and Bolam 1994). Through these close contacts, they produce large $GABA_A$-mediated inhibitory postsynaptic potentials (IPSPs) in medium spiny neurons (Plenz and Kitai 1998; Koós and Tepper 1999), and this inhibition is strong enough to delay or inhibit action potential firing in the target neuron. Other GABAergic interneurons including somatostatin-containing interneurons also receive excitatory inputs from the cortex, and they inhibit spiny projection neurons potently (Koós and Tepper 1999). Therefore, spike generation and timing of striatal projection neurons are considered to be influenced by feedforward inhibition through GABAergic interneurons.

On the other hand, collateral inhibition between striatal spiny projection neurons is believed to play an important role, especially in theoretical models of the striatum (Plenz 2003). As suggested by the "winner-take-all" model, among the striatal neurons competing with each other, a neuron that fires strongest inhibits other

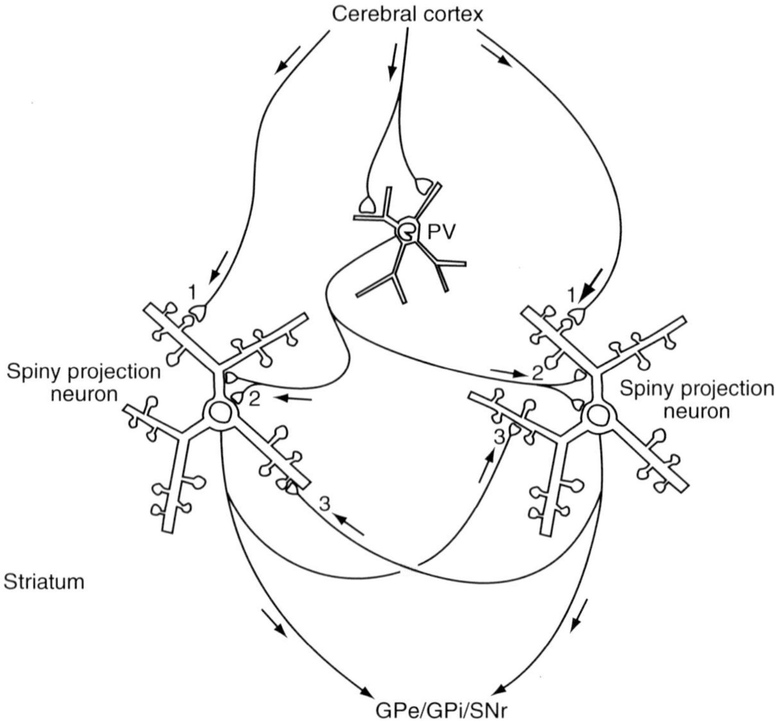

Fig. 2 A schematic diagram of the information processing in the striatum. Medium spiny projection neurons receive direct excitatory cortical inputs (**1**). Their activity is modulated by feedforward inhibition mediated by parvalbumin (PV)-containing GABAergic interneurons (**2**) and collateral inhibition between projection neurons (**3**)

surrounding neurons through mutual inhibitory connections (Wickens et al. 1991). These models are supported by neuroanatomical evidence showing collateral innervation of spiny projection neurons. Spiny projection neurons have extensive local axon collaterals that usually cover the dendritic arborization of the original neurons. The major targets of these axon collaterals are other spiny projection neurons, and most of the synapses are formed onto the dendrites and spine necks and a smaller portion onto the somata (Wilson and Groves 1980). However, physiological evidences have shown only weak functional collateral synaptic connectivity among spiny projection neurons (Jaeger et al. 1994; Koós and Tepper 1999; Czubayko and Plenz 2002; Tunstall et al. 2002). The amplitude of IPSPs is smaller than that evoked from PV-containing interneurons; the failure rate is higher (30–50%) than that for the PV-containing interneuronal synapse (<1%), and reciprocally connected pairs are rare. These physiological studies suggest that collateral inhibition between striatal spiny projection neurons controls local dendritic events, rather than operating in a winner-take-all fashion.

The present study showed that activity patterns of striatal projection neurons were enhanced after blockade of GABAergic neurotransmission in the striatum.

This is compatible with the hypothesis that GABAergic networks in the striatum modify or adjust the activity of striatal projection neurons by feedforward inhibition through interneurons and by feedback inhibition among projection neurons (Fig. 2).

The present study also showed that movement-related activity of the striatal projection neurons was greatly diminished after blockade of glutamatergic inputs. The major excitatory glutamatergic inputs to the spiny projection neurons come from the cerebral cortex and thalamic nuclei, and the synapses are formed onto the heads of the dendritic spines or dendritic shafts (Bennett and Wilson 2000). Our previous study showed that striatal projection neurons exhibit the activity specific to the cortex projecting to the recorded neurons (Takara et al. 2006). Thus, the activity of striatal projection neurons is derived mainly from cortical inputs (Fig. 2). Taken together, the activity of striatal projection neurons is (1) mainly governed by the cortical inputs and (2) modified or adjusted by GABAergic networks in the striatum.

References

Bennett BD and Bolam JP (1994) Synaptic input and output of parvalbumin-immunoreactive neurons in the neostriatum of the rat. Neuroscience 62: 707–719.

Bennett BD and Wilson CJ (2000) Synaptology and physiology of neostriatal neurones. In: Miller R and Wickens JR (eds) Brain Dynamics and the Striatal Complex. Harwood Academic Publishers, Amsterdam. pp 111–140.

Czubayko U and Plenz D (2002) Fast synaptic transmission between striatal spiny projection neurons. Proc Natl Acad Sci USA 99: 15764–15769.

Jaeger D, Kita H and Wilson CJ (1994) Surround inhibition among projection neurons is weak or nonexistent in the rat neostriatum. J Neurophysiol 72: 2555–2558.

Kawaguchi Y, Wilson CJ, Augood SJ and Emson PC (1995) Striatal interneurons: chemical, physiological and morphological characterization. Trends Neurosci 18: 527–535.

Kita H, Nambu A, Kaneda K, Tachibana Y and Takada M (2004) Role of ionotropic glutamatergic and GABAergic inputs on the firing activity of neurons in the external pallidum in awake monkeys. J Neurophysiol 92: 3069–3084.

Koós T and Tepper JM (1999) Inhibitory control of neostriatal projection neurons by GABAergic interneurons. Nat Neurosci 2: 467–472.

Mallet N, Le Moine C, Charpier S and Gonon F (2005) Feedforward inhibition of projection neurons by fast-spiking GABA interneurons in the rat striatum in vivo. J Neurosci 25: 3857–3869.

Nambu A, Tokuno H, Hamada I, Kita H, Imanishi M, Akazawa T, Ikeuchi Y and Hasegawa N (2000) Excitatory cortical inputs to pallidal neurons via the subthalamic nucleus in the monkey. J Neurophysiol 84: 289–300.

Nambu A, Kaneda K, Tokuno H and Takada M (2002) Organization of corticostriatal motor inputs in monkey putamen. J Neurophysiol 88: 1830–1842.

Plenz D (2003) When inhibition goes incognito: feedback interaction between spiny projection neurons in striatal function. Trends Neurosci 26: 436–443.

Plenz D and Kitai ST (1998) Up and down states in striatal medium spiny neurons simultaneously recorded with spontaneous activity in fast-spiking interneurons studied in cortex-striatum-substantia nigra organotypic cultures. J Neurosci 18: 266–283.

Tachibana Y, Kita H, Chiken S, Takada M and Nambu A (2008) Motor cortical control of internal pallidal activity through glutamatergic and GABAergic inputs in awake monkeys. Eur J Neurosci 27: 238–253.

Takara S, Hatanaka N, Takada M and Nambu A (2006) Activity of putaminal neurons receiving inputs from motor cortical areas in behaving monkeys. Neurosci Res 55: S247.

Tunstall MJ, Oorschot DE, Kean A and Wickens JR (2002) Inhibitory interactions between spiny projection neurons in the rat striatum. J Neurophysiol 88: 1263–1269.

Wickens JR, Alexander ME and Miller R (1991) Two dynamic modes of striatal function under dopaminergic-cholinergic control: simulation and analysis of a model. Synapse 8: 1–12.

Wilson CJ (2004) Basal ganglia. In: Shepherd GM (ed) The Synaptic Organization of the Brain, 5th Edn. Oxford University Press, New York. pp 361–413.

Wilson CJ and Groves PM (1980) Fine structure and synaptic connections of the common spiny neuron of the rat neostriatum: a study employing intracellular inject of horseradish peroxidase. J Comp Neurol 194: 599–615.

What Controls the Timing of Striatal Spiny Cell Action Potentials in the Up State?

Charles J. Wilson

Abstract The excitatory and inhibitory components of synaptic conductance in the Up state of the striatal spiny neuron were studied in cells whose voltage-sensitive ion channels were blocked by intracellular application of cesium, QX-314, and D-890. The Up state was driven by a large synaptic conductance, ranging from 13 to 46 nS. From the reversal potential of the Up and Down states, it was possible to estimate the relative contribution of the excitatory and inhibitory conductance components as 85–67% and 15–33%, respectively. The frequency spectra of the membrane fluctuations produced by excitation and inhibition were estimated by holding the cells far from or near to the reversal potential for excitatory synaptic transmission. The spectrum for excitation was dominated by slow components, interpreted as the time course of correlations among cortical neurons. The spectrum for inhibition exhibited high-frequency components in the gamma range, not present in the excitatory input. Action potentials in the Up state were triggered from fast membrane potential events, whose time course matched that of inhibitory fluctuations rather than excitatory ones. It was proposed that the timing of action potentials is controlled by disinhibition events, perhaps arising from the pallido-striatal pathway.

1 Up and Down States

In anesthetized or sleeping animals, and perhaps at other times, the membrane potential of striatal spiny neurons fluctuates between two preferred ranges. The more hyperpolarized state, which is called the Down state, is near the reversal potential for potassium and is characterized by a very low level of synaptic input, both excitatory and inhibitory (Wilson and Kawaguchi 1996). Because of the relatively

C.J. Wilson (✉)
Department of Biology, University of Texas at San Antonio, One UTSA Circle, San Antonio, TX 78249, USA
e-mail: Charles.Wilson@utsa.edu

H.J. Groenewegen et al. (eds.), *The Basal Ganglia IX*, Advances in Behavioral Biology 58, DOI 10.1007/978-1-4419-0340-2_4, © Springer Science+Business Media, LLC 2009

sparse synaptic input in the Down state, it might be expected that the input resistance of the spiny neuron would be high when hyperpolarized, and that cell would be very sensitive to increases in synaptic excitation. This does not happen, because a hyperpolarization-activated potassium current dominates the input resistance of spiny cells in the Down state, lowering the input resistance and making the cell relatively insensitive to all synaptic inputs. When synaptic input from the cerebral cortex and thalamus is strong and sustained enough to overcome the ionic conductances that maintain the Down state, the cell membrane is drawn to the depolarized Up state. The Up state is dependent on sustained excitatory synaptic activity, but the membrane potential in the Up state is less depolarized and less variable than might be expected, given the large barrage of synaptic input. Both of these deviations from expectation are caused by the action of depolarization-activated potassium currents that increase to counteract the depolarizing influence of increases in synaptic excitation, and decrease when synaptic excitation wanes momentarily. As a result, the membrane potential in the Up state is maintained in a relatively narrow range, usually slightly negative to the action potential threshold, and fluctuations in synaptic input are attenuated by voltage-dependent currents. When depolarization-activated potassium currents are blocked by intracellular application of cesium ions, the Up state becomes much more depolarized and also more variable (Wilson and Kawaguchi 1996).

Because of these depolarization-activated K^+ currents, the average membrane potential in the Up state is below the action potential threshold, and it is common for a cell to be in the Up state for long periods of time without firing. However, action potentials are triggered in the Up state in some cells at a low rate and at irregular times. Using urethane/ketamine anesthetized animals, Wickens and Wilson (1998) compared the Up states in silent spiny cells and those of neurons whose Up states occasionally evoked action potentials. They reported that silent cells have Up states approximately 10 mV more hyperpolarized than those of spontaneously active neurons recorded in the same animals. Even in the most depolarized cells, the average Up state membrane potential was subthreshold, and action potentials arose from small fluctuations in the membrane potential superimposed on the Up state.

2 What Kind of Synaptic Input Could Trigger Spiking in the Up State?

Given the action of depolarization-activated ion channels to stabilize the Up state, it is puzzling what kind of synaptic activation might be able to depolarize the spiny cell in the Up state sufficiently to trigger an action potential without evoking opposing K^+ currents that would block activation. This was addressed by Wilson (1995) in a computational study of synaptic activation in dendrites possessing depolarization-activated potassium channels. The results of that study showed that an adequate stimulus for activation of the spiny cell would be a rapid synaptic depolarizing conductance that could outrun activation of voltage-dependent K^+ currents distributed in the dendrite. To maintain firing, the input should maintain a high-frequency variability. If the membrane potential varied at sufficiently high frequency relative to the time constant of K^+ current activation, the cell could be effectively driven to fire without increasing the activation of opposing hyperpolarizing currents. Thus, the input to the striatal spiny neuron could be viewed as composed of a slow average level of synaptic input that is responsible for the Up and Down state transitions (and slow fluctuations in the Up state) and a high-frequency component that triggers action potentials. The identity of these two components has not previously been identified.

One possibility is that fast depolarizations responsible for firing in the Up state could arise by chance, given a random barrage of excitatory inputs from the cortex and elsewhere. The probability of such events depends on the total frequency of synaptic inputs participating in the cortical barrage and their individual sizes. For a truly random synaptic barrage, as would be expected from cortical neurons firing independently and randomly during their own Up states (Stern et al. 1997), this would appear as a chance coincidence of a large number of excitatory events (compared with the mean expectation). Such combinations are distributed by the binomial distribution. Because the variance of this distribution is small (relative to the mean) when the input consists of many small independent random inputs, the probability of such an event is highest if we assume that the Up state is generated by a small number of relatively large synaptic inputs. Alternatively, large sudden EPSCs could arise from momentary nonrandom synchronous barrages generated within the cortex, for example, as avalanche-like events driven through excitatory intracortical circuits (Plenz and Thiagarajan 2007). A third possibility is that the sudden depolarizations responsible for action potential generation arise from disinhibition.

3 A Role for Inhibition?

Inhibition was all but dismissed as a mechanism of spike generation in striatal neurons by Wickens and Wilson (1998) and other studies of Up and Down state transitions. Inhibition is assigned a much more important role for generating Up states in cerebral cortex neurons. In the cortex, the existence of Up states is attributed largely to a balance between synaptic excitation and inhibition (Rudolph et al. 2007). Although some evidence for an important role for voltage-sensitive dendritic currents has been presented (Waters and Helmchen 2006), there is no doubt that inhibition is a larger contributor to the Up state current balance in some cortical neurons than it is in striatal spiny cells. Measurements of the total conductance in the Up and Down states, and of the relative proportions of inhibition and excitation in the Up state of cortical cells have been made from voltage clamp data and from analysis of membrane potential variance during experimental manipulation of the Up state membrane potential (Haider et al. 2006; Rudolph et al. 2005).

Wilson and Kawaguchi (1996) showed that in cells treated intracellularly with cesium (to block potassium currents), QX-314 (to block sodium currents), and D-890 (to block calcium currents), it was possible to measure synaptically driven fluctuations in striatal spiny cell membrane potential without the interference of voltage-sensitive currents. By passing constant current, they were able to reverse the Up and Down states in three cells recorded in vivo, and they plotted the I–V curves for the Up and Down states separately. This work demonstrated that when voltage-sensitive conductances are blocked, the whole cell conductance is larger in the Up state than in the Down state as expected if the Up state is due to increased synaptic input. Wilson and Kawaguchi did not calculate the total synaptic conductance responsible for the Up state or the relative contributions of excitatory and inhibitory conductances, although their data contained the information necessary for those calculations.

4 Measurement of Excitatory and Inhibitory Conductances in the Up State

I have reanalyzed the three neurons used in the Wilson and Kawaguchi (1996) study, to obtain a measure of the total synaptic conductance in the Up state, and the contributions of excitatory and inhibitory synaptic input. The method of making this measurement is illustrated in Fig. 1. The Up and Down states were estimated

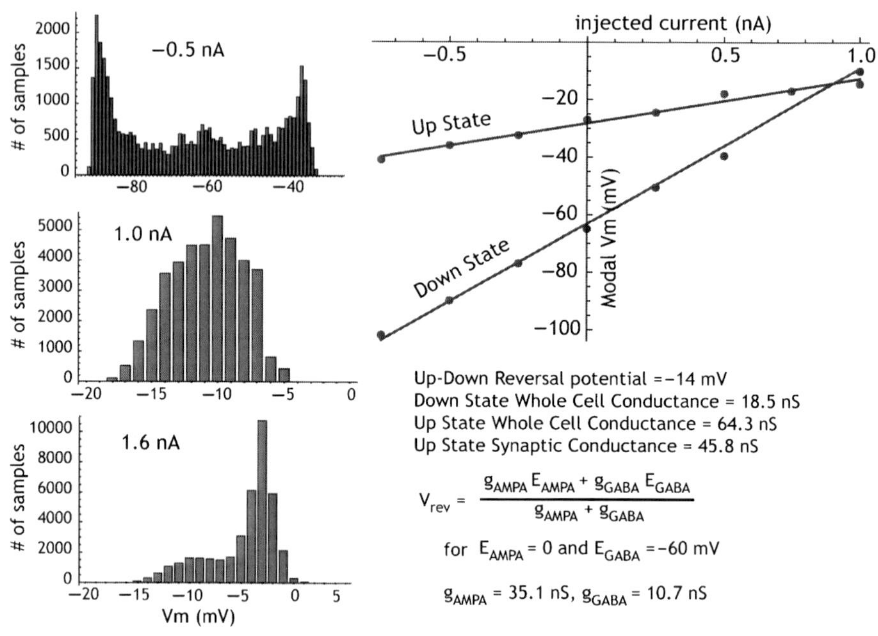

Fig. 1 All-points histograms used to measure the input resistance of a spiny neuron in the Up and Down states and the potential at which the Up and Down states reverse. From these measurements estimates of the total excitatory and inhibitory synaptic conductance in the Up state can be obtained

by modes in the all-points histogram of membrane potential fluctuations, using a variety of levels of intracellular constant current to manipulate the membrane potential in both states (Fig. 1a). The slope of the resulting steady state average current–voltage relationship (Fig. 1b) gives the whole cell input resistance in both states. As pointed out by Wilson and Kawaguchi (1996), the lower slope of the curve in the Up state indicates an overall larger conductance in that state, caused by the synaptic input. Because of the absence of most voltage-sensitive currents, the current–voltage relationship in both states is approximately linear, and a best-fitting line is fit to each state in Fig. 1b. The reciprocal of the slopes gives the whole cell conductance in each state, which for the example in the figure gives 18.5 nS in the Down state and 64.3 nS in the Up state. The difference is an estimate of the average synaptic conductance present in the Up but not the Down state. For the three cells in the sample, the conductance difference varied between 13.9 and 45.8 nS.

The synaptic conductance consists of a combination of excitatory and inhibitory inputs. Excitatory inputs arise primarily from synapses from cortical and thalamic afferents that employ glutamate as their transmitter and act largely via AMPA and NMDA receptors located on dendritic spines and distal dendritic shafts. Each spiny neuron receives about 10,000 synapses of this kind [see review in Wilson (2004)]. Because these are located on relatively distal parts of the dendritic tree, the average conductance from these synapses is probably underestimated somewhat due to current loss through the dendritic membrane conductances. Inhibitory synapses arise from collaterals of the GABAergic spiny neurons and from a small number of GABAergic interneurons distributed among the spiny cells [see review in Wilson (2007)]. Like the excitatory inputs, the recurrent mutual inhibition via collaterals of spiny cells are primarily located distally on the dendritic tree, and it is estimated that there are about 1,200–1,800 of these synapses on every spiny neuron, arising from about 500 other spiny cells. Synapses from interneurons are fewer, but end more proximally on the spiny cell and generate a much stronger synaptic conductance (Koós et al. 2004). One kind of interneuron, the fast-spiking parvalbumin-positive interneuron, makes three to six synapses on each of approximately 135–541 spiny cells it contacts. Each spiny cell receives synapses from about 4 to 27 of these cells (Koós and Tepper 1999).

The influence of synaptic location on the estimate of average Up state synaptic conductance in the data of Wilson and Kawaguchi (1996) was minimized by blockade of voltage-sensitive conductances and by the fact that Up and Down states are maintained for hundreds of milliseconds, and thus are steady state measurements relatively unaffected by membrane capacitance. Assuming that the error generated by synaptic placement is small, it is possible to use the reversal potential of Up and Down states to estimate the proportion of the total conductance contributed by excitatory and inhibitory synapses. For the three neurons in the sample, the reversal potential varied from −9 to −20 mV. For the cell in Fig. 1, the reversal potential obtained from the intersection of best-fitting lines was −14 mV. Assuming a chloride reversal potential of −60 mV (Koós et al. 2004; Gustafson et al. 2006) and a reversal potential of 0 mV for the excitatory synaptic currents, these indicate that the proportion of the total conductance contributed by inhibitory synapses in the three neurons varies from 15% to 33% of the total. These small proportions of the total current are in keeping with the conclusion of Wilson and Kawaguchi (1996) that the Up state in spiny cells is dominated by excitatory inputs and is not effectively balanced by inhibitory input. When the total conductance is taken into consideration however, these small proportions still represent a substantial inhibitory contribution to the Up state. In the cell used for the example in Fig. 1, the average inhibitory conductance in the Up state is 10.7 nS, compared with 35.1 nS of excitatory conductance.

An estimate of the total rate of synaptic input contributing to the excitatory and inhibitory components of the Up state can be obtained from the total average synaptic conductances given earlier, the time constant of decay of conductance at individual synaptic conductances, and the peak conductance associated with each kind of synapse. For AMPA-receptor-mediated synaptic conductances in the striatum, I used an estimate of 0.225 nS for the peak synaptic conductance, and instantaneous onset, and 2.5 ms for the decay time constant, obtained from voltage clamp studies of spontaneous AMPA synaptic conductances by Choi and Lovinger (1997). For $GABA_A$-receptor-mediated synaptic conductances in the striatum I used a decay time constant of 12.5 ms and peak conductances of 0.5 nS for spiny cell -> spiny cell synapses and 1.5 nS for interneuron -> spiny cell synapses, estimated by Koós et al. (2004). For a Poisson-triggered train of such synaptic conductances, the average conductance $<g>$ is given by the expression:

$$< g >= f \times g_{\text{peak}} \times \text{tau}_{\text{decay}},$$

in which f is the average frequency of synaptic input, $\text{tau}_{\text{decay}}$ is the decay time constant of synaptic conductance, and g_{peak} is the peak conductance. Thus, the frequency of input can be estimated from the average conductance. For the example in Fig. 1, the frequency of excitatory inputs is calculated to be 62,400 synapses/s. Assuming that these come from cells firing on an average 10 spikes/s during the Up state, the number of neurons contributing to the Up state is approximately 6,000, or 60% of the total excitatory innervation of the spiny neuron. This is a surprisingly large number, given modeling studies that estimate that as few as 200–300 synaptic inputs (corresponding to an average rate of 800 synapses/s) may be sufficient to trigger an Up state transition (Blackwell et al. 2003). In anesthetized animals, there may be much more input than necessary to generate and maintain the Up state. It is remarkable that such a massive average synaptic conductance produces a subthreshold depolarization, and this observation highlights the ability of the spiny neuron to suppress firing in response to relatively constant average excitation. One implication of this is that fast random fluctuations in excitatory input must be very small. The expected value of the standard deviation of the conductance generated by independent random synaptic inputs of this size and rate is approximately 1.8 nS, only about 5% of the total excitatory conductance. Of course, excitatory inputs may not be random. If excitatory cortical and/or thalamic input are responsible for generating brief fluctuation in the drive that triggers action potentials, they must do so by way of correlated firing among cells.

A similar consideration yields an estimate of only 1,712 inhibitory inputs per second if the entire conductance of 10.7 nS arose from spiny cell collaterals, and 570 synapses/s for inhibitory interneurons. Spiny cells fire sparsely during the Up state and their firing is not correlated (Stern et al. 1998), so the variance of conductance can be estimated, as mentioned earlier, to be 1.4 nS or 13% of the total inhibitory conductance. If the inhibition arose from interneurons, the standard deviation would be 2.5 nS or 23% of the total conductance. Thus, although inhibition is a much smaller component of the total synaptic conductance, random fluctuations in inhibition generated by interneurons are a more effective source of variation in synaptic input during Up states.

5 Fast Membrane Fluctuations Are Mostly Inhibitory

The fluctuations of membrane potential that are superimposed on the Up state and from which action potentials are triggered could arise from the excitatory or inhibitory inputs or both. Although a truly random barrage of excitatory inputs can not (from the earlier considerations) as effectively account for the variability of membrane potential in the Up state, nearly any kind of correlations among excitatory afferents would produce such variability, and the time course of transient membrane potential changes in the Up state would reflect the time course of correlation among inputs, rather than the time course of the individual excitatory synaptic potentials. To the extent that inhibitory synaptic potentials may generate such variability, the time course of IPSPs should be reflected in the membrane potential of the Up state. To estimate the time course of membrane potential transients generated by synaptic excitation and inhibition in the Up state, I have reanalyzed the membrane potential fluctuations in the same neurons described earlier, recorded in vivo, but with blockade of voltage-dependent conductances with intracellularly applied ion channel blockers. To obtain estimates of the variability caused by excitation and inhibition in relative isolation, membrane potential variability was compared for Up states recorded during passage of strong hyperpolarizing currents that ensured that the Up state was at least −40 mV, or during passage of depolarizing current so that the Up state was at or near the reversal potential for excitatory synapses. In the former case, the variability in membrane potential was dominated by excitation (although an influence of inhibition is still present because the Up state was not held at the GABA reversal potential). When the cell was depolarized to the vicinity of the glutamate reversal potential, the variability could be attributed to inhibition. Segments of data from Up states were extracted from traces by eye to avoid contamination with the transitions between states, and amplitude spectra of the resulting fragments of membrane potential recordings (200–600 ms in duration) were calculated and averaged.

All the three neurons in the sample showed the differences in Up state membrane spectra illustrated in Fig. 2. When excitation dominated the membrane potential fluctuations, the spectra were dominated by very low frequency components, as indicated by a nearly linear spectrum when plotted as a Bode plot (Fig. 2a). This spectrum is identical to that obtained by Stern et al. (1997) for the Up states of spiny

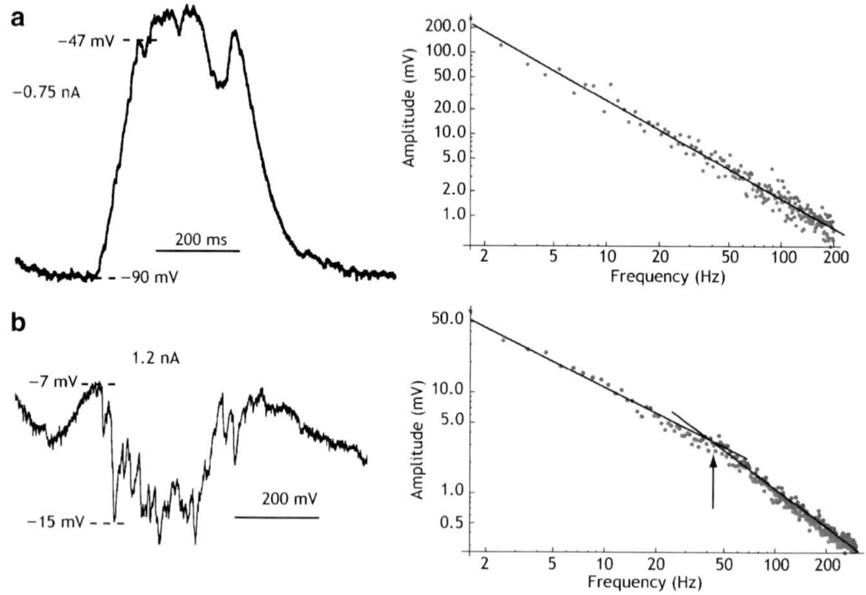

Fig. 2 The inhibitory component of the Up state contains high-frequency fluctuations. Up states recorded during passage of hyperpolarizing (**a**) or depolarizing (**b**) currents to change the relative driving forces for excitation and inhibition reveal that the inhibitory component contains proportionately more high-frequency fluctuations, whereas the excitatory component is dominated by the low-frequency process associated with the state transitions

neurons recorded without ion channel blockers. The low cutoff frequency [too low to see in Fig. 2a, but illustrated by Stern et al. (1997)] was near 1 Hz. As pointed out by Stern et al. this spectrum matches that of the Up and Down state transitions in the urethane-ketamine anesthetized animal and apparently reflects the correlation structure of excitatory inputs responsible for the generation of Up and Down states in anesthetized animals. The excitation is correlated, but with a broad correlation window in which brief transients are small and rare. The spectrum of inhibition also reflects this slow process, which is responsible for the fact that Up states average about 500 ms in duration, but the magnitude of the slow periodicity is about one fourth of that seen when excitation is dominant. In the traces dominated by inhibition, an additional cutoff frequency is seen near 50 Hz (arrow in Fig. 2b, right), which corresponded to fast fluctuations in membrane potential lasting 25 ms and less (see trace at b, left). These fast fluctuations match the time course of inhibitory synaptic potentials generated by GABAergic somatic synapses formed by the axons of fast-spiking interneurons. Thus, the variability of membrane potential in the Up state can be roughly separated into two components, one consisting of fast transients generated by the action of a small number of powerful inhibitory inputs to the spiny neuron, and slower fluctuations generated by weak correlations among excitatory inputs to the spiny cell.

These considerations inform the study of spike generation in the Up state in striatal spiny neurons in vivo. To examine the events responsible for spike generation, we compared action potentials generated naturally in the Up state and those generated by depolarizing current pulses or by antidromic activation by stimulation of the spiny cell's axon in the substantia nigra or globus pallidus. These recordings were made using sharp electrode recordings in vivo, in the absence of any ion channel blockers, and using potassium methyl sulfate as the micropipette electrolyte. This comparison is shown in Fig. 3. As expected from their remote origin, antidromically activated

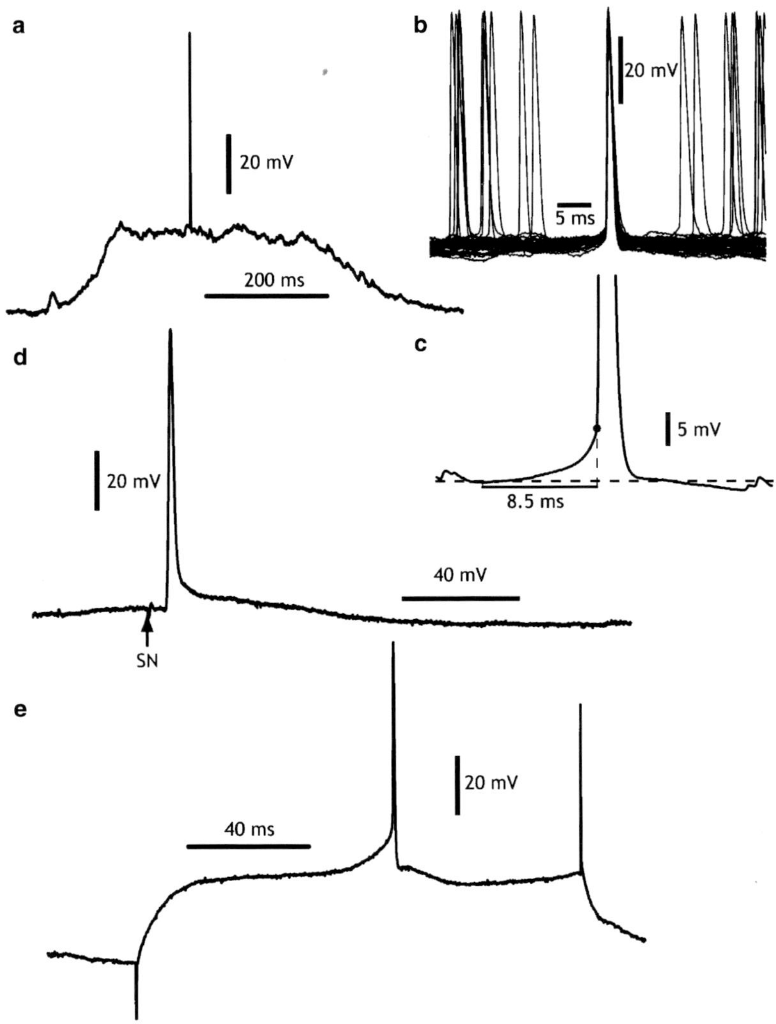

Fig. 3 Voltage transients preceding action potentials in vivo, during the Up state (**a–c**), during antidromic activation (**d**), and in response to a current step (**e**). (**a**) Action potential arising from fluctuations in membrane potential during an Up state in vivo. (**b**) Superimposed action potentials triggered in the Up state. (**c**) Average membrane potential preceding spontaneous action potentials in the Up state. (**d**) Antidromic action potential from substantia nigra (SN) stimulation in vivo. (**e**) Action potential triggered from intracellular current pulse

action potentials arose abruptly from a flat baseline potential. In contrast, action potentials generated by constant current pulses (in the absence of fast transients) arose gradually, and the trajectory to spiking was controlled by the activation of voltage-sensitive currents activated near threshold. This is the action potential trajectory expected if action potentials were generated by a near-threshold constant synaptic depolarization interacting with voltage-sensitive conductances such as those responsible for calcium currents and persistent sodium currents activated near threshold (e.g., Hernández-López et al. 1997). The average membrane potential trajectory of naturally occurring action potentials in the Up state was calculated by spike-triggered averaging of the membrane potential preceding action potentials generated in the Up state. The membrane potential trajectories preceding action potentials in the Up state were much briefer than the trajectories seen using constant current pulses, indicating that the spikes were driven by brief membrane potential transients that began 5–15 ms before the onset of the action potential (Fig. 3c), as predicted by computational studies (Wilson 1995).

These results suggest that action potentials in the Up state in anesthetized animals may be triggered by fast transients in the membrane potential caused by the offset transients of a few powerful inhibitory synapses (Koós and Tepper 1999). A likely source of such inhibitory synapses would be the axons of fast-spiking feedforward interneurons that are activated during the Up state by the same excitatory inputs that drive the Up state in the spiny cells (Plenz and Kitai 1998; Mallet et al. 2005). Although they receive synaptic excitation from the same cortical axons that innervate spiny cells, each interneuron receives a much smaller cortical innervation, so the variance in their input over time is expected to be much greater than that of the spiny neuron. This may be enough to generate the variations in synaptic inhibition shown here and to generate rapid fluctuations in membrane potential that are too fast to be dampened by corresponding changes in dendritic K^+ conductance. In addition, however, striatal GABAergic interneurons are subject to a synaptic influence not shared by the spiny neurons. GABAergic inhibitory neurons in the external segment of the globus pallidus preferentially innervate striatal interneurons (Bevan et al. 1998). Unlike neurons in the internal pallidal segment and substantia nigra pars reticulata, external pallidal neurons are often seen to show increases in firing rate in response to stimuli that also excite spiny neurons (DeLong 1971). These excitatory responses are thought to result from the powerful excitatory input to these cells arising from the subthalamic nucleus (Nambu et al. 2000). Subthalamic neurons in turn are excited by glutamatergic inputs from the cerebral cortex. If corticostriatal and cortico-subthalamic inputs are activated together, as happens when the frontal cortex is electrically stimulated, inhibition from the globus pallidus may mitigate the excitatory effect of the cortex on striatal inhibitory interneurons, but the striatal spiny cells are disinhibited at the moment that they receive cortical excitation.

This speculative circuit for the generation of action potentials in striatal spiny neurons in anesthetized or sleeping animals is shown in Fig. 4. One component of the input to the striatal neuron in this scheme is the slowly varying cortical input in the Up state, which is ineffective at triggering action potentials because it is

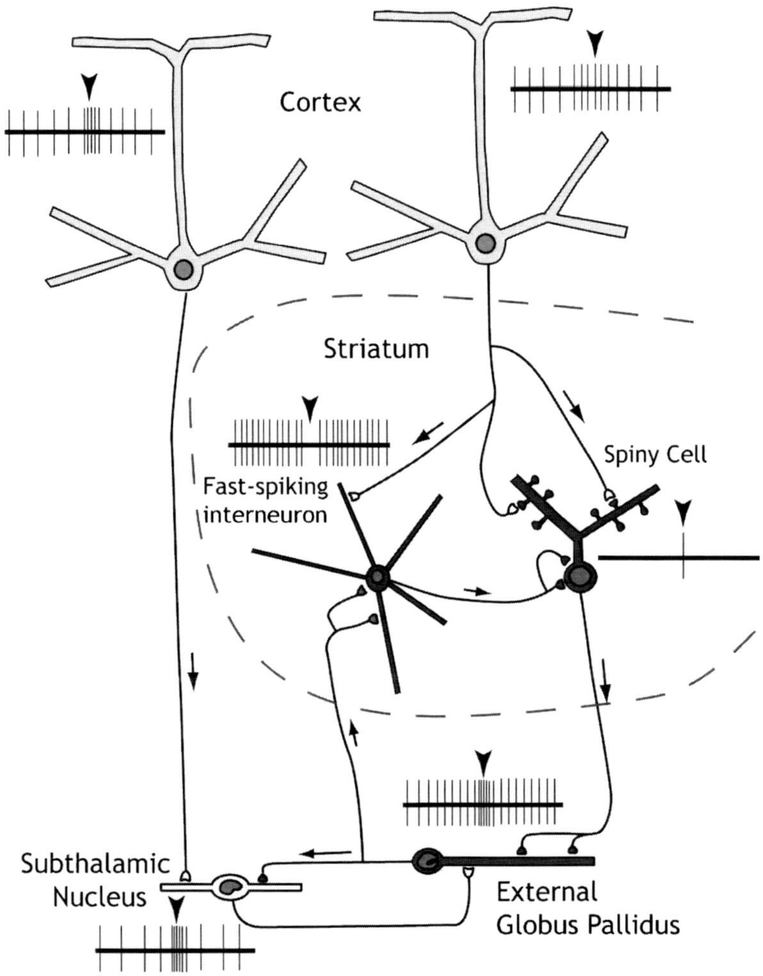

Fig. 4 A speculation on the origin of disinhibition responsible for firing of striatal spiny neurons in the Up state. Neurons of the GPe preferentially inhibit interneurons in the striatum. Their activity is controlled by excitatory input from via the cortico–subthalamo–pallidal circuit that can produce phasic increases in firing rate. Simultaneous activity in the corticostriatal and cortico–subthalamo–pallidal pathways could produce the combination of excitation and disinhibition required to generate action potentials in the spiny neuron

counteracted by the opposing effect of dendritic depolarization-activated K^+ current. The second component of the input is a rapidly varying inhibition from a small number of striatal interneurons. This input is responsible for the fast variations in membrane potential in the Up state, including fast disinhibition that can trigger action potentials. Such disinhibition events could be triggered by bursts of activity in the cortico–subthalamo–pallidal pathway.

In alert animals, striatal neurons often are maintained for long periods in a depolarized state similar to that of the Up state (Mahon et al. 2006), which correspond to similar periods of prolonged excitation in the cortex (Steriade et al. 2001; Destexhe et al. 2007). Usually, the depolarization during these Up state-like depolarizations is not as strong as that seen in anesthetized and sleeping animals, and it is also associated with sparse firing. Perhaps in this state, as in the Up state of anesthetized animals, firing of spiny neurons may depend on an interaction between cortical inputs and fast disinhibition from striatal inputs. If so, firing in striatal spiny neurons may signal the coincidence of two signals in the cortex, one directed toward the spiny neuron itself and one that excites the subthalamic nucleus and in doing so briefly relieves the spiny neuron of inhibition.

References

Bevan MD, Booth PA, Eaton SA and Bolam JP (1998) Selective innervation of neostriatal interneurons by a subclass of neuron in the globus pallidus of the rat. J Neurosci 18: 9438–9452.

Blackwell KT, Czubayko U and Plenz D (2003) Quantitative estimate of synaptic inputs to striatal neurons during up and down states in vitro. J Neurosci 23: 9123–9132.

Choi S and Lovinger DM (1997) Decreased frequency but not amplitude of quantal synaptic responses associated with expression of corticostriatal long-term depression. J Neurosci 17: 8613–8620.

DeLong MR (1971) Activity of pallidal neurons during movement. J Neurophysiol 34: 414–427.

Destexhe A, Hughes SW, Rudolph M and Crunelli V (2007) Are corticothalamic 'up' states fragments of wakefulness? Trends Neurosci 30: 334–342.

Gustafson N, Gireesh-Dharmaraj E, Czubayko U, Blackwell KT and Plenz D (2006) A comparative voltage and current-clamp analysis of feedback and feedforward synaptic transmission in the striatal microcircuit in vitro. J Neurophysiol 95: 737–752.

Haider B, Duque A, Hasenstaub AR and McCormick DA (2006) Network activity in vivo is generated through a dynamic balance of excitation and inhibition. J Neurosci 26: 4535–4545.

Hernández-López S, Bargas J, Surmeier DJ, Reyes A and Galarraga E (1997) D1 receptor activation enhances evoked discharge in neostriatal medium spiny neurons by modulating an L-type Ca^{2+} conductance. J Neurosci 17: 3334–3342.

Koós T and Tepper JM (1999) Inhibitory control of neostriatal projection neurons by GABAergic interneurons. Nat Neurosci 2: 467–472.

Koós T, Tepper JM and Wilson CJ (2004) Comparison of IPSCs evoked by spiny and fast-spiking neurons in the neostriatum. J Neurosci 24: 7916–7922.

Mahon S, Vautrelle N, Pezard L, Slaght SJ, Deniau JM, Chouvet G and Charpier S (2006) Distinct patterns of striatal medium spiny neuron activity during the natural sleep–wake cycle. J Neurosci 26: 12587–12595.

Mallet N, Le Moine C, Charpier S and Gonon F (2005) Feedforward inhibition of projection neurons by fast-spiking GABA interneurons in the rat striatum in vivo. J Neurosci 25: 3857–3869.

Nambu A, Tokuno H, Hamada I, Kita H, Imanishi M, Akazawa T, Ikeuchi Y and Hasegawa N (2000) Excitatory cortical inputs to pallidal neurons via the subthalamic nucleus in the monkey. J Neurophysiol 84: 289–300.

Plenz D and Kitai ST (1998) 'Up' and 'down' states in striatal medium spiny neurons simultaneously recorded with spontaneous activity in fast-spiking interneurons studied in cortex–striatum–substantia nigra organotypic cultures. J Neurosci 18: 266–283.

Plenz D and Thiagarajan TC (2007) The organizing principles of neuronal avalanches: cell assemblies in the cortex? Trends Neurosci 30: 101–110.

Rudolph M, Pelletier J-G, Par D and Destexhe A (2005) Characterization of synaptic conductances and integrative properties during electrically induced EEG-activated states in neocortical neurons in vivo. J Neurophysiol 94: 2805–2821.

Rudolph M, Pospischil M, Timofeev I and Destexhe A (2007) Inhibition determines membrane potential dynamics and controls action potential generation in awake and sleeping cat cortex. J Neurosci 27: 5280–5290.

Steriade M, Timofeev I and Grenier F (2001) Natural waking and sleep states: a view from inside neocortical neurons. J Neurophysiol 85: 1969–1985.

Stern EA, Kincaid AE and Wilson CJ (1997) Spontaneous subthreshold membrane potential fluctuations and action potential variability of rat corticostriatal and striatal neurons in vivo. J Neurophysiol 77: 1697–1715.

Stern EA, Jaeger D and Wilson CJ (1998) Membrane potential synchrony of simultaneously recorded striatal spiny neurons in vivo. Nature 394: 475–478.

Waters J and Helmchen F (2006) Background synaptic activity is sparse in neocortex. J Neurosci 26: 8267–8277.

Wickens JR and Wilson CJ (1998) Regulation of action-potential firing in spiny neurons of the rat neostriatum. J Neurophysiol 79: 2358–2364.

Wilson CJ (1995) Dynamic modification of dendritic cable properties and synaptic transmission by voltage-gated potassium channels. J Comput Neurosci 2: 91–115.

Wilson CJ (2004) Basal ganglia. In: Shepherd GM (ed.) The Synaptic Organization of the Brain. Oxford University Press, Oxford, pp 361–413.

Wilson CJ (2007) GABAergic inhibition in the neostriatum. Prog Brain Res 160: 91–110.

Wilson CJ and Kawaguchi Y (1996) The origins of two-state spontaneous membrane potential fluctuations of neostriatal spiny neurons. J Neurosci 16: 2397–2410.

Asymmetric Encoding of Positive and Negative Expectations by Low-Frequency Discharge Basal Ganglia Neurons

Mati Joshua, Avital Adler, and Hagai Bergman

Abstract Experimental and theoretical studies depict the basal ganglia as a reinforcement learning system where the dopaminergic neurons provide reinforcement error signal by modulation of their firing rate. However, the low tonic discharge rate of the dopaminergic neurons suggests that their capability to encode negative events by suppressing firing rate is limited. We recorded the activity of single neurons in the basal ganglia of two monkeys during the performance of probabilistic conditioning task with food, neutral and air-puff outcomes. In a related paper we analyzed the activity of five basal ganglia populations; here, we extend this to the low-frequency discharge (LFD) neurons of the main axis of the basal ganglia, that is, the striatal phasically active neurons (PANs), and the LFD neurons in the external segment of the globus pallidus (GPe). The licking and blinking behavior during the cue presentation epoch reveals that monkeys expected different probabilistic appetitive, neutral, and aversive outcomes. Nevertheless, the activity of single striatal and GPe neurons is more strongly modulated by expectation of reward than by expectation of the aversive event. The neural-behavioral asymmetry suggests that expectation of aversive events and rewards is differentially represented at many levels of the basal ganglia.

1 Introduction

Experimental and theoretical studies depict the basal ganglia as a reinforcement learning system where the dopaminergic neurons provide the reinforcement error signal by modulation of their firing rate. Previous studies in primates have shown that basal ganglia activity is modulated by expectation of rewards. Most of these studies have focused on midbrain dopaminergic neurons and striatal cholinergic

M. Joshua(✉), A. Adler, and H. Bergman
Interdisciplinary Center for Neural Computation, The Hebrew University, Jerusalem, Israel
e-mail: mati@alice.nc.huji.ac.il

H.J. Groenewegen et al. (eds.), *The Basal Ganglia IX*, Advances in Behavioral Biology 58,
DOI 10.1007/978-1-4419-0340-2_5,

interneurons [tonically active neurons (TANs), Wilson et al. (1990)]. Midbrain dopaminergic neurons have been shown to encode the mismatch in the positive domain of reinforcement; that is, they respond when conditions are better than expected (Fiorillo et al. 2003; Satoh et al. 2003; Nakahara et al. 2004; Bayer and Glimcher 2005). TANs also modulate their activity when a reward is given (Kimura et al. 1984) or expected (Graybiel et al. 1994). However, some reports have indicated that TAN modulation is invariant to reward predictability (Shimo and Hikosaka 2001; Morris et al. 2004). Finally, although to a lesser extent, several studies have demonstrated that the main axis of the basal ganglia is modulated by expectation of reward (Sato and Hikosaka 2002; Arkadir et al. 2004; Samejima et al. 2005; Darbaky et al. 2005; Pasquereau et al. 2007; Lau and Glimcher 2007).

In contrast to the extensive research on reward-related activity, only a few physiological studies have explored whether neural activity of the basal ganglia encodes the negative domain (i.e., aversive outcome or omission of rewards, which as outlined later might not be identical events). Dopamine neurons decrease their firing rate in response to reward omission (Schultz 1998; Satoh et al. 2003; Matsumoto and Hikosaka 2007). However, this suppression is limited since the firing rate is truncated at zero. In fact, other groups (Morris et al. 2004; Bayer and Glimcher 2005) have reported that the firing of dopaminergic neurons does not demonstrate incremental encoding of reward omission, and alternative encoding schemes have been proposed (Tobler et al. 2005; Bayer et al. 2007). There are even fewer studies on the responses of the basal ganglia neurons to aversive events. Although arising from slightly different experimental paradigms (instrumental vs. classical conditioning, behaving vs. anesthetized animals) and slightly different recording locations (ventral tegmental area vs. the more lateral substantia nigra pars compacta) the findings are incompatible. Some studies suggest that dopamine neurons increase their firing rate following aversive events (Guarraci and Kapp 1999; Coizet et al. 2006) whereas others have evidence of a decrease (Mirenowicz and Schultz 1996; Ungless et al. 2004). There are reports that TAN activity differentiates motivationally opposing stimuli (Ravel et al. 2003; Yamada et al. 2004), but it remains unclear whether and how TANs respond to expectation of aversion.

We designed a classical conditioning paradigm with aversive and rewarding probabilistic outcomes. Thus, symmetric manipulations of expectations of food (appetitive event) or air puff (aversive event) were built into the experimental design, allowing for comparison of neural responses to expectation of positive and negative outcomes. In parallel studies (Joshua et al. 2008; Joshua et al. 2009) we describe the activity of basal ganglia critics (midbrain dopaminergic neurons and striatal cholinergic interneurons) and high-frequency (>50 spikes/s) discharge neurons in the main axis (external and internal segments of the globus pallidus (GPe) and the substantia nigra pars reticulata). In this manuscript, we analyzed the activity of two basal ganglia populations with low-frequency (<20 spikes/s) discharge, the striatal phasically active neurons (PANs), and the low-frequency discharge (LFD) neurons in the external segment of the GPe.

2 Methods

All experimental protocols were performed in accordance with the National Institute of Health *Guide for the Care and Use of Laboratory Animals* and with Hebrew University guidelines for the use and care of laboratory animals in research, supervised by the institutional animal care and use committee. Two monkeys (L and S, *Macaque fascicularis*, female 4 kg and male 5 kg) were engaged in a probabilistic delay classical-conditioning task. At the beginning of each trial, a visual cue covering the full extent of a 17-in. computer screen was presented for 2 s. After the cue offset the monkeys received the outcome in a probabilistic manner. Images were fractal patterns constructed with the Chaos Pro 3.2 program (http://www.chaospro.de), with the same images presented during all training and recording periods. We delivered liquid food (L: 0.4 ml, 100-ms duration, S: 0.6 ml, 150 ms) as the positive reward and an air puff (L: 100-ms duration, S: 150 ms, 50–70 psi, 2 cm from eye), split and directed to both eyes, as an aversive stimulus. To enhance the monkeys' ability to discriminate trial outcomes, the beginning of the result epoch was signaled by one of three sounds that discriminated the delivery of food, the delivery of the air puff, and no outcome: that is, each possible end result was accompanied by a different sound. Sounds were normalized to the same intensity and duration. These sounds were additional to the background device sounds (air-puff solenoid and food pump). Sounds and visual cues were shuffled between monkeys. All trials were followed by a variable intertrial interval (ITI) (monkey S: 3–7 s, monkey L: 4–8 s). Because of the different probabilities and in order to equalize the average occurrence of each outcome we introduced the nondeterministic cues ($P \neq 1$ for reward or aversive) three times more than the deterministic ones. With this occurrence ratio, all trials were randomly interleaved.

3 Results

3.1 Monkey Behavior Reflects Expectation of Rewarding and Aversive Events

We recorded neuronal activity in the basal ganglia (Fig. 1a, b) during performance of a probabilistic classical conditioning task (Fig. 1c) with food or air puff as the rewarding and aversive outcomes, respectively. The two monkeys were introduced to seven different fractal visual cues, each predicting the outcome in a probabilistic manner. Three cues predicted a food outcome (reward cues) with a delivery probability of 1/3, 2/3, and 1; three cues predicted an air-puff outcome (aversive cues) with a delivery probability of 1/3, 2/3, and 1. The seventh cue (the neutral cue) was never followed by a food or air-puff outcome. The same seven fractal cues were presented to both monkeys; however, the associated outcomes were randomized

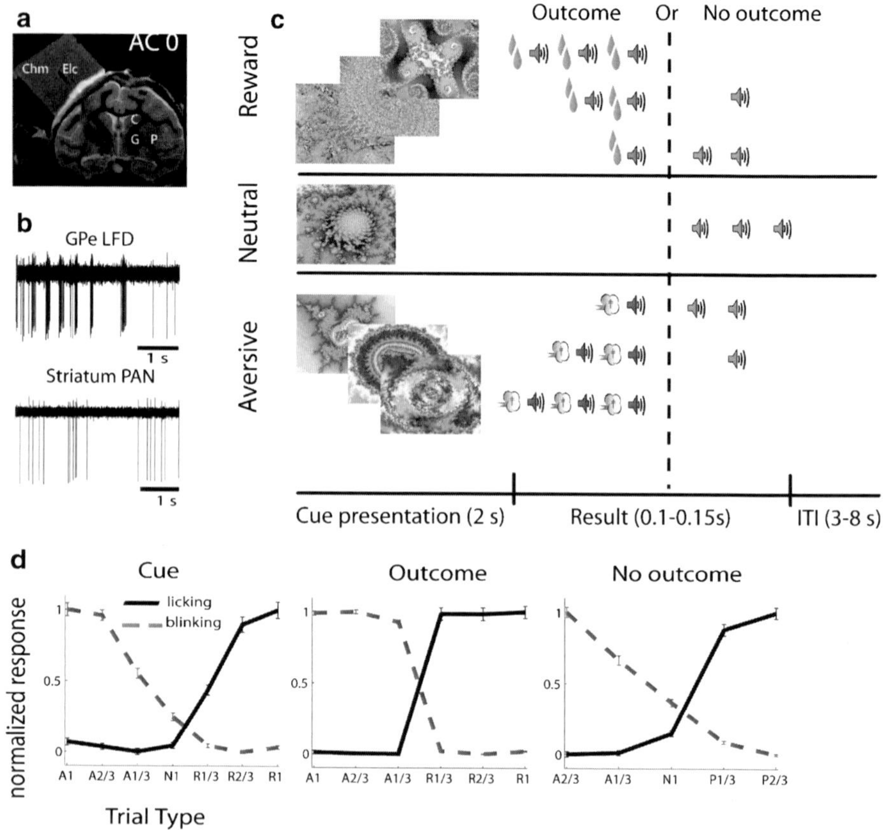

Fig. 1 MRI, electrophysiology and behavior. (**a**) MRI identification of recording coordinates. Coronal MRI image at anterior commissure level. Tungsten microelectrodes are inserted at known chamber coordinates enabling identification of the brain structures by alignment of the MRI images with the monkey atlas. Abbreviations: *C* caudate, *Chm* recording chamber (filled with 3% agar), *Elc* electrode, *G* globus pallidus, *P* putamen. (**b**) Example from the recordings of a low-frequency discharge neuron in the GPe (*top*) and a phasically activated neuron in the striatum (*down*). (**c**) Behavioral task. *Top* – reward trials; *middle* – neutral trials; *bottom* – aversive trials. (**d**) Normalized behavioral response. Licking (*black*) and blinking (*gray*) response (average ± SEM) in a time window around the behavioral event (cue: 500–0 ms before cue ending; outcome and no outcome: 0–500 ms after cue ending for blinking response and 500–1,000 ms for licking response). The responses are normalized in each epoch by the minimal and maximal values (normalized responses = [(Response – min)/(max – min)]. Abscissa: different behavioral conditions (*A* aversive, *N* neutral, *R* reward; the number is the outcome probability)

for each monkey. Cues were presented for 2 s and were immediately followed by a result epoch, which could be an outcome (food, air puff) or no outcome, according to the probabilities associated with the cue. The beginning of the result epoch was signaled by one of three sounds that discriminated the three possible events: a drop of food, an air puff, or no outcome (Fig. 1c).

We tested how conditioning affected the monkeys' behavior by monitoring licking and blinking during neural recordings. The monkeys increased their licking in response to cues predicting food but only slightly to the aversive and neutral cues. Similarly the monkeys increased their frequency of blinking to cues predicting an air puff but only slightly to reward and neutral cues (Fig. 1d, left). Moreover, the increase of blinking and licking during the cue epoch was maximal in trials where the probability of the outcome was 2/3 or 1 and smaller in trials where the probability was 1/3. When no food or air puff were delivered (no outcome epoch – the $P = 1/3$ or $P = 2/3$ trials) licking and blinking increased, respectively. Furthermore, the increase was in accordance with the previously instructed probability (Fig. 1d, right). These behavioral results indicate that the monkeys could distinguish between aversive, reward, and neutral cues and between the different outcome probabilities they were intended to signal, and that the symmetry in task design was reflected by the monkeys' behavior.

3.2 PANs and GPe LFD Activity Are Asymmetrically Modulated by Expectation of Aversive and Reward Outcomes

We recorded single unit activity from putamen PANs and from GPe LFD neurons (Fig. 1b). Cells were included in the study if they passed criteria of waveform isolation (isolation score >0.5; Joshua et al. 2007) and discharge rate stability. Of the cells that met these criteria we further analyzed only those that were recorded during at least 20 min of task performance. Finally, we grouped the neurons in the same structure recorded from both monkeys. A total of 113 neurons were recorded of which 65 neurons (38 PANs and 27 GPe LFD) passed the earlier criteria.

Figure 2a is an example of a PAN recorded during the performance of the behavioral task. The neural activity following the reward cue was larger than the activity following the neutral and aversive cues. Furthermore, the response to reward cues increased with reward probability. Population analysis shows that both GPe LFD neurons and PANs had larger responses to reward cues than to aversive cues and a larger fraction of cells responded to the reward cues (Fig. 2b, c). The population PSTH (Fig. 2b, c – left columns) is an average estimate and may be biased by a few neurons with an extreme response. However, analysis of the fraction of cells with a significant response (Fig. 2b, c – right columns) is not sensitive to the relative amplitude of the responses. We, therefore, formulated the response index as a measure of the relative differences between the neutral vs. the aversive and the reward responses of single neurons. We found that for the majority of cells the response index for the reward trials was larger than the response index for aversive trials (Fig. 3a, b). In addition, a substantial fraction of the LFD neurons in the basal ganglia showed a significant response index to reward cues, whereas only a small number of cells had a significant response index to aversive cues (Fig. 3c).

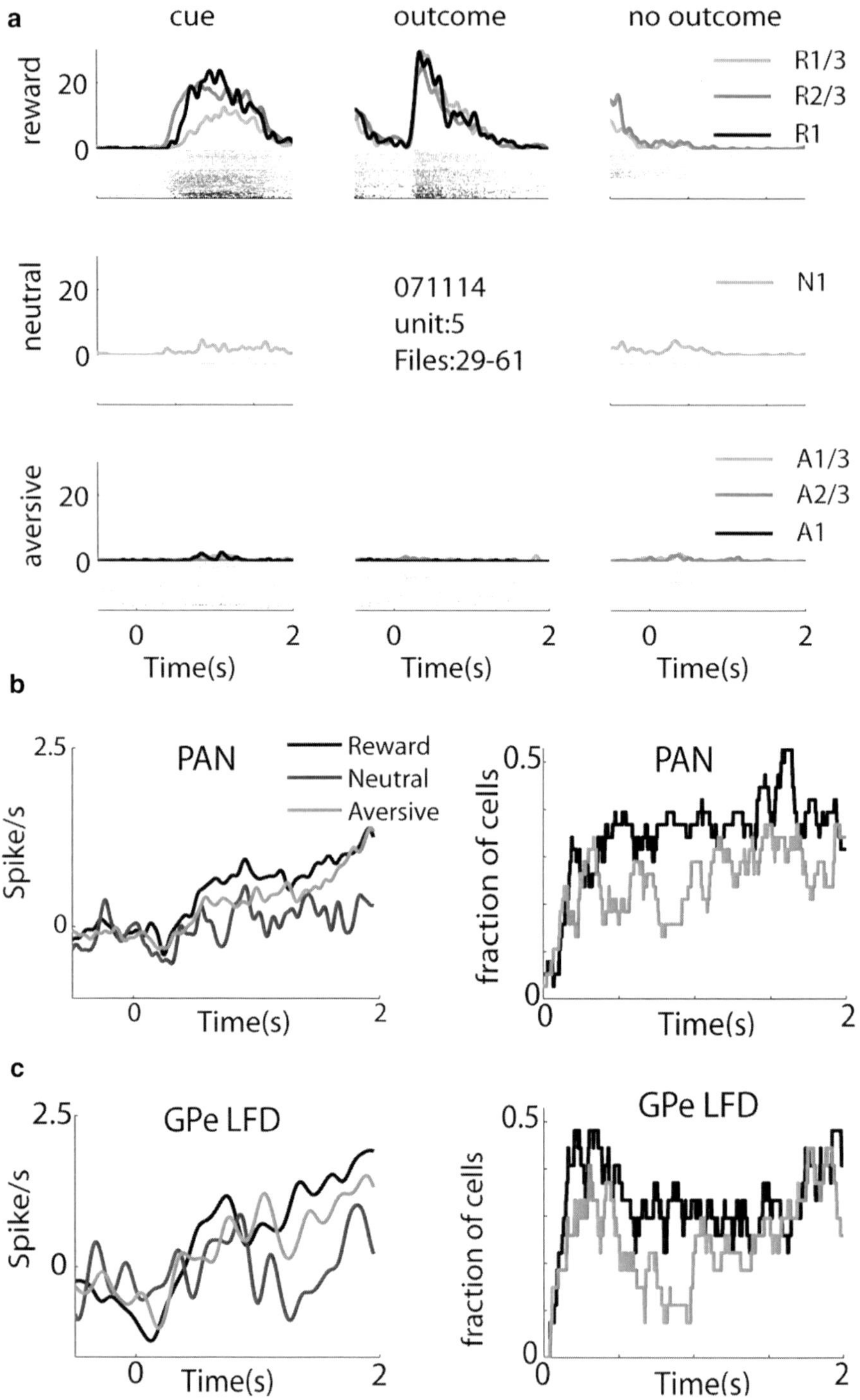
a
cue
outcome
no outcome
reward
20
0
R1/3
R2/3
R1
neutral
071114
unit:5
Files:29-61
N1
aversive
A1/3
A2/3
A1
Time(s)
b
PAN
Reward
Neutral
Aversive
Spike/s
2.5
fraction of cells
0.5
c
GPe LFD

4 Discussion

We have shown that despite the symmetry in behavior, expectation of reward – but not of an aversive event – affects rate modulations of basal ganglia LFD neurons beyond the modulation that followed a neutral cue. Asymmetry in value expectation is congruent with theories on the localization of antagonistic motivational systems (Konorski 1967). It has been shown that neural systems other than the basal ganglia, for example, the amygdala (LeDoux 2000) and the cerebellum (McCormick and Thompson 1984), are involved in aversive conditioning. In parallel studies, we showed that both the neuromodulators (TANs and SNc) and basal ganglia populations with high-frequency discharge (SNr, GPi, and GPe) encode expectation of reward but not the expectation of an aversive event (Joshua et al. 2008; Joshua et al. 2009). Here, we extend this notion and show similar results for the PANs and for the LFD neurons of the GPe.

In a previous study, Samejima et al. (2005) showed that the activity of many striatal projection neurons was selective to both values and action; however, only a few neurons were tuned to relative values or action choice. Lau and Glimcher (2007) reported that the encoding of action and outcome was carried out by largely separate populations of caudate neurons that were active after movement execution. Although these studies focus on different trial epochs they are suggestive of two different coding schemes. In our task the monkey performed actions in the cue epoch both in the reward and the aversive trials but not in the neutral trials (Fig. 1d). However, although the action dissociates neutral and aversive trials, responses to

Fig. 2 Population responses of striatal PANs and GPe LFD neurons to reward cues are larger and more common than responses to aversive cues. (**a**) Example of rasters and peristimulus time histograms (PSTHs) of a PAN aligned to behavioral events. The rows are separated according to the expected outcome. *First row*: trials with cues that predict the delivery of food. *Second row*: trials with the neutral cue (a cue always followed by no outcome). *Third row*: trials with cues that predict an air puff. Columns are aligned according to the trial epoch. *First column*: cue presentation epoch (–0.5 s to 2 s after cue onset). *Second column*: outcome epoch (–0.5 s to 2 s after delivery of food or air puff). *Third column*: trials in which no outcome was delivered; outcome omission was signaled to the monkey by the no-outcome sound (–0.5 s to 2 s after sound onset). The first 0.5 s of the second and third column overlap the last 0.5 s of the first column. Gray level codes are marked at the right side of the no-outcome rasters (*A* aversive, *N* neutral, *R* reward; the number is the outcome probability). PSTHs were constructed by summing activity across trials in 1-ms resolution and then smoothing with a Gaussian window (SD of 20 ms). (**b**) *Left column*: Population responses of PANs ($n = 38$) to behavioral cues. PSTH were smoothed with a Gaussian (SD = 20) and averaged across cells. *Black* – average responses to reward predicting cues; *gray* – neutral cue, light gray – aversive cues. *Right column*: Fraction of cells with significant (two-sigma rule) modulations of firing rate in the cue epoch. Same color coding as in the *left column*. Neutral events are not included to enable inclusion of all rewarding/aversive events in the statistical tests. The ordinate is the fraction of cells that had a significant response at each time bin (1 ms). (**c**) *Left column*: Population responses of GPe LFD neurons ($n = 27$) to behavioral cues. Smoothing with a SD = 40 ms. *Right column*: Fraction of cells with significant firing rate modulations in the cue epoch. Same analysis and gray level code as in (**b**)

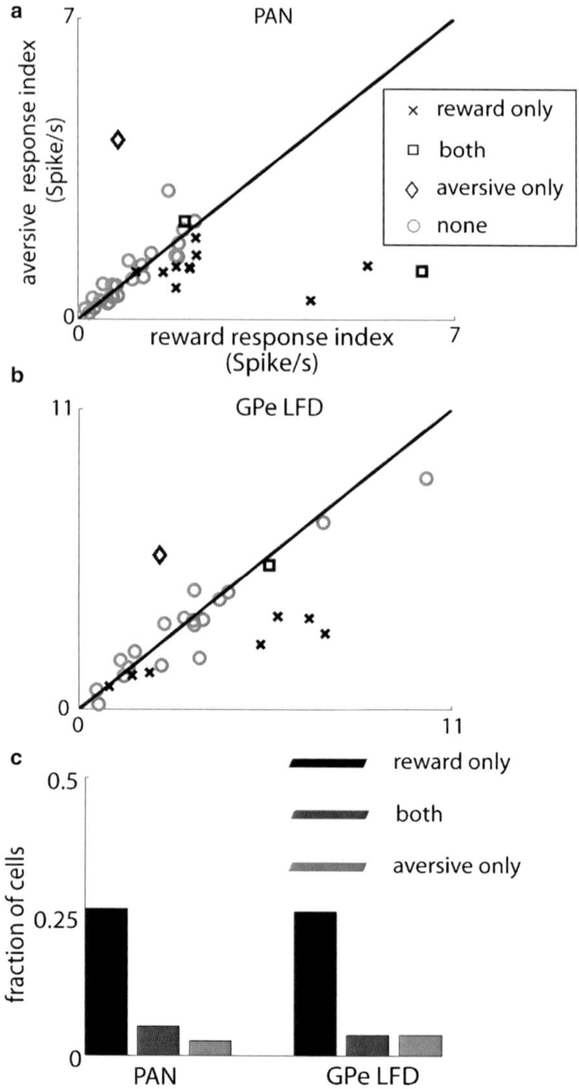

Fig. 3 Response-index analysis reveals larger responses to reward than to aversive cues. (**a**) Scatter plots comparing the response index of individual PAN to reward and aversive cues. The response index was calculated for each cell as the absolute difference between the (aversive or reward) cue-aligned PSTH and the PSTH of the neutral cue. The *black line* is the identity ($Y = X$) line. Points below this line represent cells with a response index that is larger for the reward cues than for aversive cues. The differences in the scale between nuclei reflect differences in modulation size. Significance level was $p < 0.05$. The time window used for this analysis was 0–2,000 ms from cue presentation. (**b**) Same as (**a**), for the GPe LFD population. (**c**) Summary of the fraction of cells with a significant response index ($p < 0.05$, bootstrap test)

the aversive and neutral cue were the same (Fig. 3a). This suggests that it is not action per se that is encoded in these neurons. Furthermore, in the reward trials we found considerable modulations, suggesting that it is the action value that is encoded in the striatum.

We found that the fraction of cells with a short latency response was larger for the GPe LFD than for the PANs (Fig. 2b right vs. Fig. 2c right). This difference is surprising since the striatum is the main input of the GPe. Our recordings in the striatum are limited to the putamen and it could be that the fast responses of GPe LFD are due to input from other striatum territories with faster responses.

Human decisions are not symmetric in response to negative and positive prospects (Tversky and Kahneman 1981). Here, we show that the basal ganglia encoding of the positive domain surpasses their encoding of the negative domain. We extend our report to the input stage of the basal ganglia – the striatum, as well as to a unique population of neurons in the GPe – the LFD neurons. Having two biological systems, one for the aversive domain and one for the reward domain might be the neural basis of risk-aversive, asymmetric, and nonrational human behavior (Tversky and Kahneman 1981).

References

Arkadir D, Morris G, Vaadia E and Bergman H (2004) Independent coding of movement direction and reward prediction by single pallidal neurons. J Neurosci 24: 10047–10056.

Bayer HM and Glimcher PW (2005) Midbrain dopamine neurons encode a quantitative reward prediction error signal. Neuron 47: 129–141.

Bayer HM, Lau B and Glimcher PW (2007) Statistics of midbrain dopamine neuron spike trains in the awake primate. J Neurophysiol 98: 1428–1439.

Coizet V, Dommett EJ, Redgrave P and Overton PG (2006) Nociceptive responses of midbrain dopaminergic neurones are modulated by the superior colliculus in the rat. Neuroscience 139: 1479–1493.

Darbaky Y, Baunez C, Arecchi P, Legallet E and Apicella P (2005) Reward-related neuronal activity in the subthalamic nucleus of the monkey. Neuroreport 16: 1241–1244.

Fiorillo CD, Tobler PN and Schultz W (2003) Discrete coding of reward probability and uncertainty by dopamine neurons. Science 299: 1898–1902.

Graybiel AM, Aosaki T, Flaherty AW and Kimura M (1994) The basal ganglia and adaptive motor control. Science 265: 1826–1831.

Guarraci FA and Kapp BS (1999) An electrophysiological characterization of ventral tegmental area dopaminergic neurons during differential pavlovian fear conditioning in the awake rabbit. Behav Brain Res 99: 169–179.

Joshua M, Elias S, Levine O and Bergman H (2007) Quantifying the isolation quality of extracellularly recorded action potentials. J Neurosci Methods 163: 267–282.

Joshua M, Adler A, Mitelman R, Vaadia E and Bergman H (2008) Midbrain dopaminergic neurons and striatal cholinergic interneurons encode the difference between reward and aversive events at different epochs of probabilistic classical conditioning trials. J Neurosci 28: 11673–11684.

Joshua M, Adler A, Rusin B, Vaadia E and Bergman H (2009) Encoding of probabilistic rewarding and aversive events by pallidal and nigral neurons. J. Neurophysiol 101: 758–772.

Kimura M, Rajkowski J and Evarts E (1984) Tonically discharging putamen neurons exhibit set-dependent responses. Proc Natl Acad Sci USA 81: 4998–5001.

Konorski J (1967) Integrative Activity of the Brain: An Interdisciplinary Approach. Chicago: Chicago University Press.

Lau B and Glimcher PW (2007) Action and outcome encoding in the primate caudate nucleus. J Neurosci 27: 14502–14514.

LeDoux JE (2000) The amygdala and emotion: a view through fear. In: Aggleton JP (ed) The Amygdala. Oxford: Oxford University Press, pp 289–310.

Matsumoto M and Hikosaka O (2007) Lateral habenula as a source of negative reward signals in dopamine neurons. Nature 447: 1111–1115.

McCormick DA and Thompson RF (1984) Cerebellum: essential involvement in the classically conditioned eyelid response. Science 223: 296–299.

Mirenowicz J and Schultz W (1996) Preferential activation of midbrain dopamine neurons by appetitive rather than aversive stimuli. Nature 379: 449–451.

Morris G, Arkadir D, Nevet A, Vaadia E and Bergman H (2004) Coincident but distinct messages of midbrain dopamine and striatal tonically active neurons. Neuron 43: 133–143.

Nakahara H, Itoh H, Kawagoe R, Takikawa Y and Hikosaka O (2004) Dopamine neurons can represent context-dependent prediction error. Neuron 41: 269–280.

Pasquereau B, Nadjar A, Arkadir D, Bezard E, Goillandeau M, Bioulac B, Gross CE and Boraud T (2007) Shaping of motor responses by incentive values through the basal ganglia. J Neurosci 27: 1176–1183.

Ravel S, Legallet E and Apicella P (2003) Responses of tonically active neurons in the monkey striatum discriminate between motivationally opposing stimuli. J Neurosci 23: 8489–8497.

Samejima K, Ueda Y, Doya K and Kimura M (2005) Representation of action-specific reward values in the striatum. Science 310: 1337–1340.

Sato M and Hikosaka O (2002) Role of primate substantia nigra pars reticulata in reward-oriented saccadic eye movement. J Neurosci 22: 2363–2373.

Satoh T, Nakai S, Sato T and Kimura M (2003) Correlated coding of motivation and outcome of decision by dopamine neurons. J Neurosci 23: 9913–9923.

Schultz W (1998) Predictive reward signal of dopamine neurons. J Neurophysiol 80: 1–27.

Shimo Y and Hikosaka O (2001) Role of tonically active neurons in primate caudate in reward-oriented saccadic eye movement. J Neurosci 21: 7804–7814.

Tobler PN, Fiorillo CD and Schultz W (2005) Adaptive coding of reward value by dopamine neurons. Science 307: 1642–1645.

Tversky A and Kahneman D (1981) The framing of decisions and the psychology of choice. Science 211: 453–458.

Ungless MA, Magill PJ and Bolam JP (2004) Uniform inhibition of dopamine neurons in the ventral tegmental area by aversive stimuli. Science 303: 2040–2042.

Wilson CJ, Chang HT and Kitai ST (1990) Firing patterns and synaptic potentials of identified giant aspiny interneurons in the rat neostriatum. J Neurosci 10: 508–519.

Yamada H, Matsumoto N and Kimura M (2004) Tonically active neurons in the primate caudate nucleus and putamen differentially encode instructed motivational outcomes of action. J Neurosci 24: 3500–3510.

Stimulation Effect on Neuronal Activity in the Globus Pallidus of the Behaving Macaque

Izhar Bar-Gad and Robert S. Turner

Abstract High-frequency deep brain stimulation in the globus pallidus has been shown to improve the symptoms of multiple disorders including Parkinson's disease, dystonia, and Tourette syndrome. However, the effects of pallidal stimulation on neuronal activity remain poorly understood. In this study, we used multielectrode recording and microstimulation in a normal behaving primate to study the effects of pallidal stimulation on local neuronal activity. We used multiple stimulation protocols varying in stimulus frequency and pattern. Most pallidal neurons responded to each of the protocols by altering their firing rate and pattern but the form of the response varied significantly across the neuronal population. We used principal component analysis to identify the primary features that make up those complex responses. In addition, a comparison of task-related changes in neuronal activity during different stimulation patterns revealed that the neuronal encoding of the task and the stimulation was primarily independent. These results support the notion that stimulation does not block all information flow from the basal ganglia via either inhibition or complete locking, but rather enables the partial transmission of behaviorally significant information along the cortico-basal ganglia loop.

1 Introduction

Surgical procedures involving ablation of parts of the basal ganglia have been used throughout most of the twentieth century for treating multiple disorders and primarily Parkinson's disease (Gildenberg 2003). High-frequency deep brain stimulation (HF-DBS) is replacing ablative surgeries for the treatment of

I. Bar-Gad (✉)
Gonda Brain Research Center, Goodman Faculty of Life Sciences, Bar-Ilan University, Ramat-Gan, 52900, Israel
e-mail: bargadi@mail.biu.ac.il

R.S. Turner
Department of Neurobiology & Center for the Neural Basis of Cognition, University of Pittsburgh, Pittsburgh, PA 15261, USA
e-mail: rturner@pitt.edu

H.J. Groenewegen et al. (eds.), *The Basal Ganglia IX*, Advances in Behavioral Biology 58,
DOI 10.1007/978-1-4419-0340-2_6,

Parkinson's disease (Benabid 2003), dystonia (Vidailhet et al. 2005), Tourette syndrome (Diederich et al. 2005), and other disorders associated with cortico-basal ganglia loop. In all of these disorders, the internal globus pallidus (GPi) serves as one of the key targets for stimulation and has been shown to provide a significant alleviation of the disorders' signs and symptoms.

The success of both lesions and HF-DBS in reducing disorders' signs has led many to assume that the effect of HF-DBS on the pallidal output is the same as a lesion (i.e., complete cessation of neuronal activity). This assumption was tested at the behavioral level in healthy primates (Horak and Anderson 1984a, b). These studies demonstrated significant changes in movement duration with no significant changes in reaction time during both high-frequency stimulation and lesion of the GP. The therapeutic efficacy of GPi lesions and the assumption of an inhibitory stimulation effect are also consistent with predictions of the predominant model of basal ganglia physiology, the box and arrow model (DeLong 1990; Albin et al. 1989). In that model, abnormally elevated pallidal activity leads to an inhibition of thalamocortical circuits and to hypokinetic signs. The reduction of GPi output by lesion or stimulation would thus reduce the inhibition of the cortical activity and improve normal motor activity.

Early neurophysiologic studies of the effects of DBS showed that the effect of high-frequency stimulation is indeed a complete inhibition of the neuronal activity (Boraud et al. 1996; Dostrovsky et al. 2000; Benazzouz et al. 2000; Beurrier et al. 2001). However, later studies revealed a more complex effect of stimulation. The response of pallidal neurons to microstimulation in the globus pallidus (Bar-Gad et al. 2004) and to macrostimulation in the subthalamic nucleus (STN) (Hashimoto et al. 2003) in the 1-methyl 4-phenyl 1,2,3,6-tetrahydropyridine (MPTP)-treated primate have shown that the firing rate of GP neurons does not decrease to zero and that the neuronal firing becomes partially locked to the stimulus pulses. Stimulation also leads to changes in firing pattern such as a reduction of the low-frequency oscillations characteristic of Parkinson's disease (Meissner et al. 2005; McCairn and Turner 2009). These studies, which were performed in a primate model of Parkinson's disease, addressed the effects of stimulation on pallidal activity in animals at rest.

In this study, we describe the effects of pallidal stimulation in the normal behaving primate. With this approach, we are able to explore the effects of stimulation on pallidal encoding of task-related information in conjunction with the encoding of the stimulation.

2 Methods

One male rhesus (*Macaca mulatta*) monkey (~9 kg) was involved in this study. The experimental protocol was approved by the UCSF Animal Care and Use Committee. Animals were cared for in accordance with the American Physiological Society Guiding Principles in the Care and Use of Animals (1991). The animal was trained to perform a simple two-choice reaching task with the right hand: the trial began with the placement of the hand on a central button which was followed, after a random delay, by one of the two visual cues instructing the animal to press a button located either to the left or to the right of the start position. Following a successful press of the side button, the monkey was rewarded with a food bolus. The animal achieved a stable performance level (>90% successful trials) prior to surgery.

Animals were prepared for recording and stimulation by an implantation of a recording chamber which allowed access to the left globus pallidus from a 45° lateral approach. For a detailed description of the surgical techniques see Turner and Anderson (1997). After recovery from surgery, the animal resumed the behavioral procedure and returned to the same level of task performance. During the execution of the behavioral task, four glass-coated tungsten electrodes (Alpha-Omega Engineering, Nazareth, Israel) were lowered independently using a high-resolution computer-controlled positioning system (EPS, Alpha-Omega Engineering). Electrophysiological data were acquired continuously at 20 kHz, displayed, sorted online, and saved for offline analysis (MAP, Plexon, Dallas, TX, USA).

The same electrodes were used for both recording and microstimulation (40 μA, 200 μs, biphasic – negative phase first). The stimulation produced large shock artifacts which varied over time and did not maintain their exact shape probably due to the capacitive properties of the electrode and amplification system. The stimulus artifacts were removed offline: the saturated periods (typically 0.6–1.2 ms) were replaced by null values and afterward, due to the nonstationary stimulus artifact, suppression was performed by removal of the curve produced using a nonlinear regression of a double exponential function: $a(t) = \alpha_1\beta_1 t + \alpha_2\beta_2 t + \theta$. Under the nonstationary conditions, this method produced results that were superior to simpler and faster methods such as removal of a moving mean (Bar-Gad et al. 2004).

3 Results

We recorded from one rhesus monkey performing a simple two-choice reaching task. Recordings were made using four microelectrodes in both segments of the GP during the task performance within a session. A full session consisted of 50 correct performances of the task; however, incomplete sessions of at least 20 correct trials were included in the analysis. A total of 97 recording sessions were used for the analysis which were divided to four basic types: (1) sessions with no stimulation; (2) sessions with low-frequency (10 stimulations per second) continuous stimulation; (3) sessions with high-frequency (140 stimulations per second) continuous stimulation; (4) sessions with high-frequency short bursts (200 ms bursts of 140 stimulations per second) interspersed by 500 ms of no stimulation period. The breakdown of the sessions and trials to the four types appears in Table 1.

Table 1 Number of sessions and trials across different stimulation protocols

	# Sessions	# Trials
No stimulation	37	1,504
Low-frequency stimulation	17	632
High-frequency stimulation	19	515
Burst stimulation	24	1,031

The performance of the tasks was compared between the four stimulation types. Reaction times and movement durations were compared between all pairs. The comparison showed that reaction times (i.e., the time interval between presentation of the instruction cue and onset of movement) increased significantly following burst and high-frequency stimulation protocols in the GPi. Movement durations (i.e., the time between leaving of the central button and pressing of the side button) increased significantly following all stimulation protocols in the GPe with no significant differences between the different stimulation protocols (in both cases, $p < 0.01$, one-sided ANOVA, Tukey HSD post hoc).

The effects of stimulation on single neuron firing rate and locking to stimulus varied significantly between neurons. Changes in the mean firing rate following HFS demonstrated an equal proportion of neurons increasing (23%) and decreasing (25%) their firing rate during high-frequency stimulation. Roughly half of the neurons (52%) showed no significant change in overall firing rates. The response of single neurons to the stimulation was complex and included multiple phases of excitation and inhibition. The response of the same neuron typically displayed very different modulations across different stimulation protocols. A close examination of one such neuron (Fig. 1) reveals that the neuron, which fired a basal rate of 75 spikes per second without stimulation, underwent a massive inhibition for a prolonged period (>20 ms) following low-frequency stimulation (Fig. 1a). This almost complete cessation of firing was followed by an even longer albeit milder increase in firing rate. Persistence of this response during higher frequency stimulation would have resulted in a complete cessation of neuronal firing. However, the response to high-frequency stimulation differed significantly. During continuous HF stimulation, the neuron responded with a large excitation approximately 3 ms after the stimulus which was flanked by short inhibitory periods (Fig. 1b). The stimulus artifact masked the neuronal activity within the initial 0.8 ms following the stimulus hiding any potential direct responses. The response to a repeating short burst of HF stimuli displayed time evolving dynamics during the burst itself and during the period immediately following the stimulation burst (Fig. 1c). During the burst, the neuron was inhibited for the first three stimulation pulses and initiated a partially locked firing from the fourth stimuli. Following the termination of the burst the neuron increased its rate (>150 spikes per second) immediately with a slow decay back to the baseline rate.

Stimulus-driven neuronal responses were common (Table 2) and varied significantly across the population and between stimulation conditions. Principal component analysis (PCA) was used to quantify the major elements of these responses (Fig. 2). The shape of the response that represents most of the variance within the different responses to the stimulus is described by the first principal component (PC1). The second principal component (PC2) represents the part of the response that has the most variance orthogonal to PC1 (and thus not described by it), and so forth. The response of the neurons to low-frequency stimulation was characterized primarily by an initial inhibition (20 ms) followed by a rebound excitation. The following PCs of the low-frequency stimulation describe the mid-term components (20–40 ms) with very little effects at later times (>40 ms) (Fig. 2a). The response to high-frequency stimulation was characterized by an early (1 ms) and two types

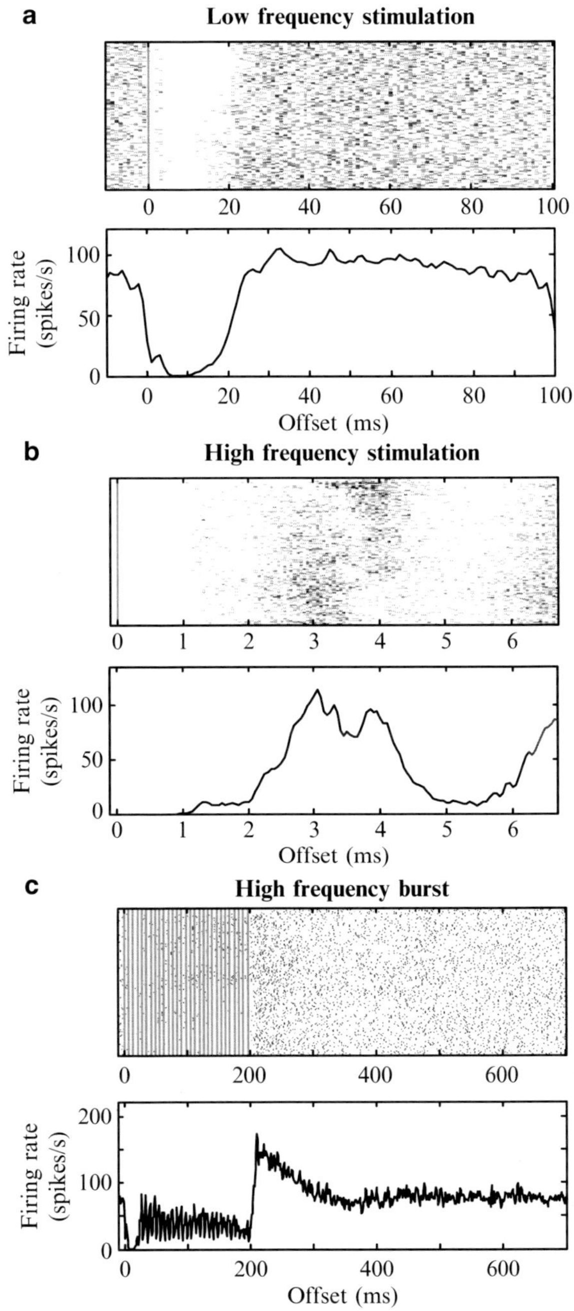

Fig. 1 Single neuron response to different pallidal stimulation protocols. Example of the response of a single pallidal neuron (T.050128.02.2a) to: (**a**) low-frequency stimulation, (**b**) high-frequency stimulation, and (**c**) burst stimulation. The response is presented via (*top*) peri-stimulus rasters (*gray lines* – stimulation times, first stimulation on top) and (*bottom*) peri-stimulus histograms. Presentation of the rasters uses grayscale encoding to enable the display of the large number of repetitions

Table 2 Locking of neurons to the stimulation across the stimulation protocols

	# Locked	# Total	%
Low-frequency stimulation	39	54	72
High-frequency stimulation	29	53	55
Burst stimulation	48	63	76

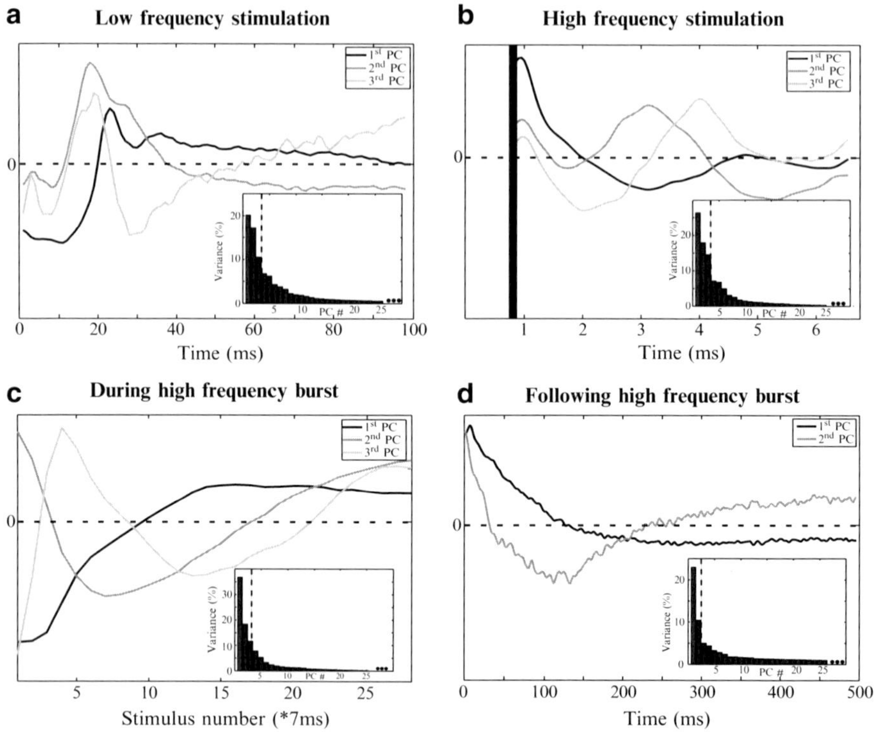

Fig. 2 Population response to different pallidal stimulation protocols. Principal components of the population's response to different stimulation protocols: (**a**) low-frequency stimulation, (**b**) high-frequency stimulation and to the two part of the burst stimulation (**c**) during the high-frequency burst and (**d**) following the high-frequency burst. The first three (**a–c**) or two (**d**) principal components are brought to demonstrate the populations response. The fraction of the variance explained by each component is brought separately for each stimulation type in the inset (*dotted line* – cutoff fraction for the components)

of late (3 and 4 ms) excitatory responses separated by periods of inhibition (Fig. 2b). Responses to burst stimulation displayed an independent response during the burst and following the burst so the PCs were calculated separately for each of these periods. The response to the stimuli within the burst maintained the same partial locking characteristics as the one seen in the high-frequency stimulation. Moreover, in addition to the locking to individual pulses, the response evolved during the burst (Fig. 2c). The change of the response within the burst is demonstrated by using bins which include complete interstimulus intervals as single bins to mask the locking to the single stimuli. The response evolves primarily during the initial ten stimuli within the burst (first PC) and includes either an increase or a decrease in the neurons' responses. Secondary changes continue after the initial ten stimuli (second and third PCs). The activity of the neurons following the end of the stimulus burst demonstrates a prolonged decay (~200 ms) back to the baseline level from an initial excitation or inhibition (Fig. 2d).

A large subset of neurons (43/84 of the neurons in the nonstimulated sessions) displayed significant task-related increases and/or decreases in firing rate in response to multiple parts of the task: cue, movement, reward, etc. The stochastic nature of the delay between the events allowed the attribution of the response to a specific event (Fig. 3a, b). Task-related changes in the neuronal activity were maintained during the stimulation sessions. The same neuron demonstrated drastically

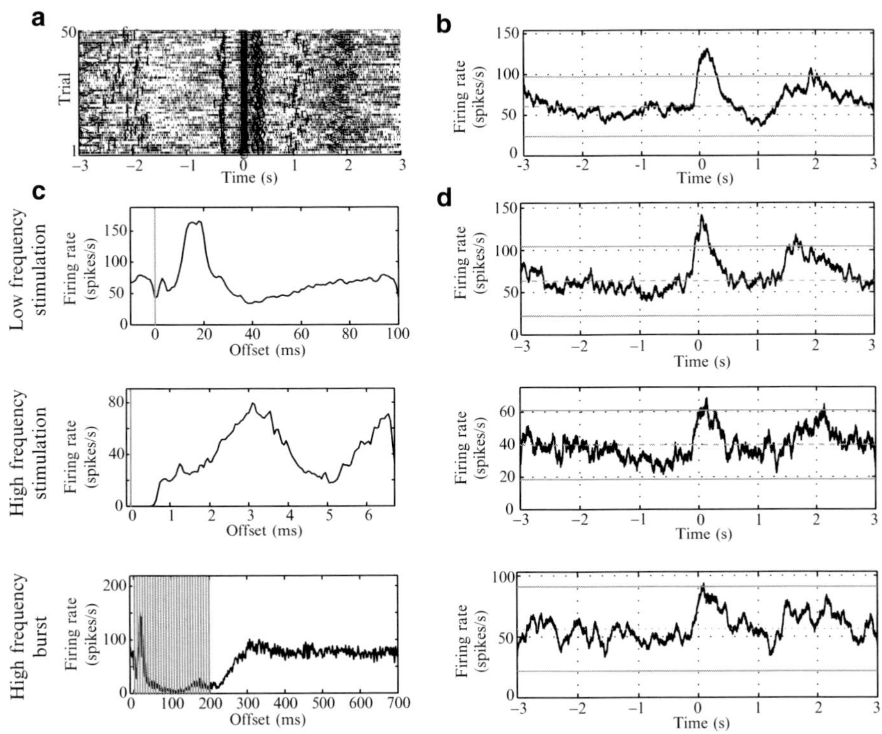

Fig. 3 Single neuron response to both stimulation and behavioral events. An example of the response of a single pallidal neuron (T.050204.03.3a) to the behavioral task is aligned to the leaving of the central button. (**a**) The raster of the trials (overlaid with the behavioral events: > – trial begin, f – fixation, t – target on, g – center leave, o – side touch, r – reward, < – trial end). (**b**) The peri-event histogram of the neuronal firing during a no-stimulation control session. (**c**) Response of the neuron to the different stimulation patterns. (**d**) The peri-event histograms of the neuronal firing of the same neuron during the stimulation sessions. (**c–d**): (*top*) low-frequency stimulation, (*center*) high-frequency stimulation, and (*bottom*) high-frequency bursts

different stimulation responses to the three different stimulation conditions (Fig. 3c). However, the neuron maintained the same basic task-related response to the "leave center button" event (Fig. 3d). The magnitude of the task-related response and the baseline firing rate surrounding it were influenced by the stimulation condition, but the general timing and form of the task-related response were preserved across stimulation conditions. The difference between task-related response and baseline rate was significant in all cases.

The response of the neuron to the combination of the stimulation and the behavioral event was analyzed by examination of the joint probably matrix of firing (Fig. 4). The product of the multiplication of firing probabilities (Fig. 4a) reflects the theoretical joint probability assuming an independent effect of stimulation (top and abscissa) and task (right and ordinate). The actual observed joint probability (Fig. 4b) closely resembled the joint distribution theoretically predicted for two independent processes.

4 Discussion

This study demonstrates that microstimulation of the globus pallidus in the normal behaving primate can alter response and movement times. The changes in behavior are accompanied by a response of most of the neurons to the stimulation by changes of the firing rate and by partial locking to the stimulation pulses. The nature of the response of each neuron to the stimulus varies and depends on the stimulation parameters. Furthermore, this response to the stimulation does not completely block the transmission of information along the cortico-basal ganglia loop.

The results of this study support the notion that neurons of the basal ganglia do not undergo a complete inhibition of their firing but rather modulate their firing transiently in response to stimulation pulses (Hashimoto et al. 2003; Bar-Gad et al. 2004; Meissner et al. 2005). Due to the prominent stimulus artifact, it is impossible in our data to analyze the first 0.6–1.2 ms of the period following the stimulus, so direct responses, such as those reported by McCairn and Turner (2009), may have been overlooked. The first few components of the locked responses seen in this study of the normal primate reveal many similar properties to the results shown in earlier studies of the MPTP-treated monkey. The low-frequency stimulation displayed a predominantly inhibitory response followed by a prolonged period of increased firing. However, the short initial excitation (3–5 ms) shown in the MPTP-treated primate was not apparent in most neurons in the normal monkey. The response to the high-frequency stimulation displayed the same increased early (3–4 ms) and late (6–7 ms) activity seen in the MPTP-treated primate but in addition also displayed a strong very early (~1 ms) excitation missing in the MPTP monkey. The results of the burst stimulation demonstrate that the effect of the stimulation on the firing rate persists for 200 ms after the end of the burst. The results demonstrate that the response of the neurons is to a large extent not dependent on the state of the animal, (i.e., dopamine levels due to input from the substantia nigra pars compacta (SNc)) but rather on the intrinsic connectivity and cellular properties of the basal

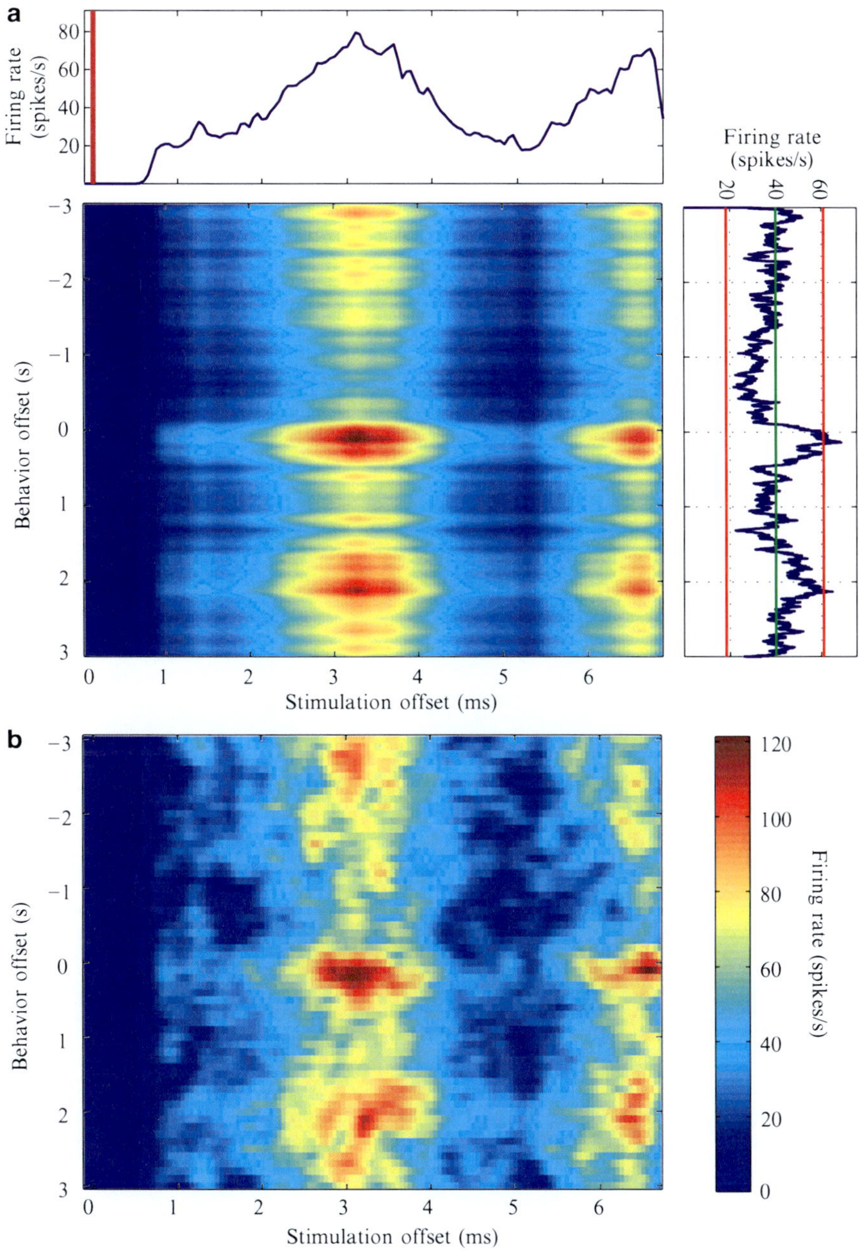

Fig. 4 Interaction of the encoding of the stimulation and the behavioral events. The response of a single pallidal neuron (same as shown in Fig. 3) to the stimulation and the behavioral event and to both together is compared. (**a**) The response of the neuron to the stimulation (peri-stimulus histogram, *top*) and the behavioral event (peri-event histogram, *right*). The multiplication of the responses results in a theoretical response matrix which assumes independents encoding of the two parameters. (**b**) The actual response of the neuron to the two events. The theoretical and actual responses to both parameters are normalized to depict the neuronal rate and color encoded

ganglia neurons. The differences observed may be not only due to the dopaminergic modulation but also due to the differences in the experimental setup.

Early approaches assumed that HF-DBS completely inhibits the output of the basal ganglia, much like the effects of pallidal ablation. This removal of the basal ganglia inhibition would enable the disinhibition of cortical activity and the execution of actions. Later studies, which have shown that the neuronal activity does not cease during stimulation, led to the concept of the formation of an artificial rhythm of neuronal activity due to the locking to the stimulus. This locked rhythm was believed to completely replace the abnormal rhythm of the parkinsonian globus pallidus. Although this effect was thought to completely abolish the normal output of information from the basal ganglia, it was seen to free thalamo-cortical function from the otherwise interfering noisy output from the parkinsonian basal ganglia (McIntyre et al. 2004; Rubin and Terman 2004). Our results hint at a third possibility: stimulation may lead to a *partial locking* of spiking activity to the stimulus leading to a reduction in the abnormal bursting and oscillating activity. However, the locking is not a simple response of one spike per stimulus pulse but rather a probabilistic response both in time to the stimulus sequence and in phase regarding the temporal offset of the spike from the stimulus pulse. This enables both the disruption of the abnormal activity and, at least partially, the transmission of behaviorally important information. This might enable a partial usage of information processed in the basal ganglia and could thus enable a better therapeutic outcome relative to ablations.

Acknowledgments We thank V. Davis for her help in acquiring and processing the data.

References

Albin RL, Young AB and Penney JB (1989) The functional anatomy of basal ganglia disorders. Trends Neurosci 12: 366–375.

Bar-Gad I, Elias S, Vaadia E and Bergman H (2004) Complex locking rather than complete cessation of neuronal activity in the globus pallidus of a 1-methyl-4-phenyl-1,2,3,6-tetrahydropyridine-treated primate in response to pallidal microstimulation. J Neurosci 24: 9410–9419.

Benabid AL (2003) Deep brain stimulation for Parkinson's disease. Curr Opin Neurobiol 13: 696–706.

Benazzouz A, Gao DM, Ni ZG, Piallat B, Bouali BR and Benabid AL (2000) Effect of high-frequency stimulation of the subthalamic nucleus on the neuronal activities of the substantia nigra pars reticulata and ventrolateral nucleus of the thalamus in the rat. Neuroscience 99: 289–295.

Beurrier C, Bioulac B, Audin J and Hammond C (2001) High-frequency stimulation produces a transient blockade of voltage-gated currents in subthalamic neurons. J Neurophysiol 85: 1351–1356.

Boraud T, Bezard E, Bioulac B and Gross C (1996) High frequency stimulation of the internal Globus Pallidus (GPi) simultaneously improves parkinsonian symptoms and reduces the firing frequency of GPi neurons in the MPTP-treated monkey. Neurosci Lett 215: 17–20.

DeLong MR (1990) Primate models of movement disorders of basal ganglia origin. Trends Neurosci 13: 281–285.

Diederich NJ, Kalteis K, Stamenkovic M, Pieri V and Alesch F (2005) Efficient internal pallidal stimulation in Gilles de la Tourette syndrome: A case report. Mov Disord 20: 1496–1499.

Dostrovsky JO, Levy R, Wu JP, Hutchison WD, Tasker RR and Lozano AM (2000) Microstimulation-induced inhibition of neuronal firing in human globus pallidus. J Neurophysiol 84: 570–574.

Gildenberg PL (2003) History repeats itself. Stereotact Funct Neurosurg 80: 61–75.

Hashimoto T, Elder CM, Okun MS, Patrick SK and Vitek JL (2003) Stimulation of the subthalamic nucleus changes the firing pattern of pallidal neurons. J Neurosci 23: 1916–1923.

Horak FB and Anderson ME (1984a) Influence of globus pallidus on arm movements in monkeys. I. Effects of kainic acid-induced lesions. J Neurophysiol 52: 290–304.

Horak FB and Anderson ME (1984b) Influence of globus pallidus on arm movements in monkeys. II. Effects of stimulation. J Neurophysiol 52: 305–322.

McCairn KW and Turner RS (2009) Deep brain stimulation of the globus pallidus internus in the parkinsonian primate: local entrainment and suppression of low-frequency oscillations. J Neurophysiol 101: 1941–1960.

McIntyre CC, Grill WM, Sherman DL and Thakor NV (2004) Cellular effects of deep brain stimulation: Model-based analysis of activation and inhibition. J Neurophysiol 91: 1457–1469.

Meissner W, Leblois A, Hansel D, Bioulac B, Gross CE, Benazzouz A and Boraud T (2005) Subthalamic high frequency stimulation resets subthalamic firing and reduces abnormal oscillations. Brain 128: 2372–2382.

Rubin JE and Terman D (2004) High frequency stimulation of the subthalamic nucleus eliminates pathological thalamic rhythmicity in a computational model. J Comput Neurosci 16: 211–235.

Turner RS and Anderson ME (1997) Pallidal discharge related to the kinematics of reaching movements in two dimensions. J Neurophysiol 77: 1051–1074.

Vidailhet M, Vercueil L, Houeto JL, Krystkowiak P, Benabid AL, Cornu P, Lagrange C, Tezenas du MS, Dormont D, Grand S, Blond S, Detante O, Pillon B, Ardouin C, Agid Y, Destee A and Pollak P (2005) Bilateral deep-brain stimulation of the globus pallidus in primary generalized dystonia. N Engl J Med 352: 459–467.

High-Frequency Stimulation of the Globus Pallidus External Segment Biases Behavior Toward Reward

Avital Adler, Mati Joshua, Inna Finkes, and Hagai Bergman

Abstract Associative learning requires the ability to learn the connections between events. Fundamental to the learning process is acquiring the expectation that certain events lead to certain outcomes. Indeed, studies have shown that when an expected outcome is changed, so does the subsequent behavior. It has also been demonstrated that basal ganglia neuronal activity is highly modulated by the expectation of future reward. However, the direct link between reward expectation, basal ganglia neuronal activity, and the final change in behavior is not clear. Electrical stimulation enables a direct manipulation of neuronal activity and has been shown to cause behavioral changes. From another perspective, high-frequency deep brain electrical stimulation (DBS) is used as treatment of advanced neurological disorders, yet the underlying mechanism is still a topic of debate.

We investigated the effect of high-frequency stimulation of the external segment of the globus pallidus, the central nucleus of the basal ganglia networks, on monkey's expectations of both reward and aversive events. We show that long-duration (30 min), high-frequency (130 Hz) stimulation changed the monkey's behavior in a classical conditioning task, enhancing its expectation of reward. The effect we observed was gradual and persisted even after stimulation had ceased, implying a plastic change. The results support the notion of asymmetric coding of the positive vs. negative domains in the basal ganglia and may suggest the GPe as another possible target for DBS therapy of depression.

1 Introduction

Expected outcomes of events have a crucial influence on decision making. Studies have shown changes in behavioral parameters preceding reward which indicate the induction of behavioral bias associated with the preferred reward (Watanabe et al. 2001; Ding

A. Adler(✉), M. Joshua, I. Finkes, and H. Bergman
The Interdisciplinary Center for Neural Computation, The Hebrew University, Jerusalem 91904 Israel
e-mail: avital.adler@gmail.com

H.J. Groenewegen et al. (eds.), *The Basal Ganglia IX*, Advances in Behavioral Biology 58,
DOI 10.1007/978-1-4419-0340-2_7,

and Hikosaka 2007; Milstein and Dorris 2007). Furthermore, basal ganglia neural activity, which is involved in the execution of goal-directed behavior, is highly modulated by expectation of reward both at the input-output structures level (Lauwereyns et al. 2002; Sato and Hikosaka 2002; Arkadir et al. 2004; Darbaky et al. 2005; Samejima et al. 2005) and at the neuromodulator level (Joshua et al. 2008). Specifically, it has been shown that changes in the activity of dopamine neurons code the discrepancy between reward occurrence and reward prediction (Hollerman and Schultz 1998).

These studies, however, shed little light on the causal relationship between neural activity and behavioral changes. One technique, which makes it possible to draw these conclusions, is to test the effects of electrical microstimulation in specific regions and specific time domains on behavior. Olds and Milner (1954) were the first to show that rats can learn to press a lever in order to receive electrical stimulation of their medial forebrain bundle. Since then, intracranial self-stimulation (ICSS) has been reported in many other animals and from a variety of structures in the brain including the substantia nigra pars compacta (SNc) and the striatum (Redgrave and Dean 1981). ICSS experiments show that the effect of local stimulation on learning and behavior can mimic that of natural reward, including the appearance of behavioral bias associated with the preferred reward.

Other studies have investigated the effects of local microstimulation on different cognitive functions. At the cortical level, Salzman et al. (1990) showed that stimulation applied to the middle temporal visual area biased the monkey's judgments toward the direction encoded by the neurons in the stimulating area. In a visually guided saccade task, microstimulation was applied to the oculomotor region of the caudate after saccades in a fixed direction. This led to facilitation of saccades (reduction of saccadic reaction time) in that direction (Nakamura and Hikosaka 2006). In yet another study, microstimulation in the caudate nucleus increased the rate of learning associations between visual images and specific movements, when it was delivered during the reinforcement period (Williams and Eskandar 2006). Finally, recent advanced studies have revealed that stimulation of even a few neurons can lead to robust behavioral effects (Brecht et al. 2004; Huber et al. 2008).

At the other end of the spectrum is the procedure of high-frequency deep brain stimulation (DBS) of different subcortical structures as treatment for neurological disorders. Generally, continuous stimulation of the subthalamic nucleus (STN) or the internal segment of the globus pallidus (GPi) is used for treatment of advanced movement disorders such as dystonia and Parkinson's disease (PD) (Benabid et al. 1998). Despite the remarkable improvement in motor symptoms, STN DBS has been shown to cause cognitive or emotional side effects such as impulsive decision making (Frank et al. 2007). Furthermore, the mechanism by which DBS brings about its therapeutic effects remains unknown (Perlmutter and Mink 2006).

Compared to extensive physiological research in the positive (appetitive) domain, research in the negative domain of associative learning is scanty. This prompted us to investigate the effects of high-frequency DBS in the central nucleus of the basal ganglia, the external segment of the globus pallidus (GPe) on the expectations of positive and negative outcomes. To this end, we devised a symmetric classical conditioning task with reward (food) and aversive (air puff) outcomes. Stimulation at high frequency and for periods of long duration (30 min) was administered while the monkey was engaged in the task. This paradigm allowed us to monitor and analyze the monkey's behavioral responses to both expectation of reward, expressed as anticipatory licking responses and expectation of aversion, expressed as blinking responses.

2 Methods

2.1 General

One male monkey (*Macaque fascicularis*, 5 kg) was used. All experimental procedures were performed in accordance with the National Institutes of Health *Guide for the Care and Use of Laboratory Animals (1966)* and Hebrew University guidelines for the use and care of laboratory animals in research, supervised by the institutional animal care and use committee.

2.2 Behavioral Task

The monkey was trained (5 months) on a probabilistic classical conditioning task. Each trial began with the presentation of a visual cue (on a 17-in. computer screen at a distance of 50 cm) for a period of 2 s. There were seven different visual cues (Fig. 1d). Three of them represented a reward outcome (with probabilities of 1, 2/3, and 1/3), three represented an aversive outcome (with probabilities of 1, 2/3, and 1/3), and one was the neutral cue. All visual cues were fractal images constructed with the Chaos Pro 3.2 program (http://www.chaospro.de). After cue presentation the monkey received an outcome in a probabilistic fashion (end-result epoch, Fig. 1d). The end result was one of the three possibilities: liquid food (0.6 mL; 150-ms duration) in the reward trials, air puff (150-ms duration; 50–70 psi; 2 cm from the eye, split and directed to both eyes) in the aversive trials, or neither in the neutral trials nor in the trials with no outcome. An auditory stimulus signaled the beginning of the end-result epoch. Each possible end result was signaled by one of the three different sounds; sounds were normalized to the same intensity and duration. In the reward and aversive trials, the auditory signal was accompanied by background sound of the devices delivering food and air puffs. Each trial was followed by an intertrial interval (ITI) of 3–7 s. The reward and aversive trials with 2/3 and 1/3 probabilities of an outcome were introduced three times more in order to maintain the same average occurrence of all possible outcomes. With this occurrence ratio, all trials were randomly interleaved. The monkey was engaged in the task continuously and, in parallel, stimulation effects were tested by alternating between stimulation and no stimulation 30-min periods (Fig. 1c).

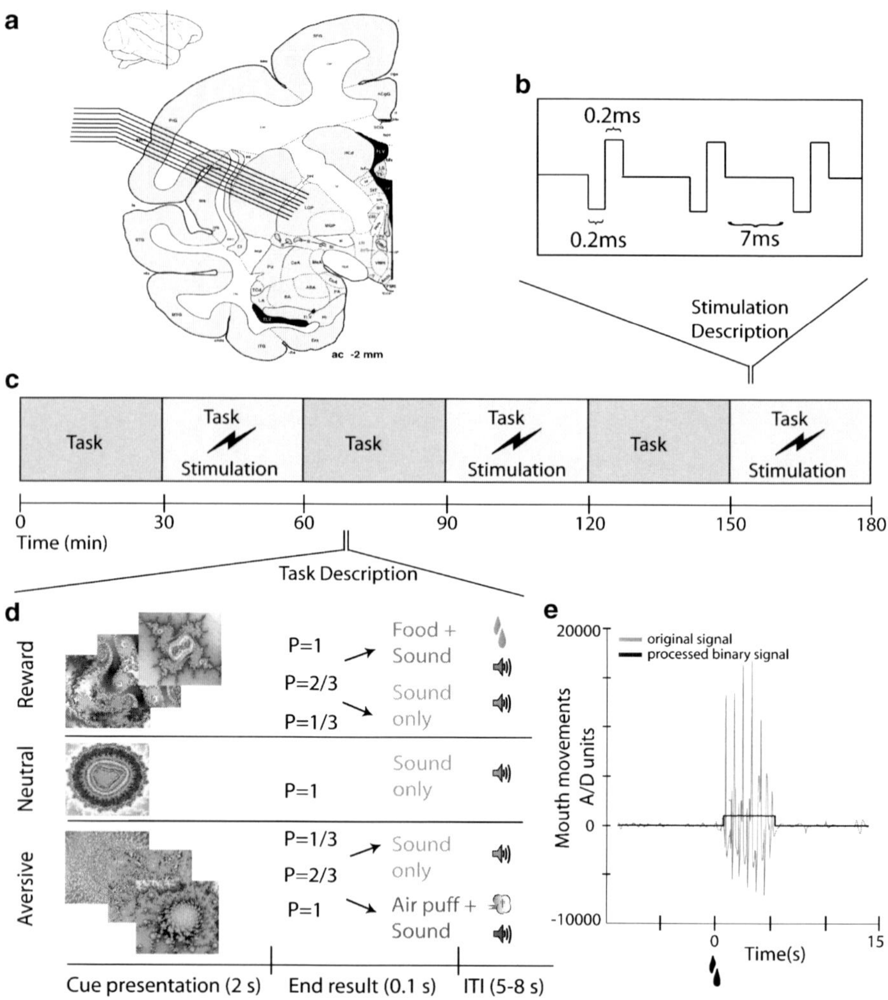

Fig. 1 Task and stimulation paradigms. (**a**) Stimulation site – A representative coronal section –2 mm from anterior commissure (Martin and Bowden 1996). Eight microelectrodes were advanced separately into the GPe. Stimulation site was 0.5-mm deep to the lateral border of the GPe. Each stimulation session had 5–8 stimulating electrodes. (**b**) Stimulation description – biphasic stimulation pulses of 40–60 μA per electrode. Pulse duration was 0.4 ms (0.2 ms each phase), interpulse interval was 7 ms, e.g., pulse frequency = 130 Hz. (**c**) Single day schedule – A single day was divided into six sessions lasting 30 minutes each, alternating between a session with stimulation and a session without stimulation. The first session (with or without stimulation) was interleaved between days. (**d**) Task description – Top – reward trials; Middle – neutral trials (no outcome); Bottom – aversive trials. (**e**) Mouth signal – an example of the original signal, mouth movements monitored using an infrared reflection detector (Dr. Bouis, Freiburg, Germany). Processed binary signal, original signal transformed to periods with licking movements and periods without licking movements. Time zero is the delivery of food reward and its auditory signal

Prior to the behavioral sessions, eight glass-coated tungsten microelectrodes with expanded exposed tip size (impedance 0.05–0.25 MΩ at 1,000 Hz), confined within a cylindrical guide (1.65 mm inner diameter) were advanced separately (EPS, Alpha-Omega Engineering, Nazareth, Israel) into the external segment of the globus pallidus (GPe, Fig. 1a). On each stimulation session, there were 3–8 stimulating electrodes (6.1 on average; high impedance electrodes or electrodes missing GPe were excluded) and biphasic stimulation pulses of 40–60 μA were conducted out of each electrode (54.75 μA per electrode on average, Fig. 1b).

2.3 Data Analysis

The monkey's behavior, licking and blinking responses, was continuously monitored. Mouth movements were monitored using an infrared reflection detector (Dr. Bouis, Freiburg, Germany). The infrared signal was filtered between 1 and 100 Hz by a band-pass four-pole Butterworth filter, and sampled at 1.56 kHz. The analog signal was transformed into binary signals with periods identified as licking and periods identified as no-licking (Fig. 1e). The blinking responses were monitored by a computerized digital video camera sampled at 50 Hz. Video analyses were constructed using home-made custom software.

Licking ratio and blinking ratio were determined as the number of trials out of all trials in a single day with licking and blinking responses, respectively. The ratios were computed separately for each event and for each epoch of the trial. A cue epoch was considered as one that had a behavioral response if the monkey made at least a single licking or blinking response in the time interval of 1.5–2 s after cue presentation. An outcome or no-outcome epoch was considered as having a licking/blinking response if the monkey made at least a single licking/blinking response in the time interval of 0.5–1 or 0–0.5 s after cue ending, respectively. These different time intervals were adjusted to the typical appearance of licking and blinking at the end-result epoch (latency of blinking responses was significantly shorter than the latency of licking responses).

The response bias (RB) is a measure of the monkey's tendency toward one behavioral response or the other (Macmillan and Creelman 2005). RB was measured for the licking responses and for the blinking responses separately

$$\mathrm{RB} = -0.5\ (Z(H) + Z(F)).$$

H denotes the ratio of correct (Hits) behavioral responses, i.e., the licking ratio in reward trials or the blinking ratio in aversive trials. F denotes the ratio of incorrect (False alarms - false positive errors) behavioral responses, i.e., the licking ratio in aversive trials or the blinking ratio in reward trials. Z is the inverse of the normal cumulative distribution function in standard deviation units.

In order to examine the change in RB, we subtracted the median RB in the prestimulation control days from the given RB. This was defined as the response bias change. The RB change during the stimulation days and the following, no-stimulation days were compared to the RB change of the prestimulation control days using a Mann-Whitney U test. Linear regression analysis was conducted on the RB change values. In all tests the significance level was 5%, unless noted otherwise.

3 Results

3.1 *High-Frequency Stimulation Changes the Monkey's Licking Response*

We monitored the monkey's licking and blinking behavior during task performance for 102 days in total. We first obtained data in 23 prestimulation control days (control 1), afterward, while the monkey was engaged in the same task, we introduced the stimulation sessions for a period of 20 days (stimulation 1). The next 22 days with task performance alone was the follow-up period (follow up). At this time we monitored the monkey's behavior for 2 days under Apomorphine (0.1 mg/kg IM) following by another 25 days of task performance. These 27 days were the control days (control 2) for the second period of stimulation which lasted for 10 days (stimulation 2).

We found that stimulation significantly ($p \ll 0.01$, Mann–Whitney U test) altered the monkey's response bias for licking behavior in all epochs of the trial and in all the three outcome probabilities both in stimulation-1 days (compared with control-1) (Fig. 2, left column) and in stimulation-2 days (compared with control-2) (Fig. 2, right column). Furthermore, we found a greater tendency to lick both at aversive and reward trials. The change in RB at cue presentation and no-outcome epochs was a result of increased licking ratios in both aversive and rewarding trials, whereas the change in RB at the outcome epoch was mainly due to larger licking ratios in aversive trials.

The effect of stimulation on RB was maintained throughout the full length of a recording day both in the trials with stimulation and in the trials without. Therefore, it cannot be attributed to mere motor effects of the stimulation. In fact the change in RB also persisted in the follow-up period ($p \ll 0.01$, Mann–Whitney U test). The change in RB tended to be stronger according to outcome probability, with the strongest change in trials involving the deterministic cue ($P = 1$ of outcome). These differences between trials with different outcome probabilities were found in the cue presentation and no-outcome epochs of the trial, though it did not reach significance (except at the no-outcome epoch in stimulation-1, $P \ll 0.01$, Kruskal-Wallis test).

There was a significant difference in RB values between the two control periods in which the medians of control-2 were lower than those of control-1 in all trial epochs (median in cue presentation, 0.61 and 0.5; in outcome epoch, 0 and −0.94; and in no-outcome epoch, 0.18 and −0.5 for control-1 and control-2, respectively; $p \ll 0.01$, Mann–Whitney U test). This difference was attributable to increased licking

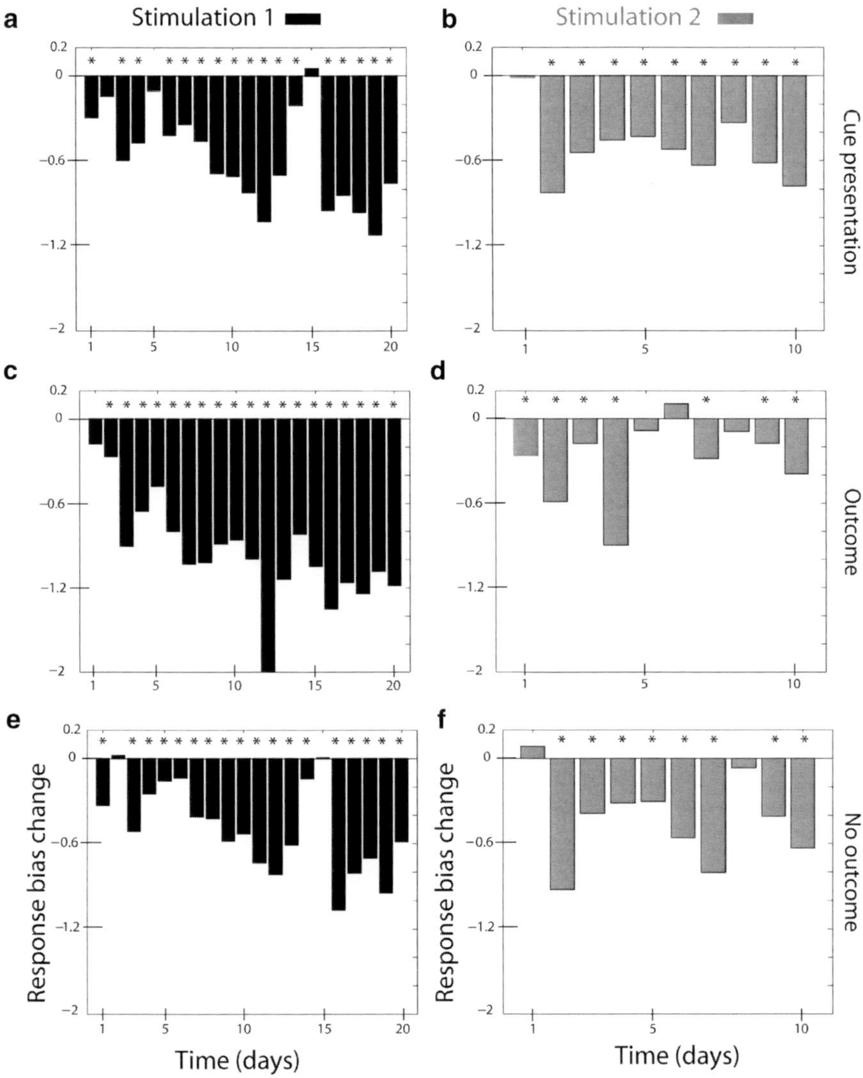

Fig. 2 Response bias change following GPe stimulation. *Bar plots* of the response bias (RB) change for each day in the stimulation-1 period (*black, left column*) and the stimulation-2 period (*gray, right column*). To calculate the RB change for a single day we subtracted the median RB in the control days (control-1 and -2 for stimulation-1 and -2, respectively) from the given RB of that day. *Asterisk* indicates a RB change significantly different from the RB change in the control periods. Values below zero indicate a greater tendency to lick on both reward and aversive trials. (**a**, **b**) RB change plot for cue presentation epoch; (**c**, **d**) RB change plot for outcome epoch; (**e**, **f**) RB change plot for no-outcome epoch

ratios, in both reward and aversive trials, in control-2 period compared with control-1. Indeed, in the follow-up period (after stimulation-1) there was a gradual increase in RB values back toward control-1 values indicating a decrease in licking ratios

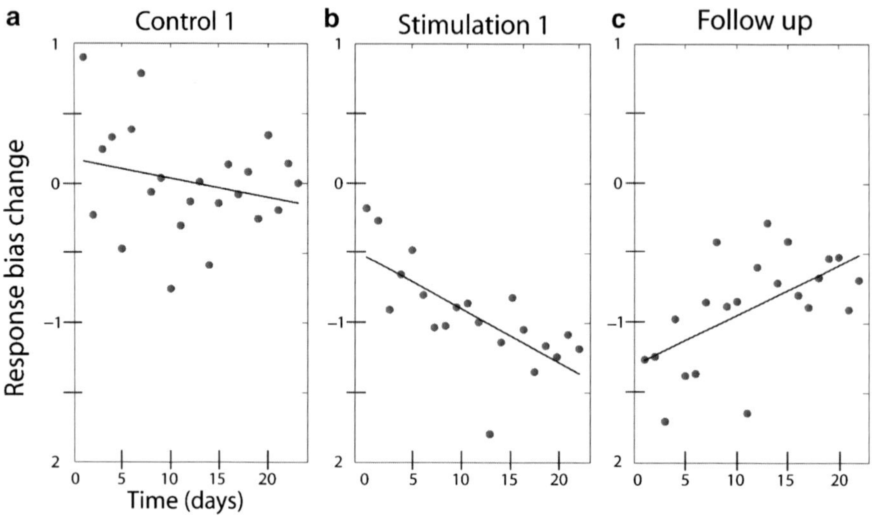

Fig. 3 Linear regression analysis of the Response Bias changes. An example of the RB change linear regression in the outcome epoch. (**a**) Control-1 period $R^2 = 0.05$, $p = 0.3$. (**b**) Stimulation-1 period $R^2 = 0.5$, $p \ll 0.001$. (**c**) Follow-up period $R^2 = 0.34$, $p \ll 0.01$

(Fig. 3c). However, in the following control-2 period this trend was attenuated and even reversed. It has been demonstrated that the stereotyped phenomena of repetitive licking behavior is correlated with the activation of the nigrostriatal system, specifically with the activation of both dopamine D1 and D2 receptors (Kelly et al. 1975). We therefore hypothesize that the apomorphine given in the first 2 days of control-2 period was partially responsible for these extreme changes in the baseline values. All the differences in stimulation-2 period mentioned above were in addition to the apomorphine-induced change. In contrast to the modulation of the monkey's licking behavior, no consistent or significant effects were found for the monkey's blinking behavior. In addition, we inspected the monkey's behavior visually during times with stimulation and without performance of the classical conditioning task. We did not observe any major effects on the monkey's behavior during stimulation alone periods.

3.2 *Progression of the Stimulation Effect*

In order to appraise the development over time of the changes in the monkey's licking behavior, we further analyzed the RB change. We examined for each day in the stimulation period (1 and 2), whether its RB change was significantly different from the RB change during the control period (1 and 2, respectively). Nonparametric confidence interval significance levels are calculated according to the data itself (control-1: 96.53% significance level and control-2: 94.78% significance level). We found that the change in RB was immediate and occurred even after a single stimulation day (see asterisks in Fig. 2).

Additionally, the values of the RB change decreased with time for the stimulation-1 period (i.e., revealing an increasing tendency to lick during the stimulation period). We found a significant correlation between the number of stimulation days and changes of the RB values in all three epochs of the trial ($p < 0.01$, linear regression). An example of the outcome epoch is given in Fig. 3b. In a complementary manner, during the follow-up period (Fig. 3c), as the number of days increased, the RB change values increased back to the control level ($p \ll 0.01$, linear regression). We did not observe these gradual changes in the stimulation-2 period, probably due to the high baseline values in the control-2 period.

4 Discussion

4.1 High-Frequency Stimulation Alters Expectation of Reward

We have shown that high-frequency stimulation of the GPe changed the monkey's behavior, increasing only its licking responses to both aversive and rewarding events while the blinking responses were unchanged. Thus, the effect of GPe stimulation on the learned associations mimicked the effect of natural (positive) reward altering the monkey's expectations. The sole enhancement of expectation of reward and not the expectation to aversion by basal ganglia stimulation may reflect the notion of localized antagonistic motivational systems (Konorski 1967). According to this view there are two neuronal networks; one for the reward domain and another for the aversive domain, whereas the BG network is biased toward encoding of the positive domain. This notion explains the differential effect of stimulation on licking vs. blinking behavior and is supported by our group's results which show that the expectation level of reward, but not aversion, is preferentially encoded in the basal ganglia (Joshua et al. 2008 and also see Chapter "*Asymmetric Encoding of Positive and Negative Expectations by Low-Frequency Discharge Basal Ganglia Neurons*" by Joshua et al. in this volume). Furthermore, it has been demonstrated that behavioral reactions reflect differences in the motivational value of the expected reward (Watanabe et al. 2001); in our experiment the RB change tended to be graded by the different outcome probabilities. Thus, basal ganglia stimulation effects are dependent on the behavioral context.

4.2 High-Frequency Stimulation is Associated with Synaptic Plasticity

We have also shown that GPe stimulation induced an effect which was gradual (additional decrease in RB change values as the cumulative stimulation days increased), long lasting (stimulation effect proceeded to the follow-up period) and not local (stimulation effect was apparent both at sessions with stimulation and sessions without stimulation in a single day). This implies that the mechanism of GPe stimulation could be related to synaptic plasticity as was suggested for the DBS mechanism in dystonia. Dystonic patients have abnormal degree of muscle contraction which is most likely a result of abnormal basal ganglia motor control (Berardelli et al. 1998). Studies have shown that improvement following DBS appears in a progressive manner and may take weeks or months to be effective (Yianni et al. 2003; Bittar et al. 2005). It was suggested that GPi DBS in these patients is associated with long-term effects on neural plasticity (Tisch et al. 2007). The precise mechanism of microstimulation that induced basal ganglia synaptic effects may be mediated through SNc innervations by pallidal neurons (Bevan et al. 1996; Paladini et al. 1999). Dopamine involvement in the modulation of corticostriatal synaptic efficacy has recently been reported (Reynolds and Wickens 2002). The reciprocal dopaminergic innervation of the pallidal complex may contribute to the GPe microstimulation-induced changes in behavior. However, other basal ganglia neurotransmitter systems, like lateral GABAergic connections between GPe neurons, can also play a role in the modification of the basal ganglia functional circuitry.

4.3 *Therapeutic Implications*

Depressive symptoms are associated with a bias toward negative prediction of future life events (Strunk et al. 2006), whereas in our experiment high-frequency GPe stimulation induced an "optimistic" bias. Mood improvements of patients with major depression (Kosel et al. 2007; Johansen-Berg et al. 2008) have been observed following high-frequency stimulation of different cortical and basal ganglia targets. Our results suggest the GPe as possible therapeutic target for depression. Moreover, as in the case with DBS treatment of motor symptoms of Parkinson's disease, future studies may reveal a differential efficacy of stimulation of different basal ganglia and cortical targets for different mood disorders. The GPe – the central nucleus of the basal ganglia network – may be added to the list of possible targets for future DBS treatment of emotional and psychiatric disorders.

References

Arkadir D, Morris G, Vaadia E and Bergman H (2004) Independent coding of movement direction and reward prediction by single pallidal neurons. J Neurosci 24: 10047–10056.

Benabid AL, Benazzouz A, Hoffmann D, Limousin P, Krack P and Pollak P (1998) Long-term electrical inhibition of deep brain targets in movement disorders. Mov Disord 13(Suppl 3): 119–125.

Berardelli A, Rothwell JC, Hallett M, Thompson PD, Manfredi M and Marsden CD (1998) The pathophysiology of primary dystonia. Brain 121: 1195–1212.

Bevan MD, Smith AD and Bolam JP (1996) The substantia nigra as a site of synaptic integration of functionally diverse information arising from the ventral pallidum and the globus pallidus in the rat. Neuroscience 75: 5–12.

Bittar RG, Yianni J, Wang S, Liu X, Nandi D, Joint C, Scott R, Bain PG, Gregory R, Stein J and Aziz TZ (2005) Deep brain stimulation for generalised dystonia and spasmodic torticollis. J Clin Neurosci 12: 12–16.

Brecht M, Schneider M, Sakmann B and Margrie TW (2004) Whisker movements evoked by stimulation of single pyramidal cells in rat motor cortex. Nature 427: 704–710.

Darbaky Y, Baunez C, Arecchi P, Legallet E and Apicella P (2005) Reward-related neuronal activity in the subthalamic nucleus of the monkey. Neuroreport 16: 1241–1244.

Ding L and Hikosaka O (2007) Temporal development of asymmetric reward-induced bias in macaques. J Neurophysiol 97: 57–61.

Frank MJ, Samanta J, Moustafa AA and Sherman SJ (2007) Hold your horses: Impulsivity, deep brain stimulation, and medication in parkinsonism. Science 318: 1309–1312.

Hollerman JR and Schultz W (1998) Dopamine neurons report an error in the temporal prediction of reward during learning. Nat Neurosci 1: 304–309.

Huber D, Petreanu L, Ghitani N, Ranade S, Hromadka T, Mainen Z and Svoboda K (2008) Sparse optical microstimulation in barrel cortex drives learned behaviour in freely moving mice. Nature 451: 61–64.

Johansen-Berg H, Gutman DA, Behrens TE, Matthews PM, Rushworth MF, Katz E, Lozano AM and Mayberg HS (2008) Anatomical connectivity of the subgenual cingulate region targeted with deep brain stimulation for treatment-resistant depression. Cereb Cortex 18: 1374–1383.

Joshua M, Adler A, Mitelman R, Vaadia E, Bergman H. (2008) Midbrain dopaminergic neurons and striatal cholinergic interneurons encode the difference between reward and aversive events at different epochs of probabilistic classical conditioning trials. J Neurosci 28: 11673–11684.

Kelly PH, Seviour PW and Iversen SD (1975) Amphetamine and apomorphine responses in the rat following 6-OHDA lesions of the nucleus accumbens septi and corpus striatum. Brain Res 94: 507–522.

Konorski J (1967) Integrative Activity of the Brain: An Interdiciplinary Approach. Chicago, IL: Chicago University Press.

Kosel M, Sturm V, Frick C, Lenartz D, Zeidler G, Brodesser D and Schlaepfer TE (2007) Mood improvement after deep brain stimulation of the internal globus pallidus for tardive dyskinesia in a patient suffering from major depression. J Psychiatr Res 41: 801–803.

Lauwereyns J, Watanabe K, Coe B and Hikosaka O (2002) A neural correlate of response bias in monkey caudate nucleus. Nature 418: 413–417.

Macmillan NA and Creelman CD (2005) Detection Theory. Mahwah, NJ: Lawrence Erlbaum Associates.

Martin RF and Bowden DM (1996) A stereotaxic template atlas of the macaque brain for digital imaging and quantitative neuroanatomy. Neuroimage 4: 119–150.

Milstein DM and Dorris MC (2007) The influence of expected value on saccadic preparation. J Neurosci 27: 4810–4818.

Nakamura K and Hikosaka O (2006) Facilitation of saccadic eye movements by postsaccadic electrical stimulation in the primate caudate. J Neurosci 26: 12885–12895.

Olds J and Milner P (1954) Positive reinforcement produced by electrical stimulation of septal area and other regions of rat brain. J Comp Physiol Psychol 47: 419–427.

Paladini CA, Celada P and Tepper JM (1999) Striatal, pallidal, and pars reticulata evoked inhibition of nigrostriatal dopaminergic neurons is mediated by GABA(A) receptors in vivo. Neuroscience 89: 799–812.

Perlmutter JS and Mink JW (2006) Deep brain stimulation. Annu Rev Neurosci 29: 229–257.

Redgrave P and Dean P (1981) Intracranial self-stimulation. Br Med Bull 37: 141–146.

Reynolds JN and Wickens JR (2002) Dopamine-dependent plasticity of corticostriatal synapses. Neural Netw 15: 507–521.

Salzman CD, Britten KH and Newsome WT (1990) Cortical microstimulation influences perceptual judgements of motion direction. Nature 346: 174–177.

Samejima K, Ueda Y, Doya K and Kimura M (2005) Representation of action-specific reward values in the striatum. Science 310: 1337–1340.

Sato M and Hikosaka O (2002) Role of primate substantia nigra pars reticulata in reward-oriented saccadic eye movement. J Neurosci 22: 2363–2373.

Strunk DR, Lopez H and DeRubeis RJ (2006) Depressive symptoms are associated with unrealistic negative predictions of future life events. Behav Res Ther 44: 861–882.

Tisch S, Rothwell JC, Bhatia KP, Quinn N, Zrinzo L, Jahanshahi M, Ashkan K, Hariz M and Limousin P (2007) Pallidal stimulation modifies after-effects of paired associative stimulation on motor cortex excitability in primary generalised dystonia. Exp Neurol 206: 80–85.

Watanabe M, Cromwell HC, Tremblay L, Hollerman JR, Hikosaka K and Schultz W (2001) Behavioral reactions reflecting differential reward expectations in monkeys. Exp Brain Res 140: 511–518.

Williams ZM and Eskandar EN (2006) Selective enhancement of associative learning by microstimulation of the anterior caudate. Nat Neurosci 9: 562–568.

Yianni J, Bain PG, Gregory RP, Nandi D, Joint C, Scott RB, Stein JF and Aziz TZ (2003) Postoperative progress of dystonia patients following globus pallidus internus deep brain stimulation. Eur J Neurol 10: 239–247.

The Subthalamic Region of Luys, Forel, and Dejerine

John S. McKenzie

Abstract The subthalamic region (ST) embraces the subthalamic nucleus (STN), zona incerta (ZI), Forel's fields (H, H1 and H2), and their nucleus (FF). JB Luys of Paris first described STN in human brain atlases, in 1865 as drawings and in 1873 in photographs of brain slices. He completely ignored ZI and Forel's fields. As an industrious neurologist and neuropathologist, he treated many mental patients by hypnosis. In his later public demonstrations, he neglected proper scientific controls, and was duped by collusive patients to believe in bizarre sensory properties and symptom transfer. In 1877 Forel produced the first adequate description of ST, in serial sections. He later specialised in neurology and psychiatry, directing the Zurich mental asylum. In combatting alcoholism as a cause of insanity he adopted the strategy of total abstinence. He was renowned also for his scientific work on ants. Dejerine and his wife Augusta Klumpke were an outstanding neurological team in Paris. In their great two-volume work on the anatomy of nervous centres, they established the detailed structure of ST, still largely accepted. Their devoted neurological work with the war wounded soldiers led to the exhausted Dejerine's death in 1917. His widow continued the work until her death in 1927.

1 Introduction

The term "subthalamus" is conveniently used for a region ventral to the dorsal thalamus, mostly rostral to the red nucleus (NR) and substantia nigra (SN). It includes the subthalamic nucleus (STN) or corpus Luysii (CL); zona incerta (ZI); nucleus of Forel's field (FF); fibre tracts of fields H, H1 and H2 of Forel (Mai et al. 1997). Deep brain stimulation of the STN-ZI area can alleviate signs of Parkinsonism (Plaha et al. 2006).

J.S. McKenzie (✉)
Department of Physiology, The University of Melbourne, VIC 3010, Australia
e-mail: jsmcke@unimelb.edu.au

H.J. Groenewegen et al. (eds.), *The Basal Ganglia IX*, Advances in Behavioral Biology 58,
DOI 10.1007/978-1-4419-0340-2_8,

The STN was first identified by Jules-Bernard Luys (1828–1897). Born and educated in Paris, he worked there for life. At the end of 1853 he was appointed as Hospital Interne. He then launched into microscopic research; his first communications to the Société de Biologie in 1855 concerned dural ossification, and multiple brain tumours (Semelaigne 1930–1932). In 1856 he won a prize from the Académie de Médecine, for the application of the microscope to pathological anatomy, diagnosis and treatment. In 1862 he was appointed to the senior rank of Médecin des Hôpitaux, practising at the Salpêtrière, then at the Charité hospital in rue des Saints-Pères. In 1864 he accepted the additional post of Director of the Ivry-sur-Seine mental hospital.

2 Research by Luys on Brain structure

2.1 Brain Atlases

In 1865 Luys published a treatise of 660 pages, *Recherches sur le système cerebro-spinal, sa structure, ses functions, et ses maladies*, accompanied by a separately bound Atlas of 40 quaint lithographic plates drawn by himself, with 80 pages of explanatory legends (Luys 1865). For this he received a prize of F2,500 from the Académie des Sciences (Semelaigne loc. cit.). He followed this with another human brain atlas, *Iconographie photographique des centres nerveux* (Luys 1873), in which pioneering photographs of thin brain slices were coupled with schemata he drew from the photos, often improving their detail and clarity.

2.2 Methods of Preparation

The 1865 Atlas was prepared from unfixed brains cut into slabs (1–2 cm thick). The slabs were then hardened in 4% aqueous chromic acid for 2–3 days. A thin shaving was removed from each face and the slabs returned to the chromic acid bath for further fixation. Excess fixative was then removed in water, and details of each surface were drawn on tracing paper. Slices as thin as 1 mm were cut from the slabs for microscopic examination (Luys 1865, pp. 1–3). For the 1873 Atlas, the brains were hardened in chromic acid before cutting into thick slabs then into 1-mm slices with an early microtome. The slices were then stained with carmine and photographed using solar illumination (Luys 1892a).

2.3 The Subthalamic Region in the Atlases

In 1865, Luys used various names for a new structure: accessory grey substance of the *locus niger* (i.e. of SN) (label 4 in his Plate 18, Fig. 2); accessory grey substance of the superior olive (i.e. of NR), in his persistent departure from naming priority

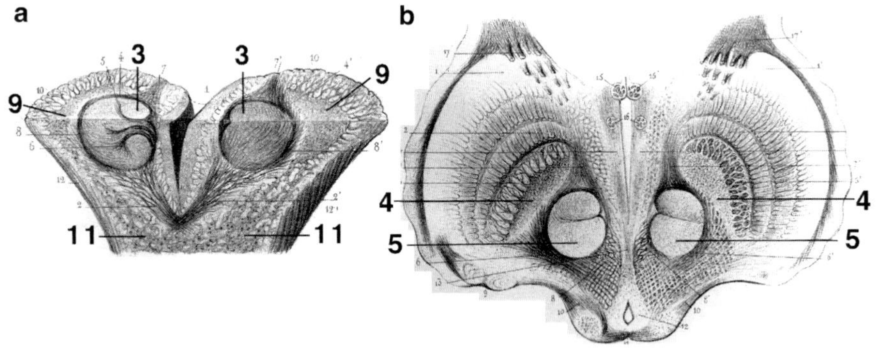

Fig. 1 (**a**) Combined vertical and horizontal cuts through midbrain. 3 – red nucleus, 9 – accessory grey nucleus (= subthalamic nucleus), 11 – substantia nigra. (Luys' Plate 13, Fig. 2). (**b**) Horizontal section through red nucleus in midbrain. 4 – substantia nigra, 5—red nucleus. (Luys' Plate 10, Fig. 3). Taken from Plates 10, 13, 18, 24 in a copy of the Luys (1865) Atlas in the author's possession

(label 19 in his Plate 24, Fig. 3); or in the full text as accessory strip (*bandelette*) of the superior olive. The new structure had the aspect of greyish matter arranged in linear form with sometimes a biconvex body. Furthermore, the tissue of the accessory strip consisted of cells and fibres that were difficult to distinguish. Figure 1a (his Plate 13) shows combined vertical and horizontal cuts through the NR region. It is stated that some superior cerebellar peduncle fibres combine to form the accessory mass of grey matter (label 9 in Fig. 1a). The grey matter of these accessory nuclei is said to be "in continuity with" the grey matter of the SN (label 11 in Fig. 1a). In shape and relations to NR and cerebral peduncles, they are remarkably similar to SN as depicted in his Plate 10, Fig. 3 (Fig. 1b).

Still more confusing is his explanation of label 6 in the legend to his Plate 18 (Fig. 2), which is stated to give details of region 4 as depicted in his Plate 10, Fig. 3 (Fig. 1b), there labelled SN: "Terminal fibres of the superior cerebellar peduncles coming either from the grey matter of the superior olive, or from that of the *locus niger*. They form a mass of conglomerated grey matter in 4" (named nucleus of the accessory grey substance of the *locus niger*). That conjoining nerve fibres could form structural anastomoses is consistent with Luys' adherence to the "reticulate" theory of nervous tissue. But his carelessness in choosing labels is not consistent with the reputation for intellectual rigour that Luys seems to have enjoyed (Parent et al. 2002, p. 287).

The Atlas of 1873 consists of photographs of whole brain thin slices coupled with schemata drawn from them. The structures of the subthalamus are defined more realistically and in better detail than in 1865, no doubt benefiting from the improved fixation procedures. The text is often a rehash of material in the earlier book, with some additions and editing. In contrast to the 1865 Atlas, the STN is rarely portrayed in biconvex form, but often as a ribbon of grey substance largely composed of grey fibres, sometimes merging into a thicker body. In the partly oblique horizontal section of his Plate 15, capturing the course of the brachium

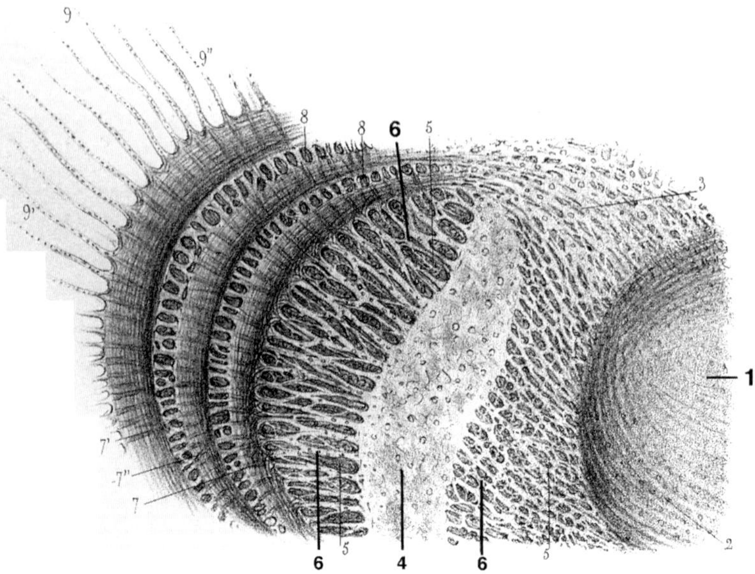

Fig. 2 Detail of region 4 of Fig. 1a. 1 – red nucleus, 4 – grey nucleus accessory to substantia nigra (= subthalamic nucleus), 6 – "terminal fibres of superior cerebellar peduncle coming from the *superior olive* (Luys' term for red nucleus) and *locus niger* (Luys' term for substantia nigra), which form a grey mass conglomerated in 4". (Luys' Plate 18, Fig. 8). Taken from Plates 10, 13, 18, 24 in a copy of the Luys (1865) Atlas in the author's possession

conjunctivum, through its decussation to the NR, the STN is clearly apparent on each side as a lentiform grey mass (see in Parent 2002).

In neither Atlas did Luys illustrate (e.g. Fig. 3) or refer to the prerubral complex of tracts and neuronal aggregates later named ZI, Fields of Forel and FF nucleus. In 1873, his Plates X and XI may include unrecognisable fragments of ZI.

Interestingly, 8 years after a letter thanking Forel for naming the STN after him, Luys (1886) published an article in his journal "L'Encéphale" in which he recapitulated the discovery of a special agglomeration of grey matter near the NR, said that all his anatomical research had been confirmed in Germany, and that Forel had confirmed his priority by naming the structure "corpus Luysii".

3 Research by Luys on Brain Functions

3.1 Neurology

After 1873 Luys practically ceased research on neuroanatomy for its own sake and began producing many reports correlating findings in the brain at autopsy with neurological signs in a large variety of patients (Semelaigne 1930–1932, pp. 90–93).

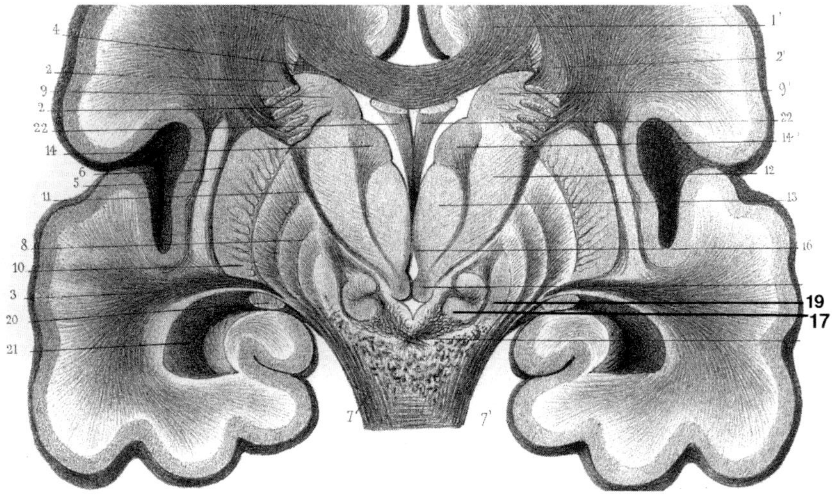

Fig. 3 Vertical brain section. 17 – red nucleus, 19 – grey matter accessory to red nucleus (= subthalamic nucleus). (Luys' Plate 24, Fig. 1). Taken from Plates 10, 13, 18, 24 in a copy of the Luys (1865) Atlas in the author's possession

He continued to issue speculative inferences on brain mechanisms, culminating in a book *The Brain and Its Functions* (1876; 3rd edition in English 1889), which was treated with reserve by other neurologists. In the preface to this book, Luys offers a reasonable and interesting plan of attack. First, he will set out the main features of brain anatomy, followed by several chapters on the subject of nerve cell properties: reactivity, storage of experienced activity and automatic activities. These bases will then be developed to encompass attention, personality, ideas, all the way to speculations on the formation of judgements. Finally, the outputs of the brain to the internal bodily processes and the external world in the form of skeleto-motor activity will be discussed. Unfortunately, his ambitious plan is characterised in its execution by many poorly supported or even completely imagined structure-function mechanisms, in an overblown metaphorical style with facile arguments by analogy. An excerpt from the chapter on the genesis and evolution of memory (p. 147) may serve to exemplify his style: "Sensory excitations, when they are diffused in the plexuses of the *sensorium* and fix themselves there in a persistent manner, do not usually remain there in the state of vague, uncertain impressions. They go further, penetrate more deeply into the recesses of cerebral life, and...". His philosophy of brain mechanisms might well be designated as "imaginative materialism".

3.2 Psychiatry

In an informative treatment of Luys's career, Parent et al. (2002) outline his work on mental patients at the Ivry-sur-Seine *maison de santé,* and emphasize the recognition he received for his work and for his books on mental illness (*Traité clinique et pratique des maladies mentales,* 693 pp., Paris, 1881; *Traitement de la folie,* 334 pp., Paris, 1893), from both his academic peers and the French Government, which made him *Chevalier* (1877) and *Officier* (1893) of the Legion of Honour.

Semelaigne, a former pupil of Luys, devotes several pages to the importance he attached to abnormal degrees or persistence of "automatism" as the "fundamental element of all types of madness" (Semelaigne, loc. cit., p. 80), an idea derived from Baillarger, a former Director of the Ivry hospice.

3.3 Hypnotism and Other Psychic Phenomena

From 1886 Luys began publishing articles on the next phase of his work, his growing interest in hysteria and hypnotism, with a note to the Société de biologie (Séance 7 august 1886) concerning purported actions of drugs at a distance on hypnotised subjects. Although he continued with more classical neurological studies, pertaining to changes in cerebral blood flow and abnormal states of excitation reinforced by cellular memory properties, there were increasing reports of bizarre findings in hysteria and hypnotism. On leaving the Salpêtrière to set up his own school at the Charité, Luys instituted clinical demonstrations following the example of Charcot at the Salpêtrière. Open to the public, they attracted a largely non-medical audience. Here he exhibited his research on hypnotherapy, as well as further curiosities such as the ability of trained hypnotised subjects to see, with allegedly hypersensitive retinas, the fields of magnetic poles (south pole red, north blue); the polarity of an electric circuit (positive red and negative blue); "emanations" from the body of normal, active human or animal subjects: red on the right side and blue on the left (Luys 1892b, 1893); the storage of emitted cerebral energy in magnetised crowns and its possible transfer (along with the mental troubles of the emitting patient) to other hypnotised subjects (Luys 1894). All these results relied on the evidence provided by trained hypnotic subjects, who where considered by Luys as veritable "living reagents" for studying these phenomena.

According to Semelaigne and other former interns at the Charité, several of these subjects were known imposters, who did not hide the fact and indeed coached each other and rehearsed their performances days in advance. Luys appears to have been quite sure of the veracity of his "reagents", although he was aware of possible suspicions of fraud by subjects or experimenters (Luys 1894).

The good faith of Luys in these activities was strongly supported by three obituarists (Ritti 1897; Beaudoin 1897; Cadet de Gassicourt 1897), two of whom criticised his excursions into these dangerous areas. Nevertheless, his good faith comes into some question in view of his last articles (Luys and David 1897a, b), in which he presented pictures claimed to be records on photographic plates of radiations emanating from the digits, the ear and the retina. He may well have been again a victim of duplicity by his assistants, but is not thereby absolved from lack of scientific rigour.

The end of the nineteenth century saw a flourishing of scientific and popular interest in hypnotic states and purported psychic phenomena, exploited by a swarm of charlatans and conjurers. Most scientists remained sceptical, but some sought to explore them as possible new discoveries, dependent on physical events such as electromagnetic waves in “the ether”, that they understood only vaguely.

4 The Subthalamus Following Luys

4.1 *Meynert*

Théodor Meynert (1833–1892) followed Luys in writing about the subthalamic region. He considered it to be part of the tegmentum, but made no significant contribution to its understanding (Meyer 1971). It was Forel and his contemporaries the Dejerines who used newer microtome and staining techniques to establish the basic structure of the whole subthalamic region, as it is now understood (Mai et al. 1997).

4.2 *Auguste Forel (1848–1931)*

Forel was born in Switzerland near Morges on Lake Geneva of French-speaking parents, in a family of French Huguenots on his mother’s side and Swiss Calvinists on his father’s (Forel 1937). A boyhood fascination with insects led him to a scientific interest in ants, and later to becoming an international expert on them. Brought up in a warm family, members of the Protestant Free Church in the Vaud Canton, the boy experienced growing scepticism of belief. At the age 16 he renounced religion, and became a non-believer for life, remaining on good terms with his mother (and later with his Christian wife).

After secondary school in Lausanne, in 1866 he began medical studies in Zurich, where, influenced by his teacher von Gudden, he decided to become a psychiatrist (Kuhlenbeck 1953). After his course in Zurich, Forel spent 7 months in 1871–1872 studying neuroanatomy under Meynert in Vienna, producing a comparative anatomical study of the thalamus (Meyer 1893). In 1873 he took a position in Munich as an assistant to von Gudden, where he produced his classical paper on the anatomy of the tegmentum in the human brain (Forel 1877) and helped von Gudden in developing the first reliable microtome, by inventing the stiffened hollow-ground blade. With it he made the first thin serial sections of the human brain (Forel 1937, p. 93). His thorough investigation of the tegmentum, with adequate, specified technique, provided new detail of tracts and nuclei in the subthalamic region. The unfortunately small images as published are yet more informative and accurate than those of Luys in 1873. Always correct, Forel named the STN *corpus Luysii* after its discoverer, but objected to his nomenclature. Forel considered the *corpus Luysii* as a cluster of neurons with associated axons, not a bundle of fibres (bandelette of Luys); for the *superior olive* of Luys he used *nucleus ruber* as all other anatomists did; in his view the Luysian body was not connected with either NR or the superior olive in the hindbrain. While working with von Gudden, he published in 1874 "The ants of Switzerland", which attracted correspondence from Charles Darwin (Forel 1968).

In 1879 Forel was appointed to direct the Burghölzli mental asylum as professor of psychiatry at the University of Zurich. Here he successfully pursued clinical neurology and psychiatry (including the use of hypnosis), some neuroanatomy, and his lifelong interest in the collection and classification of the world's ants (in 1914 his collection was the largest in existence). Many of his insane patients suffered from alcoholism, so he embraced total abstinence as a therapeutic strategy. He also undertook radical reform of the asylum's poor administration and corrupt practices.

In 1898 Forel retired from the asylum to live in the region of his childhood, and to devote himself to the scientific study of ants, travel and social works including the promotion of total abstinence from alcohol. In 1912 he suffered a stroke from which he made a partial recovery, and continued to work on his social commitments, now including pacifism, an anti-alcoholism international organisation, and democratic socialism, as best he could in the face of increasing infirmity. He died at 83 years in 1931. His memoirs, completed by 1920, were edited by his family and friends, published in 1935 in German, and translated into French and English.

4.3 Joseph-Jules Dejerine (1849–1917)

Dejerine was born in Geneva, Switzerland, of French parents, and studied medicine in Paris from 1871, where he was a student of Vulpian (Zabriskie 1953). He was a hospital interne, and received his Doctorate in 1879 for a thesis *Research on Lesions of the Nervous System in Acute Ascending Paralysis.* He then held successive posts as clinical head at the Charité hospital (1879), médecin des hôpitaux (1882), professeur agrégé (1886), consultant at the Hospital for sick children, tenured physician at the Bicêtre hospital (1887), and deputy head of the Charité medical clinic in 1893. Moving to the Salpétrière (1895), he was appointed professor of medicine at the Paris Faculty (1901). In 1907 he was appointed to the chair of internal pathology. In 1908 he entered the Académie de Médecine, and took the clinical chair of nervous diseases at the Salpêtrière in 1910 (Dictionnaire de Biographie Française 1963 fascic 57: 587).

In 1888 he married Augusta Klumpke (1859–1927), a brilliant young American who had studied medicine in Paris, and in the face of bitter male opposition was the first female *interne des hôpitaux* in France (Zabriskie 1953). The two then collaborated in many neuroanatomical and neurological studies, also publishing independently, with treatises on areas of clinical neurology. In 1917, while working to exhaustion in the treatment of wounded soldiers with nervous injuries in the Great War, Dejerine died in Paris of uraemia. After his death, his widow directed the care of paraplegics and peripheral nerve wounded at the Invalides. For her war services, she was made Chevalier, then Officer of the Legion of Honour in 1921 (André-Thomas 1928).

Mme Dejerine-Klumpke was named as collaborator on the title page of their two-volume magnum opus *Anatomie des Centres Nerveux* (Dejerine 1895–1901). In this extensive exposition, the Dejerines provided very detailed methodological information on the preparation of stained microscopic serial brain sections, cut under water, with modern illustrations in profusion of the subthalamic region and other brain structures, mainly in Weigert preparations (Fig. 4), which are largely the same as images in the standard texts of today. In the methods section, they described an adjustable projector to cast images from large, thin-mounted sections onto paper, for accurate tracing of low power microscopic detail. In their opinion this provided more satisfactory results than photography, a view confirmed by the experience of many subsequent investigators.

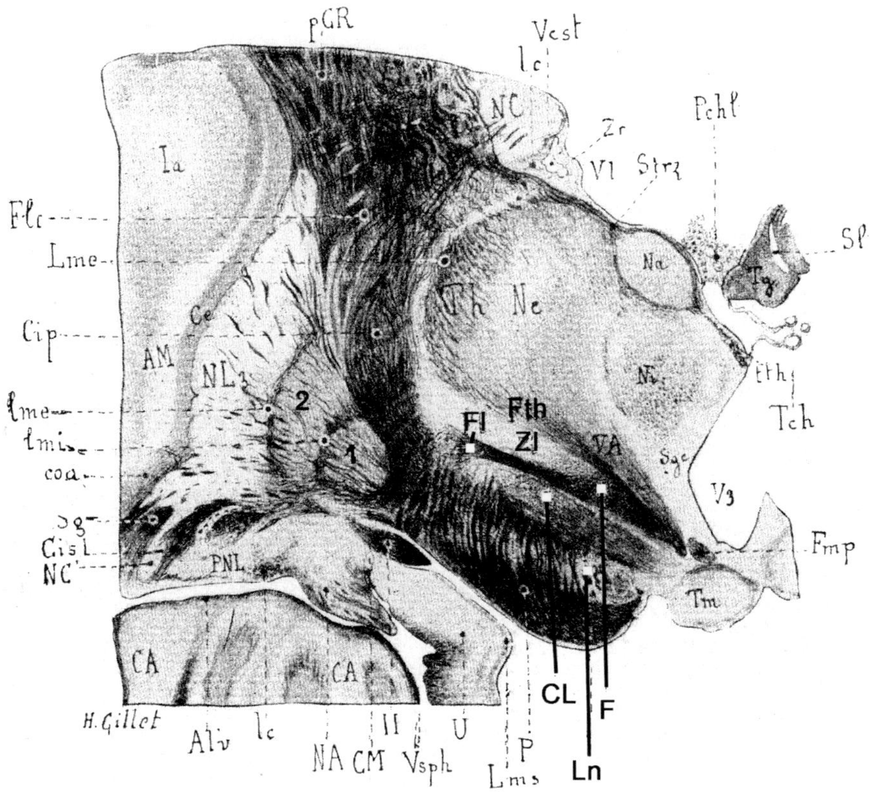

Fig. 4 Transverse section through subthalamus, from Dejerine (1895–1901) *Anatomie des centres nerveux* T. II, Fig. 324. *CL* STN (corps de Luys), *F* field H of Forel, *Fl* fasciculus lenticularis, *FTh* fasciculus thalamicus, *Ln* locus niger (substantia nigra), *NL* lenticular nucleus (1,2 – pallidum, 3 – putamen), *VA* bundle of Vic d'Azyr and *ZI* zona incerta

5 Conclusions

5.1 *Luys as Discoverer of the STN*

Luys discovered the STN (his accessory grey matter of the red nucleus) almost by accident, and apparently misinterpreted the constitution of its tissue and its connections. Although he was a relentless worker, who wrote 8 books and 105 articles and reports, he was not a rigorous thinker and had little self-critical capacity. He was more given to flights of inferential and deductive fancy than to careful testing of his imaginative explanations of brain mechanisms. Possibly, he was unduly pressed to explain the neurological and psychiatric maladies he confronted, mostly as single cases, in terms of inventive mechanistic models that seemed to him reasonable in view of existing information, particularly that provided by his own research. Nonetheless, his imaginative deductions allied to his profuse neuroanatomical illustrations, may still be found stimulating.

5.2 *Establishment of Basic Subthalamic Structure*

The principal successors to Luys in the study of the subthalamic region, Forel and the Dejerines, were content to investigate assiduously the morphological evidence without the "speculative anatomy" as practiced by Luys (Meyer 1893). In this way, they provided more solid advances in understanding the complex structure of the subthalamus, while explicitly leaving undecided questions of connectivity and its polarity to be decided by future advances in research techniques.

References

André-Thomas (1928) Necrologie. Augusta Dejerine-Klumpke (1859–1927). L'Encéphale 23: 75–88.

Beaudoin M (1897) Necrologie, M le Dr Luys. Le Progrès Méd 2: 141–142.

Cadet de Gassicourt E (1897) Osèques de M le Dr Luys. Bull Acad Med 38: 198–200.

Dejerine J (1895–1901) Anatomie des centres nerveux. Rueff, Paris.

Forel A (1877) Untersuchungen uber die Haubenregion und ihre oberen verknupfungen im Gehirne des Menschen. Arch f Psychiat 7: 393–495.

Forel A (1937) Out of my life and work (trans. B Miall). George Allen & Unwin, London.

Forel A (1968) August Forel, Briefe correspondance 1849–1927. Hans Hube, Bern.

Kuhlenbeck A (1953) August Forel (1848–1931). In Haymaker W (ed.) The founders of neurology. CC Thomas, New York, NY.

Luys J (1865) Recherches sur le système nerveux cérébro-spinal. Baillière, Paris.

Luys J (1873) Iconographie photographique du système nerveux. Baillière, Paris.

Luys J (1886) Description d'une nouvelle region de substance grise… L'Encéphale 6: 5–10.

Luys J (1889) The brain and its functions, 3rd ed. Kegan, Paul, Trench, London.

Luys J (1892a) Des procédés à employer pour l'étude anatomique et photographique du système nerveux. Ann Psychiat Hypnol 2: 129–141.
Luys J (1892b) De la visibilité, par les sujets en etat hypnotique, des effluves dégagés par les êtres vivants. C R Soc Biol Paris, séance 18 juillet 1892: 657–659.
Luys J (1893) De la visibilité directe des effluves cérébraux. C R Soc Biol Paris, séance 17 juin 1893: 638–641.
Luys (1894) De l'emmagasinement de certaines activités cerebrales dans une couronne aimantée. C R Soc Biol Paris, séance 10 février 1894: 128–130.
Luys et David (1897a) Note sur l'enregistrement photographique des effluves qui se dégagent des extrémités des doigts et du fond de l'oeil de l'être vivant... C R Soc Biol Paris, séance 29 mai 1897: 515–519.
Luys et David (1897b) Fixation par la photographie des effluves qui se dégagent de l'appareil auditif... C R Soc Biol Paris, séance 10 juillet 1897: 676–678.
Mai JK, Assheuer J, Paxinos G (1997) Atlas of the human brain. Academic, London.
Meyer A (1971) Historical aspects of cerebral anatomy. Oxford University Press, London.
Meyer Ad (1893) Neurologists and neurological laboratories. II Neurological work at Zurich. J Comp Neurol 3: 1–6.
Parent A (2002) Jules Bernard Luys and the subthalamic nucleus. Mov Disord 17: 181–185.
Parent A, Parent M, Leroux-Hugon V (2002) Jules Bernard Luys: A singular figure in 19th century neurology. Can J Neurol Sci 29: 282–288.
Plaha P, Ben-Shlomo Y, Patel NK, et al (2006) Stimulation of the caudal zona incerta is superior to stimulation of the subthalamic nucleus in improving contralateral parkinsonism. Brain 129: 1732–1747.
Ritti A (1897) Necrologie: Dr J Luys. Ann Médicopsych 6: 321–323.
Semelaigne R (1930–1932) Les pionniers de la psychiatrie française: avant et après Pinel. Baillière, Paris.
Zabriskie E G (1953) Joseph Jules Dejerine (1849–1917). In Haymaker W (ed) Op cit.

Organization of Motor Cortical Inputs to the Subthalamic Nucleus in the Monkey

Hirokazu Iwamuro, Yoshihisa Tachibana, Nobuhito Saito, and Atsushi Nambu

Abstract To investigate the motor cortical inputs to the subthalamic nucleus (STN), we examined the responses of STN neurons to electrical stimulation of the primary motor cortex (MI) and the supplementary motor area (SMA). Stimulating electrodes were chronically implanted in the orofacial, forelimb, and hindlimb regions of the MI, and the forelimb and hindlimb regions of the SMA in a macaque monkey. Cortical stimulation induced a neuronal response composed of an early excitation and a late excitation in the STN. Of the STN neurons with cortical inputs, approximately 70% responded exclusively to either MI- or SMA-stimulation and the remaining 30% responded to both MI- and SMA-stimulation. Neurons responding mainly to MI-stimulation were distributed in the dorsolateral part of the STN, while neurons responding mainly to SMA-stimulation were mainly present in the ventromedial region. Approximately 60% of STN neurons received inputs from a single somatotopical region, while the others received convergent inputs from multiple somatotopical regions. In the MI domain of the STN, neurons responding to the stimulation of the orofacial, forelimb, and hindlimb regions were represented in its lateral-most, central and medial parts, respectively. In the SMA domain, neurons receiving inputs from the forelimb region were located more medially than those receiving inputs from the hindlimb region. The present study has clearly demonstrated that the STN is somatotopically organized based on the cortical inputs from the MI and SMA despite considerable convergence.

H. Iwamuro, Y. Tachibana, and A. Nambu (✉)
Division of System Neurophysiology, National Institute for Physiological Sciences, 38 Nishigonaka, Myodaiji, Okazaki 444-8585, Japan
e-mail: nambu@nips.ac.jp

N. Saito
Department of Neurosurgery, The University of Tokyo
7-3-1 Hongo, Bunkyo-ku, Tokyo 113-8655, Japan

H.J. Groenewegen et al. (eds.), *The Basal Ganglia IX*, Advances in Behavioral Biology 58,
DOI 10.1007/978-1-4419-0340-2_9,

1 Introduction

It is essential for understanding the motor functions of the basal ganglia to investigate how the information originating from the motor cortices is processed through the cortico-basal ganglia-thalamic loop (Alexander et al. 1986; Alexander and Crutcher 1990). In the current model of the basal ganglia organization, the striatum receives direct excitatory cortical inputs and projects to the output nuclei of the basal ganglia, i.e., the internal segment of the globus pallidus (GPi) and the substantia nigra pars reticulata (SNr), via two major projection systems. One is the direct pathway that connects the striatum and the output nuclei monosynaptically. The other is the indirect pathway that connects the striatum and the output nuclei via the external segment of the globus pallidus (GPe) and the subthalamic nucleus (STN) multisynaptically (Albin et al. 1989; Alexander and Crutcher 1990). Recent studies have proposed that the STN is another input station of the basal ganglia (Mink and Thach 1993; Mink 1996; Nambu et al. 1996, 2000, 2002c; Levy et al. 1997), suggesting another pathway in the basal ganglia, cortico-STN-GPi/SNr "hyperdirect" pathway (Nambu et al. 2002c). Therefore, the STN receives cortical inputs directly through the hyperdirect pathway and indirectly through the striatum and the GPe. The first aim of the present study is to investigate how these inputs through the hyperdirect and indirect pathways integrate in the STN. Special attention was paid to the convergence of the inputs from different cortical areas, i.e., the primary motor cortex (MI) and the supplementary motor area (SMA). Another aim of the study is to establish the somatotopical map in the STN, which gives us important clues to localize the recording site during stereotaxic surgery. Previous anatomical studies have shown that cortico-STN projections are somatotopically organized (Monakow et al. 1978, Nambu et al. 1996). However, STN neurons have extensive dendrites covering large areas of the STN (Kita et al. 1983). Moreover, STN neurons have local axon collaterals that terminate neighboring neurons (Kita et al. 1983). Thus, the somatotopical map investigated electrophysiologically on the single neuronal level may be obscure, compared with the anatomical cortico-STN input map. To answer these questions, we recorded neuronal activity of STN neurons in a monkey, and examined the neuronal responses to the stimulation of the somatotopical regions in the MI and SMA.

2 Methods

A female Taiwan monkey (*Macaca cyclopis*) weighing 4.0 kg was used in this study. This study was performed in compliance with the Guide for the Use and Care of Laboratory Animals in Research of the National Institute of Physiological Sciences. After induction with ketamine hydrochloride (10 mg/kg i.m.) and xylazine hydrochloride (2 mg/kg i.m.), the monkey was anesthetized with pentobarbital sodium (25 mg/kg body weight iv) following which it underwent a surgical operation to fix its head in a stereotaxic frame attached to a monkey chair as described previously (Nambu et al. 2000, 2002a). A few days after the surgery, the MI and SMA were mapped by the intracortical microstimulation and sensory responses. According to this electrophysiological mapping, pairs of bipolar stimulating electrodes were chronically implanted into the orofacial, forelimb, and hindlimb regions of the MI and into the forelimb and hindlimb regions of the SMA. A rectangular chamber was also attached to access the STN.

After full recovery from these procedures, recording STN neuronal activity was started. A glass-coated Elgiloy-alloy microelectrode (0.7–1.5 MΩ at 1 kHz) was inserted vertically through the dura into the STN to record neuronal activity using a hydraulic microdrive in the awake state. The responses of STN neurons to cortical electrical stimulation (300-μs duration single pulse, strength of less than 0.7 mA, at 0.7 Hz) were observed by constructing peri-stimulus time histograms (PSTHs; bin width, 1 ms; sum of 100 times). This stimulating frequency was used in our previous reports (Nambu et al. 2000, 2002a), and no changes of synaptic efficacy were observed.

After the electrophysiological experiments, recording sites were examined histologically and reconstructed.

3 Results

The MI was successfully mapped in the anterior bank of the central sulcus. In this area, the orofacial, forelimb, and hindlimb regions were arranged from the lateral to medial parts along the central sulcus. The distal part of the forelimb was represented more laterally than the proximal part. In the mesial wall of the hemisphere, the forelimb region of the SMA was more rostrally located than the hindlimb region. Six pairs of stimulating electrodes were implanted into each region: the orofacial (MIo), distal and proximal forelimb (MIf) and hindlimb (MIh) regions of the MI, and the forelimb (SMAf) and hindlimb (SMAh) regions of the SMA.

We recorded activity of STN neurons that responded to the electrical stimulation of the MI and/or SMA. Some neurons responded to the stimulation of multiple cortical sites, and a total of 280 responses to the cortical stimulation were recorded from 144 STN responsive neurons. The firing rates of these responsive neurons were measured based on recordings of spontaneous discharges for 30 s before stimulation, and their average was 19.0 ± 9.9 (mean ± SD) Hz. The cortical stimulation typically induced biphasic responses, consisting of an early excitation and a late excitation, in the STN neurons (n = 225, 80%, Fig. 1a1, a2), although it occasionally induced responses with the only early excitation (n = 6, 2%, Fig. 1b) or responses with the only late excitation (n = 49, 18%, Fig. 1c). The ratio of these three response types did not differ among the cortical stimulation sites. Different response types were sometimes induced in single neurons with multiple cortical inputs.

The latencies of the early and late excitations in STN neurons evoked by the stimulation of different cortical areas are compared in Table 1. The latency of the early or late excitation evoked by the MI-stimulation was significantly shorter than that evoked by the SMA-stimulation, respectively (ANOVA with Bonferroni/Dunn post hoc tests, $P < 0.01$). Among neurons with inputs from the MI or SMA, the latencies of the excitations caused by the stimulation of different somatotopical regions were comparable. One exception was that the latency of the early excitation produced by the MIf-stimulation was shorter than that by the MIo-stimulation (ANOVA with Bonferroni/Dunn post hoc tests, $P < 0.01$).

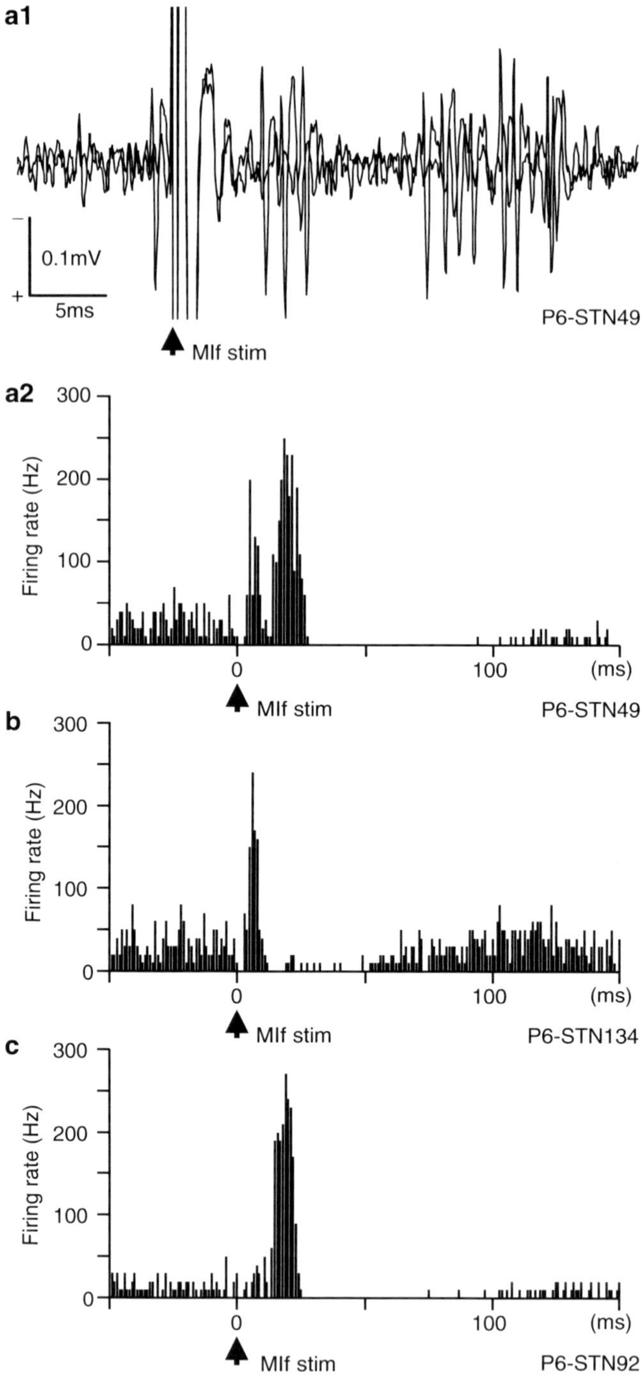

Fig. 1 Three response patterns of STN neurons evoked by cortical stimulation. (**a**) an STN neuron showing a biphasic response to MIf stimulation. (**a1**) Responses to MIf stimulation. Two traces were overlaid. (**a2**) A peri-stimulus time histogram (PSTH; 100 times; bin width, 1 ms) constructed from responses as shown in (**a1**). (**b**) A PSTH of another STN neuron showing only an early excitation to MIf stimulation. (**c**) A PSTH of another STN neuron showing only a late excitation to MIf stimulation

Table 1 Latencies of cortically evoked excitatory responses in STN neurons (means ± SD in ms)

Cortical stimulating site		Latency (ms) Early excitation	Late excitation
MI		7.0 ± 1.7*	16.7 ± 2.8**
	MIo	7.7 ± 1.8†	16.0 ± 2.8
	MIf	6.5 ± 1.7†	17.4 ± 2.9
	MIh	7.2 ± 1.5	16.2 ± 2.0
SMA		9.1 ± 1.9*	21.8 ± 3.0**
	SMAf	9.6 ± 1.7	22.1 ± 2.8
	SMAh	8.6 ± 2.0	21.4 ± 3.2

*, **, † significantly different from each other (ANOVA with Bonferroni/Dunn post hoc tests, $P < 0.01$). *o* orofacial, *f* forelimb, and *h* hindlimb

Table 2 Number of neurons with inputs from different cortical areas based on the early excitation

MI ($n = 68$)	
MIo	33
MIf	6
MIh	15
MIo+f	7
MIf+h	7
SMA ($n = 25$)	
SMAf	12
SMAh	8
SMAf+h	5
MI+SMA ($n = 37$)	
MIo+SMAf	1
MIf+SMAf	6
MIf+SMAh	5
MIh+SMAf	1
MIh+SMAh	1
MIo+f+SMAf	6
MIo+f+SMAh	1
MIf+h+SMAh	5
MIo+SMAf+h	1
MIf+SMAf+h	2
MIh+SMAf+h	1
MIf+h+SMAf+h	3
MIo+f+SMAf+h	4

o orofacial, *f* forelimb, and *h* hindlimb

One hundred and thirty STN neurons showed the early excitation evoked by cortical stimulation, and these neurons could be categorized into MI-, SMA-, and MI+SMA-recipient neurons based on the early excitation (Table 2). The majority of STN neurons (70%) received exclusive inputs from the MI or SMA. On the other hand, a substantial number of STN neurons (30%) received convergent inputs from both the MI and SMA. The STN neurons could be further categorized based on the input from different somatotopical regions (Table 2). Concerning MI- and SMA-recipient neurons, most neurons (79% of MI- and 80% of SMA-recipient neurons) received inputs from a single somatotopical region, while other neurons received convergent inputs from two somatotopical regions. However, the convergence did not occur randomly. Neurons received convergent inputs from neighboring somatotopical regions, such as MIo+f, and MIf+h, although no neurons received convergent inputs from MIo+h. Among MI+SMA-recipient neurons, some (19%) received convergent inputs from a single somatotopical region in the MI and the same somatotopical region in the SMA, such as MIf+SMAf, while the others (81%) received convergent inputs from multiple but neighboring somatotopical regions, such as MIf+SMAh and MIo+f+SMAf. Therefore, among 130 STN neurons, the majority (62%) received inputs from a single somatotopical region, while the others (38%) received convergent inputs from multiple somatotopical regions. Instead of the early excitation, STN neurons could also be categorized based on the late excitation, resulting in similar distribution (data not shown).

Locations of the STN neurons exhibiting the early excitation evoked by the stimulation of the different cortical areas are plotted in Fig. 2. The neurons that responded to the stimulation of the MI and/or SMA were distributed in the caudal portion of the STN, especially in the dorsal part. Within this area, MI-recipient neurons were mainly located in the dorsolateral part, whereas SMA-recipient neurons were predominantly located in the ventromedial part. MI+SMA-recipient neurons were found both in the MI and SMA domains. In the MI domain, MIo-recipient neurons were located in the lateral-most part, and MIh-recipient neurons were located in its medial part. MIf-recipient neurons were located in between. MIo+f-recipient neurons were intermingled with MIo- or MIf-recipient neurons, and

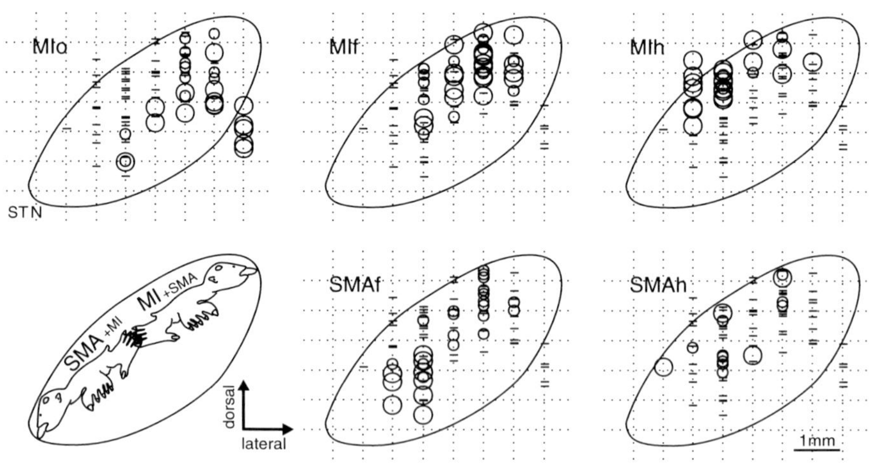

Fig. 2 Locations of STN neurons with an early excitation in the frontal section (A12) according to cortical inputs. *Large* and *small circles* mean the largest responses and the smaller responses, respectively. *Bars* mean no responses. *Lower left corner*, somatotopical organization of the STN revealed by the previous anatomical study (Nambu et al. 1996)

MIf+h-recipient neurons with MIf- or MIh-recipient neurons. In the SMA domain, SMAf-recipient neurons were located more medially than SMAh-recipient neurons. In addition to the early excitation, locations of STN neurons were also plotted based on the late excitations, resulting in the similar map (data not shown).

4 Discussion

The present study has demonstrated the following results (1) MI and SMA-stimulation induced a response composed of early and late excitations in the STN; (2) Most of STN neurons received exclusive inputs from the MI or SMA, while others received convergent inputs from the MI and SMA; (3) Cortical inputs from the MI and SMA in the STN were somatotopically organized despite considerable convergent inputs.

The common response pattern of STN neurons to the MI- or SMA-stimulation is a short-latency excitation followed by a late excitation, as reported previously (Fig. 1; Nambu et al. 2000). The early excitation evoked by the cortical stimulation is considered to be caused by the direct cortico-STN projections as investigated previously in rodents (Kitai and Deniau 1981; Ryan and Clark 1992; Fujimoto and Kita 1993; Maurice et al. 1998). On the other hand, for the late excitation in the STN, several possible origins have been proposed. One is the disinhibition in the STN through the cortico-striato-GPe-STN indirect pathway (Maurice et al. 1998; Nambu et al. 2002b). Another is an anodal-break rebound depolarization that follows the preceding inhibition induced by the cortico-STN-GPe-STN pathway (Nakanishi et al. 1987; Bevan et al. 2002). A third is a late component of a single broad excitatory response interrupted by a brief inhibition induced by the cortico-STN-GPe-STN pathway (Fujimoto and Kita 1993). The first explanation assumes that cortical inputs via the cortico-STN and cortico-striato-GPe-STN projections converge on single STN neurons. The latencies of the early and late excitations evoked by MI-stimulation in this study were very similar to those reported previously (Nambu et al. 2000), and were shorter than those by SMA-stimulation in this study (Table 1). The latency difference between MI- and SMA-stimulation-induced responses were also observed in the striatum (Nambu et al. 2002a), and may be explained by the difference of axonal conduction velocities between MI neurons projecting to the STN and SMA neurons to the STN.

Based on the present study, the majority of STN neurons received exclusive inputs from the MI or SMA, while others received convergent inputs from both the MI and SMA. Concerning somatotopical organization, the majority received inputs from a single somatotopical region, while the others received convergent inputs from multiple somatotopical regions. The convergence does not occur randomly, and obeys the following rule: the STN neurons received convergent inputs from the same or neighboring somatotopical regions (Table 2). Thus, these convergent inputs may have functional significance. In spite of such convergence, MI-recipient neurons were mainly located in the dorsolateral part of the STN, whereas SMA-recipient neurons were predominantly located in the ventromedial part in the present study (Fig. 2).

In the MI domain, MIo-, MIf- and MIh-recipient neurons were located along the dorsolateral-ventromedial axis. In the SMA domain, SMAf-recipient neurons were located more medially than SMAh-recipient neurons. Thus, the STN shows a clear somatotopical organization in both MI and SMA domains. Such somatotopical organization agrees well with our previous anatomical study (see the schema in the lower left corner of Fig. 2; Nambu et al. 1996). The cortico-STN projections from the hindlimb, forelimb, and orofacial regions of the MI are arranged from medial to lateral in the lateral part of the STN, whereas the projections from the SMA are represented in the inverse order in the medial part of the STN (Nambu et al. 1996). The anatomical study has shown that the SMA also sends minor projections to the lateral part of the STN in a somatotopical organized manner, and the MI to the medial part, supporting the present observation of STN neurons receiving convergent inputs from the same somatotopical regions of the MI and SMA.

The somatotopical organization in the STN has been also studied by observing responses to active movements and passive manipulations of individual body parts (DeLong et al. 1985; Matsumura et al. 1992; Wichmann et al. 1994). Neurons responding to these stimuli were distributed mainly in dorsolateral part of the STN. Neurons representing the hindlimb lie centrally in the mediolateral dimension of the nucleus, and neurons representing the orofacial part are located in the most lateral part. Neurons representing the forelimb tend to be sandwiched in between. This representation corresponds well to the somatotopical organization from the MI in the present study. On the other hand, activity of STN neurons in the SMA domain remains to be elucidated. The present study has also shown that the most common response pattern of STN neurons to the cortical stimulation is an early excitation followed by a late excitation. Although the origin of the late excitation is still under debate, the present study suggests that signals originating from the cortex are conveyed through the cortico-STN hyperdirect pathway, and through the cortico-striato-GPe-STN indirect or cortico-STN-GPe-STN pathway, and finally converge on a single STN neuron. Moreover, the similar somatotopical arrangement has been shown by the output map of the STN, which was studied by injecting rabies virus into the MI as a retrograde transneuronal tracer (Miyachi et al. 2006). All these studies suggest that signals originating from the cortex are processed principally in a parallel fashion and return to the original cortex through the cortico-basal ganglia-thalamic loop despite considerable convergence.

References

Albin RL, Young AB and Penney JB (1989) The functional anatomy of basal ganglia disorders. Trends Neurosci 12: 366–375.

Alexander GE and Crutcher MD (1990) Functional architecture of basal ganglia circuits: Neural substrates of parallel processing. Trends Neurosci 13: 266–271.

Alexander GE, DeLong MR and Strick PL (1986) Parallel organization of functionally segregated circuits linking basal ganglia and cortex. Ann Rev Neurosci 9: 357–381.

Bevan MD, Magill PJ, Hallworth NE, Bolam JP and Wilson CJ (2002) Regulation of the timing and pattern of action potential generation in rat subthalamic neurons in vitro by GABA-A IPSPs. J Neurophysiol 87: 1348–1362.
DeLong MR, Crutcher MD and Georgopoulos AP (1985) Primate globus pallidus and subthalamic nucleus: Functional organization. J Neurophysiol 53: 530–543.
Fujimoto K and Kita H (1993) Response characteristics of subthalamic neurons to the stimulation of the sensorimotor cortex in the rat. Brain Res 609: 185–192.
Kita H, Chang HT and Kitai ST (1983) The morphology of the intracellular labeled rat subthalamic neurons: A light microscopic analysis. J Comp Neurol 215: 245–257.
Kitai ST and Deniau JM (1981) Cortical inputs to the subthalamus: Intracellular analysis. Brain Res 214: 411–415.
Levy R, Hazrati L-N, Herrero M-T, Vila M, Hassani O-K, Mouroux M, Ruberg M, Asensi H, Agid Y, Féger J, Obeso JA, Parent A and Hirsch EC (1997) Re-evaluation of the functional anatomy of the basal ganglia in normal and parkinsonian states. Neuroscience 76: 335–343.
Matsumura M, Kojima J, Gardiner TW and Hikosaka O (1992) Visual and oculomotor functions of monkey subthalamic nucleus. J Neurophysiol 67: 1615–1632.
Maurice N, Deniau JM, Glowinski J and Thierry AM (1998) Relationships between the prefrontal cortex and the basal ganglia in the rat: Physiology of the corticosubthalamic circuits. J Neurosci 18: 9539–9546.
Mink JW (1996) The basal ganglia: Focused selection and inhibition of competing motor programs. Prog Neurobiol 50: 381–425.
Mink JW and Thach WT (1993) Basal ganglia intrinsic circuits and their role in behavior. Curr Opin Neurobiol 3: 950–957.
Miyachi S, Lu X, Imanishi M, Sawada K, Nambu A, Takada M (2006) Somatotopically arranged inputs from putamen and subthalamic nucleus to primary motor cortex. Neurosci Res 56: 300–308.
Monakow KH, Akert K and Künzle H (1978) Projections of the precentral motor cortex and other cortical areas of the frontal lobe to the subthalamic nucleus in the monkey. Exp Brain Res 33: 395–403.
Nakanishi H, Kita H and Kitai ST (1987) Electrical membrane properties of rat subthalamic neurons in an in vitro slice preparation. Brain Res 437: 35–44.
Nambu A, Takada M, Inase M and Tokuno H (1996) Dual somatotopical representations in the primate subthalamic nucleus: Evidence for ordered but reversed body-map transformations from the primary motor cortex and the supplementary motor area. J Neurosci 16: 2671–2683.
Nambu A, Tokuno H, Hamada I, Kita H, Imanishi M, Akazawa T, Ikeuchi Y and Hasegawa N (2000) Excitatory cortical inputs to pallidal neurons via the subthalamic nucleus in the monkey. J Neurophysiol 84: 289–300.
Nambu A, Kaneda K, Tokuno H and Takada M (2002a) Organization of corticostriatal motor inputs in monkey putamen. J Neurophysiol 88: 1830–1842.
Nambu A, Kita H, Akazawa T, Imanishi M and Takada M (2002b) Excitatory cortical input to the subthalamic nucleus through the cortico-striato-external pallido-subthalamic 'indirect' pathway. Abstr 32nd Ann Meeting Soc f Neurosci 358.10.
Nambu A, Tokuno H and Takada M (2002c) Functional significance of the cortico-subthalamo-pallidal 'hyperdirect' pathway. Neurosci Res 43: 111–117.
Ryan LJ, Clark KB (1992) Alteration of neural responses in the subthalamic nucleus following globus pallidus and neostriatal lesions in rats. Brain Res Bull 29: 319–327.
Wichmann T, Bergman H and DeLong MR (1994) The primate subthalamic nucleus. I. Functional properties in intact animals. J Neurophysiol 72: 494–506.

A Subpopulation of Mesencephalic Dopamine Neurons Interfaces the Shell of Nucleus Accumbens and the Dorsolateral Striatum in Rats

Yvette C. van Dongen, Bogdan P. Kolomiets, Henk J. Groenewegen, Anne-Marie Thierry, and Jean-Michel Deniau

Abstract Nigrostriatal dopaminergic neurons are usually considered to interface the ventral limbic and dorsal sensorimotor striatum, since the shell of the nucleus accumbens (Acb shell) projects to the ventral tegmental area/substantia nigra pars compacta (VTA/SNC) complex. However, both the organization of Acb shell projections to the nigrostriatal neurons innervating the sensorimotor striatum and the synaptic influence exerted by the Acb shell on these neurons remain to be determined. These questions were addressed in the rat using neuroanatomical and electrophysiological approaches.

Combined anterograde tracing from the Acb shell with retrograde tracing from the sensorimotor region of the dorsal striatum revealed that labeled fibers from the Acb shell overlap retrogradely labeled nigrostriatal neurons located in the medial SNC and the lateral VTA but avoid the nigrostriatal neurons located laterally. In addition, stimulation of the Acb-shell induced an inhibition of dopaminergic nigrostriatal neurons projecting to the sensorimotor striatal territory. In agreement with the anatomical observations, these responses were observed in nigrostriatal neurons located in the medial SNC and the lateral VTA but not in nigrostriatal neurons located laterally.

These data further establish the existence of a functional link between the Acb shell and the sensorimotor striatum via dopaminergic nigrostriatal neurons. The present study also reveals that among the dopaminergic nigrostriatal neurons innervating the sensorimotor striatal territory, only the subpopulation located in the medial SNC and lateral VTA receives an inhibitory input from the Acb shell. This indicates a functional heterogeneity within the population of dopaminergic neurons innervating a given striatal territory.

Y.C. van Dongen, B.P. Kolomiets, H.J. Groenewegen, A.–M. Thierry, and J.–M. Deniau (✉)
Institut National de la Santé et de la Recherche Médicale U 114, Chaire de Neuropharmacologie, Collège de France, 11 Place Marcelin Berthelot, 75231 Paris Cedex 05, France
e-mail: jean-michel.deniau@college-de-france.fr

H.J. Groenewegen et al. (eds.), *The Basal Ganglia IX*, Advances in Behavioral Biology 58, DOI 10.1007/978-1-4419-0340-2_10,

1 Introduction

The striatum transmits cortical signals to the basal ganglia output nuclei, i.e., the substantia nigra pars reticulata (SNR) and the internal segment of globus pallidus. Growing evidence indicates that signals originating from functionally distinct cortical areas are processed in separate striatal territories and remain segregated in the striatonigral and striatopallidal pathways, supporting the concept of parallel cortico-basal ganglia circuits (Alexander and Crutcher 1990; Groenewegen et al. 1999; Deniau and Thierry 1997; Mailly et al. 2001). Classically, the dorsal striatum is associated with sensorimotor and cognitive processes and the ventral striatum with motivation and reward.

Dopaminergic (DA) fibers from the ventral tegmental area/substantia nigra pars compacta complex (VTA/SNC) innervate both dorsal and ventral striatum. Based on observations that the ventral striatum innervates the VTA/SNC complex, Nauta et al. (1978) first proposed that DA neurons constitute an interface between the limbic and extrapyramidal motor systems. Corroborating this hypothesis, fibers from the ventral striatum were shown to establish synaptic contacts on dendrites of nigral neurons projecting to the dorsal striatum (Somogyi et al. 1981). However, the functional–anatomical properties of interactions between the limbic and sensorimotor domains of the striatum via DA nigrostriatal neurons remain to be determined.

Histochemical and anatomical studies have now established that the major component of the ventral striatum, the nucleus accumbens (Acb), comprises two main subdivisions, the core and shell that present distinctive connectional characteristics (Heimer et al. 1997). Like the dorsal striatum, the Acb core receives major inputs from the cerebral cortex, i.e., the medial and lateral prefrontal areas, and projects to the subcommissural ventral pallidum and the SNR. The Acb shell receives major inputs from hippocampus and amygdala and innervates predominantly the VTA/SNC complex in addition to the ventral pallidum (Heimer et al. 1997; Groenewegen et al. 1999; Wright et al. 1996; Montaron et al. 1996; Maurice et al. 1997, 1998). Interestingly, a recent comparative 3D analysis of the striatonigral and nigrostriatal projections in the rat revealed that each functional territory of the dorsal striatum is innervated by two main subpopulations of VTA/SNC neurons (Maurin et al. 1999). The first population, located in the SNC and denominated "proximal," occupies a position in register with the striatonigral projections in the subjacent SNR. Such a close spatial relationship suggests that this subpopulation may be involved in a reciprocal striato-nigro-striatal feedback circuit. The second population, denominated "distal," is located more medially and dorsally in the SNC and VTA, and is likely involved in nonreciprocal connections with the dorsal striatum.

The aim of the present study was to investigate to which extent the shell exerts a synaptic influence on these two subpopulations of nigrostriatal neurons which innervate the sensorimotor dorsal striatum. For this purpose injections of an anterograde tracer into the Acb shell were combined with injections of a retrograde tracer into the dorsolateral striatal territory innervated by the sensorimotor cortical areas.

In addition, the effects of electrical stimulation of the Acb shell on DA nigrostriatal neurons identified as projecting to the sensorimotor territory of the dorsolateral striatum were determined.

2 Materials and Methods

2.1 Anatomical Tracing Studies

2.1.1 Surgical Procedures

A total of 15 female Wistar rats weighing 180–240 g (Harlan/CPB, Zeist, Netherlands) were used. Animals were anesthetized with a 4:3 parts mixture of a 1% solution of ketamine (Aesco, Boxtel, Netherlands) and a 2% solution of xylazine (Bayer, Brussels, Belgium) by intramuscular injections (1 mL/kg). During surgery, anesthesia was maintained by additional doses of the same solution, while body temperature (36–37°C) was maintained by a homeothermic mat. Anesthetized animals were mounted in a stereotaxic frame. Local anesthetics were injected under the skin of the head at the site of incision. Tracers were injected using coordinates from the atlas of Paxinos and Watson (1996).

2.1.2 Tracer Injections

The anterograde tracer biotinylated dextran amine (BDA, 10,000 MW, Molecular Probes, Eugene, OR) and the retrograde tracer Fluorogold (FG, Fluorochrome, Denver, CO) were iontophoretically delivered through glass micropipettes (external diameter 3 μm for BDA and 15 μm for FG) using a positive pulsed current of 1 μA (7 s on/off) for BDA and 2.5 μA (7 s on/off) for FG, employing a constant current source (Midgard CCS-3, USA). Pipettes were filled with either 5% BDA in 0.1 M NaH_2PO_4/Na_2HPO_4 (phosphate buffer [PB], pH 7.4) or 2% FG in 0.1 M cacodylate buffer (pH 7.3). Tracer delivery lasted 2 min for BDA and 5 min for FG.

2.1.3 Histological Procedures

Following a 7-day postoperative survival period, animals were deeply re-anesthetized with sodium pentobarbital (60 mg/kg body weight, i.p.; Ceva, Paris, France) and perfused transcardially with 0.9% saline, followed by a fixative containing 4% paraformaldehyde (Merck, Darmstadt, Germany) and 0.05% glutaraldehyde (Merck-Schuchardt, Hohenbrunn, Germany) in PB (0.1 M, pH 7.4) for 15 min. Brains were post-fixed for 1.5 h and cryoprotected by storage for 18–48 h at 4°C in a mixture of 20% glycerin (Merck) and 2% dimethyl sulfoxide (DMSO;

Merck) in PB (0.1 M, pH 7.4). Horizontal or transverse 40-μm sections were cut on a sliding microtome. Sections were collected sequentially in six vials containing either PB (0.1 M, pH 7.4) for direct processing, or a mixture of glycerin/DMSO for storage at −20°C.

Sections to be double stained for BDA and FG were rinsed with PB (0.1 M, pH 7.4) followed by a rinse in 0.05 M Tris-HCl (Merck) supplemented with 0.15 M NaCl, pH 7.6 (TBS) and 0.5% Triton X-100 (TBS-Tx; Merck). They were incubated with rabbit anti-FG (Chemicon, Temecula CA; 1:5,000) overnight at 4°C. Subsequently, the sections were stained for BDA with nickel-enhanced diaminobenzidine substrate: 7.5 mg diaminobenzidine-tetrahydrochloride (DAB; Sigma, St. Louis, MO), 0.225 g nickel-ammonium sulfate (Boom, Meppel, Netherlands), 10 μL of 30% H_2O_2 in 50 mL Tris–HCl (pH 8.0) for 10–30 min.

After rinsing with TBS–Tx, sections were incubated with swine anti-rabbit serum (Dako, Glostrup, Denmark; 1:100) for 60 min at room temperature (rt). Following a rinse with TBS-Tx, the sections were incubated with rabbit peroxidase–antiperoxidase (Dako, Denmark; 1:800) for 60 min at rt. Finally, a Tris–HCl rinse (pH 7.6) was followed by staining with DAB: 1 mL DAB ([5 mg/10 mL), 9 mL Tris–HCl, pH 7.6, and 3.3 μL of 30% H_2O_2 for 15 min.

In order to determine the site of the BDA injections in relationship to either Acb shell or core, we performed a double staining for BDA and the calcium-binding protein calbindin-$D_{28\,kDA}$(CaB). The immunohistochemical staining followed exactly the procedures described by Wright et al. (1996).

Sections were mounted, dehydrated through an alcohol gradient, and coverslipped from the xylene using Entellan (Merck).

2.2 Electrophysiological Studies

Experiments were performed on 18 adult male Sprague–Dawley rats (275–300 g) anesthetized with ketamine (100 mg/kg, i.p., supplemented every hour by 50 mg/kg i.m. injections; Imalgène 500, Rhone-Mérieux, Courbevoie, France) and fixed in a stereotaxic frame (Horsley-Clark; Unimécanique, Epinay-sur-Seine, France). Body temperature was monitored rectally and maintained at 36–38°C with a homeothermic warming blanket (Harvard Apparatus, Kent, UK).

Electrical stimulation of the Acb shell (*A*: 10.4; *L*: 0.8; *H*: 7.4 mm from the cortical surface) and of the sensorimotor territory of the dorsal striatum (*A*: 9.2; *L*: 3.8; *H*: 7.4 mm from the cortical surface) were made with coaxial stainless steel electrodes (diameter 400 μm; tip–barrel distance 300 μm) positioned stereotaxically (Paxinos and Watson 1996). Electrical stimuli consisted of pulses of 0.6 ms width and 200–600 μA intensity delivered at a frequency of 1 Hz.

Single-unit activity of VTA/SNC cells ipsilateral to the stimulation sites was recorded using glass micropipettes (6–10 MΩ) filled with a 0.6 M NaCl solution containing 4% Pontamine Sky Blue. Action potentials were amplified and displayed on a memory oscilloscope. Spikes were separated from noise using a window discriminator and sampled online using a CED 1401 interface (Cambridge Electronics Design, Cambridge, UK) connected to a computer. Peristimulus time histograms were generated from 100 to 200 stimulation trials using 1-ms bins. The duration of the inhibitory response corresponds to the time during which no spike was observed. DA cells projecting to the dorsal striatum were identified using their classically defined electrophysiological characteristics: large duration spikes >2 ms, low discharge frequency <8 Hz, and latency of the antidromic spike evoked from stimulation of the dorsal striatum (Deniau et al. 1978; Guyenet and Aghajanian 1978). The antidromic spikes were characterized by their fixed latency and their collision with spontaneous discharges within an appropriate time interval. In four rats, the conduction time of the shell-VTA/SNC pathway was determined from the latency of the antidromic spikes evoked in shell cells following stimulation of the VTA/SNC complex (*A*: 3.8; *L*: 1.3; *H*: 7.9 from the cortical surface).

At the end of each recording session, the tip of the stimulating electrode was marked by a deposit of iron (15 μA anodal, 20 s) and localized in histological sections after a ferri-ferrocyanide reaction. The tip of the recording electrode was marked with Pontamine Sky Blue (8 μA cathodal, 20 min) allowing the determination of the recorded cells. Brains were removed and fixed in a 12% formalin solution, and the positions of electrodes were identified on serial frozen sections (100 μm) stained with safranin.

3 Results

3.1 Anatomical Tracing Studies

Following FG injections in the sensorimotor part of the dorsal striatum, verified by the presence of retrograde labeling in the sensorimotor cortex (Fig. 1a–d), retrogradely labeled neurons in the ventral mesencephalon were observed virtually over the entire mediolateral extent of the SNC as well as more medially in the dorsolateral part of the VTA (Fig. 1g). However, there was a clear difference in the strength of the labeling of neurons in different parts of the VTA/SNC complex. In the dorsal tier of the SNC, densely labeled neurons were intermingled with more weakly labeled cells, while in the most medial aspects of the SNC and in the VTA retrogradely labeled neurons were in general only lightly labeled. The relationships between the distribution of anterogradely labeled fibers and terminals originating in the medial Acb shell and retrogradely labeled neurons following a FG injection in the dorsal striatum are shown in Fig. 1g for a representative case. As illustrated, an extensive overlap between anterograde and retrograde labeling was observed in the dorsal parts of the medial SNC and the lateral VTA. In this region, frequent close appositions between retrogradely labeled, FG-containing neuronal cell bodies or dendrites and varicosities or boutons of anterogradely filled, BDA-containing fibers could be observed (Fig. 1e, f). By contrast in more lateral aspects of SNC, the anterogradely labeled fibers were mostly segregated from the retrogradely labeled nigrostriatal neurons, these cells occupying a ventral position with respect to the labeled fibers.

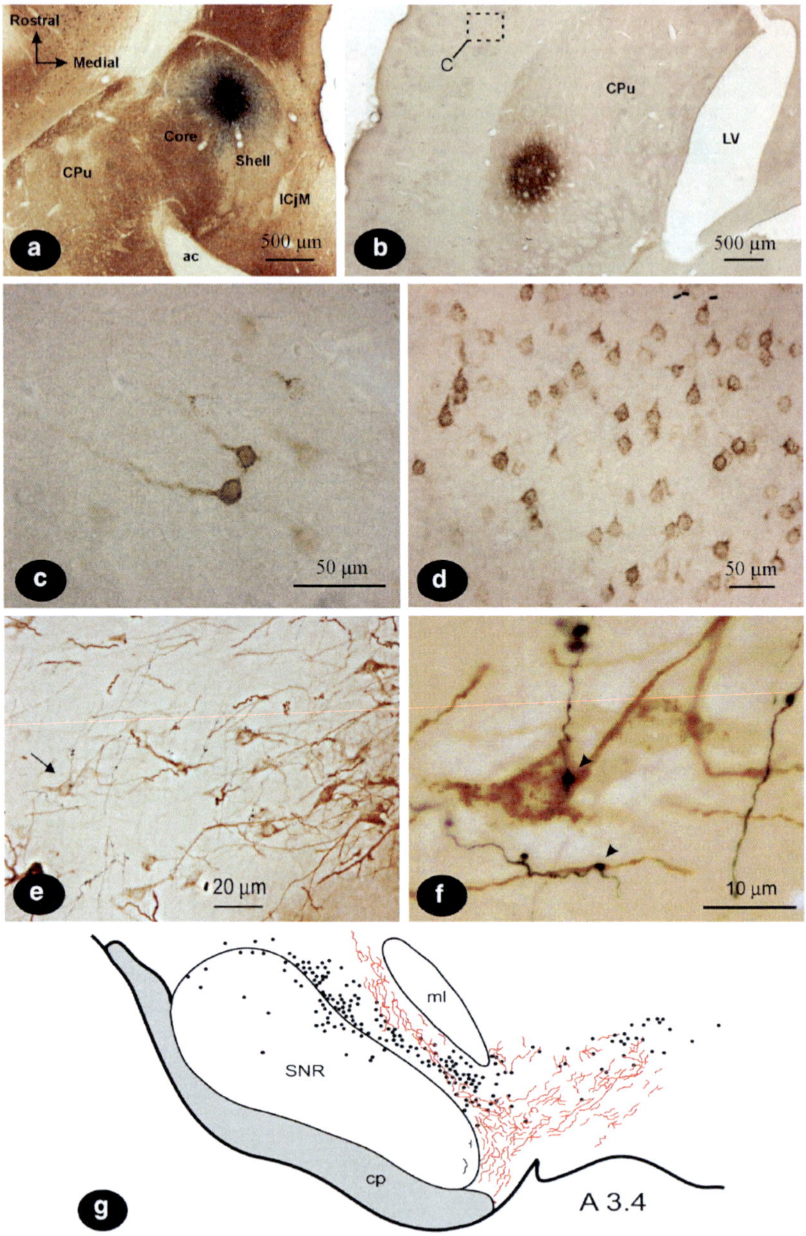

Fig. 1 Relationships in the SNC/VTA of anterogradely labeled fibers and terminals following an injection of BDA in the shell of the nucleus accumbens (**a**), and retrogradely labeled neurons after an injection of FG in the dorsolateral part of the dorsal striatum at the same side of the brain (**b**). To confirm the "sensori-motor characteristics" of the injected part of the dorsal striatum, retrogradely labeled neurons were identified in the somatosensory (**c**) and motor cortices (**d**). In (**e**) and (**f**) the close appositions of anterogradely labeled fibers and retrogradely labeled neurons in the SNC and VTA are shown. (**e**) Low-power photomicrograph of a horizontal section through the VTA and SNC showing brown-stained, retrogradely labeled FG-containing neurons and blue-black

3.2 Effect of Shell Stimulation on the Activity of VTA/SNC Dopaminergic Neurons Projecting to the Sensorimotor Territory of the Dorsal Striatum

In 14 rats, the responses evoked by electrical stimulation of the shell of the Acb were investigated in 152 cells antidromically driven from the orofacial sensorimotor territory of the dorsal striatum (Deniau et al. 1996; Deniau and Thierry 1997). These cells were characterized as DA on the basis of their long duration action potential (>2 ms), a relatively low spontaneous activity (<8 Hz), and the mean latency of their antidromic spike was 11.7 ± 0.2 ms. The cells antidromically driven from the dorsal striatum were located throughout most of the mediolateral extent of the SNC and the lateral part of the VTA. In addition, 15 other cells were antidromically driven from the shell; these cells were located more medially in the VTA.

Electrical stimulation of the shell induced in 69 of the 152 cells antidromically driven from the dorsal striatum an inhibitory response with a mean latency of 17.8 ± 0.9 ms and mean duration of 38.8 ± 3.0 ms (Fig. 2a, b). As shown in Fig. 2c, these cells were located in the medial SNC and the lateral part of the VTA mainly dorsally.

In order to determine the conduction time of the shell-VTA/SNC pathway, the latency of the antidromic responses evoked in the shell (41 cells) by stimulation of the VTA/SNC complex was determined. The mean latency of the antidromic responses was 16.7 ± 0.2 ms (range: 13– 20 ms).

4 Discussion

The present anatomical and electrophysiological data show that (1) projections of the shell of the Acb, a major component of the limbic striatum, exert an inhibitory influence on DA nigrostriatal neurons that project to the sensorimotor territory of the dorsal striatum and (2) this inhibitory influence is addressed to

Fig. 1 (continued) colored fibers BDA-containing fibers and terminals. The neuronal cell bodies and the main dendrites have a granular appearance. (**f**) High-power photomicrograph, extended focus picture of the area in (**e**) indicated with an *arrow*. Note the close appositions of two varicosities with a FG-positive cell body and dendrite (*arrowheads*). (**g**) Chartings of BDA-labeled fibers and terminals (represented in *red*) in the ventral tegmental area (VTA) and substantia nigra pars compacta (SNC) in combination with the distribution of retrogradely labeled FG-containing neurons (*black dots*). The anterograde and retrograde labeling in three adjacent sections was compiled. Note the retrograde labeling throughout the SNC, as well as in the dorsal part of VTA (rostral level) and in the substantia nigra pars reticulata (SNR). Intermingling of anterogradely labeled fibers and retrogradely labeled neurons occurs in the lateral VTA and the dorsal tier of the medial half of SNC. Abbreviations: *A* anterior, *ac* anterior commissure, *cp* cerebral peduncle, *CPu* caudate-putamen, *ICjM* major island of Calleja, *LV* lateral ventricle, *ml*, medial lemniscus

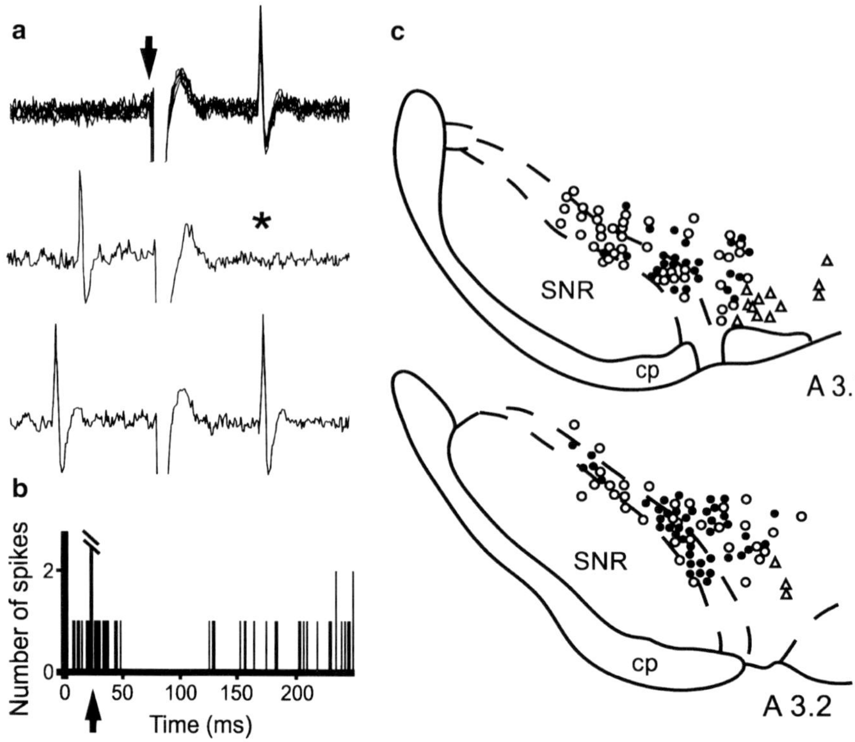

Fig. 2 Inhibitory effect evoked by stimulation of the shell of the Acb in electrophysiologically identified DA nigrostriatal neurons projecting to the sensorimotor striatal region. (**a**) Electro physiological identification of a nigrostriatal neuron by antidromic activation following an electrical stimulation applied in the orofacial sensorimotor territory of the dorsolateral striatum. *Upper trace*: fixed latency (17 ms) of the antidromic response. *Arrow* indicates the time of application of the stimulation. *Middle and lower trace*: collision test. Note the lack of the antidromic response (*star*) when the striatal stimulation is preceded by a spontaneous discharge of the neuron upon an appropriate time interval. (**b**) Inhibitory response of this nigrostriatal neuron to stimulation of the shell of Acb. *Arrow* indicates the stimulation artifact. (**c**) Distribution within the VTA/SNC complex of the neurons antidromically activated from the dorsolateral striatum and either inhibited by stimulation of the Acb shell (*filled circles*) or uninfluenced by the shell stimulation (*open circles*). *Open triangles* indicate the location of neurons antidromically activated by stimulation of the Acb shell. Abbreviations: *A* anterior, *cp* cerebral peduncle, *SNR* substantia nigra pars reticulata

the subpopulation of nigrostriatal neurons located in the medial SNC and lateral VTA and not to the nigrostriatal neurons located more laterally. These data provide the first evidence for a direct functional link between the limbic ventral striatum and the sensorimotor dorsolateral striatum and stress the functional heterogeneity of the population of DA nigrostriatal neurons innervating the sensorimotor domain of the striatum.

4.1 *Mesencephalic Dopamine Neurons as a Link Between the Shell and the Dorsolateral Striatum*

Various earlier studies indicate that mesencephalic projections from the Acb shell innervate the VTA/SNC complex (Nauta et al. 1978; Heimer et al. 1991; Ikemoto 2007). Accordingly, following injections of the anterograde tracer BDA within discrete regions of the Acb shell, labeled fibers, and terminals distributed primarily among DA neurons of both the VTA and dorsomedial SNC (Groenewegen et al. 1994). The organization of the VTA/SNC neurons that innervate distinct functional territories of the rat dorsal striatum has now been clarified (Maurin et al. 1999). This organization obeys to complex three-dimensional rules that cannot be systematized on a simple point-to-point topographical basis. As revealed by combined anterograde and retrograde tracing, each functional territory of the dorsal striatum receives an innervation from a neuronal population largely distributed along the mediolateral axis of the VTA/SNC complex. As a rule, the most dense and largest neuronal contingent lies in proximal position with respect to the corresponding striatonigral projection field, in register with the labeled striatonigral fibers. The lateral extension of this contingent of "proximal" neurons never transcends the lateral edge of the corresponding striatonigral projections. A less numerous contingent of neurons extends medially in the SNC and the adjoining VTA. This "distal" subpopulation of neurons does not follow topographical rules. Consequently, the "distal" contingent of nigrostriatal cells innervating different functional territories of the dorsal striatum intermingle within the medial SNC and lateral VTA. Interestingly, following combined injections of a retrograde tracer centered in the orofacial territory of the sensorimotor striatum and of an anterograde tracer in the Acb shell, labeled fibers were observed among retrogradely labeled neurons located in the medial SNC and the adjoining VTA corresponding primarily to the "distal" population of nigrostriatal neurons. By contrast, the retrogradely labeled neurons localized in the lateral SNC corresponding to the "proximal" subpopulation of nigrostriatal neurons was devoid of anterogradely labeled Acb shell fibers.

A further confirmation of a functional link between the Acb shell and the VTA/SNC neurons that project to the dorsal striatal sensorimotor territory is derived from the findings that an important proportion of electrophysiologically identified DA neurons antidromically activated from the orofacial striatal territory presented an inhibitory response to shell stimulation. In agreement with the anatomical data, these cells were mainly located in the medial SNC and lateral VTA suggesting that the Acb shell exerts an inhibitory influence preferentially on the "distal" subpopulation of nigrostriatal neurons. Whether indeed the shell targets this subgroup as identified by Maurin et al. (1999) awaits further investigation. The inhibitory responses evoked by Acb shell stimulation are likely monosynaptic since their latency is consistent with the conduction time of the Acb shell-VTA/SNC pathway (present study). In addition, it is known that efferent neurons from the Acb shell are GABAergic (Ferraguti et al. 1990). Finally, VTA/SNC cells antidromically driven from the dorsal striatum and inhibited by Acb shell stimulation presented the electrophysiological characteristics of dopaminergic neurons (Deniau et al. 1978; Guyenet and Aghajanian 1978). A further, rigorous test of their dopaminergic nature will be to apply apomorphine or D2 receptor agonists, known to inhibit these neurons. Taken together, these data suggest that the "distal" subpopulation of dopaminergic neurons and not the "proximal" one provides a functional link between the Acb shell and the dorsolateral striatum. These observations also support the notion that the "proximal" and "distal" subpopulations of nigrostriatal neurons constitute functionally distinct sets of neurons involved in different regulatory processes. Although the present anatomical and electrophysiological study was focused on the neurons of the VTA/SNC innervating the orofacial sensorimotor territory of the dorsal striatum, it is likely that a similar link exists between the Acb shell and the distal subpopulations of neurons innervating other sensorimotor districts of the dorsal striatum since these subpopulations share the same spatial distribution in the medial part of the VTA/SNC complex (Maurin et al. 1999).

4.2 Functional Considerations

The parallel organization of the corticostriatal circuits allows for the simultaneous processing of various kinds of sensorimotor, cognitive, and motivational information in the basal ganglia. However, for producing a coherent behavior, a communication between these specific circuits is required. In monkeys, anatomical studies have revealed that striato-nigro-striatal relationships comprise both reciprocal and nonreciprocal components suggesting that the DA system is involved in both feedback and feed-forward loops with the striatum. On the basis of the arrangement of feed-forward connections, Haber et al. (2000) have proposed that "information from the limbic system reaches the motor system through a series of connections" that form an ascending spiral from the Acb shell to the dorsolateral striatum. In this functional scheme, the limbic and motor circuits of the basal ganglia would be linked through an ordered sequential chain of events proceeding successively from the Acb shell through the cognitive processing circuits involving the Acb core and the central striatum to finally reach the motor circuits of the dorsolateral striatum. In rats, the striato-nigro-striatal relationships also comprise reciprocal and nonreciprocal components. As in the monkey brain, the organization of the nonreciprocal components of the striato-nigro-striatal circuits suggests that the different functional regions of the rodent striatum might also be interconnected by DA neurons through an ascending spiral. Indeed, as shown from the present anterograde labeling experiments and confirming earlier reports, the Acb shell innervates massively the region of VTA/SNC containing DA neurons projecting to the Acb core (Heimer et al. 1991; Maurin et al. 1999; Otake and Nakamura 2000; Ikemoto 2007). The Acb core innervates a dorsomedial region of the SNR that is overlaid by DA neurons innervating medial and central portions of the rostral striatum receiving inputs from anterior cingulate and prefrontal cortical areas (Deniau et al. 1996; Maurin et al. 1999; Groenewegen et al. 1999).

However, in contrast to monkeys in which direct limbic-motor connections seem rather limited, the present study indicates that in rats, a relatively large subpopulation of DA cells projecting to the sensorimotor territory of the dorsal striatum lies in the medial SNC and lateral VTA, a region innervated massively by the Acb shell, and receives a direct synaptic input from the Acb shell. In that respect, the Acb shell in rats appears to be in a unique position.

Yet, an important question remains: what is the functional role of this disynaptic, DA link from the Acb shell to more dorsal striatal regions? DA afferents constitute slow, modulatory inputs to a striatal network that processes information propagated by fast glutamatergic inputs from the cerebral cortex, thalamus, and amygdala. In line with this, electrophysiological data support a DA modulation of glutamatergic transmission in striatum (West et al. 2003). Furthermore, recent behavioral data strongly support a role for DA in a "spiraling" link between the ventral and dorsal striatum in the course of the development of cocaine seeking habits in rats (Belin and Everitt 2008).

4.3 Conclusion

The Acb shell receives predominant afferents from limbic structures such as the hippocampus and amygdala, and is implicated in motivation and emotional behavior. By contrast, the dorsolateral striatum, via its afferents from the sensorimotor cortical areas and its connections with the motor output pathways of the basal ganglia, is involved in executive behavior and procedural learning. Therefore, it can be proposed that the subpopulation of DA neurons that innervates the sensorimotor dorsal striatum and receives inputs from the Acb shell, by funneling information from the limbic system to the extrapyramidal motor system, provides a link through which motivational and emotional states can influence motor outcomes.

Acknowledgments Supported by a Dutch Medical Research Council program grant (NWO-ZonMW 903-42-092) and a NWO-ZonMW/INSERM travel grant (910-48-029).

References

Alexander GE and Crutcher MD (1990) Functional architecture of basal ganglia circuits: neural substrates of parallel processing. Trends Neurosci 13: 266–271.

Belin D and Everitt BJ (2008) Cocaine seeking habits depend upon dopamine-dependent connectivity linking the ventral with the dorsal striatum. Neuron 57: 432–441.

Deniau JM and Thierry AM (1997) Anatomical segregation of information processing in the rat substantia nigra pars reticulata. Adv Neurol 74: 83–96.

Deniau JM, Hammond C, Riszk A and Feger J (1978) Electrophysiological properties of identified output neurons of the rat substantia nigra (pars compacta and pars reticulata): evidences for the existence of branched neurons. Exp Brain Res 32: 409–422.

Deniau JM, Menetrey A and Charpier S (1996) The lamellar organization of the rat substantia nigra pars reticulata: Segregated patterns of striatal afferents and relationship to the topography of corticostriatal projections. Neuroscience 73: 761–781.

Ferraguti F, Zoli M, Aronsson M, Agnati LF, Goldstein M, Filer D and Fuxe K (1990) Distribution of glutamic acid decarboxylase messenger RNA-containing nerve cell populations in the male rat brain. J Chem Neuroanat 3: 377–396.

Groenewegen HJ, Berendse HW and Wouterlood FG (1994) Organization of the projections from the ventral striatopallidal system to ventral mesencephalic dopaminergic neurons. In: Percheron G and McKenzie JS (eds) The Basal Ganglia IV. Plenum, New York, NY, pp. 81–93.

Groenewegen HJ, Wright CI, Beijer AV and Voorn P (1999) Convergence and segregation of ventral striatal inputs and outputs. Ann NY Acad Sci 877: 49–63.

Guyenet PG and Aghajanian GK (1978) Antidromic identification of dopaminergic and other output neurons of the rat substantia nigra. Brain Res 150: 69–84.

Haber SN, Fudge JL and McFarland NR (2000) Striatonigrostriatal pathways in primates form an ascending spiral from the shell to the dorsolateral striatum. J Neurosci 20: 2369–2382.

Heimer L, Zahm DS, Churchill L, Kalivas PW and Wohltmann C (1991) Specificity in the projection patterns of the accumbal core and shell in the rat. Neuroscience 41: 89–125.

Heimer L, Alheid GF, de Olmos JS, Groenewegen HJ, Haber SN, Harlan RE and Zahm DS (1997) The accumbens: Beyond the Core-Shell dichotomy. J Neuropsychiatry Clin Neurosci 9: 354–381.

Ikemoto S (2007) Dopamine reward circuitry: Two projection systems from the ventral midbrain to the nucleus accumbens-olfactory tubercle complex. Brain Res Rev 56: 27–78.

Mailly P, Charpier S, Mahon S, Menetrey A, Thierry AM, Glowinski J and Deniau JM (2001) Dendritic arborizations of the rat substantia nigra pars reticulata neurons: Spatial organization and relation to the lamellar compartmentation of striato-nigral projections. J Neurosci 21: 6874–6888.

Maurice N, Deniau JM, Menetrey A, Glowinski J and Thierry AM (1997) Position of the ventral pallidum in the rat prefrontal cortex-basal ganglia circuit. Neuroscience 80: 523–534.

Maurice N, Deniau JM, Menetrey A, Glowinski J and Thierry AM (1998) Prefrontal cortex-basal ganglia circuits in the rat: Involvement of ventral pallidum and subthalamic nucleus. Synapse 29: 363–370.

Maurin Y, Banrezes B, Menetrey A, Mailly P and Deniau JM (1999) Three-dimensional distribution of nigrostriatal neurons in the rat: Relation to the topography of striatonigral projections. Neuroscience 91: 891–909.

Montaron MF, Deniau JM, Menetrey A, Glowinski J and Thierry AM (1996) Prefrontal cortex inputs of the nucleus accumbens-nigro-thalamic circuit. Neuroscience 71: 371–382.

Nauta WJ, Smith GP, Faull RL and Domesick VB (1978) Efferent connections and nigral afferents of the nucleus accumbens septi in the rat. Neuroscience 3: 385–401.

Otake K and Nakamura Y (2000) Possible pathways through which neurons of the shell of the nucleus accumbens influence the outflow of the core of the nucleus accumbens. Brain Dev 22: S17–26.

Paxinos G and Watson C (1996) The Rat Brain in Stereotaxic Coordinates. Academic, New York, NY.

Somogyi P, Bolam JP, Totterdell S and Smith AD (1981) Monosynaptic input from the nucleus accumbens-ventral striatum region to retrogradely labelled nigrostriatal neurons. Brain Res 217: 245–263.

West AR, Floresco SB, Charara A, Rosenkranz JA and Grace AA (2003) Electrophysiological interactions between striatal glutamatergic and dopaminergic systems. Ann NY Acad Sci 1003: 53–57.

Wright CI, Beijer AV and Groenewegen HJ (1996) Basal amygdaloid complex afferents to the rat nucleus accumbens are compartmentally organized. J Neurosci 16: 1877–1893.

Synchrony of the Rat Medial Prefrontal Cortex Network During Isoflurane Anaesthesia

Mathijs Stegeman, Marieke de Boer, Marcel van der Roest, and Antonius B. Mulder

Abstract In the anaesthetized rat, the membrane potential of the prefrontal cortex (PFC) fluctuates between two states of polarization: the up- and down-states. The PFC network strongly governs activity in its target areas such as various basal ganglia structures. The influence of anaesthesia on network dynamics in the prelimbic cortex will therefore have direct consequences for the activity profile of the output structures. In order to investigate the effects of isoflurane anaesthesia on network activity of the PFC, rats were anaesthetized with isoflurane of which the concentration was varied between 1.25 and 2.25%. Local EEG and single-unit activity were recorded simultaneously.

At baseline levels (1.75%), isoflurane anaesthesia induced up-state transitions in the prelimbic cortex visible in the local EEG signal. The up-state deflections in the EEG were determined as reflections of clustered firing of individual prelimbic neurons. 2.25% isoflurane strongly reduced the up-state frequency whereas 1.25% isoflurane allowed the network to shift to a continuous activity mode. These results show that neuronal activity in the medal PFC and therefore its target structures is highly dependent on the level of isoflurane anaesthesia. Considering the similarity between the synaptic targets of isoflurane by which surgical anaesthesia is maintained and the mechanisms involved in initiation, maintenance and cessation of up-states, the up- and down-state fluctuations are a direct result of anaesthesia.

1 Introduction

The prefrontal cortex (PFC) processes and integrates information from many parts of the brain. Hierarchically, it serves as an important relay station between sensory inputs and basal ganglia motor outputs, including dorsal and ventral striatal areas (Groenewegen and Uylings 2000). The PFC has a main role in the mediation and

M. Stegeman, M. de Boer, M. van der Roest, and A.B. Mulder (✉)
Department of Anatomy and Neurosciences, Cognitive Neurophysiology-CNCR,
VU University Medical Center, P.O. Box 7057, 1007 MB Amsterdam, The Netherlands
e-mail: ab.mulder@vumc.nl

H.J. Groenewegen et al. (eds.), *The Basal Ganglia IX*, Advances in Behavioral Biology 58,
DOI 10.1007/978-1-4419-0340-2_11,

guidance of complex behaviours like attention and behavioural flexibility. At a network level, the PFC like all cortical areas consists of an intricate network of pyramidal neurons and interneurons. In anaesthetized rats and cats, the membrane potential of these neurons has been shown to fluctuate between two states of polarization (Steriade et al. 1993a; Branchereau et al. 1996; Lewis and O'Donnell 2000). Transitions occur between a very negative, hyperpolarized state of the membrane, which is termed "down-state" and a more depolarized membrane potential which is near spiking threshold and is subsequently termed "up-state" (Wilson and Kawaguchi 1996). Steriade and co-workers have shown the presence of these rhythmic up- and down-state transitions in in vivo recordings of pyramidal neurons in layers III–VI of the cortical association areas 5 and 7 in anaesthetized cats (Steriade et al. 1993a–c). This rhythm represents network-generated oscillations, in which intrinsic cortico-cortical inputs are the driving force, has a frequency of 0.2–1 Hz and is distinct from the delta-wave rhythm (1–4 Hz) (Steriade et al. 1993a, b). Fast-spiking GABAergic interneurons are thought to be involved in the mediation of this cortical rhythm (Steriade et al. 2001). In this respect, thalamocortical inputs play a modulatory role; they are, however, not a prerequisite in the generation of delta waves and sleep spindles (Steriade et al. 1990, 1991, 1993b).

Neurons displaying this bimodal phenomenon have been found in many brain areas of anaesthetized and sleeping animals, including the dorsal striatum (Wilson and Groves 1981; Wilson and Kawaguchi 1996; Stern et al. 1998; Mahon et al. 2006), the nucleus accumbens (O'Donnell and Grace 1995, 1998; Goto and O'Donnell 2001a, b), the cerebellum (Loewenstein et al. 2005), (para)hippocampal regions (Isomura et al. 2006), somatosensory cortex (Luczak et al. 2007), visual cortex (Imas et al. 2005) and the PFC (Steriade et al. 1993a; Branchereau et al. 1996; Lewis and O'Donnell 2000). Furthermore, Hudetz and Imas (2007) have shown that these up-states travel across cortical areas from primary visual cortex to anterior cortices in a spatio-temporal order after flash stimuli. Excitatory recurrent feedback is typical within these neuronal populations and can act to amplify cortical feed-forward signalling (Douglas et al. 1995). Depolarization is thought to be initiated in layer V pyramidal neurons and can be self-sustaining through excitatory recurrent connections within the network propagating to other cortical layers (Sanchez-Vives and McCormick 2000) and cortical areas (Hudetz and Imas 2007).

The dynamics of the EEG signal are regulated by simultaneous activity of many neurons (Adrian and Matthews 1934). It has been shown that membrane potential transitions to the up-state on a cellular level coincide with bursts of activity seen in the local EEG signal (Amzica and Steriade 1995; Contreras and Steriade 1995). This suggests that the activity in the EEG signal is caused by synchronized transitions to the up-state of a considerable number of neurons. Intracellular recordings from pairs of spiny projection neurons in the striatum elucidated this synchrony and showed a high correlation of up- and down-state transitions between neurons, without the necessity of being connected to each other (Stern et al. 1998). Up- and down-state transitions in these striatal populations and in visual cortex (V1) neurons (Lampl et al. 1999) seem to be synchronized on a network level. Importantly, the network synchrony can only be observed in local EEG recordings, but not in global, transcranial EEG measurements (Lampl et al. 1999). Despite the synchrony of up-state transitions in the network, spike firing is not synchronized between neurons during up-states (Stern et al. 1998). This can explain that up- and down-states are observed in thc EEG signal, but the amplitude of individual up-states may differ.

O'Donnell and Grace (1995) already reported that bolus injection of chloral hydrate anaesthesia reduced the number of up-state transitions over time. However, this problem was resolved by using continuous administration of anaesthesia, and was not investigated further. Imas et al. (2005) have shown that various volatile anaesthetics modulate the spontaneous and flash-induced up-state frequency in visual cortex in a concentration-dependent manner. Since these up-state transitional waves travel over the cortex (Hudetz and Imas 2007), they might eventually reach prefrontal areas, thereby governing the activity profile of their striatal target areas during anaesthesia. In order to investigate the influence of isoflurane on network dynamics in the prelimbic cortex, the local EEG and single-unit activity were recorded in the isoflurane anaesthetized rat.

2 Materials and Methods

2.1 Surgery and Placement

Experiments were performed using five male Wistar rats (250–550 g). For surgery, the animals were initially placed in a chamber and anaesthetized with a mixture of 2.5–3.0% isoflurane (1-chloro-2,2,2-trifluoroethyl difluoromethyl ether) in O_2/N_2O. Subsequently, the anaesthetized animals were mounted in a stereotaxic frame (Kopf instruments) and placed on a heating pad. The body temperature was monitored through a rectal probe and was kept constant at 37°C. Anaesthesia was maintained continuously throughout the surgery at 1.75% isoflurane in O_2/N_2O. The skull was exposed by removal of the scalp and cleaned. Burr holes were drilled and stainless steel screws were inserted to provide adhesive stability. A four-tetrode microdrive (Neuralynx, Bozeman, USA) was placed above the medial PFC (from Bregma: AP: −3.2 mm; ML: 0.7 mm) and the tetrodes were lowered to 2.0 mm below cortical surface according to the atlas of Paxinos and Watson (2007). The microdrive was attached to the skull with stainless steel screws and dental cement. One stainless steel screw was implanted in the interparietal bone and served as ground. During the time course of the experiment, tetrodes were individually lowered to their final prelimbic targets.

2.2 Isoflurane Manipulations

First, 30 min of baseline recordings under 1.75% isoflurane were made after which the isoflurane concentration was increased to 2.25% for 15 min followed by another period of 30 min at 1.75% isoflurane. Then, isoflurane concentration was decreased to 1.25%. The minimal alveolar concentration (MAC) value for isoflurane in Wistar rats is 1.35% (Gómez de Segura et al. 1998; De Wolff et al. 1999; Criado and Gómez de Segura 2003). This MAC value produces surgical anaesthesia in 50% of subjects and is a measure of the potency of volatile anaesthetics. Since 1.25% is slightly below the isoflurane MAC value, it was maintained for short periods of time (between 3 and 10 min) not compromising general anaesthesia. Isoflurane concentration was immediately set to baseline values once the network had shifted to a continuous activity mode (see Fig. 2a). Another 30 min of 1.75% concluded the measurements.

2.3 Data Acquisition and Analysis

All recordings were amplified by a headstage amplifier (gain: 20 times) and by a subsequent preamplifier (gain: 50 times) (Plexon Inc., Dallas, USA). Final amplification was done with a variable end-stage amplifier (gain: 1–25 times). The single-unit data was band-pass filtered between 150 Hz and 9 kHz, and analog-to-digital (AD) converted at a sampling frequency of 40,000 Hz. The local EEG-signal was band-pass filtered between 0.7 Hz and 170 Hz, and AD converted at a sampling frequency of 1,000 Hz.

All data analysis was performed offline. Single units were determined and isolated by cluster-cutting analysis using Offline Sorter (Plexon Inc., Dallas, USA). Further spike firing and up-down state analysis was done using NeuroExplorer (Littleton, USA).

2.4 Histological Evaluation

At the end of the experiments rats were deeply anaesthetized (3.0% isoflurane) and small electrolytic lesions were made to mark the final placements of the tetrodes. An additional dose of 0.4 mL Nembutal (i.p.) was given after which the rats were transcardially perfused with 4% paraformaldehyde and 0.05% glutaraldehyde in 0.1 M phosphate buffer (pH 7.6). Following fixation, the brains were removed from the skull and after post-fixation for 1 day they were transferred to a 30% sucrose solution. Coronal sections (50 μm) of the PFC were cut on a freezing microtome. Subsequent Nissl staining was performed to determine the tetrode placements in the medial PFC.

3 Results

3.1 Definition of Up- and Down-States

The activity in the local medial PFC EEG signal as up- and down-states has been defined according to the parameters depicted in Fig. 1. The start of an up-state was characterized by the strong downward (negative) deflection in the EEG signal.

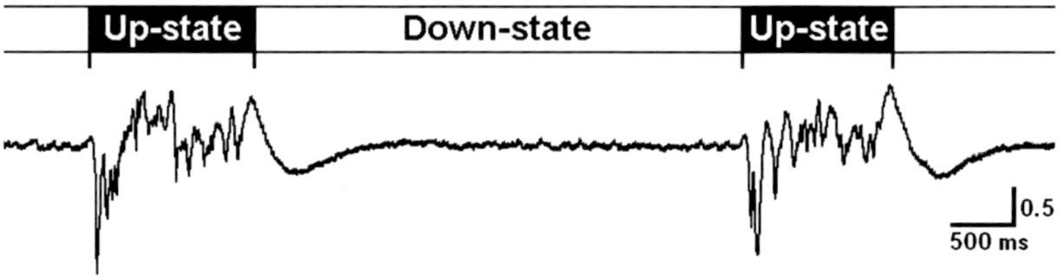

Fig. 1 Up- and down-state definition. In the local EEG, the up-state starts when there is a strong and sharp downward deflection from baseline signal. It ends at the top of the last upward deflection, which is followed by an undershoot before returning to baseline. From this top onward, the down-state starts and ends when there is a downward peak signalling the start of a new up-state

The end of an up-state was defined as the maximum of the last upward peak, which is followed by an undershoot before returning to baseline. A subsequent up-state epoch is taken into account only if its initial strong downward reflection occurs when the previous up-state signal has returned to baseline. The down-state reflects the time without clustered cortical activity, thus defining the period between up-states.

3.2 Up- and Down-State Transitions are Reversibly Influenced by Isoflurane Anaesthesia

The effect of isoflurane at different concentrations on up-state transitions was evaluated in the local EEG within the medial PFC. During the measurements, the isoflurane concentration was increased to 2.25% or decreased to 1.25% from baseline values (1.75%). Reducing the isoflurane concentration to 1.25% resulted in the disappearance of up-state transitions and caused a shift to continuous activity of the network (Fig. 2a). No individual up-states could be determined at this stage.

Increasing the amount of isoflurane from 1.75 to 2.25% elicited a strong and significant decrease in the frequency of up-state transitions (Fig. 2b; 1.75%: 0.13 ± 0.04 Hz; 2.25%: 0.03 ± 0.01 Hz; paired *t* test: $p = 0.0036$; $n = 5$). In addition, the duration of the up-states also decreased (Fig. 2c; 1.75%: 1.58 ± 0.18 s; 2.25%: 0.95 ± 0.46 s; paired *t* test: $p = 0.019$, $n = 5$). In close correlation with the up-state duration, the duration of the down-state increased drastically with increasing amounts of isoflurane (Fig. 2d; 1.75%: 6.61 ± 2.39 s; 2.25%: 46.52 ± 16.86 s; paired *t* test: $p = 0.0060$, $n = 5$). As can be seen in Fig. 2a, a considerable between-subject variation in sensitivity to the effects of isoflurane was encountered.

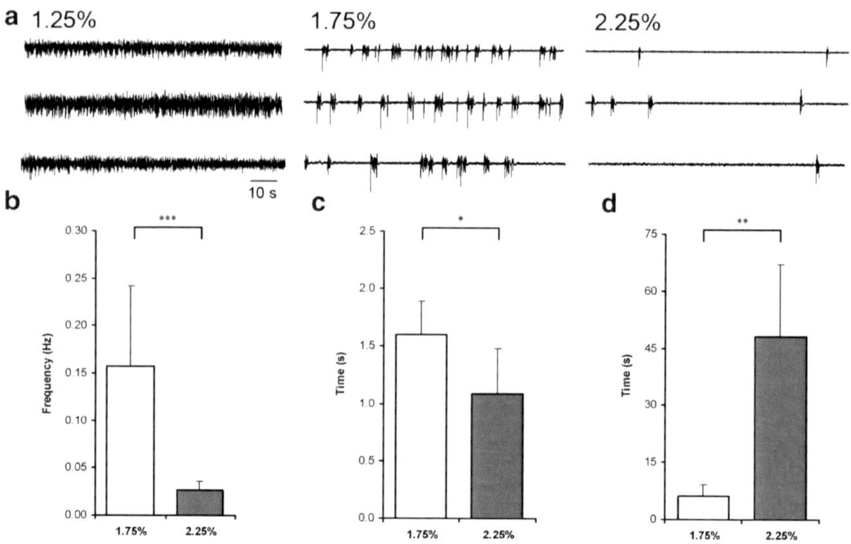

Fig. 2 Isoflurane has a concentration dependent effect on up- and down-state dynamics. (**a**) Electroencephalogram (EEG) of three rats showing the effect of different concentrations (1.25%, 1.75%, 2.25%) of isoflurane. Note that there is a general trend, although there are differences in sensitivity to the anaesthesia between animals. (**b**) The up-state frequency (mean ± SD) decreases with increasing amounts of isoflurane ($n = 5$). (**c**) The duration of the up-state (mean ± SD) also decreases with increasing amounts of isoflurane ($n = 5$). (**d**) The duration of the down-state (mean ± SD) increases strongly with increasing amounts of isoflurane ($n = 5$). *$p < 0.05$, **$p < 0.01$, ***$p < 0.005$

These results show that the frequency of up-state transitions in the prelimbic cortex is influenced by isoflurane anaesthesia in a concentration-dependent manner.

3.3 Prelimbic Neurons Fire Action Potentials Only During Up-States

Neurons in the PFC only fire action potentials when they are in the up-state mode, near firing threshold (Cowan and Wilson 1994; Branchereau et al. 1996; Lewis and O'Donnell 2000). In total 78 single units were recorded. By recording many single units simultaneously, we could show that neuronal activity was clustered during 1.75 and 2.25% of isoflurane anaesthesia (Fig. 3a). Moreover, this concerted activity was only observed during up-states whereas single units were silent during the down-state periods (Fig. 3b). At the lowest concentration of isoflurane (1.25%), no apparent up-states were detected in the local EEG and single units seemed to fire in a continuous manner.

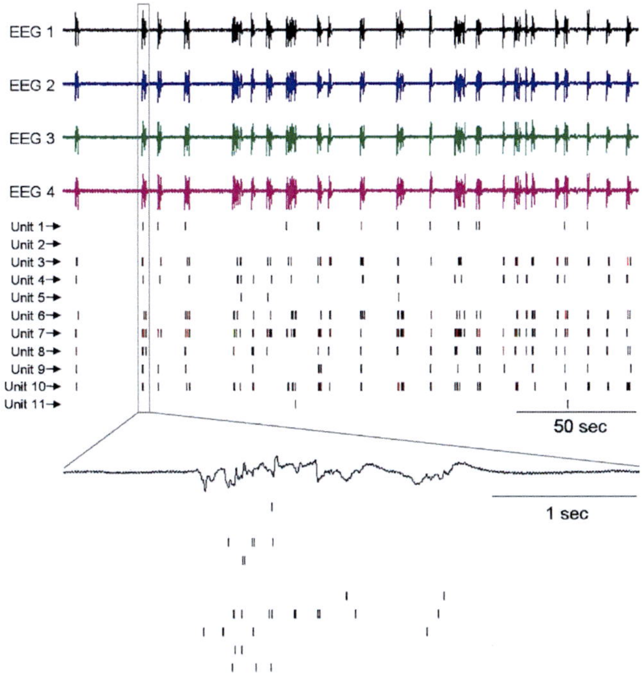

Fig. 3 Spike firing exclusively occurs during up-states. *Top panel*: The EEG signal of the four tetrodes showing up-down state transitions during 1.75% isoflurane. Note the strong similarity between recording locations. Spike-firing patterns of 11 simultaneously recorded units are depicted. Action potentials can be exclusively seen during up-states and units are completely silent during down-states. *Lower panel*: The firing of the units in relation to a single up-state. Clearly visible is the grouped firing of the units without apparent synchronicity

4 Discussion

These results show that isoflurane anaesthesia induces synchronized neuronal network activity in the prelimbic cortex resulting in up-state transitions in the local EEG signal. Conversely, the down-state represents network silence. Furthermore, it shows that the up-state deflections in the EEG are reflections of clustered firing of individual prelimbic neurons as has been shown for somatosensory cortex (Luczak et al. 2007). These results also show that neuronal activity in the medial PFC is highly dependent on the level of isoflurane anaesthesia. The window between network silence and continuous activity of the prefrontal network is small and falls between 1.25 and 2.25 volume percentage of isoflurane. In the following paragraphs, the similarity between the synaptic effects of isoflurane and the mechanisms underlying up-state fluctuations in the PFC are discussed.

4.1 Synaptic Effects of Isoflurane

The synaptic effects of isoflurane balance the dynamics of neurotransmitter release and re-uptake. The overall effect of isoflurane is the reduction of excitatory neuronal activity. This action is regulated in twofold: (1) by reducing excitatory synaptic transmission and (2) by increasing inhibitory synaptic transmission.

Presynaptically, isoflurane causes a concentration-dependent decrease in the Ca^{2+}-dependent and -independent release of glutamate from presynaptic terminals at clinically relevant concentrations (Larsen et al. 1994; MacIver et al. 1996; Larsen and Langmoen 1998; Westphalen and Hemmings 2006b). Furthermore, within the clinical concentration of isoflurane, synaptic glutamatergic transmission is further reduced by an increase in the re-uptake velocity of glutamate at the presynaptic terminal, without affecting the binding-affinity of the transporter (Larsen et al. 1997). In compliance with the boundaries set by the MAC-value, the effect on the re-uptake velocity is observed at isoflurane concentrations of 1.5 and 3.0% whereas 0.5% isoflurane was ineffective (Larsen et al. 1997). In contrast to glutamate, re-uptake kinetics of GABA is not affected (Larsen et al. 1998). An additional target of isoflurane, mediating the inhibitory effect, is the increased Ca^{2+}-dependent release of GABA from presynaptic terminals, thereby increasing the post-synaptic Cl^- conductance (Larsen et al. 1998; Nishikawa and MacIver 2001; Westphalen and Hemmings 2006a). In contrast, Westphalen and Hemmings (2006b) showed that isoflurane causes an inhibition of evoked release of GABA; an effect that is equal for glutamate release. However, the inhibition of the evoked release of glutamate is more vigorous compared to GABA release resulting in a decrease in glutamatergic signalling relative to GABAergic signalling (Westphalen and Hemmings 2006b).

At the post-synaptic site, application of isoflurane causes depression of neuronal excitability by hyperpolarization of the neurons (Berg-Johnsen and Langmoen 1990). Positive modulation of background currents through non-inactivating K^+ channels, voltage-dependent K^+ channels and Tandem Pore Domain K^+ channels also play substantial roles in decreasing neuronal excitability (Shin and Winegar 2003; Liu et al. 2004). The main post-synaptic receptor involved in the depressant actions of isoflurane on GABAergic signalling is the $GABA_A$ receptor. In vitro application of isoflurane to cultured hippocampal neurons showed a reversible increase in inward Cl^- current and causes potentiation of the GABA response duration by enhancing IPSC decay times. The strength of the $GABA_A$ receptor-mediated inhibition is concentration dependent, especially within the range of 1–1.5 times MAC (Jones et al. 1992). In addition, volatile anaesthetics potentiate kainate-type glutamate receptor-mediated currents by binding to a specific site on the GluR6-subunit (Minami et al. 1998), but do not seem to have any effect on NMDA receptor-mediated synaptic signalling in rats (Pearce et al. 1989; Verhaegen et al. 1992).

In summary, these effects on excitatory and inhibitory transmitter release and uptake, as well as the effects on receptor functioning, in combination with the actions on K^+ channels, are mainly responsible for the anaesthetic properties of isoflurane.

4.2 Isoflurane Modulates Mechanisms Underlying Up-Down-State Transitions

Our results show that isoflurane anaesthesia induces local synchronized neuronal network activity in the prelimbic cortex reflected by up-state transitions in the EEG signal. This oscillatory activity reflects the functional state of the local cortical network. The transition to and from the depolarized state is tightly regulated by GABAergic interneurons and intrinsic ionic conductances, primarily through K^+ channels (Wilson and Kawaguchi 1996; Sanchez-Vives and McCormick 2000). NMDA receptors play a role in the up-down transitions in cortical neurons (Steriade et al. 1993a; Sanchez-Vives and McCormick 2000; Milojkovic et al. 2005). Blocking of glutamate receptors, NMDA as well as AMPA, causes a severe but not complete reduction in the frequency of up-states (O'Donnell and Grace 1998; Cossart et al. 2003). Only after additional application of the $GABA_A$ antagonist picrotoxin a total cessation of up-states is apparent (Cossart et al. 2003). This indicates that the combination of active glutamate and GABA receptor activity is crucial for up-state initiation.

The membrane potential that is attained during the transition to the up-state is controlled by the voltage-dependent activation of K^+ channels. Therefore, spike firing during the up-state is determined not by the amount of excitatory input, but by influence of K^+ currents (Wilson and Kawaguchi 1996).

The transition from the up- to the down-state is proposed to be induced by increasing outward Ca^{2+}-currents and Na^+-dependent K^+ currents during sustained depolarization in the up-state, ultimately causing termination of the up-state by hyperpolarization of the network (Sanchez-Vives and McCormick 2000). In the absence of excitatory input, the down-state is maintained near the equilibrium potential for potassium by inwardly rectifying K^+ channels, but not by Cl^--currents (Wilson and Kawaguchi 1996). The prominent slow after-hyperpolarization transiently recovers. This inactivation period can be considerably long as has been shown by Kroeger and Amzica (2007). In rats, under high isoflurane (2–3%), they showed a 2 s refractory period after an up-state. Based on the present results and considering the dose-dependant action of isoflurane on K^+ currents, this refractory period fully depends on the isoflurane concentration. In fact, in the transition from 1.75 to 1.25% isoflurane, there is a gradual decrease of the up-state duration until the up-states seem to fuse. Under the lowest concentration of isoflurane (1.25%), there is no detectable refractory period in the local EEG in the prelimbic area. In vivo intracellular recordings in the prelimbic area are necessary to confirm this at the membrane potential level in single neurons.

In summary, the synaptic targets of isoflurane are similar to the signalling components underlying up-state initiation, maintenance and cessation. Therefore, the up-state–down-state phenomenon as seen in local EEG can be considered a property of anaesthesia. At the single neuron level, however, intracellular recordings in area 7 of non-anaesthetized cats and in the striatum of non-anaesthetized rats have shown up-down-state transitions in the sleeping animal during slow-wave sleep without the characteristic alternation of up- and down-states in the corresponding cortical EEG pattern (Steriade et al. 2001; Mahon et al. 2006). The local EEG effects in the anaesthetized state can be explained by the strong facilitation of up-state synchrony by anaesthesia, which is lacking in the normal sleeping animal. Whether the up-states during anaesthesia and sleep are similar or even related remains to be determined.

4.3 Final Considerations

Up- and down-state transitions remain an interesting model to study network dynamics in the anaesthetized animal. Interplay between pyramidal and interneurons, and their dependence on synaptic inputs and intracellular signalling can be studied in a controlled manner. However, since the major target areas of the PFC include several basal ganglia structures, the activity profiles of these target structures will be highly influenced by fluctuating anaesthesia concentrations. Therefore, the use of controllable volatile anaesthetics over injection anaesthesia is highly preferable for in vivo neurophysiological measurements in general and for basal ganglia research in particular.

References

Adrian ED and Matthews BHC (1934) The interpretation of potential waves in the cortex. J Physiol 81: 440–471.

Amzica F and Steriade M (1995) Short and long-range neuronal synchronization of the slow (<1 Hz) cortical oscillation. J Neurophysiol 73: 20–38.

Berg-Johnsen J and Langmoen IA (1990) Mechanisms concerned in the direct effect of isoflurane on rat hippocampal and human neocortical neurons. Brain Res 507: 28–34.

Branchereau P, van Bockstaele EJ, Chan J and Pickel VM (1996) Pyramidal neurons in rat prefrontal cortex show a complex synaptic response to single electrical stimulation of the locus coeruleus region: Evidence for antidromic activation and GABAergic inhibition using in vivo intracellular recording and electron microscopy. Synapse 22: 313–331.

Contreras D and Steriade M (1995) Cellular basis of EEG slow rhythms: A study of dynamic corticothalamic relationships. J Neurosci 15: 604–622.

Cossart R, Aronov D and Yuste R (2003) Attractor dynamics of network UP states in the neocortex. Nature 423: 283–288.

Cowan RL and Wilson CJ (1994) Spontaneous firing patterns and axonal projections of single corticostriatal neurons in the rat medial agranular cortex. J Neurophysiol 71: 17–32.

Criado AB and Gómez de Segura IA (2003) Reduction of isoflurane MAC by fentanyl or remifentanil in rats. Vet Anaesth Analg 30: 250–256.

De Wolff MH, Leather HA and Wouters PF (1999) Effects of tramadol on minimum alveolar concentration (MAC) of isoflurane in rats. Br J Anaesth 83: 780–783.

Douglas RJ, Koch C, Mahowald M, Martin KAC and Suarez HH (1995) Recurrent excitation in neocortical circuits. Science 269: 981–985.

Gómez de Segura IA, Criado AB, Santos M and Tendillo FJ (1998). Aspirin synergistically potentiates isoflurane minimum alveolar concentration reduction produced by morphine in the rat. Anesthesiology 89: 1489–1494.
Goto Y and O'Donnell P (2001a) Synchronous activity in the hippocampus and nucleus accumbens in vivo. J Neurosci 21: RC131: 1–5.
Goto Y and O'Donnell P (2001b) Network synchrony in the nucleus accumbens in vivo. J Neurosci 21: 4498–4504.
Groenewegen HJ and Uylings HB (2000) The prefrontal cortex and the integration of sensory, limbic and autonomic information. Prog Brain Res 126: 3–28.
Hudetz AG and Imas OA (2007) Burst activation of the cerebral cortex by flash stimuli during isoflurane anesthesia in rats. Anesthesiology 107: 983–991.
Imas OA, Ropella KM, Ward BD, Wood JD and Hudetz AG (2005) Volatile anesthetics enhance flash-induced gamma oscillations in rat visual cortex. Anesthesiology 102: 937–947.
Isomura Y, Sirota A, Özen S, Montgomery S, Mizuseki K, Henze DA and Buzsáki G (2006) Integration and segregation of activity in entorhinal-hippocampal subregions by neocortical slow oscillations. Neuron 52: 871–882.
Jones MV, Brooks PA and Harrison NL (1992) Enhancement of γ-aminobutyric acid-activated Cl^- currents in cultured rat hippocampal neurons by three volatile anaesthetics. J Physiol 449: 279–293.
Kroeger D and Amzica F (2007) Hypersensitivity of the anesthesia-induced comatose brain. J Neurosci 27: 10597–10607.
Lampl I, Reichova I and Ferster D (1999) Synchronous membrane potential fluctuations in neurons of the cat visual cortex. Neuron 22: 361–374.
Larsen M and Langmoen IA (1998) The effect if volatile anaesthetics on synaptic release and uptake of glutamate. Toxicol Lett 100: 59–64.
Larsen M, Grøndahl T, Haugstad T and Langmoen IA (1994) The effect of the volatile anesthetic isoflurane on the Ca^{2+}-dependent glutamate release from rat cerebral cortex. Brain Res 663: 335–337.
Larsen M, Hegstad E, Berg-Johnsen J and Langmoen IA (1997) Isoflurane increases the uptake of glutamate in synaptosomes from rat cerebral cortex. Br J Anaesth 78: 55–59.
Larsen M, Haugstad TS, Berg-Johnsen J and Langmoen IA (1998) Effect of isoflurane on release and uptake of gamma-aminobutyric acid from rat cortical synaptosomes. Br J Anaesth 80: 634–638.
Lewis BL and O'Donnell P (2000) Ventral tegmental area afferents to the prefrontal cortex maintain membrane potential 'up' states in pyramidal neurons via D_1 dopamine receptors. Cereb Cortex 10: 1168–1175.
Liu C, Au JD, Zou HL, Cotton JF and Spencer Yost C (2004) Potent activation of the tandem pore domain K^+ channel TRESK with clinical concentrations of volatile anesthetics. Anesth Analg 99: 1715–1722.
Loewenstein Y, Mahon S, Chadderton P, Kitamura K, Sompolinsky H, Yarom Y and Häusser M (2005) Bistability of cerebellar Purkinje cells modulated by sensory stimulation. Nat Neurosci 8: 202–211.
Luczak A, Bartho P, Marguet SL, Buzsaki G, Harris KD (2007) Sequential structure of neocortical spontaneous activity in vivo. Proc Natl Acad Sci USA 104: 347–352.
MacIver BM, Mikules AA, Amagasu SM and Monroe FA (1996) Volatile anesthetics depress glutamate transmission via presynaptic actions. Anesthesiology 85: 823–834.
Mahon S, Vautrelle N, Pezard L, Slaght SJ, Deniau JM, Chouvet G and Charpier S (2006) Distinct patterns of striatal medium spiny neuron activity during the natural sleep-wake cycle. J Neurosci 26: 12587–12595.
Milojkovic BA, Radojicic MS and Antic SD (2005) A strict correlation between dendritic and somatic plateau depolarizations in the rat prefrontal cortex pyramidal neurons. J Neurosci 25: 3940–3951.

Minami K, Wick M, Stern-Bach Y, Dildy-Mayfield JE, Brozowski SJ, Gonzales EL, Trudell JR and Harris RA (1998) Sites of volatile anesthetic action on kainate (glutamate receptor 6) receptors. J Biol Chem 273: 8248–8255.

Nishikawa K and MacIver MB (2001) Agent-selective effects of volatile anesthetics on $GABA_A$ receptor-mediated synaptic inhibition in hippocampal interneurons. Anesthesiology 94: 340–347.

O'Donnell P and Grace AA (1995) Synaptic interactions among excitatory afferents to nucleus accumbens neurons: Hippocampal gating of prefrontal cortical input. J Neurosci 15: 3622–3639.

O'Donnell P and Grace AA (1998) Phencyclidine interferes with the hippocampal gating of nucleus accumbens neuronal activity in vivo. Neuroscience 87: 823–830.

Paxinos G and Watson C (2007) The rat brain in stereotaxic coordinates, 6th Edition, Elsevier, Amsterdam.

Pearce RA, Stringer JL and Lothman EW (1989) Effect of volatile anesthetics on synaptic transmission in the rat hippocampus. Anesthesiology 71: 591–598.

Sanchez-Vives MV and McCormick DA (2000). Cellular and network mechanisms of rhythmic recurrent activity in neocortex. Nat Neurosci 3: 1027–1034.

Shin WJ and Winegar BD (2003) Modulation of noninactivating K^+ channels in rat cerebellar granule neurons by halothane, isoflurane and sevoflurane. Anesth Analg 96: 1340–1344.

Steriade M, Gloor P, Llinás RR, Lopes da Silva FH and Mesulam M-M (1990) Basic mechanisms of cerebral rhythmic activities. Electroencephalogr Clin Neurophysiol 76: 481–508.

Steriade M, Curró Dossi R and Nuñez A (1991) Network modulation of a slow intrinsic oscillation of cat thalamocortical neurons implicated in sleep delta waves: Cortically induced synchronization and brainstem cholinergic suppression. J Neurosci 11: 3200–3217.

Steriade M, Nuñez A and Amzica F (1993a) A novel slow (<1 Hz) oscillation of neocortical neurons in vivo: Depolarizing and hyperpolarizing components. J Neurosci 13: 3252–3265.

Steriade M, Nuñez A and Amzica F (1993b) Intracellular analysis of relations between the slow (<1 Hz) neocortical oscillation and other sleep rhythms of the electroencephalogram. J Neurosci 13: 3266–3283.

Steriade M, Contreras D, Curró Dossi R and Nuñe A (1993c) The slow (<1 Hz) oscillation in reticular thalamic and thalamocortical neurons: Scenario of sleep rhythm generation in interacting thalamic and neocortical networks. J Neurosci 13: 3284–3299.

Steriade M, Timofeev I and Grenier F (2001) Natural waking and sleep states: A view from inside neocortical neurons. J Neurophysiol 85: 1969–1985.

Stern EA, Jaeger D and Wilson CJ (1998) Membrane potential synchrony of simultaneously recorded striatal siny neurons in vivo. Nature 394: 475–478.

Verhaegen M, Todd MM and Warner DS (1992) The influence of different concentrations of volatile anesthetics on the threshold for cortical spreading depression in rats. Brain Res 581: 153–155.

Westphalen RI and Hemmings HC (2006a) Volatile anesthetic effects on glutamate versus GABA release form isolated rat cortical nerve terminals: Basal Release. J Pharmacol Exp Ther 316: 208–215.

Westphalen RI and Hemmings HC (2006b) Volatile anesthetic effects on glutamate versus GABA release form isolated rat cortical nerve terminals: 4-aminopyridine-evoked release. J Pharmacol Exp Ther 316: 216–223.

Wilson CJ and Groves PM (1981) Spontaneous firing patterns of identified spiny neurons in the rat neostriatum. Brain Res 220: 67–80.

Wilson CJ and Kawaguchi Y (1996) The origins of two-state spontaneous membrane potential fluctuations of neostriatal spiny neurons. J Neurosci 16: 2397–2410.

On the Relationships Between the Pedunculopontine Tegmental Nucleus, Corticostriatal Architecture, and the Medial Reticular Formation

David I.G. Wilson, Duncan A.A. MacLaren, and Philip Winn

Abstract Recent studies have established that the pedunculopontine tegmental nucleus (PPTg) is integrated into corticostriatal looped architecture through connections that include established basal ganglia output nuclei (pallidum, subthalamus and substantia nigra pars reticulata), thalamus and midbrain dopamine (DA) containing neurons in both the ventral tegmental area (VTA) and substantia nigra pars compacta (SNC). It is becoming apparent that the PPTg can be functionally dissociated internally. A simple dissociation is between posterior and anterior PPTg. The posterior PPTg contains a large proportion of cholinergic neurons, has polymodal sensory input that triggers very fast neuronal activity and projects preferentially to the VTA. In contrast, the anterior PPTg contains fewer cholinergic neurons, receives outflow from both corticostriatal systems and the extended amygdala and projects to the SNC. We suggest that this organization maps on to the spiral corticostriatal architecture such that the posterior PPTg interacts with ventromedial striatal systems (a proposed function of which is to integrate incentive salient stimuli to shape flexible goal-directed actions), whereas the anterior PPTg interacts with dorsolateral striatal circuits (which are thought to mediate the learning and execution of stimulus–response associations and the formation of habits). By these interactions, the PPTg en masse contributes to high-order decision making processes that shape action selection. In addition to this we also suggest that the PPTg integrates with medial reticular formation systems that operate as an immediate low-level action selection mechanism. We hypothesize that the PPTg has a pivotal position, bridging between higher order action selection mechanisms dealing with flexible learning of novel action patterns and lower level action selection processes that permit very fast responding to imperative stimuli.

D.I.G. Wilson, D.A.A. MacLaren, and P. Winn (✉)
School of Psychology, University of St. Andrews, St. Mary's Quad, South Street,
St. Andrews, Fife, Scotland KY16 9JP, UK
e-mail: pw@st-andrews.ac.uk

H.J. Groenewegen et al. (eds.), *The Basal Ganglia IX*, Advances in Behavioral Biology 58,
DOI 10.1007/978-1-4419-0340-2_12,

1 The Pedunculopontine Tegmental Nucleus

The pedunculopontine tegmental nucleus (PPTg) is a structure that is receiving increasing interest, driven most recently by demonstrations that low-frequency deep brain stimulation here has benefit in the treatment of Parkinson's disease (Plaha and Gill 2005; Stefani et al. 2007). Traditionally the PPTg has been regarded as being important in behavioural state control and locomotion. The large cholinergic neurons in the PPTg – the Ch5 group (Mesulam et al. 1983) – do indeed change their activity in characteristic ways across the sleep–wake cycle, while the idea that the PPTg was involved with locomotion came primarily from the belief that it was part of the mesencephalic locomotor region. However, contemporary studies are suggesting that the PPTg is an important member of the basal ganglia family of structures, and that it has functions that go beyond the rather automatic processes with which it has traditionally been associated – even to the extent that it appears to have functions relating to learning, memory and attention that could be regarded as cognitive. There have been several recent thorough reviews of PPTg structure and function (Mena-Segovia et al. 2004; Pahapill and Lozano 2000; Winn 2006), describing its connections with various forebrain, midbrain and brainstem structures, its mixed populations of neurons (cholinergic and non-cholinergic – most likely glutamate and GABA containing neurons) and its behavioural and psychological functions. The main purposes of this short review are (1) to highlight the possibility that anterior and posterior PPTg have distinct functions; (2) to stimulate thinking about how the PPTg relates to corticostriatal architecture (within which the basal ganglia are embedded) and the reticular formation and (3) to draw more attention to the fact that corticostriatal and basal ganglia systems that have a significant role in action selection processes cannot be divorced from those mechanisms in the brainstem that also have a role in these processes.

2 How does the PPTg Interact with the Anatomy of Corticostriatal Loops?

Current understanding of corticostriatal architecture – in which the key structures of the basal ganglia are embedded – is that there are neural circuit loops from cortex through basal ganglia to thalamus and back to cortex (cortico-basal ganglia-thalamus-cortico loops) and these are implicated in processing emotional, cognitive and motor information (Haber 2003; Joel and Weiner 2000; Parent and Hazrati 1995; Voorn et al. 2004). Three main subdivisions of these loops have been proposed based on the putative functions of the corticostriatal projections (see Fig. 1). (1) Limbic loops, implicated in processing motivational information, include projections from medial and orbital prefrontal cortex to ventromedial striatum (which includes the nucleus accumbens, ventral parts of caudate and putamen and striatal parts of the olfactory tubercle (Joel and Weiner 2000)). (2) Cognitive loops, the functions of which

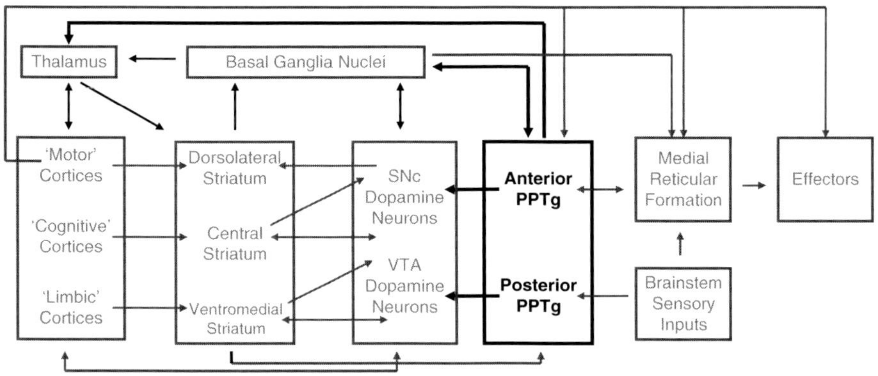

Fig. 1 Schematic representation of anatomical connections between PPTg (*bold black*) and corticostriatal architecture (cortex, striatum, basal ganglia nuclei, thalamus), midbrain dopamine neurons, reticular formation, brainstem sensory inputs and effectors. The PPTg, as a whole, has three main points of access into corticostriatal loops (*bold black arrows*): (1) via direct connections with all major basal ganglia nuclei (subthalamic nucleus, pallidum, substantia nigra pars reticulata); (2) through contact with all thalamic nuclei and (3) indirect access via midbrain dopamine neurons projections. In the latter case, posterior PPTg makes preferential projections to VTA dopamine neurons and can thus modulate neuronal activity associated with "motivational" information; anterior PPTg makes preferential projections into the SNc, thus influencing dorsolateral striatum, the recipient of information that is more "motor" than "motivational." It remains to be determined whether anterior and posterior PPTg make differential projections to basal ganglia and thalamus. Sensory input that is highly salient can reach posterior PPTg directly and can influence processing in the reticular formation which has reciprocal connections to anterior PPTg. NB: *Arrows* that extend into *boxes* indicate a specific projection to a subcomponent of the box whereas *arrows* that end outside of *boxes* indicate a projection to all structures within that *box*

include working memory, attentional set-shifting and cognitive planning, have projections from dorsolateral prefrontal cortex to the central caudate nucleus and putamen (central, dorsal striatum). (3) Motor loops, which are thought to participate in movement initiation, have projections from caudal premotor, presupplementary motor and cingulate motor cortical neurons to dorsolateral striatum.

This tripartite division of loop regions is thought to be maintained throughout each connection with basal ganglia nuclei (Haber 2003). It should be noted that information does not flow one way around the loops since most structures are connected reciprocally. Information can be integrated across loops by axons crossing over loops, by neurons converging at loop edges, by neurons connecting across hemispheres and by multiple reciprocal and feed-forward projections between cortical and thalamic neurons and between striatal and midbrain dopamine (DA) neurons (Haber 2003; Ikemoto 2007; Joel and Weiner 2000). In the latter case, striatal neurons make reciprocal connections with DA neurons and project to more ventrolaterally placed DA neurons. These DA neurons in turn project to striatal neurons that are situated more dorsolaterally in the striatum. In spirals of serial, feed-forward striatal–DA–striatal connections, striatal neurons in a particular region are in a position to influence dopamine inputs to their region and dopamine inputs to striatal neurons situated more dorsolaterally.

What information processing takes place within these loops? One prominent idea is that within each of the main loop regions (limbic, cognitive and motor) multiple parallel circuits are in competition for control of common motor resources (Prescott et al. 1999; Redgrave et al. 1999a). In the basal ganglia, conflicts between these circuits can be resolved via focused selection of the circuits associated with the appropriate action, thought or feeling and global inhibition of the circuits associated with the inappropriate ones (Mink 1996; Redgrave et al. 1999a). Thus competition amongst multiple potential actions, thoughts or feelings can be resolved to produce the "winning" appropriate goal, plan or movement. Indeed, the striatal–DA spiral architecture could enable a functional hierarchy since a cascade of neural activity from ventromedial striatum in a dorsolateral direction could facilitate conflict resolution to occur first in limbic loops, resulting in the selection of an appropriate goal, second in cognitive loops, resulting in the selection of an appropriate plan of actions to obtain that goal, and last in motor loops, facilitating the selection of the appropriate action to be implemented (Haber 2003; Joel and Weiner 2000; Redgrave et al. 1999a).

PPTg neurons can influence activity in corticostriatal pathways through multiple routes (see Fig. 1). One main way, and the focus of this review, is via direct projections from PPTg cholinergic neurons to midbrain DA neurons. A functional gradient has been found whereby cholinergic neurons in anterior PPTg project to DA neurons in the substantia nigra pars compacta (SNC) (which tend to project to more dorsal sites in the striatum) whereas cholinergic neurons in posterior PPTg project predominantly to DA neurons in the ventral tegmental area (VTA) (which target ventral striatum) (Oakman et al. 1995). A second major way that PPTg neurons could differentially affect the three regions of corticostriatal circuitry is via projections to the thalamus. Studies examining immediate early gene expression in the thalamus following PPTg stimulation have shown activation of centrolateral (CL), ventrolateral (VL) and the thalamic reticular nucleus (TRN) (Ainge et al. 2004). The two principal nuclei, VL and CL, are motor nuclei with connections to both the cortex and striatum, whereas TRN – which was activated bilaterally by PPTg stimulation – has more general functions in controlling thalamic state. Tract tracing studies have revealed that all PPTg neurons appear to project to at least one thalamic nucleus (Oakman et al. 1999) and that overlap exists between thalamic areas receiving projections from the PPTg with those thalamic areas that project to the striatum (Erro et al. 1999). Thus, different PPTg neurons can potentially influence different corticostriatal loops. Although no differential distribution of projections to thalamic nuclei across anterior–posterior sites in the PPTg have been cited, anterior portions project more densely to the thalamus than posterior portions (Erro et al. 1999). A third way in which PPTg can affect corticostriatal processing is via ascending projections to the globus pallidus and – perhaps most importantly – the subthalamic nucleus (STN). This is a key nucleus in corticostriatal architecture that appears to have a critical role in behavioural selection. Gilles and Willshaw's (1998) "braking hypothesis" suggests that the STN has a pivotal role in the process through which one of several competing actions gains control of brainstem motor machinery. Highly interconnected STN neurons are able to generate a widespread and uniform pulse in response to excitatory inputs that has the function of braking ongoing basal ganglia activity. One of the significant excitatory inputs to the STN is from the PPTg which appears to provide a combined cholinergic and glutamatergic input (Clarke et al. 1997). It is therefore possible that PPTg inputs to STN have a significant "interrupt" effect on basal ganglia processing. Finally, the PPTg is involved in corticostriatal activity on the output as well as the input side. The preceding considerations all reflect on the ability of the PPTg to change activity in various corticostriatal sites – midbrain DA neurons, thalamus and STN. But, like the substantia nigra, the PPTg both provides input to corticostriatal systems and receives output from them (Parent and Hazrati 1995). The PPTg therefore needs careful consideration both as a provider of input to corticostriatal systems and as an output station, interfacing corticostriatal output with systems even deeper in the brainstem.

3 Corticostriatal Systems, Instrumental Responding and the PPTg

Many psychological functions are thought to be mediated by corticostriatal loops and their DA afferents. These include response selection (Lyon and Robbins 1975), behavioural switching (Evenden and Carli 1985; Redgrave et al. 1999b; Robbins and Koob 1980), behavioural activation (Salamone et al. 2003), incentive salience (Berridge and Robinson 1998), hedonic attribution (Berridge 2003), reward learning (Schultz 1998) and the learning of novel actions (Redgrave and Gurney 2006). Most researchers agree that corticostriatal systems are important determinants of an animal's ability to respond appropriately in instrumental tasks, although the participation of different loops to different components of this process is under debate (Atallah et al. 2007; Berridge and Robinson 1998; Everitt and Robbins 2005; Prescott et al. 1999; Redgrave and Gurney 2006; Schultz 1998; Wickens et al. 2007; Yin and Knowlton 2006). Here we will review those studies that we believe best estimate the functional roles that the two most clearly differentiated loops ("limbic" and "motor"), through ventromedial and dorsolateral striatum, have in instrumental responding.

Lesions to the nucleus accumbens (a major component of ventromedial striatum) *spare* rats' ability to learn to approach a reward magazine upon presentation of a conditioned stimulus (Corbit et al. 2001; de Borchgrave et al. 2002; Hall et al. 2001), to perform a previously learnt instrumental response (Balleine and Killcross 1994; Cardinal and Cheung 2005; Hutcheson et al. 2001; Reading et al. 1991) and to learn a new instrumental response (Alderson et al. 2001; Cardinal and Cheung 2005; Corbit et al. 2001; de Borchgrave et al. 2002). However, such lesions can abolish the stimulatory effects of conditioned stimuli on instrumental responding (Corbit et al. 2001; de Borchgrave et al. 2002; Hall et al. 2001) and responding when a delay is introduced between action and outcome (Cardinal and Cheung 2005).

Lesions in (or inactivation of) the nucleus accumbens core also impair responding – and reacquisition of responding – for reward-related stimuli that reinforce patterns of multiple responses within second-order schedules of reinforcement (conditioned reinforcers) (Di Ciano et al. 2008; Hutcheson et al. 2001; Ito et al. 2004). Consistent with this, single neurons in the ventral striatum respond to conditioned reinforcers within a second-order schedule of reinforcement, and to a variety of conditioned stimuli distal to reward within schedules requiring multiple bar presses (Bowman et al. 1996; Shidara et al. 1998; Wilson and Bowman 2004, 2005). Together, these data suggest that the ventral striatum is required in order for an animal to use incentive salient stimuli in shaping any behaviour that requires chains or patterns of instrumental responses across long periods of time.

It is generally thought that, during instrumental learning, there is a shift from responding that is flexible, deliberative and goal-directed (action–outcome learning) to a more reflexive, involuntary habit-like response that is made without reference to an explicit goal (stimulus–response learning) (Dickinson 1985; Dickinson et al. 1995). Packard and McGaugh (1996) demonstrated that rats with lesions in the dorsal striatum were less likely to express a previously learnt stimulus-response habit. Indeed, it has been shown that damage to the dorsolateral striatum (Yin et al. 2004) and its DA inputs (Faure et al. 2005) can disrupt learning of stimulus-response habits and cause rats to revert to an action–outcome style of learning. Recent neurophysiological evidence is consistent with the hypothesis that dorsolateral circuits mediate stimulus–response learning: a proportion of dorsolateral striatal neurons exhibited stronger encoding of operant head movements once movements became habitual (Tang et al. 2007).

It has been theorized that within the basal ganglia different functional units can compete to gain access to common motor machinery (Prescott et al. 1999), and indeed there is experimental evidence that competition of this sort exists at a functional level between dorsolateral and ventromedial loops. For instance, Hernandez and colleagues found that inhibition of plasticity in ventral striatum *impaired* rats' ability to learn to lever-press for food (Hernandez et al. 2002). However, the same procedure in dorsolateral striatum *improved* instrumental learning compared to controls. They interpreted these seemingly paradoxical data as a reflection of competition between stimulus-response processing through dorsolateral striatum vs. action-outcome processing through ventral striatum. They suggested that lesions to dorsolateral striatum liberate control to ventral striatum, allowing rats to learn how to respond for the current goal; in comparison rats in the control group with sham lesions had normal-functioning plasticity in dorsolateral striatum and so striatal learning processes might be biased away from the task-related goal and towards unrelated, salient but distracting stimuli. Similarly, Atallah et al. (2007) recently demonstrated that temporary inactivation of dorsal striatum led to impaired execution of an appropriate instrumental response, but did not affect instrumental learning – reactivation of dorsal striatum after several training sessions revealed that instrumental learning had in fact taken place during the inactivation procedures and had simply been latent. The authors argued that in their task the ventral striatum aided instrumental learning (temporary inactivation of the ventral striatum impaired learning an instrumental response) and can direct dorsal striatum towards executing appropriate goal-related responses.

What functional role do afferents from midbrain DA neurons provide to these circuits? DA bursts are found following the presentation of unexpected reward-related stimuli and have been considered to encode an error in the prediction of the economic value of a stimulus (Schultz 1998; Tobler et al. 2005). However, Redgrave and Gurney have argued that these responses might in fact be present to any salient stimulus that has not yet been habituated (including conditioned stimuli) independent of their reward value. They propose that phasic bursting activity by DA neurons could interact with glutamatergic inputs of striatal neurons to strengthen active corticostriatal synapses, thereby facilitating the learning of novel actions related to the burst-inducing stimulus (Redgrave and Gurney 2006). Therefore, DA burst-firing would reinforce the selection of actions (or events) that cause important sensory events. Structures external to the basal ganglia (for example the orbitofrontal cortex and amygdala – see Schoenbaum et al. 2006) would mediate learning of the economic value of stimuli and actions to bias future action selection through the basal ganglia.

Because there are different functions for different corticostriatal loops, DA burst-firing should serve to reinforce different types of actions or events throughout sub-regions of the striatum. Ventral striatal neurons receive complex outcome and contextual sensory information via orbitofrontal, amygdala and hippocampal afferents (Grace et al. 2007; Schoenbaum et al. 2006). Thus, burst firing of DA neurons to conditioned stimuli (or any non-habituated salient stimulus) might reinforce patterns of neural activity associated with outcome-related events or contexts in the ventral striatum and thus set motivational units relevant to a particular environment. Importantly, it has been suggested that these ventral striatal neurons, via the ventral pallidum, can gate the availability of particular DA neurons that can be activated by salient events (Grace et al. 2007), a process that could also happen via feed-forward spiraling connections with DA neurons. Consequently, upon the subsequent occurrence of salient non-habituated events, such as conditioned stimuli, only gated DA neurons would burst-fire meaning that only particular patterns of connected striatal neurons would receive DA bursts. Thus specific actions could be learnt as described by Redgrave and Gurney (2006) and different goals and contexts could invoke particular patterns of actions using a particular subset of DA neuron/dorsal striatal circuits. In this way, motivational processing in the ventral striatum could modify the execution of appropriate responses across delays and with a particular level of effort (Cardinal et al. 2001; Salamone et al. 2003).

The action of DA in corticostriatal systems is also thought to be important in drug addiction, which can be viewed as a disorder of instrumental learning and responding. Everitt and Robbins (2005) have recently formulated a theory that drug addiction might reflect defects in processing the transitions between flexible, goal-directed choices for reinforcing (and possibly hedonic) drug effects towards habitual drug-taking that becomes compulsive; that is, drug addiction might be mediated by a shift from ventromedial to dorsolateral striatal circuitry control as drug use progresses. Consistent with this, repeated injections of amphetamine, a common drug of abuse, enhanced the transition from goal-directed to habitual responding (Nelson and Killcross 2006). Moreover, a unilateral lesion to the nucleus accumbens core combined with injections of a DA antagonist to the dorsolateral striatum in the opposing hemisphere impaired instrumental responding towards conditioned reinforcers associated with cocaine in the same manner as did bilateral injections of the DA antagonist in non-lesioned rats did (Belin and Everitt 2008). Because unilateral lesions had no impact on responding without injections, the authors interpreted these data as evidence of the interplay between ventral and dorsal striatum via spiraling DA connections that allows conditioned reinforcers to control instrumental responding.

These theories highlight both the differentiated functions of the corticostriatal loops and the critical role of DA within them. What role might the PPTg have in the operation of these systems? There is a significant volume of anatomical and behavioural data that strongly suggest that the PPTg has a close relationship with corticostriatal systems (for reviews, see Maskos 2008; Mena-Segovia et al. 2004, 2008). For example, excitotoxic lesions of the whole PPTg produce behavioural deficits that are similar in kind to those that follow damage to striatal systems. Lesioned rats have no difficulties with basic daily tasks – feeding, drinking, locomotion, grooming – or with tests of "emotion" such as the elevated plus maze (Walker and Winn 2007). But tasks that require responding for reward or reinforcement, and tasks with a strong associative element cause profound difficulties for PPTg lesioned rats. For example, rats with bilateral excitotoxic lesions of the whole PPTg cannot complete radial maze tasks that are also challenged by interference within corticostriatal systems (Keating and Winn 2002; Taylor et al. 2004). Likewise, rats with such lesions have difficulty learning to respond for intravenous amphetamine (though they have no difficulty in responding if trained to lever press for reward prior to lesioning) (Alderson et al. 2004). Such behavioural data indicate, at a functional level at least, a relationship between PPTg and corticostriatal systems. However, what has become more intriguing is the apparent anatomical and behavioural dissociation between anterior and posterior PPTg which suggests that different parts of the PPTg have relationships with different elements of the corticostriatal architecture.

There is both anatomical and behavioural evidence that functional differences between anterior and posterior PPTg mimic the distinction between the SNC and VTA DA systems. Anatomically, Oakman and his colleagues suggest that there is an anterior–posterior gradient through the PPTg that mirrors the mediolateral gradient amongst midbrain DA neurons (Oakman et al. 1995). In the PPTg, it is the more anterior cholinergic neurons that project to SNC while the posterior neurons tend to project to the VTA (Mena-Segovia et al. 2008). (Moreover, the adjacent Ch6 neurons in the laterodorsal tegmental nucleus, embedded in the central grey, also project to the VTA.). Taking these as a whole, the lateral neurons – those in the SNC – project predominantly into the dorsal striatum (the basal ganglia proper), while the projections to the ventral striatum come from the medially placed VTA. This anatomical difference appears to have functional significance.

Alderson and colleagues found that discriminable lesions in either the anterior or posterior portions of the PPTg – approximately corresponding to the older terms pars dissipatus and pars compactus, respectively – had clearly different behavioural effects. Lesions in the posterior PPTg had no effect on baseline levels of locomotion but altered both the locomotor response to nicotine and altered the self-administration of nicotine (an effect suggested to be the product of a change in incentive salience brought about by alteration of the firing of VTA DA neurons). In contrast, lesions in the anterior PPTg had no effect on responding to nicotine but significantly reduced baseline levels of locomotor activity (Alderson et al. 2006, 2008). These effects are consistent with the idea that SNC DA neurons are more concerned with motor events, while the DA neurons of the VTA are concerned with responses to novelty, reward and reinforcement. The reinforcing properties of nicotine for example are strongly linked to VTA DA neurons (Corrigall et al. 1994).

The inputs to anterior and posterior PPTg also appear to be different. The posterior portion – containing the bulk of the cholinergic neurons – is in receipt of sensory information. Visual, auditory and somatosensory data all appear to impact on these neurons, evoking neuronal firing with very short latencies of around 8 ms in the case of imperative auditory signals (Dormont et al. 1998; Pan and Hyland 2005). The anterior PPTg in contrast appears to be targeted by outflow from both basal ganglia and structures of the extended amygdala (Parent and Hazrati 1995; Zahm et al. 2001) (though whether these target the same neurons, or the same parts of single neurons is not clear). It thus appears that the anterior and posterior PPTg are differentially incorporated into the corticostriatal architecture: the posterior PPTg appears to be in receipt of fast sensory polymodal data that is then directed at VTA DA neurons, which could be hypothesized to inform the construction of flexible, goal-directed actions. In contrast, the anterior PPTg receives descending input from the basal ganglia and extended amygdala and has output to the DA neurons of the SNc. The confluence here of "motor" and "limbic" information suggests that the *anterior* PPTg might be a viable candidate as part of what was once thought of as a "limbic motor interface" (see Winn et al. 1997), but what precise function this might have is less clear. It is likely that the bulk of these descending inputs to PPTg is inhibitory, being GABA mediated. Understanding the ways in which these streams interact would provide a significant clue as to the nature of the feedback – is it positive feedback (maintain the current behaviour) or negative (an interrupt signal)?

4 Does PPTg Integrate Processing Between Corticostriatal and Brainstem Systems?

Output of the basal ganglia from the substantia nigra pars reticulata (SNR) produces widespread tonic inhibition not only of thalamocortical neurons but also of midbrain and mesopontine regions such as superior colliculus, periaqueductal grey, cuneiform nucleus and PPTg, structures that link basal ganglia with the reticular formation and brainstem pattern generators. It has been proposed that the basal ganglia can signal decisions on action selection to these midbrain regions in order to control different motor programs or effectors in the brainstem (Grillner et al. 2005; Hikosaka 1998, 2007; McHaffie et al. 2005). The types of behaviours controlled by central pattern generators within its control are fundamental, such as locomotion, vocalization, posture, breathing, chewing and swallowing. However, in the brainstem there are also relatively complex behaviours, such as the ability to orientate to auditory or visual stimuli, make discriminations between food types and sequence actions to hold and eat food and to groom (Humphries et al. 2007).

Indeed, Humphries et al. (2007) argue that the anatomy of the medial reticular formation (mRF) lends itself towards being a potential action selection system in the brainstem, a region into which the PPTg projects extensively. They argue that neurons are organized into clusters that are not reflective of a sensory, or motor, topography. Instead, they propose that individual clusters of mRF neurons represent single actions, since correlated activity by neurons within a cluster can recruit disparate types of muscles and movements that are idiosyncratic for a particular action. Co-activation of different clusters would trigger a coordinated behavioural response. What functional role could the PPTg play in terms of its connections to brainstem? It has been proposed that the PPTg might output to modify posture and to facilitate locomotion (Hikosaka 2007; Takakusaki et al. 2003). Thus, Hikosaka (2007) posits a ventral striatum–SNr–MLR/PPTg substrate for reward-orientated locomotion. However, as described above in Sect. 1, PPTg lesion studies suggest a distinct role in learning and not in postural control. Given this, and the possibility that learned action sequences could be carried out by mRF, it seems worth considering, albeit speculatively, that the PPTg (and in particular the anterior portion) could link action representations in the mRF (for example, a lever-press cluster) with instrumental learning procedures throughout DA-modulated corticostriatal loops. In this way basal ganglia output via the anterior PPTg could instruct particular clusters in mRF to execute particular actions.

Moreover, the PPTg clearly plays another important role in instrumental learning. PPTg activity has been shown to be essential in driving burst-firing in DA neurons (Grace et al. 2007). PPTg neurons (primarily in the posterior portion) receive polymodal sensory input and it is thought that they pass on sensory information to DA neurons related to the content, rather than the salience, of a given stimulus (Lodge and Grace 2006; Stewart and Dommett 2006). If DA neurons are activated to burst fire (if the stimulus is considered unpredictable and salient) then the stimulus information provided by the PPTg to striatum (via thalamus) can be reinforced (McHaffie et al. 2005). In this way, the information regarding stimulus content that is to be conditioned in the striatum is produced by (primarily posterior) PPTg circuitry. Without this information – for example, when the posterior PPTg is lesioned – information on the content of unpredicted, salient stimuli would be unable to induce burst firing in DA neurons, and therefore the lesioned animal would be unable to learn goal-directed behaviours.

These are two main examples of ways in which PPTg-basal ganglia and PPTg-reticular formation circuits could contribute to instrumental responding and these examples again demonstrate how anterior and posterior PPTg might influence instrumental learning in different ways. It has been posited that these systems might be part of a heterarchical layered network, such that they can work together (as outlined above), independently or compete towards action selection (Humphries et al. 2007). Thus, there may be situations where the basal ganglia may subsume the mRF and gain control such as when multiple action choices are available of approximately equal salience. This would be apparent in new environments where learning is required. In these situations, DA neurons tend to fire and perhaps this relinquishes control from brainstem circuits to that of the basal ganglia. However, it is also possible for brainstem circuits to not require the basal ganglia. Thus, the PPTg could trigger responding in mRF when an action is clearly more salient than its competitors and the environment demands it. Additionally, the brainstem could act before the basal ganglia, for instance when quick decisions are required although in these situations the basal ganglia would continue processing in parallel to enable the animal to learn how to respond more effectively in the future. The most obvious example of this multilevel approach to action selection – low level fast, higher level moderated – is the startle response (Koch et al. 1993; Koch 1999): the PPTg is known to be involved in the production of startle responses to unpredictable stimuli, but its actions can also be brought under learned control by corticostriatal systems.

5 Conclusions

In conclusion, the PPTg lies in an important position and could interface between two potential action selection systems. The internal organization of the PPTg, even at the relatively simple level of an anterior–posterior discrimination, appears to map onto the spiral nature of corticostriatal architecture and provides important drivers of both SNC and VTA DA neurons. At the same time, the PPTg is in receipt of outflow from corticostriatal systems and has strong connections with reticular formation systems that are hypothesized to be a low level, evolutionarily older, action selection system. How and when these systems co-operate or compete and the involvement of the PPTg is still unknown. However, these hypotheses provide interesting explanations of why damage to the PPTg can results in impaired instrumental learning. Due to the likely important relationship between DA burst firing and aberrant learning that may contribute to schizophrenia and drug addiction, it seems increasingly important that the PPTg could play a vital role in these abnormal learning situations as well as in normal ones.

Acknowledgments David Wilson and Duncan MacLaren are supported by Wellcome Trust project grant 081128 to PW. We wish to extend our thanks to the editors for their patience and kindness in allowing us extra time in which to complete this essay.

References

Ainge JA, Jenkins TA and Winn P (2004) Induction of c-fos in specific thalamic nuclei following stimulation of the pedunculopontine tegmental nucleus. Eur J Neurosci 20: 1827–1837.

Alderson HL, Parkinson JA, Robbins TW and Everitt BJ (2001) The effects of excitotoxic lesions of the nucleus accumbens core or shell regions on intravenous heroin self-administration in rats. Psychopharmacology (Berl) 153: 455–463.

Alderson HL, Latimer MP, Blaha CD, Phillips AG and Winn P (2004) An examination of d-amphetamine self-administration in pedunculopontine tegmental nucleus-lesioned rats. Neuroscience 125: 349–358.

Alderson HL, Latimer MP and Winn P (2006) Intravenous self-administration of nicotine is altered by lesions of the posterior, but not anterior, pedunculopontine tegmental nucleus. Eur J Neurosci 23: 2169–2175.

Alderson HL, Latimer MP and Winn P (2008) A functional dissociation of the anterior and posterior pedunculopontine tegmental nucleus: Excitotoxic lesions have differential effects on locomotion and the response to nicotine. Brain Struct Funct 213: 247–253.

Atallah HE, Lopez-Paniagua D, Rudy JW and O'Reilly RC (2007) Separate neural substrates for skill learning and performance in the ventral and dorsal striatum. Nat Neurosci 10:126–131.

Balleine B and Killcross S (1994) Effects of ibotenic acid lesions of the nucleus accumbens on instrumental action. Behav Brain Res 65: 181–193.

Belin D and Everitt BJ (2008) Cocaine seeking habits depend upon dopamine-dependent serial connectivity linking the ventral with the dorsal striatum. Neuron 57: 432–441.

Berridge KC (2003) Pleasures of the brain. Brain Cogn 52: 106–128.

Berridge KC and Robinson TE (1998) What is the role of dopamine in reward: Hedonic impact, reward learning, or incentive salience? Brain Res Brain Res Rev 28: 309–369.

Bowman EM, Aigner TG and Richmond BJ (1996) Neural signals in the monkey ventral striatum related to motivation for juice and cocaine rewards. J Neurophysiol 75: 1061–1073.

Cardinal RN and Cheung TH (2005) Nucleus accumbens core lesions retard instrumental learning and performance with delayed reinforcement in the rat. BMC Neurosci 6: 9.

Cardinal RN, Pennicott DR, Sugathapala CL, Robbins TW and Everitt, BJ (2001) Impulsive choice induced in rats by lesions of the nucleus accumbens core. Science 292: 2499–2501.

Clarke NP, Bevan MD, Cozzari C, Hartman BK and Bolam JP (1997) Glutamate-enriched cholinergic synaptic terminals in the entopeduncular nucleus and subthalamic nucleus of the rat. Neuroscience 81: 371–385.

Corbit LH, Muir JL and Balleine BW (2001) The role of the nucleus accumbens in instrumental conditioning: Evidence of a functional dissociation between accumbens core and shell. J Neurosci 21: 3251–3260.

Corrigall WA, Coen KM and Adamson KL (1994) Self-administered nicotine activates the mesolimbic dopamine system through the ventral tegmental area. Brain Res 653: 279–284.

de Borchgrave R, Rawlins JN, Dickinson A and Balleine BW (2002) Effects of cytotoxic nucleus accumbens lesions on instrumental conditioning in rats. Exp Brain Res 144: 50–68.

Di Ciano P, Robbins TW and Everitt BJ (2008) Differential effects of nucleus accumbens core, shell, or dorsal striatal inactivations on the persistence, reacquisition, or reinstatement of responding for a drug-paired conditioned reinforcer. Neuropsychopharmacology 33: 1413–1425.

Dickinson A (1985) Actions and habits: The development of behavioural autonomy. Philos Trans R Soc Lond B Biol Sci 308: 67–78.

Dickinson A, Balleine BW, Watt A, Gonzales F and Boakes RA (1995) Overtraining and the motivational control of instrumental action. Anim Learn Behav 22: 197–206.

Dormont JF, Conde H and Farin D (1998) The role of the pedunculopontine tegmental nucleus in relation to conditioned motor performance in the cat. I. Context-dependent and reinforcement-related single unit activity. Exp Brain Res 121: 401–410.

Erro E, Lanciego JL and Gimenez-Amaya JM (1999) Relationships between thalamostriatal neurons and pedunculopontine projections to the thalamus: A neuroanatomical tract-tracing study in the rat. Exp Brain Res 127: 162–170.

Evenden JL and Carli M (1985) The effects of 6-hydroxydopamine lesions of the nucleus accumbens and caudate nucleus of rats on feeding in a novel environment. Behav Brain Res 15: 63–70.
Everitt BJ and Robbins TW (2005) Neural systems of reinforcement for drug addiction: From actions to habits to compulsion. Nat Neurosci 8: 1481–1489.
Faure A, Haberland U, Conde F and El Massioui N (2005) Lesion to the nigrostriatal dopamine system disrupts stimulus–response habit formation. J Neurosci 25: 2771–2780.
Gillies AJ and Willshaw DJ (1998) A massively connected subthalamic nucleus leads to the generation of widespread pulses. Proc R Soc Lond B 265: 2101–2109.
Grace AA, Floresco SB, Goto Y and Lodge DJ (2007) Regulation of firing of dopaminergic neurons and control of goal-directed behaviors. Trends Neurosci 30: 220–227.
Grillner S, Hellgren J, Menard A, Saitoh K and Wikstrom MA (2005) Mechanisms for selection of basic motor programs – roles for the striatum and pallidum. Trends Neurosci 28: 364–370.
Haber SN (2003) The primate basal ganglia: Parallel and integrative networks. J Chem Neuroanat 26: 317–330.
Hall J, Parkinson JA, Connor TM, Dickinson A and Everitt BJ (2001) Involvement of the central nucleus of the amygdala and nucleus accumbens core in mediating pavlovian influences on instrumental behaviour. Eur J Neurosci 13: 1984–1992.
Hernandez PJ, Sadeghian K and Kelley AE (2002) Early consolidation of instrumental learning requires protein synthesis in the nucleus accumbens. Nat Neurosci 5: 1327–1331.
Hikosaka O (1998) Neural systems for control of voluntary action – a hypothesis. Adv Biophys 35: 81–102.
Hikosaka O (2007) GABAergic output of the basal ganglia. Prog Brain Res 160: 209–226.
Humphries MD, Gurney K and Prescott TJ (2007) Is there a brainstem substrate for action selection? Philos Trans R Soc Lond B Biol Sci 362: 1627–1639.
Hutcheson DM, Parkinson JA, Robbins TW and Everitt BJ (2001) The effects of nucleus accumbens core and shell lesions on intravenous heroin self-administration and the acquisition of drug-seeking behaviour under a second-order schedule of heroin reinforcement. Psychopharmacology 153: 464–472.
Ikemoto S (2007) Dopamine reward circuitry: Two projection systems from the ventral midbrain to the nucleus accumbens–olfactory tubercle complex. Brain Res Rev 56: 27–78.
Ito R, Robbins TW and Everitt BJ (2004) Differential control over cocaine-seeking behavior by nucleus accumbens core and shell. Nat Neurosci 7: 389–397.
Joel D and Weiner I (2000) The connections of the dopaminergic system with the striatum in rats and primates: An analysis with respect to the functional and compartmental organization of the striatum. Neuroscience 96: 451–474.
Keating GL and Winn P (2002) Examination of the role of the pedunculopontine tegmental nucleus in radial maze tasks with or without a delay. Neuroscience 112: 687–696.
Koch M (1999) The neurobiology of startle. Prog Neurobiol 59: 107–128.
Koch M, Kungel M and Herbert H (1993) Cholinergic neurons in the pedunculopontine tegmental nucleus are involved in the mediation of prepulse inhibition of the acoustic startle response in the rat. Exp Brain Res 97: 71–82.
Lodge DJ and Grace AA (2006) The hippocampus modulates dopamine neuron responsivity by regulating the intensity of phasic neuron activation. Neuropsychopharmacology 31: 1356–1361.
Lyon M and Robbins TW (1975) The action of central nervous system stimulant drugs: A general theory concerning amphetamine effects In: Essman W, Valzelli L (eds) Current developments in psychopharmacology. Spectrum, New York, NY.
Maskos U (2008) The cholinergic mesopontine tegmentum is a relatively neglected nicotinic master modulator of the dopaminergic system: Relevance to drugs of abuse and pathology. Br J Pharmacol 153(Suppl 1): S438–S445.
McHaffie JG, Stanford TR, Stein BE, Coizet V and Redgrave P (2005) Subcortical loops through the basal ganglia. Trends Neurosci 28: 401–407.
Mena-Segovia J, Bolam JP and Magill PJ (2004) Pedunculopontine nucleus and basal ganglia: Distant relatives or part of the same family? Trends Neurosci 27: 585–588.
Mena-Segovia J, Winn P and Bolam JP (2008) Cholinergic modulation of midbrain dopaminergic systems. Brain Res Rev 58: 265–271.

Mesulam MM, Mufson EJ, Wainer BH and Levey AI (1983) Central cholinergic pathways in the rat: An overview based on an alternative nomenclature (Ch1–Ch6). Neuroscience 10: 1185–1201.
Mink JW (1996) The basal ganglia: Focused selection and inhibition of competing motor programs. Prog Neurobiol 50: 381–425.
Nelson A, Killcross S (2006) Amphetamine exposure enhances habit formation. J Neurosci 26: 3805–3812.
Oakman SA, Faris PL, Kerr PE, Cozzari C and Hartman BK (1995) Distribution of pontomesencephalic cholinergic neurons projecting to substantia nigra differs significantly from those projecting to ventral tegmental area. J Neurosci 15: 5859–5869.
Oakman SA, Faris PL, Cozzari C and Hartman BK (1999) Characterization of the extent of pontomesencephalic cholinergic neurons' projections to the thalamus: Comparison with projections to midbrain dopaminergic groups. Neuroscience 94: 529–547.
Packard MG and McGaugh JL (1996) Inactivation of hippocampus or caudate nucleus with lidocaine differentially affects expression of place and response learning. Neurobiol Learn Mem 65: 65–72.
Pahapill PA and Lozano AM (2000) The pedunculopontine nucleus and Parkinson's disease. Brain 123: 1767–1783.
Pan WX and Hyland BI (2005) Pedunculopontine tegmental nucleus controls conditioned responses of midbrain dopamine neurons in behaving rats. J Neurosci 25: 4725–4732.
Parent A and Hazrati LN (1995) Functional anatomy of the basal ganglia. I. The cortico-basal ganglia-thalamo-cortical loop. Brain Res Brain Res Rev 20: 91–127.
Plaha P and Gill SS (2005) Bilateral deep brain stimulation of the pedunculopontine nucleus for Parkinson's disease. Neuroreport 16: 1883–1887.
Prescott TJ, Redgrave P and Gurney K (1999) Layered control architectures in robots and vertebrates. Adapt Behav 7: 99–127.
Reading PJ, Dunnett SB and Robbins TW (1991) Dissociable roles of the ventral, medial and lateral striatum on the acquisition and performance of a complex visual stimulus-response habit. Behav Brain Res 45: 147–161.
Redgrave P, Gurney K (2006) The short-latency dopamine signal: A role in discovering novel actions? Nat Rev Neurosci 7: 967–975.
Redgrave P, Prescott TJ, Gurney K (1999a) The basal ganglia: A vertebrate solution to the selection problem? Neuroscience 89: 1009–1023.
Redgrave P, Prescott TJ and Gurney K (1999b) Is the short-latency dopamine response too short to signal reward error? Trends Neurosci 22: 146–151.
Robbins TW and Koob GF (1980) Selective disruption of displacement behaviour by lesions of the mesolimbic dopamine system. Nature 285: 409–412.
Salamone JD, Correa M, Mingote S and Weber SM (2003) Nucleus accumbens dopamine and the regulation of effort in food-seeking behavior: Implications for studies of natural motivation, psychiatry, and drug abuse. J Pharmacol Exp Ther 305: 1–8.
Schoenbaum G, Roesch MR and Stalnaker TA (2006) Orbitofrontal cortex, decision-making and drug addiction. Trends Neurosci 29: 116–124.
Schultz W (1998) Predictive reward signal of dopamine neurons. J Neurophysiol 80: 1–27.
Shidara M, Aigner TG and Richmond BJ (1998) Neuronal signals in the monkey ventral striatum related to progress through a predictable series of trials. J Neurosci 18: 2613–2625.
Stefani A, Lozano AM, Peppe A, Stanzione P, Galati S, Tropepi D, Pierantozzi M, Brusa L, Scarnati E and Mazzone P (2007) Bilateral deep brain stimulation of the pedunculopontine and subthalamic nuclei in severe Parkinson's disease. Brain 130: 1596–1607.
Stewart RD and Dommett EJ (2006) Subcortical control of dopamine neurons: The good, the bad and the unexpected. Brain Res Bull 71: 1–3.
Takakusaki K, Habaguchi T, Ohtinata-Sugimoto J, Saitoh K and Sakamoto T (2003) Basal ganglia efferents to the brainstem centers controlling postural muscle tone and locomotion: A new concept for understanding motor disorders in basal ganglia dysfunction. Neuroscience 119: 293–308.

Tang C, Pawlak AP, Prokopenko V and West MO (2007) Changes in activity of the striatum during formation of a motor habit. Eur J Neurosci 25: 1212–1227.

Taylor CL, Kozak R, Latimer MP and Winn P (2004) Effects of changing reward on performance of the delayed spatial win-shift radial maze task in pedunculopontine tegmental nucleus lesioned rats. Behav Brain Res 153: 431–438.

Tobler PN, Fiorillo CD and Schultz W (2005) Adaptive coding of reward value by dopamine neurons. Science 307: 1642–1645.

Voorn P, Vanderschuren LJ, Groenewegen HJ, Robbins TW and Pennartz CM (2004) Putting a spin on the dorsal-ventral divide of the striatum. Trends Neurosci 27: 468–474.

Walker SC and Winn P (2007) An assessment of the contributions of the pedunculopontine tegmental and cuneiform nuclei to anxiety and neophobia. Neuroscience 150: 273–290.

Wickens JR, Horvitz JC, Costa RM and Killcross S (2007) Dopaminergic mechanisms in actions and habits. J Neurosci 27: 8181–8183.

Wilson DI and Bowman EM (2004) Nucleus accumbens neurons in the rat exhibit differential activity to conditioned reinforcers and primary reinforcers within a second-order schedule of saccharin reinforcement. Eur J Neurosci 20: 2777–2788.

Wilson DI and Bowman EM (2005) Rat nucleus accumbens neurons predominantly respond to the outcome-related properties of conditioned stimuli rather than their behavioral-switching properties. J Neurophysiol 94: 49–61.

Winn P (2006) How best to consider the structure and function of the pedunculopontine tegmental nucleus: Evidence from animal studies. J Neurol Sci 248: 234–250.

Winn P, Brown VJ and Inglis WL (1997) On the relationships between the striatum and the pedunculopontine tegmental nucleus. Crit Rev Neurobiol 11: 241–261.

Yin HH and Knowlton BJ (2006) The role of the basal ganglia in habit formation. Nat Rev Neurosci 7: 464–476.

Yin HH, Knowlton BJ and Balleine BW (2004) Lesions of dorsolateral striatum preserve outcome expectancy but disrupt habit formation in instrumental learning. Eur J Neurosci 19: 181–189.

Zahm DS, Williams EA, Latimer MP and Winn P (2001) Ventral mesopontine projections of the caudomedial shell of the nucleus accumbens and extended amygdala in the rat: Double dissociation by organization and development. J Comp Neurol 436: 111–125.

Microcircuits of the Pedunculopontine Nucleus

Juan Mena-Segovia and J. Paul Bolam

Abstract The notion of the pedunculopontine nucleus (PPN) is being updated in light of increasing evidence that identifies it as a key structure in the processing of exteroceptive information and its transmission to the forebrain, influencing the activity of PPN's multiple targets (from basal ganglia to the cerebral cortex). Here we review the anatomical evidence supporting the existence of a local network that forms the basis for such processing functions and propose a more complex relationship with its efferent systems. We also identify some of the critical issues that remain to be answered in order to have a comprehensive understanding of the roles of the PPN. A better insight into this connectivity will also help to understand the contribution of the PPN to the diverse systems it influences.

1 Past and Present Notions of the PPN

The pedunculopontine nucleus (PPN) is a phylogenetically old neuronal structure that maintains a close relationship with the basal ganglia, which seems to be preserved in the evolution of the brain since early chordates (Grillner et al. 2008). The PPN constitutes the main cholinergic output from the brainstem and has wide-spread projections that innervate, and influence the activity of, many other neuronal systems across the entire brain. As a part of the so-called "reticular activating system", it has long been considered that PPN's central function, along with other nuclei in the brainstem, is to convey and forward exteroceptive information to the forebrain, acting mainly as a relay structure. In this context, the PPN has been regarded as the link between the basal ganglia and the activating systems of the brainstem. Recent physiological evidence, however, shows that the PPN indeed provides an early response to sensory inputs but, more importantly, this response

J. Mena-Segovia (✉) and J.P. Bolam
MRC Antomical Neuropharmacology Unit, University of Oxford, Mansfield Road, Oxford OX1 3TH, UK
e-mail: juan.mena-segovia@pharm.ox.ac.uk

H.J. Groenewegen et al. (eds.), *The Basal Ganglia IX*, Advances in Behavioral Biology 58, DOI 10.1007/978-1-4419-0340-2_13,

might involve a degree of local processing (Pan and Hyland 2005). Unfortunately, the anatomical data supporting the idea of internal connectivity in the PPN is scarce, and therefore an anatomical basis for its role in the early processing of sensory information is lacking.

The long-held view of the PPN as a neurochemically homogeneous structure has focused research for the last two decades primarily on cholinergic neurons and their functional properties, even though they only constitute a fraction of the entire population of the PPN. There have been no reports of estimates of the proportions of cholinergic vs. non-cholinergic neurons in animals, but studies in humans show that cholinergic neurons account for less than the 50% of the total neuronal population (Manaye et al. 1999). Probably one of the reasons for this focus on cholinergic neurons is due to the particular difficulty in labelling non-cholinergic neurons to identify the markers they express. So far, there is not a reliable marker to immunolabel cell bodies of GABAergic or glutamatergic neurons in this region. Previous attempts in our laboratory using antibodies against these neurotransmitters, or their synthetic enzymes, have given inconsistent results. On the other hand, results using *in situ* hybridization (Ford et al. 1995) or post-embedding immunolabelling of axon terminals in PPN's targets (Bevan and Bolam 1995), were able to conclusively establish that the PPN contains GABAergic and glutamatergic neuronal populations[1].

It is the purpose of the present review to gather information regarding the morphological and synaptic properties of the different neuronal types in the PPN and to present a model of internal connectivity to ultimately update the current notion of the PPN.

2 Ultrastructural Analysis

Several thorough and detailed studies of the morphological properties and connectivity of the PPN were published in the late 1980s, early 1990s, and constitute most of what is now known about PPN cytoarchitecture, although some of them were exclusively centred on cholinergic neurons (Rye et al. 1987; Honda and Semba 1995; Steininger et al. 1997).

Cholinergic neurons are large cells with polygonal, triangular or fusiform shapes. The nuclei characteristically have infoldings with a large, granular nucleolus, and smooth and dispersed chromatin filaments. According to Honda and Semba (1995), half of the somatic surface of each cholinergic neuron is covered by vesicle-containing axons, some of which make synaptic contacts, whereas a quarter of the surface is occupied by glial processes, mainly arising from protoplasmic astrocytes. There is no estimate of the *total* number of neurons in the rodent PPN, although Rye and colleagues (Rye et al. 1987) estimated that the rat PPN contains approximately 1,755–1,900 cholinergic neurons. In addition, the authors acknowledge the presence of non-cholinergic neurons intermingled among cholinergic neurons, and noted that cholinergic neurons represent, on average, 45–55% of the total cell population in which they are contained. Despite this, the PPN is still often considered as an homogeneous cholinergic structure.

The soma of cholinergic neurons (20.2 ± 2.6 μm diameter) gives rise to 3–5 primary dendrites, which have an initial diameter of 3–5 μm. This diameter is maintained in the 2–3 secondary dendrites that arise 50–100 μm from the soma (Rye et al. 1987). They contain microtubules and mitochondria, and if they are over 1 μm in diameter, they also contain rough endoplasmic reticulum (Honda and Semba 1995). Typically, cholinergic dendrites have swellings that resemble axonal varicosities and are irregularly spaced. These swellings can be observed mainly in secondary dendrites, although some are present also in primary dendrites. The dendritic arbours of cholinergic neurons overlap extensively with one another, but remain relatively confined to the boundaries of the nucleus (Rye et al. 1987).

In contrast to cholinergic neurons, considerably less information is available regarding the morphological properties of non-cholinergic neurons. Even though it is known that non-cholinergic subtypes include GABAergic and glutamatergic neurons, the ultrastructural characteristics of these specific subtypes have not been studied. Thus, with the reservations that non-cholinergic neurons are an heterogeneous group, in general they are smaller than cholinergic neurons (13.1 ± 3.7 μm of diameter; Steininger et al. 1997), although a minor proportion may be similar in size to cholinergic neurons (Lavoie and Parent 1994). They have a spindle shape, their nucleus is irregular in shape, and they have smaller cytoplasmic volume but contain the same types of organelles as cholinergic neurons (Honda and Semba 1995), although in a lower proportion and more sparsely arranged (Steininger et al. 1997). The dendrites of non-cholinergic neurons are similar to the dendrites of cholinergic neurons, and occasionally afferent synaptic contacts were found at the site of the swellings.

3 Local Connectivity

The cell bodies and dendrites of cholinergic neurons were found to receive frequent synaptic contacts, occasionally multiple contacts arising from a single axon. The cell bodies receive a mean of 4.7 synaptic contacts per electron microscope section. This represents 70% more than non-cholinergic neurons, which receive a mean of 2.8 synapses per electron microscope section. Although both cholinergic and non-cholinergic neurons were contacted by similar types of axon terminals in terms of synaptic vesicle morphology and membrane specializations, non-cholinergic neurons were also contacted by terminals containing large dense-cored vesicles of 75–110 nm in diameter that were not found to contact cholinergic neurons (Honda and Semba 1995).

Cholinergic terminals were frequently observed within the boundaries of the PPN. Most of them were observed to form synaptic contacts with non-cholinergic cell bodies and dendrites, and just a minor proportion in contact with cholinergic dendrites (Honda and Semba 1995). The authors considered that the interaction of cholinergic axons with cholinergic dendrites was underestimated due to technical constraints, such as the penetration of the antibodies in myelinated cholinergic axons and the additional level of difficulty for the identification of immunoreactive elements in both pre- and postsynaptic structures.

More recently, our results from juxtacellular labelling of single cells allowed us to identify the local connectivity of cholinergic and non-cholinergic neurons in the PPN. The axons of individual cholinergic neurons have 4–6 collaterals arising in the proximity of the cell body and travelling in several directions. Most importantly, they have a large number of local axonal varicosities (59 ± 25) along all their collaterals and some of them have been identified as forming asymmetric synaptic contacts. Non-cholinergic neurons, on the other hand, have a significantly smaller number of local varicosities. These anatomical findings strongly indicate the existence of a local neuronal network driven primarily by cholinergic neurons. Indeed, the activation of acetylcholine receptors within the PPN leads to changes in activity in other regions of the brain, notably the cortex (Mena-Segovia et al. 2008b).

Further evidence supporting this idea is the physiological finding of local interconnections between PPN neurons (Garcia-Rill et al. 2007). By using *in vitro* paired recordings, the authors showed that there is a degree of interconnections between PPN neurons, and in some cases these neurons are electrically coupled through GAP junctions.

4 Local Network Model

The available information of the local connectivity in the PPN is limited and is restricted to the cholinergic neuronal population. However, the notion that the PPN is a purely relay structure is now being revised to that of a highly interconnected heterogeneous structure, the local network of which, underlies its functional properties (Androulidakis et al. 2008; Mena-Segovia et al. 2008b). The PPN thus plays a role in processing some type of exteroceptive information before transferring it to more specialized systems (e.g., midbrain dopaminergic system; Mena-Segovia et al. 2008a), a notion consistent with some of the known functions of the PPN (e.g., attentional processes; Winn 2006). In addition, it suggests that PPN might be generating a population activity that could determine the firing of its different neuronal types (Mena-Segovia, Magill and Bolam, unpublished observations), and therefore contribute to the generation of specific oscillatory activities in its targets, such as alpha (Lorincz et al. 2008), theta (Leszkowicz et al. 2007) and gamma (Steriade et al. 1991; Mena-Segovia et al. 2008b) rhythms, thus behaving as a central pattern generator (Grillner 2006).

Evidence is increasing but much work is still needed to elucidate the microcircuits of the PPN (Fig. 1). Cholinergic, glutamatergic and GABAergic neurons are the three main populations in the PPN but we do not know to what extent they interact, or to what extent they co-localize with other molecule(s) (e.g., calcium-binding proteins or neuropeptides). Indeed, the level of coexpression of the three main markers is not yet certain[1] (see discussion in Mena-Segovia et al. 2008a). Another critical point to understand the internal connectivity of the PPN is the iden-

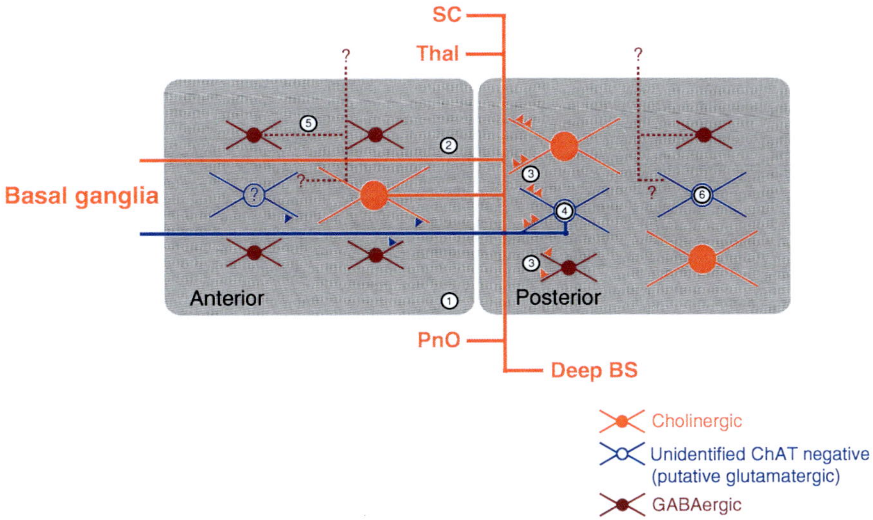

Fig. 1 Model of microcircuits in the PPN and future directions. Recent evidence obtained from the characterization of the PPN at the level of individual neurons has shown significant differences in their topography, neurochemical properties and/or connectivity. Thus, an updated model of internal connectivity in the PPN involves, first, cholinergic neurons that emit long-range axon collaterals travelling to many of the known targets of the PPN (e.g., *Thal* thalamus, *SC* superior colliculus, *PnO* nucleus pontis oralis, *BS* brainstem), and which also give rise to many local varicosities that make synaptic contacts (*triangles*) with dendrites in the PPN. Second, non-cholinergic neurons that seem to project preferentially, although not exclusively, to some of the basal ganglia nuclei, showing a minor degree of connectivity. Nevertheless, the following are some of the essential questions that still remain to be elucidated (1) whether the topographical differences between anterior and posterior PPN, in terms of neurochemistry and behaviour, also reflect distinct physiological properties and patterns of connectivity; (2) whether cholinergic neurons have a similar pattern of connectivity in the anterior, compared to the posterior, PPN; (3) definition of the post-synaptic targets of the local collaterals of cholinergic neurons (i.e., cholinergic, glutamatergic or GABAergic dendrites); (4) definition of the neurochemical nature of the neurons projecting to the basal ganglia (there are some indicators suggesting a glutamatergic nature); (5) definition of the pattern of innervation of GABAergic neurons and (6) whether subtypes of non-cholinergic neurons innervate targets other than the basal ganglia

tification of GABAergic interneurons. Although the evidence for GABAergic projection neurons is clear (Bevan and Bolam 1995), the evidence for GABAergic interneurons is rather weak as it only relies on the lack of labelling for retrograde tracers in GABAergic neurons identified by *in situ* hybridization (Ford et al. 1995). Even though the authors performed an outstanding study to identify the GABAergic neuronal population in the PPN, they could only hypothesize from their results the *possibility* that these neurons are indeed interneurons. Last, but not least, recent results of Winn and colleagues show functional differences between different regions in the PPN (Alderson et al. 2008), which may be underpinned by the differences in the distribution of different neuronal types (for a review see Mena-Segovia et al. 2008a).

In summary, PPN should be regarded as an heterogeneous structure, highly interconnected with different neuronal systems and capable of regulating its own activity. Future work should focus on elucidating the mechanisms for integration of the multiple synaptic inputs from different areas of the brain.

Note

Since the submission of this manuscript, two relevant papers characterizing non-cholinergic neurons by *in situ* hybridization have been published (Mena-Segovia et al., 2009; Wang and Morales 2009). One of them shows that GABAergic neurons account for twice the number of cholinergic neurons and are topographically distributed (Mena-Segovia et al., 2009), whereas the other one shows very low levels of co-expression of GABAergic or glutamatergic markers in cholinergic neurons (Wang and Morales 2009).

References

Alderson HL, Latimer MP and Winn P (2008) A functional dissociation of the anterior and posterior pedunculopontine tegmental nucleus: Excitotoxic lesions have differential effects on locomotion and the response to nicotine. Brain Struct Funct 213: 247–253.

Androulidakis AG, Mazzone P, Litvak V, Penny W, Dileone M, Doyle Gaynor LM, Tisch S, Di Lazzaro V and Brown P (2008) Oscillatory activity in the pedunculopontine area of patients with Parkinson's disease. Exp Neurol 211: 59–66.

Bevan MD and Bolam JP (1995) Cholinergic, GABAergic, and glutamate-enriched inputs from the mesopontine tegmentum to the subthalamic nucleus in the rat. J Neurosci 15: 7105–7120.

Ford B, Holmes CJ, Mainville L and Jones BE (1995) GABAergic neurons in the rat pontomesencephalic tegmentum: Codistribution with cholinergic and other tegmental neurons projecting to the posterior lateral hypothalamus. J Comp Neurol 363: 177–196.

Garcia-Rill E, Heister DS, Ye M, Charlesworth A and Hayar A (2007) Electrical coupling: Novel mechanism for sleep-wake control. Sleep 30: 1405–1414.

Grillner S (2006) Biological pattern generation: The cellular and computational logic of networks in motion. Neuron 52: 751–766.

Grillner S, Wallen P, Saitoh K, Kozlov A and Robertson B (2008) Neural bases of goal-directed locomotion in vertebrates – an overview. Brain Res Rev 57: 2–12.

Honda T and Semba K (1995) An ultrastructural study of cholinergic and non-cholinergic neurons in the laterodorsal and pedunculopontine tegmental nuclei in the rat. Neuroscience 68: 837–853.

Lavoie B and Parent A (1994) Pedunculopontine nucleus in the squirrel monkey: Distribution of cholinergic and monoaminergic neurons in the mesopontine tegmentum with evidence for the presence of glutamate in cholinergic neurons. J Comp Neurol 344: 190–209.

Leszkowicz E, Kusmierczak M, Matulewicz P and Trojniar W (2007) Modulation of hippocampal theta rhythm by the opioid system of the pedunculopontine tegmental nucleus. Acta Neurobiol Exp (Warshaw) 67: 447–460.

Lorincz ML, Crunelli V and Hughes SW (2008) Cellular dynamics of cholinergically induced alpha (8–13 Hz) rhythms in sensory thalamic nuclei in vitro. J Neurosci 28: 660–671.

Manaye KF, Zweig R, Wu D, Hersh LB, De Lacalle S, Saper CB and German DC (1999) Quantification of cholinergic and select non-cholinergic mesopontine neuronal populations in the human brain. Neuroscience 89: 759–770.

Mena-Segovia J, Winn P and Bolam JP (2008a) Cholinergic modulation of midbrain dopaminergic systems. Brain Res Rev 58: 265–271

Mena-Segovia J, Sims HM, Magill PJ and Bolam JP (2008b) Cholinergic brainstem neurons modulate cortical gamma activity during slow oscillations. J Physiol 586: 2947–2960

Mena-Segovia J, Micklem BR, Nair-Roberts RG, Ungless MA and Bolam JP (2009) GABAergic neuron distribution in the pedunculopontine nucleus defines functional subterritories. J Comp Neurol 515: 397–408.

Pan WX and Hyland BI (2005) Pedunculopontine tegmental nucleus controls conditioned responses of midbrain dopamine neurons in behaving rats. J Neuroscience 25: 4725–4732.

Rye DB, Saper CB, Lee HJ and Wainer BH (1987) Pedunculopontine tegmental nucleus of the rat: Cytoarchitecture, cytochemistry, and some extrapyramidal connections of the mesopontine tegmentum. J Comp Neurol 259: 483–528.

Steininger TL, Wainer BH and Rye DB (1997) Ultrastructural study of cholinergic and noncholinergic neurons in the pars compacta of the rat pedunculopontine tegmental nucleus. J Comp Neurol 382: 285–301.

Steriade M, Dossi RC, Pare D and Oakson G (1991) Fast oscillations (20–40 Hz) in thalamocortical systems and their potentiation by mesopontine cholinergic nuclei in the cat. Proc Natl Acad Sci USA 88: 4396–4400.

Wang HL and Morales M (2009) Pedunculopontine and laterodorsal tegmental nuclei contain distinct populations of cholinergic, glutamatergic and GABAergic neurons in the rat. Eur J Neurosci 29: 340–358.

Winn P (2006) How best to consider the structure and function of the pedunculopontine tegmental nucleus: Evidence from animal studies. J Neurol Sci 248: 234–250.

Part II
Basal Ganglia Model Studies

The Effects of Dopaminergic Modulation on Afferent Input Integration in the Ventral Striatal Medium Spiny Neuron

John A. Wolf, Jason T. Moyer, and Leif H. Finkel

Abstract Utilizing a highly detailed model of the medium spiny neuron and a small network of these cells, in this chapter we analyze the processing of synaptic inputs by the striatum and the modulation of this processing via D1 and D2 activation. Computational models such as the one presented here allow for an examination of the cell or network's response of an identical set of inputs under various sets of intrinsic and synaptic modulations. In the single cell model, D1 modulation leads to an increase in the output of the cell at all input frequencies, whereas D2 modulation led to a significant reduction at all of the synaptic input frequencies. These results are generally supportive of the ideas of the Albin-DeLong model, which originally suggested that there is segregation of the indirect and direct pathways in the dorsal striatum, and that D1 is excitatory and D2 is inhibitory. What has previously not been fully appreciated is the contribution and interaction of the synaptic modulations and the intrinsic modulations via DA to elicit this effect. Our model indicates that changes in the NMDA:AMPA ratio during pathological states may have an impact on normal synaptic input processing, and that lateral inhibition may have a greater effect at the network level than assumed from recent in vitro results. Elucidation of the role of dopaminergic modulation on network level interactions will play a crucial role in our understanding of information processing in the striatum.

1 Introduction

The striatum is the input structure of the basal ganglia, through which afferent input from many of the cortical and limbic regions is integrated and passed on to various output structures. The predominant cell in the striatum is the GABAergic medium

J.A. Wolf (✉), J.T. Moyer, and L.H. Finkel
Department of Neuroscience, Room 240 Skirkanich Hall, 210 S. 33rd Street, Philadelphia, PA 19104, USA
e-mail: wolfjo@upenn.edu

H.J. Groenewegen et al. (eds.), *The Basal Ganglia IX*, Advances in Behavioral Biology 58, DOI 10.1007/978-1-4419-0340-2_14,

spiny projection neuron (MSP), comprising over 95% of the striatal cells in the rodent. The processing of afferent input by these MSPs has been the focus of numerous investigations, as is the role of dopaminergic (DA) modulation in affecting both the integration of inputs and the output of the MSPs. Since these cells have a bimodal membrane potential under anesthesia or in vitro, one of the critical questions about them has been whether they can maintain their depolarized state via intrinsic mechanisms without any further synaptic input. With a modeling approach, we have addressed many of these aspects by looking not only at modeled current injection, but also at how much synaptic input is required to keep the cell in the depolarized state. We have also examined the questions of whether these cells are intrinsically bistable with or without DA modulation, and how changes in the ratios of the glutamatergic conductances of the cell affect its ability to entrain to afferent input.

2 Striatal Anatomy

The striatum includes the structures of the caudate, putamen, and nucleus accumbens (Nacb), and is a subset of the subcortical interconnected nuclei that make up the basal ganglia (Wilson 1998). In mammals, the Nacb is a part of the ventral striatum (VS), which is comprised of the Nacb and the olfactory tubercle. This division of the striatum into dorsal and ventral sections has been challenged recently, as there are few neurochemical or structural boundaries that conform to a single division (Voorn et al. 2004). Generally, however, the dorsal striatum processes more input from the motor regions of the cortex, while the ventral striatum processes input predominantly from the limbic and cognitive areas of the brain (Alexander et al. 1986). Disorders localized to the dorsal striatum predominantly (but not exclusively) lead to motoric dysfunction, while pathology in the ventral striatum is hypothesized to disrupt more cognitive aspects of brain function (Graybiel 1995).

2.1 *Anatomy of the Nucleus Accumbens*

The Nacb receives inputs from all the major limbic and cognitive areas of the brain, as well as the thalamus (Finch 1996) (Fig. 1). These inputs overlap in various fashions throughout the Nacb, with their relative weights changing throughout not only the dorsal–ventral axis, but also the anterior-posterior axis as well (Groenewegen et al. 1999). One prominent area of overlap in the rat is of hippocampal and prefrontal cortical afferents, each of which provides a major input into the Nacb (Lopes da Silva et al. 1984; Groenewegen et al. 1987; Vertes 2004). Based on the arrangement of inputs and a number of functional differences, the Nacb has been divided into two sections – the core, which is crossed by the anterior commissure, and the shell, the more ventromedial portion of the Nacb that surrounds the core (Groenewegen et al. 1999). These areas may process different information based on their anatomical

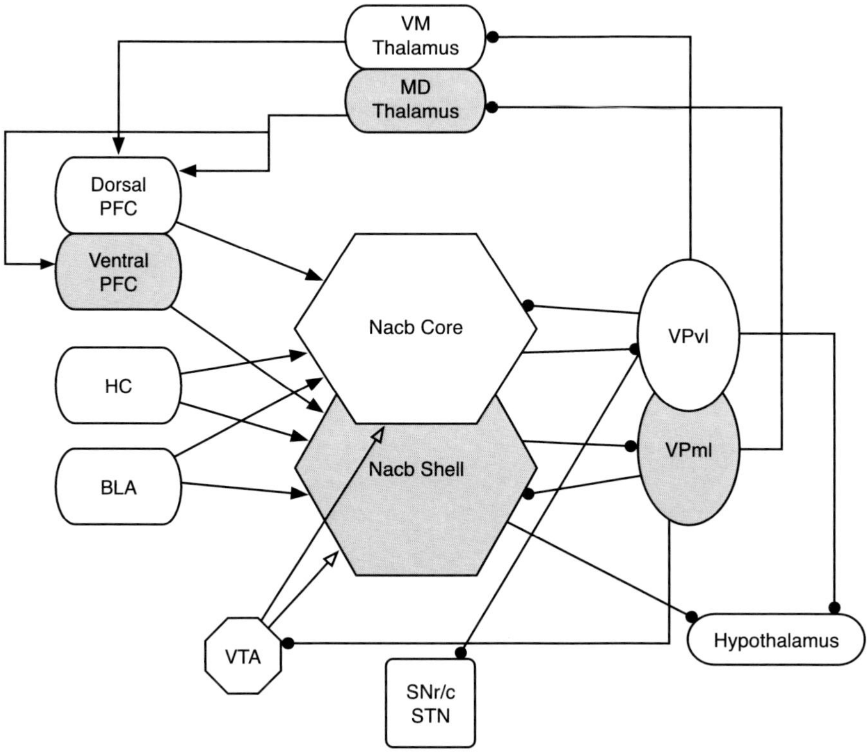

Fig. 1 Afferent information flow to the Nacb core and shell and their outputs. The more dorsal cortico-striatal-thalamocortical loop (*white*) goes through the core, while the more ventral pathway is through the shell (*gray*). *Black arrows* – glutamatergic, *open arrows* – neuromodulatory, *closed circles* – GABAergic. Abbreviations: *PFC* prefrontal cortex, *HC* hippocampus, *BLA* basolateral amygdala, *VTA* ventral tegmental area, *SNr/c* substantia nigra reticulata or compacta, *STN* subthalamic nucleus, *VP* ventral pallidum, *vl* ventrolateral, *ml* mediolateral, *vm* ventromedial, and *md* mediodorsal

inputs, but have been demonstrated to communicate with each other as well (Voorn et al. 2004).

The predominant cell in the striatum is the GABAergic MSP, so called for the large number of spines on each dendrite. The other 3–5% of cells in the striatum are GABAergic and cholinergic interneurons, each of which plays an important role in striatal function (Wilson 1998; Wang et al. 2006). While each MSP cell has a small soma (10–15 μm), it has a large dendritic structure (200–300 μm) and a branching axonal segment. The branching axonal arbor can be larger than the dendritic arbor, with one portion of the axonal segment projecting to the efferent structures.

Medium spiny cells from the Nacb project predominantly to the ventral pallidum, with a subset of MSP cells projecting to the lateral hypothalamus (Heimer et al. 1991). The information flow to the ventral pallidum, also a predominantly

GABAergic inhibitory structure, leads to a release of inhibition of the thalamic nuclei to which they project (O' Donnell et al. 1997). These nuclei in turn project back to the prefrontal areas from which some of the afferent Nacb input arose, thus closing a long cortico-striatal-cortical loop (Fig. 1). This corticostriatal feedback loop has been proposed as one of the dominant cognitive pathways in the brain (Lawrence et al. 2000; Houk 2005). The output to the LH has been linked to feeding and other forms of behavioral output via release of inhibition (Kelley 2004).

3 Functional Role of MSP Cells and DA Modulation

A tremendous amount of research has been devoted to deciphering what the neurons of the striatum, and in particular the Nacb, encode from their afferent input (for review, see Pennartz et al. 1994). As reward-related plasticity is at least partially mediated in the Nacb, cellular activity related to both natural (food) and drug rewards has been extensively investigated (Carelli 2002; Kelley 2004). It has been demonstrated with a number of paradigms that appetitive behavior and reinforcement are predominantly encoded in the ventral portion of the striatum (Kelley 2004). Some of the strongest evidence supports the hypothesis that the Nacb encodes the motivational significance of cues, as well as the association of a cue with aversive and rewarding outcomes. Neurons in the Nacb will respond to primary rewarding or aversive stimuli, and then develop responses to predictive cues associated with them (Roitman et al. 2005). It has also been demonstrated that cells in the dorsal striatum appear to respond to reversal learning prior to the prefrontal cortex (PFC), giving support to the primacy of the striatal responses in guiding reward-related behavior (Pasupathy and Miller 2005).

Along with their location at the interface of the limbic system and motor output, one of the most intriguing aspects of this area is the large innervation of the ventral striatum by projections from the dopaminergic (DA) nuclei of the ventral tegmental area (VTA) of the midbrain (Mogenson et al. 1980; Swanson 1982). Phasic DA signaling from these afferent areas has been proposed to be an error prediction signal about rewards, with theorists suggesting that this represents a signal that can be utilized for a type of temporal delay learning (Schultz and Dickinson 2000; Daw and Doya 2006). Recent findings from Goto and Grace (2005) have also demonstrated the possible role of DA modulation on the balance of hippocampus (HC) and PFC inputs and synaptic plasticity in the ventral striatum. They found that the tonic DA levels appeared to selectively modulate the PFC inputs via D2 receptors, while the phasic DA facilitated the HC input via D1 receptors, increasing HC control of the Nacb (Goto and Grace 2005). Activation of D2 receptors led to the suppression of the PFC inputs and affected the set-shifting behaviors tested experimentally. DA was therefore suggested to maintain the balance of the limbic and cortical input to the Nacb.

3.1 Intrinsic Properties and Membrane Behavior of MSP Neurons

Among the most intriguing aspects of the MSP neurons of the Nacb are their intrinsic membrane properties and their behavior in response to synaptic input. These cells have a large dominant inwardly rectifying potassium current with a low reversal potential (–90 mV) that is open at rest (Uchimura et al. 1989), and lead to a very low resting membrane potential (~ –85 mV) (O' Donnell and Grace 1993). Since this current is inwardly rectifying, it is open fully at or below the K+ reversal potential but shuts down as the membrane potential rises above the K+ reversal potential. These kinetics are due to a blockade of the ionic current by intracellular Mg^{2+} or polyamines as the membrane depolarizes (Lu 2004). When the MSP cells are depolarized, another set of potassium currents dominates the behavior of the membrane. These are the A-type potassium currents, of which there are two in the MSP cells. One is a fast, noninactivating current (K_{Af}) and the other is a slower inactivating current (K_{As}). These currents begin to open at depolarized membrane potentials prior to spike threshold and remain open during repetitive firing.

These potassium currents largely mediate two defining electrophysiological characteristics of the MSP cells – a depolarizing ramp preceding the first spike in response to current injection and a bimodal membrane potential in response to sustained synaptic input. Potassium inward-rectifying (K_{IR}) currents act to clamp the Vm near resting level but close rapidly once the membrane depolarizes, while K_{As} opposes depolarizing inputs near firing threshold, delaying the generation of the first action potential in a train and leading to a long depolarizing ramp in response to current pulses. The transitions between rest, which has been termed the "down" state, and the state just prior to action potential threshold, which has been termed the "up" state, lead to the characteristic bimodality of the Vm under certain input conditions. This bimodality has also been termed "bistability" due to the apparent inability of the cell to remain in between these two states. However, the recordings that demonstrate this effect in the cell are all performed either under anesthesia or in slices in vitro. Anesthetics tend to synchronize the afferents of the striatum, and could therefore account for a large portion of the changes in afferent input that lead to these state transitions (Callaway and Henriksen 1992; Contreras et al. 1997; Mahon et al. 2001). Recordings in awake cats have also shown that throughout the neocortex and thalamus cells remain depolarized during brain activated states (waking and REM sleep) without transitions to the down state, further establishing the notion that up and down states (or the slow oscillation as it was first described by Steriade et al. 1993) are characteristic of slow wave sleep and anesthesia. Recently, an awake head-fixed intracellular rat preparation was utilized to demonstrate that MSP cell Vm rarely enters the "down" state in the awake animal, but that under slow-wave sleep conditions the Vm becomes bimodal as in the anesthetized condition (Mahon et al. 2006).

Despite the fact that transitions to the "down" state may not occur in the awake animal, much has been inferred from the transition of these cells between states.

There are many models of how the striatum functions that are dependent on the current MSP membrane state and how fast the cell can integrate inputs at a given time. Some models suggest that the modulation by dopamine (DA) of the currents that dominate the state transitions is one of the primary ways that DA exerts its effects in the striatum (Nicola et al. 2000; Murer et al. 2002; Gruber et al. 2003). These models suggest that in the down state DA might help to further filter out uncoordinated inputs, while the up-state DA may make the cell more prone to fire in response to less coordinated input. In this manner the cell might switch from being a coincidence detector to an integrator of relevant information. Several models of striatal/basal ganglia function have also been built on the idea that striatal MSPs are inherently bistable, or become bistable after D1 modulation. Two of these models posit the basal ganglia as an action-selection mechanism, in which striatal MSP bistability enables the cortico-basal ganglionic loop to function as a pattern detector (Beiser and Houk 1998) or enhances the duration and intensity of striatal activity (Gruber et al. 2003).

4 Computational Model

In order to address how the MSP cell integrates inputs under various conditions, we constructed a computational model of the MSP neuron. The 189-compartment Nacb MSP cell was developed using the NEURON simulation environment (Hines and Carnevale 1997), and includes most of the known ionic currents in these cells. The model is the first MSP model to individually include all reported species of calcium and calcium-dependent potassium currents as well as recently updated parameters for several potassium currents. Whenever feasible, channels were taken directly from published reports in accumbens or striatum and utilized in the model with minimal adjustments.

The cell was tuned solely by balancing the maximum conductance levels of intrinsic currents against each other to match in vitro adult rat accumbens current injection data. Briefly, the cell has two sodium currents – fast (NaF) and persistent (NaP) – and six different potassium currents: inward-rectifying (K_{IR}), slow A-type (K_{As}), fast A-type (K_{Af}), 4-AP resistant-persistent (K_{RP}), small-conductance calcium-dependent (SK), and large-conductance calcium-dependent (BK). We also incorporated six calcium currents: N-, Q-, R-, T-, $Ca_v1.2$ (high-voltage activated) L- and $Ca_v1.3$ (low-voltage activated) L-types (for review, see Catterall 2000).

Due to the lack of knowledge of the true concentration of various calcium channels in the model, we developed two tunings of the model. One model used the prediction of dendritic calcium conductances from the measured dissociated somatic concentrations (Churchill and Macvicar 1998). The second tuning was an attempt to maximize the number and concentration of calcium channels in the model, while maintaining the model's match of the current injection data as closely as possible. The rationale behind this "high calcium" tuning is that these channels are heavily modulated by DA and have therefore been predicted by many to be one of the major ways in which DA can modulate the intrinsic activity of the MSP cell. We therefore wanted to include the maximal amount of current possible to test theories predicting DA modulation of excitability and bistability.

The model receives synaptic input from simulated spike trains via NMDA, AMPA, and GABA-A receptors. Glutamatergic synapses in MSP cells are generally located in spines, which are primarily located in the mid-to-distal dendrites (Wilson 1992). Accordingly, we distributed our glutamatergic synapses in the dendrites only – AMPA and NMDA channels were co-localized one-to-one at each synapse and received the same inputs. AMPA, NMDA, and GABA synaptic strengths were set to match published values (Dalby and Mody 2003; Myme et al. 2003; Nusser et al. 1998). Previous studies have indicated that GABA synapses are located throughout MSP cells in the striatum, but with higher concentration near the soma (Fujiyama et al. 2000), and were modeled as such. Although feed-forward and lateral inhibitory input from GABAergic neurons is likely different from the glutamatergic input, for simplicity the GABAergic input trains were modeled as duplicates of the glutamatergic inputs.

During simulations, each synapse (AMPA/NMDA or GABA) received an independent spike train generated using MATLAB (Mathworks, Natick, MA). State transitions were generated using this method solely by changing the mean frequency of inputs to the cell between 3 Hz/synapse and 7.5 Hz/synapse. The ratio of glutamatergic inputs to GABA inputs was held constant at roughly 1:1 for all simulations (Blackwell et al. 2003). NMDA:AMPA ratios were set by changing the total conductance of each synapse, starting with a baseline of 0.5 (Myme et al. 2003). However, since published values of NMDA:AMPA vary widely (0.5–4; Myme et al. 2003) and may change following dopaminergic modulation (Flores-Hernandez et al. 2002) or cocaine addiction (Thomas et al. 2001), this represented a baseline condition for upwards modulation and should be regarded as a conservative estimate of the normal ratio.

4.1 Current Injection Responses

Results from both of the tunings of the models were generally similar (Moyer et al. 2007). The cell was tuned in each case so that the model closely matched the *I–V* relationship present in the recorded cell. The firing rate in relation to the current injection was also taken into account in the final tunings, as can be seen in the response to current injection in the model. After tuning the model by adjusting maximal current conductance for the channels in each compartment, the model cell closely matched whole cell recordings from an adult rat Nacb slice preparation during multiple current injections (Fig. 2).

DA modulation of the intrinsic properties was modeled by changing the maximal conductance of specific channels using values predicted by the literature after an extensive review (see Moyer et al. 2007). One of the notable conclusions of this review was that although there were areas where the literature showed disagreement

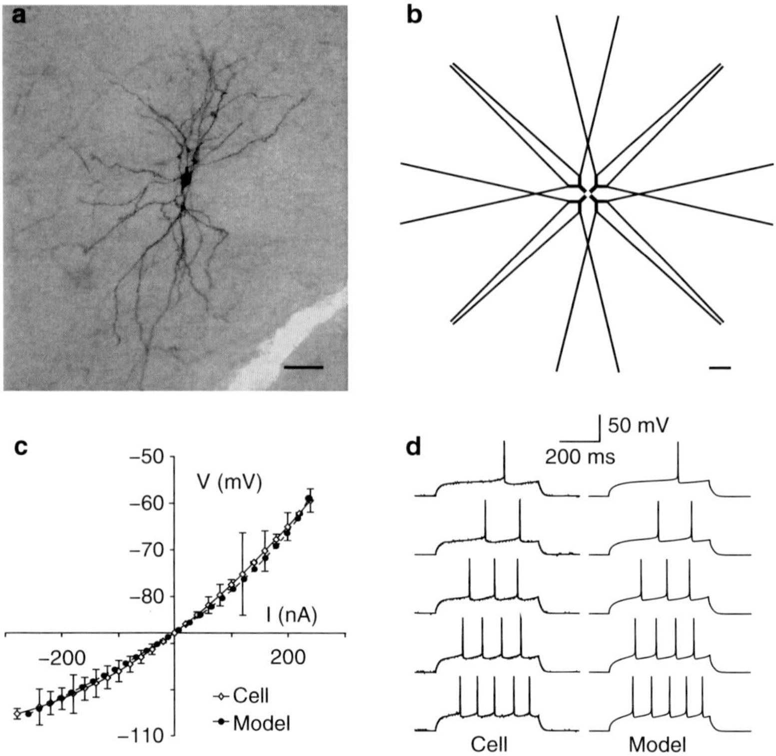

Fig. 2 Comparison of nucleus accumbens MSP cell and model. (**a**) An example of an MSP cell filled with neurobiotin from which whole-cell recordings were performed (bar = 50 μm). (**b**) Morphology of the MSP cell model. Dendritic length and diameter have been increased in order to compensate for spine membrane area (bar = 50 μm). (**c**) Comparison of the current–voltage relationships of the model and the mean of 7 NAcb MSP cells; voltages were recorded 450 ms into a 500 ms current pulse. The model's response (*squares*) is within the SD of the cell's responses (*circles*; *bars* = SD). The model cell was matched to the mean resting membrane potential of −87.75 mV. (**d**) In vitro (*left*) and model (*right*) response to five current injections resulting in spike frequencies of 2–10 Hz [modified from Wolf et al. (2005), Copyright 2005 by the Society for Neuroscience]

about the extent to which a particular channel modulated the MSP cell properties, there was a surprising amount of agreement about which channels were modulated and in what direction D1 or D2 modulation exerted its effects on the conductance. Briefly, the channels modulated by D1 in the model were a 5% reduction in the NaF current, a 50% reduction in the CaP and CaQ maximal conductances, an 80% reduction of CaN, a 10-mV negative shift in the Cav1.3 activation curve (leading to earlier and greater activation), a doubling of the Cav1.2 conductance, and a 25% increase in the K_{IR} current. The intrinsic properties modulated for the D2 expressing cells were modeled by increasing the Na+ current by 10%, decreasing the CaV1.3 by 25%, and increasing the K_{As} conductance by 10%.

When DA modulation was applied to the cell, the response of the cell changed, but in a manner that was not predicted by many of the standard models of DA modulation of MSP cells (Fig. 3). Under current injection, when the cell was modulated via D1 there was a change in the slope of the response. Therefore, at some injection

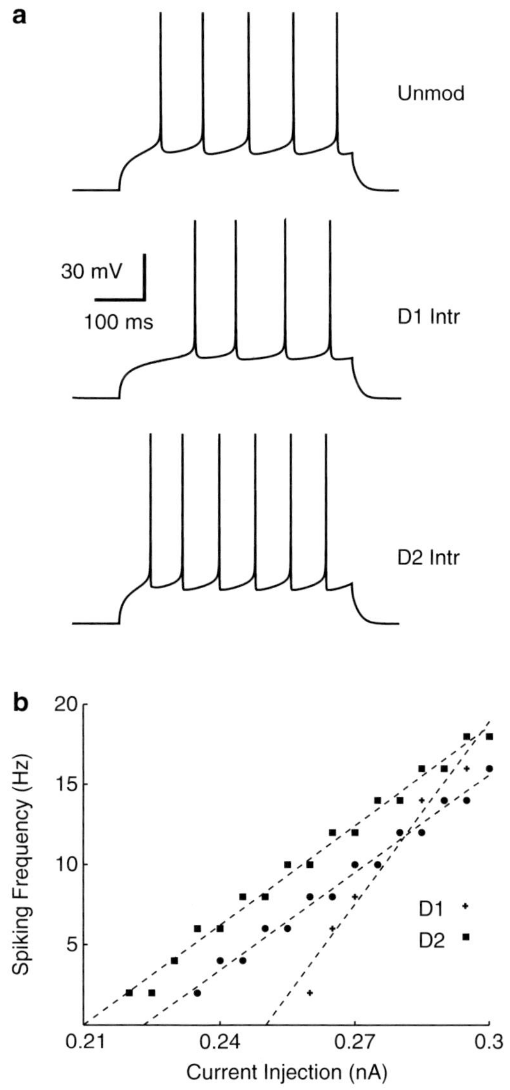

Fig. 3 (**a**) Model's response to 0.271-nA current injection in unmodulated state, and following D1R-mediated modulation and D2R-mediated modulation of intrinsic channels. (**b**) Spiking frequency vs. current injection for unmodulated (*circles*), D1 intrinsic (*pluses*), and D2 (*squares*) intrinsic mediated modulation of model currents [modified from Moyer et al. (2007), used with permission]

levels there was less of a response to the input, while at greater levels of current injection the D1 modulation increased the response. This was dependent on the tuning of the cell. When D2 modulation was tested, there was a consistent but moderate increase in response to the current injection under all levels of current injection, which is contrary to some published reports. However, this was predominantly dependent on the 10% increase in the Na+ current, which is not a well-defined modulation and for which further in vitro data would be important.

4.2 Synaptic Input Responses

One of the primary motivations to build a biophysically detailed model is to test the comparison of the responses of current injection to that of synaptic inputs, the conditions under which real cells continually operate. This also allows for the visualization of the effects of both intrinsic and synaptic changes on the behavior of the cell under various synaptic input conditions, including oscillatory input and the neuromodulation of synaptic currents. One of the most powerful aspects of this technique is that it allows for the examination of the response to the exact same set of synaptic inputs under various sets of intrinsic modulations.

One of the most striking aspects of the response to a set of inputs designed to drive the cell into and out of the "up" state was the realization that very few inputs were required at any given time in order for the cell to reach the up state, and then to keep it there to fire action potentials. Prior to running these experiments, we presumed that since there were so many spines in these cells, and such a strong K_{IR} expression throughout the cell, that many of the synapses would need to be active in order to bring the cell to the "up" state and to fire action potentials. However, when inputs of approximately 1,000 Hz in total and above were fed into the cell, it easily transitioned to the up state and began firing (Fig. 4). This suggests that only 1% (~100) of the 10–15,000 spines of an MSP cell need to be active if the afferent cells are firing in the theta (4–12 Hz) frequency band. If the cell can be fired so easily by a small subsection of its inputs, this suggests that there are many ensembles of cortical input that can potentially be represented by the cell (see below).

Another interesting aspect of the visualization of the model in response to the synaptic input was the individual responses of the dendrites to these afferent inputs. We assumed that many of the dendrites would have to be active and depolarized in order for the soma to make the transition to the depolarized state, but it was readily apparent from observing the dendritic membrane potentials that only 25% of the dendrites had to be depolarized for the soma to make this transition. During the transition, the rest of the dendrites then followed the soma in unison to the depolarized state, at which point they began to move independently in voltage space again (Wolf et al. 2005).

The NMDA and AMPA currents dominated different aspects of the transitions between the "up" and "down" states of the model cell's Vm. Due to the differences in the NMDA and AMPA current kinetics, the AMPA currents lead the cell into the

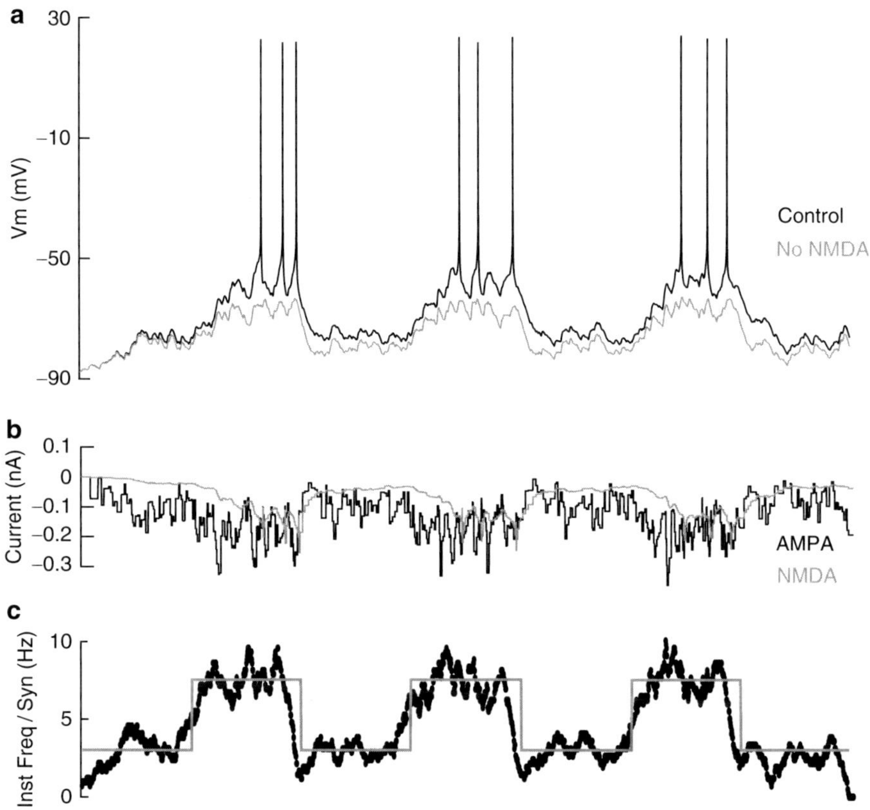

Fig. 4 Response of the model to changes in synaptic input and influence of different synaptic currents. GABAergic and glutamatergic synaptic input to the model was simulated as independent spike trains distributed throughout the cell. State transitions in the model Vm (**a**; *black trace*) were generated by changing the mean frequency of synaptic input from 504 Hz (3 Hz/synapse) to 1,260 Hz (7.5 Hz/synapse) (**c**; *gray line*). Blockade of NMDA currents prevents the model from fully reaching the up state (**a**; *gray trace*). Synaptic input patterns are the same for both control and no-NMDA cases. (**b**) Whole cell synaptic currents underlying the model's activity: AMPA dominates the transition into the up state. NMDA lags the AMPA current and is highly prominent in the transition to the down state. (**c**) Mean instantaneous frequency per synapse is plotted over the simulation time (*gray*). A sliding window filter (50 ms wide) of the instantaneous frequency at all synapses combined is plotted in *black*. Note that the somatic membrane potential matches this in many respects, with a slight delay for dendritic processing of the inputs [modified from Wolf et al. (2005), Copyright 2005 by the Society for Neuroscience]

depolarized state, while the long time constant of the off kinetics of the NMDA current (320 ms at 22°C) are dominant in the transition to the hyperpolarized state. The necessity of the NMDA current to the maintenance of the depolarized state is evident when removed (Fig. 4), which is also apparent in vitro (Vergara et al. 2003). The balance of the NMDA and AMPA currents is also crucial to the ability to entrain properly to oscillatory input (see below). The GABA-A current appeared to have very little role in these transitions. This is probably due to the fact that the GABA synaptic inputs were not independently modeled, but were copies of the excitatory input sent to different synapses. We are currently investigating the role of the GABAergic inputs on these transitions and oscillatory entrainment when they are explicitly modeled as either feed-forward (from interneurons) or lateral inhibition (from other MSP cells).

The flexibility of the model allows for the examination of the response to synaptic input during the modulation of either the intrinsic properties or the synaptic properties, or a combination of both. D1 modulation of the synaptic inputs was modeled by increasing the conductance of the NMDA receptor by 30%, while D2 modulation reduced the AMPA conductance by 20%. DA modulation of synaptic properties appears to have a more profound effect on MSP function when compared to DA modulation of intrinsic properties alone. When synaptic input to the cell was modeled along with the modulation of just the intrinsic properties (but no synaptic modulations), the results were much the same as the current injection: a change in slope for the D1 modulation, with an overall increase in response to the D2 modulation (Fig. 5a). However, when the synaptic modulations were added along with the intrinsic modulations, there was a significant change in the response with both the D1 and the D2 modulations. The D1 modulation now led to an increase in slope *and* an increase in the output of the cell at all input frequencies. The D2 modulation led to a significant reduction at all of the synaptic input frequencies (Fig. 5b). These results are generally supportive of the ideas of the Albin-DeLong model, which originally suggested that there is segregation of the indirect and direct pathways in the dorsal striatum, and that D1 is excitatory and D2 is inhibitory (Albin et al. 1989). What has previously not been fully appreciated

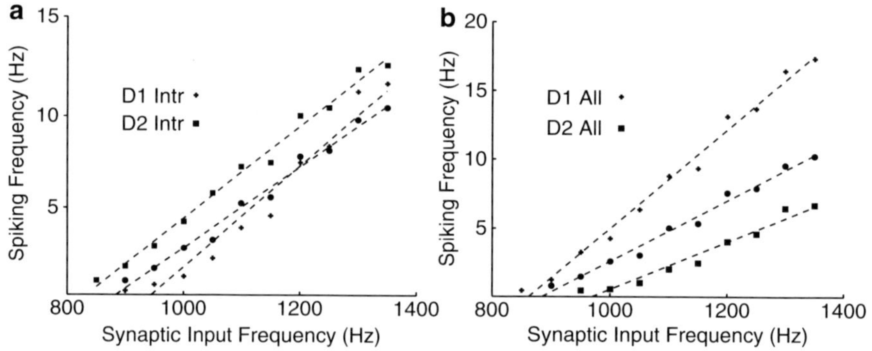

Fig. 5 (**a**) Spiking frequency vs. synaptic input frequency for unmodulated (*circles*), D1R (*pluses* – D1 Intr), and D2R (*squares* – D2 Intr) mediated modulation of intrinsic properties only. (**b**) D1R-modulation of both intrinsic and synaptic properties (D1 All) leads to excitation at all input levels (*pluses*). D2R-modulation of both intrinsic and synaptic properties (D2 All) leads to inhibition at all input levels (*squares*). The effects of synaptic modulations counteract the effects of intrinsic modulations on excitability in both the D1 All and D2 All conditions, and the resulting effects in these conditions agree with previous proposals regarding DA's effects on MSP neuron excitability [modified from (Moyer et al. (2007), used with permission]

is the contribution and interaction of the synaptic modulations and the intrinsic modulations via DA to elicit this effect.

5 Bistability or Bimodality?

One of the most interesting aspects of the MSP neurons and the previous descriptions of their behavior is their tendency to remain in one of the two states in both the anesthetized and the in vitro slice preparation: the "up" state or the "down" state. Because of the bimodal nature of the Vm during the anesthetized preparation, some have proposed that the cells themselves are "bistable." This proposal requires that the cells prefer to remain in the depolarized or hyperpolarized state, but not in between, and that by definition there is a bifurcation in the membrane state. This mechanism has been proposed to arise from the intrinsic properties of the cells, including the inward-rectifying potassium current modulating the down state, and a combination of A-type K+ currents and calcium currents modulating the depolarized state. The NMDA currents in the up-state have also been proposed to mediate the depolarized state due to the long tau-off time constants.

We tested this hypothesis in a number of ways, primarily under synaptic input conditions, to see whether the cells could indeed maintain only two states under realistic input conditions and DA modulation. One of the most straightforward ways to examine this is to linearly increase the number of inputs that the cell receives, and observe the Vm of the cell and the cell output under different modulation conditions and NMDA:AMPA ratios. One of the current hypotheses was that modulation via the D1 receptor might induce this bistable state via an increase in the K_{IR} current in the down state, and modulation of calcium currents in the up state. When we tested this in the full biophysical model, it was readily apparent that the cell did not exhibit nonlinearity under these conditions, much less true bistability. The cell easily maintained the intermediate voltages between states over extended periods of time, suggesting that bistability was not induced via D1 modulation (Fig. 6).

We then examined the extremes of the NMDA:AMPA ratio to see what would be required to get the model to behave in a bistable manner. An NMDA:AMPA ratio of 2.5:1 was required to achieve a nonlinear response to input. Based on the work from other laboratories, this ratio appears to be outside of the physiological range of this ratio in the striatum, and in the model induced a nonphysiological rate of firing in the depolarized state (Myme et al. 2003; Moyer et al. 2007). The results from the modulated and unmodulated version of the model suggest that bistability between membrane states is not the main mechanism via which the striatum is processing input, and that the bimodal nature of the membrane potential under anesthesia is probably due to the synchronization of cortical inputs during the slow-wave oscillation (Steriade et al. 1993). This argument was recently further strengthened by a report that severe state changes are not present in the awake rat, and that the cells maintained a state between the up and down states in this preparation (Mahon et al. 2006).

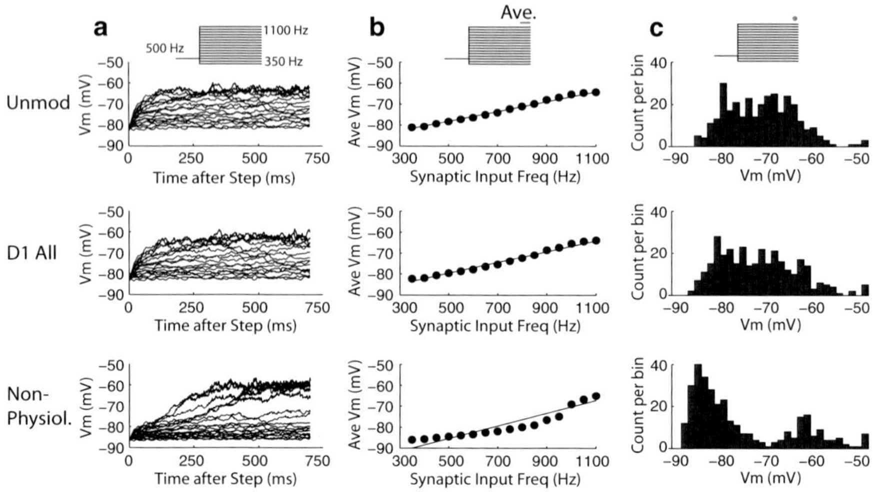

Fig. 6 Effects of D1R-mediated modulation on nonlinearity, hysteresis, and negative slope conductance in the model. (**a**) Averaged response (18 trials) of the model to constant frequency synaptic input in the unmodulated state (Unmod) and following D1R-mediated modulation of both intrinsic and synaptic channels (D1 All). In each trial, the model received synaptic input at 500 Hz for 200 ms, and then the input was stepped to a new frequency and held for 700 ms (see *inset*). The input steps were linearly spaced between 350 and 1,100 Hz. (**b**) Averaged membrane potential (18 trials) vs. synaptic input frequency. Each point represents the average of the last 50 ms of the corresponding trace in Fig. 3a, with a least squares linear fit drawn through the data for comparison. Neither the unmodulated, the D1 Intrinsic, nor the D1 All states exhibited nonlinearity. Upon dramatically increasing the NMDA conductance (500% of baseline), the model exhibited some nonlinearity (nonphysiol), however, this may be a nonphysiological level of NMDA. (**c**) Membrane potential distribution in all trials 670 ms after the step in synaptic input frequency in the unmodulated (Unmod) condition and following D1R-modulation of intrinsic and synaptic channels (D1 All). The distribution was unimodal in these conditions, but upon dramatically increasing NMDA, the distribution became bimodal (nonphysiol) [modified from Moyer et al. (2007), used with permission]

5.1 *Effect of NMDA:AMPA Ratio Changes*

As well as being modulated via D1 receptors, the NMDA current and the ratio of this current to the AMPA current in individual synapses is also modified under a number of other conditions. Cocaine exposure appears to increase the NMDA current temporarily (Thomas et al. 2001), while a decrease in the NMDA current has been proposed as one of the mechanisms underlying schizophrenia (Coyle 2006). Changes in the NMDA:AMPA ratio between different classes of inputs to the MSP cell have also recently been noted, and may be a mechanism whereby one afferent area temporarily exerts greater control over the membrane potential and output of the cell. We therefore investigated with the model how changes in the NMDA:AMPA ratio affected the behavior of the cell in response to synaptic input.

When altering this ratio, we soon discovered that the increase in the NMDA current had a profound impact on how the cell responds to both small and large changes in the input to the cell. When we modeled the changes in the cell's membrane state by using synaptic input ramps, the cell demonstrated an increased tendency to stay in its current state, otherwise known as hysteresis. This hysteresis of the membrane is evident in the increased time that the cell takes to transition to the up state, and then as the input is removed, the increased time that the cell dwells in the depolarized state before returning to the hyperpolarized state (Fig. 7).

While it affects the transitions between states, an even more interesting aspect of how the NMDA:AMPA ratio changes the response to afferent input is that it affects the cell's response to even the smallest changes in oscillatory input. Much of the ventral striatum receives input from the HC, where cells are entrained to the theta oscillation during navigation and exploration. When the synaptic input from these HC cells reaches the MSP cell, it can entrain the firing of the MSP cell to the oscillation in the afferent structure. When a very small number of inputs (10–20% of the total afferent input to the cell at any given time) were synchronized to an afferent theta oscillation, the MSP cell firing was then also entrained to theta (Fig. 8). Considering that only around 1% of the cell's total synapses have to be active to maintain the up state, this means that a surprisingly small 10–20 out of 10–15,000 spines have to be entrained to the afferent oscillation in order to entrain the cell.

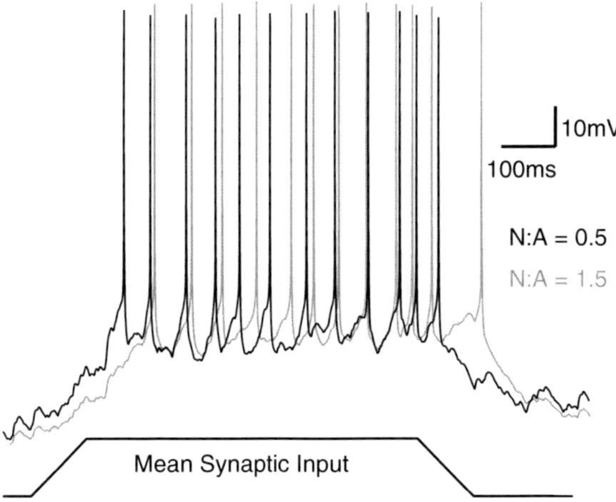

Fig. 7 Increasing the NMDA:AMPA (N:A) ratio contributes to hysteresis in the MSP model membrane potential during synaptic input ramps. An N:A of 1.5 (*gray trace*) results in a delayed and extended up state compared to a 0.5 N:A ratio (*black trace*) in a synaptic input ramp experiment. The same input stimulus consisting of a ramp up, plateau, and ramp down was used for both traces and is indicated at the *bottom* [modified from Wolf et al. (2005), Copyright 2005 by the Society for Neuroscience]

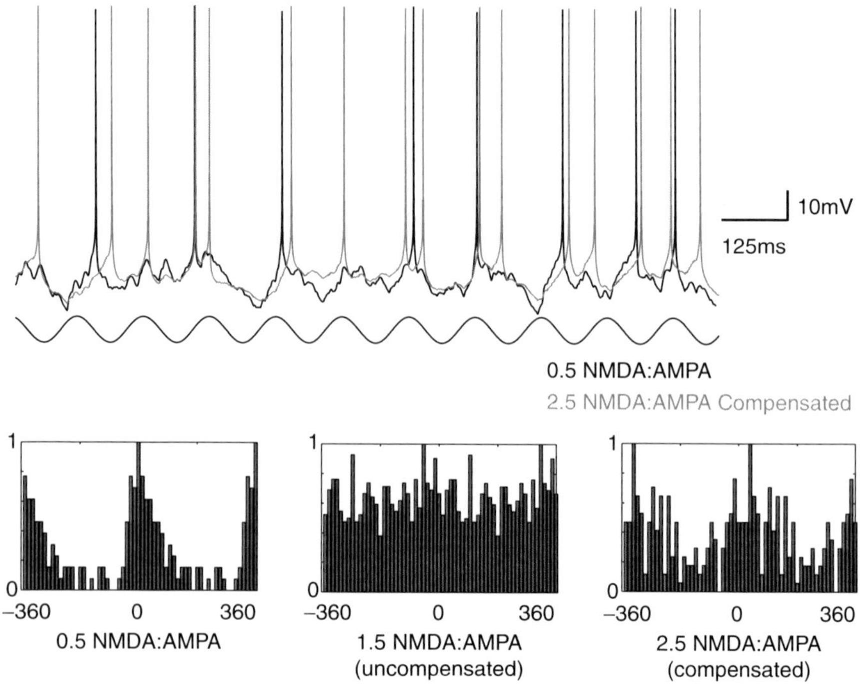

Fig. 8 Increasing the NMDA:AMPA (N:A) ratio degrades the cell's ability to entrain to oscillatory input. (*Top*) At lower values of N:A (0.5), the cell spikes regularly in response to 8-Hz oscillatory input (*black trace*). At a higher N:A ratio (2.5), with weights increased to give approximately one spike per period, the cell does not entrain as regularly to the input. (*Bottom*) PSTHs of MSP spikes relative to peak of sine wave input are plotted. For clarity, the PSTH is centered at 0° and plotted in both directions. At 0.5 N:A ratios, a clear peak is evident at 8-Hz oscillatory input, indicating that the cell is entraining well to the input. Increased values of N:A that are not compensated for spike output lead to loss of entrainment (uncompensated condition). With overall synaptic weight increased to give approximately one spike per cycle, as the N:A ratio is increased, the cell's response to the oscillatory input becomes more variable (2.5 compensated condition). This is evidenced by the loss of a clear peak and spurious spikes throughout the phase cycle [modified from Wolf et al. (2005), Copyright 2005 by the Society for Neuroscience]

Changing the NMDA:AMPA ratio has a profound input on how well the cell can perform the above entrainment. When the NMDA:AMPA ratio is increased above 1.5, the cell can no longer properly entrain the afferent oscillation, even when the total synaptic input current is compensated for this ratio shift (Fig. 8). We and others have speculated that entrainment to afferent oscillations is important to striatal function (Decoteau et al. 2007). This aberrant effect of the NMDA:AMPA ratio on the ability of the cells to entrain to afferent oscillations may be playing a role in the disruption of the normal processes of the striatum under pathological conditions.

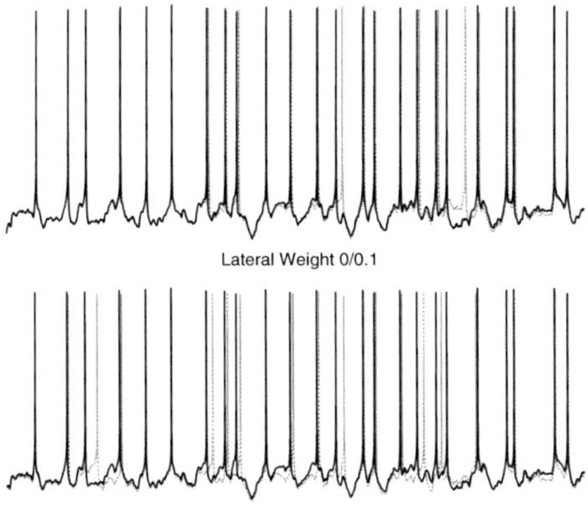

Fig. 9 Example of voltage traces from a cell in the middle of a 100-cell MSP network under two different all-to-all lateral inhibition strengths. In *black* is the voltage when the cell has no lateral inhibition. As the lateral inhibition is turned on, at even 0.1× of the reported conductance value, spike timing can be dramatically affected. Increasing the weight to the reported lateral inhibition conductance strength leads to timing changes in almost all of the action potentials, which may affect such processes as spike-timing-dependent plasticity

5.2 *Local Inhibition*

An important aspect of the striatal network is the local GABAergic inhibition that each MSP cell receives, both feed-forward input from interneurons and lateral inhibition from MSP cells. Unfortunately, the effect on our single cell model from the GABAergic input was relatively insignificant, probably because the inputs mimic the glutamatergic input too closely. The next level of the network we are building will allow us to address a number of important issues related to local GABAergic signaling in the striatum, including the role of feed-forward and lateral inhibition. A preliminary model built to examine lateral inhibition with a 100-cell all-to-all connected network demonstrates that even at very low weights of lateral connectivity, the inhibition has an effect on spike timing. When comparing the response of the cell to the same input train with and without lateral inhibition, even at GABAergic conductance values one-tenth of those predicted by the literature, the output of the cell is clearly different (Fig. 9). This in turn could affect aspects of synaptic plasticity dependent on spike timing (STDP). This suggests that even very small weights of inhibitory connections between MSP cells may be having a profound effect on information processing in the striatum.

6 Afferent Ensembles and Integration

The Nacb can detect the co-occurrence of patterns of input that relate content, emotion, and working memory due to its inputs from afferent areas that represent these aspects of cognition. One of the major aspects of the MSP cell's function is the ability to integrate these inputs faithfully. We have examined both the integration of synaptic input and the DA modulation of this integration in the current model, and demonstrated that the cell is capable of accurately representing windows of synaptic input that range from 10 to 60 ms. This ability to faithfully represent the input stream arriving from many different areas may be one of the hallmark features of the MSP cell's information processing abilities.

The results from the model also suggest that there are an enormous number of afferent ensembles that can be represented in a single cell's synaptic input. We have demonstrated that the MSP neuron can be driven into the depolarized state easily by a small number of inputs. Even using the most minimal calculation of afferent input dimensions, the cell can represent 100 different afferent ensembles of input. If we assume that a cell can participate in more than one ensemble, this number increases dramatically. What role would a cell that can accurately reproduce the inputs it is receiving from multiple afferent input ensembles in its membrane state have in the brain? Afferent ensembles may be the "snapshots" of particular cortical states, and the MSP cell may be particularly good at examining the local features of this larger cortical representation. These units may collectively be acting as a form of dimensionality reducer (Bar–Gad et al. 2003), which may be useful if a particular state needs to be reinforced or labeled, which is one hypothesized role of DA and synaptic plasticity in the striatum.

7 Implications of the Model for Disease States

The striatum appears to be exquisitely sensitive to disruptions of both DA modulation and shifts in the ratios of the synaptic input conductances. Aberrations in these aspects of afferent information processing may have implications for the pathology underlying many disease states. The ventral striatum has often been implicated in schizophrenia because of the large concentration of D2 receptors in this area, at which virtually all of the pharmacological agents that can improve positive schizophrenic symptoms act. These agents may be regulating corticostriatal activity presynaptically, which may help to overcome the hypofrontality associated with the disease. These facts supported the DA hypothesis of schizophrenia, which suggests that an overabundance of DA in the striatal area leads to the symptoms of the disease (Meltzer and Stahl 1976). More recent hypotheses suggest that a larger cortical-level disruption involving primarily a glutamatergic disruption may underlie the disease, with dopaminergic receptor activity as a compensatory reaction to that primary dysfunction (Coyle 2006). A reduction in the amount of NMDA current, as demonstrated above, would have drastic effects on the ability of the cell to actively process inputs. We speculate that a compensatory increase in total current would be necessary to maintain the cells firing in this condition. This in turn would lead to a reduction in the number of afferent inputs that could be appropriately represented by each cell. Any change in the glutamatergic receptors in the striatum may lead to a loss of the ability to properly entrain to afferent oscillations as well.

A combination of hypofrontality and increased DA in the striatum suggests that D2 receptor blockers may be acting to remove suppression of the PFC inputs presynaptically to the striatum (Bamford et al. 2004). This may act against both the hypofrontal nature of the disease, and any potential decrease in input strength from alterations in the glutamatergic receptors.

The Nacb has also been particularly implicated in human addiction and animal models of addiction and relapse. Addiction has recently been described as the pathology of motivation and choice, where the entire pathway of goal-directed behavior undergoes pathological modification (Kalivas and Volkow 2005). Many drugs of abuse function via increasing the amount of DA release or blocking its reuptake. This has led to the prominent hypothesis that drugs of abuse "hijack" the natural reward system of the brain during the addictive process (Kauer and Malenka 2007). Under this model, while DA release is the initiator of plasticity that leads to addiction, it is the glutamatergic interaction between the PFC and the Nacb that is the final common pathway that mediates drug seeking and relapse. Our model indicates that changes in the NMDA:AMPA ratio during addiction may be having an impact on synaptic inputs in such a manner that the drug-related contexts and other stimuli may come to dominate the normal assembly representation that is taking place in the striatum. We suggest that further elucidation of the local network mechanisms in the striatum, and the effects of DA modulation on these interactions, will lead to greater insight into the role that the striatum is playing in the pathology of these devastating diseases.

References

Albin RL, Young AB and Penney JB (1989) The functional anatomy of basal ganglia disorders. Trends Neurosci 12: 366–375.

Alexander GE, DeLong MR and Strick PL (1986) Parallel organization of functionally segregated circuits linking basal ganglia and cortex. Annu Rev Neurosci 9: 357–381.

Bamford NS, Zhang H, Schmitz Y, Wu NP, Cepeda C, Levine MS, Schmauss C, Zakharenko SS, Zablow L and Sulzer D (2004) Heterosynaptic dopamine neurotransmission selects sets of corticostriatal terminals. Neuron 42: 653–663.

Bar-Gad I, Morris G and Bergman H (2003) Information processing, dimensionality reduction and reinforcement learning in the basal ganglia. Prog Neurobiol 71: 439–473.

Beiser DG and Houk JC (1998) Model of cortical-basal ganglionic processing: Encoding the serial order of sensory events. J Neurophysiol 79: 3168–3188.

Blackwell KT, Czubayko U and Plenz D (2003) Quantitative estimate of synaptic inputs to striatal neurons during up and down states in vitro. J Neurosci 23: 9123–9132.

Callaway CW and Henriksen SJ (1992) Neuronal firing in the nucleus accumbens is associated with the level of cortical arousal. Neuroscience 51: 547–553.

Carelli RM (2002) Nucleus accumbens cell firing during goal-directed behaviors for cocaine vs. 'natural' reinforcement. Physiol Behav 76: 379–387.

Catterall WA (2000) Structure and regulation of voltage-gated Ca2+ channels. Ann Rev Cell Dev Biol 16: 521–555.

Churchill D and Macvicar BA (1998) Biophysical and pharmacological characterization of voltage-dependent Ca2+ channels in neurons isolated from rat nucleus accumbens. J Neurophysiol 79: 635–647.

Contreras D, Durmuller N and Steriade M (1997) Plateau potentials in cat neocortical association cells in vivo: Synaptic control of dendritic excitability. Eur J Neurosci 9: 2588–2595.
Coyle JT (2006) Glutamate and schizophrenia: Beyond the dopamine hypothesis. Cell Mol Neurobiol 26: 365–384.
Dalby NO and Mody I (2003) Activation of NMDA receptors in rat dentate gyrus granule cells by spontaneous and evoked transmitter release. J Neurophysiol 90: 786–797.
Daw ND and Doya K (2006) The computational neurobiology of learning and reward. Curr Opin Neurobiol 16: 199–204.
Decoteau WE, Thorn C, Gibson DJ, Courtemanche R, Mitra P, Kubota Y and Graybiel AM (2007) Learning-related coordination of striatal and hippocampal theta rhythms during acquisition of a procedural maze task. Proc Natl Acad Sci USA 104: 5644–5649.
Finch DM (1996) Neurophysiology of converging synaptic inputs from the rat prefrontal cortex, amygdala, midline thalamus, and hippocampal formation onto single neurons of the caudate/putamen and nucleus accumbens. Hippocampus 6: 495–512.
Flores-Hernandez J, Cepeda C, Hernandez-Echeagaray E, Calvert CR, Jokel ES, Fienberg AA, Greengard P and Levine MS (2002) Dopamine enhancement of NMDA currents in dissociated medium-sized striatal neurons: Role of D1 receptors and DARPP-32. J Neurophysiol 88: 3010–3020.
Fujiyama F, Fritschy JM, Stephenson FA and Bolam JP (2000) Synaptic localization of GABA(A) receptor subunits in the striatum of the rat. J Comp Neurol 416: 158–172.
Goto Y and Grace AA (2005) Dopaminergic modulation of limbic and cortical drive of nucleus accumbens in goal-directed behavior. Nat Neurosci 8: 805–812.
Graybiel AM (1995) The basal ganglia. Trends Neurosci 18: 60–62.
Groenewegen HJ, Vermeulen-Van der Zee E, te Kortschot A and Witter MP (1987) Organization of the projections from the subiculum to the ventral striatum in the rat. A study using anterograde transport of Phaseolus vulgaris leucoagglutinin. Neuroscience 23: 103–120.
Groenewegen HJ, Wright CI, Beijer AV and Voorn P (1999) Convergence and segregation of ventral striatal inputs and outputs. Ann NY Acad Sci 877: 49–63.
Gruber AJ, Solla SA, Surmeier DJ and Houk JC (2003) Modulation of striatal single units by expected reward: A spiny neuron model displaying dopamine-induced bistability. J Neurophysiol 90: 1095–1114.
Heimer L, Zahm DS, Churchill L, Kalivas PW and Wohltmann C (1991) Specificity in the projection patterns of accumbal core and shell in the rat. Neuroscience 41: 89–125.
Hines ML and Carnevale NT (1997) The NEURON simulation environment. Neural Comput 9: 1179–1209.
Houk JC (2005) Agents of the mind. Biol Cybern 92: 427–437.
Kalivas PW and Volkow ND (2005) The neural basis of addiction: A pathology of motivation and choice. Am J Psychiatry 162: 1403–1413.
Kauer JA and Malenka RC (2007) Synaptic plasticity and addiction. Nat Rev Neurosci 8: 844–858.
Kelley AE (2004) Memory and addiction: Shared neural circuitry and molecular mechanisms. Neuron 44: 161–179.
Lawrence AD, Watkins LH, Sahakian BJ, Hodges JR and Robbins TW (2000) Visual object and visuospatial cognition in Huntington' s disease: Implications for information processing in corticostriatal circuits. Brain 123: 1349–1364.
Lopes da Silva FH, Arnolds DE and Neijt HC (1984) A functional link between the limbic cortex and ventral striatum: Physiology of the subiculum accumbens pathway. Exp Brain Res 55: 205–214.
Lu Z (2004) Mechanism of rectification in inward-rectifier K+ channels. Annu Rev Physiol 66: 103–129.
Mahon S, Deniau JM and Charpier S (2001) Relationship between EEG potentials and intracellular activity of striatal and cortico-striatal neurons: An in vivo study under different anesthetics. Cereb Cortex 11: 360–373.

Mahon S, Vautrelle N, Pezard L, Slaght SJ, Deniau JM, Chouvet G and Charpier S (2006) Distinct patterns of striatal medium spiny neuron activity during the natural sleep-wake cycle. J Neurosci 26: 12587–12595.

Meltzer HY and Stahl SM (1976) The dopamine hypothesis of schizophrenia: A review. Schizophr Bull 2: 19–76.

Mogenson GJ, Jones DL and Yim CY (1980) From motivation to action: Functional interface between the limbic system and the motor system. Prog Neurobiol 14: 69–97.

Moyer JT, Wolf JA and Finkel LH (2007) Effects of dopaminergic modulation on the integrative properties of the ventral striatal medium spiny neuron. J Neurophysiol 98: 3731–3748.

Murer MG, Tseng KY, Kasanetz F, Belluscio M and Riquelme LA (2002) Brain oscillations, medium spiny neurons, and dopamine. Cell Mol Neurobiol 22: 611–632.

Myme CI, Sugino K, Turrigiano GG and Nelson SB (2003) The NMDA-to-AMPA ratio at synapses onto layer 2/3 pyramidal neurons is conserved across prefrontal and visual cortices. J Neurophysiol 90: 771–779.

Nicola SM, Surmeier J and Malenka RC (2000) Dopaminergic modulation of neuronal excitability in the striatum and nucleus accumbens. Annu Rev Neurosci 23: 185–215.

Nusser Z, Hajos N, Somogyi P and Mody I (1998) Increased number of synaptic GABA(A) receptors underlies potentiation at hippocampal inhibitory synapses. Nature 395: 172–177.

O' Donnell P and Grace AA (1993) Physiological and morphological properties of accumbens core and shell neurons recorded in vitro. Synapse 13: 135–160.

O' Donnell P, Lavin A, Enquist LW, Grace AA and Card JP (1997) Interconnected parallel circuits between rat nucleus accumbens and thalamus revealed by retrograde transynaptic transport of pseudorabies virus. J Neurosci 17: 2143–2167.

Pasupathy A and Miller EK (2005) Different time courses of learning-related activity in the prefrontal cortex and striatum. Nature 433: 873–876.

Pennartz CM, Groenewegen HJ and Lopes da Silva FH (1994) The nucleus accumbens as a complex of functionally distinct neuronal ensembles: An integration of behavioural, electrophysiological and anatomical data. Prog Neurobiol 42: 719–761.

Roitman MF, Wheeler RA and Carelli RM (2005) Nucleus accumbens neurons are innately tuned for rewarding and aversive taste stimuli, encode their predictors, and are linked to motor output. Neuron 45: 587–597.

Schultz W and Dickinson A (2000) Neuronal coding of prediction errors. Annu Rev Neurosci 23: 473–500.

Steriade M, Contreras D, Curro Dossi R and Nunez A (1993) The slow (<1 Hz) oscillation in reticular thalamic and thalamocortical neurons: Scenario of sleep rhythm generation in interacting thalamic and neocortical networks. J Neurosci 13: 3284–3299.

Swanson LW (1982) The projections of the ventral tegmental area and adjacent regions: A combined fluorescent retrograde tracer and immunofluorescence study in the rat. Brain Res Bull 9: 321–353.

Thomas MJ, Beurrier C, Bonci A and Malenka RC (2001) Long-term depression in the nucleus accumbens: A neural correlate of behavioral sensitization to cocaine. Nat Neurosci 4: 1217–1223.

Uchimura N, Higashi H and Nishi S (1989) Membrane properties and synaptic responses of the guinea pig nucleus accumbens neurons in vitro. J Neurophysiol 61: 769–779.

Vergara R, Rick C, Hernandez-Lopez S, Laville JA, Guzman JN, Galarraga E, Surmeier DJ and Bargas J (2003) Spontaneous voltage oscillations in striatal projection neurons in a rat corticostriatal slice. J Physiol 553: 169–182.

Vertes RP (2004) Differential projections of the infralimbic and prelimbic cortex in the rat. Synapse 51: 32–58.

Voorn P, Vanderschuren LJ, Groenewegen HJ, Robbins TW and Pennartz CM (2004) Putting a spin on the dorsal-ventral divide of the striatum. Trends Neurosci 27: 468–474.

Wang Z, Kai L, Day M, Ronesi J, Yin HH, Ding J, Tkatch T, Lovinger DM and Surmeier DJ (2006) Dopaminergic control of corticostriatal long-term synaptic depression in medium spiny neurons is mediated by cholinergic interneurons. Neuron 50: 443–452.

Wilson CJ (1992) Dendritic morphology, inward rectification, and the functional properties of neostriatal neurons. In: McKenna T (ed) Single Neuron Computation. Academic, New York, NY, pp. 141–171.

Wilson CJ (1998) Basal ganglia. In: Shepherd GM (ed) Synaptic Organization of the Brain. Oxford University Press, Oxford, NY, pp. 329–375.

Wolf JA, Moyer JT, Lazarewicz MT, Contreras D, Benoit-Marand M, O' Donnell P and Finkel LH (2005) NMDA/AMPA ratio impacts state transitions and entrainment to oscillations in a computational model of the nucleus accumbens medium spiny projection neuron. J Neurosci 25: 9080–9095.

A Spiking Neuron Model of the Basal Ganglia Circuitry that Can Generate Behavioral Variability

Osamu Shouno, Johane Takeuchi, and Hiroshi Tsujino

Abstract The cortical-basal ganglia circuitry has been implicated in action selection, action initiation, and generation of behavioral variability. However, underlying mechanisms for these functions still remain unresolved. In this paper, we propose a new spiking neuron model for the basal ganglia circuitry that includes different functions for the direct and indirect pathways: the indirect pathway selects an action to be executed and then the direct pathway determines the timing of the selected action. Computer simulations demonstrate that the basal ganglia circuitry supports action selection and self-timed initiation of actions. The circuitry can also generate trial-by-trial variations in the selection and timing of actions for exploration. Finally, these variations can optimally and independently be tuned by the strength of corticostriatal synapses for exploitation.

1 Introduction

Animals can adaptively shape their behavior to pursue their goals. In a delayed-response task, an animal can choose an appropriate movement and initiate it at the right time in response to external events that inform the animal how and when to react and which rewards to expect (Hollerman et al. 1998; Pasupathy and Miller 2005; Turner and Anderson 2005). To perform these behaviors, animals not only react to stimuli but also flexibly use cognitive resources, such as working memory and/or internal triggers for action, to comply with task requirements (Hollerman et al. 1998; Lee and Assad 2003; Pasupathy and Miller 2005; Turner and Anderson 2005; Maimon and Assad 2006). However, the question remains: how does the brain realize these flexible behaviors?

The basal ganglia are a strong candidate for one of the structures underlying flexible behavior. The striatum, the major input station of the basal ganglia, is

O. Shouno (✉), J. Takeuchi, and H. Tsujino
Honda Research Institute Japan Co., Ltd, 8-1 Honcho, Wako-shi, Saitama 351-0188, Japan
e-mail: shouno@jp.honda-ri.com

H.J. Groenewegen et al. (eds.), *The Basal Ganglia IX*, Advances in Behavioral Biology 58, DOI 10.1007/978-1-4419-0340-2_15, © Springer Science+Business Media, LLC 2009

thought to be important for learning and acting on associations between external/internal events and rewarded actions. Striatal neurons are differentially activated during individual task events (Hollerman et al. 1998), and neuronal activation changes distinctively when specific behavioral tasks are being learned (Tremblay et al. 1998; Pasupathy and Miller 2005). Differential activation of striatal neurons is related to biased choices of movements (Samejima et al. 2005), implying action selection mechanisms downstream of the striatum. Additionally, striatal and pallidal neurons generally have phasic, movement-related activity that is comparable for externally triggered, reactive movements and internally triggered, self-timed movements (Lee and Assad 2003; Turner and Anderson 2005), suggesting common circuits and mechanisms control movement initiation in the basal ganglia. However, how a candidate action is selected and how the action is initiated in the basal ganglia remain open questions. An animal performs the task differently each time in order to learn the appropriate choice and timing of an action by trial and error. This generates both good and bad variations in behavior, which enables good variations to be reinforced. Although the basal ganglia are implicated in the generation of the behavioral variability that is necessary for trial-and-error learning (Barnes et al. 2005; Olveczky et al. 2005; Kao and Brainard 2006; Tumer and Brainard 2007), how the basal ganglia generate variability in behavior still remains unresolved.

In this paper, we propose a new model for the basal ganglia circuitry with different functions for the direct and indirect pathways of the basal ganglia. First, the indirect pathway selects an action to be executed, and then the direct pathway determines the timing of the selected action. The model also assumes that dynamic interactions between the external segment of the globus pallidus (GPe) and subthalamic nucleus (STN) in the indirect pathway provide a source of behavioral variation. We have constructed a spiking neuron model of the basal ganglia circuitry and tested these hypotheses by computer simulations. These simulations demonstrate that the basal ganglia circuitry supports action selection and initiation of self-timed actions. We also find that the network can generate trial-by-trial variations in the selection and timing of action, and these variations can be optimally and independently tuned by the strength of corticostriatal synapses.

2 Computational Hypotheses for Basal Ganglia Function

Our basic ideas on the function of the basal ganglia are as follows:

1. The multiple loops (channels) of the cortical-basal ganglia circuitry are specialized for both selection and initiation of action. The selection of an action and timing of the action are accomplished by computational processes intrinsic to the basal ganglia.
2. Corticostriatal inputs modulate the selection and timing of an action, and learned changes of corticostriatal synaptic strength achieve adaptive tuning of behavior.

In natural settings, a number of events can happen simultaneously. Brain cognitive systems are continuously processing a plethora of information in parallel about multiple features of an event, including the present perception as well as memories. Convergence of corticostriatal projections from different cortical areas indicates that each channel of the basal ganglia coincidentally receives multiple types of information, some of which is important for achieving the goal. Experimental evidence indicates that the basal ganglia learn faster than the frontal cortex (Pasupathy and Miller 2005). Therefore, it is unlikely that incoming corticostriatal information is already selective for a given channel before learning. Rather, striatal neurons must learn significant information that predicts rewarded actions and their timing.

3. The selection of an action and the timing of the selected action are carried out by the indirect and direct pathways, respectively.

Because action selection and action execution are sequential rather than simultaneous processes, we assume that there is one distinct neural substrate for each. Traditionally, the direct pathway is thought to control the initiation of actions through a disinhibition mechanism. Neural activity in the striatum inhibits neurons in the substantia nigra pars reticulata (SNr) and the internal segment of the GP (GPi), thereby reducing their activity below the threshold for initiation of an action. We suggest that the action selection process occurs in the GPe of the indirect pathway, resulting in sustained enhancement (or reduction) of GPe activity that subsequently reduces (or increases) SNr/GPi activity, and thus modulates the threshold level of action initiation. This idea is consistent with the prediction that action selection is realized downstream of the striatum (Samejima et al. 2005) and it is also supported by neural recordings in the GPe and GPi (Arkadir et al. 2004; Pasquereau et al. 2007). This modulation may determine whether striatal medium spiny neuron activity induces an action. Thus, a sustained reduction assists striatal medium spiny neurons in inducing an action even when activity of these neurons is insufficient to induce an action. In contrast, sustained enhancement prevents striatal neural activity from inducing an action, even when their activity should be sufficient to induce an action.

We propose several biologically plausible mechanisms to substantiate these ideas:

(a) A selection mechanism is implemented by mutual inhibition between populations of neurons in the GPe, based on anatomical findings of local axonal collaterals of GABAergic neurons in the GPe (Kita and Kitai 1994; Bevan et al. 1998; Sato et al. 2000).
(b) Because task-related sustained activity is observed in both the GPe and subthalamic nucleus (STN) (Hikosaka et al. 2000; Arkadir et al. 2004), we propose that this task-related, sustained activity depends largely on electrophysiological properties of neurons in the GPe and STN. As shown in physiological experiments and computer simulations, reciprocally connected GPe inhibitory neurons and excitatory STN neurons spontaneously generate self-sustained pacemaker activity in both the GPe and STN (Plenz and Kitai 1999; Terman et al. 2002). Thus, we hypothesize that multiple, interacting GPe–STN loops

dynamically generate chaotic, self-sustained bursting activity. Rebound depolarization of STN neurons is a key mechanism for bursting activity (Plenz and Kitai 1999). Enhanced inhibition from GPe neurons induces rebound depolarization through activity-dependent modulation of GPe-to-STN synapses, via an unknown mechanism (Hallworth and Bevan 2005). Based on these findings, we propose that corticosubthalamic inputs, which convey motivation-related information, excite GPe neurons via excitatory projections from STN neurons. This strengthens GPe-to-STN inhibitory projections and consequently facilitates self-sustained burst activity. Corticosubthalamic inputs are implemented as Poisson spike trains whose mean firing rates are transiently increased. The activity-dependent strengthening of GPe-to-STN inhibitory projections is modeled by short-term facilitation for the sake of simplicity.

(c) We assumed that medium spiny neurons are usually silent even if corticostriatal input exceeds their threshold, because corticostriatal inputs simultaneously excite striatal fast-spiking GABAergic interneurons, which fire more readily than spiny neurons and exert tonic inhibition on medium spiny neurons. A subset of GPe neurons projects to fast-spiking GABAergic interneurons in the striatum (Bevan et al. 1998). We propose that subpopulations of GPe neurons with chaotic self-sustained burst activity disinhibit striatal medium spiny neurons via pallidostriatal projections. Therefore, the indirect pathway controls the temporal sequences of windows through which corticostriatal inputs shape the activation of striatal medium spiny neurons.

3 Simulation Results

Our model was designed for binary selection of a channel and output timing. The model consists of two channels that represent two brain areas, the basal ganglia and the superior colliculus (Fig. 1). Each channel is selective for a distinct action. The basal ganglia network contains both direct and indirect pathways and consists of four nuclei, namely the striatum, GPe, STN, and SNr. In the striatum, there are two types of neurons, medium spiny output neurons (MS) and fast-spiking GABAergic interneurons (FS). Striatal neurons can be subdivided into direct pathway neurons (MSd and FSd) and indirect pathway neurons (MSi and FSi). Each nucleus contains competing neural populations selective for a distinct action, and selectivity is preserved throughout the processing stages. For the sake of simplicity, we did not include the thalamocortical pathway in our model.

It has been proposed that the basal ganglia promote behavioral variability or bias behavioral fluctuations in the optimum direction according to reinforcement signals during trial-and-error learning (Doya and Sejnowski 1995; Barnes et al. 2005; Kao et al. 2005; Olveczky et al. 2005; Tumer and Brainard 2007). We assumed that dopamine- and activity-dependent synaptic plasticity of corticostriatal inputs (Reynolds et al. 2001) modulates the balance between these two opposing basal ganglia effects. A shift from initial “exploration,” in which behavior is variable, to

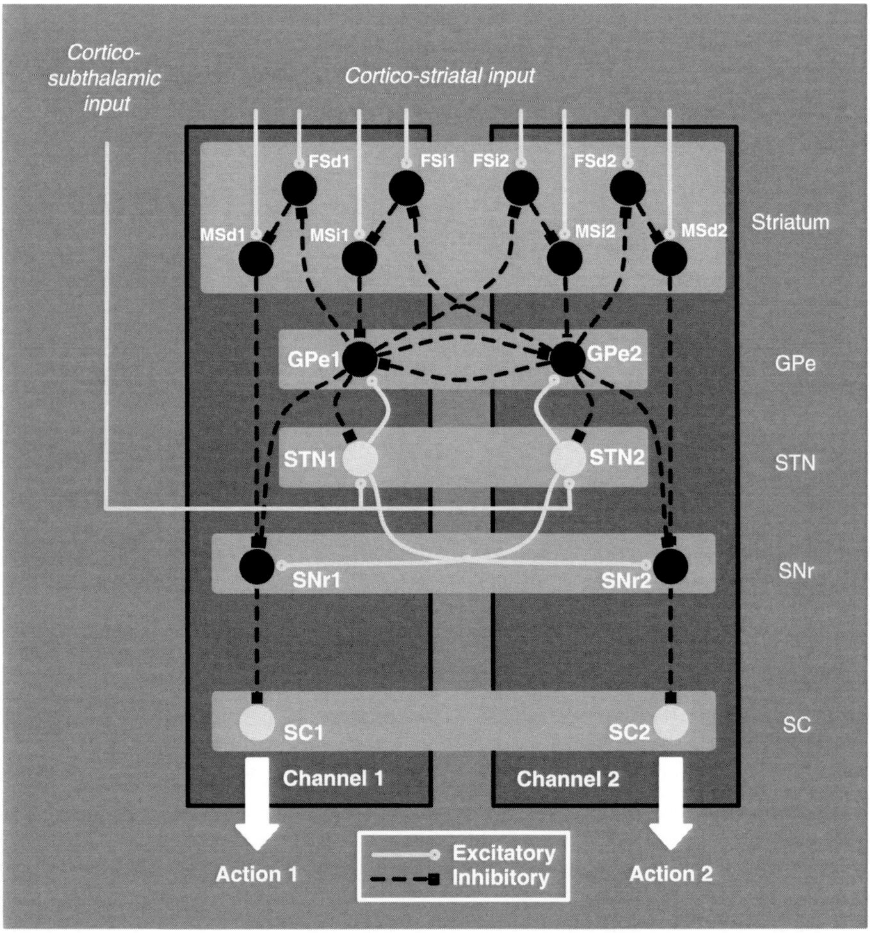

Fig. 1 Schematic model architecture. *Circles* represent populations of neurons (*light gray*, excitatory neurons; *dark gray*, inhibitory neurons). Each type of neuron has characteristic membrane dynamics. Each population is simulated by noisy spiking neurons using NEST simulation software (http://www.nest-initiative.org/)

"exploitation," in which successful behaviors are reproduced consistently, is realized through learning. We examined whether the model retains these required capacities for both phases of learning.

3.1 *Selection and Timing Mechanisms for Exploration in the Basal Ganglia Network*

We first considered the behavior of our model in the early stages of trial-and-error learning. According to the hypothesis that the striatum learns earlier than the cortex, it is difficult to assume that before learning there are spatiotemporal patterns of corticostriatal inputs that are appropriately tailored to accomplish basal ganglia functions. Hence, we first demonstrate a single-trial simulation of the model under conditions where the corticostriatal inputs are not selective for each channel of the basal ganglia and do not change abruptly reflecting immediate external events that trigger actions. The corticostriatal inputs are modeled as Poisson spike trains whose mean firing rates are constant in time and identical between channels (Fig. 2a). We assumed that the corticosubthalamic inputs reflect an increase in motivation. These inputs are also modeled as Poisson spike trains whose mean firing rates are similar for both channels and transiently increased when a task starts. Before a task begins, neurons in the GPe, STN, and SNr exhibit spontaneous firing (Fig. 2b, c). Due to the inhibition from the SNr, superior colliculus neurons are silent despite excitatory inputs that exceed their threshold. In the striatum, fast-spiking interneurons are continuously active in response to corticostriatal inputs. These fast-spiking interneurons silence medium spiny neurons even though their corticostriatal inputs exceed their firing threshold. As corticosubthalamic input increases, activity in the STN and GPe increases and induces self-sustained bursts (Fig. 2b, c, P1). Competition between populations of GPe neurons leads to the selection of channel 2. GPe2 and STN2 neurons continue bursting, while activity for those neurons in channel 1 decreases (Fig. 2b, c, P2). Consequently, SNr2 neuron activity is reduced and SNr1 activity is enhanced. As activity in the GPe increases, fast-spiking neurons in both channels start to exhibit occasional silent periods that correspond to bursts of activity in innervating GPe neurons (Fig. 2b, c, P1). As a result, medium spiny neurons are released from fast-spiking neuron inhibition and exhibit long-lasting, low-frequency discharges, which are characteristic of these neurons (Wilson 1998). After selection of GPe2, the activity of fast-spiking neurons starts to differ between channels; FSd2 neurons in the direct pathway exhibit silent periods, while FSd1 neurons return to a continuous discharge pattern without pause (Fig. 2b, c, P2). Fast-spiking neurons in the indirect pathway show the opposite activity pattern because FSi neurons receive projections from GPe neurons of a distinct channel. Consequently, medium spiny neurons exhibit a corresponding pattern of activity (i.e., MSi1 and MSd2 remain active). When the overlap between burst periods of GPe2 neurons innervating FSd2 is sufficiently long, overall MSd2 neurons become more active, shut down SNr2 and release SC2 from inhibition (Fig. 2b, c, P3). When SC2 fires a strong burst, it triggers a motor action.

Under the same input conditions, selection of a channel shows trial-by-trial randomness and selection probabilities are approximately equal (Figs. 2d and 3a), indicating that the model can select a channel probabilistically. Moreover, the action execution time, defined by the SC burst onset, varies markedly across trials (Fig. 2e). These results demonstrate that the basal ganglia circuitry supports action selection and the initiation of self-timed actions. Furthermore, a mechanism that generates behavioral variability, which is crucial for exploration, is realized by our model.

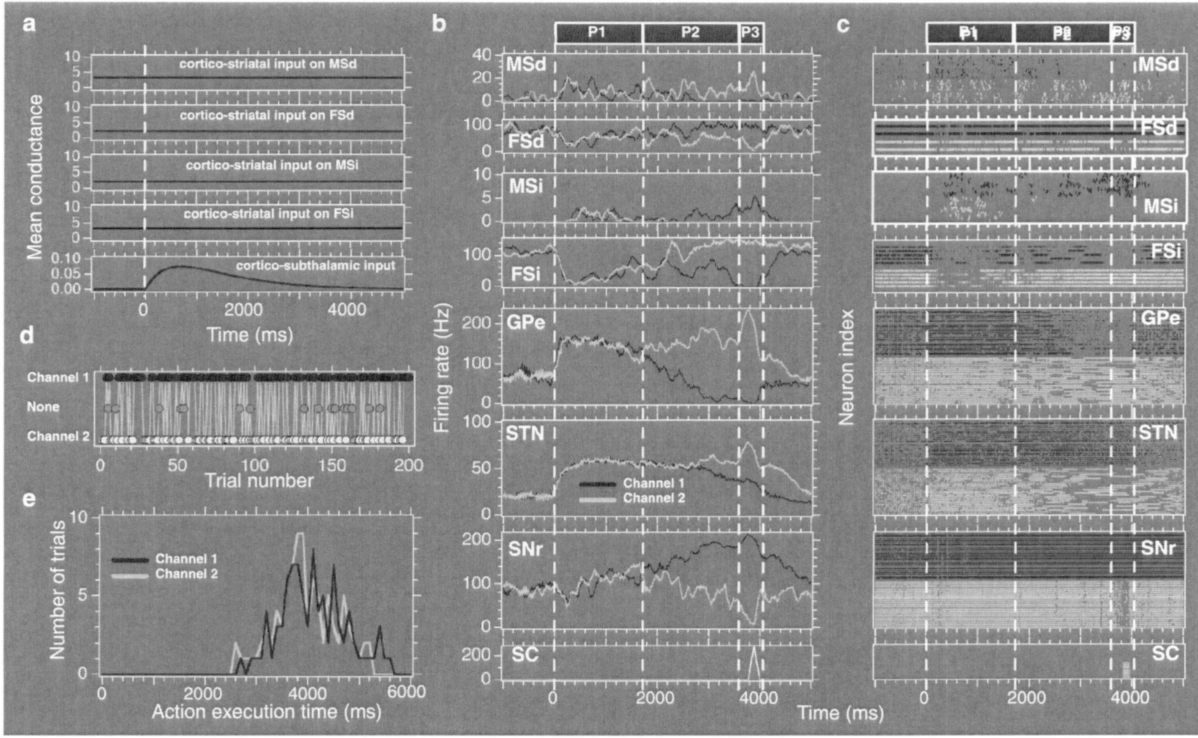

Fig. 2 Performance of the network model when corticostriatal inputs do not have specific spatiotemporal patterns. (**a**) Mean conductance of the excitatory input to the network. Each channel receives corticostriatal and corticosubthalamic inputs with common statistical properties. (**b**) Population firing rates for MSd, FSd, MSi, FSi, GPe, STN, SNr, and SC. *Dark gray*, channel 1; *Light gray*, channel 2. (**c**) Corresponding raster plots for the population firing rates. *Dark gray*, channel 1; *Light gray*, channel 2. (**d**) Record of individual choices of trials. (**e**) Distribution of action execution times for each channel. $t = 0$ corresponds to the onset of the task. *MSd* Medium spiny output neuron in the direct pathway, *FSd* fast-spiking GABAergic interneurons in the direct pathway, *MSi* medium spiny output neuron in the indirect pathway, *FSd* fast-spiking GABAergic interneurons in the indirect pathway, *GPe* external segment of globus pallidus, *STN* subthalamic nucleus, *SNr* substantia nigra pars reticulata, *SC* superior colliculus

3.2 Modulation of Selection Probability and Timing for Exploitation

Next, we considered the behavior of our model in the later stages of trial-and-error learning. We explored how corticostriatal synaptic efficacy affects the selection probability and action execution time. We have shown that MSi neurons in the channel that is not selected exhibit higher activity during a task than those in the selected channel. Furthermore, only MSd neurons in the selected channel exhibit increased activity around the time of action execution (see Fig. 2 for example). If the trial shown in Fig. 2 leads to success, the active corticostriatal synapses on MSi1 and MSd2 neurons are potentiated as a result of dopamine- and activity-dependent plasticity. These changes can result in two types of spatiotemporal modulation of corticostriatal inputs: (1) an increased difference in synaptic efficacy of MSi neurons between channels and (2) strengthening of the active inputs on MSd2 neurons when an action is executed (Fig. 3b, top panel). We independently examined the effect of these types of modulation on spatiotemporal patterns of corticostriatal inputs.

First, as the difference in synaptic efficacy of the corticostriatal inputs between MSi1 and MSi2 neurons increases, the selection is biased toward a channel with lower efficacy (Fig. 3a) due to differential activation of MSi neurons, which provides

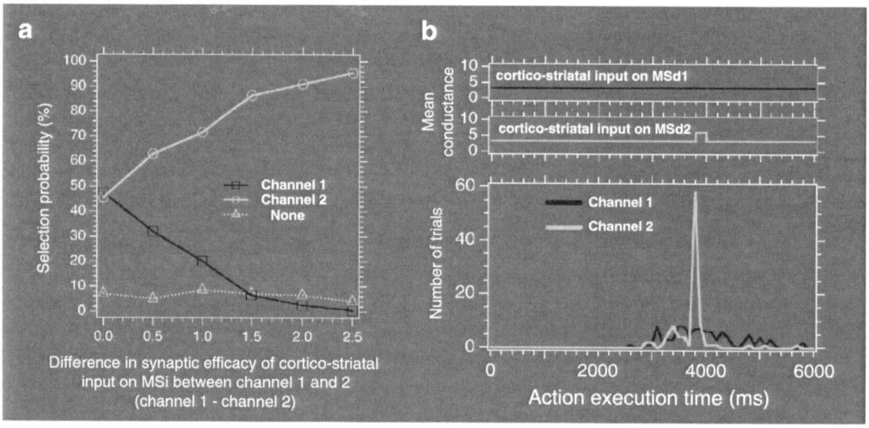

Fig. 3 (**a**) Selection probability is dependent on interchannel differences in the synaptic efficacy of corticostriatal inputs to MSi neurons. The zero point of the horizontal axis corresponds to the conditions described in Fig. 2. (**b**) Phasic modulation of the synaptic efficacy of corticostriatal inputs to MSd can reduce the variability of action execution time. *Top* and *middle panels*: Time course of the synaptic efficacy of corticostriatal synaptic inputs to MSd1 and MSd2 neurons, respectively. Other conditions are identical to those described in Fig. 2. *Bottom panel*: Histograms of action execution time. Compare the sharp distribution of the action execution time for channel 2 with the phasic increase of synaptic efficacy. In contrast, channel 1 has a broad distribution of action execution time and lacks temporal modulation. *MSd* Medium spiny output neuron in the direct pathway and *MSi* medium spiny output neuron in the indirect pathway

unbalanced inhibition to GPe during competition. These results demonstrate that the selection probability can be modulated and biased by changes in the efficacy of corticostriatal synapses onto medium spiny neurons in the indirect pathway.

Second, a phasic increase in the synaptic efficacy of corticostriatal inputs onto MSd2 at a fixed time reduces the variability of action execution timing by channel 2. The distribution of the action execution time is biased toward the time of increased synaptic efficacy, while execution times for channel 1 are unaffected (Fig. 3b). These results suggest that changes in the synaptic efficacy of corticostriatal synapses onto medium spiny neurons in the direct pathway can tune the timing of self-timed actions, from a variable to a fixed execution time.

Taken together, these simulation results show that our basal ganglia model has the capacity for trial-and-error learning because it promotes variability in behavior for exploration and then reduces variability to shape the optimal behavior for exploitation.

4 Discussion

In this paper, we present a spiking neuron model of the basal ganglia that describes the computational processes of action selection and self-timed initiation of the selected action. Action selection is accomplished by the indirect pathway, whereas timing of the selected action is determined by the direct pathway of the basal ganglia. In our model, both processes are accomplished even in early phases of learning when there are no specific spatiotemporal patterns for corticostriatal inputs, consistent with physiological observations (Pasupathy and Miller 2005) and requirements for trial-and-error learning. Notably, due to chaotic, self-sustained burst activity in GPe and STN neurons, both action selection and timing of the selected action are probabilistic, realizing variability in behavior that is required for trial-and-error learning. Moreover, behavioral variability can be tuned by changes in the synaptic strength of corticostriatal inputs. These results are consistent with the idea that cortical-basal ganglia loops underlie the explore–exploit behavior necessary for trial-and-error learning (Doya and Sejnowski 1995; Barnes et al. 2005; Kao et al. 2005; Olveczky et al. 2005; Tumer and Brainard 2007).

Acknowledgments We thank Drs. Marc-Oliver Gewaltig, Ursula Koerner, and Prof. Edgar Koerner for advice and encouragement, and Profs. Kenji Doya and Tomoki Fukai for valuable comments.

References

Arkadir D, Morris G, Vaadia E and Bergman H (2004) Independent coding of movement direction and reward prediction by single pallidal neurons. J Neurosci 24: 10047–10056.

Barnes TD, Kubota Y, Hu D, Jin DZ and Graybiel AM (2005) Activity of striatal neurons reflects dynamic encoding and recoding of procedural memories. Nature 437: 1158–1161.

Bevan MD, Booth PAC, Eaton SA and Bolam JP (1998) Selective innervation of neostriatal interneurons by a subclass of neuron in the globus pallidus of the rat. J Neurosci 18: 9438–9452.
Doya K and Sejnowski TJ (1995) A novel reinforcement model of birdsong vocalization learning. In: Tesauro G, Touretzky DS and Leen TK (eds.) Advances in Neural Information Processing Systems, Vol. 7, MIT, Cambridge, MA, pp. 101–108.
Hallworth NE and Bevan MD (2005) Globus pallidus neurons dynamically regulate the activity pattern of subthalamic nucleus neurons through the frequency-dependent activation of postsynaptic GABAA and GABAB receptors. J Neurosci 25: 6304–6315.
Hikosaka O, Takikawa Y and Kawagoe R (2000) Role of the basal ganglia in the control of purposive saccadic eye movements. Physiol Rev 80: 953–978.
Hollerman JR, Tremblay L and Schultz W (1998) Influence of reward expectation on behavior-related neuronal activity in primate striatum. J Neurophysiol 80: 947–963.
Kao MH, Brainard MS (2006) Lesions of an avian basal ganglia circuit prevent context-dependent changes to song variability. J Neurophysiol 96: 1441–1455.
Kao MH, Doupe AJ and Brainard MS (2005) Contributions of an avian basal ganglia-forebrain circuit to real-time modulation of song. Nature 433: 638–643.
Kita H and Kitai ST (1994) The morphology of globus pallidus projection neurons in the rat: An intracellular staining study. Brain Res 636: 308–319.
Lee IH and Assad JA (2003) Putaminal activity for simple reactions or self-timed movements. J Neurophysiol 89:2528–2537.
Maimon G and Assad JA (2006) A cognitive signal for the proactive timing of action in macaque LIP. Nat Neurosci 9:948–955.
Olveczky BP, Andalman AS and Fee MS (2005) Vocal experimentation in the juvenile songbird requires a basal ganglia circuit. PLoS Biol 3: e153.
Pasquereau B, Nadjar A, Arkadir D, Bezard E, Goillandeau M, Bioulac B, Gross CE and Boraud T (2007) Shaping of motor responses by incentive values through the basal ganglia. J Neurosci 27: 1176–1183.
Pasupathy A and Miller EK (2005) Different time courses of learning-related activity in the prefrontal cortex and striatum. Nature 433: 873–876.
Plenz D and Kitai S (1999) A basal ganglia pacemaker formed by the subthalamic nucleus and external globus pallidus. Nature 400: 677–682.
Reynolds JN, Hyland BI and Wickens JR (2001) A cellular mechanism of reward-related learning. Nature 413: 67–70.
Samejima K, Ueda Y, Doya K and Kimura M (2005) Representation of action-specific reward values in the striatum. Science 310: 1337–1340.
Sato F, Lavallée P, Lévesque M and Parent A (2000) Single-axon tracing study of neurons of the external segment of the globus pallidus in primate. J Comp Neurol 417: 17–31.
Terman D, Rubin JE, Yew AC and Wilson CJ (2002) Activity patterns in a model for the subthalamopallidal network of the basal ganglia. J Neurosci 22: 2963–2976.
Tremblay L, Hollerman JR and Schultz W (1998) Modifications of reward expectation-related neuronal activity during learning in primate striatum. J Neurophysiol 80: 964–977.
Tumer EC and Brainard MS (2007) Performance variability enables adaptive plasticity of 'crystallized' adult birdsong. Nature 450: 1240–1244.
Turner RS and Anderson ME (2005) Context-dependent modulation of movement-related discharge in the primate globus pallidus. J Neurosci 25: 2965–2976.
Wilson CJ (1998) Basal ganglia. In: Shepherd GM (ed.) The Synaptic Organization of the Brain, fourth edition. Oxford University Press, New York, NY, pp. 329–375.

Learning with an Asymmetric Teacher: Asymmetric Dopamine-Like Response Can Be Used as an Error Signal for Reinforcement Learning

Rea Mitelman, Mati Joshua, and Hagai Bergman

Abstract According to many computational models, dopamine (DA) neurons in the basal ganglia (BG) play a major role in reinforcement learning. DA signal is proportional to the difference between actual and predicted reward, hence it could serve as the error signal in a temporal difference (TD) learning algorithm implemented in the BG. Indeed, a proportional increase in the firing rate of DA neurons in states with higher values than expected (the positive domain of the error signal) has been found experimentally. However, many studies indicate that DA neurons do not decrease their firing rate symmetrically in the negative domain. Some studies report a smaller gain relative to the positive domain, whereas others report a decrease in the firing to a constant level.

Our work focuses on using such an asymmetric error signal in a TD-like computational algorithm. We simulated a probabilistic classical conditioning task in which the agent sequentially received stimuli with different probabilities of reward or aversion. The algorithm calculated the value of each of the stimuli, using an asymmetric TD signal. This was done by manipulating the negative domain of the error signal function by decreasing its slope by a multiplicative factor smaller than one, fixing its negative values to a constant negative level, or fixing them to zero.

We show that learning can be achieved when the negative domain of the error signal function is either constant negative or with reduced gain, although it is slower than with a symmetric error signal. However, learning cannot be achieved with a constant value of zero for the negative domain of the error signal function. We examined learning by comparing the values the algorithm assigned to the stimuli with those assigned by a symmetric TD algorithm. These values had a non-linear concave trend, with higher calculated values than those assigned by the symmetric TD algorithm.

This suggests that the DA asymmetric signal could be used as an error signal in a TD algorithm as implemented in the BG and that DA asymmetric coding does not

R. Mitelman (✉), M. Joshua, and H. Bergman
Department of Physiology, Hadassah Medical School and the Interdisciplinary Center for Neural Computation, The Hebrew University, Jerusalem, Israel
e-mail: ream@alice.nc.huji.ac.il

H.J. Groenewegen et al. (eds.), *The Basal Ganglia IX*, Advances in Behavioral Biology 58, DOI 10.1007/978-1-4419-0340-2_16, © Springer Science+Business Media, LLC 2009

require a complementary BG modulatory error signal. We further hypothesize that the concave values of the error signal curve could thus lead to some aspects of irrational human behavior.

1 Introduction

1.1 Reinforcement Learning and the Basal Ganglia

The neural network of the basal ganglia can be portrayed as comprised of two subsystems: a main axis and neuron modulators that modulate activity along the main axis (Haber and Gdowski 2004). The main axis includes the input nuclei, the striatum, and subthalamic nucleus, which project both directly and indirectly (via the external segment of the globus pallidus) to the output nuclei of the basal ganglia: the internal segment of the globus pallidus and the substantia nigra pars reticulata. The main elements of the neuron-modulator subsystem are the dopaminergic neurons of the substantia nigra pars compacta and the cholinergic interneurons of the striatum.

Previous studies have shown that activity in the basal ganglia is altered by expectation of reward. A key response is observed in the dopaminergic neurons (a neuron-modulator subsystem), which increase their firing rate when conditions are better than expected, and decrease their firing rate below their background discharge rate (and even pause) when conditions are worse than predicted (Schultz et al. 1997). This was well demonstrated in deterministic and stochastic classical (Pavlovian) conditioning tasks (Mirenowicz and Schultz 1994; Fiorillo et al. 2003). In such tasks, a sensory stimulus with no intrinsic reward value, the conditioned stimulus (CS), is associated with a stimulus with intrinsic value, the unconditioned stimulus (US), such as an appetitive stimulus (Pavlov 1927). The animal learns to associate the CS with the US, and shifts its behavior accordingly. When a novel US is presented to the animal, there is a transient increase in the firing of the dopaminergic neurons. However, after a CS–US pairing, the reaction of the dopaminergic cells shifts, and they increase their firing in response to the CS. In the stochastic form of this task, one of a set of CSs is presented at each trial, and the US is given at a constant probability (per CS). In this case, after learning, the increase in DA firing for the CS is proportional to its probability of reward. Furthermore, immediately after the US is presented, a second increase of dopaminergic firing appears which is reversely proportional to the predicted probability of receiving reward. On the trials of omitted US, a decrease of dopaminergic firing appears (Mirenowicz and Schultz 1994; Hollerman and Schultz 1998; Satoh et al. 2003).

It has been suggested that the dopaminergic signal plays the role of an error signal in a temporal difference (TD) algorithm implemented in the basal ganglia to address the reinforcement learning problem (Schultz et al. 1997; Sutton and Barto 1998). In the TD algorithm, an agent shifts between different states and assigns values to them according to expected rewards. In the simplest form of TD learning, the agent remembers only the current value of each state (without an explicit memory of the entire history), and updates it according to an error signal. This error signal can be easily understood as a measurement of surprise, or deviation between expected and perceived reward.

In the simple example of the stochastic classical conditioning task, with reward as a US, the error signal of the TD algorithm (after convergence) is indeed very similar to the observed DA signal after learning. When presented with a reward predicting CS, the error signal is positive and proportional to the probability of reward attached to this CS. At the time of the US the error signal depends on whether or not the rewarding US was actually given, due to the stochastic nature of this task. In the case of presented US, the error signal is positive but inversely proportional to CS reward prediction (i.e., higher probability of reward leads to lower error signal). Furthermore, when the US is omitted, the error signal is negative, and its absolute value is proportional to CS reward prediction (i.e., higher probability of reward leads to a more negative error signal).

1.2 Dopaminergic Signal Is Truncated at Zero

As mentioned above, the error signal in TD algorithm can be both positive and negative. For the dopamine discharge rate to play the role of an error signal, we have to look at its deviation from the basal firing rate, since a firing rate is always positive. However, dopaminergic neurons have a relatively low basal firing rate (~5 Hz), which sets the lower boundary of the deviation. On the other hand, the deviation of the peak in the response to unpredicted reward is known to be much higher than 5 Hz. This limits the ability of the DA signal to play the role of error signal in the negative domain; i.e., the error signal for states with values smaller than predicted.

As mentioned earlier, reward omission causes a transient decrease in the firing rate of the dopaminergic neurons, but different studies have found dissimilar levels of this decrease. Satoh et al. (2003) report the decrease of DA firing rate to be inversely proportional to reward prediction, but with a lower gain than for the positive domain. Other groups have found no significant trend for the decrease (Morris et al. 2004; Bayer and Glimcher 2005). They suggested therefore that the reaction of the dopaminergic neurons in the negative domain is characterized by a decrease to a constant level, regardless of reward prediction.

A possible solution to the asymmetric range for negative and positive TD coding by discharge rate of DA neurons is the observation that the negative domain of the error signal is encoded by the length of the decrease in firing rate. In other words, the positive domain is encoded by the height or the integral of the short peak of increased firing rate, whereas the negative domain is encoded by the duration of the decrease of firing rate (Bayer et al. 2007). However, this has not been observed, at least in the peri-stimulus time histograms (PSTH) analysis of other groups (Fiorillo et al. 2003; Morris et al. 2004).

Both representations for the DA discharge rate signal in the negative domain (different gain and a negative plateau) are different from the symmetric error signal of the TD algorithm. Nevertheless, the theoretical significance of this difference was not addressed so far. Here we examine how DA-like asymmetric error signals affect learning: does learning take place at all; if so, what values will be learned and how long does it take for the learning to converge? We also looked at the extreme case in which the negative domain of the error signal was set to zero, equivalent to no change in dopaminergic firing from the basal level for negative surprises.

2 Methods

We modeled a discrete time, stochastic version of the classical conditioning task, based on the TD(0) algorithm (Schultz et al. 1997; Sutton and Barto 1998). An agent randomly received one of a set of 19 stimuli which could be followed by an aversive outcome (9 stimuli with probabilities of 1/9 to 1), a reward outcome (another 9 stimuli with probabilities of 1/9 to 1), or a neutral outcome (a single stimulus with a probability of 0 for an outcome). The stimuli were given at $t = 1$, the outcome was given two time steps after the stimulus, and the following time step was a terminal step, after which the agent returned to the initial state for a new trial. The agent had no explicit memory of the previous trials, and their only trace was the updated states.

The model parameters were defined as follows: $V(s,t)$ was the value of the state (s,t), i.e., following t time steps after the agent was presented with stimulus s. Since the first step did not include any stimulus presentation, it was defined as $V(0,0)$. The algorithm was initiated with a value of 0 for all states. For simplicity, when we refer to the value of a stimulus s, we mean the value of the stimulus when it was presented, i.e., $V(s,1)$. The future discount rate, γ, determines to what extent does the error signal influenced by future states' values. It can vary continuously between 1 – no discount in the influence of future states and 0 - complete discount.

After each time step, the value of the previous TD-error was calculated

$$\delta(t) = r(t) + \gamma V(t+1,s) - V(t,s) \quad (1)$$

where $r(t)$ is the reinforcement at time t (+1 for reward, –1 for aversion, 0 for no outcome). Note that reward omission (no outcome) has a value of zero in our model, and not a negative value as in models of asymmetric tasks (e.g., tasks with only appetitive or omission outcomes).

The state value was updated according to the TD error:

$$V(t,s) \leftarrow V(t,s) + \alpha \cdot F(\delta(t)) \quad (2)$$

where α is the step-size parameter controlling the speed of learning and F is the function which differed from one model to another.

In the original TD(0) model:

$$F(\delta(t)) = \delta(t) \tag{3}$$

For decreased gain of the negative domain we used:

$$F(\delta(t)) = \begin{cases} \delta(t) & \text{if } (\delta)0 \\ d \cdot \delta(t) & \text{otherwise} \end{cases} \tag{3.1}$$

where d is the slope of the negative domain of the error signal, $0 < d \leq 1$ (Fig. 1a), and $d = 1$ is identical to the regular TD(0) algorithm (3).

For constant negative slope we used:

$$F(\delta(t)) = \begin{cases} \delta(t) & \text{if } (\delta)0 \\ -k & \text{otherwise} \end{cases} \tag{3.2}$$

where $0 < k < 1$ is the constant value of the negative error signal. We present here $k = 0.2$ (Fig. 2a), however, changing k values does not have qualitative effects on our results (data not shown).

Finally, we simulated the extreme case in which the negative domain of the error signal is set to 0:

$$F(\delta(t)) = \begin{cases} \delta(t) & \text{if } (\delta)0 \\ 0 & \text{otherwise} \end{cases} \tag{3.3}$$

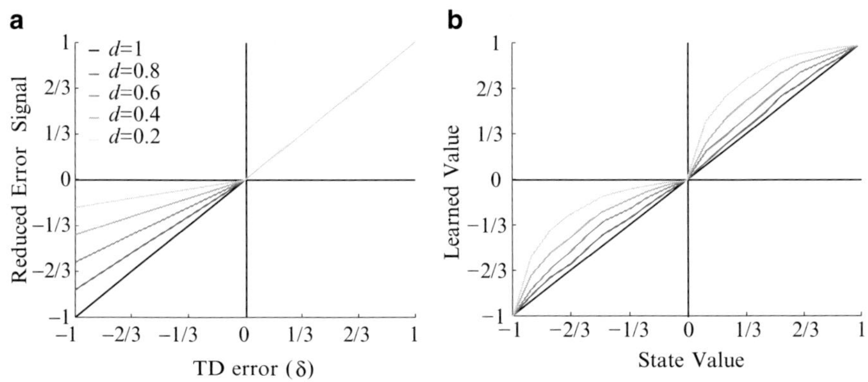

Fig. 1 Biologically inspired TD learning with lower gain for the negative domain. (**a**) The DA-like error signal with a lower gain in the negative domain, as a function of the error signal of the symmetric model. Different values of d are used as decreased gain, from $d = 1$, which is identical to the symmetric TD and below (*gray level* coding). (**b**) Values of the different stimuli, as learned by the algorithm, as a function of their values according to the symmetric model, for different values of d. The *gray levels* are identical to (**a**), namely each learned values curve is the outcome of an error signal in the same *gray level*

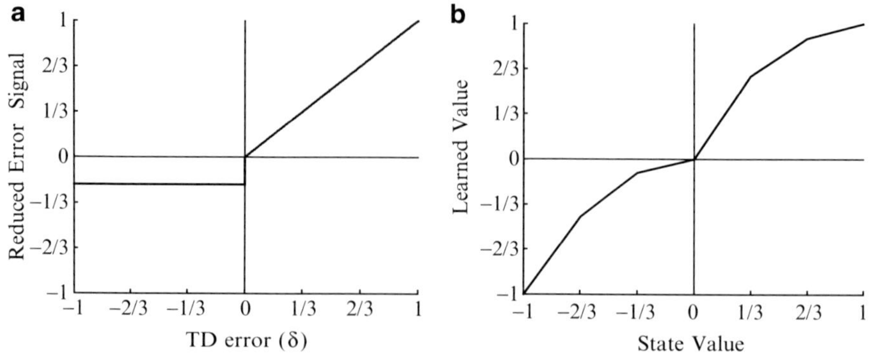

Fig. 2 Biologically inspired TD learning with a constant negative value for the negative domain. (**a**) The DA-like error signal as a function of the error signal of the symmetric model. (**b**) Values of the different stimuli, as learned by the algorithm, as a function of their values according to the symmetric model

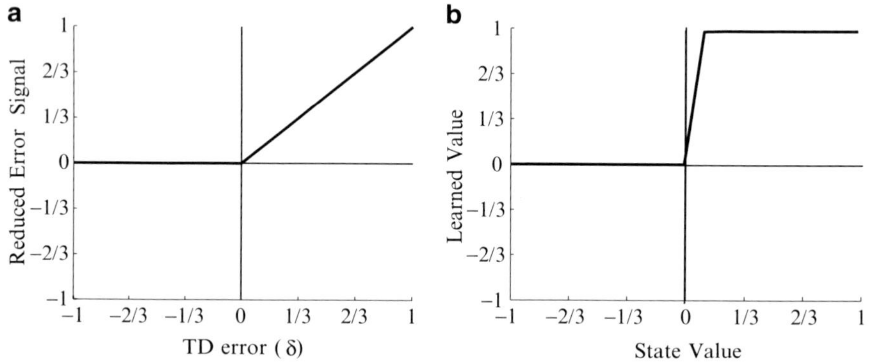

Fig. 3 The extreme case of TD learning with zero values for the negative domain. (**a**) The DA-like error signal as a function of the error signal of the symmetric model. Note that this is identical to Fig. 1b with $d = 0$. (**b**) Values of the different stimuli, as learned by the algorithm, as a function of their value according to the symmetric mode. Note that all stimuli were assigned the learned value of either zero or one

This is identical to both $d = 0$ for equation (3.1) and $k = 0$ for equation (3.2), and presented in Fig. 3a.

We used a high value of $\gamma = 0.99$ (i.e., minimal discounting of future events); however, we obtained the same qualitative values for lower values of the temporal discounting factor (data not shown). The learning rate constant (2) was selected to have a value of $\alpha = 0.1$. To smooth the learning curve we repeated and averaged the learning results. The algorithm was run until convergence was obtained (i.e., constant values for the averaged learning curve). To assess the learning speed we used $t_{1/2}$, i.e., the number of iterations it took the algorithm to reach half of the final value (Fig. 4).

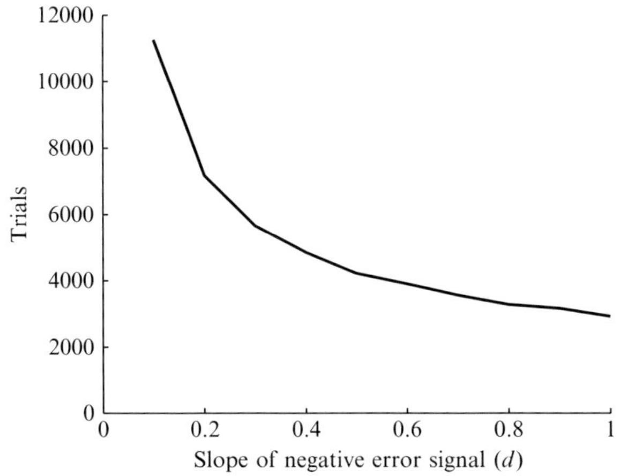

Fig. 4 Learning duration for the biologically inspired error signal with lower gain for the negative domain. Learning duration is calculated by the number of iterations necessary to reach half the final value $t_{1/2}$, as a function of the slope of the negative domain, d. Note that since convergence was ill-defined for $d = 0$, as the algorithm did not learn the values of the different probabilities, and convergence duration is presented starting at the lower positive value of d we simulated ($d = 0.1$)

3 Results

3.1 Error Signal with Lower Gain for the Negative Domain

First, we examined an error signal which has a lower gain in the negative domain (Fig. 1a). The algorithm managed to converge for all values of d, i.e., the achieved learned values encode the different stimuli they are associated with. These values form a monotonously increasing function with a concave trend in comparison to the values learned by the symmetric TD algorithm (Fig. 1b). This trend increases with the decrease in the value of d (i.e., with greater deviation from the symmetric TD function), for both rewarded and aversive stimuli. This was the consequence of using an error signal whose negative domain was equal to or greater than the error signal of the symmetric TD [$F(\delta) \geq \delta$ for $\delta < 0$, Equation 3.1]. This asymmetric error signal was used for all stimuli, causing a concave trend for both the rewarded and aversive stimuli.

Next, we examined the effect of changing d on the convergence duration (Fig. 4), by continuously shifting it between $d = 1$ (identical to symmetric TD) and $d = 0.1$. Convergence duration decreases the closer the algorithm is to the symmetric TD.

3.2 Error Signal with a Constant Negative Value for the Negative Domain

Next, we addressed the second possibility for the dopaminergic signal (Equation 3.2), which is a constant negative error signal value for the negative domain. It turns out that the consequences of this model (Fig. 2b) were not different qualitatively from the previous one (Fig. 1b), despite the fact that the negative domain of the error signal was fixed. The algorithm manages to assign different values to the different stimuli despite the identical negative correction they received. We equalized the number of presentations of the stimuli; thus, the stimulus with probability $p = 8/9$ for aversive outcome (A8/9) received negative corrections of the stimulus value more frequently than the stimulus with probability $p = 1/9$ for aversive outcome (A1/9). Hence, despite the same magnitude of negative correction, A8/9 had a final learned value which was more negative than the learned value of A1/9, and the algorithm managed to encode different probabilities in the learned values. Similarly, stimuli with a high probability of reward received the constant negative correction less frequently than stimuli with a low probability of reward.

3.3 Error Signal Set to Zero for the Negative Domain

Finally, in the extreme case of an error signal which was set to zero for all negative corrections, we obtained a constant learned value of zero for all aversive stimuli, and a constant learned value of one for all rewarded stimuli (Fig. 3b). In other words, the algorithm could not learn the outcome probability, only the sign of the outcome. This is because all values of $F(\delta)$ were nonnegative, so all corrections of the learned value were done by increasing it. For the aversive stimuli, there was never a positive correction (initial values of all stimuli are set to zero and the outcome of an aversive trial can equal -1 or 0 only); hence they did not change their initial value.

4 Discussion

In this chapter, we show that dopamine discharge rate alone, despite its asymmetric properties, can be used as an error signal in a TD algorithm implemented in the BG. This is true as long as it shows any trend of decrease in firing rate below the baseline level for negative errors. Without such a decrease, our modeling results show that the values of CSs which differ in their probabilities for future outcome cannot be learned.

Learning with asymmetric error signal is slower than learning with symmetric TD and less linear with expected values. Such nonlinearity, if used for decision making (for example, in a more complicated task of operant conditioning, when reward depends on some action of the animal), can be observed as irrational behavior. For instance, a stimulus with a reward at $p = 2/3$ would be worth twice as much as a stimulus with a reward at $p = 1/3$ (i.e., it will receive twice the value) for rational behavior. However, the nonlinear, concave curve of the learned value function of a reinforcement learning model with a biologically driven asymmetric error signal assigns values that are much closer to each other for these two stimuli (Figs. 1b and 2b). The greatest differences in the learned values are between the neutral (indicating no outcome) and a slight indication of future reward. On the other hand, in the negative domain, the learned values are biased toward a more optimistic estimation of the CS predicting future aversive events.

Irrationality also characterize human decision making (Tversky and Kahneman 1981). According to this, the positive domain of the curve of subjective value (as a function of the objective curve) is characterized by a nonlinear, concave, monotonous increase. This is very similar to the positive domain of the learned values achieved by biologically driven error signals, as mentioned above. It is possible that one of the reasons for human irrational behavior – at least when it comes to rewarded stimuli – is the asymmetric nature of the dopaminergic signal. However, Kahneman and Tversky's prospect theory asserts that the prospect of a loss has a greater impact on decision making than does the prospect of an equivalent gain. Thus, a full model of human irrational behavior necessitates the incorporation of other neural systems in addition to the basal ganglia and the dopamine systems. Indeed, a recent work on responses of different nuclei in the basal ganglia to rewarded and aversive stimuli (Joshua et al. 2008; see Chapter "Asymmetric Encoding of Positive and Negative Expectations by Low-Frequency Discharge Basal Ganglia Neurons" by Joshua et al. in this volume) shows that the dopaminergic signal does not encode the error for aversive stimuli. This suggests that learning the values of aversive stimuli takes place outside the basal ganglia.

The slower learning achieved by the asymmetric TD may raise the question of its relevance to learning in the nervous system. However, the question of learning speed arises even for the symmetric TD algorithm as a model of learning in animals, since animals usually perform faster than predicted by the TD algorithm (Kawato and Samejima 2007). Unlike computer models, animal learning probably uses a multitude of algorithms (e.g., declarative explicit learning paradigms) in parallel with the reinforcement (procedural) learning algorithm studied here (Kawato and Samejima 2007). Nevertheless, theoretical models enable us to test the boundaries of the learning algorithms employed in the neural networks of the basal ganglia. The ability to achieve informative reinforcement learning, despite the asymmetric encoding of the negative and the positive domains of a dopamine-like error signal, suggests that the basal ganglia reinforcement learning networks can be modulated by a single critic – the dopaminergic system. Future physiological studies of the encoding schemes and the relative contribution of other basal ganglia critics – such as the striatal cholinergic interneurons and dorsal raphe serotonergic neurons – should shed more light on actual learning algorithms of the basal ganglia and the brain in general.

References

Bayer HM and Glimcher PW (2005) Midbrain dopamine neurons encode a quantitative reward prediction error signal. Neuron 47: 129–141

Bayer HM, Lau B and Glimcher PW (2007) Statistics of midbrain dopamine neuron spike trains in the awake primate. J Neurophysiol 98: 1428–1439

Fiorillo CD, Tobler PN and Schultz W (2003) Discrete coding of reward probability and uncertainty by dopamine neurons. Science 299: 1898–1902

Haber SN and Gdowski MJ (2004) The basal ganglia. In Paxinos G and Mai JK (eds) The Human Nervous System. Elsevier, London, pp. 676–738.

Hollerman JR and Schultz W (1998) Dopamine neurons report an error in the temporal prediction of reward during learning. Nat Neurosci 1: 304–309

Joshua M, Adler A, Mitelman R, Vaadia E and Bergman H (2008) Midbrain dopaminergic neurons and striatal cholinergic interneurons encode the difference between reward and aversive events at different epochs of probabilistic classical conditioning trials. J Neurosci 28: 11673–11684

Kawato M and Samejima K (2007) Efficient reinforcement learning: Computational theories, neuroscience and robotics. Curr Opin Neurobiol 17: 205–212

Mirenowicz J and Schultz W (1994) Importance of unpredictability for reward responses in primate dopamine neurons. J Neurophysiol 72:1024–1027

Morris G, Arkadir D, Nevet A, Vaadia E and Bergman H (2004) Coincident but distinct messages of midbrain dopamine and striatal tonically active neurons. Neuron 43: 133–143

Pavlov IP (1927) Conditioned Reflexes. Oxford University Press, Oxford

Satoh T, Nakai S, Sato T and Kimura M (2003) Correlated coding of motivation and outcome of decision by dopamine neurons. J Neurosci 23: 9913–9923

Schultz W, Dayan P and Montague PR (1997) A neural substrate of prediction and reward. Science 275: 1593–1599

Sutton RS and Barto AG (1998) Reinforcement Learning: An Introduction. MIT, Cambridge, MA

Tversky A and Kahneman D (1981) The framing of decisions and the psychology of choice. Science 211: 453–458

A Theoretical Information Processing-Based Approach to Basal Ganglia Function

Mandar Jog and Dorian Aur

Abstract This chapter presents an information processing approach to basal ganglia function. Concepts that build the understanding of information processing utilize published experimental work that shows how information within a network can be represented even as far down as the ionic flux. Ionic flux that results in and from action potentials is shown to be a carrier of information flow within a neuronal network. This flux can organize during behavior and can represent behavioral changes that occur within a striatal network during the acquisition of a learning task. The present chapter further develops the idea of this adaptation to the abrupt behavioral learning seen in striatal neurons. A hypothesis that then attempts to synthesize these ideas into a more system level understanding of basal ganglia function and dysfunction in Parkinson's disease is put forward.

1 Introduction

The repertoire of movements that is achieved by humans in their lifetime is enormous. While we are able to perform gentle tasks that require finesse and detail such as painting or sculpture, we use the same motor system also to perform manually intense and forceful tasks from chopping wood to lifting weights. Indeed, it is possible for the system to not only know how to do this, but also know how to rapidly switch between activities within a matter of milliseconds. One can gently crack an egg and instantaneously switch to chopping onions while making an omelet without any effort whatsoever. The problems that face us from a scientific viewpoint in understanding how the brain achieves this phenomenal flexibility of the output are vast. All subdisciplines of neuroscience interested in motor control have invested heavily in trying to essentially solve this very problem. These fields include human

M. Jog (✉) and D. Aur
Department of Clinical Neurological Sciences, University of Western Ontario,
Room A10-026, 339 Windermere Road, London, ON, Canada N6A 5A5
e-mail: mandar.jog@lhsc.on.ca

H.J. Groenewegen et al. (eds.), *The Basal Ganglia IX*, Advances in Behavioral Biology 58,
DOI 10.1007/978-1-4419-0340-2_17,

kinesiology and kinematics, functional neuroimaging in humans, electrophysiology and electrical recordings within various parts of the nervous system, some branches of neurobiology and, in the past few decades, mathematics and computational neuroscience. However, despite the increasing knowledge of the neuroanatomy, chemistry, and physiology of the nervous system, the issues of motor control and essentially the question of how inputs are transformed into performing the desired output, with control, remain essentially elusive to us.

Biologists and computational neuroscientists have adapted the computer science and engineering terminology of "motor program," "input–output," "information flow" among others in their daily speaking about motor control. Apart from providing a general concept or construct of what we may want to communicate, these words have little meaning since there is no conceptual understanding of what a motor program actually is within the nervous system. Even though we know that the system must obviously contain information, again the nature of this at the biological level is a mystery. And if information is gathered, stored, retrieved, and indeed communicated between networks of neurons, how is this done? Finally, despite the approach of control theory and the sensory input–output paradigms that have been modeled by many, how sensory–motor integration actually occurs and a solution is registered and provided, also is essentially unknown.

Solving these problems is not trivial and there are no complete answers currently available. Fundamental neurobiological understanding and the approach of cataloging every cellular mechanism is a never-ending task. However, increasing our knowledge at this level does not help in increasing system level understanding of large-scale behavior. Human kinematic experimentation, where direct observation of motion can be accurately quantified, tells us about how we move, but it is not possible to concurrently study the underlying biological processes that underlie the observations. In addition, one has to also explain how the system fails to perform in disorders such as Parkinson's disease. How does one approach this problem with any hope?

The answers may lie in a theoretically based framework that assumes that there are basic and fundamental rules that govern how neurons process information. These rules are common to neurons in general. How individual neurons perform under these rules may not only depend on their intrinsic properties but also depend on the received inputs. Alterations in every aspect of the neuron, ranging from its internal chemistry to its axo-dendritic connectivity, may be guided by these principles. At a larger scale, the organization of groups of neurons in certain anatomical structures with probabilistic connectivity patterns that change over time is also determined to allow neural efficiency and reduce entropy within the system, while allowing certain redundancies. Finally, having built this connectivity, observed neurobiological and neurochemical characteristics of individual cells, networks, and larger structures are only emergent properties of the system that may adapt abnormally when disease affects the system.

In this chapter, we will present some novel informational theoretical ideas and concepts, and suggest a different approach toward how neurons process information.

2 The Functional Perspective

The basal ganglia (BG) appear to be well suited for taking large quantities of functionally important information including context, cognition, and sensory input, and performing an integrative function to provide a particular motor output. The current theories of BG functioning can be summarized from a motor control and computational perspective. An initial concept was based upon the fact that BG structures were involved in performing routine automatic tasks (Alexander and Crutcher 1990). This was followed by associations of the basal ganglia to a slightly more complex function such as generation of sequences (Jenkins et al. 1994). Then came the hypothesis that the BG had relationship to task learning through their dopaminergic system (Mirenowicz and Schultz 1996; Schultz et al. 1993; Schultz et al. 1995). Dopamine was shown to be responsible for reinforcement of the cortical inputs coming into the basal ganglia. This led to the idea of reinforcement learning based on the actor-critic architecture. According to this concept, the output of an actor, supposedly implemented by the striatal matrix, is corrected by a critic signal, possibly delivered by the dopaminergic system, such that the reinforcement given in a task is maximized.

Based on animal recordings, many functions were subsequently assigned to the basal ganglia in terms of what they might be "focusing" on, including planning, initiation, control, and execution of movement. However, the numbers of synapses involved in the basal ganglia loop make it difficult for them to actually perform any task directly. Subsequently, the functional models were conceptually further advanced by the "action selection" framework. Following these ideas, the basal ganglia were felt to be responsible predominantly for the "focusing" or selection of desired motor programs (Mink 1996). Motor programs that were beneficial for the motor task were enhanced and the unnecessary competing motor programs were effectively inhibited. The basal ganglia thus reinforced cortical "programs" in a context-dependent manner.

Nevertheless, recent understanding and theoretical work has begun to cast doubt that the basal ganglia have anything to do with actual coding of any movement parameters. It is very likely that the basal ganglia operate in a multidimensional space that may have little to do with three-dimensional action space and indeed do not provide anything close to an actual action solution. In this respect, the output of the basal ganglia is more akin to generating a "map" within the cortical regions of what the boundary conditions that exist in the imagined task space might be. In a way, the concept is similar to the idea of dimensionality reduction that has been hypothesized by Bar-Gad and Bergman (2001). According to these authors, the cortical and other inputs into the basal ganglia are funneled through, based on the anatomy of the structures, to define the restraints within which the task needs to be performed. This signal is then transferred to the cerebral cortex where a somewhat lower level, more somatotopically organized construct of the action plan may reside in a distributed network. Sensory inputs and motor commands interact directly at the cortical level without actually requiring basal ganglia involvement in a routine motor task. However, the representation within the cortex is still unlikely to be in three-dimensional action space and would continue to be abstract.

3 The Neurophysiologic Basis

New concepts need to be formulated that "bridge" the basic neurobiological framework all the way to the actual observed behavior. Although this theoretical approach is completely a work in progress, a certain level of development has been achieved by our group. At a network level, our prior work has shown that during recording electrical activity of a large number of striatal neurons that are responding during a motor learning task in a rodent, a change in responsivity is seen (Jog et al. 1999). During this task of learning how to select an appropriate arm of a T-maze to obtain reward, the firing patterns of the striatal neurons were recorded over a period of 10–12 days. The pattern of activity within the striatal network changed from being a fairly random firing pattern to one that showed that the animal had some concept of what the task was all about. Before a trial began, neurons had built some anticipatory activity for what is going to happen, i.e., a choice. Subsequently they reduced their firing during the actual performance of the task, and finally the response rebuilt at the end of the task if the choice was correct (Jog et al. 1999) Further analysis shows that not only the neuronal firing is reorganized when an animal learns a task, but it is also possible for the network to partially unlearn and relearn when re-exposed to the task (Barnes et al. 2005). Network level analysis has shown that at least at the striatal level, learning a motor task produces a form of reorganization that may be related to an alteration of connectivity between neurons that make up the striatal neuronal network. This analysis revealed to us important aspects of habit formation.

The above-described analysis did not describe in any further detail how the information might be getting reorganized. Specifically, there was no exploration in terms of the nature of information flow within the system, how it was built within the network. The work in our laboratory in the past 3 years has been devoted to tackle these very issues. We therefore computed, using the same dataset, an analysis of alteration of the mutual information (MI) between neurons that spike in the network (Jog et al. 2007). This revealed an intriguing quality of how the system learns information. Contrary to the expectation that the build up of such information would be gradual, we found that this occurred abruptly within the striatal network. This build up of information corresponded to the time that the animal had acquired the task and had begun to perform it correctly. Further analysis of this striatal learning was carried out in order to address what was actually happening at the level of the neuronal network. Again, electrophysiological data was recorded from rodents learning a T-maze task (Jog et al. 1999). The sum of the mean neural activities, computed for each animal, showed that motor task learning occurred at the level of striatal neurons: neuronal activity recorded on day 1 was concurrent mostly with the auditory cue, given in the center of the maze, as opposed to the activity recorded on day 12 that was concentrated at the starting point of the maze and at the left and right endpoints where reward was located. Changes in neuronal activities that occurred while animals learned the T-maze task were also analyzed by computing the spike directivity (Aur et al. 2005; Aur and Jog 2006) and ensemble mutual information (EMI) (Jog et al. 2007). Changes in the neuronal activity during the T-maze task were accompanied by changes in the organization of the calculated directionality of the ionic charge flow (Aur and Jog 2007a). From the point of view of MI analysis, the T-maze learning showed three distinct phases (1) an early learning phase with low MI values, (2) a task acquisition phase showing a rapid increase of MI, and (3) an overtraining phase in which MI values varied around a high mean value. These results indicated that the learning in the striatal network at the neuronal level is not gradual but has a pronounced nonlinear aspect (Jog et al. 2007).

At that stage of conceptualization, it was becoming clear that one of the functions of the basal ganglia, as indicated in our recordings in the striatum, was to build some representation of the task from incoming cortical information. The striatum was somehow representing some components of the task, presumably the most salient aspects and was probably not coding for the actual motion itself. In addition, it was also interesting to note that in terms of information within the striatal network, the change in the EMI was not a linear phenomenon but an abrupt one. However, these approaches still remain at a substantially large scale at which it is impossible to determine the actual processes underlying the carriage of information within the network. To tackle this problem, we turned to an even deeper level of charges and ionic flux, the skeleton underneath the electrochemical properties of neuronal function.

4 The Informational Theoretical Construct

Understanding how neurons represent, process, and manipulate information is one of the main goals of neuroscience. These issues are fundamentally abstract and information theory plays a key role in formalizing and addressing them. However, application of an information theory to experimental data is problematic. A variety of innovative analytical techniques can be used. Our hypothesis is that information is transferred within ionic flow in spiking activities. Statistically, more information is carried when an event is less likely to happen than when it actually occurs. This interpretation relates the event probability to spike occurrence. This means that low levels of information are coded by ionic fluxes during the neuron's interspike interval. Therefore, the transient state described by an action potential (AP) occurrence results in substantial ionic flux that transfers high levels of information. Since an AP is generated by ionic fluxes, it is expected that a relation between ionic flux and information transfer could be established if one can provide an estimation of the ionic flow during AP.

4.1 Ionic Mechanisms

Each signal source or AP in the neuron can be thought of as being generated by charge flow that includes sodium, potassium, chloride, and calcium. The significant increase in the ionic conductance during the AP is generated by an abrupt change in the permeability characteristics of the membrane that for short term becomes more permeable to Na^+ ions than to K^+ ions. The generation of an AP is then the result of an increase in the ionic conductance of the membrane and the activation of voltage sensitive ion channels.

The propagation of excitation along the dendrites and axon of a neuron can be described by a partial differential equation

$$C_{\mathrm{m}}\frac{\partial \mathrm{V(r,t)}}{\partial \mathrm{t}}=\alpha(r)\frac{\partial^2 V(r,t)}{\partial r^2}+\beta(r)\frac{\partial V(r,t)}{\partial r}-I_{\mathrm{ionic}}+I_{\mathrm{in}}(r,t) \quad (1)$$

C_{m} is the membrane capacitance and $I_{\mathrm{in}}(r, t)$ represents the injected current density along the neuron's membrane (Toth and Crunelli 1999). The coefficients α and β depend on the dendrite's shape and account for eccentricity in the dendrite cross section and I_{ionic} is the sum of all ionic current densities including I_{leak} (2)

$$I_{\mathrm{ionic}}=I_{\mathrm{Na}}+I_{\mathrm{K}}+I_{\mathrm{leak}} \quad (2)$$

In this equation the currents I_{Na} and I_{K} are determined by the ionic flux of Na^+ and K^+, respectively. After membrane depolarization, Na^+ permeability increases even more; therefore, an avalanche of Na^+ is produced and, consequently, Na^+ ions invade the cell. However, the process is transient and is accompanied by an increase in the already high K^+ permeability and by Na^+ current inactivation. Immediately following an AP there is a time interval during which no stimulus can generate a second AP potential and this time interval is called the refractory period.

Groups of several ionic charges are therefore responsible for this AP mechanism. A detailed analysis of individual action potentials recorded from a neuron on the basis of charge movement model could provide valuable insights regarding electrical flow of information within each neuron and across chains or circuits of neurons. This can also be achieved in vitro using slice recordings and high-resolution techniques employing an ultra small array of electrodes to observe the electrical activity of neurons controlling the synaptic input of individual neurons (Spencer et al. 2004). In vivo techniques such as two-photon microscopy (Helmchen et al. 1999), surface imaging, and directional analysis in tetrode recordings (Takahashi et al. 2003) have been used to capture electrical flow during an AP. However, such systems cannot determine any impact of behavior on the electrical flow in the neurons. An analysis of electrical flow in an in vivo system that utilizes behaving animals would therefore be extremely useful in order to understand pattern changes within APs during behavior.

Therefore, we proposed a novel approach to obtain a representation for electrical flow of charges during each AP, based upon an independent component analysis technique. Having separated the multiple signal sources into putative individual neurons using established spike sorting techniques (Jog et al. 2002), our method advanced the understanding of the properties of individual APs assigned to one putative neuron. The method uses the four-dimensional views of every AP generated by tetrode recordings and provides details regarding the electrical pattern of activation within each spike (Aur and Jog 2006).

4.2 Application to Behavior

Utilizing the fact that charge flow may be important in the computation of information that is exchanged by neurons, the next task was to develop an understanding of how this might be useful in the real-world setting of behavior. In order to address this issue, albeit indirectly, we turned to the computation of directivity of charge flow that was occurring within the striatal neuronal network. In the analysis of directivity, we used the idea that during an AP, we can study charge flow that occurs within the network and that this is reflected by the extracellular electrical activity. The directionality of this charge flow can be computed due to the fact that we use a tetrode (a four-tip recording electrode). This allows us to mathematically localize in the three-dimensional space where the charges might be flowing. Such directionality of charge flux can be computed for every spike from every striatal cell that we are recording from. In a series of theoretical papers, we showed that this is possible in a model (Aur et al. 2005; Aur and Jog 2006).

The computational details of the directivity calculations for each spike recorded and the resulting charge flow were then performed using already published techniques (Aur et al. 2005). These calculations were performed in "tetrode space" (Aur et al. 2005). In short, the following steps were utilized. The charge flow trajectory was calculated for each and every spike by using the point charge model and the triangulation method. Estimation of spike directivity in "tetrode space" was then achieved. Therefore, the computed directivity was not in real space but in a transformed space where the tetrode is a frame of reference. This is what is meant by "tetrode space." The trajectory in "tetrode space" was analyzed for every spike using singular value decomposition (SVD) in order to find the best linear approximation of the spike direction (Stewart 1993). A model of the spatial signal distribution was obtained by utilizing the largest singular value and the corresponding right singular vector that represented direction cosines of the best linear approximation in "tetrode space" (Aur et al. 2005). This SVD technique generated three cosines (v_1,v_2,v_3), one for each axis. Higher singular values of the decomposition indicate dimensions with higher energy within the data.

We then separately estimated the probability distribution for each of the three cosines. This calculation was performed for specific time periods for every tetrode in all animals during free exploration and for all trials per tetrode for each neuron for the entire period of the T-maze task. This level of spatial randomness within the spike train directivity was quantified by analyzing the values of cosine angles using the Shannon information theory (Shannon and Weaver 1949). Shannon introduced the notion of the entropy for a random discrete variable x, as the average of the quantity of information brought by it. Shannon information entropy is a function of the probability distribution p

$$H_S(p) = -\sum p(x_i)\log p(x_i) \tag{3}$$

Shannon entropy H_S is a measure of uncertainty about the outcome of the random variables and the value H_S does not depend on the state values $x_1, x_2, \ldots, x_n$.

Applying this spike directivity analysis to the actual animal recordings in the same learning behavior as the T-maze above, we were able to show that spike directivity changes in time are in a determined relationship with behavioral events on the T-maze. Based on these analyses correlated with the behavioral data, we demonstrated that the cosine of the angle between two spike directivity vectors is related to the "semantic" distance between neuronal spikes. This semantic distance measures the similarity of meaning between two spikes from the same neuron. Therefore, predictability of the upcoming behavior is an inherent property of "expert" neurons as revealed in spike directivity changes. These analyses of spike directivity in recorded cells demonstrate the link between changes in neuronal activity and semantics of behavior at a detailed level that cannot be achieved by using spike timing paradigms (Aur and Jog 2007b).

Interestingly, this ability to alter the directionality of the charge flow was only seen in the time period that the tone (signaling the turn direction) and the turning of the animal began. The beginning and the end of the task appeared to be not relevant to the animal. However, the time between the tone delivery and the beginning of the turn was the most salient part of the task. An incorrect choice would have meant a wrong motor act of turning to the opposite side and then potentially not getting the reward. Hence, after learning the striatal recordings continued to show a directionality of charge flux predicting the choice the animal was making. But when the animal had overlearnt the task, the directionality was really an indication of the learning that had occurred in the input structure which is the cortex. Hence, the striatal neuronal activity that we were recording was really activity from the cerebral cortex that was reflecting what part of the task was the most salient. The actual act of turning was left to the other structures to do. This is the first time that the salience of the actual basal ganglia output, and the fact that it can "teach" a component of a motor task to the efferent structures, has been demonstrated.

5 Can We Make Any Real Sense of This?

Concluding the above paragraphs, we have shown that the basal ganglia activity, at least at the level of the main input nucleus, the striatum, does organize itself, that it does this rather abruptly and, finally, that this indeed appears to actually reflect in its activity what the basal ganglia may be submitting to the cerebral cortex. With this framework in mind, we need to go back to a conceptual level with the basal ganglia as the teacher. In order to understand this, we will utilize the concept of an "action plan," i.e., a spatiotemporal solution that incorporates the salient features or boundary conditions of a sensorimotor output. This action plan is probably still not in the three-dimensional action space, but in a higher order, multidimensional space, albeit less than what the basal ganglia probably worked upon. Hence, the internal constraints are mapped onto this action plan. This dimensionally reduced action plan can be further simplified through the brainstem, with the cerebellum providing appropriate gain, accuracy, and further refining the salient components. After the brainstem centers, the action plan is ultimately directed toward the spinal cord. At that level, an additional set of transformations takes place that convert the gradually reduced dimensionality of the action plan into actual joint torques and angles of motion. In this way, the multidimensional, sensorimotor integrative space is gradually transformed into a Euclidean three-dimensional space.

Therefore, it is possible that evolution, genetic, and environmental influences have come up with a substantially complex core set of action plans that the motor cortical descending projection system can deal with. In addition, the basal ganglia system, through continuous adaptation adds and modifies these plans, as hinted by the results presented above. When a demand is placed on the system, the cortex executes an action plan fairly quickly. The action and plan can therefore be thought of as a connection of a variety of different cortical areas such as the prefrontal cortex, premotor cortex, motor cortex, sensory cortex, association cortex, etc., with varying number of neurons involved in each of these areas, connected through a common representation.

In this context, the basal ganglia can be seen as having two main functions (see Fig. 1). The first function is termed a "teacher function," the second the "reference function." Thus, as already referred to above, the basal ganglia-thalamic system performs the saliency extraction task that is vital in generating efficient boundary conditions, albeit in the action plan space and not directly in the task performance space. This occurs in novel situations, where novelty may be internally or externally driven. Gradually and over time, depending on the number of nonlinearities within the demand, an efficient action plan is put together. In the mean time, approximate action plans are executed and their impact is assessed by the sensory system. This information is communicated to all levels of the system including directly to the cerebral cortex and the basal ganglia via the thalamic relay nuclei.

As this action plan is being put together, the basal ganglia continue to receive information flowing through its nuclei. Gradually, this information begins to represent the action plan and the basal ganglia saliency output comes closer and closer to the action plan within the cerebral cortex. When this happens, the basal ganglia can be seen as representing a "reference function." In the reference function mode, the basal ganglia do not any more modify the cortical action plans substantially. The role of teacher, at least in the context of the current action plan, is then over. Thus the two functions, the teacher and the reference, are concurrent and work dynamically. The mode with which the basal ganglia and cortex interact with each other is therefore determined by the requirement of the system. Other networks may simultaneously be reorganizing other action plans (Almeida et al. 2003; Johnson et al. 2003; Abbruzzese and Berardelli 2003; Berardelli et al. 2001).

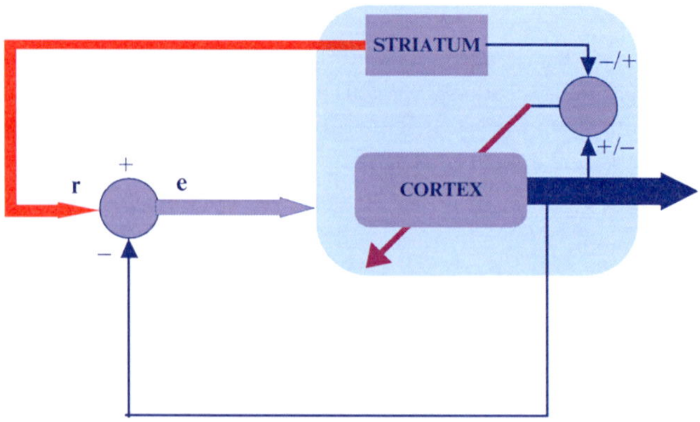

Fig. 1 The schematic representation of a classical dynamical system where the striatum processes information. Feedback from the cortex is compared with the reference **r** (in *red* color) provided by the striatum and defines the error **e**. The teacher function where the striatum changes the cortical plans substantially is schematically represented in *magenta*

Finally, our discussion would not be complete, in the context of this volume, unless we can put forward what may happen in Parkinson's disease (PD). In Parkinson's disease, the important and primary pathophysiological phenomenon is the loss of the dopamine input from the nigrostriatal system. This deficiency develops over time and hence the deficits that occur in PD have to be explained in terms of this gradual development. In addition, PD patients do not simply have a deficiency of learning new tasks but their predominant dysfunction is one of an inability to perform already learnt tasks. In this scenario, the basal ganglia output gradually changes over time. While most well-engrained tasks do not require to be modified, the basal ganglia start to do exactly this. However, much of it may be related to an expansion of the boundary conditions that do not actually alter the resident cortical action plan, but simply switch it from an automatic mode to one that now has to listen to an increasing basal ganglia output when it was used to not having to at all. Thus the basal ganglia gradually and intermittently switch between the reference and teacher functions, and this switch is unpredictable. A loss of dopamine, which normally would not have been useful in the truly automatic task, is now felt by the system as enhancing the noise that is being submitted from the basal ganglia. The information that is then provided to the cerebral cortex may in fact be corrupted so that the cortex is led to believe that its selected action plan may in fact be wrong. In this way, the basal ganglia could then be seen as a bad teacher.

One can conceptualize that over a period of time, abnormal teacher function of the basal ganglia could gradually corrupt to a certain extent existing action plans in the cerebral cortex. Not only the movement plans will now be slowed down because of noisy reference function of the basal ganglia, but indeed the cortical action plans will also primarily be interrupted or corrupted so that it takes much longer for initiation and execution of the action plan. These phenomena can be seen in patients with Parkinson's disease. Additionally, it is possible that the basal ganglia output could lock cortical systems in an oscillatory mode as well, producing the classic symptoms of tremor. These features can be modeled fairly well computationally.

In conclusion, we have discussed above the integration of a structural and a functional model of the basal ganglia. Incoming internally and externally driven behavioral action plans might exist in the cerebral cortex. These have been built over years through inputs that are sent from the cortical systems through the basal ganglia. The basal ganglia process this information and over time filter out the most nonsalient features of the task as demonstrated in our data. The processing of this information occurs by building up informational entropy within the network of neurons and probably in every structure. Ionic flux is one of the vehicles by which this informational entropy is represented within the network. During APs, this entropy is dynamically altered and the ionic flux resulting from the interspike collection of information entropy is transferred across to the next and neighboring neurons within the network. This ionic or charge flux has specific directionality that is altered by the input and output connectivities of the network. In the basal ganglia, this directional change of ionic flux represents the reorganization at a fundamental level, the informational entropy of the network. This represents an alteration of how the network communicates when the system learns a behaviorally related task and, further, that ionic flux represents a carrier with which this communication can be understood.

As the appropriate saliencies are acquired by the network, and this can occur rather abruptly (Jog et al. 2007), dopamine and other neuromodulators modify the generated network state. In the cortico-basal-ganglia system, the basal ganglia receive information from the cerebral cortex, gradually build a representation of the action solution, and through ionic flux communicate this to the cortex in the conceptual "teacher" and "reference" modes. The entire system is extremely plastic and action plans may be modified over time. In diseases such as PD, the system aberrantly and unpredictably switches between the teacher and the reference functions and many of the cortical representations are slowed as more and more cortical areas are included within the action plan space. The system becomes inefficient and fails. Dopamine has an important role in enhancing the salient features and inhibiting the others. When absent, all throughput is homogenized, while dopamine replacement immediately highlights the saliency making the basal ganglia output appear much less noisy to the cortex, and enhancing motor output.

The basal ganglia are well situated anatomically to transform such contextually dependent incoming sensory information into a learned motor output or action plan. The basal ganglia anatomical machinery is designed to provide in an iterative process such as the mapping of patterns through the basal ganglia circuits to generate action solutions. The basal ganglia can teach the cerebral cortex for over a period of time to modulate these action plans and when diseases such as Parkinson's disease or Huntington's disease disrupt the circuitry, an abnormal amount of information, either increase or decrease, or simply corrupt information, is provided to the cortical action plans. The latter over a period of time, gradually get either temporarily or permanently disrupted.

References

Abbruzzese G and Berardelli A (2003) Sensorimotor integration in movement disorders. Mov Disord 18: 231–240.

Alexander GE and Crutcher MD (1990) Functional architecture of basal ganglia circuits: neural substrates of parallel processing. Trends Neurosci 13: 266–271.

Almeida QJ, Wishart LR and Lee TD (2003) Disruptive influences of a cued voluntary shift on coordinated movement in Parkinson's disease. Neuropsychologia 41: 442–452.

Aur D and Jog MS (2006) Building spike representation in tetrodes. J Neurosci Methods 157: 364–373.

Aur D and Jog MS (2007a) Neuronal spatial learning, Neural Process Lett 25: 31–47.

Aur D and Jog MS (2007b) Reading the neural code: What do spikes mean for behavior? Nat Precedings. http://dx.doi.org/10.1038/npre.2007.61.1

Aur D, Connolly CI and Jog MS (2005) Computing spike directivity with tetrodes. J Neurosci Methods 149: 57–63.

Bar-Gad I and Bergman H (2001) Stepping out of the box: Information processing in the neural networks of the basal ganglia. Curr Opin Neurobiol 11: 689–695.

Barnes TD, Kubota Y, Hu D, Jin DZ and Graybiel AM (2005) Activity of striatal neurons reflects dynamic encoding and recoding of procedural memories. Nature 437: 1158–1161.

Berardelli A, Rothwell JC, Thompson PD, Hallett M (2001) Pathophysiology of bradykinesia in Parkinson's disease. Brain 124: 2131–2146.

Helmchen F, Svoboda K and Tank DW (1999) In vivo dendritic calcium dynamics in deep-layer cortical pyramidal neurons. Nat Neurosci 2: 989–996.

Jenkins IH, Brooks DJ, Nixon PD, Frackowiak RSJ and Passingham RE (1994) Motor sequence learning: A study with positron emission tomography. J Neurosci 14: 3775–3790.

Jog MS, Kubota YK, Connolly CI, Hillegaart V and Graybiel AM (1999) Building neural representations of habits. Science 286: 1745–1749.

Jog MS, Connolly CI, Kubota Y, Iyengar DR, Garrido L, Harlan R and Graybiel AM (2002) Tetrode technology: Advances in implantable hardware, neuroimaging, and data analysis techniques. J Neurosci Methods 117: 141–152.

Jog MS, Aur D, Connolly C (2007) Is there a tipping point in neuronal ensemble during learning? Neurosci Lett 412: 39–44.

Johnson AM, Vernon PA, Almeida QJ, Grantier LL and Jog MS (2003) A role of the basal ganglia in movement: The effect of precues on discrete bi-directional movements in Parkinson's disease. Motor Control 7: 71–81.

Mink JW (1996) The basal ganglia: Focused selection and inhibition of competing motor programs. Prog Neurobiol 50: 318–425.

Mirenowicz J and Schultz W (1996) Preferential activation of midbrain dopamine neurons by appetitive rather than aversive stimuli. Nature 379: 449–451.

Schultz W, Apicella P and Ljungberg T (1993) Responses of monkey dopamine neurons to reward and conditioned stimuli during successive steps of learning a delayed response task. J Neurosci 13: 900–913.

Schultz W, Romo R, Ljungberg T, Mirenowicz J, Hollerman JR and Dickinson A (1995) Reward-related signals carried by dopamine neurons. In: Houk JC, Davis JL and Beiser DG (Eds) Models of Information Processing in the Basal Ganglia. MIT, Cambridge, MA, pp. 233–248.

Shannon CE and Weaver W (1949) The Mathematical Theory of Communication. University of Illinois Press, Champaign, IL.

Spencer L, Smith JW and Otis TS (2004) An ultra small array of electrodes for stimulating multiple inputs into a single neuron. J Neurosci Methods 133: 109–114.

Stewart GW 1993) On the early history of the singular value decomposition. SIAM Rev 35: 551.

Takahashi S, Anzai Y and Sakurai Y (2003) A new approach to spike sorting for multi-neuronal activities recorded with a tetrode /how ICA can be practical. Neurosci Res 46: 265–272.

Toth TI and Crunelli V (1999) Solution of the nerve cable equation using Chebyshev approximations. J Neurosci Methods 87: 119–136.

Part III
Pharmacological and Receptor Studies in the Basal Ganglia

The Cellular Localisation of $GABA_A$ and Glycine Receptors in the Human Basal Ganglia

Henry J. Waldvogel, Kristin Baer, Ray T. Gilbert, Weiping Gai, Mark I. Rees, and Richard L.M. Faull

Abstract We have investigated the cellular localisation of $GABA_A$ ($GABA_AR$) and glycine (GLYR) receptors in the human basal ganglia using immunohistochemical techniques and light and confocal laser scanning microscopy. $GABA_AR$ were most highly expressed on GABAergic striatal interneurons (α_1, $\beta_{2,3}$, γ_2 subunits), cholinergic interneurons (α_3), and striatal projection neurons (α_2, α_3, $\beta_{2,3}$, γ_2 subunits). GLYR were present mainly on ChAT and a subset of parvalbumin striatal interneurons. The neurons of the globus pallidus (GP) showed high levels of α_1, α_3, $\beta_{2,3}$, γ_2 subunits (no α_2) whereas GLYRs were only distributed on a subpopulation of pallidal neurons. In addition, GLYRs selectively stained neurons in the intermedullary laminae of the GP. Neurons in the SNr and SNc were labelled with GLYR but had different $GABA_AR$ subunit configurations. SNr neurons expressed α_1, α_3, $\beta_{2,3}$, γ_2 (no α_2) subunits and SNc neurons expressed mainly $GABA_AR$ $\alpha_3\gamma_2$ subunits. These results demonstrate that in the basal ganglia, neurons are generally associated with one of four different $GABA_A$ receptor configurations. This suggests that throughout the basal ganglia GABA acts via $GABA_A$ receptors with various subunit combinations and that glycine acts through glycine receptors on neurons of the SNr and SNc and groups of interneurons scattered throughout the striatum and GP.

1 Introduction

The basal ganglia of the human brain are a group of nuclei involved in the complexity of motor and mood control. They include the striatum, the external (GPe) and internal (GPi) segments of the globus pallidus (GP), and the pars compacta (SNc) and the pars reticulata (SNr) of the substantia nigra (SN). GABA is the major neurotransmitter in the basal ganglia. It is present in the striatopallidal (GABA-enkephalin,

H.J. Waldvogel (✉), K. Baer, R.T. Gilbert, W. Gai, M.I. Rees, and R.L.M. Faull
Department of Anatomy with Radiology, Faculty of Medical and Health Sciences, University of Auckland, Private Bag, 92019, Auckland, New Zealand
e-mail: h.waldvogel@auckland.ac.nz

H.J. Groenewegen et al. (eds.), *The Basal Ganglia IX*, Advances in Behavioral Biology 58,
DOI 10.1007/978-1-4419-0340-2_18, © Springer Science+Business Media, LLC 2009

GABA-substance P), the striatonigral (GABA-substance P), and the pallidonigral (GABA-parvalbumin) projections, and is also the neurotransmitter for striatal interneurons, pallidal neurons, and SNr projection neurons (Graybiel 1995; Hornykiewicz 2001; Smith et al. 1998). In addition, rodent studies have implicated glycine as an inhibitory neurotransmitter in the basal ganglia (Darstein et al. 1997, 2000). $GABA_A$ and glycine receptors are chloride ion channels that facilitate inhibitory neurotransmission in the mammalian central nervous system and are comprised of combinations of subunits from a variety of different subunit classes ($GABA_A$- α_{1-6}, β_{1-3}, γ_{1-3}, δ, ε, π; glycine- α_{1-3}, β) (Bernard et al. 1987; Rajendra et al. 1997). We have localised $GABA_A$ and glycine receptor subunits with detailed immunohistochemical techniques to further investigate the distribution of inhibitory receptors in the human basal ganglia.

2 Methods

2.1 *Brain Tissue*

The human brain tissue for this study was obtained from the Neurological Foundation of New Zealand Human Brain Bank (Department of Anatomy with Radiology, University of Auckland). The University of Auckland Human Participants Ethics Committee approved the protocols used in these studies and all tissues were obtained with full consent of the families. Brain tissue was obtained from neurologically normal cases, with a range of ages (48–83 years), with no history of neurological disease and no evidence of neuropathology and had a *post-mortem* interval between 5 and 23 h after death. For the immunohistochemical studies, the human brains were processed as previously described (Waldvogel et al. 2004, 2006, 2007). In brief, the human brains were fixed by perfusion through the basilar and internal carotid arteries, first with phosphate-buffered saline (PBS) with 1% sodium nitrite, followed by 15% formalin in 0.1 M phosphate buffer, pH 7.4. After perfusion, blocks from the basal ganglia were carefully dissected out and kept in the same fixative for 24 h. The tissue blocks were cryoprotected in 30% sucrose in 0.1 M phosphate buffer with 0.1% Na-azide and then quickly frozen on dry ice and stored at −80°C. The blocks were sectioned on a freezing microtome at 50–70 μm and the sections stored in PBS with 0.1% sodium azide (PBS-azide).

2.2 *Immunohistochemical Procedures*

2.2.1 Primary Antibodies

A series of antibodies were used to detect $GABA_A$ receptor subunits and glycine receptors and to identify the various cell phenotypes in the basal ganglia. (1) The monoclonal antibody bd24 directed against an extracellular epitope of the α_1

subunit of the $GABA_AR$ and the monoclonal antibody bd17 against the $\beta_{2,3}$ subunits of the $GABA_AR$ (H. Mohler and J.-M. Fritschy, Institute of Pharmacology and Toxicology, University of Zurich, Switzerland) used at a dilution of 1:20,000 (Benke et al. 1991). (2) Three polyclonal guinea pig antibodies directed against the α_2, α_3, or γ_2 subunit of the $GABA_AR$ were used at dilutions of 1:10,000, 1:5,000, and 1:2,000, respectively (Benke et al., 1996; Fritschy and Mohler 1995), as well as a polyclonal rabbit anti-$GABA_AR$ α_3-subunit antiserum, used at 1:1,000 (Alomone Labs Ltd, Israel). (3) Polyclonal rabbit anti-tyrosine hydroxylase antiserum used at a dilution of 1:1,000 (Protos-Biotech, New York, USA). (4) Polyclonal rabbit anti-parvalbumin antiserum used at a dilution of 1:5,000 (SWANT, Switzerland; PV-28). (5) Antibodies used to localise glycine receptors were the monoclonal antibody Mab4a (Synaptic Systems, Germany), which recognizes the human glycine receptor [48 kDa glycine receptor α_1 subunit and 58 kDa glycine receptor β subunit (Pfeiffer et al. 1984; Schroder et al. 1991)]; and a rabbit polyclonal antibody (RGLYR) raised against the N terminus of the human GLYR α_1 subunit with cross reactivity to the GLYR α_2 subunit. All antibodies were dissolved in immunobuffer consisting of 1% goat serum in PBS with 0.2% Triton-X and 0.4% Thimerosol (Sigma).

2.2.2 Single Immunoperoxidase Labelling

Adjacent series of sections were selected and processed free-floating in tissue culture wells using standard immunohistochemical procedures. Sections were washed in PBS and 0.2% Triton-X (PBS-triton) and where appropriately pre-treated for antigen retrieval using standard protocols (Fritschy et al. 1998; Waldvogel et al. 1999) before being processed for immunohistochemistry. Briefly, sections for antigen retrieval were incubated overnight in 0.1 M sodium citrate buffer, pH 4.5, microwaved in a 650-W microwave oven for 30 s and allowed to cool before washing (3 × 15 min) in PBS-triton. The sections were then incubated for 20 min in 50% methanol and 1% H_2O_2, washed (3 × 15 min) in PBS-triton, incubated in primary antibodies for 2–3 days on a shaker at 4°C, washed and incubated overnight in species-specific biotinylated secondary antibodies followed by 4 h at room temperature in ExtrAvidin™, 1:1,000 (Sigma), or streptavidin peroxidase complex 1:1,000 (Chemicon). The sections were reacted in 0.05% 3,3-diaminobenzidine tetrahydrochloride (DAB; Sigma) and 0.01% H_2O_2 in 0.1 M phosphate buffer, pH 7.4, for 15–30 min to produce a brown reaction product. A nickel-intensified procedure was also used in which 0.4% nickel ammonium sulphate was added to the DAB solution to produce a blue-black reaction product (Adams 1981). The sections were mounted on gelatine chrom-alum coated slides, dehydrated through a graded alcohol series to xylene, and coverslipped with DPX (BDH, Poole, England, UK).

Control sections were routinely processed to determine non-specific staining using the same immunohistochemical procedures detailed above except that the primary antibody was omitted from the procedure or blocking peptide was preincubated with the primary antibodies. In addition, some sections were Nissl stained with cresyl violet according to standard techniques.

Table 1 Distribution of $GABA_A$ receptor subunits α_1, α_2, α_3, $\beta_{2,3}$, γ_2 and glycine receptors (GLYR) on the different cell types in the various nuclei of the basal ganglia

	α_1	α_2	α_3	$\beta_{2,3}$	γ_2	GlyR
Striatal projection neurons (type 4)	–	++++	++	++++	+	–
Striatal GABA interneurons (types1-3)	++++	–	+	++++	+++	+
Striatal ChAT interneurons	–	–	+++	–	–	++++
Globus pallidus	++++	–	+++	++++	+++	+
SNc	+	–	+++	+	+++	++++
SNr	++++	–	+++	++++	+++	++++
Intermedullary lamina (MML and LML)	++	–	–	++	+	++++

– Not detected, + low levels of IR, ++ moderate levels of IR, +++ high levels of IR, ++++ highest levels of IR

2.2.3 Immunofluorescent Double Labelling

Double immunofluorescent labelling was carried out to determine which cell types in the various nuclei of the basal ganglia were associated with the various $GABA_A$ and glycine receptors (Table 1). Sections were incubated in a cocktail of monoclonal and polyclonal primary antibodies followed by secondary antibodies linked to different fluorophores. The various receptor antibodies were co-incubated with various cell phenotype markers, that is, parvalbumin, calbindin, somatostatin, neuropeptide Y, ChAT, calretinin, and TH. The sections were prepared using essentially the same procedure as the single-labelled sections. The sections were incubated in primary antibodies for 2–3 days on a shaker at 4°C, washed, and then incubated in species-specific fluorescent secondary antibody directly linked to AlexaFluor 488, AlexaFlour 594 (1:400), or CY3 (1:500). To detect for any cross reactivity in the double labelling experiments, each primary antibody in the pair was incubated with both secondary antibodies to be used in the double label. Control sections where the primary antibody was omitted showed no immunoreactivity. The sections were washed, mounted on slides with antifade mounting media, and viewed with epifluorescence and by confocal laser scanning microscopy.

3 Results

3.1 Striatum

The regional and cellular localisation of five $GABA_A$ receptor subunits (α_1, α_2, α_3, $\beta_{2,3}$, and γ_2) and glycine receptors was investigated in the human basal ganglia (Fig. 1). The results demonstrated that $GABA_A$ receptors in the striatum showed considerable

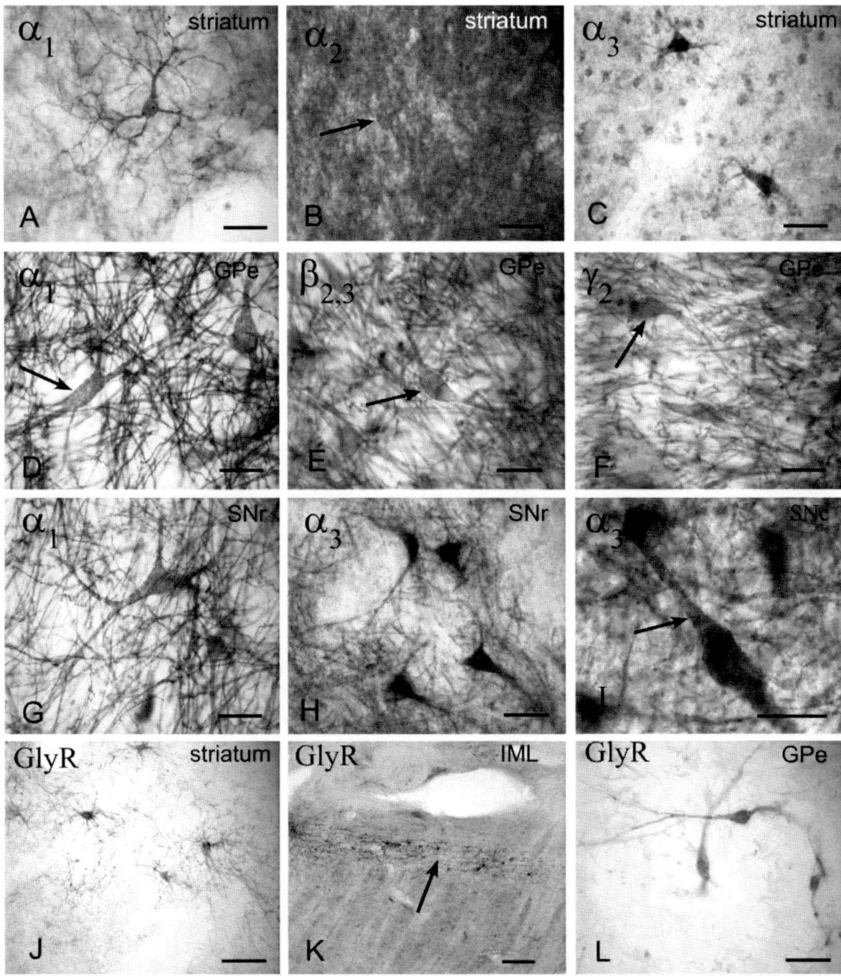

Fig. 1 Photomicrographs showing examples of $GABA_A$ receptor subunit and glycine receptor labelling in (*arrow*) representative sections of basal ganglia regions. (**A–C**), $GABA_A$ receptor labelling in the human striatum: (**A**) Medium-sized $GABA_A$ receptor α_1-subunit immunoreactive neuron; (**B**) Dense $GABA_A$ receptor α_2-subunit immunoreactive neuropil with a cell outlined by receptor labelling (*arrow*); (**C**) $GABA_A$ receptor α_3-subunit labelling of large intensely immunoreactive neurons and several medium-sized neuronal cell bodies with lower labelling density. (**D**, **E**), $GABA_A$ receptor labelling in the globus pallidus: (**D**), $GABA_A$ receptor α_1-subunit immunoreactive neuronal cell bodies (*arrow*) amongst a dense network of immunoreactive dendrites; (**E**), $GABA_A$ receptor $\beta_{2,3}$-subunit immunoreactive neurons and dendrites; (**F**), $GABA_A$ receptor γ_2-subunit immunoreactive neurons (*arrow*) and dendrites. (**G–I**), $GABA_A$ receptor labelling in the substantia nigra: (**G**), α_1-subunit immunoreactive neuron in the SNr; (**H**), α_3-subunit immunoreactive neurons in the SNr; (I) α_3-subunit immunoreactive pigmented neuron in the SNc (*arrow*). (**J-L**), Glycine receptor labelling in; the striatum (**J**), IML [*arrow*, (**K**)], and globus pallidus (**L**). Scale bars (**A–H**), $L = 50$ μm, $H = 500$ μm, $J = 100$ μm, $K = 200$ μm

subunit heterogeneity in their regional distribution and cellular localisation. At the regional level, high densities of $GABA_A$ receptors containing the α_2, α_3, $\beta_{2,3}$, γ_2 subunits and a lack of glycine receptors were found in the striosome compartment while lower densities of $GABA_AR$ containing the α_1, α_2, α_3, $\beta_{2,3}$, and γ_2 subunits as well as low levels of glycine receptors were found in the matrix compartment. Our results in the human brain at the cellular level have shown a complex and intricate morphological distribution of $GABA_A$ and glycine receptors throughout the various nuclei of the basal ganglia (Waldvogel et al. 1999, 2004, 2007, 2008) which agree remarkably well with those observed in previous animal studies (Boyes and Bolam 2007; Fujiyama et al. 2000). As detailed in Fig. 2, six different types of neurons were identified in the striatum on the basis of $GABA_A$ receptor subunit configuration, cellular morphology, and chemical neuroanatomy; these included five types of

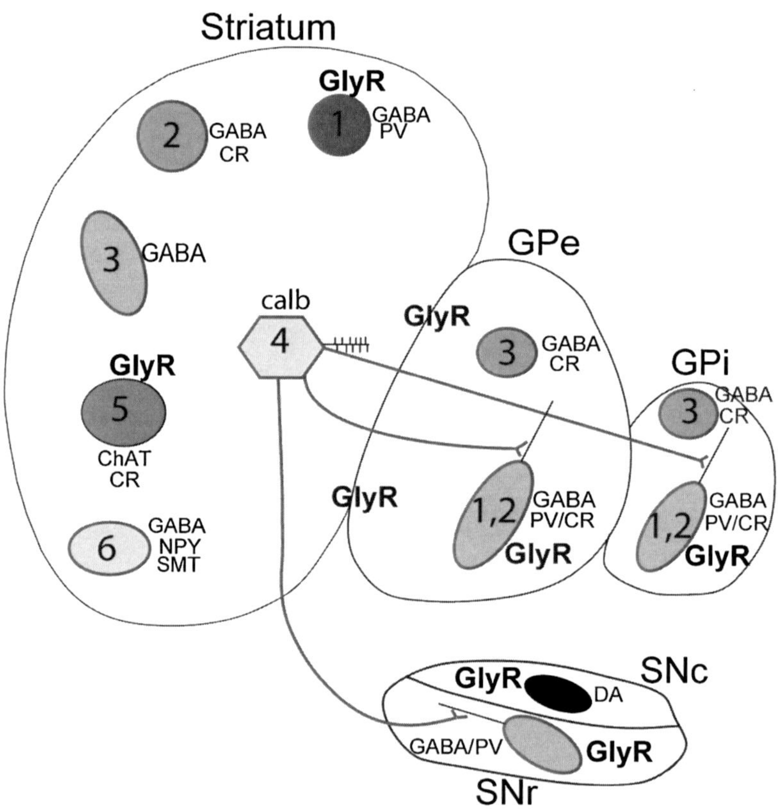

Fig. 2 Summary diagram showing the chemical phenotypic characteristics and glycine receptors on the various cell types in the striatum, globus pallidus, and substantia nigra of the human basal ganglia (see text for GABAA receptor subunit localisation on each of the cell types). *ChAT* Choline acetyltransferase, *calb* calbindin, *CR* calretinin, *DA* dopamine, *GABA* γ-aminobutyric acid, *GLYR* glycine receptors, *NPY* neuropeptide Y, *PV* parvalbumin, *SMT* somatostatin

interneurons (types, 1–3, 5, 6) and one type of projection neuron (type 4). Of these, three types (types 1, 2, 3) were GABAergic interneurons which displayed $\alpha_1,\beta_{2,3},\gamma_2$-subunit immunoreactivity. Of these three, the most numerous (60%) were medium-sized parvalbumin positive neurons (type 1; Fig. 1A), followed by medium-large-sized aspiny neurons immunoreactive for calretinin (type 2), and third, large sparsely spiny neurons immunoreactive for α_1, α_3, $\beta_{2,3}$, γ_2 subunits (type 3). Type 4 neurons were medium spiny projection neurons immunoreactive for α_2, α_3, $\beta_{2,3}$, γ_2 subunits, i.e. their cell soma contained mainly α_2 and α_3 subunits but lacked α_1 subunits (Fig. 1B). The type 5 neurons were the cholinergic interneurons that were immunoreactive for ChAT and α_3 subunit (Fig. 1C), and type 6 neuropeptide Y positive neurons showed no GABA$_A$ receptor subunit immunoreactivity.

Glycine receptors were found on striatal cell types 1, 3, and 5 (Fig. 1J).

3.2 *Globus Pallidus*

The globus pallidus contained neurons that were (1) large-sized and immunoreactive for α_1, α_3, $\beta_{2,3}$, γ_2 subunits (Fig. 1D–F), and immunoreactive for parvalbumin alone (type 1), or for both parvalbumin and calretinin (type 2). (2) Type 3 neurons were medium-sized and immunoreactive for calretinin alone and α_1 $\beta_{2,3}$, γ_2 subunits.

Glycine receptors were often observed on many of the cell types in the external and internal segments of the globus pallidus (Fig. 1I).

3.3 *Lateral Medullary Lamina and Medial Medullary Lamina*

The neurons of the lateral medullary lamina (LML) intervening between the putamen and GPe, and the Medial Medullary Lamina (MML) intervening between the GPe and GPi (Mai et al. 2003; Woolsey 2003) show high concentrations of glycine receptors (Fig. 1K) and GABA$_A$ receptor subunits $\alpha_1\beta_{2,3},\gamma_2$ were also present.

3.4 *Substantia Nigra*

3.4.1 Substantia Nigra Pars Compacta

In the SNc, the majority of the pigmented neurons were largely devoid of intense dendritic and neuronal α_1 and $\beta_{2,3}$ subunit-IR, but were mainly labelled with the α_3 (Fig. 1J) and γ_2 subunits. However, a subpopulation of pigmented neurons (6.5%) also contained α_1 and $\beta_{2,3}$ subunits.

In addition, approximately 80% of the pigmented neurons were immunoreactive for glycine receptors.

3.4.2 Substantia Nigra Pars Reticulata

The non-pigmented neurons of the SNr showed parvalbumin immunoreactivity and a $GABA_A$ receptor subunit configuration of α_1, α_3, $\beta_{2,3}$, and γ_2 subunits (Fig. 1G) similar to the majority of the large parvalbumin positive neurons of the globus pallidus. By contrast approximately 40% of the SNr neurons were immunoreactive for glycine receptors.

4 Discussion

4.1 Striatum

Our studies on the regional and cellular localisation of $GABA_A$ receptor subunits and glycine receptors indicate a diversity of inhibitory receptor types in the striatum. First, at a regional level the various $GABA_A$ receptor subunits and glycine receptor show a heterogeneous distribution that follows the well-established striosome/matrix neurochemical subdivisions of the striatum (Graybiel 1983). The highest density of $GABA_A$ receptor subunits was found in the striosome compartment where glycine receptors were not present, whereas the matrix compartment contained lower levels of the $GABA_AR$ subunits and low levels of glycine receptors. The significance of this difference in the two compartments is still unclear, although it may be related to the concept that the striosomes and matrix represent two different functional domains within the striatum with the matrix carrying mainly sensory-motor information and the striosomes carrying mainly limbic associated information. It is well established that the subunit configuration of $GABA_A$ receptors determines their pharmacological and physiological properties (Barnard et al. 1987), and our results suggesting that different configurations or densities of receptors occur in the striosomes and matrix compartments indicate that different GABAergic and glycinergic receptor processing may operate in these two neurochemical compartments. In addition, the receptor combinations found on individual neurons would determine the receptor's physiological and pharmacological properties and govern their responses to activation by GABA (Barnard et al. 1984). The glycine receptor is also comprised of different subunits which may assemble into various combinations (Colquhoun and Sivilotti 2004).

At the cellular level, six types of striatal neurons were identified which included one group of projection neurons (type 4 medium spiny neurons) and five groups of striatal interneurons (types 1, 2, 3, 5, 6). See also Waldvogel et al. (1999) for details.

The medium spiny projection neurons of the striatum were labelled with calbindin and $GABA_AR$ α_2, α_3, $\beta_{2,3}$, γ_2 subunits (*type 4*). Although calbindin labels only the cell soma and proximal dendrites, it is assumed that these subunits also label the distal dendrites and spines. The medium spiny neurons which make up the vast majority of all striatal cells receive a variety of GABAergic inputs (Bolam et al. 2000) which would act on $GABA_A$ receptors with a different configuration to those on the striatal interneurons. In particular, they would include an α_2 subunit in their receptor configuration instead of an α_1 subunit which is prominent on GABAergic interneurons.

Of the striatal interneurons, three types (types 1, 2, and 3) were GAD-positive interneurons with $GABA_AR$ containing α_1, $\beta_{2,3}$, γ_2 subunits. The most intense staining of this subunit combination occurred on the *type 1* parvalbumin positive neurons, the *type 2* calretinin neurons, as well as on large pallidal-like *type 3* neurons. The parvalbumin positive neurons are the fast-spiking neurons of the striatum that inhibit medium spiny neurons (Koos and Tepper 1999). The inhibitory input onto the parvalbumin neurons has been shown to come from the globus pallidus (Bolam et al. 2000) which could act through the $GABA_A$ receptors (α_1, $\beta_{2,3}$, γ_2 subunits) coating these neurons. In addition, a subgroup of these parvalbumin positive interneurons was also labelled with glycine receptors, which suggests that they are influenced by an as yet unknown source of glycine. The *type 5* cholinergic interneurons were mainly labelled with $GABA_A$ α_3 subunits as well as highly expressing glycine receptors. This agrees well with animal studies which report that cholinergic neurons are associated with $GABA_A$ α_3 subunits (Gao et al. 1993) as well as with glycine receptors (Darstein et al. 2000). Previous studies suggest that activation of α_3 subunit containing $GABA_A$ receptors and glycine receptors can influence cholinergic neuron function which is in turn implicated in the regulation of dopaminergic mechanisms in the striatum (Darstein et al. 2000; Pisani et al. 2007). On the *type 6* somatostatin/NPY/NOS interneurons of the striatum, we were unable to show any labelling of the five major $GABA_A$ receptor subunits or glycine receptor. The receptor types mediating inhibition on this type of interneuron is inhibited are still unclear.

4.2 Globus Pallidus

In the globus pallidus, three types of neurons were identified; large parvalbumin and calretinin positive projection neurons (type 1 and 2 GP neurons) and a population of medium-sized calretinin positive neurons (type 3 GP neurons) which all have $GABA_AR$ assembled from the α_1, α_3, $\beta_{2,3}$, and γ_2 subunits (no α_2 subunit was detected in our studies). These subunits have also been detected in the GP in animal studies (Fritschy and Mohler 1995; Pirker et al. 2000), and other studies indicate that α_1 and α_3 subunits can be found alone or in combination in receptor complexes (Benke et al. 2004). Therefore, receptors formed from a complement of α_1, α_3, $\beta_{2,3}$, and γ_2 subunits may combine into various configurations including $\alpha_1\beta_{2,3}\gamma_2$, α_3,

$\beta_{2,3}\gamma_2$ or possibly $\alpha_1\alpha_3\beta_{2,3}\gamma_2$ receptors on pallidal neurons. Additionally glycine receptors were found on a subgroup of neurons within the globus pallidus which showed varying morphology; whether they are interneurons or projection neurons and if they colocalise with $GABA_AR$ containing neurons still needs further investigation.

An interesting finding in our recent studies was that neurons which form a thin "shell-like" layer between the putamen and globus pallidus (LML) and between the GPe and GPi (MML) show high densities of glycine receptor immunoreactivity (Fig. 1K) that were co-localised with cholinergic and parvalbumin positive neurons (see also Waldvogel et al. 2007). However, it is unclear whether these neurons contain both $GABA_A$ receptors and glycine receptor types. The possibility arises that this particular group of neurons in the medullary laminae encapsulates the globus pallidus and may serve a distinctive inhibitory function in the circuitry of the basal ganglia. They may well serve to modulate neuronal function in the globus pallidus in a manner which resembles the GABAergic modulation of the thalamic nuclei by the thalamic reticular nucleus (McCormick et al. 1997). The source of glycine in the striatum and globus pallidus is still unknown, but is highly likely to be derived from brain stem pathways which are enriched with glycine (Rajendra et al. 1997) or possibly generated locally by glycine secreting neurons or glia. In addition, glycine may not be the only agonist involved in activation of these glycine receptors, as the amino acid taurine may also be an endogenous ligand for glycine receptors (Ericson et al. 2006).

4.3 Substantia Nigra

Glycine receptor immunoreactivity was localized to the majority of pigmented neurons in the SNc and approximately 40% of the neurons in the SNr, see also Waldvogel et al. (2008) for details. However, the neurons of the SNc and SNr had different $GABA_AR$ configurations. SNr neurons had α_1, α_3, $\beta_{2,3}$, γ_2 (no α_2) subunits and showed similar subunit configurations to those of the globus pallidus. By contrast SNc neurons expressed mainly $GABA_AR$ α_3, γ_2 subunits, which are a unique subtype of receptor in the basal ganglia. The possibility exists that in the SNc other subunits may also be present in $GABA_A$ receptor complexes such as β_1 or a_4 in addition to α_3 and γ_2 subunits.

These results demonstrate that in the basal ganglia, neurons were generally associated with one of the four different $GABA_A$ receptor configurations: (1) α_1, α_3, $\beta_{2,3}$, γ_2 subunits are associated with striatal GABAergic interneurons and projection neurons of the GP and the SNr; (2) α_2, α_3, $\beta_{2,3}$, γ_2 on striatal projection neurons; (3) α_3 subunits on ChAT neurons; and (4) α_3, γ_2 on SNc neurons. Glycine receptors were found on striatal interneurons, pallidal and nigral neurons, and neurons of the intermedullary laminae.

4.4 Functional Considerations

GABA is the major transmitter used by the vast majority of cells in the basal ganglia (Smith et al. 2001). Recent studies on the effects of GABA in the monkey and rat striatum demonstrates that GABA acting through $GABA_A$R strongly inhibits firing rates of striatal, pallidal, and SNr neurons (Windels and Kiyatkin 2006). In the striatum, it limits synchronous striatal and cortical activity (Darbin and Wichmann 2008; Koos and Tepper 1999), and in the GP and SNr it reduces the firing rates of projection neurons. The results also show that in the substantia nigra there is marked $GABA_A$ receptor subunit heterogeneity in the SNr compared to the SNc suggesting that GABA exerts quite different effects on pars compacta and pars reticulata neurons in the human SN via $GABA_A$ receptors of different subunit configurations. In addition, glycine also plays an as yet undefined inhibitory role in the substantia nigra.

These findings provide new information for understanding the complexity of GABAergic and glycinergic functions in the human basal ganglia. This will translate into a better knowledge of inhibitory mechanisms in normal brain function and their involvement in neurodegenerative diseases such as Huntington's disease which results from the selective degeneration of inhibitory projection neurons in the striatum of the human basal ganglia.

Acknowledgements This work was supported by grants from the Neurological Foundation of New Zealand and the Health Research Council of New Zealand. KB is grateful for support from the British Royal Society. We thank the Neurological Foundation of New Zealand Human Brain Bank for providing the human brain tissue used in these studies. We also thank the Biomedical Imaging Research Unit (BIRU) in the Department of Anatomy with Radiology University of Auckland for expert assistance and use of their facilities.

References

Adams JC (1981) Heavy metal intensification of DAB-based HRP reaction product. J Histochem Cytochem 29: 775.

Barnard EA, Stephenson FA, Sigel E, Mamalaki C, Bilbe G, Constanti A, Smart TG and Brown DA (1984) Structure and properties of the brain GABA/benzodiazepine receptor complex. Adv Exp Med Biol 175: 235–254.

Barnard EA, Darlison MG and Seeburg P (1987) Molecular biology of the GABAA receptor: The receptor/channel superfamily. Trends Neurosci 10: 502–509.

Benke D, Mertens S, Trzeciak A, Gillessen D and Mohler H (1991) GABAA receptors display association of gamma 2-subunit with alpha 1- and beta 2/3-subunits. J Biol Chem 266: 4478–4483.

Benke D, Honer M, Michel C and Mohler H (1996) Gaba(a) receptor subtypes differentiated by their gamma-subunit variants – prevalence, pharmacology and subunit architecture. Neuropharmacology 35: 1413–1423.

Benke D, Fakitsas P, Roggenmoser C, Michel C, Rudolph U and Mohler H (2004) Analysis of the presence and abundance of GABA(A) receptors containing two different types of alpha subunits in murine brain using point-mutated alpha subunits. J Biol Chem 279: 43654–43660.

Bolam JP, Hanley JJ, Booth PA and Bevan MD (2000) Synaptic organisation of the basal ganglia. J Anat 196: 527–542.

Boyes J and Bolam JP (2007) Localization of GABA receptors in the basal ganglia. Prog Brain Res 160: 229–243.

Colquhoun D and Sivilotti LG (2004) Function and structure in glycine receptors and some of their relatives. Trends Neurosci 27: 337–344.

Darbin O and Wichmann T (2008) Effects of striatal GABA-A receptor blockade on striatal and cortical activity in monkeys. J Neurophysiol. 99:1294–1305.

Darstein M, Loschmann PA, Knorle R and Feuerstein TJ (1997) Strychnine-sensitive glycine receptors inducing [3H]-acetylcholine release in rat caudatoputamen: A new site of action of ethanol? Naun Schmied Arch Pharmacol 356: 738–745.

Darstein M, Landwehrmeyer GB, Kling C, Becker CM and Feuerstein TJ (2000) Strychnine-sensitive glycine receptors in rat caudatoputamen are expressed by cholinergic interneurons. Neuroscience 96: 33–39.

Ericson M, Molander A, Stomberg R and Soderpalm B (2006) Taurine elevates dopamine levels in the rat nucleus accumbens; antagonism by strychnine. Eur J Neurosci 23: 3225–3229.

Fritschy JM and Mohler H (1995) GABA(A)-receptor heterogeneity in the adult rat brain - Differential regional and cellular distribution of seven major subunits. J Comp Neurol 359: 154–194.

Fritschy JM, Weinmann O, Wenzel A and Benke D (1998) Synapse-specific localization of NMDA and GABA(A) receptor subunits revealed by antigen-retrieval immunohistochemistry. J Comp Neurol 390: 194–210.

Fujiyama F, Fritschy JM, Stephenson FA and Bolam JP (2000) Synaptic localization of GABA(A) receptor subunits in the striatum of the rat. J Comp Neurol 416: 158–172.

Gao B, Fritschy JM, Benke D and Mohler H (1993) Neuron-specific expression of GABAA-receptor subtypes: Differential association of the a1 and a3-subunits with serotonergic and GABAergic neurons. Neuroscience 54: 881–892.

Graybiel AM (1983) Compartmental organization of the mammalian striatum. Progr Brain Res 58: 247–256.

Graybiel AM (1995) The basal ganglia. Trends Neurosci 18: 60–62.

Hornykiewicz O (2001) Chemical neuroanatomy of the basal ganglia - normal and in Parkinson's disease. J Chem Neuroanat 22: 3–12.

Koos T and Tepper JM (1999) Inhibitory control of neostriatal projection neurons by GABAergic interneurons. Nat Neurosci 2: 467–472.

Mai J, Paxinos G and Assheuer J (2003) Atlas of the Human Brain. Academic, New York, NY.

McCormick DA, Huguenard JR, Bal T and Pape HC (1997) In Steriade M, Jones EG and McCormick D (Eds) Thalamus. Oxford, Elsevier.

Pfeiffer F, Simler R, Grenningloh G and Betz H (1984) Monoclonal antibodies and peptide mapping reveal structural similarities between the subunits of the glycine receptor of rat spinal cord. Proc Natl Acad Sci USA 81: 7224–7227.

Pirker S, Schwarzer C, Wieselthaler A, Sieghart W and Sperk G (2000) GABA(A) receptors: Immunocytochemical distribution of 13 subunits in the adult rat brain. Neuroscience 101: 815–850.

Pisani A, Bernardi G, Ding J and Surmeier DJ (2007) Re-emergence of striatal cholinergic interneurons in movement disorders. Trends Neurosci 30: 545–553.

Rajendra S, Lynch JW and Schofield PR (1997) The glycine receptor. Pharmacol Ther 73: 121–146.

Schroder S, Hoch W, Becker CM, Grenningloh G and Betz H (1991) Mapping of antigenic epitopes on the alpha 1 subunit of the inhibitory glycine receptor. Biochem 30: 42–47.

Smith Y, Bevan MD, Shink E and Bolam JP (1998) Microcircuitry of the direct and indirect pathways of the basal ganglia. Neuroscience 86: 353–387.

Smith Y, Charara A, Paquet M, Kieval JZ, Pare JF, Hanson JE, Hubert GW, Kuwajima M and Levey AI (2001) Ionotropic and metabotropic GABA and glutamate receptors in primate basal ganglia. J Chem Neuroanat 22:13–42.

Waldvogel HJ, Kubota Y, Fritschy JM, Mohler H and Faull RLM (1999) Regional and cellular localisation of GABA(A) receptor subunits in the human basal ganglia: An autoradiographic and immunohistochemical study. J Comp Neurol 415: 313–340.

Waldvogel HJ, Billinton A, White JH, Emson PC and Faull RL (2004) Comparative cellular distribution of GABAA and GABAB receptors in the human basal ganglia: Immunohistochemical colocalization of the alpha 1 subunit of the GABAA receptor, and the GABABR1 and GABABR2 receptor subunits. J Comp Neurol 470: 339–356.

Waldvogel HJ, Curtis MA, Baer K, Rees MI and Faull RL (2006) Immunohistochemical staining of post-mortem adult human brain sections. Nat Protoc 1: 2719–2732.

Waldvogel HJ, Baer K, Allen KL, Rees MI and Faull RL (2007) Glycine receptors in the striatum, globus pallidus, and substantia nigra of the human brain: An immunohistochemical study. J Comp Neurol 502: 1012–1029.

Waldvogel HJ, Baer K, Gai WP, Gilbert RT, Rees MI, Mohler H and Faull RL (2008) Differential localization of GABA(A) receptor subunits within the substantia nigra of the human brain: An immunohistochemical study. J Comp Neurol 506: 912–929.

Windels F and Kiyatkin EA (2006) GABAergic mechanisms in regulating the activity state of substantia nigra pars reticulata neurons. Neuroscience 140: 1289–1299.

Woolsey T (2003) In Woolsey T, Hanaway J, and Gado M (Eds) The Brain Atlas: A Visual Guide to the Central Nervous System, 2nd ed. Wiley, Hoboken, NJ.

Comparative Ultrastructural Analysis of D1 and D5 Dopamine Receptor Distribution in the Substantia Nigra and Globus Pallidus of Monkeys

Michele A. Kliem, Jean-Francois Pare, Zafar U. Khan, Thomas Wichmann, and Yoland Smith

Abstract Dopamine acts through the D1-like (D1, D5) and D2-like (D2, D3, D4) receptor families. Various studies have shown a preponderance of presynaptic dopamine D1 receptors on axons and terminals in the internal globus pallidus (GPi) and substantia nigra reticulata (SNr), but little is known about D5 receptors distribution in these brain regions. In order to further characterize the potential targets whereby dopamine could mediate its effects in basal ganglia output nuclei, we undertook a comparative electron microscopic analysis of D1 and D5 receptors immunoreactivity in the GPi and SNr of rhesus monkeys. At the light microscopic level, D1 receptor labeling was confined to small punctate elements, while D5 receptor immunoreactivity was predominantly expressed in cellular and dendritic processes throughout the SNr and GPi. At the electron microscopic level, 90% of D1 receptor labeling was found in unmyelinated axons or putative GABAergic terminals in both basal ganglia output nuclei. In contrast, D5 receptor labeling showed a different pattern of distribution. Although the majority (65–75%) of D5 receptor immunoreactivity was also found in unmyelinated axons and terminals in GPi and SNr, significant D5 receptor immunolabeling was also located in dendritic and glial processes. Immunogold studies showed that about 50% of D1 receptor immunoreactivity in axons was bound to the plasma membrane providing functional sites for D1 receptor-mediated effects on transmitter release in GPi and SNr. These findings provide evidence for the existence of extrastriatal pre- and postsynaptic targets through which dopamine and drugs acting at D1-like receptors may regulate the basal ganglia outflow and possibly exert some of their anti-parkinsonian effects.

M.A. Kliem (✉), J.-F. Pare, Z.U. Khan, T. Wichmann, and Y. Smith
Yerkes National Primate Research Center, Emory University, 954 Gatewood Road, Atlanta, GA 30322, USA
e-mail: mkliem@emory.edu

H.J. Groenewegen et al. (eds.), *The Basal Ganglia IX*, Advances in Behavioral Biology 58, DOI 10.1007/978-1-4419-0340-2_19,

1 Introduction

Dopamine is an important modulator of neuronal activity in the basal ganglia circuitry. The most prominent dopaminergic projection originates in the substantia nigra pars compacta (SNc) and terminates in the striatum (Bernheimer et al. 1973; Hornykiewicz and Kish 1987). Degeneration of this pathway is known to contribute to the development of parkinsonism. However, dopamine also reaches basal ganglia areas outside of the striatum, including the internal pallidal segment (GPi; Smith et al. 1989; Pifl et al. 1990; Schneider and Rothblat 1991; Whone et al. 2003) and the substantia nigra pars reticulata (SNr; Bernheimer et al. 1973; Geffen et al. 1976; Cheramy et al. 1981). It is, therefore, possible that dopamine loss at these sites may also contribute to the development of parkinsonism. In GPi, dopamine is released from terminals of direct axonal projections from the SNc which, in monkeys, are in large part separate from the nigrostriatal projection (Smith et al. 1989; Jan et al. 2000). By contrast, the dopamine supply to the SNr is through release from dendrites of SNc neurons (Björklund and Lindvall 1975; Nieoullon et al. 1978; Arsenault et al. 1988).

The physiological actions of dopamine are mediated through two families of metabotropic receptors, D1-like receptors (D1LRs; Clark and White 1987; Neve 1997) and D2-like receptors (Neve 1997). D1LRs are strongly expressed in the monkey GPi and SNr (Richfield et al. 1987; Besson et al. 1988). D1 receptors are predominately presynaptic in axons and axon terminals of striatopallidal and striatonigral projection neurons (Barone et al. 1987; Fremeau et al. 1991; Mengod et al. 1991; Levey et al. 1993; Yung et al. 1995), where they regulate GABA release (Kliem et al. 2007). The exact location and function of D5 receptors have not been extensively studied, but qualitative immunohistochemical observations revealed that these receptors are expressed at pre- and postsynaptic sites in the rat SNr (Khan et al. 2000). Electrophysiologic studies have shown D5 receptor-mediated modulation of neuronal activity in other basal ganglia nuclei (Yan and Surmeier 1997; Baufreton et al. 2003).

In a recent electrophysiologic study in monkeys, we demonstrated that local microinfusions of a D1LR antagonist increased neuronal discharge rates in GPi, suggesting that these receptors are tonically occupied by endogenous dopamine (Kliem et al. 2007). In contrast, infusions of a D1LR agonist significantly reduced the neuronal firing rate in both GPi and SNr. Similar agonist injections in parkinsonian animals resulted in the same effects, demonstrating that D1LRs are functionally active in the dopamine-depleted state (unpublished data). In order to find out whether loss of nigral dopamine neurons influences the location of D1LRs in basal ganglia output nuclei, we compared the cellular and ultrastructural location of D1 and D5 receptor immunoreactivity in the GPi and SNr of normal and parkinsonian monkeys.

2 Materials and Methods

2.1 Animals

Brain tissue from 13 (D1, $n = 11$; D5, $n = 7$) Rhesus monkeys (*Macacca mulatta*, 3–10 kg) was used for this study. All experiments were performed in accordance with the NIH Guide for the Care and Use of Laboratory Animals, and the PHS Policy on Humane Care and Use of Laboratory Animals (amended 2002), and were approved by the Institutional Animal Care and Use Committee at Emory University.

2.2 MPTP Administration and Behavioral Assessment

Five of the animals received the neurotoxin 1-methyl-4-phenyl-1,2,3,6-tetrahydropyridine (MPTP; Sigma) i.m. (0.25 mg/kg) twice weekly until a stable parkinsonism was reached. Two of these animals had previously received two MPTP injections into the right carotid artery (0.5–0.7 mg/kg per injection). A parkinsonian rating scale, changes in spontaneous cage behavior, and automated activity counting procedures (using an infrared beam system) were used to document the degree of MPTP-induced motor disability and its stability. All MPTP-treated animals showed stable parkinsonian motor signs, including bradykinesia, rigidity, and postural instability.

2.3 Tissue Preparation

The animals were killed with an overdose of pentobarbital, followed by transcardiac perfusion with paraformaldehyde (4%) and glutaraldehyde (0.1%). The brains were then removed from the skull and cut in 10-mm-thick blocks containing GPi or SNr. Tissue processed for immunocytochemistry was cut in 60-μm sections with a vibratome, rinsed in phosphate-buffered saline (PBS; 0.01 M, pH 7.4), incubated with 1% sodium borohydride solution in PBS, rinsed in PBS, exposed to cyroprotectant, frozen at –80°C, thawed and rinsed in PBS.

2.4 Primary Antisera

The specific monoclonal D1 receptor antibodies used in this study (1:75; Sigma-Aldrich, St. Louis, MO; Levey et al. 1993) were raised in rats against 97 amino acids in the carboxy terminus of the D1 receptor, while the selective D5 receptor antibodies used in this study (1:500; Khan et al. 2000) recognize ten different amino acids in the carboxy terminus of the D5 receptor.

2.5 Immunoperoxidase Procedure

The sections were first exposed for 1 h at 4°C to a solution containing normal goat serum (10%) and bovine serum albumin (1%) to block nonspecific binding. They were then incubated for 48 h in primary antibody solutions (rat anti-D1 antibodies or rabbit anti-D5 antibodies). Following this, they were incubated in a solution containing biotinylated secondary antibodies [goat anti-rat (D1) antibodies or anti–rabbit (D5) antibodies, Vector Labs, Burlingame, CA; 1:200] followed by the avidin-biotin complex solution (Vectastain Standard Kit, Vector Labs; 1:100). Following rinses in PBS and Tris buffer (0.5 M, pH 7.6) they were exposed to a 10-min incubation in a solution containing imidazole (0.01 M; Fisher Scientific, Hampton, NH), hydrogen peroxide (0.006%) and 3–3′-diaminobenzidine tetrahydrochloride (0.025%; Sigma-Aldrich).

After rinsing in PBS and phosphate buffer (PB; 0.1 M, pH 7.4), the sections were postfixed in osmium tetroxide (1%), followed by washes in PB. Then, a series of incubations in increasing concentrations of alcohol followed by propylene oxide were performed to dehydrate the tissue. Sections were exposed to uranyl acetate (1%) in a 70% alcohol solution to enhance the contrast under the microscope. Finally, the sections were embedded in epoxy resin (Durcupan, ACM; Fluka, Ft. Washington, PA), mounted on microscope slides and placed in the oven (60°C) for 2 days.

Blocks of GPi and SNr tissue were cut from the microscope slides and glued onto resin blocks. Ultrathin sections (60 nm) were cut with an ultramicrotome (Leica Ultracut T2, Nussloch, Germany) and mounted onto Pioloform-coated single copper grids, and stained with lead citrate.

2.6 Immunogold Procedure

After they were treated with a solution containing milk (5%) and bovine serum albumin (BSA, 1%) in PBS for 1 h, followed by rinses in Tris-buffered saline-gelatin, sections were transferred to wells containing milk (1%), BSA (1%), and the D1 receptor primary antibody (see above). Next, the sections were rinsed and exposed to goat anti-rat IgGs (1:100) conjugated to 1.4-nm gold particles. Afterward, they were transferred to the HQ kit (Nanoprobes) for silver intensification of the gold particles (5–10 min). The sections were then rinsed with aqueous sodium acetate buffer (2%) and PB, and then treated with osmium tetroxide (0.5% in PB, 0.1 M, pH 7.4). The remainder of the tissue preparation steps was similar to those described for the immunoperoxidase reaction.

2.7 Ultrastructural Analysis

Randomly encountered labeled elements in the immunoperoxidase- or immunogold-processed tissue were photographed under the electron microscope (Zeiss EM 10C, Thornwood, NY) at 16,000–25,000× using an electronic camera (DualView 300 W; Gatan, Pleasanton, CA), which was controlled by Digital Micrograph Software (Gatan, Inc., Warrendale, PA; v. 3.6.5). The micrographs were adjusted for brightness and contrast, if needed, with Digital Micrograph or Photoshop software (Adobe Systems, San Jose, CA) to improve the quality of the images for analysis.

For the analysis of immunoperoxidase data, immunoreactive elements from a series of 30–50 electron micrographs were categorized as axons, terminals, dendrites, or glial processes. The relative proportion of these elements was calculated and expressed as a percent of total labeled elements in the tissue areas that were examined. The mean percentages and standard deviations (SD) of labeled elements in GPi and SNr of normal and MPTP-treated monkeys were calculated.

For the analysis of immunogold data, silver-intensified gold particles were categorized as plasma membrane-bound or intracellular depending on whether they were in contact with the plasma membrane.

3 Results

3.1 Ultrastructural Localization of D1 Receptor Immunoreactivity in GPi and SNr

At the light microscopic level, the whole extent of the neuropil in GPi and SNr was enriched in D1-receptor-immunoreactive aggregates of small punctate structures, while cell bodies were devoid of immunoreactivity. At the electron microscopic level, the D1 receptor immunoreactivity was mostly located in unmyelinated, preterminal axons, accounting for 90.4 ± 2.2% and 89 ± 5.2% of all labeled elements in GPi and SNr, respectively, in normal monkeys (Figs. 1 and 3). The pattern was the same in MPTP-treated parkinsonian monkeys in which the relative proportion of labeled axonal processes was 90.2 ± 0.4% and 88.8 ± 5.9% of total labeled elements in GPi and SNr, respectively.

In immunogold-labeled tissue from three normal monkeys (Fig. 1), 47.7 ± 1.2% of total gold particles in GPi and 47.8 ± 2.1% of gold labeling in SNr was bound to the axonal plasma membrane. A significant proportion of gold particles were also bound to axonal plasma membranes in GPi (62.1%) and SNr (60.1%) of a parkinsonian monkey.

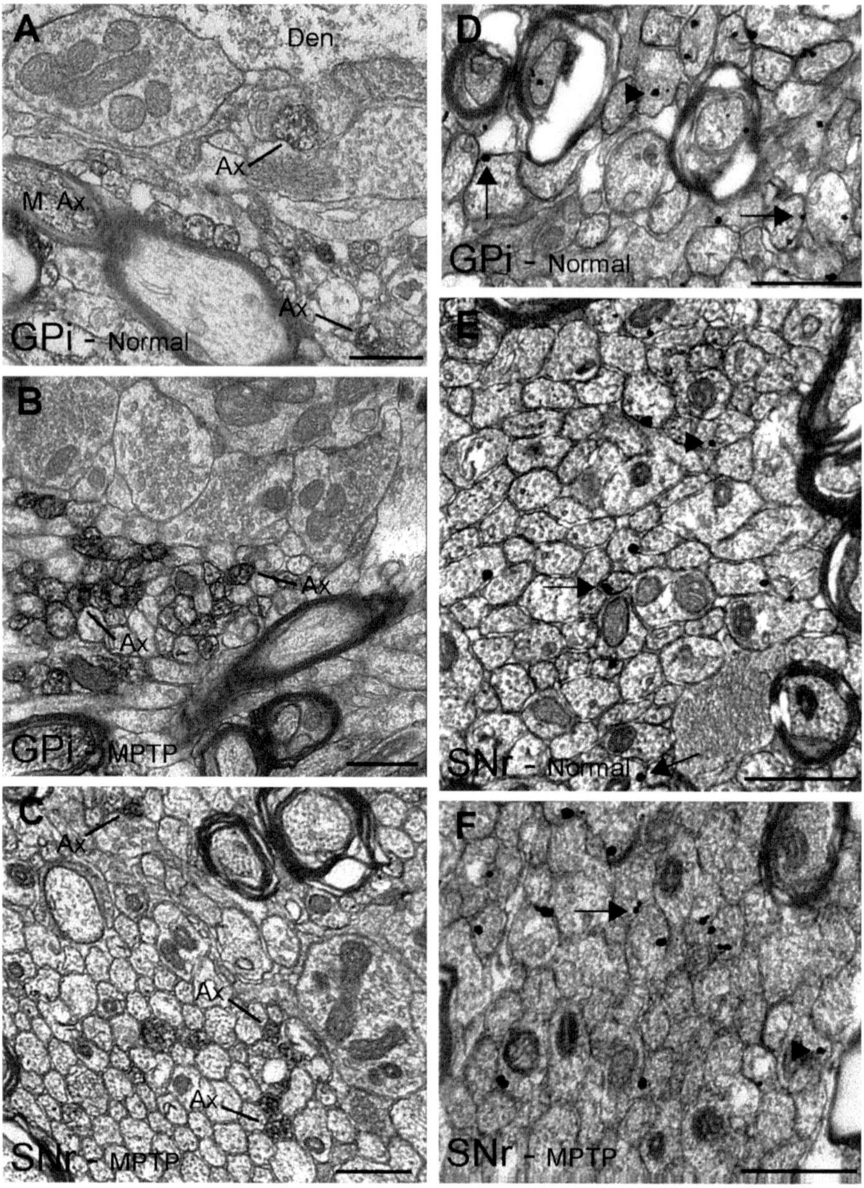

Fig. 1 Localization of D1 receptors in GPi and SNr of normal and MPTP-treated monkeys. D1 receptor-immunolabeled unmyelinated axons (Ax) are shown in the GPi and SNr of normal and MPTP-treated monkeys (**A–C**). An immunoreactive myelinated axon (M.Ax) is also depicted in GPi (**A**). D1 receptor immunogold labeling is apposed to the plasma membrane [*arrows* in (**D–F**)] or intracellular [*arrowheads* in (**D–F**)] in unmyelinated axons that travel through GPi and SNr of normal and MPTP-treated monkeys (**D–F**). Scale bars: 0.5 μm

3.2 Ultrastructural Localization of D5 Receptor Immunoreactivity in GPi and SNr

Overall, the pattern of D5 receptor labeling in GPi and SNr resembled that described for D1 receptors (Figs. 2 and 3). We found that 72.0 ± 8.5% of D5 receptor labeling in GPi and 67.0 ± 3.7% of labeling in SNr was expressed in unmyelinated axons. In MPTP-treated monkeys, 74.3 ± 17.6% and 74.8 ± 5.8% of D5 receptor immunoreactivity was found in unmyelinated axons in GPi and SNr, respectively.

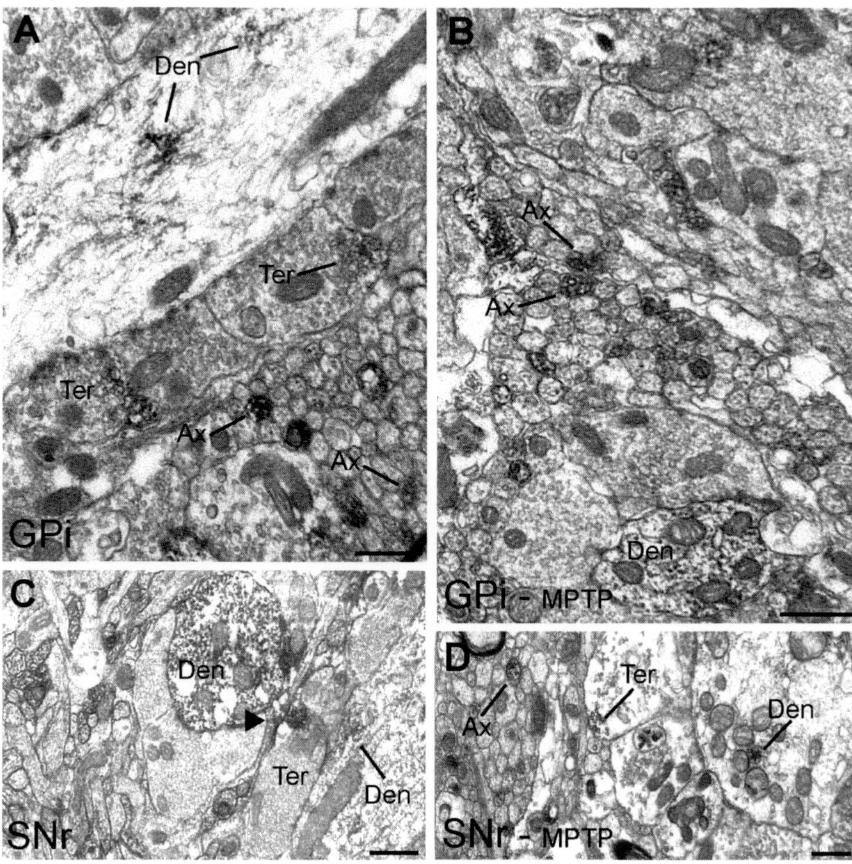

Fig. 2 Localization of D5 receptor immunoreactivity in GPi and SNr in normal and MPTP-treated monkeys. (**A–D**) Illustrate D5 receptor immunolabeling in dendrites (Den), unmyelinated axons (Ax) and terminals (Ter) in GPi (**A**, **B**) and SNr (**C**, **D**) of normal and MPTP-treated monkeys. Occasional glial labeling is also depicted in the SNr [*arrowhead* in (**C**)]. Scale bars: 0.5 μm

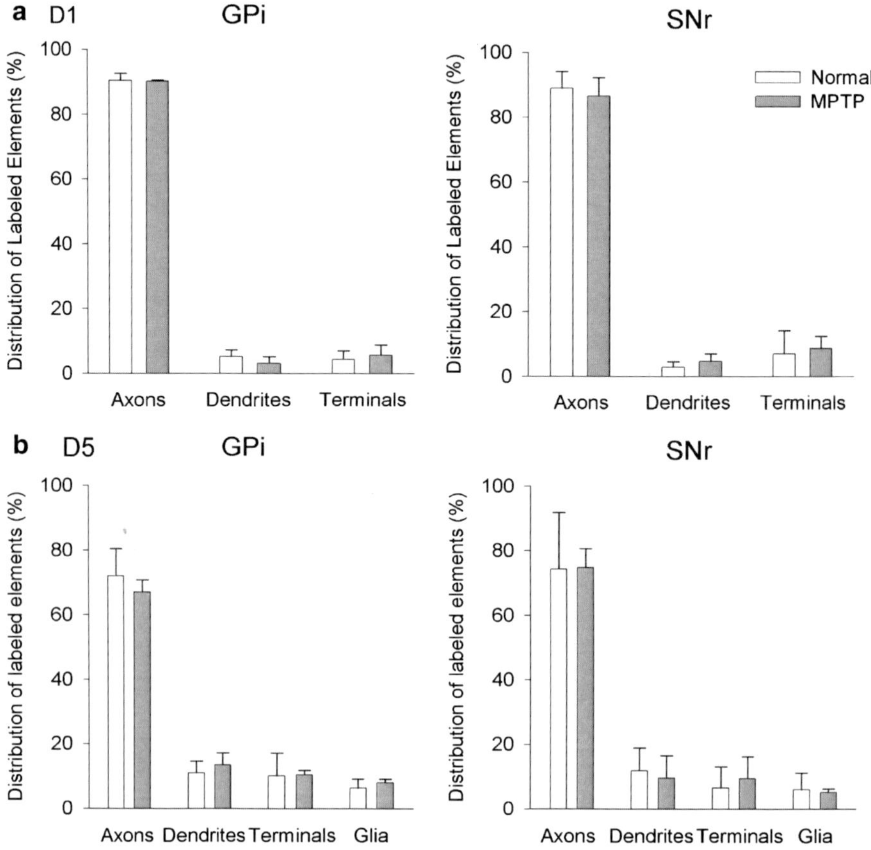

Fig. 3 Comparative distribution of D1 (**a**) and D5 (**b**) receptors in axons, dendrites, terminals and glia (**b** only) in GPi and SNr of normal and MPTP-treated monkeys. D1 receptors were almost exclusively found in unmyelinated axonal segments in the SNr [$n = 4$; surface area (SA), 2,936.4 μm^2 in normal; $n = 3$; SA, 2,202.3 μm^2 in MPTP-treated monkeys]. A similar pattern of labeling was found in GPi ($n = 3$ in normal; $n = 3$ in MPTP-treated monkeys; SA, 2,202.3 μm^2). The majority of D5 receptor immunoreactivity was confined to unmyelinated axons in GPi ($n = 3$ in normal; $n = 3$ in MPTP-treated monkeys; SA, 3,670.5 μm^2) and SNr ($n = 3$ in normal; $n = 3$ in MPTP-treated monkeys; SA, 3,450.3 and 3,670.5 μm^2, respectively). Data are expressed as percentages of immunolabeled elements. Each *bar* represents the mean ± SD

However, significant D5 receptor immunoreactivity was also detected in dendrites, accounting for 11.0 ± 3.6% of labeling in GPi and 12.0 ± 7.0% in SNr in normal monkeys. In parkinsonian monkeys, 13.5 ± 3.6% of all D5 receptor immunoreactive elements in GPi were in dendrites, while dendritic labeling accounted for 9.7 ± 6.9% of the labeled structures in the SNr.

4 Discussion

We found that D1 and D5 receptors are predominately located in preterminal axons and axon terminals in the GPi and SNr of normal and MPTP-treated parkinsonian monkeys. Our findings also revealed that D5, but not D1, receptors were significantly expressed postsynaptically in dendrites of GPi and SNr neurons in normal and parkinsonian monkeys. The prominent presynaptic location of D1LRs, presumably on GABAergic terminals of axons originating in the striatum, provides the endogenous dopamine system a means to influence the activity of the basal ganglia output nuclei through presynaptic regulation of striatofugal GABAergic transmission. Pre- and postsynaptic D1LRs may also provide potential targets for therapeutic agents with D1LR activity in Parkinson's disease.

4.1 Localization of D1/D5 Receptors in GPi and SNr

As in previous studies in rats, monkeys, and humans (Levey et al. 1991; Gnanalingham et al. 1993; Yung et al. 1995; Caille et al. 1996; Betarbet and Greenamyre 2004), we found that D1 receptors are strongly expressed in unmyelinated axons and putative GABAergic axon terminals in GPi and SNr of normal and MPTP-treated monkeys. In immunogold studies, most gold particles were located on axonal plasma membranes in GPi and SNr of normal and parkinsonian monkeys. Therefore, endogenous dopamine or drugs acting at these sites may act to increase GABA efflux and reduce discharge of basal ganglia output neurons in normal and pathological conditions.

Previous qualitative light microscopy studies using in situ hybridization and immunohistochemical techniques demonstrated D5 receptor immunoreactivity on cell bodies and dendrites of GPi and SNr neurons in monkeys (Bergson et al. 1995; Choi et al. 1995; Ciliax et al. 2000), or axons and dendrites in the rat SNr (Khan et al. 2000). Data from our study are in line with those of Kahn et al. (2000), using the same antibodies; D5 receptor immunoreactivity being expressed either presynaptically, in unmyelinated axons and axon terminals, or postsynaptically in dendrites in both basal ganglia output nuclei of normal and MPTP-treated monkeys.

4.2 Functional Consequences of Dopamine D1/D5 Receptor Activation

It is difficult to distinguish pharmacologically between D1 and D5 receptors, and effects of activation of these receptors in GPi and SNr have not been studied separately. It is known, however, that D1LR activation (i.e., combined activation of D1 and D5 receptors) enhances GABA efflux in the rodent and monkey basal ganglia output nuclei (Floran et al. 1990; Aceves et al. 1995; Timmerman and Westerink 1995; Ferre et al. 1996; Rosales et al. 1997; Trevitt et al. 2002; Kliem et al. 2007). It is, therefore, expected that activation of D1LRs at these sites increases GABA release from terminals of the striato-GPi and striatonigral projections, thereby reducing neuronal firing in SNr and GPi. We, indeed, demonstrated recently that administration of the selective dopaminergic D1LR agonist, SKF82958, significantly reduced neuronal discharge in GPi and SNr of normal monkeys (Kliem et al. 2007), as was previously shown for neurons in the rat SNr and entopeduncular nucleus (Timmerman and Abercrombie 1996; Radnikow and Misgeld 1998; Floran et al. 2002; Windels and Kiyatkin 2006).

Comparatively, little is known about potential changes of D1LR function in the parkinsonian state. Studies using rat brain slices from dopamine-depleted animals have shown that D1LR effects on GABAergic transmission in the entopeduncular nucleus are reduced compared to the normal state (Floran et al. 1990; Aceves et al. 1995). In vivo electrophysiologic studies showed that systemically administered D1LR agonists reduce firing rates of SNr neurons in dopamine-depleted rats (Waszczak et al. 1984; Weick and Walters 1987). We have also recently demonstrated that local activation of D1LRs significantly inhibits GPi and SNr cells in MPTP-treated monkeys (Kliem et al. unpublished data), suggesting that D1LR-mediated effects are retained after dopamine depletion, consistent with the ultrastructural data presented in this study. Despite significant dopamine depletion and clear behavioral signs of parkinsonism, no significant change was found in the location or function of D1LRs in GPi and SNr of MPTP-treated monkeys (Kliem et al. unpublished data). At first glance, these observations appear to be in contrast with findings from the striatum showing that dopamine depletion induces significant changes in binding, affinity, and, presumably, function of D1LR in striatal projection neurons (Gagnon et al. 1990; Graham et al. 1993; Blanchet et al. 1996; Betarbet and Greenamyre 2004). However, it is noteworthy in our study that neither attempt at characterizing the pharmacological properties of D1LRs, provides any direct evidence for absolute decrease or increase in the total number of D1 or D5 receptor immunoreactive elements in basal ganglia output nuclei of parkinsonian monkeys.

4.3 Dopamine D1/D5 Receptor Activation May Influence Behavior

Studies in the rat have demonstrated that D1LR blockade in the SNr impairs motor activity and increases EMG activity and rigidity (Hemsley and Crocker 2001; Trevitt et al. 2001; Bergquist et al. 2003), suggesting that dendritic dopamine release in the nigra plays a role in regulating basal ganglia-related motor behaviors. Thus far, there is no clear evidence for a functionally significant dopaminergic tone in the monkey SNr, although our previous experiments have demonstrated that D1LR may be tonically activated by ambient dopamine in GPi. Degeneration of the SNc-GPi projection may also contribute to the development of some of the abnormal behaviors associated with parkinsonism because lower dopamine concentrations have been reported in GPi of parkinsonian patients (Bernheimer et al. 1973) and MPTP-treated monkeys (Pifl et al. 1992). Systemic administration of drugs with D1LR activity improves motor symptoms of parkinsonism in MPTP-treated monkeys (Blanchet et al. 1996; Goulet and Madras 2000). Based on findings presented in this and in our functional studies (Kliem et al. unpublished; Kliem et al. 2007), it is reasonable to believe that some of these behavioral effects may be mediated through modulation of GPi neuronal activity. Due to the limited knowledge about the localization and function of D5 receptors and the overall greater number of D1 receptors in the brain, most physiological and behavioral effects of mixed D1/D5-receptor agonist in the brain have been attributed to activation of D1 receptors. However, because dopamine has a higher affinity for the D5 receptor subtype, combined with our findings showing that D5 receptors are located to subserve pre- and postsynaptic regulation of neuronal activity in basal ganglia output nuclei, we should consider that D5 receptors may have a more significant role in normal and pathological basal ganglia functions than previously thought (Grandy et al. 1991; Sunahara et al. 1991; Tiberi et al. 1991).

Recently it has become possible to study the functions of D1 and D5 receptors separately, by genetic inactivation of one of the two receptor subtypes. These studies suggested that D1 receptor stimulation may facilitate movement, while D5 receptor stimulation may have an opposite effect on motor activity. For instance, intracerebroventricular administration of an antisense oligodeoxynucleotide against the D5 receptor in dopamine-depleted rats blocks D1LR agonist-induced contralateral turning (Dziewczapolski et al. 1998). This is further supported by the observation that D5 receptor knock-out mice are hyperactive (Sibley 1999), whereas D1 receptor knock-out mice show reduced motor activity (Centonze et al. 2003). Hopefully, the future development of selective D1 or D5 receptor agonists and antagonists will open up the possibility to clearly assess the functional roles of each of these receptors individually in normal monkeys or in animal models of parkinsonism.

4.4 Functional Significance of D1LRs-Mediated Effects in Parkinsonism

Our study demonstrates that the overall pattern of D1LR expression in both basal ganglia output nuclei is unchanged after dopamine depletion in MPTP-treated monkeys, suggesting that D1LR-mediated effects may not be significantly affected in Parkinson's disease, and therefore, remain potential therapeutic targets. Dopamine replacement therapies aimed at the SN, indeed, support this view. For instance, glia-derived neurotrophic factor infusions (Gerhardt et al. 1999) and fetal mesencephalic tissue grafts (Starr et al. 1999) in the SN ameliorate parkinsonian motor signs in adult MPTP-treated monkeys, underlining the importance of dendritic dopamine release in mediating parkinsonism in primates. Our results suggest that enhancing dopamine release in GPi in parkinsonism may also be a useful therapeutic strategy for parkinsonism. Therapies that exploit the prokinetic effects of D1 receptor activation, and possibly D5 receptor antagonism, could also be particularly useful.

References

Aceves J, Floran B, Sierra A and Mariscal S (1995) D-1 receptor mediated modulation of the release of gamma-aminobutyric acid by endogenous dopamine in the basal ganglia of the rat. Prog Neuropsychopharmacol Biol Psychiatry 19: 727–739.

Arsenault MY, Parent A, Seguela P and Descarries L (1988) Distribution and morphological characteristics of dopamine-immunoreactive neurons in the midbrain of the squirrel monkey (*Saimiri sciureus*). J Comp Neurol 267: 489–506.

Barone P, Bankiewicz KS, Corsini GU, Kopin IJ and Chase TN (1987) Dopaminergic mechanisms in hemiparkinsonian monkeys. Neurology 37: 1592–1595.

Baufreton J, Garret M, Rivera A, de la Calle A, Gonon F, Dufy B, Bioulac B and Taupignon A (2003) D5 (not D1) dopamine receptors potentiate burst-firing in neurons of the subthalamic nucleus by modulating an L-type calcium conductance. J Neurosci 23: 816–825.

Bergquist F, Shahabi HN and Nissbrandt H (2003) Somatodendritic dopamine release in rat substantia nigra influences motor performance on the accelerating rod. Brain Res 973: 81–91.

Bergson C, Mrzljak L, Smiley JF, Pappy M, Levenson R and Goldman-Rakic PS (1995) Regional, cellular, and subcellular variations in the distribution of D1 and D5 dopamine receptors in primate brain. J Neurosci 15: 7821–7836.

Bernheimer H, Birkmayer W, Hornykiewicz O, Jellinger K and Seitelberger F (1973) Brain dopamine and the syndromes of Parkinson and Huntington. J Neurol Sci 20: 415–455.

Besson MJ, Graybiel AM and Nastuk MA (1988) [3H]SCH 23390 binding to D1 dopamine receptors in the basal ganglia of the cat and primate: Delineation of striosomal compartments and pallidal and nigral subdivisions. Neuroscience 26: 101–119.

Betarbet R and Greenamyre JT (2004) Regulation of dopamine receptor and neuropeptide expression in the basal ganglia of monkeys treated with MPTP. Exp Neurol 189: 393–403.

Björklund A and Lindvall O (1975) Dopamine in dendrites of substantia nigra neurons: Suggestions for a role in dendritic terminals. Brain Res 83: 531–537.

Blanchet PJ, Grondin R, Bedard PJ, Shiosaki K and Britton DR (1996) Dopamine D1 receptor desensitization profile in MPTP-lesioned primates. Eur J Pharmacol 309: 13–20.

Caille I, Dumartin B and Bloch B (1996) Ultrastructural localization of D1 dopamine receptor immunoreactivity in rat striatonigral neurons and its relation with dopaminergic innervation. Brain Res 730: 17–31.

Centonze D, Grande C, Saulle E, Martin AB, Gubellini P, Pavon N, Pisani A, Bernardi G, Moratalla R and Calabresi P (2003) Distinct roles of D1 and D5 dopamine receptors in motor activity and striatal synaptic plasticity. J Neurosci 23: 8506–8512.

Cheramy A, Leviel V and Glowinski J (1981) Dendritic release of dopamine in the substantia nigra. Nature 289: 537–542.

Choi WS, Machida CA and Ronnekleiv OK (1995) Distribution of dopamine D1, D2, and D5 receptor mRNAs in the monkey brain: Ribonuclease protection assay analysis. Brain Res 31: 86–94.

Ciliax BJ, Nash N, Heilman C, Sunahara R, Hartney A, Tiberi M, Rye DB, Caron MG, Niznik HB and Levey AI (2000) Dopamine D(5) receptor immunolocalization in rat and monkey brain. Synapse 37: 125–145.

Clark D and White FJ (1987) D1 dopamine receptor – the search for a function: A critical evaluation of the D1/D2 dopamine receptor classification and its functional implications. Synapse 1: 347–388.

Dziewczapolski G, Menalled LB, Garcia MC, Mora MA, Gershanik OS and Rubinstein M (1998) Opposite roles of D1 and D5 dopamine receptors in locomotion revealed by selective antisense oligonucleotides. Neuroreport 9: 1–5.

Ferre S, O'Connor WT, Svenningsson P, Björklund L, Lindberg J, Tinner B, Stromberg I, Goldstein M, Ogren SO, Ungerstedt U, Fredholm BB and Fuxe K (1996) Dopamine D1 receptor-mediated facilitation of GABAergic neurotransmission in the rat strioentopenduncular pathway and its modulation by adenosine A1 receptor-mediated mechanisms. Eur J Neurosci 8: 1545–1553.

Floran B, Aceves J, Sierra A and Martinez-Fong D (1990) Activation of D1 dopamine receptors stimulates the release of GABA in the basal ganglia of the rat. Neurosci Lett 116: 136–140.

Floran B, Barajas C, Floran L, Erlij D and Aceves J (2002) Adenosine A1 receptors control dopamine D1-dependent [(3)H]GABA release in slices of substantia nigra pars reticulata and motor behavior in the rat. Neuroscience 115: 743–751.

Fremeau RT, Jr., Duncan GE, Fornaretto MG, Dearry A, Gingrich JA, Breese GR and Caron MG (1991) Localization of D1 dopamine receptor mRNA in brain supports a role in cognitive, affective, and neuroendocrine aspects of dopaminergic neurotransmission. Proc Natl Acad Sci USA 88: 3772–3776.

Gagnon C, Bedard PJ and Di Paolo T (1990) Effect of chronic treatment of MPTP monkeys with dopamine D-1 and/or D-2 receptor agonists. Eur J Pharmacol 178: 115–120.

Geffen LB, Jessell TM, Cuello AC and Iversen LL (1976) Release of dopamine from dendrites in rat substantia nigra. Nature 260: 258–260.

Gerhardt GA, Cass WA, Huettl P, Brock S, Zhang Z and Gash DM (1999) GDNF improves dopamine function in the substantia nigra but not the putamen of unilateral MPTP-lesioned rhesus monkeys. Brain Res 817: 163–171.

Gnanalingham KK, Smith LA, Hunter AJ, Jenner P and Marsden CD (1993) Alterations in striatal and extrastriatal D-1 and D-2 dopamine receptors in the MPTP-treated common marmoset: An autoradiographic study. Synapse 14: 184–194.

Goulet M and Madras BK (2000) D(1) dopamine receptor agonists are more effective in alleviating advanced than mild parkinsonism in 1-methyl-4-phenyl-1,2,3, 6-tetrahydropyridine-treated monkeys. J Pharmacol Exp Ther 292: 714–724.

Graham WC, Sambrook MA and Crossman AR (1993) Differential effect of chronic dopaminergic treatment on dopamine D1 and D2 receptors in the monkey brain in MPTP-induced parkinsonism. Brain Res 602: 290–303.

Grandy DK, Zhang Y, Bouvier C, Zhou QY, Johnson RA, Allen L, Buck K, Bunzow JR, Salon J and Civelli O (1991) Multiple human-D5 dopamine receptor genes - a functional receptor and 2 pseudogenes. Proc Natl Acad Sci USA 88: 9175–9179.

Hemsley KM and Crocker AD (2001) Changes in muscle tone are regulated by D1 and D2 dopamine receptors in the ventral striatum and D1 receptors in the substantia nigra. Neuropsychopharmacology 25: 514–526.

Hornykiewicz O and Kish SJ (1987) Biochemical pathophysiology of Parkinson's disease. Adv Neurol 45: 19–34.

Jan C, Francois C, Tande D, Yelnik J, Tremblay L, Agid Y and Hirsch E (2000) Dopaminergic innervation of the pallidum in the normal state, in MPTP-treated monkeys and in parkinsonian patients. Eur J Neurosci 12: 4525–4535.

Khan ZU, Gutierrez A, Martin R, Penafiel A, Rivera A and de la Calle A (2000) Dopamine D5 receptors of rat and human brain. Neuroscience 100: 689–699.

Kliem MA, Maidment NT, Ackerson LC, Chen S, Smith Y and Wichmann T (2007) Activation of nigral and pallidal dopamine D1-like receptors modulates basal ganglia outflow in monkeys. J Neurophysiol 89: 1489–1500.

Levey AI, Kitt CA, Simonds WF, Price DL and Brann MR (1991) Identification and localization of muscarinic acetylcholine receptor proteins in brain with subtype-specific antibodies. J Neurosci 11: 3218–3226.

Levey AI, Hersch SM, Rye DB, Sunahara RK, Niznik HB, Kitt CA, Price DL, Maggio R, Brann MR and Ciliax BJ (1993) Localization of D1 and D2 dopamine receptors in brain with subtype-specific antibodies. Proc Natl Acad Sci USA 90: 8861–8865.

Mengod G, Vilaro MT, Niznik HB, Sunahara RK, Seeman P, O'Dowd BF and Palacios JM (1991) Visualization of a dopamine D1 receptor mRNA in human and rat brain. Brain Res 10: 185–191.

Neve KA (1997) The Dopamine Receptors. Totowa, NJ: Humana Press.

Nieoullon A, Cheramy A and Glowinski J (1978) Release of dopamine evoked by electrical stimulation of the motor and visual areas of the cerebral cortex in both caudate nuclei and in the substantia nigra in the cat. Brain Res 145: 69–83.

Pifl C, Bertel O, Schingitz G and Hornykiewicz O (1990) Extrastriatal dopamine in symptomatic and asymptomatic Rhesus monkeys treated with 1-methyl-4-phenyl-1,2,3,6-tetrahydropyridine (MPTP). Neurochem Int 17: 263–270.

Pifl C, Schingnitz G and Hornykiewicz O (1992) Striatal and non-striatal neurotransmitter changes in MPTP-parkinsonism in rhesus monkey: The symptomatic versus the asymptomatic condition. Neurochem Int 20(Suppl): 295S–297S.

Radnikow G and Misgeld U (1998) Dopamine D1 receptors facilitate GABAA synaptic currents in the rat substantia nigra pars reticulata. J Neurosci 18: 2009–2016.

Richfield EK, Young AB and Penney JB (1987) Comparative distribution of dopamine D-1 and D-2 receptors in the basal ganglia of turtles, pigeons, rats, cats, and monkeys. J Comp Neurol 262: 446–463.

Rosales MG, Martinez-Fong D, Morales R, Nunez A, Flores G, Gongora-Alfaro JL, Floran B and Aceves J (1997) Reciprocal interaction between glutamate and dopamine in the pars reticulata of the rat substantia nigra: A microdialysis study. Neuroscience 80: 803–810.

Schneider JS and Rothblat DS (1991) Neurochemical evaluation of the striatum in symptomatic and recovered MPTP-treated cats. Neuroscience 44: 421–429.

Sibley DR (1999) New insights into dopaminergic receptor function using antisense and genetically altered animals. Annu Rev Pharmacol Toxicol 39: 313–341.

Smith Y, Lavoie B, Dumas J and Parent A (1989) Evidence for a distinct nigropallidal dopaminergic projection in the squirrel monkey. Brain Res 482: 381–386.

Starr PA, Wichmann T, van Horne C and Bakay RA (1999) Intranigral transplantation of fetal substantia nigra allograft in the hemiparkinsonian rhesus monkey. Cell Transplant 8: 37–45.

Sunahara RK, Guan HC, Odowd BF, Seeman P, Laurier LG, Ng G, George SR, Torchia J, Vantol HHM and Niznik HB (1991) Cloning of the gene for a human dopamine D5-Receptor with higher affinity for dopamine than D1. Nature 350: 614–619.

Tiberi M, Jarvie KR, Silvia C, Falardeau P, Gingrich JA, Godinot N, Bertrand L, Yangfeng TL, Fremeau RT and Caron MG (1991) Cloning, molecular characterization, and chromosomal assignment of a gene encoding a 2nd D1 dopamine receptor subtype - differential expression pattern in rat-brain compared with the D1a receptor. Proc Natl Acad Sci USA 88: 7491–7495.

Timmerman W and Abercrombie ED (1996) Amphetamine-induced release of dendritic dopamine in substantia nigra pars reticulata: D1-mediated behavioral and electrophysiological effects. Synapse 23: 280–291.

Timmerman W and Westerink BH (1995) Extracellular gamma-aminobutyric acid in the substantia nigra reticulata measured by microdialysis in awake rats: Effects of various stimulants. Neurosci Lett 197: 21–24.

Trevitt JT, Carlson BB, Nowend K and Salamone JD (2001) Substantia nigra pars reticulata is a highly potent site of action for the behavioral effects of the D1 antagonist SCH 23390 in the rat. Psychopharmacology 156: 32–41.

Trevitt T, Carlson B, Correa M, Keene A, Morales M and Salamone D (2002) Interactions between dopamine D1 receptors and gamma-aminobutyric acid mechanisms in substantia nigra pars reticulata of the rat: Neurochemical and behavioral studies. Psychopharmacology 159: 229–237.

Waszczak BL, Lee EK, Tamminga CA and Walters JR (1984) Effect of dopamine system activation on substantia nigra pars reticulata output neurons: Variable single-unit responses in normal rats and inhibition in 6-hydroxydopamine-lesioned rats. J Neurosci 4: 2369–2375.

Weick BG and Walters JR (1987) Effects of D1 and D2 dopamine receptor stimulation on the activity of substantia nigra pars reticulata neurons in 6-hydroxydopamine lesioned rats: D1/D2 coactivation induces potentiated responses. Brain Res 405: 234–246.

Whone AL, Moore RY, Piccini PP and Brooks DJ (2003) Plasticity of the nigropallidal pathway in Parkinson's disease. Ann Neurol 53: 206–213.

Windels F and Kiyatkin EA (2006) Dopamine action in the substantia nigra pars reticulata: Iontophoretic studies in awake, unrestrained rats. Eur J Neurosci 24: 1385–1394.

Yan Z and Surmeier DJ (1997) D5 dopamine receptors enhance Zn2+-sensitive GABA(A) currents in striatal cholinergic interneurons through a PKA/PP1 cascade. Neuron 19: 1115–1126.

Yung KK, Bolam JP, Smith AD, Hersch SM, Ciliax BJ and Levey AI (1995) Immunocytochemical localization of D1 and D2 dopamine receptors in the basal ganglia of the rat: Light and electron microscopy. Neuroscience 65: 709–730.

Motor-Skill Learning in a Novel Running-Wheel Paradigm: Long-Term Memory Consolidated by D1 Receptors in the Striatum

Ingo Willuhn and Heinz Steiner

Abstract The sensorimotor striatum mediates procedural learning. Our previous studies revealed molecular changes in the sensorimotor striatum associated with motor-skill learning in a running-wheel task. In the present series of studies, we developed a novel test to measure the motor skill learned during running-wheel training and investigated mechanisms of wheel-skill learning in the striatum. Our results show that even a short wheel training (two daily sessions) produces robust long-term memory that lasts for months. Studies using systemic and intrastriatal D1 dopamine receptor antagonism in conjunction with cocaine treatment indicate that the acquisition of this wheel skill is dependent on optimal D1 receptor signaling in the striatum. Moreover, these studies demonstrate that striatal D1 receptors are critical for the formation of late, but not early, long-term skill memory. Further studies show that striatal processing after the training mediates consolidation of this long-term skill memory, as this memory formation was disrupted by posttrial drug infusions into the striatum. Interestingly, pretrial administration of cocaine prevented this posttrial interference, suggesting that cocaine stabilizes processes of memory formation, possibly by enhancing learning-related molecular changes. Together, these findings demonstrate that this running-wheel paradigm is an efficient model to investigate mechanisms of procedural learning and memory formation.

1 Introduction

The dorsal striatum has long been implicated in various forms of procedural learning, including habit (S-R) learning and motor-skill learning (e.g., Graybiel 1995; Knowlton et al. 1996; Packard and Knowlton 2002). For example, lesions or transient inactivation of the sensorimotor striatum have been shown to disrupt instrumental

I. Willuhn and H. Steiner(✉)
Department of Cellular and Molecular Pharmacology, Rosalind Franklin University of Medicine and Science, The Chicago Medical School, North Chicago, IL 60064, USA
e-mail: Heinz.Steiner@rosalindfranklin.edu

H.J. Groenewegen et al. (eds.), *The Basal Ganglia IX*, Advances in Behavioral Biology 58, DOI 10.1007/978-1-4419-0340-2_20,

learning in maze and lever-press tasks (e.g., Packard and McGaugh 1992; Yin et al. 2004; Featherstone and McDonald 2005). Similarly, motor-skill learning, another form of procedural learning (Squire 1987), is also dependent on normal function of the sensorimotor striatum (e.g., Ogura et al. 2005; Akita et al. 2006; Dang et al. 2006). The cellular and molecular processes underlying such learning, however, remain poorly understood. Evidence indicates that striatal dopamine regulates such processes. For example, dopamine receptor agonists infused into the striatum modified procedural maze learning (Packard and White 1991; Packard et al. 1994). Conversely, skill learning in a rotarod task was impaired by moderate striatal dopamine depletion (Ogura et al. 2005; Akita et al. 2006).

Recent interest in the regulation of procedural learning by dopamine was sparked by the proposal that addiction to psychostimulant drugs, which often are indirect dopamine receptor agonists, may involve aberrant procedural learning as a consequence of excessive dopamine receptor stimulation and ensuing molecular adaptations in striatal neurons (White 1996; Berke and Hyman 2000; Everitt et al. 2001). This notion is based in part on findings showing that psychostimulants often affect the same molecular mechanisms that are known to mediate learning and memory (Berke and Hyman 2000; Nestler 2001; Kelley 2004). For example, molecular changes induced by psychostimulants such as cocaine and amphetamine include altered expression of genes that encode transcription factors such as c-*fos* and *zif 268* (e.g., Harlan and Garcia 1998; Steiner and Gerfen 1998; Yano and Steiner 2007) and synaptic scaffolding proteins such as *homer* (e.g., Swanson et al. 2001; Yano and Steiner 2005). The former are implicated in various forms of neuroplasticity, including memory consolidation (Stork and Welzl 1999; Davis et al. 2003), while the latter are thought to directly regulate synaptic efficacy (Xiao et al. 2000; Thomas 2002), presumably a mechanism of long-term memory formation.

We have recently employed c-*fos* and *homer 1a* as markers to map motor-skill learning-associated molecular changes in the striatum and determine their modification by cocaine (Willuhn et al. 2003; Willuhn and Steiner 2005, 2006). In these studies, we used a novel motor-skill learning paradigm in which rats learn to run on a running wheel, a motor skill that rats acquire within one to two trials (Willuhn and Steiner 2006). Our results showed that this wheel-skill learning is associated with transiently enhanced c-*fos* and *homer 1a* inducibility in specific parts of the sensorimotor striatum (Willuhn and Steiner 2005, 2006). Importantly, these learning-related molecular changes were abnormally enhanced when the wheel-skill training occurred under the influence of cocaine (Willuhn et al. 2003; Willuhn and Steiner 2005, 2006).

In our earlier studies, we assessed wheel-skill learning by measuring performance errors during the training (Willuhn and Steiner 2006). While our results demonstrated that rats learn to run within one to two sessions, it also became clear that differences in general activity between groups, for example, induced by psychostimulants, complicated interpretation of differences in error counts because these counts were to some degree dependent on activity levels. We, therefore, developed a novel skill test that allows assessment of skill performance before and after the training (Willuhn and Steiner 2008). The following paragraphs describe this test and summarize the results of a series of studies that investigated the role of the sensorimotor striatum in this motor-skill learning.

2 Motor-Skill Learning in the Running Wheel

2.1 Wheel-Skill Training and Test

Training and testing were performed with standard running wheels for rats (Wahmann Company, Baltimore, MD, USA). These wheels consisted of a rotating metal chamber with a wire mesh floor (diameter, 35 cm; width, 11 cm), attached to a stationary metal wall with an access opening that could be closed. During the training session the rat was free to run, but could not leave the wheel. A mechanical counter recorded full wheel revolutions. In these studies, rats were trained once a day (40- or 60-min session) on 2 consecutive days.

Our previous findings showed that, in the beginning of the running-wheel training, rats are unable to run with an appropriate speed to remain at the bottom of the wheel (Willuhn and Steiner 2006). The rat often moves too fast or too slow relative to the speed of the wheel. Consequently, because of the momentum of the wheel, the rat/wheel "swings." During continued wheel training, the rat learns to minimize wheel swinging by adjusting its movements to control/balance the wheel ("wheel skill"; Willuhn and Steiner 2006). The wheel-skill test assesses this skill by measuring the rat's ability to stop the wheel from swinging (Willuhn and Steiner 2008).

The wheel skill is tested one day before the first training session (pretest) and repeatedly after the last training session. One day before the pretest the animal is habituated to the test/training room and wheel by placing it on a locked wheel for 1 h. On the test day, the rat is again first placed inside the locked wheel for a 2-min habituation period. The wheel with the rat is then gently rotated by 90° (rat's head up) (Fig. 1a). Upon release, the wheel swings back and forth until the rat stops the swinging by counterbalancing. Behavior on the wheel is videotaped for analysis. Wheel-skill performance is measured by counting the number of interruptions by the rat's body (minus tail) of a horizontal plane at one-third of the wheel diameter above the wheel bottom (marked by a line on the video monitor; see Fig. 1a), until the animal fails to interrupt for 3 s. In these studies, two trials per test (intertrial interval, 1 min) were performed, and the scores were averaged for a measure of performance error. The treatment groups were balanced with respect to pretest scores.

2.2 Practice is Essential for Wheel-Skill Learning

We first determined the relationship between running/practicing and wheel-skill acquisition by varying the running opportunity. Rats that trained on a running wheel (2 days, 60-min sessions) were compared with controls that were confined

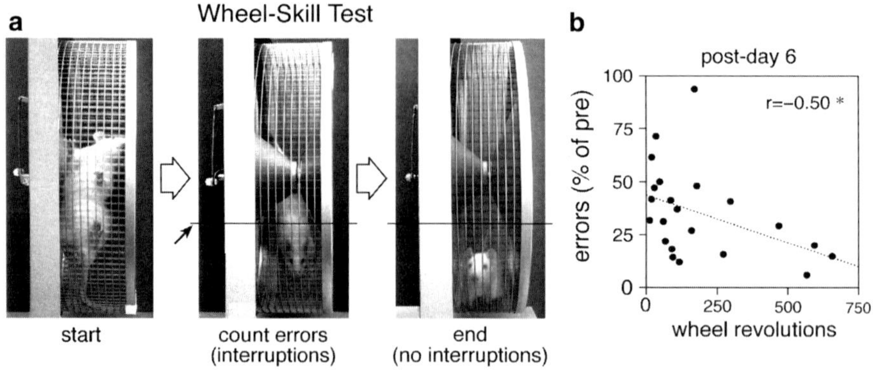

Fig. 1 Motor-skill learning on the running wheel. (**a**) Video stills depict a rat during the wheel-skill test. To start the test, swinging of the wheel is induced by the experimenter. Thus, the wheel with the rat is rotated by 90° (*left*) and released. The number of interruptions of a line marked on the video monitor (*small arrow*) by the rat's body minus tail (*middle*) is counted until the rat fails to interrupt for 3 s (*right*) (see Willuhn and Steiner 2008). These interruptions (swings) serve as an index of performance error. This test is administered before and after the running-wheel training. During the training, the rat learns to control/balance the wheel in order to reduce such swinging (wheel skill). (**b**) Scatter plot depicts, for individual animals, the relationship between the total number of wheel revolutions (practicing) during a 2-day training and the number of performance errors (swings) committed during a skill test 6 days after the training (expressed in percent of pretraining errors); $*p < 0.05$

to an identical but immobilized (locked) wheel. Thus, these controls could freely move about the wheel but could not run (turn the wheel). The wheel skill was tested before and 1 and 6 days after the training. Our results demonstrated that the running group displayed a significant training effect ($p < 0.001$, Friedman test), with a 60–80% reduction in the number of performance errors (swings) on both posttraining days 1 and 6 (Willuhn and Steiner 2008). In contrast, animals that spent the same amount of time on a locked wheel showed no improvement in their wheel skill ($p > 0.1$).

To further describe the relationship between practicing and wheel-skill learning, we compared the amount of running (wheel revolutions) during a 2-day training with skill improvement on postdays 1 and 6, in individual animals. This analysis showed a significant negative correlation between the total number of wheel revolutions during the training and the number of performance errors on both postday 1 ($r = -0.61, p < 0.01$, Spearman rank correlation) and postday 6 ($r = -0.50, p < 0.05$; Fig. 1b) (Willuhn and Steiner 2008). Thus, these results demonstrate that skill learning is directly related to the amount of practicing.

2.3 *Wheel-Skill Memory Lasts for Months*

In another experiment, rats were trained 2 days (60-min sessions) and then tested every 2 weeks for up to 4 months. Results showed that these rats retained their

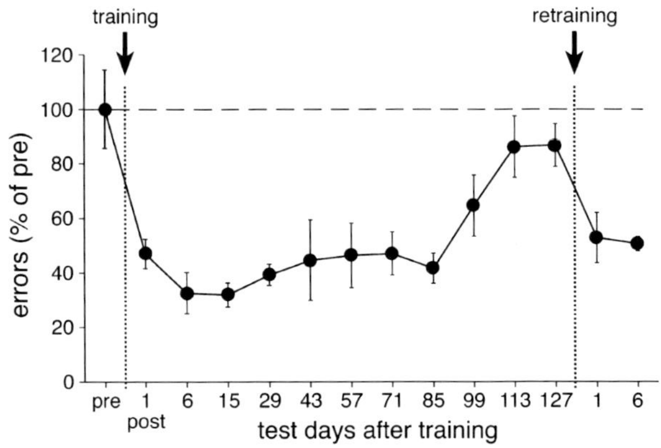

Fig. 2 Wheel-skill memory is long-lasting. The number (mean ± SEM) of performance errors (in percent of pretraining values) is given for rats tested before (pre) and repeatedly after (post) a 2-day training (60-min sessions) (n= 6). These rats were retrained in two daily sessions at 130–131 days after the initial training

improved wheel skill for approximately 80–90 days and then lost this skill (Fig. 2). A 2-day retraining at 130–131 days after the initial training reinstated this wheel skill, showing that this skill loss reflected a loss of motor memory/retrieval rather than possible age-dependent motor or other impairments. These findings demonstrate that a skill training as short as two daily sessions is sufficient to produce very robust and long-lasting long-term memory.

2.4 *Effects of Cocaine and D1 Receptor Stimulation in the Striatum on Wheel-Skill Learning*

As mentioned earlier, our previous studies showed an effect of cocaine on skill learning-associated gene regulation in the striatum (see earlier). Effects of psychostimulants on striatal gene regulation are principally mediated by D1 dopamine receptors (e.g., Graybiel et al. 1990; Steiner and Gerfen 1995; Drago et al. 1996; Moratalla et al. 1996). Moreover, recent findings indicate a role for D1 receptors in procedural learning (e.g., Eyny and Horvitz 2003; Hale and Crowe 2003). In further studies, we thus investigated whether cocaine and/or striatal D1 receptor antagonism modified wheel-skill learning.

In a first study, we investigated the effects of systemic D1 receptor antagonism, with and without cocaine, on wheel-skill learning (2 days, 40-min sessions) (Willuhn and Steiner 2008) (Fig. 3). The D1 receptor antagonist SCH-23390 (0, 3, 10 μg/kg, i.p.) and/or cocaine (25 mg/kg, i.p.) were administered before each training session. Importantly, these SCH-23390 doses did not attenuate wheel running, either alone or in conjunction with cocaine (Willuhn and Steiner 2008). The wheel skill was evaluated on days 1, 6, 18, and 26 after the training. Statistical analysis of the test scores over time for individual groups showed that training effects varied

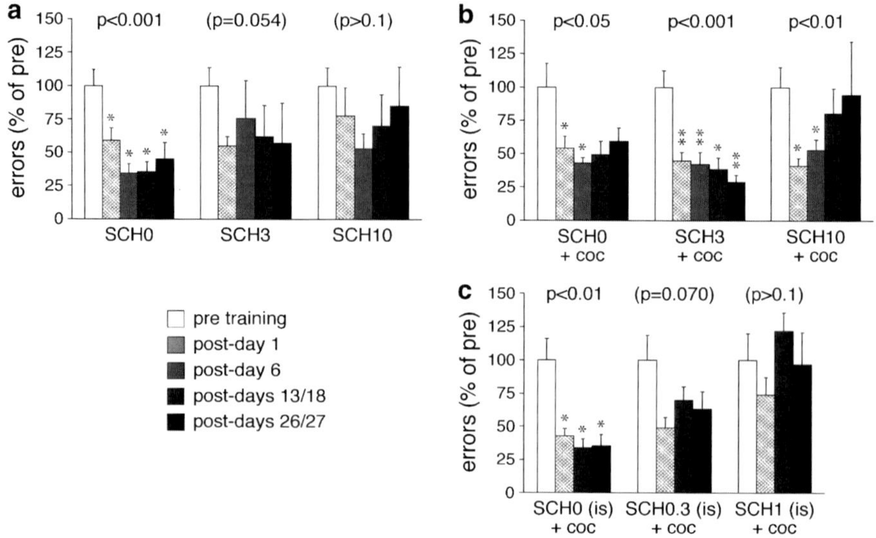

Fig. 3 Effects of systemic and intrastriatal D1 receptor blockade and/or cocaine during running-wheel training on motor-skill learning. (**a**, **b**) The number (mean ± SEM) of performance errors (in percent of pretest values) before (pre) and 1, 6, 18, and 26 days after (post) a 2-day running-wheel training (40-min sessions) are given for groups of rats (n= 7–9) that received a systemic (i.p.) injection of vehicle (SCH0), 3 μg/kg (SCH3), or 10 μg/kg (SCH10) of the D1 receptor antagonist SCH-23390, either alone (**a**) or followed by an injection of cocaine (25 mg/kg) (**b**), before each training session. (**c**) The number (mean ± SEM) of errors (in percent of pretest values) before (pre) and 1, 13, and 27 days after (post) a 2-day training are shown for groups of rats (n= 8–9) that received an intrastriatal (is) infusion of vehicle (SCH0), 0.3 μg (SCH0.3), or 1 μg (SCH1) of SCH-23390, followed by an injection of cocaine (25 mg/kg), before each training session. The p value for the overall training effect (Friedman test) in individual groups is also shown (**$p < 0.01$, *$p < 0.05$, vs. pretest values)

depending on the drug treatments (Fig. 3a, b). Vehicle only-treated rats (SCH0) displayed a highly significant training effect ($p < 0.001$, Friedman test), with an improved wheel skill present at all time points up to 26 days after the training ($p < 0.05$, Wilcoxon test) (Fig. 3a). The D1 receptor antagonist attenuated wheel-skill learning in a dose-dependent manner. Thus, no significant effect of the training on skill performance was found in rats that received 3 μg/kg (SCH3; $p = 0.054$) or 10 μg/kg of SCH-23390 (SCH10; $p > 0.1$) before each training session, but the group receiving the lower dose did show a trend for improvement (Fig. 3a).

Rats that received cocaine alone before each training session (SCH0 + coc) only showed a marginal training effect ($p < 0.05$), with significant wheel-skill improvement present until day 6 after the training ($p < 0.05$; Fig. 3b) (Willuhn and Steiner 2008). Thus, cocaine attenuated wheel-skill learning to some degree. The effects of SCH-23390 were biphasic when the D1 receptor antagonist was administered in conjunction with cocaine. Animals treated with the lower dose of SCH-23390 (3 μg/kg) plus cocaine (SCH3 + coc) displayed a highly significant training effect on wheel-skill performance ($p < 0.001$), with an improved skill that was stable throughout the posttraining period (postdays 1–26, $p < 0.05$; Fig. 3b). In contrast, rats treated with the higher dose (10 μg/kg) plus cocaine (SCH10 + coc) showed a significant training effect ($p < 0.01$), but this effect did not last. While these rats did improve to a similar degree on days 1–6 after the training ($p < 0.05$), they then lost their improved skill over time (postdays 18 and 26, $p > 0.05$; Fig. 3b). Thus, when given with the lower dose of SCH-23390, cocaine improved both early (postdays 1–6) and late long-term skill memory (postdays 18–26), whereas when given with the higher dose of SCH-23390, this psychostimulant treatment only reversed the inhibition of early, but not late, long-term memory formation (Fig. 3b). Together, these findings indicate that cocaine affects motor-skill learning via a D1 receptor-dependent mechanism.

In a further study, we investigated whether D1 receptor stimulation in the striatum is necessary for wheel-skill learning (2-day training, 40-min sessions) (Willuhn and Steiner 2008; Fig. 3c). The D1 receptor antagonist SCH-23390 (0, 0.3, and 1 μg) was infused into the dorsolateral striatum (targeting the area that showed wheel learning-associated gene regulation in our previous studies; see earlier) in conjunction with systemic cocaine. Statistical analysis revealed that the animals treated with cocaine only displayed a significant training effect ($p < 0.01$), with reduced error scores present on days 1, 13, and 27 after the training ($p < 0.05$) (Fig. 3c). In contrast, the group that received the lower dose of SCH-23390 (0.3 μg) showed only a trend for improved skill performance ($p = 0.070$), with near-normal improvement on postday 1 (Fig. 3c), and the higher dose of SCH-23390 (1 μg) completely prevented this motor learning ($p > 0.1$; Fig. 3c). Again, no effect of the D1 receptor antagonist on the amount of running during the training was observed (Willuhn and Steiner 2008).

In summary, our findings indicate that both decreasing dopamine tone at the D1 receptor (by the D1 receptor antagonist SCH-23390) and abnormally increasing dopamine tone (by the indirect dopamine receptor agonist cocaine) during the training impaired wheel-skill learning. However, combining a low dose of the D1 receptor antagonist with cocaine restored skill learning. These findings indicate that optimal D1 receptor stimulation is necessary for this motor-skill learning. Moreover, our results with intrastriatal D1 receptor antagonism demonstrate that D1 receptors in the striatum are critical for the acquisition of long-term skill memory.

2.5 *Critical Role for Striatum in Wheel-Skill Consolidation and Stabilizing Effects of Cocaine*

Similar to declarative memory, a critical step in the formation of long-term procedural memory is the consolidation of the memory trace, which involves neuronal processing after the training ("off-line" processing) (Robertson et al. 2004; Krakauer and Shadmehr 2006). This memory stabilization converts the memory from a labile state that can easily be disrupted by various manipulations to a more stable state that is resistant to interference. Such memory stabilization is also necessary for the formation of enduring motor skills (e.g., Brashers-Krug et al. 1996; Luft et al. 2004). We next investigated whether the sensorimotor striatum participates in such skill consolidation (off-line processing). Moreover, given the role of dopamine in skill learning and the finding that the indirect dopamine receptor agonist cocaine modified acquisition of wheel-skill memory (see earlier), we also assessed the effects of cocaine on consolidation of this skill memory (Willuhn and Steiner 2009).

Rats were trained on a running wheel for 2 days (40 min/day). They received an injection of vehicle or cocaine (25 mg/kg) before (pretrial) and intrastriatal infusion of 2% lidocaine (1 μl) or saline, or a sham infusion immediately after (posttrial) each training session. Rats that received vehicle and a posttrial sham infusion improved in their wheel-skill performance ($p < 0.001$), with significantly fewer performance errors after the training in all posttraining tests ($p < 0.05$) (Fig. 4a; Willuhn and Steiner 2009). In contrast, animals that received posttrial infusion of either saline or lidocaine into the striatum displayed no significant overall training effect ($p > 0.05$) (Fig. 4a). However, both groups showed a tendency for an improved skill performance (saline, $p = 0.054$; lidocaine, $p = 0.080$), which was mainly due to the performance on postday 1 when both groups displayed fewer errors ($p < 0.01$ vs. pretest), comparable with the sham controls. These results show that off-line processing in the striatum is critical for skill-memory consolidation and that this process is very labile, as it is easily disrupted by posttrial interference. Importantly, these findings again demonstrated a dissociation in mechanisms mediating consolidation of late-stage vs. early long-term skill memory, as the former but not the latter was disrupted by posttrial striatal interference (see later).

Cocaine administered before the training session abolished the effects of posttrial interference on wheel-skill consolidation (Willuhn and Steiner 2009). Controls

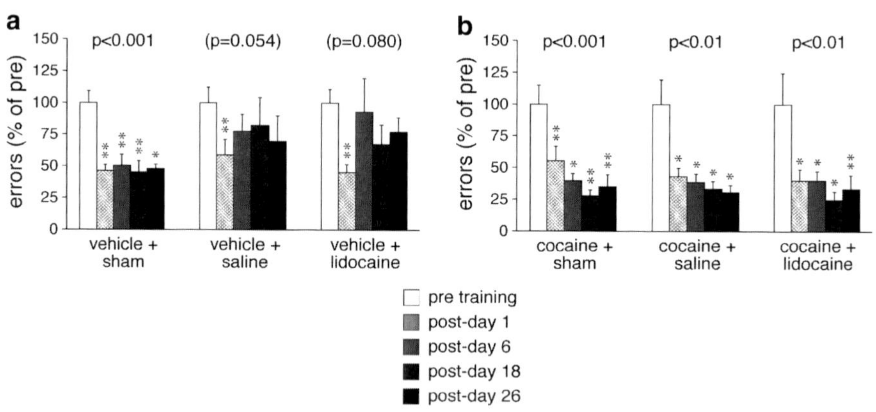

Fig. 4 Effects of posttrial intrastriatal interference with and without cocaine on motor-skill learning. (**a**, **b**) The number (mean ± SEM) of performance errors (in percent of pretest values) committed before (pre) and 1, 6, 18, and 26 days after (post) a 2-day running-wheel training are given for rats (n= 8–9) that received an injection of vehicle (**a**) or cocaine (25 mg/kg) (**b**) before (pretrial) plus a bilateral intrastriatal infusion of lidocaine (2%, 1 μl each side) or saline, or a sham infusion after (posttrial) each training session. The p value for the overall training effect (Friedman test) in individual groups is also shown (**$p < 0.01$, *$p < 0.05$, vs. pretest values)

that received pretrial cocaine and a posttrial sham infusion improved significantly in their skill performance ($p < 0.001$), with fewer performance errors up to 26 days after training ($p < 0.05$) (Fig. 4b). A similar effect was found with either posttrial intrastriatal infusion, as animals that received pretrial cocaine plus posttrial infusion of saline ($p < 0.01$) or lidocaine ($p < 0.01$) displayed significant skill improvement after the training and retained this skill for at least 4 weeks ($p < 0.05$) (Fig. 4b). Thus, pretrial administration of cocaine prevented the disruption of wheel-skill consolidation by posttrial intrastriatal infusion of saline or lidocaine.

In summary, our results demonstrate that the sensorimotor striatum plays a critical role in the consolidation of long-term motor-skill memory. Specifically, these results show that intrastriatal drug infusions following the training session interfered with the consolidation of late-stage (posttraining days 6–26), but not early (postday 1), long-term skill memory, indicating a dissociation between these two stages of long-term skill memory also for consolidation. Cocaine administered before the training session prevented this posttrial disruption of skill-memory consolidation. These findings suggest that cocaine stabilizes processes of consolidation, thus rendering them resistant to interference.

3 Discussion

3.1 *Wheel-Skill Learning*

In our studies (Willuhn et al. 2003; Willuhn and Steiner 2005, 2006, 2008, 2009), we have used a novel running-wheel paradigm to investigate the neuronal mechanisms of procedural learning. In this paradigm, rats learn to perform a motor skill while running on a wheel. Our findings show that in the beginning of the training rats are unable to run on the wheel without causing the wheel to swing (Willuhn and Steiner 2006). During continued training, they learn to control the wheel by adjusting their movements to stabilize the wheel and avoid swinging (wheel skill). As is typical for motor-skill learning (cf. Costa et al. 2004), rats learn fast within and between the first training sessions (Willuhn and Steiner 2006, 2008).

In this task, wheel swings serve as a measure of performance error for the assessment of skill learning. In our first study using this measure (Willuhn and Steiner 2006), wheel swings were counted during the training sessions to determine learning. However, because the number of swings during the training is directly related to overall activity levels, differences in the amount of running between experimental groups (e.g., induced by a drug treatment; Willuhn and Steiner 2006) complicate interpretation of differences in such swing counts. We, therefore, developed a novel wheel-skill test that can be administered before and after the training, thus allowing assessment of skill learning independent of the training situation. As a further advantage, this test can be performed repeatedly after the training to determine the duration of this motor memory. Our results obtained with this test demonstrate that, as expected, wheel-skill learning is dependent on factors such as the amount of running (practicing) during the training and the training duration (number of sessions) (Willuhn and Steiner 2008). These tests also showed that this wheel-skill memory lasts for months after the training (long-term memory).

Studying motor-skill learning in this running-wheel paradigm has a number of advantages over other learning paradigms used to investigate processes of procedural learning. For example, in contrast to other learning paradigms, wheel-skill learning is not dependent on potentially confounding motivational manipulations such as electric shocks (e.g., used for avoidance learning), food deprivation (e.g., in instrumental learning), or forced locomotion (e.g., in skill learning on rotarod), because rats voluntarily run on wheels (Sherwin 1998). Also, our results demonstrate that, unlike in many other procedural learning paradigms, long training periods are not necessary for robust motor-memory formation; even a short training (one to two daily sessions) was sufficient to produce stable long-term memory. Finally, the separation of training and testing permits the assessment of drug effects on skill memory in a drug-free state, thus avoiding confounds by drug effects on the tested motor response. In summary, our findings demonstrate that this running-wheel paradigm is an efficient and simple tool to investigate procedural learning and memory formation.

3.2 Role of Striatal D1 Receptors and Effects of Cocaine on Skill Consolidation

Our earlier reviewed studies show that cocaine modified motor-skill learning/consolidation, at least in part, by altering D1 receptor stimulation in the striatum. Overall, these findings are consistent with those of previous studies showing effects of psychostimulants and striatal dopamine receptor stimulation on acquisition and/or consolidation of other forms of procedural learning (see Willuhn and Steiner 2008 for discussion). Our findings are, to our knowledge, the first to demonstrate cocaine/D1 receptor regulation of motor-skill learning.

An important finding of our studies is that cocaine prevented disruption of memory consolidation by posttrial intrastriatal interference. Several mechanisms by which cocaine may achieve this effect are conceivable (see Willuhn and Steiner 2009 for discussion). We speculate that cocaine facilitates memory consolidation by affecting molecular mechanisms of synapse modification necessary for long-term memory. Our previous molecular findings demonstrated that cocaine enhanced skill learning-associated changes in gene regulation in the sensorimotor striatum. Specifically, animals trained under the influence of cocaine in the running-wheel task showed enhanced inducibility of genes encoding transcription factors such as c-*fos* and *zif 268* and molecules that directly regulate synaptic plasticity (Homer) in the striatum, selectively during the learning phase (Willuhn and Steiner 2005, 2006). The induction of such genes during learning is thought to be an initial step in a molecular cascade that ultimately produces changes in structure and efficacy of synapses to mediate long-term memory consolidation (Bailey and Kandel 1993; Moser 1999; Stork and Welzl 1999; Izquierdo et al. 2006). In support of this notion, our most recent results show that blockade of striatal D1 receptors prevents both these learning-associated molecular changes in the striatum (Willuhn and Steiner, in preparation) as well as wheel-skill learning (Willuhn and Steiner 2008). Thus, cocaine may stabilize motor-memory consolidation by enhancing D1 receptor-mediated molecular processes in the striatum, thereby reducing their susceptibility to interference. Abnormal stabilization of motor-memory consolidation by cocaine may contribute to aberrant procedural learning proposed to play a role in psychostimulant addiction (Berke and Hyman 2000; Everitt et al. 2001).

3.3 Dissociation of Early- and Late-Stage Long-Term Skill Memory

Another important finding of our studies is the dissociation of early- (1–6 days) vs. late-stage (>6–18 days) long-term skill-memory formation with respect to their D1 receptor dependence and susceptibility to intrastriatal interference. Short-term memory is considered to last ~3–6 h (Izquierdo et al. 2006), while memory 1–6 days after the training is early long-term memory. Our results indicate that the formation of early- and late-stage (>6 days) long-term memory is mediated by differential processes (Willuhn and Steiner 2008). Late-stage, but not early, wheel-skill memory was attenuated when animals were treated with a high enough dose of the D1 receptor antagonist SCH-23390 before the training, irrespective of whether they also received cocaine. In contrast, while such D1 receptor antagonism alone (SCH-23390 only) also attenuated early wheel-skill memory (1–6 days), this effect was reversed by coadministration of cocaine. Thus, these rats demonstrated apparently normal wheel-skill improvement on posttraining days 1–6, but then lost their wheel skill by day 18 after the training (Willuhn and Steiner 2008). A similar effect was seen with intrastriatal administration of 0.3 μg of SCH-23390. These and similar findings in the posttrial interference studies (Willuhn and Steiner 2009) indicate that while consolidation of late-stage long-term skill memory is critically dependent on striatal D1 receptor signaling, early long-term skill-memory formation is less so. It is presently unclear whether increased dopamine tone at D1 (or other) dopamine receptors or other neurochemical effects of cocaine, in the striatum or other structures, mediated this rescue of early long-term memory by cocaine. However, our findings demonstrate the critical importance of D1 receptor-mediated processes in the striatum for normal consolidation of late long-term memory for this motor skill. Future studies will have to elucidate the differential mechanisms that mediate early- vs. late-stage long-term skill-memory formation.

Acknowledgments This work was supported by USPHS grant DA011261. We would like to thank Joel Beverley for excellent technical assistance.

References

Akita H, Ogata M, Jitsuki S, Ogura T, Oh-Nishi A, Hoka S and Saji M (2006) Nigral injection of antisense oligonucleotides to synaptotagmin I using HVJ-liposome vectors causes disruption of dopamine release in the striatum and impaired skill learning. Brain Res 1095: 178–189.

Bailey CH and Kandel ER (1993) Structural changes accompanying memory storage. Annu Rev Physiol 55: 397–426.

Berke JD and Hyman SE (2000) Addiction, dopamine, and the molecular mechanisms of memory. Neuron 25: 515–532.

Brashers-Krug T, Shadmehr R and Bizzi E (1996) Consolidation in human motor memory. Nature 382: 252–255.

Costa RM, Cohen D and Nicolelis MA (2004) Differential corticostriatal plasticity during fast and slow motor skill learning in mice. Curr Biol 14: 1124–1134.

Dang MT, Yokoi F, Yin HH, Lovinger DM, Wang Y and Li Y (2006) Disrupted motor learning and long-term synaptic plasticity in mice lacking NMDAR1 in the striatum. Proc Natl Acad Sci USA 103: 15254–15259.

Davis S, Bozon B and Laroche S (2003) How necessary is the activation of the immediate early gene *zif268* in synaptic plasticity and learning? Behav Brain Res 142: 17–30.

Drago J, Gerfen CR, Westphal H and Steiner H (1996) D1 dopamine receptor-deficient mouse: Cocaine-induced regulation of immediate-early gene and substance P expression in the striatum. Neuroscience 74: 813–823.

Everitt BJ, Dickinson A and Robbins TW (2001) The neuropsychological basis of addictive behaviour. Brain Res Rev 36: 129–138.

Eyny YS and Horvitz JC (2003) Opposing roles of D1 and D2 receptors in appetitive conditioning. J Neurosci 23: 1584–1587.

Featherstone RE and McDonald RJ (2005) Lesions of the dorsolateral striatum impair the acquisition of a simplified stimulus-response dependent conditional discrimination task. Neuroscience 136: 387–395.

Graybiel AM (1995) Building action repertoires: Memory and learning functions of the basal ganglia. Curr Opin Neurobiol 5: 733–741.

Graybiel AM, Moratalla R and Robertson HA (1990) Amphetamine and cocaine induce drug-specific activation of the *c-fos* gene in striosome-matrix compartments and limbic subdivisions of the striatum. Proc Natl Acad Sci USA 87: 6912–6916.

Hale MW and Crowe SF (2003) Facilitation and disruption of memory for the passive avoidance task in the day-old chick using dopamine D1 receptor compounds. Behav Pharmacol 14: 525–532.

Harlan RE and Garcia MM (1998) Drugs of abuse and immediate-early genes in the forebrain. Mol Neurobiol 16: 221–267.

Izquierdo I, Bevilaqua LR, Rossato JI, Bonini JS, Medina JH and Cammarota M (2006) Different molecular cascades in different sites of the brain control memory consolidation. Trends Neurosci 29: 496–505.

Kelley AE (2004) Memory and addiction: Shared neural circuitry and molecular mechanisms. Neuron 44: 161–179.

Knowlton BJ, Mangels JA and Squire LR (1996) A neostriatal habit learning system in humans. Science 273: 1399–1402.

Krakauer JW and Shadmehr R (2006) Consolidation of motor memory. Trends Neurosci 29: 58–64.

Luft AR, Buitrago MM, Ringer T, Dichgans J and Schulz JB (2004) Motor skill learning depends on protein synthesis in motor cortex after training. J Neurosci 24: 6515–6520.

Moratalla R, Xu M, Tonegawa S and Graybiel AM (1996) Cellular responses to psychomotor stimulant and neuroleptic drugs are abnormal in mice lacking the D1 dopamine receptor. Proc Natl Acad Sci USA 93: 14928–14933.

Moser MB (1999) Making more synapses: A way to store information? Cell Mol Life Sci 55: 593–600.

Nestler EJ (2001) Molecular basis of long-term plasticity underlying addiction. Nat Rev Neurosci 2: 119–128.

Ogura T, Ogata M, Akita H, Jitsuki S, Akiba L, Noda K, Hoka S and Saji M (2005) Impaired acquisition of skilled behavior in rotarod task by moderate depletion of striatal dopamine in a pre-symptomatic stage model of Parkinson's disease. Neurosci Res 51: 299–308.

Packard MG and Knowlton BJ (2002) Learning and memory functions of the basal ganglia. Annu Rev Neurosci 25: 563–593.

Packard MG and McGaugh JL (1992) Double dissociation of fornix and caudate nucleus lesions on acquisition of two water maze tasks: Further evidence for multiple memory systems. Behav Neurosci 106: 439–446.

Packard MG and White NM (1991) Dissociation of hippocampus and caudate nucleus memory systems by posttraining intracerebral injection of dopamine agonists. Behav Neurosci 105: 295–306.

Packard MG, Cahill L and McGaugh JL (1994) Amygdala modulation of hippocampal-dependent and caudate nucleus-dependent memory processes. Proc Natl Acad Sci USA 91: 8477–8481.

Robertson EM, Pascual-Leone A and Miall RC (2004) Current concepts in procedural consolidation. Nat Rev Neurosci 5: 576–582.

Sherwin CM (1998) Voluntary wheel running: A review and novel interpretation. Anim Behav 56: 11–27.

Squire LR (1987) Memory and Brain. Oxford: Oxford University Press.

Steiner H and Gerfen CR (1995) Dynorphin opioid inhibition of cocaine-induced, D1 dopamine receptor-mediated immediate-early gene expression in the striatum. J Comp Neurol 353: 200–212.

Steiner H and Gerfen CR (1998) Role of dynorphin and enkephalin in the regulation of striatal output pathways and behavior. Exp Brain Res 123: 60–76.

Stork O and Welzl H (1999) Memory formation and the regulation of gene expression. Cell Mol Life Sci 55: 575–592.

Swanson CJ, Baker DA, Carson D, Worley PF and Kalivas PW (2001) Repeated cocaine administration attenuates group I metabotropic glutamate receptor-mediated glutamate release and behavioral activation: A potential role for Homer. J Neurosci 21: 9043–9052.

Thomas U (2002) Modulation of synaptic signalling complexes by Homer proteins. J Neurochem 81: 407–413.

White NM (1996) Addictive drugs as reinforcers: Multiple partial actions on memory systems. Addiction 91: 921–949.

Willuhn I and Steiner H (2005) Motor learning-related gene regulation in the striatum: Effects of cocaine. In: The Basal Ganglia VIII (Bolam JP, Ingham CA, Magill PJ, eds), pp 197–207. New York: Plenum Press.

Willuhn I and Steiner H (2006) Motor-skill learning-associated gene regulation in the striatum: Effects of cocaine. Neuropsychopharmacology 31: 2669–2682.

Willuhn I and Steiner H (2008) Motor-skill learning in a novel running-wheel task is dependent on D1 dopamine receptors in the striatum. Neuroscience 153: 249–258.

Willuhn I and Steiner H (2009) Skill-memory consolidation in the striatum: Critical for late but not early long-term memory and stabilized by cocaine. Behav Brain Res 199: 103–107.

Willuhn I, Sun W and Steiner H (2003) Topography of cocaine-induced gene regulation in the rat striatum: Relationship to cortical inputs and role of behavioural context. Eur J Neurosci 17: 1053–1066.

Xiao B, Tu JC and Worley PF (2000) Homer: A link between neural activity and glutamate receptor function. Curr Opin Neurobiol 10: 370–374.

Yano M and Steiner H (2005) Methylphenidate (Ritalin) induces Homer 1a and zif 268 expression in specific corticostriatal circuits. Neuroscience 132: 855–865.

Yano M and Steiner H (2007) Methylphenidate and cocaine: The same effects on gene regulation? Trends Pharmacol Sci 28: 588–596.

Yin HH, Knowlton BJ and Balleine BW (2004) Lesions of dorsolateral striatum preserve outcome expectancy but disrupt habit formation in instrumental learning. Eur J Neurosci 19: 181–189.

Discriminative Stimulus- vs. Conditioned Reinforcer-Induced Reinstatement of Drug-Seeking Behavior and *arc* mRNA Expression in Dorsolateral Striatum

Matthew D. Riedy, Raymond P. Kesner, Glen R. Hanson, and Kristen A. Keefe

Abstract Drug seeking in the context of stimuli associated with drug use contributes to recidivism in drug-addicted people. Although numerous brain areas have been implicated in mediating the effects of drug-conditioned cues on drug-seeking behavior, the role of the dorsolateral striatum in such cue-induced reinstatement is less clear. Particularly, the neuronal populations in which plasticity occurs and in which subsequent activity produces the effects of these cues on drug-seeking behavior are not known. Cellular compartment analysis of temporal activity by fluorescence in situ hybridization (catFISH) identifies whether the same neurons are activated by temporally discrete events. We applied this approach to brains from animals trained to both initiate cocaine self-administration in response to a discriminative stimulus and to reinstate/maintain responding to gain access to a conditioned reinforcer to examine whether the same neuronal ensembles are activated in the dorsolateral striatum by these two types of cues. CatFISH experiments did not reveal significant increases in the number of cells with *arc* mRNA expression in animals reinstating drug seeking relative to control animals exposed to the same stimuli. However, there was a significant correlation between *arc* mRNA expression and cue-induced operant responding on the test day, suggesting that the dorsolateral striatum in general, and *arc* expression in particular, may play a role in cue-induced reinstatement of drug-seeking behavior.

1 Introduction

Drug addiction is defined as uncontrollable, compulsive drug seeking and use in the face of negative consequences (http://www.nida.nih.gov/Published_Articles/Essence.html). Such compulsive behavior underlies the severe personal and societal costs associated with drug abuse and addiction. Current theories of drug addiction posit that addiction

M.D. Riedy, R.P. Kesner, G.R. Hanson, and K.A. Keefe (✉)
Department of Pharmacology & Toxicology, University of Utah,
Salt Lake City, UT 84112, USA
e-mail: k.keefe@utah.edu

H.J. Groenewegen et al. (eds.), *The Basal Ganglia IX*, Advances in Behavioral Biology 58,
DOI 10.1007/978-1-4419-0340-2_21,

reflects the abnormal association of previously neutral stimuli with drug-seeking and -taking behaviors and the subsequent ability of these stimuli to elicit such behaviors (Hyman and Malenka 2001; Everitt and Robbins 2005). Significant research has examined the neural circuits underlying the ability of conditioned reinforcers (CR) to maintain or reinforce ongoing drug-seeking behavior; much less is known about the neural circuits mediating the ability of discriminative stimuli (DS) to elicit drug-seeking behavior. Furthermore, determining the extent to which these two types of drug-associated stimuli engage the same or distinct neuronal populations within specific brain regions in an individual remains to be accomplished and is critical for fully understanding and managing the role of conditioned stimuli in addictive behaviors (Di Ciano and Everitt 2003).

1.1 Neural Substrates Mediating the Effects of Conditioned Stimuli on Reinstatement

Numerous studies have delineated brain areas involved in the ability of CRs to maintain drug-seeking behavior. These studies have provided substantial evidence for an important role of limbic brain areas including basolateral amygdala (BLA) (Kantak et al. 2002b; McLaughlin and See 2003), nucleus accumbens core (NAc) (Fuchs et al. 2004a; Ito et al. 2004), ventral tegmental area (VTA) (Di Ciano and Everitt 2004), and frontal cortex (Hutcheson and Everitt 2003; McLaughlin and See 2003; Fuchs et al. 2004b) in the ability of these CRs to sustain drug-seeking behavior. Taken together, these data suggest a circuit involving the BLA, prefrontal cortex, NAc, and VTA subserving the ability of CRs to maintain drug-seeking behavior (See 2005).

The role of the dorsal striatum in the reinstatement/maintenance of drug-seeking behavior by conditioned stimuli is less clear. Theoretically, the dorsal striatum should be involved, as it mediates stimulus-response learning/habit formation (Packard and Knowlton 2002; Yin et al. 2004). In support of this hypothesis, contingent, but not noncontingent, presentation of a CR increases extracellular dopamine (DA) levels in dorsal striatum, but not in nucleus accumbens (Ito et al. 2000, 2002). Also, blockade of dopamine or glutamate receptors in dorsolateral (DL) striatum blocks maintenance of drug-seeking behavior by response-contingent presentation of a CR (Vanderschuren et al. 2005; Alleweireldt et al. 2006). Finally, chemical inactivation of the DL striatum blocked the enhancement of reinstatement induced by contingent presentation of a CR after a period of abstinence (Fuchs et al. 2006). Conversely, lidocaine-induced inactivation of the dorsal, caudal striatum did not disrupt maintenance of drug-seeking behavior by a CR on a second-order schedule of reinforcement (Kantak et al. 2002a). Likewise, contingent presentation of a CR in the presence of a DS predictive of drug availability did not alter Fos expression in the dorsomedial (DM) or DL striatum, nor did noncontingent presentation of a CR that elicited drug-seeking behavior (Neisewander et al. 2000). Interestingly, the effect of noncontingent presentation of a CR on *zif268* expression in the dorsal striatum was not reported in the paper examining the expression of this plasticity-related gene throughout the brain in response to such stimulus presentation (Thomas et al. 2003). Therefore, the role of dorsal striatum in cue-induced reinstatement or maintenance of drug seeking by CRs and DS remains to be established.

1.2 Cellular Compartment Analysis of Temporal Activity by Fluorescence In Situ Hybridization

Studies have examined the extent to which particular brain regions are activated by drug-associated cues by assessing the expression of immediate early genes (IEGs) in animals exposed to such cues (Neisewander et al. 2000; Thomas et al. 2003). However, the results of these studies are not always consistent with the results of lesion/inactivation studies. Furthermore, given the more extended survival times used (Neisewander et al. 2000; Thomas et al. 2003), the specificity of the IEG changes in relation to the presentation of the drug-associated cues and drug-seeking behavior may not always be clear. Third, the effects of drug-associated cues on the expression of *arc* (activity regulated, cytoskeleton-associated protein), an effector IEG thought to be critically involved in synaptic modification underlying learning and memory (Tzingounis and Nicoll 2006), have not been previously reported. Finally, as reviewed by Guzowski et al. (2005), previous analyses assume that the involvement of a given brain region in a behavior is reflected by a greater number of cells with IEG expression. Coding of information in the brain, however, may be realized as much by changes in the populations of neurons within a structure that show activation as by the absolute number of neurons activated. To more fully evaluate neural substrates underlying the ability of DS and CRs to elicit and maintain drug-seeking behavior, it is useful to have a measure of IEG activation that is more temporally constrained to the cue and the behavior and that can assess whether the same populations of neurons are activated by different stimuli both relative to control conditions and to each other.

Cellular compartment analysis of temporal activity by fluorescence in situ hybridization (CatFISH) represents an approach for examining which neuronal populations show activation in response to temporally discrete events (Guzowski et al. 2001a, 2005). To date, however, this approach has not been applied to questions related to cue-induced changes in drug-seeking behavior. With catFISH it is possible to see the initiation of gene transcription for several IEGs within minutes of exposure to a discrete stimulus, allowing one to draw stronger conclusions about the relation between the stimulus and the neuronal activation (Guzowski et al. 1999; Vazdarjanova et al. 2002). In addition, the known kinetics of *arc* mRNA trafficking from the nucleus to the cytoplasm allow one to determine whether specific neurons were activated by stimuli impinging upon the animal at particular times relative to sacrifice (Guzowski et al. 1999; Daberkow et al. 2007).

The goal of the present work therefore was to determine whether rats could be trained to self-administer cocaine in association with a DS and a CR (dual DS/CR paradigm) and to reinstate and maintain cocaine-seeking behavior in response to contingent and noncontingent presentation of the CR and DS, respectively. We then applied the catFISH approach to analyze the activation of *arc* mRNA in phenotypically identified neurons of the DL striatum to determine whether the same neuronal ensembles were activated in rats reinstating and maintaining cocaine-seeking behavior in the context of temporally separated presentations of a DS and a CR.

2 Materials and Methods

2.1 Subjects

Adult male Sprague–Dawley rats (~300 g, Charles River Labs) were individually housed on a 10/14-h light/dark cycle with free access to food and water. After arrival, rats were food restricted for 2 days and then placed in operant conditioning chambers housed within light- and sound-attenuating boxes (Coulbourn Instruments). Rats were trained to lever press (>300 in 12 h) on a free operant schedule (FR1) to obtain food reward (45-mg pellet, Noyes Research Diets) during the dark phase of their light/dark cycle. The lever-pressing behavior was recorded during all sessions via Graphic State Software (Coulbourn Instruments).

Rats completing food training were separated into experimental and control groups. Five cohorts were run consisting of one experimental animal (EXP), a stimulus novelty control (NC), a stimulus control (SC), and a drug control (DC). An additional cohort consisted of entirely EXP animals. The NC animals were placed in operant chambers during training, but not exposed to the DS or CR or given any infusions. The SC rats were exposed to the DS and CR by being yoked to the EXP rats, but received saline infusions (50 μL 0.9% sterile saline/2-s infusion) instead of cocaine. The DC rats received infusions of cocaine (0.25-mg cocaine-HCl/50-μL 0.9% sterile saline/2-s infusion) yoked to those of the EXP animals, but were not exposed to the DS or CR during training. All control rats received identical extinction training and exposure to the stimuli on the reinstatement test day, being yoked to the EXP animal. Finally, age-matched, caged control animals (CC) were maintained in home cages for the same number of days as each cohort undergoing training and extinction and were sacrificed on the reinstatement test day immediately prior to each EXP animal.

2.2 Surgical Procedure

Rats (EXP, SC, DC) were implanted with chronic jugular vein catheters under ketamine/xylazine anesthesia (90/10 mg/kg, i.p.) using sterile techniques as previously described (Fuchs et al. 2004a) using 22-gauge guide cannulae (Plastics One, Inc.), silastic tubing (0.64-mm i.d., 1.19-mm o.d.; Dow Corning), and Prolite polypropylene monofilament mesh (Atrium Medical Corp.). Catheterized rats recovered from surgery for 4 days, while they adjusted to a reverse light/dark cycle and had free access to food and water. On the fifth day postsurgery, rats were food restricted for 1 day (1/2 their normal food intake), and on the sixth day cocaine self-administration training began.

2.3 *Cocaine Self-Administration Training*

Self-administration experiments were conducted using operant conditioning chambers containing two levers, a white house light located on the rear wall of the chamber, tricolor LEDs located above both the right and left levers, and a speaker mounted just outside the chamber. All rats were run during the dark phase of their day/night cycle. Each session lasted 4 h and 35 min and began with a 5-min delay. Rats then engaged in either a 2-h DS or CR training session followed by a 30-min, nonsignaled time-out period, and then the other (CR or DS) 2-h session. Therefore, each rat underwent 2 h of DS and 2 h of CR training each day, with session order counterbalanced across animals during training. Rats in the NC group were removed from their home cages and placed in dark operant chambers with levers available.

2.3.1 DS Self-Administration Training

During phase 1, the DS (white house light) was activated until the rat pressed the active lever. Upon lever press, the DS was extinguished and the rat received an infusion of cocaine (0.25 mg/50 μL/2-s i.v. infusion) followed by a 5-min, nonsignaled time-out period. At the end of the time-out period, the DS was turned on again and remained on until the rat engaged in another lever press resulting in the same cocaine infusion/time-out period. When a rat engaged in ≥10 operant responses resulting in drug infusions within one session, it advanced to phase 2, in which the duration of the DS was decreased to 2 min and the time-out period was changed to a variable interval, 8-min schedule (VI8; range 6–10 min). A lever press occurring during the 2-min DS presentation resulted in extinguishing of the DS, cocaine infusion, and entry into the nonsignaled, VI8 time-out period. Failure to respond during DS presentation resulted in entry into the VI8 time-out period.

2.3.2 CR Self-Administration Training

Ten milliseconds after a rat engaged in an active-lever press, it received a cocaine infusion (0.25 mg/50 μL/2-s i.v. infusion) paired with illumination of the tricolor LEDs and the sounding of a pure tone (2 kHz, 78 dB) stimulus.

The light/tone stimulus persisted for 18 s after the infusion. After the CR was extinguished, the rat entered a 20-s, nonsignaled time-out period. After the time-out period, an active lever press again resulted in a cocaine infusion/CR pairing/time-out sequence on an FR1 schedule. Once a rat had self-administered ≥10 cocaine doses in a single training session, response requirements were increased to an FR3 schedule. Pursuant to ≥10 doses at FR3, the schedule increased to FR5.

2.3.3 Extinction Training

When EXP rats achieved criterion for stable cocaine self-administration, defined as a coefficient of variation [CV = (SD/mean) × 100] for cocaine infusions of ≤15 across three consecutive sessions [average CV was 5.3 ± 1.8 (mean ± SEM, $n = 6$)], all rats from that cohort entered the extinction phase. During extinction, all rats were placed in the operant chambers for 14 daily 4.5-h sessions. Levers were always available and presses recorded, but pressing had no consequences. The DS and CR were not given to any animals during extinction training.

2.3.4 Reinstatement Test Day

After the 14th extinction session animals remained in the chamber overnight to avoid the effects of handling on *arc* mRNA expression. The next day, when the EXP animal spontaneously engaged in an active lever press, the CR was presented for 5 s. For the next 5 min, the CR was presented for 2 s contingent on active lever pressing (FR1). At the end of the 5-min CR period, there was a 20-min, nonsignaled time-out period. Then, the DS was illuminated for 2 min, extinguished for 1 min, reilluminated for 2 min, and finally extinguished. Each cohort of rats (EXP, NC, SC, DC), other than the caged controls, received the same stimulus exposure as the EXP animal during the reinstatement test. After the final exposure to the DS, all rats were removed from their chambers, sacrificed immediately via exposure to CO_2, and decapitated. Brains were removed and frozen in 2-methyl butane chilled on dry ice.

2.4 *In Situ Hybridization*

Brains were sectioned coronally at 12 μm and thaw-mounted onto Superfrost Plus (VWR International) microscope slides. Slides from all animals to be directly compared were processed together as previously described (Daberkow et al. 2007). Briefly, sections were postfixed, delipidated, and then hybridized with digoxigenin- [(DIG), for *arc* mRNA] and fluorescein- [(FITC), for *preproenkephalin* (*PPE*) mRNA] labeled, full-length, antisense ribonucleotide probes. The probes were transcribed in vitro from cDNAs using DIG- and FITC-UTP RNA labeling kits and T7 RNA polymerase (Roche Applied Science) and purified on G-50 spin columns (GE Healthcare). The *PPE* and *arc* cRNA probes were visualized with peroxidase-coupled, anti-FITC and anti-DIG antibodies (Roche Applied Science) and Cy5 and Cy3 Tyramide Signal Amplification kits, respectively. Finally, Sytox Green (Invitrogen/Molecular Probes) nuclear counterstain was applied to the slides prior to permanent mounting in an antifade reagent (Invitrogen/Molecular Probes).

2.5 Imaging and Cellular Compartment Analysis of Gene Expression

The DL striatum was visualized using a sequential scanning confocal microscope (Olympus FVX; IX70 microscope) and an automated stage (Prior ProScan II). The DL striatum was imaged with a 60×, 1.2-NA oil-immersion objective (plan APO), and a 3 × 3 montage of nine adjacent 235 μm × 235 μm fields was compiled. Each field consisted of ten 1-μm thick optical sections, with each optical section scanned sequentially by 488, 543, and 633-nm lasers to resolve Sytox Green (nuclear stain), Cy3 (*arc* mRNA), and Cy5 (*PPE* mRNA) fluorescence, respectively. Montages were taken from DL striata of two anterior sections (+0.70 mm from Bregma) for each animal.

An experimenter blinded to the treatment condition of each animal from which images were collected performed cellular compartment analysis (CatFISH), counting the number of neurons in each field with 1–2 intranuclear foci of *arc* mRNA expression, cytoplasmic *arc* mRNA expression, or both intranuclear and cytoplasmic *arc* mRNA expression, as previously defined (Daberkow et al. 2007). All *arc* mRNA-positive cells were phenotypically identified by the presence or absence of *PPE* mRNA expression. Cell counts were summed across the nine fields in each montage and averaged across the two sections from each animal.

3 Results

3.1 Rats Can be Trained to Self-Administer Cocaine Under the Dual DS/CR Paradigm

Fifteen EXP animals entered cocaine self-administration training under the dual DS/CR paradigm. Of those 15, five animals failed to reach the criterion due to blocked catheters. The remaining ten reached the criterion after 24.9 ± 2.6 (mean ± SEM) consecutive daily sessions.

3.2 Rats Trained Under Dual Paradigm Conditions Exhibit Cue-Induced Reinstatement of Drug-Seeking Behavior

Cue-induced reinstatement of drug-seeking behavior was defined as ≥1 lever press per minute during the CR and DS segments of the test-day session. Of the ten rats reaching the criterion on the dual training paradigm, six reinstated drug-seeking behavior in response to both the DS and CR on the test day (Fig. 1). A two-factor ANOVA (training group × test-day segment) revealed significant main effects for training group ($F_{3,14} = 8.5$, $p = 0.01$) and test-day segment ($F_{2,28} = 3.5$, $p = 0.04$), as well as a significant interaction ($F_{6,28} = 2.5$, $p = 0.04$). Post hoc *t* tests revealed significant increases in EXP lever pressing during the CR ($p = 0.02$) and DS ($p = 0.01$) segments of the reinstatement test, relative to the last extinction session. Post hoc *t* tests also revealed significant differences between the EXP group and all other groups during the CR (all $p < 0.02$) and DS (all $p < 0.01$) segments.

3.3 Cellular Compartment Analysis of arc mRNA Expression in Neurons of Dorsolateral Striatum

A two-factor ANOVA (training group × phenotype) on five EXP animals vs. CCs revealed a significant interaction between treatment group and neuronal phenotype ($F_{1,8} = 9.3$, $p = 0.02$; Fig. 2a) and a trend toward a main effect for the treatment group ($F_{1,8} = 4.4$, $p = 0.07$). Post hoc analysis via *t* test confirmed that there was a

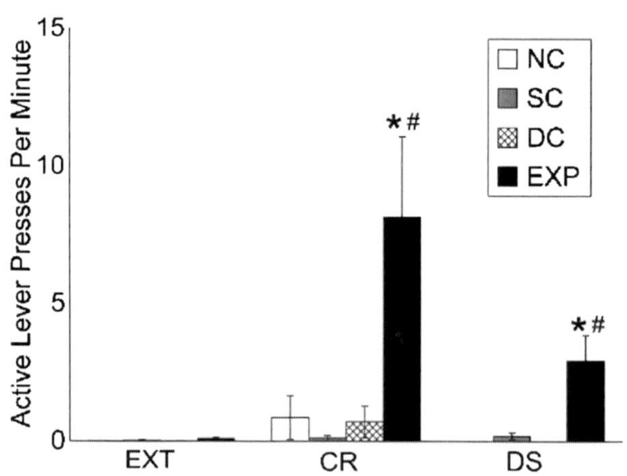

Fig. 1 Active lever pressing (presses per minute) during the final extinction session (EXT) and during the conditioned reinforcer (CR) and discriminative stimulus (DS) segments of the reinstatement test. Stimulus novelty control (NC), stimulus control (SC), and drug control (DC) rats received CR and DS presentations yoked to their respective experimental (EXP) animal. See Sect. 2.1 for training group details. *Significantly different from EXT session, $p < 0.05$. #Significantly different from other control groups, $p < 0.05$

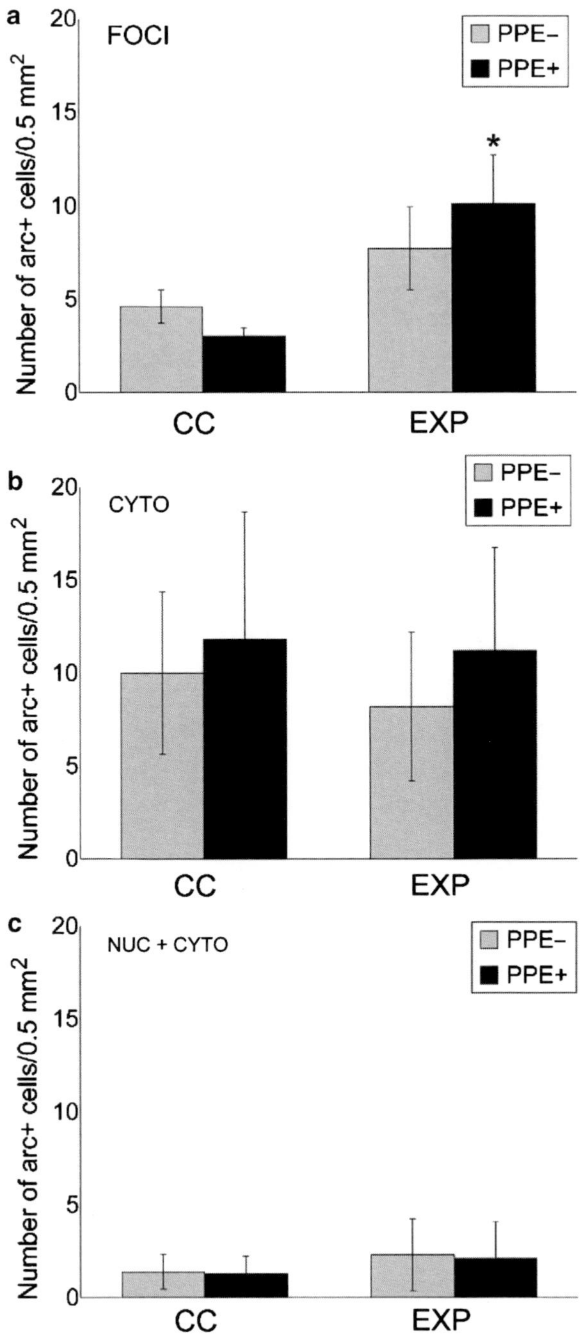

Fig. 2 The number of *preproenkephalin* mRNA-positive (*PPE*+) and *PPE*-negative (*PPE*−) cells/0.5 mm^2 area of DL striatum with *arc* mRNA expression in (**a**) 1–2 discrete foci in the nucleus (FOCI), (**b**) only in the cytoplasm (CYTO), or (**c**) in both the nucleus and cytoplasm (NUC + CYTO). *CC* caged control rats, *EXP* experimental rats. *Significantly different from PPE+ neurons in CC, $p < 0.05$

significant (t = 2.70, p = 0.03) increase in the number of *PPE*+ neurons with intranuclear foci of *arc* mRNA expression in the EXP group. A similar analysis of the number of cells with cytoplasmic expression of *arc* mRNA (Fig. 2b) revealed no significant main effects for the treatment group ($F_{1,8}$ = 0.03, p = 0.87) or phenotype ($F_{1,8}$ = 1.3, p = 0.28) and no significant interaction ($F_{1,8}$ = 0.1, p = 0.78). Likewise, analysis of the number of cells with dual nuclear and cytoplasmic expression of *arc* mRNA (Fig. 2c) revealed no significant main effects for the treatment group ($F_{1,8}$ = 0.2, p = 0.71) or phenotype ($F_{1,8}$ = 1.2, p = 0.31) and no interaction ($F_{1,8}$ = 0.1, p = 0.72).

The aforementioned data suggested that the DS predictive of cocaine availability for the EXP animals was activating *arc* mRNA induction in neurons of the DL striatum. To examine whether this increase reflected the associative history between the DS and cocaine availability, we compared the number of cells with intranuclear foci of *arc* mRNA expression in the EXP group with the number of cells in the NC, SC, and DC groups. These animals received the same exposure to the DS on the reinstatement test day; however, the stimuli had no associative value for these animals. Preliminary analysis (n = 3 per group) failed to reveal a significant main effect of training group or phenotype, or a significant interaction for the number of cells with intranuclear foci of *arc* mRNA expression (Fig. 3a; training group, $F_{4,10}$ = 1.6, p = 0.25; phenotype, $F_{1,10}$ = 0.4, p = 0.57; interaction, $F_{4,10}$ = 1.5, p = 0.26), cytoplasmic *arc* mRNA expression (Fig. 3b: training group, $F_{4,10}$ = 0.5, p = 0.72; phenotype, $F_{1,10}$ = 3.9, p = 0.08; interaction, $F_{4,10}$ = 0.2, p = 0.91), or combined nuclear and cytoplasmic *arc* mRNA expression (Fig. 3c: training group, $F_{4,10}$ = 0.3, p = 0.85; phenotype, $F_{1,10}$ = 0.2, p = 0.68; interaction, $F_{4,10}$ = 0.2, p = 0.94).

3.4 Arc mRNA Expression Was Correlated with Cue-Induced Reinstatement of Drug-Seeking Behavior

In the EXP group, there was no significant correlation between the number of cells with intranuclear foci of *arc* mRNA expression and lever pressing per minute induced by the DS exposure 5 min prior to sacrifice (R^2 = 0.35, p = 0.29). Likewise, there was no significant correlation between the number of cells with cytoplasmic expression of *arc* mRNA and the CR-associated lever pressing 25–30 min prior to sacrifice (R^2 = 0.56, p = 0.15) or between the number of cells with combined intranuclear and cytoplasmic *arc* mRNA expression and lever pressing per minute maintained by the CR presented contingent on lever pressing 25–30 min prior to sacrifice (R^2 = 0.62, p = 0.11). There was, however, a significant correlation between the total number of cue-induced lever presses per minute on the reinstatement test day and the total number of cells in DL striatum with cytoplasmic *arc* mRNA expression (R^2 = 0.78, p = 0.05; Fig. 4). Overall, the total number of cells with cytoplasmic *arc* mRNA expression were highly correlated with the total number of cells with dual nuclear plus cytoplasmic label (R^2 = 0.92, p = 0.01). Also, the number of *PPE*+ and *PPE*− cells with cytoplasmic and nuclear plus cytoplasmic

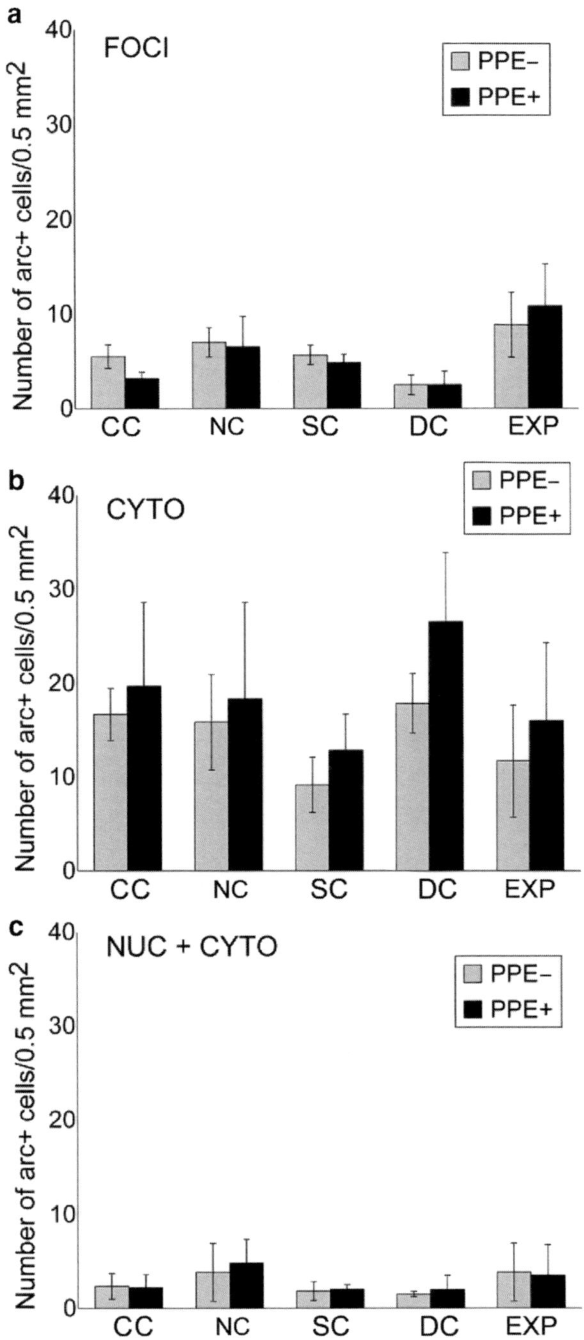

Fig. 3 The number of *preproenkephalin* mRNA-positive (*PPE*+) and *PPE*-negative (*PPE*−) cells/0.5 mm^2 area of DL striatum with *arc* mRNA expression in (**a**) 1–2 discrete foci in the nucleus (FOCI), (**b**) only in the cytoplasm (CYTO), or (**c**) in both the nucleus and cytoplasm (NUC + CYTO). *CC* caged control rats, *NC* stimulus novelty control, *SC* stimulus control, *DC* drug control, *EXP* experimental rats

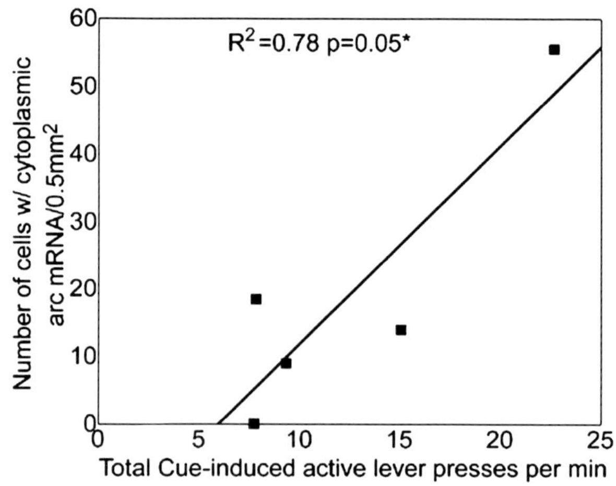

Fig. 4 Correlation between the total number of cue-induced lever presses by experimental rats and the number of cells with cytoplasmic *arc* mRNA expression (CYTO)/0.5 mm^2 area during the reinstatement test analyzed in DL striatum. *Significant correlation, $p < 0.05$

arc mRNA expression were highly correlated with the total number of cells in each category, as well as with each other for each type of labeling (all $p < 0.0005$), suggesting that there were not notable differences between the expression patterns in the *PPE*+ and *PPE*– neurons.

4 Discussion

The purpose of these experiments was to determine the feasibility of training rats to self-administer cocaine in a dual-paradigm model in order to utilize the strengths of the CatFISH approach to identify neural populations activated during reinstatement of drug seeking by a CR and a DS. The data indicate that only ~40% of rats learn to self-administer and reinstate under the dual-training paradigm. Furthermore, there were no significant increases in the number of cells with intranuclear foci, dual nuclear and cytoplasmic, or cytoplasmic *arc* mRNA expression in the DL striatum of the EXP group relative to other groups exposed to the same stimuli but for whom the stimuli held no relationship to cocaine exposure. The results of this study suggest that this approach may not be fruitful in identifying neuronal ensembles in DL striatum activated by drug-conditioned cues. Correlational analysis did, however, reveal a significant correlation between *arc* mRNA expression in DL striatum and cue-induced lever pressing during the reinstatement test, suggesting that DL striatum in general, and *arc* expression in DL striatum in particular, may play a role in cue-induced reinstatement of drug-seeking behavior.

As noted earlier, it was possible to train rats to self-administer cocaine in a dual DS/CR paradigm. However, the percentage of rats successfully completing training and reinstating lever pressing to both cues was only 40%. We think that two factors likely contribute to the high failure rate. First, the number of days required for rats to reach the criterion was relatively high. Consequently, one-third of rats entering training failed to reach the criterion due to obstruction of the catheter lines after relatively long periods of training. The second factor contributing to the high failure rate in the dual DS/CR paradigm is likely the general difficulty of the behavioral task. Some rats that completed training reinstated only to one cue. In the results presented earlier, such animals were excluded from analysis. Therefore, although both DS and CRs likely contribute to the reinstatement and maintenance of drug-seeking behavior in addicted individuals (Di Ciano et al. 2003), the ability to use an animal model to systematically establish and examine the specific roles that each type of stimulus might play within an individual seems limited.

The second goal of this work was to identify the neuronal ensembles activated, in the same rat, by a CR and DS. Previous studies (Guzowski et al. 1999; Daberkow et al. 2007) have shown that cells with cytoplasmic *arc* mRNA are activated 15–30 min prior to sacrifice (i.e., by the CR in the present study), whereas cells with intranuclear foci of *arc* mRNA are activated ~5 min prior to sacrifice (i.e., by the DS in the present study). Cells with both intranuclear and cytoplasmic expression were likely activated throughout the 30-min period prior to sacrifice (which would reflect activation by both the CR and DS aspects of the reinstatement session). Reinstatement of cocaine-seeking in the EXP rats by the CR and DS was not associated with an increase in the number of cells in DL striatum with *arc* mRNA expression in any of these cellular compartments. There may be several reasons for the apparent lack of activation of *arc* expression. First, the DL striatum may not be involved in cue-induced reinstatement of cocaine-seeking behavior. Studies examining cue-induced increases in IEG expression have reported effects in BLA, prelimbic cortex, and NAc (Neisewander et al. 2000; Thomas et al. 2003), but to date no one has reported changes in the expression of any IEG in dorsal striatum. However, contingent presentation of a CR is associated with increased extracellular dopamine levels in the dorsal striatum (Ito et al. 2002). Furthermore, See et al. (2007) have suggested that inactivation of DL striatum attenuates enhanced cue-induced reinstatement of cocaine-seeking behavior after abstinence. A second reason is that *arc* mRNA may not be induced during reactivation/reconsolidation of neural circuits already modified. However, recent data showing that similar proportions of neurons in hippocampus show *arc* mRNA induction when animals are repeatedly exposed to the same environment argue against this possibility (Guzowski et al. 2006). Finally, previous work (Guzowski et al. 2001a; Daberkow et al. 2007) demonstrates that *arc* mRNA is induced in brain regions of rats engaged in a behavioral task regardless of whether that brain region is involved in the learning and memory processes underlying the task. Therefore, in the present study, exposure of all EXP and control groups to the stimuli on the test day may have induced expression of *arc* mRNA in the DL striatum of control animals as well, even though the stimuli held no associative meaning for those animals. Furthermore, all animals, including the caged controls, were tested during the night (active) portion of their day–night cycle, raising the possibility that all rats may have been behaviorally active and thus had more basal *arc* mRNA expression in the striatum. Such elevated levels might make it difficult to detect an increase in the number of *arc* mRNA-positive neurons specifically induced by the cue-induced drug seeking. Consistent with this possibility is the observation that the number of cells with cytoplasmic *arc* mRNA expression in the caged controls in the present study is greater than the number of cells with cytoplasmic *arc* mRNA expression that we previously reported (Daberkow et al. 2007).

Previous studies (Guzowski et al. 2001b; Daberkow et al. 2007) suggest that the correlation between the degree of *arc* mRNA expression and behavioral performance may be a better metric for identifying the involvement of a particular brain region in a particular task. Interestingly, despite the lack of an apparent increase in the number of *arc* mRNA-positive neurons in the DL striatum of EXP animals, there was a significant correlation between the number of cells with *arc* mRNA expression in the cytoplasm and total cue-induced lever pressing engaged in by the EXP animals during the reinstatement test. This correlation suggests that DL striatum may be involved in cocaine-seeking behavior induced by exposure of rats to drug-associated cues. What this relationship reflects is unclear at present. Given the lack of correlation between *arc* mRNA in specific cellular compartments and lever pressing during specific segments of the reinstatement test, the correlation between overall lever pressing during the entire reinstatement period and the number of neurons with *arc* mRNA in the cytoplasm may not reflect specific learned associations between cues and operant responding. Rather, the correlation may simply reflect the degree of motor activity induced by the drug-associated cues and the role of DL striatum in the execution of lever-pressing behavior. In this regard, it is interesting that extracellular DA increases in DL striatum only when animals engage in lever pressing to obtain exposure to a conditioned stimulus, and not when animals are given noncontingent presentation of the stimulus in the absence of lever pressing (Ito et al. 2002).

In conclusion, the dual self-administration training model discussed here was developed and implemented to take advantage of the strengths of the catFISH approach – namely the ability to identify neurons activated by temporally discrete environmental events – to answer the question of whether different neuronal ensembles in DL striatum were participating in reinstatement of drug-seeking behavior induced by two distinct forms of associative learning. The catFISH analysis failed to detect cue-related changes in *arc* mRNA expression in DL striatum. However, additional analysis revealed a correlation between cue-induced operant responding and the number of cells with *arc* mRNA expression in the cytoplasm, implicating DL striatum in cue-induced reinstatement of cocaine-seeking behavior.

References

Alleweireldt AT, Hobbs RJ, Taylor AR and Neisewander JL (2006) Effects of SCH-23390 infused into the amygdala or adjacent cortex and basal ganglia on cocaine seeking and self-administration in rats. Neuropsychopharmacology 31: 363–374.

Daberkow DP, Riedy MD, Kesner RP and Keefe KA (2007) Arc mRNA induction in striatal efferent neurons associated with response learning. Eur J Neurosci 26: 228–241.

Di Ciano P and Everitt BJ (2003) Differential control over drug-seeking behavior by drug-associated conditioned reinforcers and discriminative stimuli predictive of drug availability. Behav Neurosci 117: 952–960.

Di Ciano P and Everitt BJ (2004) Contribution of the ventral tegmental area to cocaine-seeking maintained by a drug-paired conditioned stimulus in rats. Eur J Neurosci 19: 1661–1667.

Everitt BJ and Robbins TW (2005) Neural systems of reinforcement for drug addiction: From actions to habits to compulsion. Nat Neurosci 8: 1481–1489.

Fuchs RA, Evans KA, Parker MC and See RE (2004a) Differential involvement of the core and shell subregions of the nucleus accumbens in conditioned cue-induced reinstatement of cocaine seeking in rats. Psychopharmacology 176: 459–465.

Fuchs RA, Evans KA, Parker MC and See RE (2004b) Differential involvement of orbitofrontal cortex subregions in conditioned cue-induced and cocaine-primed reinstatement of cocaine seeking in rats. J Neurosci 24: 6600–6610.

Fuchs RA, Branham RK and See RE (2006) Different neural substrates mediate cocaine seeking after abstinence versus extinction training: A critical role for the dorsolateral caudate-putamen. J Neurosci 26: 3584–3588.

Guzowski JF, McNaughton BL, Barnes CA and Worley PF (1999) Environment-specific expression of the immediate-early gene arc in hippocampal neuronal ensembles. Nat Neurosci 2: 1120–1124.

Guzowski JF, McNaughton BL, Barnes CA and Worley PF (2001a) Imaging neural activity with temporal and cellular resolution using fish. Curr Opin Neurobiol 11: 579–584.

Guzowski JF, Setlow B, Wagner EK and McGaugh JL (2001b) Experience-dependent gene expression in the rat hippocampus after spatial learning: A comparison of the immediate-early genes *arc*, *c-fos*, and *zif268*. J Neurosci 21: 5089–5098.

Guzowski JF, Timlin JA, Roysam B, McNaughton BL, Worley PF and Barnes CA (2005) Mapping behaviorally relevant neural circuits with immediate-early gene expression. Curr Opin Neurobiol 15: 599–606.

Guzowski JF, Miyashita T, Chawla MK, Sanderson J, Maes LI, Houston FP, Lipa P, McNaughton BL, Worley PF and Barnes CA (2006) Recent behavioral history modifies coupling between cell activity and arc gene transcription in hippocampal ca1 neurons. Proc Natl Acad Sci USA 103: 1077–1082.

Hutcheson DM and Everitt BJ (2003) The effects of selective orbitofrontal cortex lesions on the acquisition and performance of cue-controlled cocaine seeking in rats. Ann NY Acad Sci 1003: 410–411.

Hyman SE and Malenka RC (2001) Addiction and the brain: The neurobiology of compulsion and its persistence. Nat Rev Neurosci 2: 695–703.

Ito R, Dalley JW, Howes SR, Robbins TW and Everitt BJ (2000) Dissociation in conditioned dopamine release in the nucleus accumbens core and shell in response to cocaine cues and during cocaine-seeking behavior in rats. J Neurosci 20: 7489–7495.

Ito R, Dalley JW, Robbins TW and Everitt BJ (2002) Dopamine release in the dorsal striatum during cocaine-seeking behavior under the control of a drug-associated cue. J Neurosci 22: 6247–6253.

Ito R, Robbins TW and Everitt BJ (2004) Differential control over cocaine-seeking behavior by nucleus accumbens core and shell. Nat Neurosci 7: 389–397.

Kantak KM, Black Y, Valencia E, Green-Jordan K and Eichenbaum HB (2002a) Stimulus-response functions of the lateral dorsal striatum and regulation of behavior studied in a cocaine maintenance/cue reinstatement model in rats. Psychopharmacology 161: 278–287.

Kantak KM, Black Y, Valencia E, Green-Jordan K and Eichenbaum HB (2002b) Dissociable effects of lidocaine inactivation of the rostral and caudal basolateral amygdala on the maintenance and reinstatement of cocaine-seeking behavior in rats. J Neurosci 22: 1126–1136.

McLaughlin J and See RE (2003) Selective inactivation of the dorsomedial prefrontal cortex and the basolateral amygdala attenuates conditioned-cued reinstatement of extinguished cocaine-seeking behavior in rats. Psychopharmacology 168: 57–65.

Neisewander JL, Baker DA, Fuchs RA, Tran-Nguyen LT, Palmer A and Marshall JF (2000) Fos protein expression and cocaine-seeking behavior in rats after exposure to a cocaine self-administration environment. J Neurosci 20: 798–805.

Packard MG and Knowlton BJ (2002) Learning and memory functions of the basal ganglia. Annu Rev Neurosci 25: 563–593.

See RE (2005) Neural substrates of cocaine-cue associations that trigger relapse. Eur J Pharmacol 526: 140–146.

See RE, Elliot JC and Feltenstein MW (2007) The role of dorsal vs ventral striatal pathways in cocaine-seeking behavior after prolonged abstinence in rats. Psychopharmacology 194: 321–331.

Thomas KL, Arroyo M and Everitt BJ (2003) Induction of the learning and plasticity-associated gene *zif268* following exposure to a discrete cocaine-associated stimulus. Eur J Neurosci 17: 1964–1972.

Tzingounis AV and Nicoll RA (2006) Arc/arg3.1: Linking gene expression to synaptic plasticity and memory. Neuron 52: 403–407.

Vanderschuren LJ, Di Ciano P and Everitt BJ (2005) Involvement of the dorsal striatum in cue-controlled cocaine seeking. J Neurosci 25: 8665–8670.

Vazdarjanova A, McNaughton BL, Barnes CA, Worley PF and Guzowski JF (2002) Experience-dependent coincident expression of the effector immediate-early genes arc and homer 1a in hippocampal and neocortical neuronal networks. J Neurosci 22: 10067–10071.

Yin HH, Knowlton BJ and Balleine BW (2004) Lesions of dorsolateral striatum preserve outcome expectancy but disrupt habit formation in instrumental learning. Eur J Neurosci 19: 181–189.

Preferential Modulation of the GABAergic vs. Dopaminergic Function in the Substantia Nigra by 5-HT$_{2C}$ Receptor

Giuseppe Di Giovanni, Vincenzo Di Matteo, Massimo Pierucci, and Ennio Esposito

Abstract Serotonin (5-HT) is intimately involved in the modulation of the basal ganglia circuitry and in its pathologies. The 5-HT pivotal role is supported by anatomical evidence demonstrating a large serotonergic innervation throughout the basal ganglia, with the highest concentration of this indole in the substantia nigra (SN). Among all the 5-HT receptors present in the SN, the 5-HT$_{2C}$ receptor subtype seems to be one of the principal receptors through which 5-HT exerts its function. In this chapter, we present in vivo electrophysiology and microdialysis evidence showing that the selective activation of 5-HT$_{2C}$ receptors does not affect dopaminergic function whereas it has a profound impact on GABAergic function in the substantia nigra pars reticulata (SNr). 5-HT excites the neurons of the SNr by acting on 5-HT$_{2C}$ receptors, and this control seems to be phasic rather than tonic in nature. Consequently, activation of 5-HT$_{2C}$ receptors boosts the concentration of GABA in the SNr, likely increasing GABA somatodendritic release from SNr neurons and from other GABA-containing neurons projecting to the SNr as well. Therefore, drugs acting on 5-HT$_{2C}$ receptors may provide a novel non-dopaminergic target for improving therapies for some basal ganglia disorders such as Parkinson's disease.

1 Introduction

Several recent studies have emphasized the involvement of serotonergic pathways in the central nervous system (CNS) in the modulation of the basal ganglia and in the pathophysiology of human involuntary movement disorders (Nicholson and Brotchie 2002; Di Matteo et al. 2008). This pivotal role played by 5-HT is supported by

G. Di Giovanni (✉)
Dipartimento di Medicina Sperimentale, Sezione di Fisiologia Umana,
"G. Pagano," Università degli Studi di Palermo, Corso Tuköry 129, 90134 Palermo, Italy
e-mail: g.digiovanni@unipa.it

V. Di Matteo, M. Pierucci, and E. Esposito
Istituto di Ricerche Farmacologiche "Mario Negri", Consorzio Mario Negri Sud, Santa Maria Imbaro (CH), Italy

H.J. Groenewegen et al. (eds.), *The Basal Ganglia IX*, Advances in Behavioral Biology 58,
DOI 10.1007/978-1-4419-0340-2_22,

anatomical evidence demonstrating a dense serotonergic innervation of the basal ganglia (Soubrié et al. 1984; Mori et al. 1987; Moukhles et al. 1997; Blackburn 2004). There is no doubt that a gradual degeneration of serotonergic (5-HT) neurons occurs in the raphe nuclei in Parkinson's disease (PD). In fact, 5-HT is reduced in ascending pathways; post-mortem examinations have shown a reduction of up to 50% of 5-HT in some areas of the cortex and the basal ganglia, and antemortem studies have shown a reduction of cerebrospinal fluid levels of the major 5-HT metabolite 5-hydroxyindolacetic acid (5-HIAA) in PD patients (Chase 1974; Birkmayer and Riederer 1986). Among all the 5-HT receptors present in the basal ganglia, perturbation of 5-HT_{2C} receptor signalling seems to be one of the principal pathological events that occurs in PD and related movement disorders. However, its relative contribution to the pathophysiology of these diseases has not been fully elucidated (Di Giovanni et al. 2006a, b).

A vast amount of research has led to the discovery and characterisation of a plethora of 5-HT receptor subtypes. At present, seven classes of 5-HT receptors (5-HT_{1-7} receptors) have been identified, which comprise at least a total of 15 subtypes (Hoyer et al. 2002). The 5-HT_2 receptors form a closely related subgroup of G-protein-coupled receptors (GPCR) and show the typical heptahelical structure of an integral membrane protein monomer. They are currently classified as 5-HT_{2A}, 5-HT_{2B} and 5-HT_{2C} subtypes (Hoyer et al. 2002; Di Giovanni et al. 2006b). Evidence from a variety of sources points to the involvement of the 5-HT_{2C} receptor in several important physiological and pathological non-motor processes, such as ingestive behaviour, brain development, depression, anxiety and schizophrenia, suggesting that the 5-HT_{2C} receptor is a promising target for the development of novel psychotropic drugs (Di Giovanni et al. 2002, 2006b).

2 Serotonin$_{2C}$ Receptor Distribution Within the Basal Ganglia Nuclei

The basal ganglia circuitry receives a dense 5-HT innervation, mainly from the rostral mesencephalic serotonergic group. The dorsal raphe nucleus (DRN) sends projections to the corpus striatum, the globus pallidus (GP), the subthalamic nucleus (STN), the substantia nigra (SN) and the pedunculopontine nucleus (PPN), and it provides most of the serotonergic innervation of the frontal cortex, including the motor and prefrontal cortices. Among these regions, the SN receives the densest 5-HT innervation (Fibiger and Miller 1977; Corvaja et al. 1993), which is distributed homogeneously across the rostro-caudal levels of the nucleus. 5-HT-containing raphe cells send projections to both DAergic and GABAergic cells of SN but in a higher density to the pars reticulata than to the pars compacta (Moukhles et al. 1997).

Studies of 5-HT_{2C} receptor distribution in the basal ganglia have revealed marked differences in the distribution among subregions of the basal ganglia (Mengod et al. 1990; Hoffman and Mezey 1989; Eberle-Wang et al. 1997). In the striatum, 5-HT_{2C} receptors are expressed by medium-sized efferent neurons but not by large cholinergic interneurons or glial cells. 5-HT_{2C} receptor mRNA does not show a preferential co-localization with neuropeptide markers for either the striatonigral pathway (substance P and dynorphin) or the striato-pallidal projections (enkephalin).

This would seem to indicate that 5-HT_{2C} activation is involved in both striatal output pathways. However, 5-HT_{2C} receptor mRNA is almost exclusively localized in the patch compartment areas (Ward and Dorsa 1996), revealing a preferential modulation of the projection to the substantia nigra, pars compacta (SNc). In the striatal target areas, high levels of 5-HT_{2C} receptor mRNA are present in the entopeduncular nucleus (EPN) while no mRNA is found in the GP, an area where 5-HT_{2C} receptor binding sites have been detected. The STN contains large numbers of cells that express a high level of 5-HT_{2C} receptor mRNA (Eberle-Wang et al. 1997).

2.1 Serotonin$_{2c}$ Receptor Distribution Within the Substantia Nigra

It has been calculated that the density of 5-HT-immunoreactive varicosities in the substantia nigra, pars reticulata (SNr) is in the order of 9×10^6 per mm^3, higher than the 6×10^6 varicosities per mm^3 in the SNc (Moukhles et al. 1997). Moreover, virtually all 5-HT varicosities form synaptic specialization in the SNr, whereas only 50% do so in the SNc (Moukhles et al. 1997). This picture gives an idea of the great influence that 5-HT could exert on the activity of SN neurons. Despite the homogenous 5-HT-innervation of the SN, 5-HT_{2C} receptors are expressed only in about half of the neuronal population, and in both nuclear subdivisions the density is markedly higher caudally than rostrally. Receptor expression is confined to GABAergic neurons and does not involve DAergic cell bodies (Pasqualetti et al. 1999; Di Giovanni et al. 2001). The distribution of 5-HT_{2C} receptor mRNA-expressing cells does not present any obvious similarity with the lamellar distribution of neurons projecting to different sets of target structures (Deniau and Chevalier 1992).

2.2 Serotonin$_{2c}$ Receptor Regulation in PD

In situ hybridization and autoradiographic studies in lesioned rats and postmortem brain tissue from patients with PD have revealed differential 5-HT_{2C} receptor regulation within the basal ganglia nuclei. In the striatum, 5-HT_{2C} receptors seem to be either down-regulated (Numan et al. 1995; Zhang et al. 2007) or uninfluenced by DA depletion (Fox and Brotchie 2000a; Basura and Walker 1999). In post-mortem brain tissue of PD patients, Fox and Brotchie (2000b) found normal 5-HT_{2C} binding levels in the external GP. In a study in the 6-hydroxydopamine (6-OHDA)-lesioned rat, there were no apparent changes in 5-HT_{2C} receptor mRNAs in the STN (Zhang et al. 2007). By contrast, 5-HT_{2C} receptor binding was doubled in the SNr of patients with PD, strictly in accordance with the 50% increase in 5-HT_{2C} receptor binding observed in 6-OHDA-lesioned rats (Fox and Brotchie 2000b).

The earlier data highlight a selective change in 5-HT_{2C} receptor activity following DA depletion only in the output regions of the basal ganglia. 5-HT_{2C} receptor up-regulation might be compensatory, being a consequence of a decreased level of 5-HT in these nuclei. The 5-HT_{2C} receptor may therefore play a role in the neuronal mechanisms involved in PD (Fox and Brotchie 2000b). In line with this notion, systemic administration of the selective 5-HT_{2C} antagonist SB 206553 was shown to enhance the anti-Parkinsonian effect of dopamine D1 and D2 agonists in 6-OHDA-lesioned rats (Fox et al. 1998; Fox and Brotchie 2000a), suggesting that the use of a 5-HT_{2C} receptor antagonist in combination with a dopamine receptor agonist may reduce the reliance upon dopamine replacement therapies.

3 Serotonin$_{2C}$ Modulation of the Substantia Nigra

3.1 Electrophysiological Data

3.1.1 Substantia Nigra Pars Compacta

Administration of mCPP (5–320 μg/kg, i.v.) and MK 212 (5–320 μg/kg, i.v.), two mixed 5-$HT_{2A/2C}$ agonists, induced a slight decrease in the basal firing rate, which was not statistically significant, and had no effect on the bursting activity of SNc DA neurons (Di Giovanni et al. 2000). Administration of the most selective 5-HT_{2C} receptor agonist available so far, RO 60–0175, either intravenously (5–640 μ/kg) (Invernizzi et al. 2007) or microiontophoretically (20–60 nA) did not cause any change in the basal firing rate and bursting activity of SNc DA neurons (Fig. 1). The representative rate histograms are reported in Fig. 1a, c, which show the typical lack of effect of RO 60–0175. Some DA neurons showed a slight inhibitory response to RO 60–0175, but the overall effect was not statistically significant. In addition, the mixed 5-$HT_{2C/2B}$ receptor antagonist SB 206553 caused only a slight increase in the basal activity of DA neurons in the SNc (Di Giovanni et al. 1999). Furthermore, SB 242084 (160–640 μg/kg, i.v.) (Di Matteo et al. 1999) and SB 243213, two new potent selective 5-HT_{2C} antagonists (160–320–640 μg/kg, i.v.), did not cause any significant change in the basal firing rate and firing pattern of the DA neurons recorded. Consistently, it has been reported that SB 243213 does not significantly alter the basal firing rate or pattern of SNc neurons, while its repeated administration modifies only the pattern of discharge but not the number of spontaneously active SNc DA cells (Blackburn et al. 2002). Likewise, both acute and chronic (21-day) administration of a novel 5-HT_{2C} receptor-selective agonist WAY-163909 (1–10 mg/kg) did not significantly alter the number of spontaneously active DA neurons compared with vehicle-treated animals in the SNc (Marquis et al. 2007).

3.1.2 Substantia Nigra Pars Reticulata

Contrary to the lack of effect of mCPP on SNc DA neurons, intravenous administration (5–320 mg/kg) or microiontophoretic application (10–40 nA) of mCPP caused a dose-dependent increase in the basal firing rate of a subpopulation of neurons in the SNr (50%). Pretreatment with the selective 5-HT_{2C} receptor antagonist

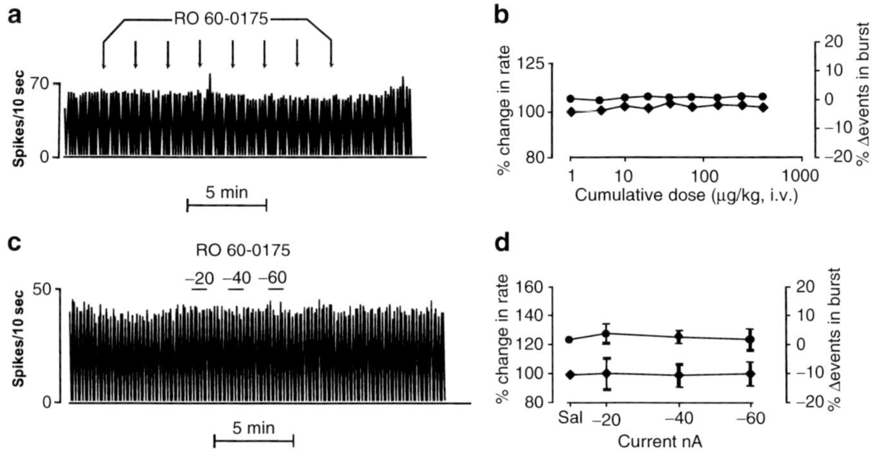

Fig. 1 Effect of systemic and microiontophoretic RO 60–0175 administration on the firing rate of dopaminergic SNc neurons. (**a**) Representative rate histograms showing the effects elicited by i.v. administration of cumulative doses of RO 60–0175 (5, 10, 20, 40, 80, 160, 320, 640 μg/kg, at *arrows*). (**b**) Cumulative dose-response curves showing the mean percentage changes (±S.E.M.) in firing rate (*filled diamond*) and the %Δ event in burst (*filled circle*) of SNc neurons elicited by administration of RO 60–0175. (**c**) Typical rate histograms showing the effect elicited by locally applied RO 60–0175. Numbers above each bar indicate the ejecting currents in nA. (**d**) Current–response curves showing the mean percentage of change (±S.E.M.) in firing rate (*filled diamond*) and the %Δ event in burst (*filled circle*), elicited by microintophoretic application of RO 60–0175 on SNc neurons

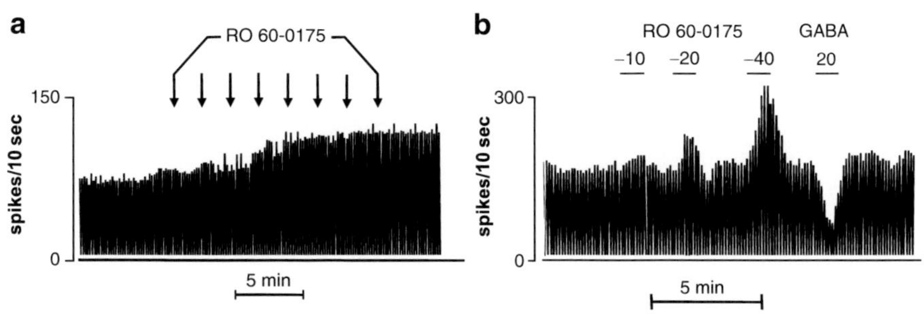

Fig. 2 Effect of systemic and local administration of RO 60–0175 on the firing rate of SNr neurons. (**a**) Representative rate histograms showing the effects elicited by i.v. administration of cumulative doses of RO 60–0175 (5, 10, 20, 40, 80, 160, 320, 640 μg/kg, at *arrows*). (**b**) Typical rate histograms showing the excitatory effect elicited by locally applied RO 60–0175. Numbers above each bar indicate the ejecting currents in nA

SB 242084 (200 mg/kg, i.v.) completely blocked the excitatory effect of i.v. and microiontophoretically applied mCPP (Di Giovanni et al. 2000). The selective stimulation of 5-HT_{2C} receptors within the SNr through RO 60–0175 treatment (5–640 μ/kg, i.v.) caused a dose-dependent excitation of about 30% of the SNr GABAergic neurons recorded (Fig. 2a) (Invernizzi et al. 2007).

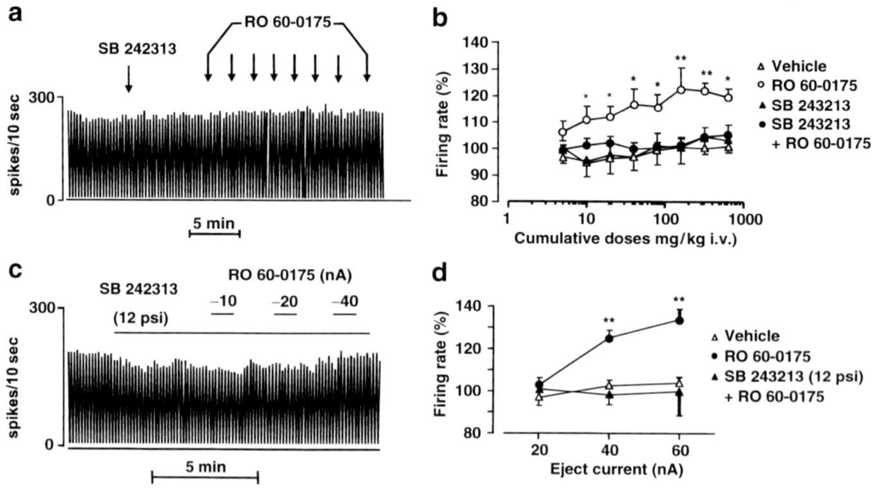

Fig. 3 Blockade by SB 243213 of the excitatory actions of RO 60–0175 on the firing rate of SNr neurons. (**a**) Representative rate histogram showing that i.v. SB 243213 (300 μg/kg) prevents the excitatory effect of RO 60–0175 (5, 10, 20, 40, 80, 160, 320, 640 μg/kg, at *arrows*). (**b**) Cumulative dose-response curves showing the mean percentage of change (±S.E.M.) in firing rate of SNr neurons elicited by administration of RO 60–0175 alone and after pretreatment with SB 243213. Statistical analysis evidenced a significant difference between excitatory ($n = 5$) effects compared with the control group ($n = 5$) (one-way ANOVA; $*p < 0.05$, $**p < 0.01$ vs. vehicle group; Fisher's PLSD). Pretreatment with SB 243213 abolished stimulatory (one-way ANOVA; $*p < 0.05$, $**p < 0.01$ RO 60–0175 vs. SB 243213 + RO 60–0175; Fisher's PLSD) effects of RO 60–0175 on SNr neurons. (**c**) Representative rate histogram showing that local co-administration of SB 242313 by micropressure ejection (12 psi; ejection time indicated by *largest horizontal bar*) prevented the excitatory effects elicited by locally applied RO 60–0175 ($n = 6$) on the firing rate of SNr neurons. Numbers above each small horizontal bar indicate the ejecting currents in nA. (**d**) Current-response curves showing the effects of local administration of RO 60–0175 alone or SB 243213 + RO 60–0175 ($n = 6$) on SNr neurons. Micropressure ejection of SB 243213 significantly counteracted the stimulatory effects exerted by locally applied RO 60–0175 on the firing rate of SNr neurons (one-way ANOVA; $**p < 0.01$ vs. SB 243213 + RO 60–0175; Fisher's PLSD) (modified from Invernizzi et al. 2007, with permission from Elsevier; © 2007)

The increase reached a peak of 22.6 ± 7.9% (above baseline) at a cumulative dose of 160 μg/kg (Fig. 3b). Moreover, either an inhibition or no response was seen in the other 70% of cells recorded, although the overall effect was not statistically significant. When RO 60–0175 was locally applied by microiontophoresis, this caused excitation in the majority of SNr neurons tested (50%) (Figs. 2b and 3d). Systemic and microiontophoretic administration of RO 60–0175 did not cause any significant change in the firing pattern of SNr neurons (Invernizzi et al. 2007). Furthermore, no significant associations were found among waveform, basal firing rate and pattern, location of SNr neurons and the effects elicited by systemic administration of RO 60–0175. The excitatory effect of both systemic and microiontophoretic RO 60–0175 administration was completely blocked by the selective 5-HT_{2C} antagonist SB 243213 administered either as pretreatment i.v. (300 μg/kg) (Fig. 3a, b) or when co-applied by micropressure ejection (12 psi) (Fig. 3c, d).

These data clearly indicate an involvement of 5-HT_{2C} receptor subtype in the effects elicited by RO 60–0175 on SNr neurons (Invernizzi et al. 2007).

3.2 Neurochemical Data

3.2.1 Striatum

Consistent with the earlier-described electrophysiological data, it has been demonstrated that the mixed 5-$HT_{2A/2C}$ agonists mCPP (1 mg/kg i.p.) and MK 212 (1 mg/kg i.p.) do not significantly affect in vivo DA release in the striatum (Di Giovanni et al. 2000). Moreover, in other studies the mixed 5-$HT_{2C/2B}$ receptor antagonist SB 206553 (5 mg/kg i.p.) caused only a slight increase in striatal DA release (Di Giovanni et al. 1999) while the selective antagonists SB 242084 (Di Matteo et al. 1999) and SB 243213 (0.5 mg/kg, i.p) did not modify DA release and DOPAC efflux in the striatum. Consistent with these observations, the selective 5-HT_{2C} receptor agonist RO 60–0175 (1 mg/kg, i.p.) failed to change DA release, whereas it decreased DOPAC outflow by a maximum of 27.79 ± 5% (below baseline) 160 min after injection in the striatum (Di Matteo et al. 1999). In spite of the lack of a local effect on DA cell bodies, 5-HT_{2C} receptors located on the DA nerve terminals in target regions seem to play a role in the regulation of DA release. Indeed, striatal infusion of the antagonist SB 206553 (0.1–100 μM) increased DA efflux in the striatum, and this effect was attenuated by systemic administration of the agonist mCPP (1.0 mg/kg i.p.) (Alex et al. 2005).

3.2.2 Substantia Nigra Pars Reticulata

Again consistent with the electrophysiological data, both systemic (1 mg/kg) and intranigral (1 μM) administration of the selective 5-HT_{2C} agonist RO 60–0175, markedly increased extracellular GABA levels in the SNr (Fig. 4a, b). RO 60–0175 (1 mg/kg) boosted GABA concentration by about 100% at 20 min, returning to baseline levels at 40 min and subsequently reaching a maximum increase 60–80 min after injection (200% above baseline levels). RO 60–0175 (1 μM) infused through the probe into the SNr significantly raised extracellular GABA levels in this nucleus. Although the infusion was continued for 2 h, the increase of GABA was short-lasting being significant 20 min after the beginning of the infusion of the drug (Fig. 4b). Systemic administration selective 5-HT_{2C} antagonist SB 243213 (0.3 mg/kg, s.c.), which did not cause any effect by itself, completely prevented the stimulatory effect of systemic RO 60–0175 (1 mg/kg, s.c.) on GABA release (Fig. 4c). Moreover, the infusion of 1 μM SB 243213 into the SNr antagonized the effect of 1 mg/kg RO 60–0175 only in the increase of extracellular GABA at 80 min and facilitates the increase at 20 min of extracellular GABA (Fig. 4d). 1 μM SB 243213 alone does not modify GABA levels (Invernizzi et al. 2007).

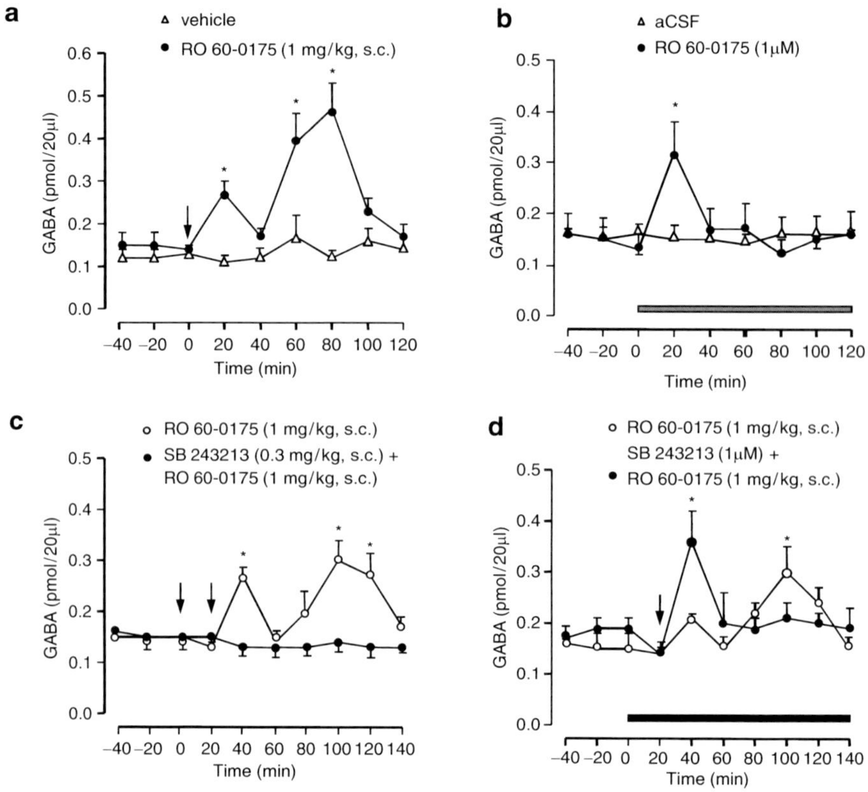

Fig. 4 Time course of the effects of 5-HT$_{2C}$ pharmacological manipulation on extracellular levels of GABA in the SNr. (**a**) Extracellular levels of GABA in the SNr of rats given 1 mg/kg, s.c. (n = 7) RO 60–0175 (*filled circle*) or vehicle (Δ, n = 6). *Arrow* indicates the injection of RO 60–0175 or vehicle. Data represent mean ± S.E.M. *p < 0.05 vs. vehicle (Tukey-Kramer's test). (**b**) Extracellular levels of GABA in the SNr of rats given 1 μM (n = 5) RO 60–0175 (filled circle) or aCSF (Δ, n = 5) through the probe. Data represent mean ± S.E.M. *Horizontal bar* indicates the duration of RO 60–0175 infusion. *p < 0.05 vs. aCSF (Tukey-Kramer's test). (**c**) SB 243213 (*filled circle*, 0.3 mg/kg, s.c. n = 5) prevents the stimulatory effect of RO 60–0175 (*open circle*) on extracellular GABA in the SNr. Rats were injected subcutaneously with SB 243213 (*first arrow*) 20 min before 1-mg/kg Ro 60–0175. (**d**) SB 243213 (*filled circle*, 1 μM, n = 5) was infused through the probe 20 min before subcutaneous RO 60–0175 administration (1 mg/kg, s.c.) and the infusion maintained for 140 min. Data represent mean ± S.E.M. *p< 0.05 vs. vehicle (Tukey-Kramer's test) (modified from Invernizzi et al. 2007, with permission from Elsevier; © 2007)

4 Conclusions

A series of studies carried out in our laboratory have clearly shown that 5-HT$_{2C}$ receptors play a prominent role in the control of mesocorticolimbic DA-mediated functions. Serotonin exerts both tonic and phasic control of DA-containing neurons in the ventral tegmental area (VTA), inhibiting their firing and consequently the release of DA in the nucleus accumbens (Di Matteo et al. 2001). Despite the evidence of strong 5-HT$_{2C}$ control of the mesocorticolimbic system, the in vivo data presented here clearly indicate a lack of effect on the nigrostriatal DA system. This noteworthy preferential control over the VTA compared with the nigral DA pathway by the 5-HT$_{2C}$ receptor subtype may provide the rationale for the use of 5-HT$_{2C}$ receptor agonists as antipsychotics that could improve the mood disorder and cognitive impairments associated with schizophrenia without producing extrapyramidal side effects (Marquis et al. 2007).

In the SNr the scenario is completely different. 5-HT$_{2C}$ receptor activation seems to exert a strong control over the activity of SNr GABAergic neurons and on nigral GABA levels (Di Giovanni et al. 2001; Invernizzi et al. 2007). The fact that the selective 5-HT$_{2C}$ antagonist SB 243213 does not cause any change in the basal activity of SNr neurons suggests that 5-HT$_{2C}$ receptors do not exert any tonic control upon these neurons. Systemic administration of the selective 5-HT$_{2C}$ receptor agonist RO 60–0175 caused a dose-dependent excitation of about one-third of the SNr neurons recorded. The excitatory effects of RO 60–0175 were completely prevented by pre-treatment with SB 243213, thus indicating that the effects were mediated by 5-HT$_{2C}$ receptors. It is conceivable that the prevalent effect of 5-HT$_{2C}$ receptor activation in the SNr is stimulatory, in as much as the direct application of RO 60–0175 in the SNr by microiontophoresis caused excitation in about half of the neurons tested (25 out of 53). The excitatory effects elicited by local application of RO 60–0175 were completely prevented by micropressure ejection of SB 243213 directly into the SNr. Moreover, it is important to note that a subpopulation of SNr neurons was not affected by RO 60–0175, thus confirming neuroanatomical data showing that about half of the SNr neurons do not express 5-HT$_{2C}$ receptors (Eberle-Wang et al. 1997).

The present findings confirm and extend previous in vitro and in vivo data showing that 5-HT exerts a direct excitatory effect on GABAergic neurons in the SNr by acting on 5-HT$_{2C}$ receptors (Rick et al. 1995; Stanford and Lacey 1996; Di Giovanni et al. 2001). Specifically, we showed that the 5-HT$_{2C}$ receptor agonist RO 60–0175, administered subcutaneously at a dose of 1 mg/kg, markedly increased extracellular GABA levels in the SNr. The effects of RO 60–0175 on GABA release could partly be due to the firing-dependent release of the neurotransmitter from GABAergic neurons of the SNr. It is interesting to note that RO 60–0175 caused a biphasic increase in GABA release, with an early peak at 20 min, a return to almost baseline level at 40 min and a second peak 80 min after systemic administration of the drugs. It is tempting to speculate that the multiple sources of GABA released in the SNr are at the basis of the biphasic effect elicited by RO 60–0175 on nigral GABA outflow. Thus, it is conceivable that the initial increase in GABA release induced by the 5-HT$_{2C}$ agonist would trigger some compensatory inhibitory mechanism that would bring GABA output back almost to baseline levels. It is likely that the second wave of GABA release would occur when this compensatory mechanism subsides and/or is overwhelmed by the stimulatory action of RO 60–0175 on 5-HT$_{2C}$ receptors. The fact that systemic administration of the 5-HT$_{2C}$ receptor agonist SB 243213 completely antagonized the RO 60–0175-induced rise of extracellular GABA at 20 and 80 min strongly supports the role of 5-HT$_{2C}$ receptors in these effects.

It was also found that the infusion of RO 60–0175 through the probe significantly increased extracellular GABA levels in the SNr. This suggests that the effect of RO 60–0175 on extracellular GABA is at least partly related to a direct action of the drugs on 5-HT_{2C} receptors in the SNr. It is likely that GABA-containing afferents to the SNr contribute to GABA release induced by systemic RO 60–0175. Interestingly, the local administration of 1-μM SB 243213 directly into the SNr through the microdialysis probe blocked only the increase of extracellular GABA at 80 min and facilitates the increase at 20 min. This suggests that the control exerted by 5-HT_{2C} receptors on extracellular GABA in the SNr is likely to involve both intra- and extra-nigral components. Therefore, these results might be due to an opposite control exerted by nigral and extranigral 5-HT_{2C} receptors on extracellular GABA.

In conclusion, selective activation of 5-HT_{2C} receptors seems not to affect nigrostriatal DAergic functions whereas it has a profound impact on GABAergic mechanisms in the SNr. From the earlier discussion it seems clear that the use of drugs acting at the 5-HT_{2C} receptor might be a promising avenue in the treatment of PD, drug-related motor disturbances and the depressive symptoms often associated with these neurological disorders.

Acknowledgements This work was supported in part by Ateneo di Palermo research funding, project ORPA068JJ5 (coordinator G. Di Giovanni).

References

Alex KD, Yavanian GJ, McFarlane HG, Pluto CP and Pehek EA (2005) Modulation of dopamine release by striatal 5-HT2C receptors. Synapse 55: 242–251.

Basura GJ and Walker PD (1999) Serotonin 2A receptor mRNA levels in the neonatal dopamine-depleted rat striatum remain upregulated following suppression of serotonin hyperinnervation. Brain Res Dev Brain Res 116: 111–117.

Birkmayer W and Riederer P (1986) Biological aspects of depression in Parkinson's disease. Psychopathology 19: 58–61.

Blackburn TP (2004) Serotonergic agents and Parkinson's disease. Drug Discov Today Ther Strateg 1:35–41.

Blackburn TP, Minabe Y, Middlemiss DN, Shirayama Y, Hashimoto K and Ashby Jr CR (2002) Effect of acute and chronic administration of the selective 5-HT2C receptor antagonist SB-243213 on midbrain dopamine neurons in the rat: an *in vivo* extracellular single cell study. *Synapse* 46: 129–139.

Chase TN (1974) Serotonergic-dopaminergic interactions and extrapyramidal function. Adv Biochem Psychopharmacol 11:377–385.

Corvaja N, Doucet G and Bolam JP (1993) Ultrastructure and synaptic targets of the raphe-nigral projection in the rat. Neuroscience 55: 417–427.

Deniau JM and Chevalier G (1992) The lamellar organization of the rat substantia nigra pars reticulata: distribution of projection neurons. Neuroscience 46: 361–377.

Di Giovanni G, De Deurwaerdére P, Di Mascio M, Di Matteo V, Esposito E and Spampinato U (1999) Selective blockade of serotonin-2C/2B receptors enhances mesolimbic and mesostriatal dopaminergic function: a combined in vivo electrophysiological and microdialysis study. Neuroscience 91: 587–597.

Di Giovanni G, Di Matteo V, Di Mascio M and Esposito E (2000) Preferential modulation of mesolimbic vs. nigrostriatal dopaminergic function by serotonin(2C/2B) receptor agonists: a combined in vivo electrophysiological and microdialysis study. Synapse 35: 53–61.

Di Giovanni G, Di Matteo V, La Grutta V and Esposito E (2001) *m*-Chlorophenylpiperazine excites non-dopaminergic neurons in the rat substantia nigra and ventral tegmental area by activating serotonin-2C receptors. Neuroscience 103: 111–116.

Di Giovanni G, Di Matteo V and Esposito E (2002) Serotonin/dopamine interaction-focus on 5-HT2C receptor, a new target of psychotropic drugs. Indian J Exp Biol 40: 1344–1352.

Di Giovanni G, Di Matteo V, Pierucci, Benigno A and Esposito E (2006a) Serotonin involvement in the basal ganglia pathophysiology: could the 5-HT2C receptor be a new target for therapeutic strategies? Curr Med Chem 13: 3069–3081.

Di Giovanni G, Di Matteo V, Pierucci M, Benigno A and Esposito E (2006b) Central serotonin2C receptor: from physiology to pathology. Curr Top Med Chem 6: 1909–1925.

Di Matteo V, Di Giovanni G, Di Mascio M and Esposito E (1999) SB 242084, a selective serotonin2C receptor antagonist, increases dopaminergic transmission in the mesolimbic system. Neuropharmacology 38: 1195–1205.

Di Matteo V, De Blasi A, Di Giulio C and Esposito E (2001) Role of 5-HT(2C) receptors in the control of central dopamine function. Trends Pharmacol Sci 22: 229–232.

Di Matteo V, Pierucci M, Esposito E, Crescimanno G, Benigno A and Di Giovanni G (2008) Serotonin modulation of the basal ganglia circuitry: therapeutic implication for Parkinson's disease and other motor disorders. Prog Brain Res 172: 423–463.

Eberle-Wang K, Mikeladze Z, Uryu K and Chesselet MF (1997) Pattern of expression of the serotonin2C receptor messenger RNA in the basal ganglia of adult rats. J Comp Neurol 384: 233–247.

Fibiger HC and Miller JJ (1977) An anatomical and electrophysiological investigation of the serotonergic projection from the dorsal raphe nucleus to the substantia nigra in the rat. Neuroscience 2: 975–987.

Fox SH and Brotchie JM (2000a) 5-HT(2C) receptor antagonists enhance the behavioural response to dopamine D(1) receptor agonists in the 6-hydroxydopamine-lesioned rat. Eur J Pharmacol 398: 59–64.

Fox SH and Brotchie JM (2000b) 5-HT2C receptor binding is increased in the substantia nigra pars reticulata in Parkinson's disease. Mov Disord 15: 1064–1069.

Fox SH, Moser B and Brotchie JM (1998) Behavioral effects of 5-HT2C receptor antagonism in the substantia nigra zona reticulata of the 6-hydroxydopamine-lesioned rat model of Parkinson's disease. Exp Neurol 151: 35–49.

Hoffman BJ and Mezey E (1989) Distribution of serotonin 5-HT1C receptor mRNA in adult rat brain. FEBS Lett 247: 453–462.

Hoyer D, Hannon JP and Martin GR (2002) Molecular, pharmacological and functional diversity of 5-HT receptors. Pharmacol Biochem Behav 71: 533–554.

Invernizzi RW, Pierucci M, Calcagno E, Di Giovanni G, Di Matteo V, Benigno A and Esposito E (2007) Selective activation of 5-HT(2C) receptors stimulates GABA-ergic function in the rat substantia nigra pars reticulata: a combined in vivo electrophysiological and neurochemical study. Neuroscience 144: 1523–1535.

Marquis KL, Sabb AL, Logue SF, Brennan JA, Piesla MJ, Comery TA, Grauer SM, Ashby CR Jr, Nguyen HQ, Dawson LA, Barrett JE, Stack G, Meltzer HY, Harrison BL and Rosenzweig-Lipson S (2007) WAY-163909 [(7bR,10aR)-1,2,3,4,8,9,10,10a-octahydro-7bH-cyclopenta-[b][1,4]diazepino[6,7,1hi]indole]: a novel 5-hydroxytryptamine 2C receptor-selective agonist with preclinical antipsychotic-like activity. J Pharmacol Exp Ther 320: 486–496.

Mengod G, Nguyen H, Le H, Waeber C, Lübbert H and Palacios JM (1990) The distribution and cellular localization of the serotonin 1C receptor mRNA in the rodent brain examined by in situ hybridization histochemistry. Comparison with receptor binding distribution. Neuroscience 35: 577–591.

Mori S, Matsuura T, Takino T and Sano Y (1987) Light and electron microscopic immunohistochemical studies of serotonin nerve fibers in the substantia nigra of the rat, cat and monkey. Anat Embryol 176: 13–18.

Moukhles H, Bosler O, Bolam JP, Vallée A, Umbriaco D, Geffard M and Doucet G (1997) Quantitative and morphometric data indicate precise cellular interactions between serotonin and postsynaptic targets in rat substantia nigra. Neuroscience 76: 1159–1171.

Nicholson SL and Brotchie JM (2002) 5-Hydroxytryptamine (5-HT, serotonin) and Parkinson's disease – opportunities for novel therapeutics to reduce the problems of levodopa therapy. Eur J Neurol 9: 1–6.

Numan S, Lundgren KH, Wright DE, Herman JP and Seroogy KB (1995) Increased expression of 5HT2 receptor mRNA in rat striatum following 6-OHDA lesions of the adult nigrostriatal pathway. Brain Res Mol Brain Res 29: 391–396.

Pasqualetti M, Ori M, Castagna M, Marazziti D, Marazziti D, Cassano GB and Nardi I (1999) Distribution and cellular localization of the serotonin type 2C receptor messenger RNA in human brain. Neuroscience 92: 601–611.

Rick CE, Stamford IM and Lacey MG (1995) Excitation of rat substantia nigra pars reticulata neurons by 5-hydroxytryptamine in vitro: evidence for a direct action mediated by 5-hydroxytryptamine2C receptors. Neuroscience 69: 903–913.

Soubrié P, Reisine TD and Glowinski J (1984) Functional aspects of serotonin transmission in the basal ganglia: a review and an in vivo approach using the push-pull cannula technique. Neuroscience 13: 605–625.

Stanford IM and Lacey MG (1996) Differential actions of serotonin, mediated by 5-HT1B and 5-HT2C receptors, on GABA-mediated synaptic input to rat substantia nigra pars reticulata neurons in vitro. J Neurosci 16: 7566–7573.

Ward RP and Dorsa DM (1996) Colocalization of serotonin receptor subtypes 5-HT2A, 5-HT2C, and 5-HT6 with neuropeptides in rat striatum. J Comp Neurol 370: 405–414.

Zhang X, Andren PE and Svenningsson P (2007) Changes on 5-HT2 receptor mRNAs in striatum and subthalamic nucleus in Parkinson's disease model. Physiol Behav 92: 29–33.

Blockade of GABA Transporter (GAT-1) Modulates the GABAergic Transmission in the Rat Globus Pallidus

Xiao-Tao Jin, Jean-Francois Paré, and Yoland Smith

Abstract Previous studies from our laboratory have demonstrated that application of SKF 89976A, a selective inhibitor of the GABA transporter GAT-1, reduces the activity of pallidal neurons in monkeys. However, the functional role of GAT-1 on GABAergic synaptic transmission in the globus pallidus (GP) is poorly understood. In the present study, we applied the whole-cell patch clamp recording technique to study the effects of blockade of GAT-1 on $GABA_A$ receptor-mediated inhibitory postsynaptic currents (IPSCs) recorded from rat GP slice preparations. Under maximal striatal stimulation (15–20 V) in parasagittal slices, SKF 89976A (10 μM) significantly prolonged the decay time, without significant effect on the amplitude, of IPSC. In contrast, SKF 89976A increased the amplitude, but did not prolong the decay time, of IPSCs under minimal striatal stimulation (2–5 V). We did not find any significant effect of SKF 89976A on IPSCs evoked locally from GP coronal slices. Furthermore, neither the amplitude nor the frequency of miniature IPSCs were changed following bath application of SKF 89976A. These results demonstrate that GABA reuptake through GAT-1 plays a major activity-dependent role in regulating GABAergic transmission at striatopallidal synapses in the GP.

1 Introduction

GABA is the main inhibitory neurotransmitter in the mammalian brain that mediates its effects through activation of $GABA_A$, $GABA_B$, and $GABA_C$ receptors. GABA released from presynaptic terminals is removed from the vicinity of the synaptic cleft by GABA transporters (GATs), and this action is believed to be a key event in terminating inhibitory transmission. Another critical function of GATs is the regulation of GABA spillover to neighboring synapses, thereby modulating synaptic

X.-T. Jin (✉), J.-F. Paré, and Y. Smith
Division of Neuroscience, Yerkes National Primate Research Center and Department of Neurology, Emory University, 954, Gatewood Road Northeast, Atlanta, GA 30322, USA
e-mail: Xjin@rmy.emory.edu

H.J. Groenewegen et al. (eds.), *The Basal Ganglia IX*, Advances in Behavioral Biology 58, DOI 10.1007/978-1-4419-0340-2_23, © Springer Science+Business Media, LLC 2009

cross talk [for review, see Borden (1996)]. GATs are also involved in maintaining a low extracellular GABA concentration throughout the brain, preventing excessive tonic activation of synaptic and extrasynaptic receptors. To date, five different high-affinity GATs have been cloned. Of these, VGAT is a vesicular transporter while GAT-1, BGT-1, GAT-2, and GAT-3 are members of the large family of 12-trans-membrane spanning Na^+/Cl^- coupled transporters [for review, see Borden (1996)]. The role of the GAT-1 transporter has been studied in various brain regions. For instance, blockade of GAT-1 with selective antagonists prolongs inhibitory GABAergic synaptic responses in the hippocampus (Roepstorff and Lambert 1992, 1994; Thompson and Gähwiler 1992; Isaccson et al. 1993; Draguhn and Heinemann 1996; Overstreet et al. 2000; Overstreet and Westbrook 2003) and neocortex (Keros and Hablitz 2005).

The globus pallidus (GP) [external globus pallidus (GPe) in primates] plays a central integrative role in the basal ganglia circuitry (Plenz and Kitai 1999; Bevan et al. 2002). It receives GABAergic inputs from the striatum (Str) and a major glutamatergic innervation from the subthalamic nucleus (STN). In turn, the GP sends GABAergic projections back to the STN and other basal ganglia nuclei.

We have shown that GAT-1 is expressed on glial processes and unmyelinated axons in the monkey pallidum and that intrapallidal application of SKF 89976A, a selective GAT-1 inhibitor, reduces the activity of external and internal globus pallidus (GPe and GPi) neurons in awake monkey (Galvan et al. 2005). Apart from this recent work, very little is known about the functions of GATs in the basal ganglia circuitry. In this study, we used electron microscopic immunocytochemistry and whole cell patch-clamp recording to determine the localization of GAT-1 immunoreactivity and elucidate the role of this transporter in regulating inhibitory synaptic transmission in the rat GP.

2 Material and Methods

2.1 Electron Microscopic Immunocytochemistry

Three 17–20-days-old Sprague–Dawley rats were used in this study. Rats were perfusion-fixed with 500 ml of cold oxygenated Ringer's solution followed by 4% paraformaldehyde and 0.1% glutaraldehyde in phosphate buffer (PB) (0.1 M, pH 7.4). The brains were then cut in 60-μm-thick sections with a vibrating microtome and processed for the immunohistochemical localization of GAT-1 at the electron microscopic level according to procedures described in detail in our previous study (Galvan et al. 2005).

A commercially available GAT-1 antiserum (Chemicon, Temecula, CA) was used. Sections processed for immunoperoxidase were stained with diaminobenzidine (DAB) using the avidin-biotin-peroxidase method (ABC, Vector Labs, Burlingame, CA, USA). After immunostaining, the sections were processed for electron microscopy and ultrathin sections of GP tissue were prepared.

The ultrathin sections collected from immunostained GP were scanned in the electron microscope for the presence of immunoreactive elements. Labeled structures were photographed and characterized on the basis of ultrastructural features described in Peters et al. (1991). The total number of labeled elements from a series of ultrathin sections collected from blocks of GP tissue of the three young (P17-20) rats were tabulated and expressed as relative percentages of immunoreactive elements.

2.2 *Whole-Cell Patch Clamp Recording*

All electrophysiological experiments were performed on slices from 17 to 20-days-old Sprague–Dawley rats (22 rats, six slices/rat) (Charles River Laboratories, Wilmington, MA). After decapitation, brains were removed and submerged in ice-cold oxygenated sucrose buffer containing (in mM): 233.4 sucrose, 20 glucose, 47.3 $NaHCO_3$, 3 KCl, 1.9 $MgSO_4$, 1.2 KH_2PO_4, and 2 $CaCl_2$. Parasagittal and coronal slices (300 μm in thickness) were made on a Vibratome 3000 (The Vibratome Company, St. Louis, MO) in ice-cold oxygenated sucrose buffer. The slices were stored at room temperature in a chamber containing artificial cerebrospinal fluid (ACSF) (in mM): 124 NaCl, 2.5 KCl, 1.3 $MgSO_4$, 1.0 NaH_2PO_4, 2.0 $CaCl_2$, 20 glucose, and 26 $NaHCO_3$, at pH 7.3–7.4 with 95% O_2, 5% CO_2 bubbling through it. The osmolarity of the ACSF was ~310 mOsm.

Whole cell patch-clamp recordings were performed as described previously (Jin et al. 2006; Jin and Smith 2007). During the recording, the slice was maintained fully submerged in the recording chamber and perfused with oxygenated ACSF (~3 ml/min). The ACSF was heated to 32–35°C by an in-line heater (Warner Instruments, Hamden, CT). GP neurons were visualized by IR-differential interference contrast microscopy (BX51WI) using a 40× water immersion objective (Olympus, Pittsburgh, PA). Electrodes were pulled from borosilicate glass on a vertical patch pipette puller (Narishige, Tokyo, Japan) to have resistance in the range of 3–5 MΩ when filled with the internal solution containing (in mM): 124 Cs-methanesulfonate, 11 KCl, 2 $MgCl_2$, 10 HEPES, 2 Na_2ATP, 0.3 GTP, and 5*N*-(2,6-dimethylphenylcarbamoylm-ethyl) triethylammonium bromide (QX314) (pH 7.4 and 300–310 mOsm). Tight-seal (>1 GΩ) whole-cell recording was obtained from the cell body of GP neurons. Series resistance was regularly monitored during recording, and cells were rejected if the series resistance changed by >20%. Neurons were voltage-clamped at a holding potential of −10 mV and whole-cell membrane currents were recorded with a Patch-Clamp PC-501A (Warner Instruments, Hamden, CT). To isolate $GABA_A$ receptor-mediated IPSCs, CNQX (50 μM) and D-AP5 (50 μM) were added to the ACSF. Bipolar matrix stimulating electrodes (FHC, Bowdoinham, ME) were placed in the striatum close to the GP. Evoked inhibitory postsynaptic currents (IPSCs) in GP neurons were recorded by stimulation of striatum with single pulses (0.15–0.2 ms) that ranged from minimal (2–5 V) to maximal stimulation (15–20 V), delivered once every 20 s. To record miniature IPSCs (mIPSCs), pipettes contained the following (in mM): 125 KCl, 10 NaCl, 1 $CaCl_2$, 2 $MgCl_2$, 10BAPTA, 10 HEPES, 2 Na_2-ATP,

0.3 GTP, and 0.5% biocytin (pH 7.4 and 300–310 mOsm). The mIPSCs were recorded at a holding potential of –60 mV in the presence of tetrodotoxin (1 μM TTX), CNQX (50 μM), and D-AP5 (50 μM).

All drugs were purchased from Toris Cookson (Ellisville, MO). Signals were filtered at 5 kHz and digitized with a Digidata 1322A and analyzed off-line using pClamp 9. The mIPSCs were detected and analyzed using Mini Analysis software (Synaptosoft, Fort Lee, NJ). All group data were expressed as means ± SEM. Statistical significance was assessed by Student's *t* test.

3 Results

3.1 Expression of GAT-1 in the GP

In the first set of experiments, we used a specific commercially availabe GAT-1 antiserum to immunohistochemically localize GAT-1 immunoreactivity at both light and electron microscopic level. Consistent with our monkey data (Galvan et al. 2005), GAT-1 was heavily expressed in glial processes and preterminal axons in the rat GP (Fig. 1).

3.2 Effects of GAT-1 Transporter Inhibitor on Evoked IPSCs in the GP

3.2.1 Parasagittal Striatopallidal Slice Preparation

Previous single-cell filling studies in rats and monkeys showed that striatal axons travel to the GP in the parasagittal plane (Kawaguchi et al. 1990; Parent et al. 1995). Therefore, to reduce the damage of striatopallidal axons, parasagittal slices were used to record IPSCs evoked by striatal stimulation in the rat GP (Jin and Smith 2007).

In hippocampal slices, GAT-1 transporter blockade prolongs the decay time, but does not change the amplitude of large evoked IPSCs (eIPSCs) (Roepstorff and Lambert 1992, 1994; Thompson and Gahwiler 1992; Isaacson et al. 1993; Draguhn and Heinemann 1996). On the basis of these findings, we first tested the effects of GAT-1 inhibitor on striatal eIPSCs in rat GP neurons using maximal striatal stimulation (15–20 V). Consistent with the hippocampal results, bath application of SKF 89976A (10 μM), a selective GAT-1 inhibitor, prolonged the decay (148.1 ± 6.1% of control; $P < 0.005$), but had no effect on the amplitude (104 ± 4% of control; $P > 0.5$) of eIPSCs in six cells tested (Fig. 2a, c).

We then examined whether blockade of GAT-1 also affects IPSCs evoked by minimal striatal stimulation (2–5 V). Under these conditions, bath application of

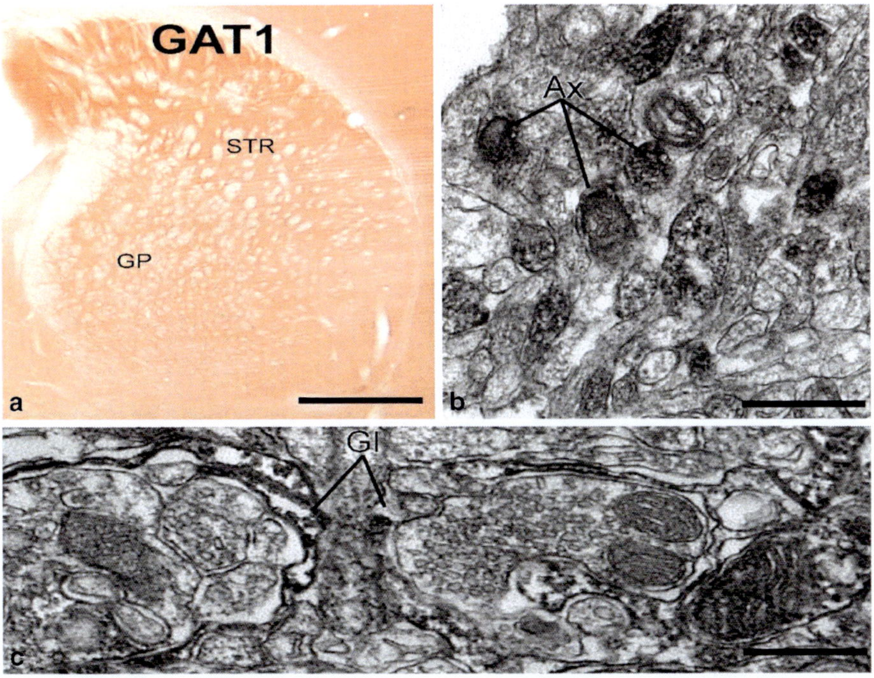

Fig. 1 Expression of GAT-1 in the globus pallidus of a 19-days-old rat. (**a**) Light photomicrographs of GAT-1 immunoreactivity in the rat striatopallidal complex. (**b**) GAT-1-immunolabeled unmyelinated axons (Ax). (**c**) GAT-1-immunolabeled glial process. Scale bars: (**a**) 1 mm and (**b**, **c**) 0.5 μm

SKF-89976A did not have a significant effect on the decay (103 ± 5.3% of control; $P > 0.5$), but increased the amplitude (154 ± 11.7% of control; $P < 0.005$) of eIPSCs in five cells recorded (Fig. 2b, d). These results demonstrate that blockade of GAT-1 has activity-dependent regulatory effects on GABAergic striatopallidal transmission in the rat GP.

3.2.2 Coronal Striatopallidal Slice Preparation

Since some of the GP neurons project back to the striatum and have local axon collaterals in the GP (Kita and Kitai 1991; Bevan et al. 1998; Smith et al. 1998; Kita et al. 1999; Kita and Kita 2001; Sadek et al. 2007), electrical stimulation of the striatum could activate both striatopallidal fibers and intrinsic collaterals of pallidostriatal axons in the rat GP. Thus, the effect of SKF 89976A on IPSCs evoked by striatal stimulation described in the previous experiment could involve activation of local axon collaterals of GP neurons. To test this possibility, we recorded IPSCs evoked locally in coronal slices of the GP, a section plane that should maintain

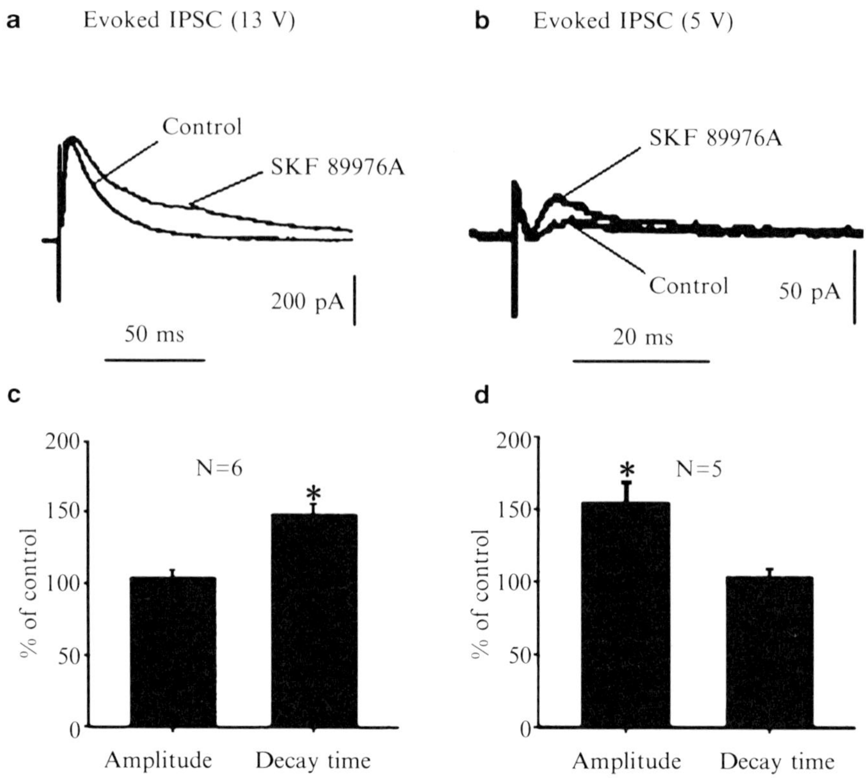

Fig. 2 Effects of GAT-1 inhibitor on GABAergic synaptic transmission in rat GP using parasagittal slices of the striatopallidal complex. (**a**) Bath application SKF 89976A (10 μM) prolongs the decay, but does not change the amplitude of IPSCs evoked in GP neurons by maximal striatal stimulation. (**b**) Summary bar graph showing the effect of SKF 89976A on the amplitude and decay time of striatal-evoked IPSCs as percent of control ± SEM (*asterisk*: significantly different from control, $P < 0.005$). (**c**) Bath application SKF 89976A increases the amplitude, but does not affect the decay of IPSCs evoked by low-intensity stimulation of the striatum. (**d**) Summary bar graph showing the effect of SKF 89976A on the amplitude and decay time of IPSCs evoked by low-intensity striatal stimulation as percent of control ± SEM (*asterisk*: significantly different from control, $P < 0.005$). Evoked IPSCs in GP neurons were recorded following high [(**a**) 13 V] and low [(**b**) 4 V] single pulse stimulation of the striatum delivered once every 20 s at a holding potential of −10 mV in the presence of 50 μM CNQX and 50 μM D-AP5

intrapallidal local axon collaterals communication, but significantly damage the striatopallidal pathway. Six neurons were recorded at the stimulation range between 7 and 15 V in these sections. Bath application of SKF 89976A (10 μM) did not have any significant effect on either the decay (103.8 ± 4% of control; $P = 0.39$) or the amplitude (97.6 ± 6.5% of control; $P = 0.73$) of eIPSCs by intrapallidal stimulation

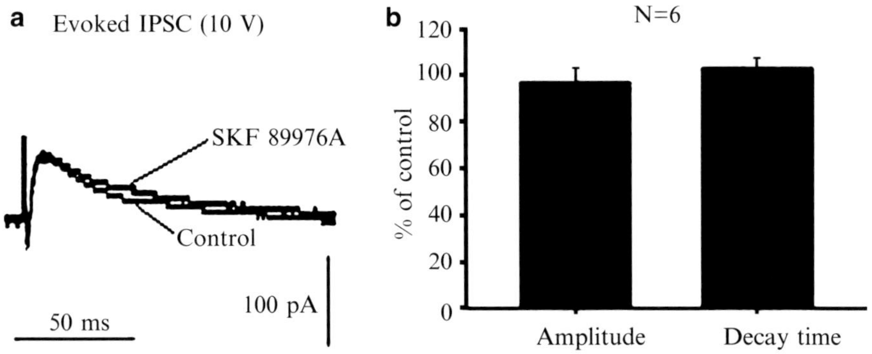

Fig. 3 Effects of GAT-1 inhibitor on GABAergic synaptic transmission in rat GP using coronal slices of the striatopallidal complex. (**a**) Bath application of SKF 89976A (10 μM) did not affect the decay time and amplitude of eIPSCs following local stimulation of GP. (**b**) Summary bar graph showing that the IPSCs decay time and amplitude evoked by GP stimulation were not changed by SKF 89976. Evoked IPSCs in GP neurons were induced by local single pulse (10 V) stimulation of GP

(Fig. 3a, b). Together with the data from the previous experiment, these results suggest that GAT-1 plays an important regulatory role of GABAergic striatopallidal transmission, but does not mediate any significant effect on intrapallidal GABAergic communication through local axon collaterals of GP neurons.

3.3 *Blockade of GAT-1 Has No Effect on mIPSCs*

The effects of SKF 89976A on the striatal eIPSCs in GP parasagittal slices suggest that blockade of GAT-1 modulates the striatopallidal GABAergic synaptic transmission under conditions that recruit a large number of striatopallidal fibers. To test whether blockade of GAT-1 transporter also modulates the GABAergic transmission at individual release sites, we recorded miniature inhibitory postsynaptic currents (mIPSCs) following bath application of SKF 89976A. In contrast to the effect of GAT-1 antagonist on eIPSCs from GP parasagittal slices, we found no significant effect of SKF 89976A on mIPSC amplitude (98.5 ± 4.6% of control; $P = 0.73$), decay time (95 ± 5.3% of controls; $P = 0.46$), and frequency (100.6 ± 6% of control; $P = 0.9$) (Fig. 4a, b). These results are consistent with previous studies demonstrating that blockade of GAT-1 does not have a significant effect on mIPSCs in hippocampal slices (Thompson and Gähwiler 1992; Isaacson et al. 1993; Overstreet and Westbrook 2003).

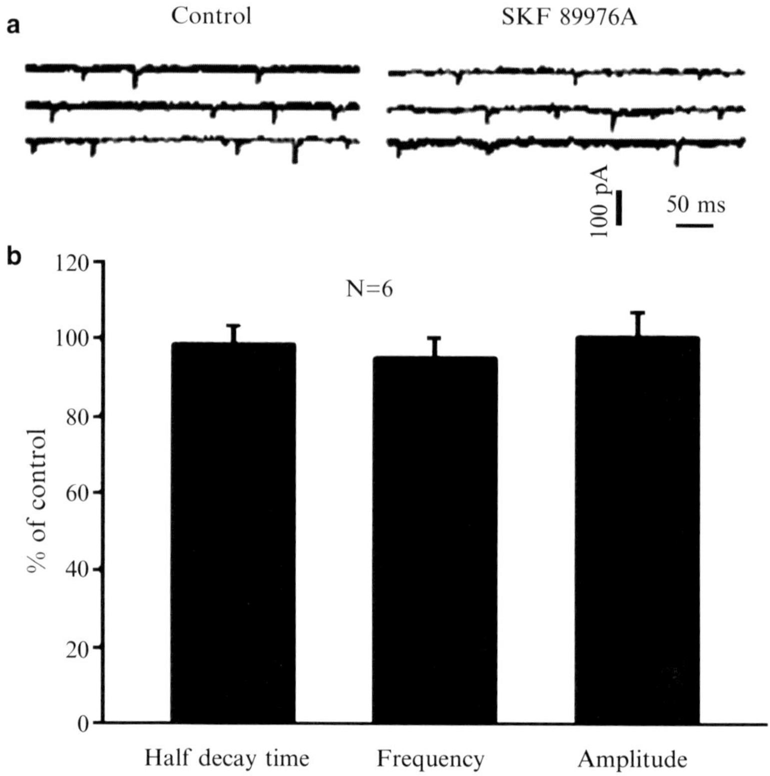

Fig. 4 Effects of GAT-1 inhibitor on tetrodotoxin (TTX)-resistant GABAergic synaptic transmission in rat GP. (**a**) Examples - mIPSCs recorded in control condition (*left*) or in the presence of 10 μM SKF 89976A (*right*). (**b**) Bar graph showing mIPSC half decay time; frequency and amplitude (percent of control ± SEM) were not affected by SKF 89976A application. The mIPSCs were recorded from parasagittal slices of rat GP at a holding potential of −60 mV in the presence of 50 μM CNQX and 50 μM D-AP5 and 1 μM TTX

4 Discussion

The data presented in this study demonstrate that application of GAT-1 inhibitor prolongs the decay time of large striatal eIPSCs, while it increases the amplitude of small eIPSCs in rat GP neurons. These observations provide the first evidence that GAT-1 plays a dual, activity-dependent regulatory role on striatopallidal GABAergic synaptic transmission in the GP.

Several studies have demonstrated that bath application of GAT-1 inhibitors, either NO711 or tiagabine, dramatically prolonged the decay of large eIPSCs without changing IPSC amplitude in hippocampal neurons (Roepstorff and Lambert 1992, 1994; Thompson and Gahwiler 1992; Isaacson et al. 1993; Draguhn and Heinemann 1996;

Overstreet et al. 2000). In constrast, small eIPSCs were unaffected by GAT-1 blockade in this brain region (Isaacson et al. 1993; Roepstorff and Lambert 1994). In the present study, we used SKF 89976A as GAT-1 blocker because the results from a previous study in our laboratory showed that this inhibitor reduced the activity of GPe and GPi neurons in monkeys (Galvan et al. 2005). In line with results reported from hippocampal slices using other GAT-1 blockers, our data demonstrate that bath application of SKF 89976A increased the decay, but not the amplitude, of large eIPSCs in the rat GP. However, SKF 98876A application was found to increase the amplitude, without any significant effect on the decay time, of small eIPSCs in the rat GP. Similar findings have been reported in the hippocampus (Vivar and Gutiérrez 2005). Although it remains to be clarified, there are various mechanisms through which the blockade of GAT-1 may increase GABAergic transmission in the rat GP. One possibility is that under the condition of minimal stimulation, a large pool of unbound $GABA_A$ receptors can be activated by synaptically released GABA, resulting in larger eIPSCs after increasing GABA level through GAT-1 blockade. This result supports our previous hypothesis that increasing the activity of postsynaptic $GABA_A$ receptors may be responsible for the inhibitory effect of SKF 89976 on the activity of GP neurons (Galvan et al. 2005). Together, our data indicate that the functional role of GAT-1 on GABAergic transmission in GP depends on the strength of striatopallidal fiber activity.

As mentioned earlier, electrical stimulation in the striatum could activate both striatopallidal and pallidostriatal fibers, which have local axon collaterals in the GP (Kita and Kitai 1991, 2001; Bevan et al. 1998; Smith et al. 1998; Kita et al. 1999; Sadek et al. 2007). Thus, the effect of GAT-1 blockade on eIPSCs recorded in GP after striatal stimulation could be the result of striatopallidal regulation or modulation of transmission at local intrapallidal axon collaterals of pallidostriatal neurons. This possibility can be tested by dual recording of pair of functionally connected GP neurons. However, due to the low probability of finding such pairs in normal GP (Oorschot 1996; Sadek et al. 2005), we recorded IPSCs evoked locally from rat GP stimulation in coronal slices of the striatopallidal complex, assuming that in such preparations IPSCs are mainly generated locally by axon collaterals of GP neurons. Under these conditions, bath application of GAT-1 inhibitor did not affect the amplitude or decay of eIPSCs, suggesting that GAT-1 has a more significant effect on the modulation of the striatopalli dal synaptic transmission than on local intrinsic GABAergic microcircuitry of GP neurons.

Similarly, we did not detect any significant effect of SKF 89976A on mIPSCs. This result is different from a previous study showing that tigabine, a GAT-1 selective blocker, prolonged the decay and reduced the frequency of mIPSCs in rat GP neurons (Chen and Yung 2003). It is not clear why these two highly potent and specific drugs have different effects on mIPSCs in the rat GP since their potency and specificty are very close [for review, see Borden (1996)]. A recent study has demonstrated that rat GP neurons can be separated into two functional domains: the neurons in the inner region of GP receiving inputs from more than twice as many GP boutons than neurons in the outer region (Sadek et al. 2007). Thus, the action potential-independent spontaneous release of GABA should be significantly higher in the inner than in the outer GP regions. Because most of our recordings were made from the outer GP region close to the striatal border, a differential location of recording sites in the GP between the two studies might explain why we did not find any significant effect of GAT-1 blockade on mIPSCs. Based on light microscopic localization of GAT-1 in the GP (Fig. 1a), it is unlikely that a differential expression level of GAT-1 in the inner GP region compared with that in the outer GP region underlies these physiological effects. On the other hand, it is noteworthy that our findings are in line with previous studies showing that application of GAT-1 inhibitors has no effects on the amplitude and decay time of mIPSCs in slices of rat hippocampus (Thompson and Gahwiler 1992; Overstreet and Westbrook 2003). These results are further supported by the evidence that the amplitude and decay time of mIPSCs are not changed in GAT-1-deficient mice (Jensen et al. 2003). In agreement with these studies, our results support the idea that GAT-1 transporter modulates the GABAergic synaptic transmission at striatopallidal synapses during intense activation of presynaptic afferents, but plays a minor role in regulating spontaneous GABA release at individual synapses.

Acknowledgments This work was supported by a grant from NIH to YS (R01 NS 042937) and the Yerkes base grant (RR00165).

References

Bevan MD, Booth PA, Eaton SA and Bolam JP (1998) Selective innervation of neostriatal interneurons by a subclass of neuron in the globus pallidus of the rat. J Neurosci 18: 9438–9452.

Bevan MD, Magill P, Terman D, Bolam JP and Wilson CJ (2002) Move to the rhythm: oscillations in the subthalamic nucleus-external globus pallidus network. Trends Neurosci 25: 525–531.

Borden LA (1996) GABA transporter heterogeneity: pharmacology and cellular localization. Neurochem Int 29: 335–356.

Chen L and Yung WH (2003) Effects of the GABA-uptake inhibitor tiagabine in rat globus pallidus. Exp Brain Res 152: 263–269.

Draguhn A and Heinemann U (1996) Different mechanisms regulate IPSC kinetics in early postnatal and juvenile hippocampal granule cells. J Neurophysiol 76: 3983–3993.

Galvan A, Villalba RM, West SM, Maidment NT, Ackerson LC, Smith Y and Wichmann T (2005) GABAergic modulation of the activity of globus pallidus neurons in primates: in vivo analysis of the functions of GABA receptors and GABA transporters. J Neurophysiol 94: 990–1000.

Isaacson JS, Solis JM and Nicoll RA (1993) Local and diffuse synaptic actions of GABA in the hippocampus. Neuron 10: 165–175.

Jensen K, Chiu CS, Sokolova I, Lester HA and Mody I (2003) GABA transporter-1 (GAT1)-deficient mice: differential tonic activation of GABAA versus GABAB receptors in the hippocampus. J Neurophysiol 90: 2696–2701.

Jin XT, Pare JF, Raju DV and Smith Y (2006) Localization and function of pre- and postsynaptic kainate receptors in the rat globus pallidus. Eur J Neurosci 23: 374–386.

Jin XT and Smith Y (2007) Activation of presynaptic kainate receptors suppresses GABAergic synaptic transmission in the rat globus pallidus. Neuroscience 149: 338–349.

Kawaguchi Y, Wilson CJ and Emson PC (1990) Projection subtypes of rat neostriatal matrix cells revealed by intracellular injection of biocytin. J Neurosci 10: 3421–3438.

Keros S and Hablitz JJ (2005) Subtype-specific GABA transporter antagonists synergistically modulate phasic and tonic GABAA conductances in rat neocortex. J Neurophysiol 94: 2073–2085.

Kita H and Kitai ST (1991) Intracellular study of rat globus pallidus neurons: membrane properties and responses to neostriatal, subthalamic and nigral stimulation. Brain Res 564: 296–305.

Kita H and Kita T (2001) Number, origins, and chemical types of rat pallidostriatal projection neurons. J Comp Neurol 437: 438–448.

Kita H, Tokuno H and Nambu A (1999) Monkey globus pallidus external segment neurons projecting to the neostriatum. NeuroReport 10: 1467–1472.

Oorschot DE (1996) Total number of neurons in the neostriatal, pallidal, subthalamic, and substantia nigral nuclei of the rat basal ganglia: a stereological study using the cavalieri and optical disector methods. J Comp Neurol 366: 580–599.

Overstreet LS, Jones MV and Westbrook GL (2000) Slow desensitization regulate the availability of synaptic $GABA_A$ receptors. J Neurosci 20: 7914–7921.

Overstreet LS and Westbrook GL (2003) Synapse density regulates independence at unitary inhibitory synapses. J Neurosci 23: 2618–2626.

Parent A, Charara A and Pinault D (1995) Single striatofugal axons arborizing in both pallidal segments and in the substantia nigra in primates. Brain Res 698: 280–284.

Peters A, Palay SL and Webster HF (1991) The Fine Structure of the Nervous System: Neurons and Their Supporting Cells (3rd edition). New York: Oxford Press.

Plenz D and Kitai ST (1999) A basal ganglia pacemaker formed by the subthalamic nucleus and external globus pallidus. Nature 400: 677–682.

Roepstorff A and Lambert JD (1992) Comparison of the effect of the GABA uptake blockers, tiagabine and nipecotic acid, on inhibitory synaptic efficacy in hippocampal CA1 neurons. Neurosci Lett 146: 131–134.

Roepstorff A and Lambert JD (1994) Factors contributing to the decay of the stimulus-evoked IPSC in rat hippocampal CA1 neurons. J Neurophysiol 72: 2911–2926.

Sadek AR, Magill PJ and Bolam JP (2005) Local connectivity between neurons of the globus pallidus. In: The Basal Ganglia VIII. Bolam JP, Ingham CA and Magill PJ (eds). New York: Springer, pp 611–619.

Sadek AR, Magill PJ and Polam JP (2007) A single-cell analysis of intrinsic connectivity in the rat globus pallidus. J Neurosci 27: 6352–6362.

Smith Y, Bevan MD, Shink E and Bolam JP (1998) Microcircuitry of the direct and indirect pathways of the basal ganglia. Neuroscience 86: 353–387.

Thompson SM and Gähwiler BH (1992) Effects of the GABA uptake inhibitor tiagabine on inhibitory synaptic potentials in rat hippocampal slice cultures. J Neurophysiol 67: 1698–1701.

Vivar C and Gutiérrez R (2005) Blockade of the membranal GABA transporter potentiates GABAergic responses evoked in pyramidal cells by mossy fiber activation after seizures. Hippocampus 15: 281–284.

Nitric Oxide Modulation of the Dopaminergic Nigrostriatal System: Focus on Nicotine Action

Vincenzo Di Matteo, Massimo Pierucci, Arcangelo Benigno, Ennio Esposito, Giuseppe Crescimanno, Maurizio Casarrubea, and Giuseppe Di Giovanni

Abstract Nitric oxide (NO) signalling plays an important role in the integration of information processed by the basal ganglia nuclei. Accordingly, considerable evidence has emerged indicating a role for NO in pathophysiological conditions such as Parkinson's disease (PD), schizophrenia and drug addiction. To further investigate the NO modulation of dopaminergic function in the basal ganglia circuitry, in this study we used in vivo electrophysiology and microdialysis in freely-moving rats. Pharmacological manipulation of the NO system did not cause any significant changes either in the basal firing rate and bursting activity of the dopamine (DA) neurons in the substantia nigra pars compacta (SNc) or in DA release in the striatum. In contrast, the disruption of endogenous NO tone was able to counteract the phasic dopaminergic activation induced by nicotine treatment in both experimental approaches. These results further support the possibility that nicotine acts via a NO mechanism and suggest a possible state-dependent facilitatory control of NO on the nigrostriatal DA pathway. Thus, NO selectively modulates the DA exocytosis associated with increased DA functiov n.

1 Introduction

Nitric oxide (NO) has been associated with a variety of physiological and pathological processes in the human body since it was identified as a novel signal molecule by Furchgott and Zawadzki (1980). NO is synthesized from L-arginine by a nitric oxide synthase (NOS) using nicotinamide adenine dinucleotide phosphate (NADPH) and molecular oxygen (Bian and Murad 2003). To date, three isoforms

V. Di Matteo, M. Pierucci, and E. Esposito
Istituto di Ricerche Farmacologiche "MArio Negri", Consorzio Mario Negri Sud, S. Maria Imbaro (Ch), Italy

A. Benigno, G. Crescimanno, M. Casarrubea, and G. Di Giovanni (✉)
Dipartimento di Medicina Sperimentale, Sezione di Fisiologia Umana, "G. Pagano," Università degli Studi di Palermo, Corso Tuköry 129, 90134 Palermo, Italy
e-mail: g.digiovanni@unipa.it

H.J. Groenewegen et al. (eds.), *The Basal Ganglia IX*, Advances in Behavioral Biology 58, DOI 10.1007/978-1-4419-0340-2_24, © Springer Science+Business Media, LLC 2009

of NOS, that is, neuronal NOS (nNOS), endothelial NOS (eNOS), and inducible NOS (iNOS), have been identified. While nNOS and eNOS are constitutively expressed, the expression of iNOS is induced through the inflammatory response process of cells to infections or injuries (Dawson and Dawson 1998). The characteristics of neurotransmitter NO are that (a) it is synthesized postsynaptically, (b) it is not stored in vesicles being a diffusible gas, (c) it does not act at conventional receptors on the surface of adjacent neurons and (d) it can act as a retrograde messenger diffusing to the presynaptic terminal. Based on this evidence we can affirm that NO in the nervous system works as an unorthodox neurotransmitter. A major biochemical function of NO is to activate the soluble form of guanylyl cyclase (sGC), inducing the accumulation of cyclic guanosine monophosphate (cGMP) in target cells. Cyclic-GMP subsequently acts via protein kinases, phosphodiesterases, and perhaps directly on ion channels (Arnold et al. 1977; Bredt et al. 1990). Furthermore, NO can exert its biological effects through other mechanisms, such as modulating the function of monoamine transporters and S-nitrosylation of receptors. It has been demonstrated that NO can S-nitrosylate the NMDA receptor leading to its down-regulation (Choi and Lipton 2000).

2 Nitric Oxide Distribution in the Basal Ganglia

Nitric oxide signalling plays an important role in controlling motor behaviour modulating the integration of information processed by the basal ganglia nuclei. Most likely, it interacts with dopaminergic (DAergic), serotonergic, cholinergic and glutamatergic neurotransmission at different levels of these nuclei. Consistently, mice mutants for nNOS have altered locomotor abilities, and rats and mice treated with various NOS inhibitors show problems with fine motor control. NO, furthermore, antagonizes the increase in locomotor activity found after DA agonist administration. The pharmacological blocking of nNOS decreases locomotion and induces catalepsy in different animal species [see Del Bel et al. (2007) for a review]. Although NOS neurons are present throughout all basal ganglia nuclei and in other regions involved in motor control such as the motor cortices and the pedunculopontine tegmental nucleus (PPTg), their concentration varies significantly (Bredt et al. 1990; Del Bel et al. 2007; Egberongbe et al. 1994; Eve et al. 1998; Garthwaite and Boulton 1995; Leontovich et al. 2004; Nisbet et al. 1994; Vincent and Kimura 1992).

In comparison with other brain centres, the substantia nigra (SN) may be considered a rather NOS cell-poor nucleus (Johnson and Ma 1993; González-Hernández et al. 2000). Some studies even failed to detect any NADPH-diaphorase (NADPH-d) (+) cells in the SNc (Govsa and Kayalioglu 1999). However, a double-labelling study showed that in general few small fusiform neurons (not more than 1%) express NOS on the dorsal border of the rostrolateral part of the SNc, where it touches the zona incerta pars ventralis. These cells appear to be tyrosine hydroxylase (TH)-(+) DA neurons (González-Hernández et al. 2000). By contrast, Del Bel and colleagues have revealed in the SNc a population of NOS neurons that have almost the same density as those present in the ventral pallidum and nucleus accumbens (Del Bel and Guimarães 2000; Gomes and Del Bel 2003). The scenario appears different in the substantia nigra, pars reticulata (SNr); in fact, many NADPH-d (+) dendrites and axon-like processes are present, some of which have close relationship with vessels (Govsa and Kayalioglu 1999). The origin of these dendrites and axon-like processes may be local, constituted by an ample intrinsic subpopulation of SNr γ-aminobutyric acid (GABA)/NOS neurons, and extrinsic, constituted by medium to large cholinergic/NOS neurons of the PPTg and laterodorsal tegmental nucleus (LDTg) (Gonzáles-Hernández and Rodríguez 2000). Within the SNr two types of GABA cells co-express NOS: one is represented by large cells in the rostrolateral part of the SNr (rlSNr) containing parvalbumine (PV) and NOS (rlSNr) and another population is constituted by small GABA cells located in the rostromedial portion of the SNr (González-Hernández and Rodríguez 2000).

The NOS expression in the striatum has been more extensively studied (Bernácer et al. 2005; Egberongbe et al. 1994; Eve et al. 1998; Govsa and Kayalioglu 1999; Johannes et al. 2003; Leontovich et al. 2004; Nisbet et al. 1994; Vincent and Kimura 1992). NOS neurons are one type of the four different classes of interneurons present in the striatum, and in the human they populate the entire striatum, including the tail and head of the caudate nucleus. They represent 1–2% of all striatal cells and are spiny, co-expressing NK1 receptors, somatostatin and neuropeptide Y (Vincent and Kimura 1992). The striatal nitrergic neurons are essentially small in diameter (12–25 μm), slender, bipolar and fusiform with a long dendrite. The neuronal population of the human striatum seems to be very heterogeneous. In fact, up to 12 different subtypes of NOS neurons have been described in humans and, strikingly, one of them is a large reticular NOS cell that resembles the characteristics of a projecting neuron (Johannes et al. 2003). This efferent nitrergic cell was demonstrated to project to the insular cortex. Thus, at least some of the NOS reticular neurons of the human striatum have direct cortical projections, though the existence of their axon collaterals in striatal tissues close to the maternal cells also demonstrates that they influence surrounding cells (Leontovich et al. 2004). Moreover, the matrix is the striatal compartment with the densest NOS neuronal population that tends to be located at the boundaries between the striosomes and the matrix, as well as at the boundaries between the core and the peripheral region of the striosomes (Bernácer et al. 2005). This finding has an important functional implication, because the NOS neurons that occur at the edges between the two compartments are thought to mediate interactions between the medium spiny neurons (MSNs) of the matrix and the striosomes. Despite the recent advances, the role of NOS interneurons in the striatum is still not clear. Among others their postulated functions are (a) to control local blood flow in the striatum by releasing NO acting directly on sGC in the vascular smooth muscle and causing vasodilatation and (b) to produce NO that acts as a neurotransmitter modulating striatal discharge and plasticity, either through direct interactions with ligand-gated channels or by influencing surrounding striatal MSNs via the stimulation of second messenger systems (West and Grace 2002).

Messenger-RNA expression studies have revealed a scattered subpopulation of NOS neurons in the internal segment and medial medullary lamina (MML) of the globus pallidus (GP) and in almost all neurons of the subthalamic nucleus (STN), which are presumed to be glutamatergic and excitatory. Moreover, it is important to note that NOS neurons have not been detected in the external GP (GPe). This may have functional implications, since it is the internal segment of the GP (GPi) and not the GPe that relays the basal ganglia output to the motor nuclei of the thalamus. In the GP, NO is probably co-localized with GABA, which has previously been shown to be the neurotransmitter of virtually all pallidal neurons (Nisbet et al. 1994; Eve et al. 1998).

3 Involvement of NO in Neurodegeneration of Dopaminergic Nigrostriatal System

Nitric oxide is a Janus-faced molecule and, notwithstanding, the exact role (i.e. neuroprotective vs. neurotoxic) it plays in neurodegenerative disorders is still ambiguous. Substantial evidence demonstrates a causative role for NO in the degeneration of DAergic neurons of the nigrostriatal pathway in PD (Esposito et al. 2007; Di Matteo et al. 2006; Duncan and Heales 2005; Zhang et al. 2006; Eve et al. 1998; Nisbet et al. 1994; Hunot et al. 1996). Consistently, NOS polymorphisms increased expression of astroglial iNOS, and nNOS have been reported in PD patients and in different toxin-induced experimental models of PD. Evidence in animal models and in PD patients suggests that NOS is up-regulated in the basal ganglia nuclei and enhanced NO formation takes place after partial injury of the nigrostriatal DAergic system. The exact mechanisms of the NO contribution to neurodegenerative diseases are not completely understood. Multiple lines of evidence indicate that NO is associated with excitotoxicity, DNA damage and protein modifications, which are common pathogenic mechanisms involved in multiple neurodegenerative diseases (Del Bel et al. 2007). Nevertheless, NO produced by either nNOS or iNOS plays an important role in DA degeneration. iNOS once induced remains active for several hours to days and produces NO in 1,000-fold greater quantities than the constitutive enzyme nNOS. A robust increase in iNOS mRNA levels has been observed after lipopolysaccharide (LPS), 1-methyl 4-phenyl 1,2,3,6-tetrahydropyridine (MPTP) or 6-hydroxydopamine (6-OHDA) injection in striatum and SN (Dawson and Dawson 1998; Bian and Murad 2003). Damage to striatal DAergic fibres seems to be mainly mediated by NO produced by nNOS, while the damage to nigral DAergic neurons is largely inflicted by NO generated by iNOS. Evidence from human post-mortem studies has revealed an increase of NOS mRNA expression only in MML of the GP and in dorsal STN. Instead, a reduction has been shown to occur in the striatum although it was not statistically significant (Eve et al. 1998). Such altered activity of NOS neurons of the MML and STN may play a role in the compensatory up-regulation of nigrostriatal DAergic neurotransmission in PD, but might also exert an excitotoxic effect on striatal neurons and nigrostriatal terminals. In animal 6-OHDA-models of PD nNOS expression is reduced while a proportion of nNOS nerve fibres in the striatum is apparently lost following DAergic deafferentation, resulting in a 50% decrease in NOS activity, and depression of the NO-cGMP pathway (De Vente et al. 2000; Sancesario et al. 2004). In contrast, Gomes et al. (2003) showed that 6-OHDA lesion induced a significant increase in NOS cell numbers in the ipsilateral dorsal striatum while a decrease was seen in the ipsilateral SNc and contralateral nucleus accumbens.

4 Nitric Oxide Modulation of the Activity of Dopaminergic Nigrostriatal System

Recently, after we had investigated the role of NO in MPP^+ (Di Matteo et al. 2006) and 6-OHDA-induced nigral neurodegeneration (Di Matteo et al. 2009), we have been studying the NO modulation of the activity of the DAergic nigrostriatal system. Although there have been numerous neurochemical studies (see West et al. 2002) the manipulation of NO within the SNc has not been investigated under normal conditions using an electrophysiological approach. Indeed, only the results of modification of tonic striatal NO tone are available at the moment (West and Grace 2000). In particular, the effect of striatal NO has been investigated on the responsiveness of SNc DA neurons to the intermittent electrical stimulation of the striatum and orbital prefrontal cortex (oPFC). Increasing NO tone in the striatum was able to counteract the decrease in firing rate of DA cells observed in control animals during intermittent stimulation. Additionally, removal of NO tone increased the proportion of DA neurons responding to striatal stimulation and increased the prevalence of the initial inhibitory responses (West and Grace 2000). Thus, it has been proposed that NO may play a pivotal role in controlling the delicate homeostatic processes that normally provide stability to the DA-nigral system. Indeed, it may be capable of dynamically regulating the relative phasic DA responsivity via its action on tonic DA levels, in a manner dependent on the arousal state of the animal. Under rest, NO produced by glutamatergic activation of NOS interneurons might increase DA release either by intensifying glutamate (GLU) release or by influencing the activity of DA transporter (DAT), decreasing DA uptake and possibly causing a DA reverse release. This increase in tonic DA would down-modulate spike-dependent phasic dopamine release via stimulation of the very sensitive DA autoreceptors present on DA terminals. In contrast, during behavioural arousal, NO exerts an opposite effect on tonic extracellular DA levels that seems to be concentration-dependent. The strong production of NO, caused by intense glutamatergic corticostriatal transmission, would result in the inhibition of NMDA receptor function and produce less inhibition of phasic DA release via disinhibition of the DA autoreceptors (West et al. 2002). As striatal NO controls DA concentration mutually striatal NO is also under a DAergic influence (Sammut et al. 2006, 2007); indeed both electrical and chemical stimulation of the SNc elicited a robust surge in striatal NO efflux. This release seems to be neuronally dependent being blocked by pre-treatment with nNOS inhibitors and also evoked only by high-frequency stimulation that resembles the natural burst firing of DA SNc neurons. This last piece of evidence indicates that NO efflux occurs only when DA transmission is phasically increased and suggests that information transmitted via the nigrostriatal pathway during DA cell burst firing may be processed and/or amplified by NOS interneurons (Sammut et al. 2006, 2007). Dopamine within the striatum could directly modulate NO efflux, exciting NOS interneurons through the activation of $DA_{1/5}$ receptors present on their somas and increasing the release of NO (Sammut et al. 2006). Alternatively, DA modulates striatal NO levels via D_2 receptors in an opposing manner. This inhibitory control seems to be indirect and it is plausible that D_2 receptors are in fact presynaptically localized on GLU and acetylcholine (ACh) fibres impinging on NOS interneurons (Sammut et al. 2007).

5 NO/DA Interaction: Focus on Nicotine Effect

5.1 *Experimental Data*

To further investigate NO modulation of the nigrostriatal system, in this study we used in vivo electrophysiology and microdialysis in the rat. Extracellular single-unit recordings coupled with microiontophoresis were performed from putative DAergic-containing neurons in the SNc; local DA and its metabolite dihydroxyphenylacetic acid (DOPAC) levels were studied in the striatum by in vivo microdialysis in freely-moving rats [for more technical details, see Di Giovanni et al. (1999)].

Dopaminergic and non-dopaminergic (presumably GABA-ergic) neurons in the SNc and SNr, respectively, were identified on the basis of their established anatomical and electrophysiological characteristics i.e. waveform, firing rate and pattern (Invernizzi et al. 2007; Di Giovanni et al. 1999).

Systemic administration of two NOS inhibitors, *N*-ω-nitro-L-arginine methyl ester (L-NAME, 50 mg/kg, i.p.) and 7-nitroindazole (7-NI, 50 mg/kg, i.p.) (Fig. 1a, b), did not cause any significant change in the basal firing rate and bursting activity of the DA neurons in the SNc recorded ($n = 5$ for each drug). The effects of the drug treatment were evaluated for at least 20 min or more when it was possible, at 5-min intervals. Accordingly, treatment with the NO precursor L-arginine (L-ARG, 50 mg/kg, i.p.; $n = 5$) and the NO releaser molsidomine (MOL, 50 mg/kg, i.p.; $n = 5$) did not produce any significant modification of the neuronal discharge (Fig. 1c). Moreover, the modification of NO levels within the SNc, by L-NAME (−20 to −60 nA, pH 6.5; $n = 5$) or L-arginine (L-ARG, +40 nA for 5 min, pH 6; $n = 5$) microiontophoretic application, did not produce any modification in the discharge of SNc DA neurons (Fig. 1d). Consistent results with the extracellular recordings were obtained by neurochemical approach. In fact, 7-NI or L-NAME (50 mg/kg, i.p.; $n = 5$ each) treatment did not modify the DA release in the striatum, although DOPAC efflux was significantly reduced by the 7-NI treatment (−38.5 ± 4.8) (Fig. 2a, b).

Furthermore, neither MOL (50 mg/kg, i.p.; $n = 5$) nor L-ARG (50 mg/kg, i.p.; $n = 5$) modified DA release or DOPAC efflux in the striatum (data not shown). In contrast to this lack of effect under normal conditions, the inhibition of NO system was able to counteract the typical excitatory effect induced by treatment with nicotine on nigrostriatal function (Figs. 2 and 3). Pre-treatment with 7-NI or L-NAME

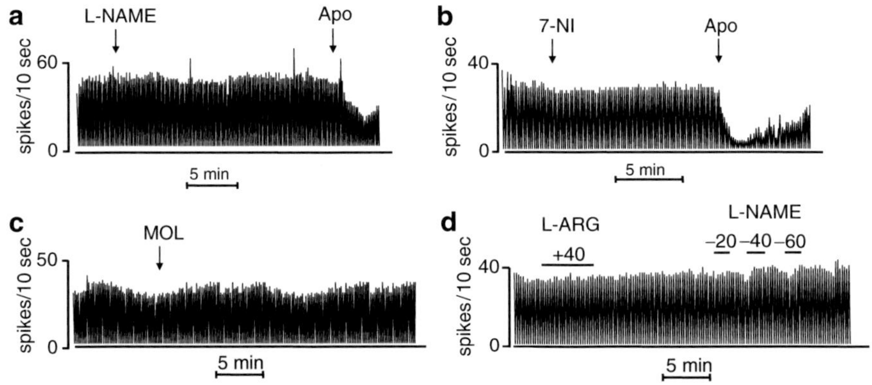

Fig. 1 Effect of systemic and local manipulation of NO signalling on the firing rate of DAergic SNc neurons. Representative rate histograms showing the effects elicited by i.p. administration (at *arrows*) of L-NAME (50 mg/kg) (**a**), 7-NI (50 mg/kg) (**b**), MOL (50 mg/kg) (**c**). APO, apomorphine administration (10 µg/kg, i.v., at *arrow*). (**d**) Typical rate histogram showing the effects elicited by locally applied ARG and L-NAME. Numbers above each bar indicate the ejecting currents in nA; the length of the bars is proportional to the time of the drug application

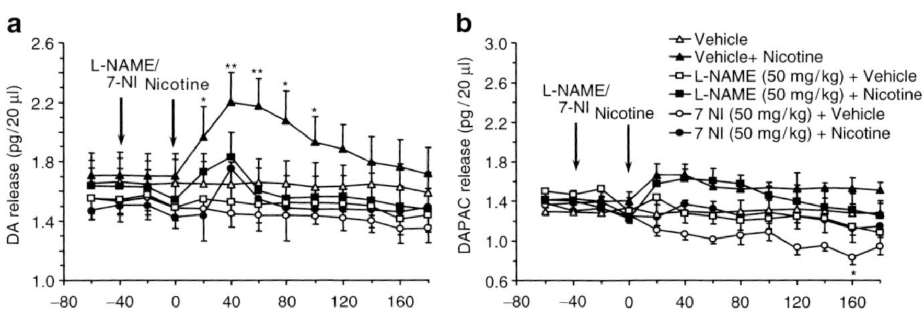

Fig. 2 Time course of the effects of 7-NI (open circle, 50 mg/kg, i.p., $n = 5$), L-NAME (open square, 50 mg/kg, i.p., $n = 5$) and nicotine (filled triangle, 1 mg/kg, i.p., $n = 5$) on extracellular DA and DOPAC levels in the striatum (**a**, **b**). Control groups treated with vehicle (inverted open triangle). The drugs were administered at the time indicated by *vertical arrows*. Each data point represents mean ± SEM absolute levels of DA, without considering probe recovery. Statistical analysis shows a significant effect of nicotine (one-way ANOVA; $p < 0.01$) as compared with the control group. 7-NI or L-NAME prevented nicotine-induced increase in DA release [two-way ANOVA; $*p < 0.05$, $**p < 0.01$ vehicle + nicotine as compared with 7-NI + nicotine (closed circle, $n = 5$) or L-NAME + nicotine (closed square, $n = 5$) by Tukey–Kramer's post hoc test]. In the striatum all pharmacological treatments failed to change DOPAC release, although 7-NI at 160 min reduced it significantly (one-way ANOVA; $p < 0.05$)

(50 mg/kg, i.p.) blocked the dose-dependent increase of the firing rate and the bursting activity of DA neurons in the SNc induced by acute i.v. injections of nicotine (25–400 μ/kg) (Fig. 3b–d).

As shown by the dose–response curve reported in Fig. 3d, nicotine reached its maximal effect (+93 ± 19%, above baseline) at the cumulative dose of 775 μg/kg that was statistically significant compared with the groups treated with the vehicles of 7-NI and L-NAME ($n = 5$). Pre-treatment with either 7-NI or L-NAME (50 mg/kg, i.p.), completely prevented nicotine-induced increase in DA firing rate and burst firing (the maximum effect was +34 ± 18% and +26 ± 8% after 7-NI and L-NAME treatment, respectively; Fig. 3d). In addition, the same pre-treatment with 7-NI or L-NAME (50 mg/kg, i.p.) prevented the enhancement in DA release elicited by acute nicotine (1 mg/kg, i.p.; $n = 5$) in the corpus striatum as well (Fig. 2a). In conclusion, inhibition of NOS by 7-NI or L-NAME was capable of counteracting the activation of nigral DA neurons caused by acute administration of nicotine.

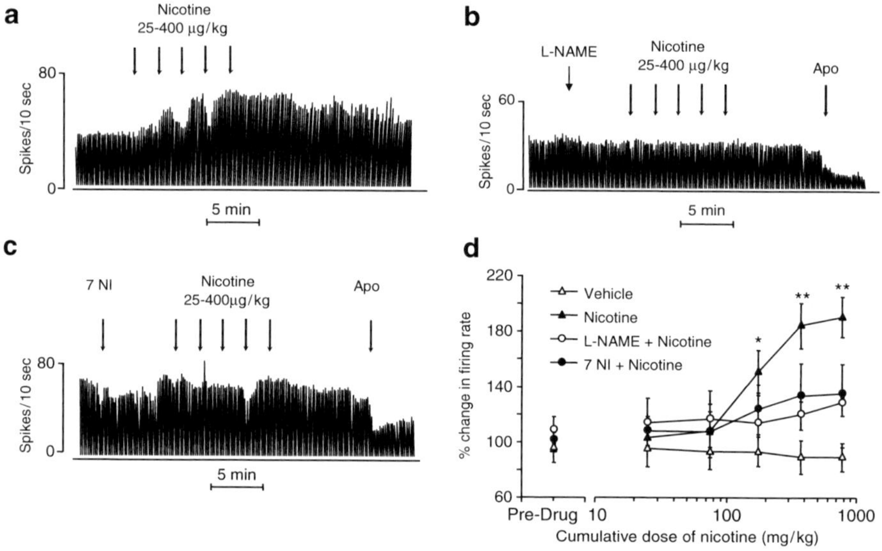

Fig. 3 Blockade by 7-NI and L-NAME of the excitatory actions of nicotine on the firing rate of SNc neurons. (**a**) Representative rate histogram showing the i.v. effect of nicotine (25, 50, 100, 200 and 400 μg/kg, at *arrows*). (**b**, **c**) Representative rate histograms showing that pre-treatment with 7-NI and L-NAME (50 mg/kg, i.p.) prevents the excitatory effect of nicotine. (**d**) Cumulative dose–response curve showing the mean percentage change (±S.E.M.) in firing rate after nicotine, vehicle, 7-NI + nicotine and L-NAME + nicotine. Statistical analysis revealed a significant effect of nicotine (one-way ANOVA; $p < 0.01$; $n = 7$) compared with the group treated with the vehicle of 7-NI and L-NAME ($n = 5$). Pre-treatment with 7-NI or L-NAME (50 mg/kg, i.p.), which did not cause any significant effect by itself (one-way ANOVA; $p < 0.29$; $n = 6$), completely prevented nicotine-induced increase in DA firing rate (two-way ANOVA; $^{*}p < 0.05$; $^{**}p < 0.01$ by Tukey–Kramer post hoc test)

5.2 Discussion

An important finding of our new work is that NO does not control DA tonic neurotransmission, in agreement with other investigations (Cox and Johnson 1998; Nowak et al. 2002; Schilström et al. 2004; Campos et al. 2006). In fact, neither the disruption of NO levels by treatment with NOS inhibitors, 7-NI and L-NAME, nor the NO elevation by L-ARG and MOL treatment was able to produce any changes in the firing rate and burst firing of SNc DA neurons. This lack of effect was obtained by general and local application of the drugs by microiontophoresis revealing that, probably, under tonic DAergic activation nigral NO does not play a relevant physiological role. Our findings are consistent with the evidence that L-NAME and L-ARG treatment did not modify the firing discharge of the ventral tegmental area (VTA) DAergic neurons in vivo (Schilström et al. 2004) and of both VTA and SNc neurons in vitro (Cox and Johnson 1998; Schilström et al. 2004). Furthermore, our neurochemical data are consistent with the electrophysiological recordings, likewise showing that endogenous nitrergic tone does not influence basal striatal DA release. The absence of effect of NO on DA function might be contingent upon confounding factors e.g. drug type-dependent effects, background redox state of the brain tissue, different NO regulation in different striatal subregions, etc. Nevertheless, we are confident in excluding limitations of the techniques since using a similar approach we have found effects of NO manipulation on the firing activity of other nuclei of the basal ganglia (Di Giovanni et al. 2003, 2006). Despite the inability of NO to modify basal DA function, the inhibition of NOS completely counteracted the stimulation of DA outflow induced by nicotine. We used nicotine because it is well known to be able to activate the nigrostriatal system at both nigral and striatal levels as well as the mesocorticolimbic system (Di Matteo et al. 2007). Consistent with previous studies, nicotine caused a robust phasic surge in nigral DAergic function due to combinatory effects. Exogenous nicotine increases the DA SNc activity in part via a direct activation of α_7-subunit containing nicotinic ACh receptors (α_7-nAChRs) present on the soma and axon of DA neurons. Simultaneously, nicotine causes a persistent depression of the inhibitory GABAergic inputs and a potentiation of glutamatergic afferents to the SNc (Di Matteo et al. 2007). The net effect is a shift toward excitation of the dopamine reward system following nicotine exposure.

It is worth highlighting that this intense nicotine DAergic activation is mediated almost completely by NO. Indeed, disrupting NO endogenous tone reduced nicotine-induced DA release by $\approx$60% and contextually the SNc DA neuron firing rate is similarly decreased. The mechanism by which inhibition of NOS influences the nicotine-induced activation of DA cells in the SNc observed in vivo is far from simple. It is unlikely that a direct synaptic effect on DA neurons is involved, since it has been shown that L-NAME does not alter the firing rate induced by nicotine in vitro and does not affect nicotine-induced inward currents in the VTA (Schilström et al. 2004). This assumption agrees with the anatomical evidence that only few DA neurons in the SNc express NOS machinery. Nicotine increases both firing and burst rate of DA neurons reducing their inhibitory drive and enhancing the excitatory influence on the SNc. NO might be involved in both effects modulating GLU and GABA release in an opposite way. The phasic DA activation boosts striatal NO tone (Sammut et al. 2006, 2007) that in turn excites MSNs as well as excites SNc DA cells by disinhibiting SNr neurons and permitting the nicotine effect. At the same time, the inhibitory striatal inputs to the SNc, those that seem not to be dependent on NO, are active as well, limiting the excitatory nicotine effect. Furthermore, due to the sustained DA nicotine overflow higher levels of striatal NO eventually may result in an inhibition of the function of NMDA receptors and a decrease of the activity of MSNs projecting to the SNr/SNc (Choi and Lipton 2000; Di Giovanni et al. 2003), thus reducing the activity of indirect excitatory striatonigral pathway.

It is possible that removal of endogenous NO tone by 7-NI or L-NAME treatment might decrease the indirect excitatory pathway through the SNr balancing the direct inhibitory one, reducing GLU release and leading the SNc neurons to a hypo-functional state. Nicotine in this condition might be unable to exert its excitatory effect.

Therefore, we propose that the degree of activity of nigrostriatal DA neurons may constitute a key factor for the expression of the NO/DA interaction, in that enhanced DA synthesis and/or release would be required to permit the occurrence of a NO modulatory control. In line with this hypothesis is the evidence that striatal NO increases only when SNc neurons fire at high frequency and in burst (Sammut et al. 2006, 2007). Further, it has been shown that the permissive role played by NO in phasic DA activation is not exclusive for nicotine and is applicable to other excitatory stimuli, both in physiological (West and Grace 2000) and in pharmacological terms (Tayfun Uzbay and Oglesby 2001). Thus, NO seems to have a more general role in controlling the DA brain reward and motivation circuitries even being implicated in the placebo effect (Fricchione and Stefano 2005). Accordingly, NO system appears to be crucial in the development of dependence on different drugs of abuse such as nicotine, morphine, cocaine and alcohol. In line with this, NOS inhibitors have been shown to be able to attenuate the development and expression of the abstinence syndrome for such psychostimulants (Tayfun Uzbay and Oglesby 2001).

6 Conclusions

In summary, our neurochemical and electrophysiological results, in combination with previous findings reviewed here, further demonstrate that NO is involved in both physiological and pathophysiological processes in the nigrostriatal system. Noticeably, our evidence indicates that endogenous NO positively modulates the efflux of DA in the striatum only when DA transmission is increased above basal levels. The determination of the exact mechanisms underlying this interaction may, in the near future, open new insights for the understanding of the functional role of the central NO system within the basal ganglia. Furthermore, these results suggest that the NO system represents an important therapeutic target for the development of agents for preventing or reducing DA neurodegeneration and in general for those conditions with an impairment of DA functions. In particular, we propose that NOS inhibitors might facilitate tobacco smoking cessation blocking the hedonic nicotine increase in DA release. Our findings are further supported by the recent evidence that bupropion, one of the effective treatments available for the cessation of smoking, seems to act inhibiting the L-arginine–NO–cGMP signaling pathway rather than having a direct effect on the DA system acting as a DAT reuptake inhibitor (Dhir and Kulkarni 2007). The challenge for pharmaceutical research now is to achieve selective inhibition of NOS isoforms, a situation complicated by the possibility that NOS inhibitors can indiscriminately affect beneficial and pathological NO signalling pathways.

Acknowledgements This study was supported in part by Ateneo di Palermo research funding project ORPA068JJ5 (coordinator G. Di Giovanni).

References

Arnold WP, Mittal CK, Katsuki S and Murad F (1977) Nitric oxide activates guanylate cyclase and increases guanosine 3′:5′-cyclic monophosphate levels in various tissue preparations. Proc Natl Acad Sci USA 74: 3203–3207.

Bernácer J, Prensa L and Giménez-Amaya JM (2005) Morphological features, distribution and compartmental organization of the nicotinamide adenine dinucleotide phosphate reduced-diaphorase interneurons in the human striatum. J Comp Neurol 489: 311–327.

Bian K and Murad F (2003) Nitric oxide (NO)-biogeneration, regulation, and relevance to human diseases. Front Biosci 8: 264–278

Bredt DS, Hwang PM and Snyder SH (1990) Localization of nitric oxide synthase indicating a neural role for nitric oxide. Nature 347: 768–770.

Campos F, Alfonso M, Vidal L, Faro LR and Durán R (2006) Mediation of glutamatergic receptors and nitric oxide on striatal dopamine release evoked by anatoxin-a. An in vivo microdialysis study. Eur J Pharmacol 548: 90–98.

Choi YB and Lipton SA (2000) Redox modulation of the NMDA receptor. Cell Mol Life Sci 57: 1535–1541.

Cox BA and Johnson SW (1998) Nitric oxide facilitates *N*-methyl-D-aspartate-induced burst firing in dopamine neurons from rat midbrain slices. Neurosci Lett 255: 131–134.

Dawson VL and Dawson TM (1998) Nitric oxide in neurodegeneration. Prog Brain Res 118: 215–229

Del Bel EA and Guimarães FS (2000) Sub-chronic inhibition of nitric-oxide synthesis modifies haloperidol-induced catalepsy and the number of NADPH-diaphorase neurons in mice. Psychopharmacology 147: 356–361.

Del Bel EA, Bermúdez-Echeverry M, Salum C and Raisman-Vozari R (2007) Nitric oxide system and basal ganglia physiopathology. In: Di Giovanni G (ed) The Basal Ganglia Pathophysiology: Recent Advances. Transworld Research Network, Kerala, India, pp. 129–158.

De Vente J, Markerink-van Ittersum M, van Abeelen J, Emson PC, Axer H and Steinbusch HW (2000) NO-mediated cGMP synthesis in cholinergic neurons in the rat forebrain: effects of lesioning dopaminergic or serotonergic pathways on nNOS and cGMP synthesis. Eur J Neurosci 12: 507–519.

Dhir A and Kulkarni SK (2007) Involvement of nitric oxide (NO) signaling pathway in the antidepressant action of bupropion, a dopamine reuptake inhibitor. Eur J Pharmacol 568: 177–185.

Di Giovanni G, De Deurwaerdére P, Di Mascio M, Di Matteo V, Esposito E and Spampinato U (1999) Selective blockade of serotonin-2C/2B receptors enhances mesolimbic and mesostriatal dopaminergic function: a combined in vivo electrophysiological and microdialysis study. Neuroscience 91: 587–597.

Di Giovanni G, Ferraro G, Sardo P Galati S, Esposito E and La Grutta V (2003) Nitric oxide modulates striatal neuronal activity via soluble guanylyl cyclase: an in vivo microiontophoretic study in rats. Synapse 48: 100–107.

Di Giovanni G, Ferraro G, Sardo P, Di Maio R, Carletti F and La Grutta V (2006) Microiontophoretic evidence that nitric oxide alters spontaneous activity of the substantia nigra pars reticulata neurons in the rat. Acta Physiol 188 (Suppl 652): P184.

Di Matteo V, Benigno A, Pierucci M, Giuliano DA, Crescimanno G, Esposito E and Di Giovanni G (2006) 7-Nitroindazole protects striatal dopaminergic neurons against MPP+-induced degeneration: an in vivo microdialysis study. Ann N Y Acad Sci 1089: 462–471.

Di Matteo V, Pierucci M, Di Giovanni G Benigno A and Esposito E (2007) The neurobiological bases for the pharmacotherapy of nicotine addiction. Curr Pharm Des 13: 1269–1284.

Di Matteo V, Pierucci M, Benigno A, Crescimanno G, Esposito E and Di Giovanni G (2009) Involvement of nitric oxide in 6-OHDA-induced neurodegeneration: an ex vivo study. Ann NY Acad Sci 1155: 309–315.

Duncan JA and Heales RJS (2005) Nitric oxide and neurological disorders. Mol Aspects Med 26: 67–96.

Egberongbe YI, Gentleman SM, Falkai P, Bogerts B, Polak JM and Roberts GW (1994) The distribution of nitric oxide synthase immunoreactivity in the human brain. Neuroscience 59: 561–578.

Esposito E, Di Matteo V and Di Giovanni G (2007) Death in the substantia nigra: a motor tragedy. Expert Rev Neurother 7: 7677–7697.

Eve DJ, Nisbet AP, Kingsbury AE, Hewson EL, Daniel SE, Lees AJ, Marsden CD and Foster OJ (1998) Basal ganglia neuronal nitric oxide synthase mRNA expression in Parkinson's disease. Brain Res Mol Brain Res 63: 62–71.

Fricchione G and Stefano GB (2005) Placebo neural systems: nitric oxide, morphine and the dopamine brain reward and motivation circuitries. Med Sci Monit 11: 54–65.

Furchgott RF and Zawadski J (1980) The obligatory role of the endothelium in the relaxation of arterial smooth-muscle by acetylcholine. Nature 288: 373–376.

Garthwaite J and Boulton CL (1995) Nitric oxide signalling in the central nervous system. Annu Rev Physiol 57: 683–706.

Gomes MZ and Del Bel EA (2003) Effects of electrolytic and 6-hydroxydopamine lesions of rat nigrostriatal pathway on nitric oxide synthase and nicotinamide adenine dinucleotide phosphate diaphorase. Brain Res Bull 62: 107–115.

González-Hernández T and Rodríguez M (2000) Compartmental organization and chemical profile of dopaminergic and GABAergic neurons in the substantia nigra of the rat. J Comp Neurol 421: 107–135.

Govsa F and Kayalioglu G (1999) Relationship between nicotinamide adenine dinucleotide phosphate-diaphorase-reactive neurons and blood vessels in basal ganglia. Neuroscience 93: 1335–1337.

Hunot S, Boissière F, Faucheux B, Brugg B, Mouatt-Prigent A, Agid Y and Hirsch EC (1996) Nitric oxide synthase and neuronal vulnerability in Parkinson's disease. Neuroscience 72: 355–363.

Johannes S, Reif A, Senitz D, Riederer P and Lauer M (2003) NADPH-diaphorase staining reveals new types of interneurons in human putamen. Brain Res 980: 92–99.

Johnson MD and Ma PM (1993) Localization of NADPH diaphorase activity in monoaminergic neurons of the rat brain. J Comp Neurol 332: 391–406.

Leontovich TA, Mukhina YK and Fedorov AA (2004) Neurons of the basal ganglia of the human brain (striatum and basolateral amygdala) expressing the enzyme NADPH-d. Neurosci Behav Physiol 34: 277–286.

Nisbet AP, Foster OJ, Kingsbury A, Lees AJ and Marsden CD (1994) Nitric oxide synthase mRNA expression in human subthalamic nucleus, striatum and globus pallidus: implications for basal ganglia function. Brain Res Mol Brain Res 22: 329–332.
Nowak P, Brus R, Oświecimska J, Sokoła A and Kostrzewa RM (2002) 7-Nitroindazole enhances amphetamine-evoked dopamine release in rat striatum. An in vivo microdialysis and voltammetric study. J Physiol Pharmacol 53: 251–263.
Sammut S, Dec A, Mitchell D, Linardakis J, Ortiguela M and West AR (2006) Phasic dopaminergic transmission increases NO efflux in the rat dorsal striatum via a neuronal NOS and a dopamine D(1/5) receptor-dependent mechanism. Neuropsychopharmacology 31: 493–505.
Sammut S, Bray KE and West AR (2007) Dopamine D2 receptor-dependent modulation of striatal NO synthase activity. Psychopharmacology 191: 793–803.
Sancesario G, Giorgi M, D'Angelo V, Modica A, Martorana A, Morello M, Bengtson CP and Bernardi G (2004) Down-regulation of nitrergic transmission in the rat striatum after chronic nigrostriatal deafferentation. Eur J Neurosci 20: 989–1000.
Schilström B, Mameli-Engvall M, Rawal N, Grillner P, Jardemark K and Svensson TH (2004) Nitric oxide is involved in nicotine-induced burst firing of rat ventral tegmental area dopamine neurons. Neuroscience 125: 957–964.
Tayfun Uzbay I and Oglesby MW (2001) Nitric oxide and substance dependence. Neurosci Biobehav Rev 25: 43–52.
Vincent SR and Kimura H (1992) Histochemical mapping of nitric oxide synthase in the rat brain. Neuroscience 46: 755–784.
West AR and Grace AA (2000) Striatal nitric oxide signalling regulates the neuronal activity of midbrain dopamine neurons in vivo. J Neurophysiol 83: 1796–1808.
West AR, Galloway MP and Grace AA (2002) Regulation of striatal dopamine neurotransmission by nitric oxide: effector pathways and signalling mechanisms. Synapse 44: 227–245.
Zhang L, Dawson VL and Dawson TM (2006) Role of nitric oxide in Parkinson's disease. Pharmacol Ther 109: 33–41.

Regulation of Dopamine Release by Striatal Acetylcholine and Nicotine Is via Distinct Nicotinic Acetylcholine Receptors in Dorsal vs. Ventral Striatum

Richard Exley, Michael A. Clements, and Stephanie J. Cragg

Abstract Striatal dopamine (DA) neurotransmission plays a fundamental role in the reinforcing and ultimately addictive effects of nicotine. Both nicotine and endogenous acetylcholine (ACh) regulate striatal DA release via β2 subunit-containing (β2*) nicotinic acetylcholine receptors (nAChRs) on striatal axons. The subfamily of β2*-nAChRs responsible for these potent synaptic effects could offer a molecular target for therapeutic strategies in nicotine addiction. We explored the role of the α6β2*-nAChRs in the nucleus accumbens (NAc) and caudate-putamen (CPu) by observing action potential-dependent DA release from synapses in real time using fast-scan cyclic voltammetry at carbon-fibre microelectrodes in mouse striatal slices. We show that α6β2*-nAChRs dominate in the control of DA release by ACh and nicotine in NAc but have a more minor role in CPu alongside other β2*-nAChRs (e.g. α4*). These data offer new insights to suggest striatal α6*-nAChRs as a molecular target for a therapeutic strategy for nicotine addiction.

1 Introduction

Striatal dopamine (DA) neurotransmission plays key roles in signalling information about natural as well as artificial reinforcers like addictive drugs (including nicotine) (Imperato et al. 1986; Corrigall et al. 1992; Nisell et al. 1994; Schultz 2002; Wise 2004). In particular, mesostriatal dopaminergic neurons signal reward-predicting stimuli and receipt of unpredicted rewards by brief bursts of high-frequency neuron activity (Schultz 1986, 2002). Striatal nicotinic acetylcholine receptors (nAChRs) operate a powerful and complex neuromodulatory control over the

R. Exley (✉), M.A. Clements, and S.J. Cragg
Department of Physiology, Anatomy and Genetics, University of Oxford, Oxford, OX1 3PT, UK
e-mail: richard.exley@dpag.ox.ac.uk

H.J. Groenewegen et al. (eds.), *The Basal Ganglia IX*, Advances in Behavioral Biology 58, DOI 10.1007/978-1-4419-0340-2_25, © Springer Science+Business Media, LLC 2009

dynamic probability of DA release probability during such bursts as well as non-burst activity (Zhou et al. 2001; Rice and Cragg 2004; Zhang and Sulzer 2004). Tonic levels of striatal ACh at β2-subunit-containing (β2*)-nAChRs promote DA release by individual action potentials but, owing to accompanying short-term synaptic depression, minimize the release of DA by subsequent action potentials in high-frequency bursts (Rice and Cragg 2004; Cragg 2006). Nicotine, at concentrations seen in smokers, desensitizes tonically active β2*-nAChRs (Pidoplichko et al. 1997; Mansvelder et al. 2002); in striatum, this reduces initial DA release probability, removes short-term depression and consequently facilitates release by bursts (Zhou et al. 2001; Rice and Cragg 2004; Zhang and Sulzer 2004).

Nicotine also has effects at the somatodendritic level on DA neurons and inputs. Through a complex activation and desensitization of α7 and β2*-nAChRs, respectively, in midbrain, nicotine increases DA neuron excitability (Grenhoff et al. 1986; Mansvelder et al. 2002; Schilstrom et al. 2003). Together these somatodendritic and axonal actions offer a two-step mechanism (Exley and Cragg 2008) through which nicotine can promote how burst activity in DA neurons facilitates DA-dependent reinforcement processing that ultimately contributes to addiction.

The neurochemical as well as corresponding reinforcing effects of nicotine depend critically on β2*-nAChRs (Picciotto et al. 1998). However, β2*-nAChRs comprise a diverse family expressed widely throughout the brain. The identity of the corresponding α-subunits of the specific nAChRs that participate in the effects of nicotine within striatum is unresolved. Rodent striatal DA axon terminals contain β2 subunits co-expressed predominantly with α4, or α6, α5 and also β3 subunits within at least three types of heteromeric pentamers, including α4β2, α6β2β3 and α6α4β2β3 (Charpantier et al. 1998; Klink et al. 2001; Grady et al. 2002; Zoli et al. 2002; Champtiaux et al. 2003; Cui et al. 2003; Salminen et al. 2004; Quik et al. 2005; Quik and McIntosh 2006; Exley and Cragg 2008). Unlike other subunits, expression of the α6 subunit is relatively restricted to catecholaminergic (and some visual system) neurons (Le Novere et al. 1996; Quik et al. 2001, 2002). Moreover, after somatic expression of α6 mRNA in VTA/SN DA neurons (Azam et al. 2002, 2007), α6*-nAChRs may account for up to 40% of β2*-nAChRs on DA axons in striatum (in rat) (Quik et al. 2002; Zoli et al. 2002). Given their restricted localization to striatal DA axons, α6*-nAChRs are attracting attention as promising targets for selective pharmacotherapies in dopaminergic disorders including nicotine addiction and Parkinson's disease (Quik and McIntosh 2006).

Here, by detecting dopamine release in real time using fast-scan cyclic voltammetry (FCV) at carbon-fibre microelectrodes in striatal slices, we explored directly the role of striatal α6β2*-nAChRs in the dynamic control by endogenous ACh and nicotine of DA release probability during burst and non-burst activity. We reveal a role for α6β2*-nAChRs in the dynamic control of striatal DA release probability that differs markedly between the NAc and CPu to which different aspects of motivational and sensorimotor function are attributed.

2 Materials and Methods

2.1 *Slice Preparation and Voltammetry*

Coronal striatal slices, 300 μm thick, were prepared from brains of C57/Bl6j mice (20–30 g; Charles River, UK) using previously described methods (Rice and Cragg 2004; Exley et al. 2008). Extracellular DA concentration ($[DA]_o$) was monitored at 32°C in bicarbonate-buffered artificial cerebrospinal fluid (containing 2.4 mM Ca^{2+}) using FCV with 7-μm carbon-fibre microelectrodes (tip length ~50–100 μm, fabricated in-house) and a Millar Voltammeter (PD Systems, UK) as described previously (Rice and Cragg 2004; Exley et al. 2008). In brief, the scanning voltage was a triangular waveform (−0.7 V to +1.3 V range vs. Ag/AgCl) at a scan rate of 800 V/s and sampling frequency of 8 Hz. The evoked current signal was attributed to DA by comparison of the potentials for peak oxidation and reduction currents with those of DA in calibration media (+500–600 and −200 mV vs. Ag/AgCl, respectively). Electrodes were calibrated in 2 μM DA in experimental media.

2.2 *Electrical Stimulation*

DA release was evoked by a surface, bipolar concentric electrode (25 μm diameter Pt/Ir; FHC, USA) ~100 μm from the recording electrode (Cragg and Greenfield 1997; Cragg 2003; Rice and Cragg 2004). Stimulus pulses (200 μs duration) were generated out-of-phase with FCV scans at currents (0.5–0.7 mA) that generate maximal DA release with a single pulse. Release is Ca^{2+}-dependent and TTX-sensitive (i.e. action potential-dependent) (Cragg and Greenfield 1997; Cragg 2003).

Evoked DA release on the timescale of these experiments is not modulated by glutamate or GABA at ionotropic receptors (Avshalumov et al. 2003; Cragg 2003) or DA D_2 receptors (Cragg 2003). However, a basal tone of endogenous ACh via presynaptic nAChRs on DA axons does govern DA release (Zhou et al. 2001; Rice and Cragg 2004; Zhang and Sulzer 2004). This background of endogenous ACh is provided by local cholinergic interneurons, which are spontaneously active in striatal slice preparations as they are in vivo (Aosaki et al. 1994; Bennett and Wilson 1999) and provide a dense network of ACh release sites (Contant et al. 1996; Descarries and Mechawar 2000; Zhou et al. 2003). ACh release evoked by the stimulus does not appear to contribute to these effects: similar results were seen in pilot studies in parasaggital slices in which DA release was elicited by remote pathway stimulation (up to 1,000 μm ventrocaudal from the recording electrode) (Rice and Cragg 2004).

2.3 Experimental Design and Analysis

Stimuli were repeated at 3-min intervals to ensure consistent release. Stimuli consisted of either single pulse (1 pulse, p) or bursts of pulses (2–5 p) at a physiological range of DA neuron firing frequencies, which included 'tonic' rates (1–10 Hz) and 'phasic' burst frequencies (25–100 Hz) that mimic the firing patterns of DA neurons that accompany 'reward-related' events in vivo e.g. the presentation of primary reward, a conditioned reinforcer or reward predictor (Schultz 1986; Hyland et al. 2002; Bayer and Glimcher 2005). The term 'burst sensitivity' refers to the relative sensitivity of DA release to a burst vs. a single pulse and denotes the ratio of release by a 4p burst vs. 1p. All data are means ± s.e.m. and the sample size, *n*, is the number of observations. Comparisons for differences in means were assessed by one- or two-way ANOVA, post hoc multiple comparison *t* tests (Bonferroni) or unpaired *t* tests.

3 Results

3.1 Identification of Different nAChR Subtypes in CPu and NAc

As shown previously (Zhou et al. 2001; Rice and Cragg 2004; Exley et al. 2008) nicotine (500 nM) desensitizes striatal nAChRs and thus reverses actions of endogenous striatal ACh. However, due to the non-selective action of nicotine it is *all* nAChRs that are desensitized. By using the α6-selective nAChR antagonist, α-Ctx-MII (Whiteaker et al. 2000; Champtiaux et al. 2002; Nicke et al. 2004) we can explore the role of specific receptor subtypes within dorsal (CPu) and ventral (NAc) striatum (Fig. 1). In both CPu and NAc, in control conditions, $[DA]_o$ evoked by a four-pulse burst was slightly but significantly greater than release by a single pulse (Fig. 1a, c). Application of α-CtxMII (30 nM) to both regions reduced $[DA]_o$ evoked by a single pulse and increased the sensitivity of DA release to burst vs. non-burst stimuli compared with controls (Fig. 1b, d). In NAc, however, subsequent co-application of DHβE in the continued presence of α-CtxMII did not further modify release by one pulse or four pulses (Fig. 1c, d). By contrast, in CPu the co-application of DHβE and α-CtxMII further reduced $[DA]_o$ evoked by 1p compared with α-CtxMII and also enhanced $[DA]_o$ evoked by a 4p/100 Hz burst compared with α-CtxMII (and compared with drug-free controls) (Fig. 1a, b).

3.2 α6β2*-nAChRs Can Account for All Frequency Filtering of DA Release by nAChRs in NAc but Not CPu

To explore the component role of α6*-nAChRs in the regulation of DA signals evoked by reward-related frequencies vs. non-reward-related frequencies, we used

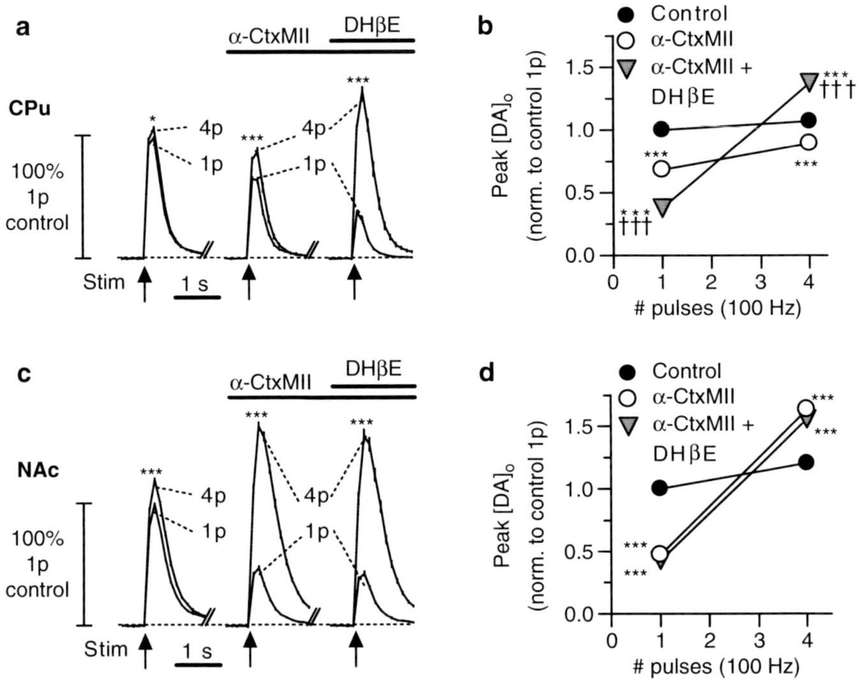

Fig. 1 α-CtxMII reveals a dominant control of dopamine release by α6β2*-nAChRs in NAc but not CPu. (**a**, **c**) Averaged profiles of $[DA]_o$ following stimuli (*arrows*) of one or four pulses (*p*) at 100 Hz in (*left*) control conditions, (*center*) α-CtxMII (30 nM) or (*right*) α-CtxMII plus DHβE (1 μM) normalized to $[DA]_o$ released by 1p in controls in (**a**) CPu and (**c**) NAc. One-way ANOVA for number of pulses, $*P < 0.05$, $***P < 0.001$, $n = 18–63$. (**b**, **d**) Mean peak $[DA]_o$ ± s.e.m. vs. number of pulses (1p or 4p, 100 Hz) in controls (*filled circles*), α-CtxMII (*unfilled*), or α-CtxMII plus DHβE (*triangles*) normalized to $[DA]_o$ released by 1p in controls in (**b**) CPu and (**d**) NAc. Significance of post hoc *t* tests indicated by asterisks for comparisons vs. controls ($*P < 0.05$, $***P < 0.001$) and by cruciform symbols for comparisons vs. α-CtxMII (†††$P < 0.001$). Reproduced with permission from Exley et al. (2008)

five pulse trains at frequencies ranging from 1 to 100 Hz. These changes in dopaminergic firing frequency within a burst are critical to the processing of reward-related information: Typically, low-frequency activity (0.5–5 Hz) is associated with non-reward-related activity, while high frequencies (20–100 Hz) signal reward-related information (Schultz 1986, 2002; Hyland et al. 2002). We have previously shown that nAChRs critically govern how DA is released by different frequencies of presynaptic activity and that nAChR desensitization by nicotine changes this frequency filtering (Rice and Cragg 2004; Zhang and Sulzer 2004). In control conditions, $[DA]_o$ varied significantly with frequency according to a slight inverted-U relationship in both CPu (Fig. 2a, b) and NAc (Fig. 2c, d). In CPu, the addition of α-CtxMII slightly reduced $[DA]_o$ at all frequencies compared with that in the controls. However, the ratio of $[DA]_o$ released by 100 Hz compared

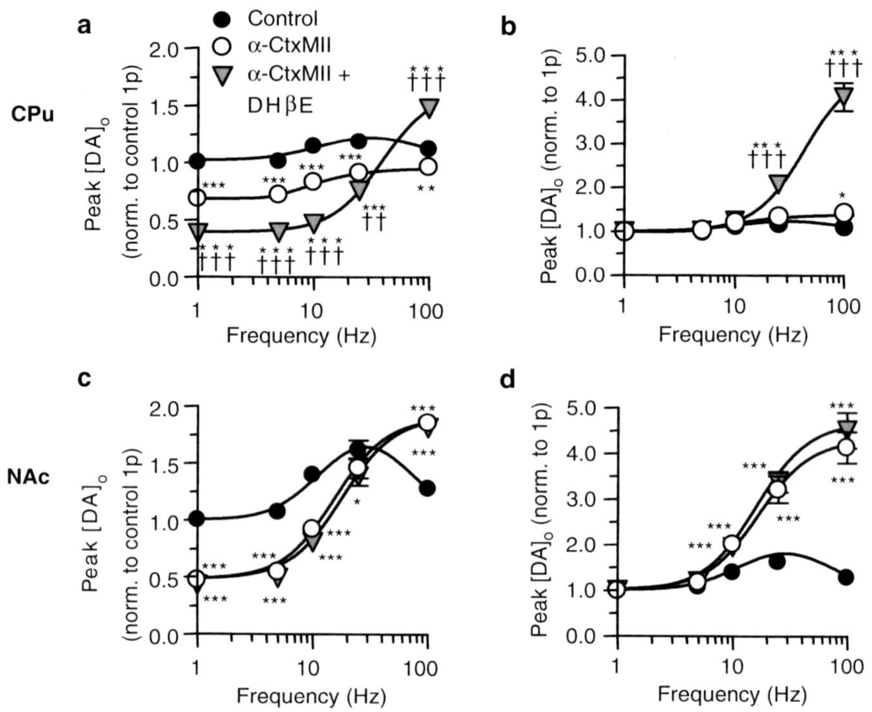

Fig. 2 α6β2*-nAChRs can account for all frequency filtering of dopamine release by nAChRs in NAc but not CPu. (**a–d**) Mean peak $[DA]_o$ ± s.e.m. vs. frequency during five pulse trains (1–100 Hz) in controls (*filled circles*), α-CtxMII (*unfilled*), or α-CtxMII plus DHβE (*triangles*) normalized to $[DA]_o$ released by (**a**, **c**) 1p in controls or (**b**, **d**) 1p in each drug condition in (**a**, **b**) CPu (n = 12–63) and (**c**, **d**) NAc (n = 9–52). Significant effects of frequency: One-way ANOVAs $P < 0.001$. Curve fits in this frequency range are modified Gaussian (controls, $R^2 > 0.9$) or sigmoidal curves ($R^2 > 0.99$). Significance of post hoc t tests is indicated by asterisks for comparisons vs. controls (*$P < 0.05$, **$P < 0.01$, ***$P < 0.001$) and by cruciform for comparisons for DHβE vs. α-CtxMII (††$P < 0.01$, †††$P < 0.001$). Reproduced with permission from Exley et al. (2008)

to a single pulse was enhanced in the presence of α-CtxMII compared with controls (Fig. 2b). With co-application of DHβE and α-CtxMII in CPu, $[DA]_o$ evoked by low-stimulation frequencies (≤25 Hz) was reduced whilst $[DA]_o$ evoked by higher frequencies (>25 Hz) was enhanced.

In NAc by contrast, the addition of α-CtxMII alone reduced $[DA]_o$ at low frequencies (≤10 Hz) compared with controls and increased $[DA]_o$ by frequencies >25 Hz compared with controls (Fig. 2c, d). This was not further modified by the addition of DHβE to α-CtxMII (Fig. 2c, d). Thus, selective block of α6*-nAChRs alone enhanced the absolute concentration range (Fig. 2c) and frequency-sensitive contrast of evoked $[DA]_o$ in NAc compared with controls (Fig. 2d).

3.3 Regional Differences in α6-nAChR Function Are Not Due to Confounding Differences in Endogenous ACh Tone or DA Release Probability

These data suggest that striatal ACh and nicotine regulate DA release probability primarily via α6β2*nAChRs in NAc and via a more mixed population including non-α6, β2*-nAChR subtypes in CPu. This greater role for α6*-nAChRs in NAc could be due to a difference in the functional nAChRs present in axon terminals in NAc and CPu. However, differences in key driving forces that contribute to DA release probability e.g. ACh tone could produce these functional differences with a more uniform arrangement of nAChR subtypes in NAc and CPu. For example, heightened ACh tone at striatal nAChRs (or DA release probability per se) in one region e.g. CPu may still be sufficient to maintain DA release probability after selective blockade by α-CtxMII via cholinergic excitation at the remaining α-CtxMII-resistant (non-α6*) nAChRs (or depolarization due to other driving factors). This may then mask any apparent influence of the α6* subpopulation.

To eliminate these confounding explanations, we first used the AChE inhibitor, ambenonium, to enhance extracellular ACh (Zhang et al. 2004) to test for different (lower) ACh tone in NAc vs. CPu. Markers of striatal cholinergic innervation (choline acetyltransferase and AChE) show dense and patchy distributions within both dorsal and ventral striatum (Graybiel et al. 1986; Phelps and Vaughn 1986; Zahm and Brog 1992; Holt et al. 1997) but there is not a consensus from microdialysis reports for the comparative ACh tone (Parada et al. 1997; Ichikawa et al. 2002). However, ambenonium in striatum has been reported to increase single-pulse-evoked DA release at low concentrations (~10^{-9} to 10^{-8} M) but suppress DA release at higher concentrations (>10^{-8} M) when ACh reaches concentrations sufficient to desensitize nAChRs (Zhang et al. 2004). Thus, where ACh tone at nAChRs is highest, ambenonium (at higher concentrations) should more readily induce ACh-dependent nAChR desensitization. We explored DA release by a single pulse (1p) and by a four-pulse burst (100 Hz) in the presence of increasing concentrations of ambenonium (0.1 nM–10 μM). Ambenonium did not potentiate evoked $[DA]_o$ at any concentration in either CPu or NAc and concentration-dependently suppressed $[DA]_o$ evoked by 1p in a manner that was similar in CPu and NAc (Fig. 3a). Suppression of 1p release by ambenonium in either region was accompanied by corresponding enhancements in the ratio of $[DA]_o$ evoked by burst (4p, 100 Hz) vs. 1p in a similar manner in CPu and NAc (Fig. 3a). These data suggest no underlying differences in ACh tone in NAc and CPu.

Second, we compared whether other driving forces on release probability of DA differ between CPu and NAc by assessing the dependence of DA release evoked

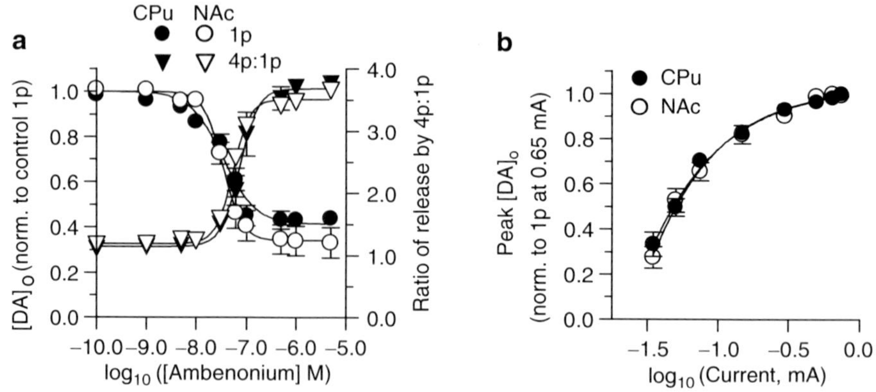

Fig. 3 No apparent variation in endogenous ACh tone and DA release probability in NAc vs. CPu. (**a**) *Circles*, Mean peak $[DA]_o$ ± s.e.m. evoked by 1p as a function of concentration of the AChE inhibitor ambenonium infilled CPu (*circles*, $n = 9$) and NAc (*open circles*, $n = 9$). $[DA]_o$ are normalized to control 1p. *Triangles*, Corresponding co-variation in ratio of release by a burst (4p, 100 Hz) vs. a single pulse (*right-hand y-axis*) as a function of ambenonium concentration in CPu (*filled*) and NAc (*unfilled*). Curve fits are sigmoidal ($R^2 = 0.8$–0.98, maximum constrained to 100%): For 1p data, curve fits and IC50s are not significantly different (32 nM, CPu; 34 nM, NAc). (**b**) Mean peak $[DA]_o$ ± s.e.m. evoked by 1p of varying stimulation current in CPu (*filled circles*, n = 11–98) and NAc (*open circles*, n = 11–100). $[DA]_o$ are normalized to 1p release at perimaximal stimulation current, I_{max} (I_{max} = 0.65 mA). Curve fits are exponential rise to maximum ($R^2 > 0.99$, maximum constrained to 100%; comparison of curves: $P < 0.05$). Reproduced with permission from Exley et al. (2008)

by a single pulse on stimulation current. Reductions in current (from I= 650 to 35 μA) gradually reduced $[DA]_o$ to zero (or below limit of detection) and in a manner that was similar in CPu and NAc (Fig. 3b). These data indicate that the greater role of α6*-nAChRs in NAc is not due to differences in the net driving force on DA release probability.

4 Discussion

4.1 β2-nAChRs and Their Subtypes: A Filter on DA Release Probability in CPu and NAc*

At concentrations seen in the plasma of tobacco smokers, nicotine via desensitization of striatal β2*-nAChRs (Zhou et al. 2001) enhances DA release in response to high-frequency, reward-related bursts (Rice and Cragg 2004; Zhang and Sulzer 2004) and thus may enhance the 'salience' or processing of reward-related information. Here, we reveal that α6β2*-nAChRs are the receptors most responsible for these effects in the NAc, but not in the CPu. These data offer a potential cell-selective target for future therapy in nicotine addiction.

4.2 Dominant Role for α6 Subunit in NAc but Not CPu

The effects of the α6*-nAChR antagonist, α-CtxMII, revealed a function for α6β2*-nAChRs in the regulation of DA release by endogenous striatal ACh in CPu and NAc, but with a striking dominance in the NAc. In CPu, the effects of α-CtxMII were a modest fraction of those seen with nicotine or the nAChR antagonist DHβE. α-CtxMII slightly diminished DA release probability by single pulses, marginally relieved short-term depression during bursts and correspondingly, only modestly enhanced the sensitivity of DA release to presynaptic frequency. In NAc by contrast, the effects of α-CtxMII on DA release probability and burst and frequency sensitivity were dramatic and indistinguishable from DHβE. These observations suggest that the β2*-nAChRs activated by endogenous ACh (and desensitized by nicotine) in CPu include a small population of α6β2*-nAChRs and other non-α6,β2*-nAChR populations, whereas in NAc these receptors are predominantly α6β2*-nAChRs.

The most simple explanation for the regional distinction in α6*-nAChR function revealed here is that a greater proportion of β2*-nAChRs activated by endogenous ACh (and desensitized by nicotine) contain the α6-subunit (i.e. α4α6β2β3- or α6β2β3-nAChRs) in NAc than in CPu. We found no evidence for competing hypotheses that the more dominant role of α6β2*-nAChRs in NAc could be otherwise explained by variations in ACh tone or in the sensitivity of DA axon membranes to depolarization.

This powerful role for α6β2*-nAChRs is in keeping with ligand binding and immunoprecipitation studies that indicate that α6-subunits may exist in up to 40% of β2*-nAChRs in dopaminergic axon terminals in rat CPu (Zoli et al. 2002) including α6β2β3 or α4α6β2β3 nAChRs (Zoli et al. 2002; Champtiaux et al. 2003; Salminen et al. 2004). Nicotine-evoked release of [^{3}H]-DA from rodent striatal synaptosomes and slices is also partially inhibited by α-CtxMII (Kulak et al. 1997; Kaiser et al. 1998; Grady et al. 2002; Champtiaux et al. 2003; Salminen et al. 2004). This work is the first to be able to reveal this dominance of α6*-nAChR function in NAc. It is unique in that our approach has access to the nAChRs that are tonically activated by physiological levels of endogenous ACh. It is these receptors that regulate release probability of endogenous DA by action potentials and that will be desensitized by nicotine during smoking. Previous studies have been limited to probe the nAChRs available to mediate an agonist secretagogue effect of nicotine on the release of exogenous DA. Whilst these data do not preclude the existence of non-α6,β2*-nAChRs in DA axons in NAc, it suggests that if present, these subtype populations are relatively redundant compared with α6β2*-nAChRs in the functional nAChR regulation of action potential-evoked DA release.

While these regional differences primarily identify a nAChR that offers a potential means for differentially modifying Ach–DA and nicotine–DA interactions in ventral vs. dorsal striatum, it is possible that these data might also impact on how synchronized *pauses* in cholinergic neurons that occur in response to reward-related cues (Morris et al. 2004) govern striatal dopamine function (Cragg 2006). Differences (albeit currently unresolved) in affinity for ACh of each receptor type might have some impact on the efficacy or speed (and therefore time window) with which a fall and rise in striatal ACh tone at nAChRs would be detected and impact on striatal DA release following a synchronized pause in ACh neurons.

5 Conclusions

Striatal β2*-nAChRs play a key role in the powerful effects of nicotine in governing how DA neuron activity is relayed into DA release. Thus nicotine, via nAChR desensitization, can promote DA signalling by high-frequency, reward-related activity (Rice and Cragg 2004; Zhang and Sulzer 2004) and thus participate in the signalling and learning of reward-related information, which ultimately participates in nicotine dependence. The current data now indicate that α6*-nAChRs play a key component function in this dynamic frequency filtering of DA release probability by endogenous ACh and nicotine. Furthermore, these data reveal for the first time that α6*-nAChRs dominate in the effects of nicotine in the NAc. As a consequence, targeting of α6β2* or non-α6β2* nAChRs could differentially modulate the components of behavioral reinforcement and sensorimotor function with which ventral and dorsal striatum are differently associated. The discrete localization within the brain of α6*-nAChRs to catecholaminergic neurons, in conjunction with these key distinctions in α6*-nAChR control of DA neurotransmission between different DA systems, suggests that α6*-nAChRs may offer a powerful molecular target for a highly selective future pharmacotherapeutic or genetic strategy to combat nicotine addiction.

Acknowledgements The authors acknowledge support from the Parkinson's Disease Society (UK), MJ Fox Foundation, the Biotechnology and Biological Sciences Research Council (UK), and Eli Lilly UK, the Paton Fellowship (University of Oxford).

References

Aosaki T, Tsubokawa H, Ishida A, Watanabe K, Graybiel AM and Kimura M (1994) Responses of tonically active neurons in the primate's striatum undergo systematic changes during behavioral sensorimotor conditioning. J Neurosci 14: 3969–3984.

Avshalumov MV, Chen BT, Marshall SP, Pena DM and Rice ME (2003) Glutamate-dependent inhibition of dopamine release in striatum is mediated by a new diffusible messenger, H_2O_2. J Neurosci 23: 2744–2750.

Azam L, Winzer-Serhan UH, Chen Y and Leslie FM (2002) Expression of neuronal nicotinic acetylcholine receptor subunit mRNAs within midbrain dopamine neurons. J Comp Neurol 444: 260–274.

Azam L, Chen Y and Leslie FM (2007) Developmental regulation of nicotinic acetylcholine receptors within midbrain dopamine neurons. Neuroscience 144: 1347–1360.

Bayer HM and Glimcher PW (2005) Midbrain dopamine neurons encode a quantitative reward prediction error signal. Neuron 47: 129–141.

Bennett BD and Wilson CJ (1999) Spontaneous activity of neostriatal cholinergic interneurons in vitro. J Neurosci 19: 5586–5596.

Champtiaux N, Han ZY, Bessis A, Rossi FM, Zoli M, Marubio L, McIntosh JM and Changeux JP (2002) Distribution and pharmacology of alpha 6-containing nicotinic acetylcholine receptors analyzed with mutant mice. J Neurosci 22: 1208–1217.

Champtiaux N, Gotti C, Cordero-Erausquin M, David DJ, Przybylski C, Lena C, Clementi F, Moretti M, Rossi FM, Le Novere N, McIntosh JM, Gardier AM and Changeux JP (2003) Subunit composition of functional nicotinic receptors in dopaminergic neurons investigated with knock-out mice. J Neurosci 23: 7820–7829.

Charpantier E, Barneoud P, Moser P, Besnard F and Sgard F (1998) Nicotinic acetylcholine subunit mRNA expression in dopaminergic neurons of the rat substantia nigra and ventral tegmental area. Neuroreport 9: 3097–3101.

Contant C, Umbriaco D, Garcia S, Watkins KC and Descarries L (1996) Ultrastructural characterization of the acetylcholine innervation in adult rat neostriatum. Neuroscience 71: 937–947.

Corrigall WA, Franklin KB, Coen KM and Clarke PB (1992) The mesolimbic dopaminergic system is implicated in the reinforcing effects of nicotine. Psychopharmacology 107: 285–289.

Cragg SJ (2003) Variable dopamine release probability and short-term plasticity between functional domains of the primate striatum. J Neurosci 23: 4378–4385.

Cragg SJ (2006) Meaningful silences: how dopamine listens to the ACh pause. Trends Neurosci 29: 125–131.

Cragg SJ and Greenfield SA (1997) Differential autoreceptor control of somatodendritic and axon terminal dopamine release in substantia nigra, ventral tegmental area, and striatum. J Neurosci 17: 5738–5746.

Cui C, Booker TK, Allen RS, Grady SR, Whiteaker P, Marks MJ, Salminen O, Tritto T, Butt CM, Allen WR, Stitzel JA, McIntosh JM, Boulter J, Collins AC and Heinemann SF (2003) The beta3 nicotinic receptor subunit: a component of alpha-conotoxin MII-binding nicotinic acetylcholine receptors that modulate dopamine release and related behaviors. J Neurosci 23: 11045–11053.

Descarries L and Mechawar N (2000) Ultrastructural evidence for diffuse transmission by monoamine and acetylcholine neurons of the central nervous system. Prog Brain Res 125: 27–47.

Exley R and Cragg SJ (2008) Presynaptic nicotinic receptors: a dynamic and diverse cholinergic filter of striatal dopamine neurotransmission. Br J Pharmacol 153 Suppl 1: S283–S297.

Exley R, Clements MA, Hartung H, McIntosh JM and Cragg SJ (2008) Alpha6-containing nicotinic acetylcholine receptors dominate the nicotine control of dopamine neurotransmission in nucleus accumbens. Neuropsychopharmacology 33: 2158–2166.

Grady SR, Murphy KL, Cao J, Marks MJ, McIntosh JM and Collins AC (2002) Characterization of nicotinic agonist-induced [(3)H]dopamine release from synaptosomes prepared from four mouse brain regions. J Pharmacol Exp Ther 301: 651–660.

Graybiel AM, Baughman RW and Eckenstein F (1986) Cholinergic neuropil of the striatum observes striosomal boundaries. Nature 323: 625–627.

Grenhoff J, Aston-Jones G and Svensson TH (1986) Nicotinic effects on the firing pattern of midbrain dopamine neurons. Acta Physiol Scand 128: 351–358.

Holt DJ, Graybiel AM and Saper CB (1997) Neurochemical architecture of the human striatum. J Comp Neurol 384: 1–25.

Hyland BI, Reynolds JN, Hay J, Perk CG and Miller R (2002) Firing modes of midbrain dopamine cells in the freely moving rat. Neuroscience 114: 475–492.

Ichikawa J, Dai J, O'Laughlin IA, Fowler WL and Meltzer HY (2002) Atypical, but not typical, antipsychotic drugs increase cortical acetylcholine release without an effect in the nucleus accumbens or striatum. Neuropsychopharmacology 26: 325–339.
Imperato A, Mulas A and Di Chiara G (1986) Nicotine preferentially stimulates dopamine release in the limbic system of freely moving rats. Eur J Pharmacol 132: 337–338.
Kaiser SA, Soliakov L, Harvey SC, Luetje CW and Wonnacott S (1998) Differential inhibition by alpha-conotoxin-MII of the nicotinic stimulation of [3H]dopamine release from rat striatal synaptosomes and slices. J Neurochem 70: 1069–1076.
Klink R, de Kerchove d'Exaerde A, Zoli M and Changeux JP (2001) Molecular and physiological diversity of nicotinic acetylcholine receptors in the midbrain dopaminergic nuclei. J Neurosci 21: 1452–1463.
Kulak JM, Nguyen TA, Olivera BM and McIntosh JM (1997) Alpha-conotoxin MII blocks nicotine-stimulated dopamine release in rat striatal synaptosomes. J Neurosci 17: 5263–5270.
Le Novere N, Zoli M and Changeux JP (1996) Neuronal nicotinic receptor alpha 6 subunit mRNA is selectively concentrated in catecholaminergic nuclei of the rat brain. Eur J Neurosci 8: 2428–2439.
Mansvelder HD, Keath JR and McGehee DS (2002) Synaptic mechanisms underlie nicotine-induced excitability of brain reward areas. Neuron 33: 905–919.
Morris G, Arkadir D, Nevet A, Vaadia E and Bergman H (2004) Coincident but distinct messages of midbrain dopamine and striatal tonically active neurons. Neuron 43: 133–143.
Nicke A, Wonnacott S and Lewis RJ (2004) Alpha-conotoxins as tools for the elucidation of structure and function of neuronal nicotinic acetylcholine receptor subtypes. Eur J Biochem 271: 2305–2319.
Nisell M, Nomikos GG and Svensson TH (1994) Systemic nicotine-induced dopamine release in the rat nucleus accumbens is regulated by nicotinic receptors in the ventral tegmental area. Synapse 16: 36–44.
Parada MA, Hernandez L, Puig de Parada M, Rada P and Murzi E (1997) Selective action of acute systemic clozapine on acetylcholine release in the rat prefrontal cortex by reference to the nucleus accumbens and striatum. J Pharmacol Exp Ther 281: 582–588.
Phelps PE and Vaughn JE (1986) Immunocytochemical localization of choline acetyltransferase in rat ventral striatum: a light and electron microscopic study. J Neurocytol 15: 595–617.
Picciotto MR, Zoli M, Rimondini R, Lena C, Marubio LM, Pich EM, Fuxe K and Changeux JP (1998) Acetylcholine receptors containing the beta2 subunit are involved in the reinforcing properties of nicotine. Nature 391: 173–177.
Pidoplichko VI, DeBiasi M, Williams JT and Dani JA (1997) Nicotine activates and desensitizes midbrain dopamine neurons. Nature 390: 401–404.
Quik M and McIntosh JM (2006) Striatal $\alpha6^*$ nicotinic acetylcholine receptors: potential targets for Parkinson's disease therapy. J Pharmacol Exp Ther 316: 481–489.
Quik M, Polonskaya Y, Kulak JM and McIntosh JM (2001) Vulnerability of 125I-α-conotoxin MII binding sites to nigrostriatal damage in monkey. J Neurosci 21: 5494–5500.
Quik M, Polonskaya Y, McIntosh JM and Kulak JM (2002) Differential nicotinic receptor expression in monkey basal ganglia: effects of nigrostriatal damage. Neuroscience 112: 619–630.
Quik M, Vailati S, Bordia T, Kulak JM, Fan H, McIntosh JM, Clementi F and Gotti C (2005) Subunit composition of nicotinic receptors in monkey striatum: effect of treatments with 1-methyl-4-phenyl-1,2,3,6-tetrahydropyridine or L-DOPA. Mol Pharmacol 67: 32–41.
Rice ME and Cragg SJ (2004) Nicotine amplifies reward-related dopamine signals in striatum. Nat Neurosci 7: 583–584.
Salminen O, Murphy KL, McIntosh JM, Drago J, Marks MJ, Collins AC and Grady SR (2004) Subunit composition and pharmacology of two classes of striatal presynaptic nicotinic acetylcholine receptors mediating dopamine release in mice. Mol Pharmacol 65: 1526–1535.
Schilstrom B, Rawal N, Mameli-Engvall M, Nomikos GG and Svensson TH (2003) Dual effects of nicotine on dopamine neurons mediated by different nicotinic receptor subtypes. Int J Neuropsychopharmacol 6: 1–11.

Schultz W (1986) Responses of midbrain dopamine neurons to behavioral trigger stimuli in the monkey. J Neurophysiol 56: 1439–1461.
Schultz W (2002) Getting formal with dopamine and reward. Neuron 36: 241–263.
Whiteaker P, McIntosh JM, Luo S, Collins AC and Marks MJ (2000) 125I-α-conotoxin MII identifies a novel nicotinic acetylcholine receptor population in mouse brain. Mol Pharmacol 57: 913–925.
Wise RA (2004) Dopamine, learning and motivation. Nat Rev Neurosci 5: 483–494.
Zahm DS and Brog JS (1992) On the significance of subterritories in the "accumbens" part of the rat ventral striatum. Neuroscience 50: 751–767.
Zhang H and Sulzer D (2004) Frequency-dependent modulation of dopamine release by nicotine. Nat Neurosci 7: 581–582.
Zhang L, Zhou FM and Dani JA (2004) Cholinergic drugs for Alzheimer's disease enhance in vitro dopamine release. Mol Pharmacol 66: 538–544.
Zhou FM, Liang Y and Dani JA (2001) Endogenous nicotinic cholinergic activity regulates dopamine release in the striatum. Nat Neurosci 4: 1224–1229.
Zhou FM, Wilson C and Dani JA (2003) Muscarinic and nicotinic cholinergic mechanisms in the mesostriatal dopamine systems. Neuroscientist 9: 23–36.
Zoli M, Moretti M, Zanardi A, McIntosh JM, Clementi F and Gotti C (2002) Identification of the nicotinic receptor subtypes expressed on dopaminergic terminals in the rat striatum. J Neurosci 22: 8785–8789.

Nitrergic Tone Influences Activity of Both Ventral Striatum Projection Neurons and Interneurons

Sarah Jane French and Henrike Hartung

Abstract Nitric oxide (NO) is a well-established striatal neuromodulator, effecting both the activity and electrical coupling of striatal projection neurons. The NO-producing interneurons within the striatum are altered in schizophrenia brain tissue, and they may be key to the pathophysiology and future treatment of schizophrenia. We investigated in vivo the effect of locally applied NO-active drugs on the firing rate of electrophysiologically and anatomically identified, medium-sized densely spiny neurons and interneurons in the ventral striatum.

Juxtacellular recording and labelling experiments were performed on ventral striatal neurons during prefrontal cortex electrical stimulation. A NO donor, precursor or scavenger were applied microiontophoretically and single unit responses were recorded; after labelling, neurons were examined morphologically to determine neuronal type.

Correlation of electrophysiological and anatomical findings revealed four drug response profiles and four types of neurons. The nitrergic modulation of ventral striatal neurons is neuronal-type specific and may be effector-mechanism dependent, and it is involved in the gating of cortically driven ventral striatal output and the temporal and spatial synchrony of the striatal networks.

1 Introduction

The ventral striatum (VST), which includes the nucleus accumbens, plays an essential role in incentive reward responding/motivational behaviour and is implicated in pathologies such as schizophrenia and addiction. A striatal neuromodulator that may be altered in schizophrenia is nitric oxide (NO); the NO-generating neurons show abnormal morphology and reduced number (Lauer et al. 2005), and neuronal

---

S.J. French (✉) and H. Hartung
Department of Pharmacology, University of Oxford, Mansfield Road, Oxford, OX1 3QT, UK
e-mail: sarah.french@pharm.ox.ac.uk

H.J. Groenewegen et al. (eds.), *The Basal Ganglia IX*, Advances in Behavioral Biology 58, DOI 10.1007/978-1-4419-0340-2_26, © Springer Science+Business Media, LLC 2009

NO synthase (nNOS) promoter polymorphisms are linked with the disease (Reif et al. 2006).

NO is key to the normal function of the striatum; it is critical to the synaptic modulation of gap junctions following the activation of the glutamatergic corticostriatal fibres (O'Donnell and Grace 1997), as well as exerting a powerful tonic modulatory influence over the membrane activity of the striatal medium-sized densely spiny neurons (MSNs) via the activation of guanylate cyclase and stimulation of cGMP production (West and Grace 2004). NO modulation of striatal interneurons may also act to regulate the functioning of individual MSNs and may co-ordinate the MSNs to form neuronal ensembles (Pennartz et al. 1994). There is a well-described network of nNOS-immunopositive interneurons (French et al. 2005) in the VST. The influence of nitrergic tone from and on these and other interneurons, as well as on the MSNs, is of critical importance to our understanding of the modulation that occurs at both the single cell and microcircuit levels in the striatum and what may therefore occur in the diseased state.

2 Materials and Methods

2.1 Animal Preparation

Experimental procedures were performed on male Wistar rats, 210–280 g (Harlan), and were conducted in accordance with the Animals (Scientific Procedures) Act of 1986 (UK). Anesthesia was induced with and maintained with urethane (1.3 g kg^{-1}, i.p.; ethyl carbamate; Sigma, UK) and supplemental doses of ketamine (30 mg kg^{-1}, i.p.; Ketaset; Willows Francis) and xylazine (3 mg kg^{-1}, i.p.; Rompun; Bayer). The animals were then placed in a stereotaxic frame (Kopf). Anesthesia levels were assessed by examination of the electrocorticogram (ECoG) and supplemental doses of the ketamine/xylazine mixture were given when necessary. The ECoG was recorded via two 1-mm diameter steel screws juxtaposed to the dura mater; raw ECoG was amplified (2,000×; NL104 AC preamplifier; Digitimer) before acquisition. Two discrete craniotomies (3–4 mm^2) were performed above the right prefrontal cortex (PFC) and VST and the dura mater was removed for insertion of stimulating and recording electrodes, respectively.

2.2 Microelectrodes and PFC Activation

The VST neurons act to integrate a large number of cortical inputs, thus continuous stimulation of the PFC was applied during the entire duration of all experiments using a platinum bipolar twisted electrode (outer diameter 0.075 mm, Plastics One). Current pulses were timed using a Master 8 stimulator (Intracel) and generated with a constant-current stimulus isolation unit (A360, WPI). A 500–700 μA square-wave pulse was delivered at 1 Hz.

The multibarrel microiontophoretic pipette and extracellular recording/labelling microelectrode assembly was cemented together in parallel with the microelectrode tip protruding 15 μm beyond the tip of the multibarrel pipette. The extracellular recording microelectrode was filled with saline solution (0.5 M NaCl) and neurobiotin (1.5% w/v; Vector Labs). One barrel of the multibarrel pipette was used for automatic balancing and was filled with 2 M NaCl. The other barrel of the microiontophoretic pipette was filled with a solution of one of the following drugs (all purchased from Sigma): the NO donor 3-morpholinosydnonimine hydrochloride (SIN-1), 40 mM in water, pH 4.5; the NO precursor hydroxy-L-arginine (H-Arg), 50 mM in water, pH 4.5; the NO scavenger 2-(4-carboxyphenyl)-4,5-dihydro-4,4,5,5-tetramemethyl-1H-imidazol-1-yl-oxy-3-oxide (C-PTIO), 50 mM in water, pH 4.5. The resistance of the recording microelectrode was 15–25 MΩ.

2.3 Electrophysiological Recordings and Microiontophoresis

Extracellular recordings of action potentials of VST neurons were made from 110 cells in 35 rats. Electrode signals were amplified (10×) through the active bridge circuitry of a Neurodata amplifier (Cygnus Technologies), AC-coupled and amplified a further 100× (NL106 AC-DC Amp; Digitimer), before being filtered between 0.3 and 5 kHz (NL125; Digitimer). Spikes were approximately 1 mV in amplitude and always exhibited an initial positive deflection. As verified in ECoG recordings, neuronal activity was recorded during slow-wave activity.

To determine the role of NO in VST cell firing, microiontophoretic techniques were used on 99 cells. SIN-1 and H-Arg were retained in the microiontophoretic pipette using a negative current of 10 nA and expelled using a positive current of 10–100 or 5–50 nA, respectively, with a Neurophore microiontophoresis BH-2 system (Medical Sciences Corp); C-PTIO was held with positive current and expelled with negative current (5–50 nA).

All biopotentials were digitized online using a Micro1401 Analog–Digital converter (CED) and a PC running Spike2 acquisition and analysis software (version 5; CED). Unit activity and the ECoG were sampled at 20 kHz. Baseline mean firing rates were calculated from 5 min immediately prior to drug ejection. Over the same period the average waveform was plotted, and the average action potential (AP) duration was determined. All data were expressed as spike events in 6-s bins. Frequency distribution histograms were generated automatically with a custom CED Spike 2 script; the script calculates the mean baseline firing rate of the 6-s bins and the mean ± [2 × standard deviation (SD)]. These levels were superimposed upon the histogram and the cell was considered to be significantly activated/inhibited if the firing frequency in three consecutive bins was beyond mean ± 2SD of the baseline (illustrated in Fig. 1a–c).

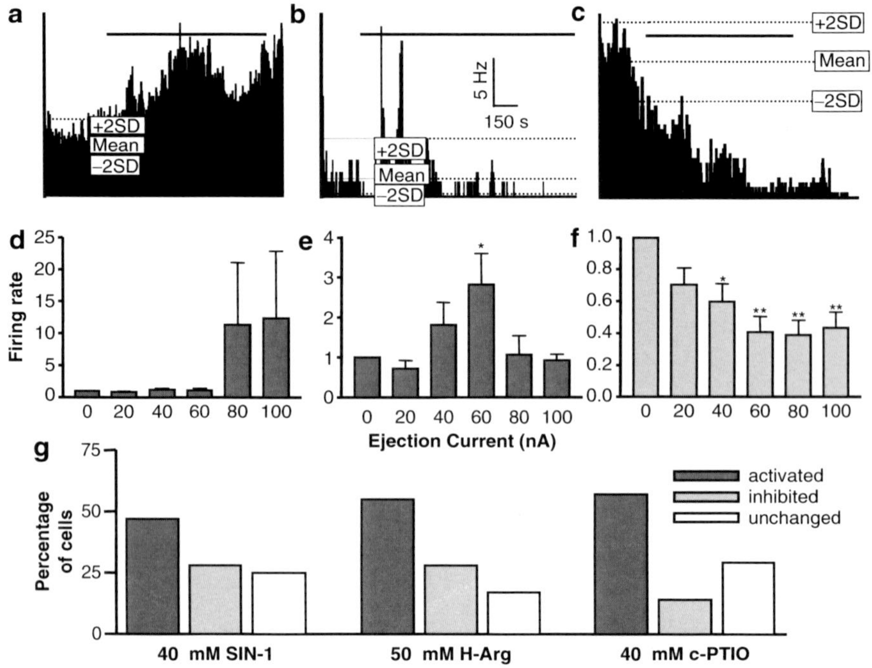

Fig. 1 A comparison of the neuronal responses to the NO donor SIN-1. Representative individual neurons were significantly activated in a more-linear current-dependant manner (**a**), or in a bell-shaped manner (**b**) or inhibited (**c**). Normalised current response graphs for all neurons activated by SIN-1 activation [n = 27 + 4; (**d**, **e**)] and all neurons inhibited by SIN-1 [n = 19; (**f**)]: The application of SIN-1 at 60 nA significantly increased the firing rates of neurons that responded with a bell-shaped curve ($F = 3.1$, $P < 0.05$, ANOVA; 20 vs. 60 nA, $P < 0.05$, Tukey's post hoc test). The application of SIN-1 at currents ranging from 40 to 100 nA significantly reduced the firing rates of inhibited neurons ($F = 6.5$, $P < 0.0001$, ANOVA; 0 vs. 40 nA, $P < 0.05$, 0 vs. 60/80/100 nA, $P < 0.001$, Tukey's post hoc test). (**g**) A comparison of the neuronal responses to the NO donor SIN-1, precursor H-Arg and scavenger c-PTIO

2.4 *Juxtacellular Labelling and Histochemistry of Single Neurons*

To identify the locations and morphological properties of recorded units, neurons were then labelled with neurobiotin by the juxtacellular method (Pinault 1996). After the recording and labelling sessions, the animals were given a lethal dose of ketamine (150 mg kg^{-1}) and perfused as described previously (French and Totterdell 2002). The fixed brains were parasagittally sectioned at 70 μm on a vibrating Microtome (VT1000M; Leica). The biotin was revealed and the tissue was processed for microscopy by methods described previously (French et al. 2002). After examination under a light microscope, reconstructions of neural structures were performed with the aid of a *camera lucida*. In addition, to create light micrograph montages with enhanced depth of field, several micrographs were taken at sequential focal depths and then processed in Adobe Photoshop software.

3 Results

3.1 Effects of the NO Manipulation VST Neurons

The responses of 67 neurons to application of the NO donor SIN-1 were recorded (Fig. 1). Of these neurons, 31 were activated (46%; Fig. 1a, d). Nineteen of the 67 recorded cells (28%) were significantly inhibited by the microiontophoresis of SIN-1; these cells decreased their firing rate to either no activity (six cells) or to 55–4% of their baseline activity (Fig. 1c, f). A further 17 cells did not respond significantly to the application of SIN-1. In responsive cells, firing rates significantly altered within 5.3 ± 4 min of the onset of the drug ejection period and were maintained for 6.4 ± 5.5 min. As can be seen in Fig. 1d–f, the excitatory or inhibitory effect of SIN-1 was related to the amount of current applied. Four of the 31 activated cells responded in a bell-shaped fashion (Fig. 1b, e); they were activated at intermediate currents only and inhibited at higher currents.

To confirm the activation of MSNs by NO and to demonstrate that stimulating the natural release pathway of NO that involves NOS could also lead to a similar effect to that seen with SIN-1 we undertook a second series of experiments with local application of the NO precursor H-arginine (H-Arg). H-Arg is a substrate for the NOS enzyme in NOS-containing interneurons. The responses of 18 cells to H-Arg were examined in the VST. Ten of the 18 neurons (55%) were significantly activated by the drug. Seven of these ten neurons were activated at the highest currents applied, in a similar manner to that seen with SIN-1. The other three activated cells displayed a bell-shaped response to the drug. Five of the 18 examined neurons (27%) were significantly inhibited by the drug. Three cells (17%) did not display a significant response to the drug (data summarised in Fig. 1g). The timescale of the responses to H-Arg were approximately twice as long as that displayed by neurons responding to SIN-1.

As there is likely to be an endogenous nitrergic tone within the striatum (West and Galloway 1997), we also tested for nitrergic tone with an NO scavenger (c-PTIO) and looked for the effects on the VST neurons (Fig. 1g). Eight of the 14 recorded cells (57%) were activated by the drug. Of these, fivc cells were activated at the highest currents applied. The increase in firing rate under application of the NO scavenger was in the range between 192 and 423% of the baseline activity. Three of the eight activated neurons exhibited a bell-shaped response to the drug. Two neurons were significantly inhibited by the drug; they decreased their firing rate to 15.98% and 22.29% of the initial baseline firing rate. Four neurons did not show a significant response to the NO scavenger. The time scale was similar to that observed for H-Arg with a significant effect seen by 10.7 ± 10.3 min, which on an average lasted for 11.5 ± 8.3 min, with the majority of neurons also not returning towards baseline activity rates after the ejection current was stopped.

3.2 Anatomical Identification of Neurons

The 20 VST neurons successfully labelled and reconstructed fell into four anatomical classes: MSNs, giant aspiny/sparsely spiny interneurons, medium-sized sparsely spiny interneurons and medium-sized aspiny interneurons.

The labelled MSNs showed the morphology typical of this class of striatal neurons (Fig. 2a–c). Of the 12 labelled MSNs, 6 received the NO donor, 3 responded with a significant increase in their firing rate and 3 significantly decreased their firing rate.

The two giant sparsely spiny labelled neurons had very similar morphology; one was within the most ventral aspect of the VST and the other within the ventral core of the nucleus accumbens. The giant sparsely spiny neuron (cell body approx. 36 μm length) from the ventral core is illustrated in Fig. 2d. The density of spines was much lower than those of the MSNs (Fig. 2g). This neuron responded to SIN-1 by increasing its firing rate at the highest currents applied.

The three medium-sized sparsely spiny interneurons were of varied morphologies, with soma sizes between 18 and 33 μm diameter (Fig. 2e). The response of these three medium-sized sparsely spiny interneurons to increased VST nitrergic tone was as follows: one significantly increasing its firing rate to H-Arg, one significantly decreasing its firing rate to SIN-1 and one did not respond to SIN-1 application.

All the medium-sized aspiny interneurons that we recorded and labelled had the shorter AP durations ≤ 1.4 ms. These neurons had medium- to large-sized soma (15–18 μm diameter) and varicose non-spiny dendrites. The application of SIN-1 to all the labelled medium-sized aspiny interneurons resulted in an increase of firing rate in a bell-shaped manner.

In the current study 20 VST neurons had short AP durations, ranging between 1.0 and 1.4 ms, and 90 neurons had an AP duration greater than 1.5 ms (Fig. 3); these AP durations had a Gaussian distribution. Other than the medium-sized aspiny interneurons, all the interneurons had AP durations between 1.9 and 2.1 ms (when using minimum AP amplitude of 0.9 mV). These latter values fall within the distribution shown for the MSNs.

4 Discussion

4.1 Modulation of Striatal Activity by Alterations of Nitrergic Tone

This study investigated the impact of alterations in nitrergic tone on the activity of different populations of VST neurons with ongoing PFC stimulation. A major finding of our study is that, by either increasing or decreasing the nitrergic tone, cells are

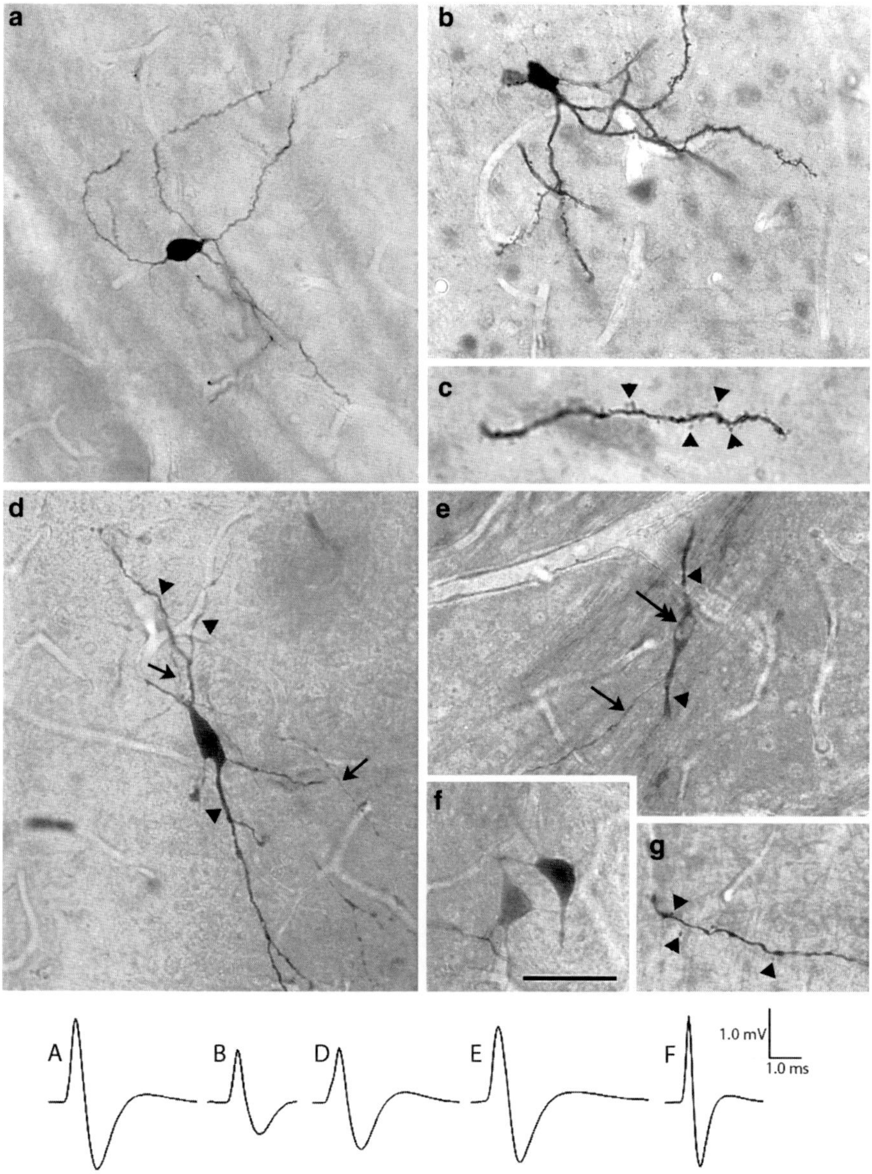

Fig. 2 Light level photomicrographs of juxtacellularly labelled VST neurons and their action potential waveforms. (**a**, **b**) MSNs with torturous densely spiny dendrites: the piece of dendrite is shown at higher magnification in (**c**), and spines are indicated by *arrowheads*. (**d**) Giant sparsely spiny neuron, clearly demonstrating dendritic (*arrowheads*) and axonal (*arrows*) labelling. (**e**) Labelled sparsely spiny medium-sized interneuron: the neuron is lightly labelled but the dendrites (*arrowheads*) and axon (*arrow*) can be clearly seen; the lighter labelling also allows the indented nucleus to be visualised (*double arrow*). (**f**) Medium-aspiny interneurons with varicose dendrites: note the short duration of this neurons waveform. (**g**) A high-power photomicrograph of a distal sparsely spiny dendrite of the cell in (**d**) – spines indicated by *arrowheads*. Scale bar = 50 μm [apart from (**c**) and (**g**) = 20 μm]

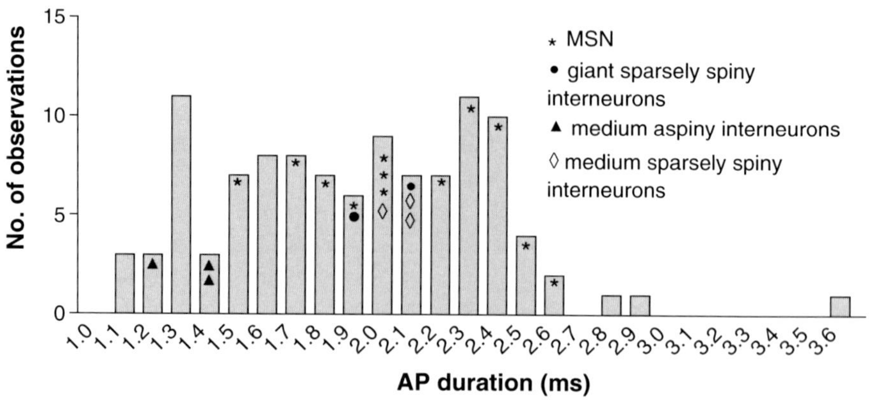

Fig. 3 A frequency distribution histogram of the AP durations of all recorded neurons illustrating the Gaussian distribution. Superimposed upon the distribution are the AP duration positions of the anatomically identified interneurons and MSNs. Note the clustering of the medium-sized aspiny interneurons at shorter AP durations and also the distribution of the other interneuron classes within that of the MSNs

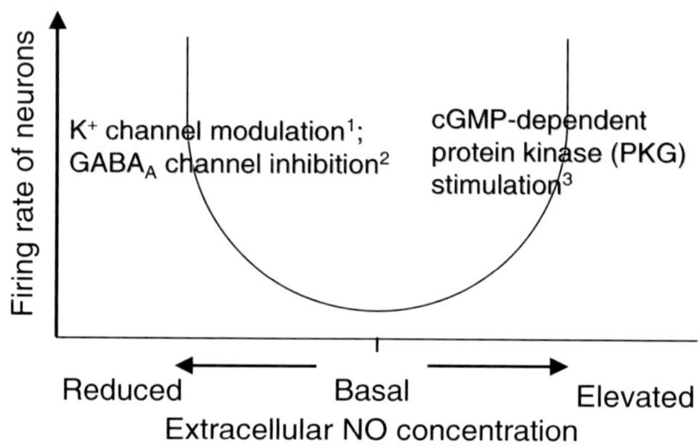

Fig. 4 A hypothetical schematic of the biphasic response of striatal neurons to alterations in the basal concentration of nitric oxide, with suggested effector mechanisms (Fagni and Bockaert 1996; Robello et al. 1996; Garthwaite and Boulton 1995)

predominantly excited. Moreover, within the cells that responded by increasing their firing rate, two patterns of response were seen; the majority of cells respond at only the highest microiontophoresis currents, whilst a small proportion respond only at intermediate currents.

Our findings demonstrate that enhancing striatal NO tone significantly increased the firing activity of 41, and inhibited 24, of the 85 neurons investigated. Reducing the nitrergic tone also significantly excited 8 of the 14 neurons tested. Given that most neurons tested were excited by increasing nitrergic tone, it is perhaps surprising that scavenging NO also excited most neurons tested. Thus, we hypothesise that the VST neurons respond to an alteration of nitrergic tone in a U-shaped manner (Fig. 4). NO may employ other effector mechanisms e.g. the inhibition of chloride influx through $GABA_A$ receptors (Robello et al. 1996) or the modulation of potassium channels (Fagni and Bockaert 1996), both of which if reduced would increase neuronal excitability.

Previous studies have investigated the action of nitrergic drugs on the activity of striatal neurons (Di Giovanni et al. 2003; Sardo et al. 2003; West and Grace 2004). One aspect of our study that is entirely novel is the correlation of single-cell anatomical properties with drug effect. In addition, no study has specifically examined the effect on striatal interneurons.

4.2 Nitrergic Modulation of VST Interneurons

We morphologically identified two giant sparsely spiny interneurons characteristic of the cholinergic interneurons (Bolam et al. 1984) that form 1–2% of the total neuronal population. In agreement with in vitro intracellular studies that reported that striatal cholinergic interneurons were excited by hydroxylamine (NO donor) and SIN-1 (Centonze et al. 2001), the current study provides evidence that these giant sparsely/aspiny putative cholinergic neurons are excited by a NO donor.

We labelled three medium-sized sparsely spiny interneurons with soma sizes and dendritic and axonal arbours consistent with those of NADPH-D-positive neurons (Kawaguchi 1993) and somatostatin-positive neurons (Kawaguchi and Kubota 1996), which are known to be those that co-localise NOS (Vincent et al. 1983). A striking feature of these neurons is that the axonal arborisation is not dense within the region of the dendritic field, but is extensive and the dendrites are relatively unbranched for long distances. The presence of sparse spines on these labelled neurons argues against them being calretinin interneurons, which are devoid of spines (Bennett and Bolam 1993). The altered firing rate of these putative NOS-containing interneurons in response to nitrergic drugs is not unexpected, as anatomical studies have already demonstrated a NOS–NOS network of neurons (French et al. 2005). This network is proposed to widely propagate a hippocampally derived spatial signal received by individual NOS-containing neurons potentially synchronising and recruiting striatal MSNs into a functional ensemble.

The modulation of PV interneurons is of critical importance to our understanding of the feedforward inhibitory role this class of interneurons may play in striatal output synchronisation (Mallet et al. 2005) and striatal synaptic plasticity control (Pennartz et al. 1993). We labelled three neurons that are morphologically similar to those described to contain PV immunoreactivity (Kita et al. 1990). Of particular note is their aspiny varicose dendrites and medium- to large-sized soma. The soma size of our medium aspiny neurons are too large to be those belonging to the calretinin class. The PV neurons are also known to be of the FSI class (Kawaguchi 1993). The nitrergic effect on these FSIs was of particular interest as they increased their firing only at intermediate currents, thus they are recruited to fire when the NO tone is initially increased, but stop firing when it is higher. These cells have dense but spatially limited axonal arbour; the increased NO would thus lead to a subsequent but limited-duration inhibitory influence on the MSNs within their strict vicinity.

The influence of nitrergic signalling in the striatum is functionally relevant to the flow of information through the MSNs output neurons, as well as the specific gating of signal transduction across corticostriatal synapses. This modulation may have fundamentally different influences on the spatial information processing and co-ordinated network synchrony depending on the neuron subtype involved. This ability to either increase or decrease the cell activity may represent the ability of NO to balance output activity such that neurons within a restricted region of the VST can act as a group or ensemble.

These findings also have implications for our understanding of the processes that may be dysfunctional in the VST of patients with schizophrenia. These neurons may normally generate a balanced gating of VST excitation by cortical glutamatergic afferents, contributing to the temporal and spatial regulation of information processing, and so their dysfunction may explain why the cognitive and motivational functions and the reward processes attributed to VST are disrupted in this disorder.

References

Bennett BD and Bolam JP (1993) Characterization of calretinin-immunoreactive structures in the striatum of the rat. Brain Res 609: 137–148.

Bolam JP, Wainer BH and Smith AD (1984) Characterization of cholinergic neurons in the rat neostriatum. A combination of choline acetyltransferase immunocytochemistry, Golgi-impregnation and electron microscopy. Neuroscience 12: 711–718.

Centonze D, Pisani A, Bonsi P, Giacomini P, Bernardi G and Calabresi P (2001) Stimulation of nitric oxide-cGMP pathway excites striatal cholinergic interneurons via protein kinase G activation. J Neurosci 21: 1393–1400.

Di Giovanni G, Ferraro G, Sardo P, Galati S, Esposito E and La Grutta V (2003) Nitric oxide modulates striatal neuronal activity via soluble guanylyl cyclase: an in vivo microiontophoretic study in rats. Synapse 48: 100–107.

Fagni L and Bockaert J (1996) Effects of nitric oxide on glutamate-gated channels and other ionic channels. J Chem Neuroanat 10: 231–240.

French SJ and Totterdell S (2002) Hippocampal and prefrontal cortical inputs monosynaptically converge with individual projection neurons of the nucleus accumbens. J Comp Neurol 446: 151–165.

French SJ, van Dongen YC, Groenewegen HJ and Totterdell S (2002) Synaptic convergence of hippocampal and prefrontal cortical afferents to the ventral striatum in rat. In: Nicholson LFB and Faull RL (eds) The Basal Ganglia VII. Kluwer Academic/Plenum, New York, pp 399–408.

French SJ, Ritson GP, Hidaka S and Totterdell S (2005) Nucleus accumbens nitric oxide immunoreactive interneurons receive nitric oxide and ventral subicular afferents in rats. Neuroscience 135: 121–131.

Garthwaite J and Boulton CL (1995) Nitric oxide signaling in the central nervous system. Annu Rev Physiol 57: 683–706.

Kawaguchi Y (1993) Physiological, morphological, and histochemical characterization of three classes of interneurons in rat neostriatum. J Neurosci 13: 4908–4923.

Kawaguchi Y and Kubota Y (1996) Physiological and morphological identification of somatostatin- or vasoactive intestinal polypeptide-containing cells among GABAergic cell subtypes in rat frontal cortex. J Neurosci 16: 2701–2715.

Kita H, Kosaka T and Heizmann CW (1990) Parvalbumin-immunoreactive neurons in the rat neostriatum: a light and electron microscopic study. Brain Res 536: 1–15.

Lauer M, Johannes S, Fritzen S, Senitz D, Riederer P and Reif A (2005) Morphological abnormalities in nitric-oxide-synthase-positive striatal interneurons of schizophrenic patients. Neuropsychobiology 52: 111–117.

Mallet N, Le Moine C, Charpier S and Gonon F (2005) Feedforward inhibition of projection neurons by fast-spiking GABA interneurons in the rat striatum in vivo. J Neurosci 25: 3857–3869.

O'Donnell P and Grace AA (1997) Cortical afferents modulate striatal gap junction permeability via nitric oxide. Neuroscience 76: 1–5.

Pennartz CM, Ameerun RF, Groenewegen HJ and Lopes da Silva FH (1993) Synaptic plasticity in an in vitro slice preparation of the rat nucleus accumbens. Eur J Neurosci 5: 107–117.

Pennartz CM, Groenewegen HJ and Lopes da Silva FH (1994) The nucleus accumbens as a complex of functionally distinct neuronal ensembles: an integration of behavioural, electrophysiological and anatomical data. Prog Neurobiol 42: 719–761.

Pinault D (1996) A novel single-cell staining procedure performed in vivo under electrophysiological control: morpho-functional features of juxtacellularly labeled thalamic cells and other central neurons with biocytin or neurobiotin. J Neurosci Methods 65: 113–136.

Reif A, Herterich S, Strobel A, Ehlis AC, Saur D, Jacob CP, Wienker T, Topner T, Fritzen S, Walter U, Schmitt A, Fallgatter AJ and Lesch KP (2006) A neuronal nitric oxide synthase (NOS-I) haplotype associated with schizophrenia modifies prefrontal cortex function. Mol Psychiatry 11: 286–300.

Robello M, Amico C, Bucossi G, Cupello A, Rapallino MV and Thellung S (1996) Nitric oxide and GABAA receptor function in the rat cerebral cortex and cerebellar granule cells. Neuroscience 74: 99–105.

Sardo P, Ferraro G, Di Giovanni G and La Grutta V (2003) Nitric oxide-induced inhibition on striatal cells and excitation on globus pallidus neurons: a microiontophoretic study in the rat. Neurosci Lett 343: 101–104.

Vincent SR, Johansson O, Hokfelt T, Skirboll L, Elde RP, Terenius L, Kimmel J and Goldstein M (1983) NADPH-diaphorase: a selective histochemical marker for striatal neurons containing both somatostatin- and avian pancreatic polypeptide (APP)-like immunoreactivities. J Comp Neurol 217: 252–263.

West AR and Galloway MP (1997) Endogenous nitric oxide facilitates striatal dopamine and glutamate efflux in vivo: role of ionotropic glutamate receptor-dependent mechanisms. Neuropharmacology 36: 1571–1581.

West AR and Grace AA (2004) The nitric oxide-guanylyl cyclase signaling pathway modulates membrane activity states and electrophysiological properties of striatal medium spiny neurons recorded in vivo. J Neurosci 24: 1924–1935.

Part IV
Basal Ganglia Disorders: Animal Studies

Kainic Acid-Induced Cell Proliferation in the Striatum Is Not Estrogen Dependent

Magda Giordano and Daniela Cano-Sotomayor

Abstract In the past decade it has become increasingly clear that the adult central nervous system has the ability to generate new neuronal cells. In particular, the subgranular layer (SGL) in the hippocampal dentate gyrus and the subventricular zone (SVZ) lining the lateral ventricles appear to be specially suited for this purpose. In these areas neurogenesis is observed due to the persistence of neuronal progenitors throughout life. The neuroprotective and neurogenic properties of estrogen have been demonstrated in vitro and in vivo. In this study we examined the role of estradiol benzoate (0, 25, and 100 μg per rat) in neuronal proliferation in the kainic acid (KA) rat model of Huntington's disease. Cell proliferation and phenotype of newborn cells were assessed using antibodies against bromodeoxyuridine (BrdU), doublecortin, and glial fibrillary acidic protein in the adult female rat striatum, SVZ of the lateral ventricle, and corpus callosum, at different antero-posterior levels. The statistical analysis for the number of BrdU-positive cells in these cerebral regions showed that estradiol did not increase cellular proliferation beyond that induced by the KA lesion and that ovariectomy did not inhibit the proliferative response.

1 Introduction

The presence of undifferentiated cells with the capacity to proliferate and to attain a glial or neuronal fate, known as stem cells, in the SVZ and the SGL of the hippocampal dentate gyrus of the adult mammalian nervous system has been verified repeatedly in the last decade (Reynolds and Weiss 1992; Alvarez-Buylla et al. 2002; Lledo et al. 2006; Doetsch et al. 1999). In addition to the interest in the understanding of adult neurogenesis per se, these findings have been greeted with the expectation of finding a source of new cells to replace those lost due to aging and disease. To reach a therapeutic goal, however, information about the factors that control proliferation,

M. Giordano (✉) and D. Cano-Sotomayor
Instituto de Neurobiología, Universidad Nacional Autónoma de México,
Juriquilla, Qro. 76230, México
e-mail: giordano@servidor.unam.mx

H.J. Groenewegen et al. (eds.), *The Basal Ganglia IX*, Advances in Behavioral Biology 58, 351
DOI 10.1007/978-1-4419-0340-2_27, © Springer Science+Business Media, LLC 2009

guide the migration, and allow the integration of these new cells into the existing networks is clearly needed. In this study, we sought to determine the role of a particular hormone, estrogen, which has been linked to neuroprotection and neurogenesis, in the cell proliferation induced by kainic acid (KA) lesions of the adult female striatum. A brief description of the organization of the SVZ in the adult and the identity of the neural stem cells will be presented, followed by an account of the role of estrogen in adult neurogenesis and by a report of our results.

1.1 Organization of the Subventricular Zone in the Adult Mammalian Nervous System

Within the SVZ there is a population of multipotent cells that can give rise to new neurons or glial cells. Three cell types have been described in mammals: type A cells, which are migrating cells with an elongated cell body with one or two processes and express markers such as polysialic acid-neuronal cell adhesion molecule (PSA-NCAM); type B cells, which are astrocytes and have irregular nuclei, abundant intermediate filaments, and stain darkly with antiglial fibrillary acidic protein (GFAP) antibodies; and type C cells, which are proliferative cells (Garcia-Verdugo et al. 2002). The dominant hypothesis is that astrocytes in the adult SVZ act as slow-dividing neural stem cells, capable of generating a progeny of neuroblast precursors (Lledo et al. 2006). These cells migrate toward the olfactory bulb along the rostral migratory stream (RMS), where they mature into olfactory inhibitory interneurons (granule and periglomerular cells) (Lledo et al. 2006). It has been shown that striatal radial glia give rise to SVZ astrocytes that continue to generate neurons in the adult brain (Merkle et al. 2004). In addition, a recent study has suggested that SVZ neural stem cells are restricted in the types of neurons they can generate, that is, postnatal stem cells in different regions produce different types of neurons, even when heterotopically grafted or grown in culture. The potential of postnatal neural stem cells appears to be determined by a spatial code, a mosaic, raising the possibility that the activity of stem cells may be regionally modulated in order to regulate the production of different types of neurons (Merkle et al. 2007).

1.2 The Role of Estrogen in Adult Neurogenesis

The production, migration, development, and survival of newly generated neurons in adults are regulated by a variety of factors (Gheusi and Lledo 2007). Among these factors estrogen has been shown to modulate neurogenesis in various types of experiments. Murashov et al. (2004) found that 17β estradiol accelerated both differentiation and maturation of neuronal cells in a substrate-dependent manner.

In mouse embryonic stem cells the combination of fibronectin coating with 17β estradiol increased the number of generated neurons. Both types of estrogen nuclear receptors (ERα and ERβ) have been found in embryonic and adult rat neural stem cells, and 17β estradiol was shown to increase the proliferation of embryonic although not adult neural stem cells (Brännvall et al. 2002).

In vivo it has been found that ovariectomy in rats diminished the number of cells labeled with the thymidine analog bromodeoxyuridine (BrdU) in the dentate gyrus and that this effect was reversed by estrogen replacement (Tanapat et al. 1999). A low dose of KA (7 mg/kg b.w.) injected in the morning of proestrus or in ovariectomized rats resulted in significant loss of hilar dentate neurons in the hippocampus. In contrast KA administration in the morning of estrus did not induce loss of hilar neurons. Estrogen prevented hilar neuronal loss when injected either 24 h before or simultaneously with KA and not when it was injected 24 h after the administration of the toxin (Azcoitia et al. 1999). In the female prairie vole, an increase in BrdU-labeled cells in the anterior portion of the SVZ was observed after induction into estrus by male exposure. Ovariectomized females exposed to a male did not show an increase in serum estrogen or BrdU labeling in the RMS. Ovariectomized females injected with estrogen (1 μg) had more BrdU-labeled cells in the RMS than vehicle-injected females (Smith et al. 2001).

In the dentate gyrus of adult female meadow voles, and of adult female rats, estradiol benzoate (EB) increased (4 h), and decreased (48 h) the number of new cells (Ormerod et al. 2001, 2003). The decrement appeared to be related to EB-stimulated adrenal activity. On the basis of these results, the authors concluded that estradiol regulates cell proliferation within the dentate gyrus in a time-dependent manner (Ormerod et al. 2003). Other factors have been shown to regulate estrogen effects on cell proliferation; for example, Banasr et al. (2001) found that serotonin mediates estrogen stimulation of cell proliferation in the adult female dentate gyrus.

In addition to its effects on cell proliferation, estrogen has important effects on cell survival, axonal sprouting, regenerative responses, and enhanced synaptic transmission (Garcia-Segura et al. 2001). Ciriza et al. (2004) found that 17β estradiol prevented neuronal loss in the hilus of the hippocampus after KA administration (7 mg/kg b.w.) and also prevented reactive gliosis. In contrast, selective estrogen receptor modulators (SERMs) had a similar neuroprotective effect, but did not decrease reactive gliosis suggesting that estradiol acts through different mechanisms than SERMs.

The mechanisms underlying the neuroprotective and proliferating effects of estrogen have yet to be thoroughly understood. Some of those effects may be mediated by the classically defined nuclear estrogen receptors, while others may involve unidentified membrane receptors, direct modulation of neurotransmitter receptor function, or antioxidant effects (in Garcia-Segura et al. 2001). Among the nongenomic effects of estrogen that have been described are the activation of extracellular signal-regulated kinases (ERK) and phospoinositol-3-kinase (PI3K)-Akt pathways in cortical and hippocampal cells (Brann et al. 2007). The possible sources of estrogen for the central nervous system are the circulation, the local conversion of peripheral testosterone by reactive astroglia, and the local conversion of locally formed testosterone by reactive astroglia (Garcia-Segura et al. 2001). Also endogenous net production of estrogen from cholesterol has been observed after stimulation of hippocampal neurons with *N*-methyl-D-aspartate (Hojo et al. 2004 cited in Brann et al. 2007). Thus, endogenous production of estrogen by brain cells may exert a local effect on synaptic plasticity (Brann et al. 2007).

Injury and disease can also promote cell proliferation (Taupin 2005). In the subependymal layer (SEL) adjacent to the caudate nucleus in the human brain of patients suffering from Huntington's disease (HD), an increase in the number of proliferating cells was observed in comparison with control tissue. The number of proliferating cells was correlated to the pathological grade and the number of CAG trinucleotide repeat length in the *IT15* gene (Curtis et al. 2003). In an animal model of HD, Tattersfield et al. (2004) observed increases in SVZ neurogenesis, leading to migration of neuroblasts to damaged areas of the striatum. Using the same animal model, Collin et al. (2005) have found that the newly generated neurons survive up to 6 weeks, express the mature neuronal marker NeuN, and also markers of striatal medium spiny neurons and interneurons. In addition, the magnitude of neurogenesis correlated with the size of the striatal damage. Arvidsson et al. (2002) also found increased cell proliferation in the SVZ after a transient middle cerebral artery occlusion in adult rats.

2 Methods

In this study we sought to determine whether estrogen administration to ovariectomized female rats had an influence on the number of newly formed cells after intrastriatal administration of KA. Twenty female Sprague–Dawley rats (200–300 g body weight) were kept under a 10:14 h light/dark schedule with free access to food and water. Animals were ovariectomized under anesthesia (with ketamine, 70 mg/kg, and xylazine, 6 mg/kg) and were given 7 days to recover. On the eighth day, 25 or 100 μg/rat s.c. β-estradiol 3-benzoate (EB; Sigma Chemical Co.) or vehicle were administered. On the same day, kainic acid (KA) (5 nmol/0.5 μl) was unilaterally infused into the striatum (AP +1.2, ML +2.6, DV −4.5 from Bregma). In the contralateral side, only the empty cannula was inserted into the striatum, and no KA was infused. After the lesion all animals received 0.5 ml of ampicillin and 0.1 ml of diazepam (5 mg/ml). On days 3 and 5 after the lesion, the animals received EB again; on days 4 and 5, the mitogenic marker 5′bromo-2′deoxyuridine (BrdU) (200 mg/kg, i.p.) (Sigma Chemical, St. Louis, MO, USA) in 0.9% NaCl was administered. On day 7 the animals were euthanized with sodium pentobarbital and intracardially perfused using 4% paraformaldehyde in phosphate buffer (0.1 M, pH 7.4); the brain was removed and left in 30% sucrose until ready to section.

Coronal sections (40 μm) were processed for indirect immunofluorescence using an antibody against BrdU (mouse anti-BrdU, BD Bioscience; 1:250). Briefly, the sections were incubated in the primary antibody for 16 h at 4°C, after blocking and several rinses; these were incubated for 2 h with a secondary antibody goat-anti-mouse conjugated with the fluorochrome Cy™3 (Jackson Immunoresearch, 1:1,000). Some sections were processed for double immunofluorescence against BrdU and glial acidic fibrillary protein (GFAP) (rabbit anti-DAKO Z0334; 1:1,000), or BrdU and doublecortin (Dcx) (guinea pig-anti-doublecortin Chemicon AB5910; 1:500). The secondary antibodies were a goat-anti-rabbit IgG (H + L) FITC conjugated for GFAP (ZYMED) (1:500) and a Cy™2-conjugated Affini-Pure goat-anti-guinea pig IgG (H + L) for Dcx (Jackson ImmunoResearch) (1:100). The sections were mounted with glycerol, observed under a confocal microscope (Nikon Eclipse E-600), and captured by means of simple PCI (C/Imaging Systems, Compix Inc., Cranberry Township, PA, USA).

To quantify the number of BrdU-positive nuclei in the striatum and corpus callosum, three different coronal antero-posterior levels were selected; these were 2.04, 1.08, and –0.12 mm from Bregma. At each level three different fields were quantified per hemisphere. For the SVZ of the lateral ventricle, a similar procedure was used. However, it was not possible to distinguish one nucleus from the next, and therefore only the percentage of the SVZ of the lateral ventricle occupied by BrdU-positive nuclei was estimated. Confocal z-series were obtained by collecting 15 optical sections of the same field (1 μm apart). Images of a series were then projected onto a single image using the Simple PCI software.

3 Results

The results of the quantification of BrdU-positive nuclei in the lesioned striatum of ovariectomized females treated with EB were evaluated using a three-way analysis of variance (group × hemisphere × level). A significant effect of hemisphere (KA lesioned vs. nonlesioned) was found [$F(1,13) = 34.36$, $p < 0.05$]; no other effect was significant. As can be seen in Fig. 1, the various EB treatments neither increased nor decreased the number of BrdU-positive nuclei after the KA lesion. Similar effects were observed for the corpus callosum, and there was only a significant effect of hemisphere [$F(1,14) = 31.27$, $p < 0.05$]. Similarly, the qualitative observations of the percentage of the perimeter of the SVZ occupied by BrdU-positive nuclei showed no differences among groups. Differences between hemispheres were not as marked as for the number of BrdU-positive nuclei. At the second anterior-posterior level, there was a slightly higher percentage of the SVZ with BrdU-positive staining in the lesioned hemisphere than in the control hemisphere. However, the most marked difference was found between the anterior level and the other two levels. That is, a higher percentage of the SVZ showed BrdU-positive staining in the most anterior section studied than in the other two more posterior sections.

The double-immunostaining study showed both GFAP and Dcx-positive cells in the lesioned striatum and in the corpus callosum ipsilateral to the lesion in all groups. Few cells showed overlap of both markers, BrdU and GFAP or BrdU and Dcx (see Fig. 2).

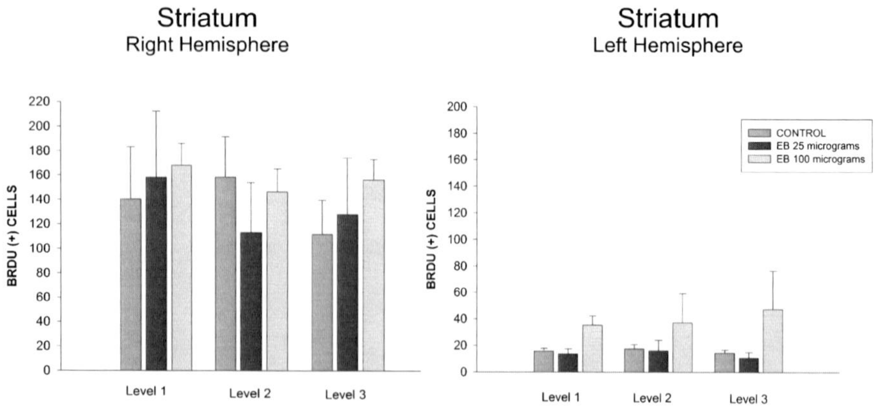

Fig. 1 Kainic acid (KA) striatal administration increases cell proliferation in the striatum. Graphs show mean ± SEM of the total number of BrdU-positive cells in the striatum after estrogen treatment. *Left panel* shows data obtained from the lesioned hemisphere, and *right panel* shows data from the sham hemisphere. BrdU-positive cells were quantified at three different levels with respect to the site of KA administration. Significant differences were observed between sham and lesioned hemispheres. Estrogen treatment had no effect on the number of BrdU-positive cells in the striatum

4 Discussion

These results corroborate what has been published before regarding the effects of striatal excitotoxic lesions in male animals, that is, an increase in BrdU-positive cells in the striatum. Treatment with EB did not significantly increase the number of BrdU-positive cells either in the striatum or in the corpus callosum. Adult neurogenesis appears to be affected by fluctuating estrogen levels and thus has been considered to possess both neuroprotective and neurogenic properties as has been documented by some authors (Ciriza et al. 2004; Taupin 2005). However, this is still a matter of debate. Our results indicate that exogenous estrogen in ovariectomized females under the conditions of administration employed in this study does not increase cell proliferation beyond that induced by an excitotoxic lesion in the striatum. These results also show that ovariectomy does not inhibit KA-induced cell proliferation in the brain. It appears that estrogen, at least estrogen from the circulation, is not a requisite for this response.

The effects of estradiol on cell proliferation and cell survival appear to depend on various factors, such as presence of progesterone, species, sex, and most importantly the length and time of exposure (Galea et al. 2006). In adult neural stem cells EB did not increase proliferation (Brännvall et al. 2002). Also, the effects of EB on cell proliferation appear to be region specific. For instance, EB decreased the density of newly generated cells in the accessory olfactory bulb, but did not induce changes in proliferation in the main olfactory bulb (Hoyk et al. 2006).

The fate of the BrdU-positive cells induced by KA administration in this study is not clear since there was no significant overlap either with a glial marker (GFAP) or with a marker for immature neurons (Dcx). Our evaluation of the type of cells that were proliferating was not exhaustive. However, it appears that the proliferating

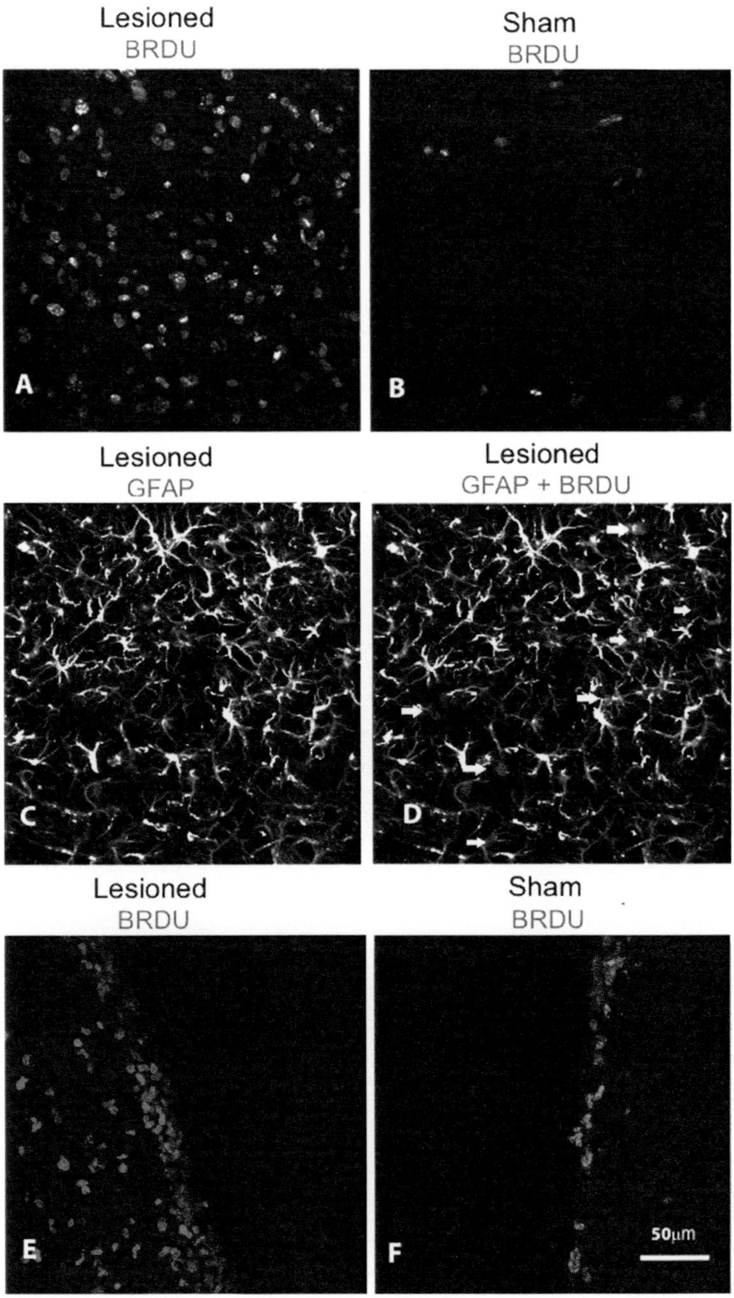

Fig. 2 BrdU-positive cells in the kainic acid lesioned striatum. (**A**) and (**B**) are confocal microscopy images showing BrdU-immunofluorescence in the striatum ipsilateral (**A**) and contralateral (**B**) to the insult at 7 days after KA administration. On the contralateral side BrdU-immunoreactivity is restricted to isolated cells, whereas abundant BrdU cells are present in the ipsilateral striatum. (**C**) and (**D**) are confocal images of GFAP-immunofluorescence (**C**) and GFAP + BrdU double-immunofluorescence (**D**). In (**D**) BrdU-positive nuclei are indicated by *arrows*; only a few cells show coexpression of both markers. (**E**) and (**F**) are confocal images showing BrdU-positive cells adjacent to the lateral ventricle. (**E**) shows the ipsilateral (lesioned) hemisphere and (**F**) shows the contralateral (sham) hemisphere. Bar = 50 μm

cells are other than astrocytes proliferating as a result of the excitotoxic lesion. It has been proposed that in the striatum progenitor cells proliferate in response to excitotoxic insults and that they may migrate into the lesioned striatum and replace lost neurons (Collin et al. 2005; Tattersfield et al. 2004). Although this may very well be the case, it is also important to consider the fact that the excitotoxic lesion may interfere with the migration of progenitor cells through the RMS and into the olfactory bulb. The presence of proliferating cells within the lesioned striatum could be the result of them becoming "trapped" due to changes in the cytoarchitecture induced by the lesion. One way to assess if this is true would be to quantify the number of proliferating cells arriving into the olfactory bulb ipsilateral and contralateral to the lesion. It is interesting to note that the anterior SVZ was the area with greatest BrdU-positive labeling. In the newborn's brain, the RMS, including the anterior SVZ, has been shown to be predominantly neurogenic, while the SVZ located in the more posterior portion of the lateral ventricle is primarily gliogenic (Martoncikova et al. 2006).

In conclusion, unilateral KA administration into the striatum results in cell proliferation ipsilateral to the lesion; estrogen does not increase cell proliferation beyond that induced by the excitotoxic lesion, and ovariectomy does not inhibit the proliferative response.

Acknowledgments The authors wish to thank E.N. Hernández, Soledad Mendoza, and Ulises G. Pacheco for technical assistance. This work was supported by grants IN225305 from PAPIIT-DGAPA and 46161-M, 41477-Q from CONACyT.

References

Alvarez-Buylla A, Seri B and Doetsch F (2002) Identification of neural stem cells in the adult vertebrate brain. Brain Res Bull 57: 751–758.

Arvidsson A, Collin T, Kirik D, Kokaia Z and Lindvall O (2002) Neuronal replacement from endogenous precursors in the adult brain after stroke. Nat Med 8: 963–970.

Azcoitia I, Fernandez Galaz C, Sierra A and Garcia-Segura LM (1999) Gonadal hormones affect neuronal vulnerability to excitotoxin-induced degeneration. J Neurocytol 28: 699–710.

Banasr M, Hery M, Brezun JM and Daszuta A (2001) Serotonin mediates oestrogen stimulation of cell proliferation in the adult dentate gyrus. Eur J Neurosci 14: 1417–1424.

Brann DW, Dhandapani K, Wakade C, Mahesh VB and Kahn MM (2007) Neurotrophic and neuroprotective actions of estrogen: basic mechanisms and clinical implications. Steroids 72: 381–405.

Brännvall K, Korhonen L and Lindholm D (2002) Estrogen-receptor-dependent regulation of neural stem cell proliferation and differentiation. Mol Cell Neurosci 21: 512–520.

Ciriza I, Carrero P, Azcoitia I, Lundeen SG and Garcia-Segura LM (2004) Selective estrogen receptor modulators protect hippocampal neurons from kainic acid excitotoxicity: differences with the effect of estradiol. J Neurobiol 61: 209–221.

Collin T, Arvidsson A, Kokaia Z and Lindvall O (2005) Quantitative analysis of the generation of different striatal neuronal subtypes in the adult brain following excitotoxic injury. Exp Neurol 195: 71–80.

Curtis MA, Penney EB, Pearson AG, van Roon-Mom, Butterworth NJ, Dragunow M, Connor B and Faull RLM (2003) Increased cell proliferation and neurogenesis in the adult human Huntington's disease brain. Proc Natl Acad Sci USA 100: 9023–9027.

Doetsch F, Caille I, Lim DA, Garcia-Verdugo JM and Alvarez-Buylla A (1999) Subventricular zone astrocytes are neural stem cells in the adult mammalian brain. Cell 97: 703–716.

Galea LA, Spritzer MD, Barker JM and Pawluski JL (2006) Gonadal hormone modulation of hippocampal neurogenesis in the adult. Hippocampus 16: 225–232.

Garcia-Segura LM, Azcoitia I and DonCarlos LL (2001) Neuroprotection by estradiol. Prog Neurobiol 63: 29–60.

Garcia-Verdugo JM, Ferron S, Flames N, Collado L, Desfilis E and Font E (2002) The proliferative ventricular zone in adult vertebrates: a comparative study using reptiles, birds, and mammals. Brain Res Bull 57: 765–775.

Gheusi G and Lledo PM (2007) Control of early events in olfactory processing by adult neurogenesis. Chem Senses 32: 397–409.

Hojo Y, Hattori TA, Enami T, Furukawa A, Suzuki K, Ishii HT, Mukai H, Morrison JH, Janssen WG, Kominami S, Harada N, Kimoto T and Kawato S (2004) Adult male hippocampus syntesizes estradiol from pregnenolone by cytochromes P45017α and P450 aromatase localized in neurons. Proc Natl Acad Sci USA 101: 865–870.

Hoyk Z, Varga C and Parducz A (2006) Estrogen-induced region specific decrease in the density of 5-bromo-2-deoxyuridine-labeled cells in the olfactory bulb of adult female rats. Neuroscience 141: 1919–1924.

Lledo PM, Alonso M and Grubb MS (2006) Adult neurogenesis and functional plasticity in neuronal circuits. Nat Rev Neurosci 7: 179–193.

Martoncikova M, Racekova E and Orendacova J (2006) The number of proliferating cells in the rostral migratory stream of rat during the first postnatal month. Cell Mol Neurobiol 26: 1453–1461.

Merkle FT, Tramontin AD, Garcia-Verdugo JM and Alvarez-Buylla A (2004) Radial glia give rise to adult neural stem cells in the subventricular zone. Proc Natl Acad Sci USA 101: 17528–17532.

Merkle FT, Mirzadeh Z and Alvarez-Buylla A (2007) Mosaic organization of neural stem cells in the adult brain. Science 317: 381–384.

Murashov AK, Pak ES, Hendricks WA and Tatko LM (2004) 17β-Estradiol enhances neuronal differentiation of mouse embryonic stem cells. FEBS Lett 569: 165–168.

Ormerod BK and Galea LAM (2001) Reproductive status regulates cell proliferation within the dentate gyrus of the adult female meadow vole: a possible regulatory role for estradiol. Neurosci 2: 169–179.

Ormerod BK, Lee TT and Galea LA (2003) Estradiol initially enhances but subsequently suppresses (via adrenal steroids) granule cell proliferation in the dentate gyrus of adult female rats. J Neurobiol 55: 247–260.

Reynolds BA and Weiss S (1992) Generation of neurons and astrocytes from isolated cells of the adult mammalian central nervous system. Science 255: 1707–1710.

Smith MT, Pencea V, Wang Z, Luskin MB and Insem TR (2001) Increased number of BrdU-labeled neurons in the rostral migratory stream of the estrous prairie vole. Horm Behav 39: 11–21.

Tanapat P, Hastings NB, Reeves AJ and Gould E (1999) Estrogen stimulates a transient increase in the number of new neurons in the dentate gyrus of the adult female rat. J Neurosci 19: 5792–5801.

Tattersfield AS, Croon RJ, Liu YW, Kells AP, Faull RL and Connor B (2004) Neurogenesis in the striatum of the quinolinic acid lesion model of Huntington's disease. Neuroscience 127: 319–332.

Taupin P (2005) Adult neurogenesis in the mammalian central nervous system: functionality and potential clinical interest. Med Sci Monit 11: 247–252.

Striatal Dopaminergic Denervation and Spine Loss in MPTP-Treated Monkeys

Rosa M. Villalba, Heyne Lee, Dinesh Raju, and Yoland Smith

Abstract Dopamine (DA) plays a critical role in regulating spine density on medium-sized spiny neurons (MSNs) in the striatum. Quantitative analysis of Golgi-impregnated MSNs in striatal sections at the level of the anterior commissure (commissural striatum) from MPTP-treated monkeys with complete DA denervation showed a significant reduction in spine density in both the caudate nucleus and putamen compared with that in controls. Similar analysis in partially DA-denervated striata revealed that the loss of spines was tightly correlated with the relative degree of dopamine depletion. Quantitative electron microscopy (EM) analyses showed a significant decrease in the density of both D1-immunolabeled and immunonegative spines in the commissural putamen of MPTP-treated monkeys, suggesting that DA depletion causes a reduction of spines on both direct and indirect pathway MSNs in the striatum of parkinsonian monkeys. Both the degree of striatal dopamine loss and extent of spine degeneration are correlated with low level of calbindin expression in striatal subregions, suggesting an important role of calcium in striatal pathogenesis of Parkinson's disease. This is further supported by electron microscopic data showing that calbindin-containing spines are less sensitive to degeneration in the caudate nucleus of partially depleted MPTP-treated monkeys. These findings provide strong evidence that striatal spine loss is an early sign of parkinsonism pathogenesis that is tightly correlated with the degree of striatal dopamine denervation and affects both direct and indirect striatofugal pathway neurons predominantly in calbindin-poor striatal subregions of MPTP-treated monkeys.

1 Introduction

Dopamine (DA) plays a critical role in regulating spine density on striatal medium-sized spiny neurons (MSN) (Arbuthnott et al. 2000; Robinson and Kolb 1999). A significant reduction in spine density has been demonstrated in postmortem tissue

R.M. Villalba (✉), H. Lee, D. Raju, and Y. Smith
Yerkes National Primate Research Center, Emory University, 954 Gatewood Road NE, Atlanta, GA 30322, USA
e-mail: rvillal@emory.edu

H.J. Groenewegen et al. (eds.), *The Basal Ganglia IX*, Advances in Behavioral Biology 58, DOI 10.1007/978-1-4419-0340-2_28,

from human parkinsonians (Stephens et al. 2005; Zaja-Milatovic et al. 2005) and in 6-hydroxydopamine (6-OHDA) rodent model of Parkinson's disease (PD) (Ingham et al. 1998). The role of DA in modulating striatal spine morphogenesis is also depicted by the significant increase in the number of spines on MSNs in animals treated with psychostimulants (Robinson and Kolb 1999; Kolb et al. 2003; Li et al. 2003; Norrholm et al. 2003). Rodent models of PD, characterized by a substantial loss of striatal dopamine, show a significant reduction in overall spine density (Ingham et al. 1993). Therefore, dopamine is a key transmitter that plays an important role in regulating the growth, maintenance, and plasticity of dendritic spines in the striatum.

Although the etiology of the degenerative process that underlies clinical deterioration in PD remains unknown, there is clear evidence that striatal dopaminergic denervation precedes nigral neuronal loss (Bernheimer et al. 1973; Herkenham et al. 1991) suggesting that striatal dysfunction in dopamine transmission and, likely, MSNs pathology are early steps toward nigrostriatal dopaminergic degeneration in PD.

In this study, we took advantage of the progressive degenerative process induced by low doses of MPTP to assess the effects of partial or severe dopaminergic depletion on spine loss in MSNs of specific functional striatal subregions in monkeys.

Findings of these studies have been presented in abstract forms (Villalba et al. 2006, 2007).

2 Materials and Methods

2.1 *Animals and Tissue Preparation*

In total, six control and six MPTP-treated adult Rhesus monkeys (*Macaca mulatta*) (Yerkes National Primate Research Center colony) were used in this study (2–8 years old). The housing, feeding, and experimental conditions used in these studies were in line with those of the National Institutes of Health guidelines and approved by Emory University Institutional Animal Care and Use Committee.

2.1.1 MPTP Injections and Parkinsonism

In four monkeys, MPTP was injected unilaterally through the left carotid (total dose 2–3 mg/kg; Sigma-Aldrich, St-Louis, MO), while the two other animals received intramuscular injections of MPTP. Following these injections, behavioral changes and parkinsonian motor signs were measured with quantitative methods routinely used in our laboratory to assess parkinsonian behaviors in MPTP-treated monkeys (Bergman et al. 1990; Wichmann et al. 2001; Wichman and Soares 2006; Soares et al. 2004; Bogenpohl et al. 2007). The four monkeys that received intracarotid MPTP injections developed obvious signs of parkinsonism on the side of the body contralateral to the injected carotid (the ipsilateral side was not significantly affected). One of the systemically injected monkeys that received a total dose of 9.3 mg/kg MPTP over a 6 months period developed stable bilateral parkinsonian motor signs (bradykinesia, rigidity, flexed limb and body posture). The other animal that was systemically administered 4.3 mg/kg MPTP over 2 months did not develop any significant motor impairment. The monkeys were perfused 6–18 months following the last MPTP injection.

2.1.2 Animal Perfusion

Animals were deeply anesthetized with an overdose of pentobarbital (100 mg/kg, iv) and perfused transcardially with cold oxygenated Ringer's solution, followed by fixative containing 4% paraformaldehyde and 0.1% glutaraldehyde in phosphate buffer (PB; 0.1 M, pH 7.4). After perfusion, the brains were cut into 10 mm-thick blocks and tissue sections (60 μm thick) were obtained with a Vibratome.

2.2 *Golgi Impregnation*

The Golgi impregnation technique was used on 60 μm thick Vibratome striatal coronal sections from normal and MPTP-treated monkeys. The sections were immersed in 1% osmium tetroxide (30 min) followed by subsequent overnight incubations in 3.5% potassium dichromate and 2% silver nitrate. Golgi-impregnated sections were then dehydrated in graded ethyl alcohol, embedded in resin (Durcupan ACM; Fluka, Buchs, Switzerland), mounted, and coverslipped. The resin was polymerized at 60°C (48 h).

2.3 *Immunocytochemistry*

2.3.1 Primary Antibodies

In this study we used well-characterized commercially available antibodies. The source and dilution of these antibodies were: (1) Mouse monoclonal anti-TH antibody (dilution 1:1,000; Chemicon, Temecula, CA); (2) Mouse monoclonal anti-CB antibody (dilution: 1:1,000; Sigma, St Louis, MO); (3) Rat monoclonal anti-D1 antibody (dilution: 1:75; Sigma, St Louis, MO).

2.3.2 Immunoperoxidase Labeling for Light and Electron Microscopy

The avidin–biotin complex method (ABC, Vectastain Standard kit) and the chromogen diaminobenzidine (DAB) were used for the peroxidase reaction. For light microscopy (LM), the commissural striatal sections were mounted onto gelatin-coated slides, dehydrated, and then coverslipped with Permount. The tissue was examined with a Leica DMRB microscope (Leica Microsystems, Inc., Bannockburn, IL) and images were taken with a CCD camera (Leica DC 500; Leica IM50 software). For electron microscopy (EM), samples of tissue were cut from the sections, mounted onto resin blocks, and cut into 60 nm sections with an ultramicrotome (Leica Ultracut T2). The 60 nm sections were examined on the Zeiss EM-10C electron microscope. Electron micrographs were taken at 25,000X magnification and saved with a CCD camera (DualView 300 W; DigitalMicrograph software, version 3.10.1; Gatan, Inc., Pleasanton, CA).

2.4 *Data Analysis*

2.4.1 Quantitative Analysis of Dendritic Spine Density in Golgi-Impregnated Neurons

In commissural striatal sections from control and DA-depleted animals, the longest dendrites from randomly selected MSNs were quantitatively analyzed using the 100X immersion oil objective and a computer-assisted tracing system (Neurolucida v7; MicrobrightField Inc., USA). In partially DA-depleted animals, the longest dendrite was quantitatively analyzed from impregnated MSNs randomly selected from specific striatal areas characterized by different levels of TH immunoreactivity.

2.4.2 Immunolabeled Spine Analysis

Blocks of tissue were prepared from commissural putamen sections immunostained with D1 and CB antibodies in control and MPTP-treated parkinsonian monkeys. These blocks were coded and mixed to avoid any bias in the analysis of the tissue. To minimize false negatives, only ultrathin sections from the most superficial sections of blocks were scanned in the electron microscope. A total of 100 electron micrographs per monkey were taken at a magnification of 25,000X with each electron micrograph representing an area of approximately 12.56 μm^2. All labeled and unlabeled dendritic spines with identifiable asymmetric synapses were quantified.

2.4.3 Statistical Analysis and Photomicrograph Production

Statistics were performed using the SigmaStat 2.03 software. The data from each control and MPTP-treated monkeys were analyzed for variance (one-way ANOVA) to determine the differences among animals in each group. All data were expressed as mean ± SEM and compared by *t* test to determine statistical differences between control and MPTP-treated monkeys. Pictures were digitally acquired and imported in TIFF format into Adobe Photoshop software (version 9.0; Adobe System, Inc., San Jose, CA). Some images were pseudocolored using the NIH Image program *ImageJ*. Micrographs were then compiled into figures in Adobe Illustrator 12.0.

3 Results

3.1 *Striatal Spine Loss: Golgi Analysis*

3.1.1 Severely Dopamine-Depleted Striatum

In a first series of experiments, striatal spine loss was examined in the dorsal striatum of four severely dopamine-depleted monkeys that displayed significant parkinsonian motor signs. The whole extent of the caudate nucleus and putamen was almost completely devoid of TH innervation in these animals (data not shown). Commissural striatal sections were Golgi-impregnated and used for quantitative measurements of total spine density on MSNs from both controls (Fig. 1a–c) and MPTP-treated (Fig. 1d–f) monkeys. Quantitative spine density was measured from the longest dendrites of randomly encountered Golgi-impregnated MSNs. The analysis of 25–30 dendrites in both control and MPTP-treated animals revealed a significant reduction (38–40%; Student's t test, $P < 0.05$) in spine density in MSNs from MPTP-treated monkeys (Fig. 2).

3.1.2 Partially Dopamine-Depleted Striatum

In a third group of monkeys, we tested the effects of partial dopamine depletion on striatal spine loss. The striatal tissue used for this part of the study was collected from either the contralateral side of the four intracarotid MPTP-treated monkeys or from the monkey that received systemic MPTP, but did not develop parkinsonian motor signs. The heterogeneous and progressive rostrocaudal and mediolateral gradient in dopamine denervation of the striatum allowed us to assess the degree of spine loss in subregions of the commissural caudate nucleus and putamen showing differential degrees of dopamine denervation (Fig. 3a, b). Pseudocolored images *(ImageJ)* of TH-immunostained striatal sections from partially dopamine-depleted MPTP-treated monkeys are shown in Fig. 3b. The spine density in MSNs from the severely dopamine-depleted lateral sector (region 1 in Fig. 3b) was significantly lower than in the less dopamine-denervated medial regions (region 3 in Fig. 3b). Overall, the density of spines was tightly correlated with the relative abundance of TH-immunoreactive fibers in these animals (Fig. 3).

3.2 *Density of D1-Immunoreactive Spines in the Striatum*

To determine if the loss of dendritic spines in dopamine-denervated striatum affects both direct and indirect pathway neurons, we measured the relative density of total spines and D1-containing spines in the commissural putamen of three control and three severely dopamine-depleted MPTP-treated monkeys (Fig. 4). As shown in

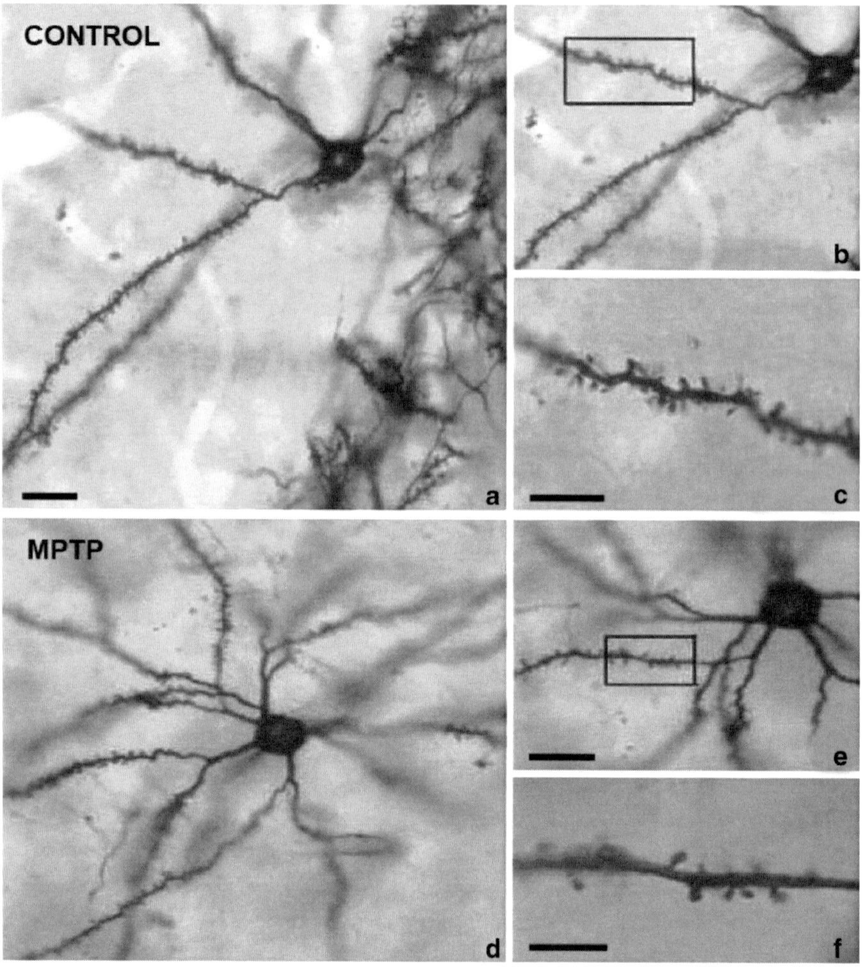

Fig. 1 Golgi-impregnated medium-sized spiny neurons (MSNs) in the commissural putamen of control (**a–c**) and MPTP-treated monkeys (**d–f**). (**c**) and (**f**): High-magnification images of dendrites boxed in (**b**) and (**e**), respectively. Scale bar in (**a**) [valid for (**b**), (**d**), (**e**)] = 25 μm. Scale bars in (**c**) and (**f**) = 5 μm

previous rodent studies (Hersch et al. 1995; Lei et al. 2004), D1 immunoreactivity was strongly expressed in dendrites and spines of MSNs in control monkeys (Fig. 4a). On an average, the density of total spines and D1-immunoreactive spines was significantly (Student's *t* test; $P < 0.05$) decreased by 30% and 33%, respectively, in the MPTP-treated striatum (Fig. 4c, d). Similarly, the density of D1-immunonegative spines was decreased by 27% in the dopamine-depleted putamen (Fig. 4e) suggesting that both D1-immunoreactive and D1-negative spines are lost following striatal dopamine denervation in MPTP-treated monkeys. These data are further supported

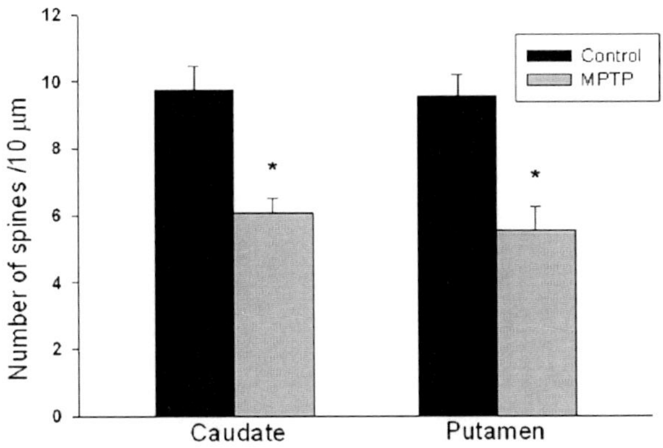

Fig. 2 Histograms showing the quantitative analysis of spine density on medium-sized spiny neurons (MSNs) in the commissural caudate and putamen of controls and severely DA-depleted MPTP-treated monkeys. Spine density (spine number per 10 μm) was measured along the length of the longest dendrite of Golgi-impregnated MSNs. Statistical analysis (*t* test) showed a significant difference (*) in the mean density (±SEM) of spines between control (caudate: $n = 25$; putamen: $n = 26$) and MPTP-treated (caudate: $n = 27$; putamen: $n = 30$) groups ($P < 0.001$). The differences in the mean densities of spines in control animals were not statistically different (ANOVA analysis, $P = 0.86$). There was no significant interindividual difference in spine density between animals in control (one way ANOVAs, $P = 0.155$; $N = 3$) or the MPTP group (one-way ANOVA, $P = 0.340$; $N = 5$). *n* number of dendrites analyzed, *N* number of animals used

by quantitative measurements showing no significant difference in the relative percentages of D1-immunoreactive spines between control and MPTP-treated monkeys (50.3 ± 1.3% in control vs. 48.4 ± 2.9% in MPTP-treated cases) (Fig. 4f).

3.3 Calbindin Immunolabeling in the Striatum

Recent studies have highlighted the importance of calcium dysregulation in striatal spine loss following dopamine depletion in rodents (Day et al. 2006). If this is also the case in MPTP-treated monkeys, striatal MSNs that express calcium binding proteins may be selectively less susceptible to spine degeneration than other striatal neurons. To address this issue, we correlated the progressive pattern of TH denervation in partially dopamine-depleted animals (Fig. 5a) with the distribution of CB immunoreactivity in normal monkeys (Fig. 5b). We found that the extent of DA denervation and corresponding striatal spine loss were, indeed, tightly correlated with the relatively low level of CB immunoreactivity. For instance, the dorsolateral striatal sector of the commissural striatum, which expressed low CB immunoreactivity, showed the highest degree of dopamine denervation and the most significant spine loss, whereas more medial striatal regions enriched in CB immunoreactivity showed a significantly

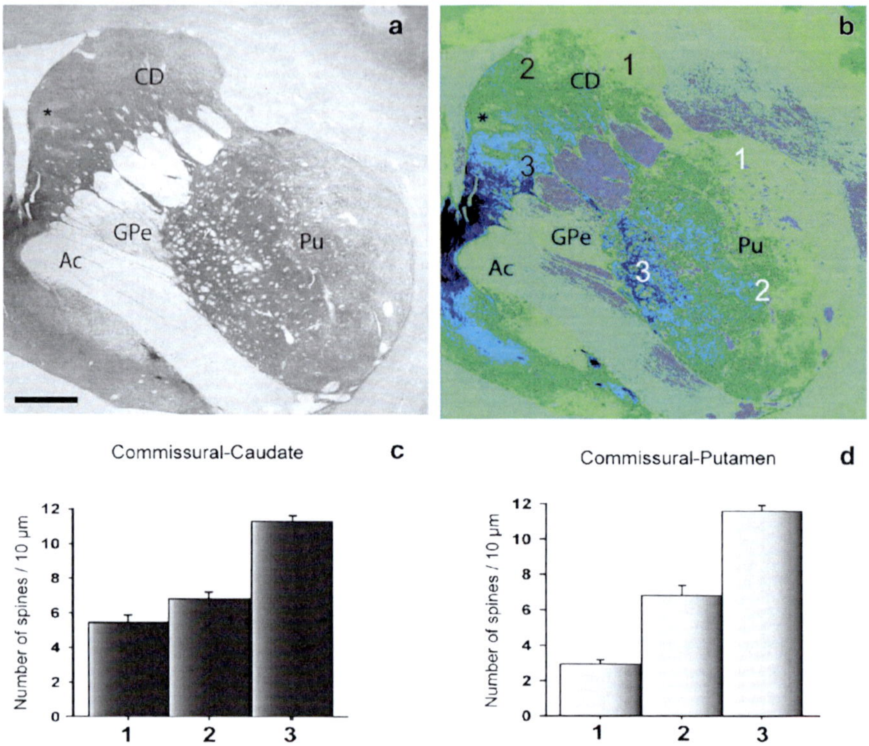

Fig. 3 (**a**) A tyrosine hydroxylase-immunoreactive (TH-IR) coronal section at the level of the commissural striatum of a partially dopamine-denervated MPTP-treated monkey. (**b**) *Black and white* pseudocolored image (NIH *ImageJ* program) of TH-immunostained sections shown in (**a**) to better illustrate the mediolateral gradient in TH denervation in the caudate nucleus and putamen. (**c–d**) Histograms showing the quantitative analysis of the dendritic spine density of Golgi-impregnated neurons from striatal areas with different degrees of TH-IR depletion. Each bar represents the mean spine density (±SEM) per 10 μm of dendritic length from at least ten neurons in the striatal areas labeled with the corresponding number and color in (**b**). Abbreviations: *Ac* anterior commissure, *CD* caudate, *Pu* putamen, *GPe* globus pallidus, external segment. Asterisks indicate dopamine-denervated striatal patch-like areas in the caudate nucleus. Scale bar in (**a**) = 1 mm [valid for (**b**)]

lower degree of dopamine denervation and a more modest level of spine pathogenesis (Fig. 5a, b).

To further demonstrate the neuroprotective effects of CB toward spine loss, we quantified the relative change in the density of CB-immunoreactive spines between control and severely dopamine-denervated MPTP-treated monkeys (Fig. 5c–f). Quantitative EM immunocytochemical analyses (Fig. 5e) showed a significant increase (Chi-square; $P = 0.005$) in the density of CB-immunoreactive spines relative to unlabeled spines in the ventral part of the caudate nucleus of MPTP-treated monkeys compared with controls, suggesting a neuroprotective role of CB against spine loss in this region. However, the ventral putamen, one of the striatal sectors

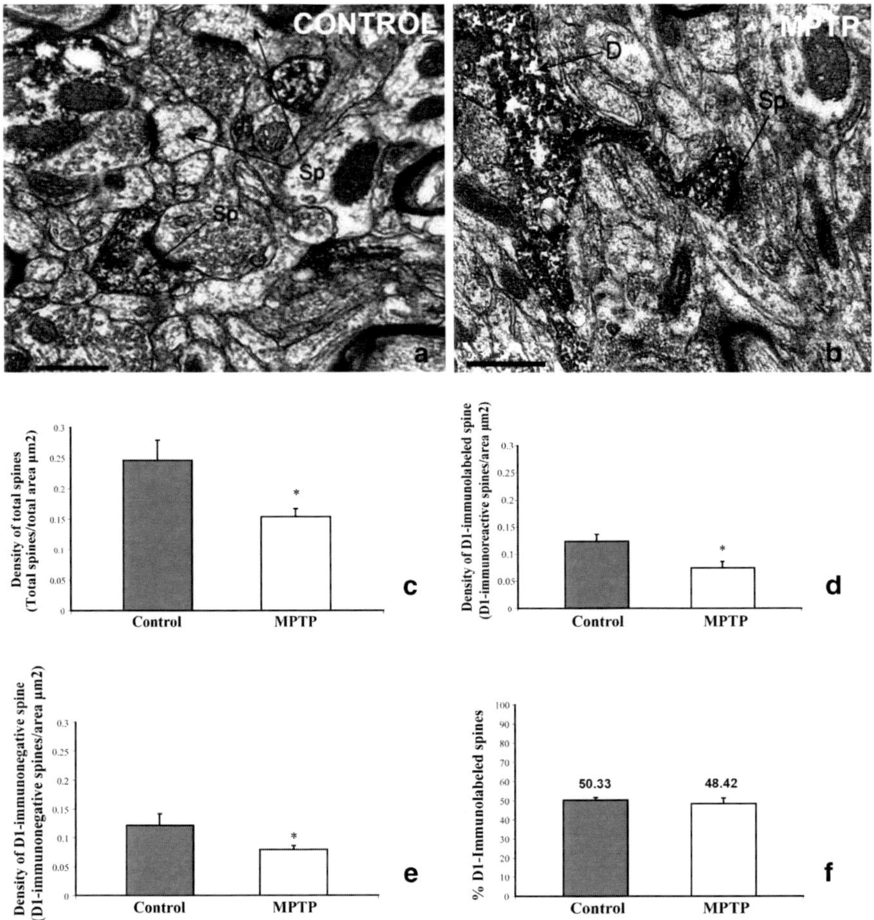

Fig. 4 (**a**, **b**) Electron micrographs of D1-immunolabeled spines in the commissural putamen from control (**a**) and MPTP-treated monkeys (**b**). (**c**) Histograms showing a significant decrease in the density of total spines in the commissural putamen of MPTP-treated monkeys compared with control monkeys (total spine density: control = 0.23 ± 0.03; MPTP = 0.16 ± 0.01, *t* test, $P < 0.05$). (**d**, **e**) Histograms showing the density of D1-immunoreactive (**d**) and D1-immunonegative spines (**e**) in the commissural putamen of control and MPTP-treated monkeys. A significant reduction in both D1-labeled (control = 0.12 ± 0.01; MPTP = 0.08 ± 0.01; *t* test, $P < 0.05$) and unlabeled (control = 0.11 ± 0.02; MPTP = 0.08 ± 0.002; *t* test, $P < 0.05$) spines was found in MPTP-treated monkeys. (**f**) No significant difference was found in the relative percentage of D1-immunoreactive spines in the commissural putamen of control animals compared with MPTP-treated monkeys. Three control and three MPTP-treated monkeys were used in this analysis. A total of 600 electron micrographs were analyzed (300 in control and 300 in MPTP-treated monkeys). Average area/picture = 12.56 µm^2. Abbreviations: *Sp* spine, *D* dendrite. Scale bar in (**a**) and (**b**): 0.5 µm

most enriched in CB immunoreactivity, showed a decrease in the ratio of calbindin-labeled to unlabeled spines, while no significant difference was observed in the dorsal caudate and dorsal putamen (Fig. 5e, f).

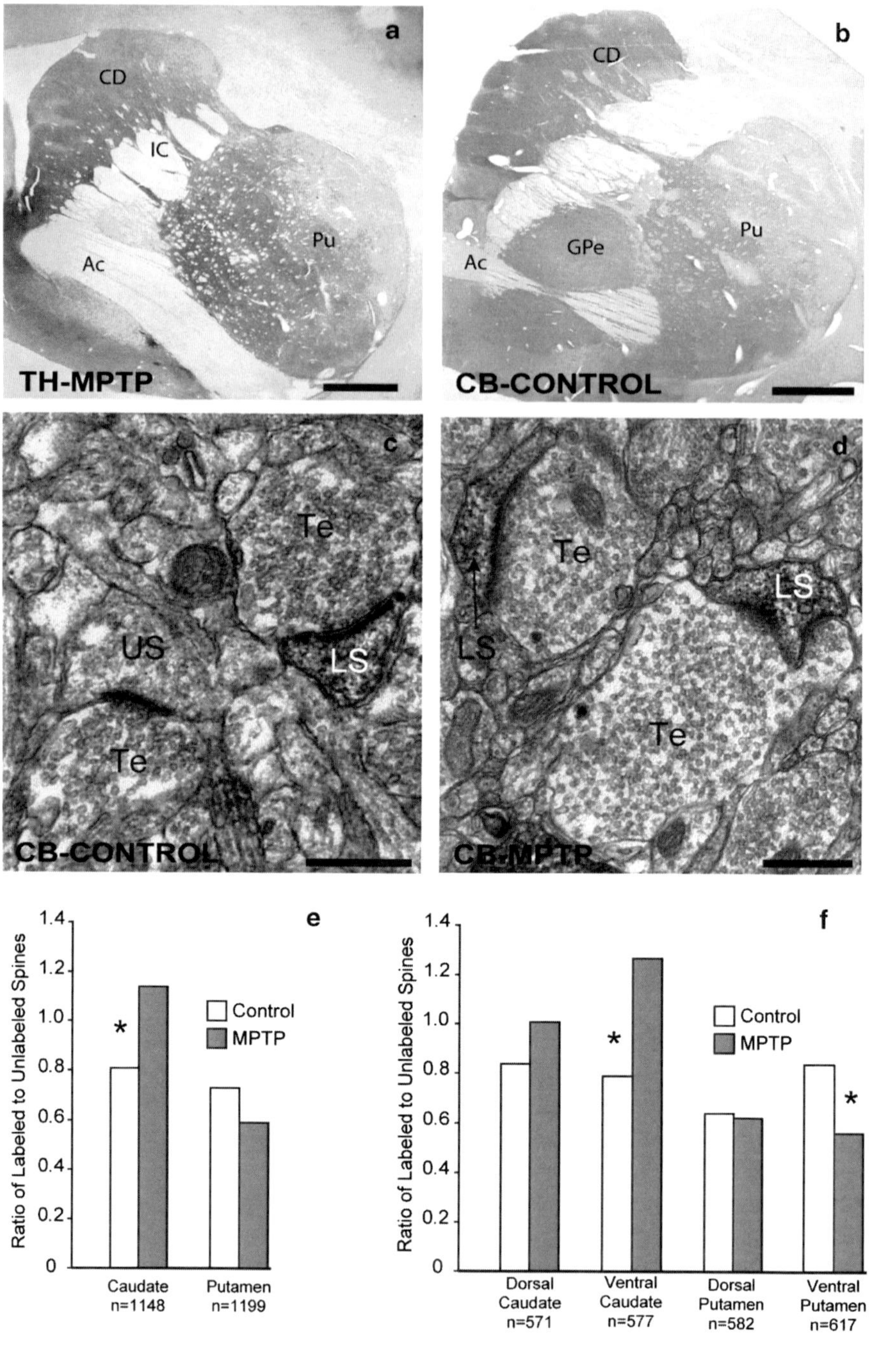
a
CD
IC
Pu
Ac
TH-MPTP
b
CD
Pu
GPe
Ac
CB-CONTROL
c
Te
US
LS
Te
CB-CONTROL
d
Te
LS
LS
Te
CB-MPTP
e
Ratio of Labeled to Unlabeled Spines
1.4
1.2
1.0
0.8
0.6
0.4
0.2
0
*
Control
MPTP
Caudate
n=1148
Putamen
n=1199
f
Ratio of Labeled to Unlabeled Spines
Control
MPTP
*
*
Dorsal
Caudate
n=571
Ventral
Caudate
n=577
Dorsal
Putamen
n=582
Ventral
Putamen
n=617

4 Discussion

The present study provides further evidence for significant striatal spine pathology in parkinsonism. In line with recent human parkinsonian data (Zaja-Milatovic et al. 2005), a dramatic spine loss was found in the striatum of MPTP-treated monkeys. Our findings also demonstrate that striatal spine loss is an early pathogenic feature of parkinsonism, tightly linked with the severity of dopamine depletion, which occurs in both symptomatic parkinsonian monkeys and animals with partial striatal dopamine depletion that do not show any signs of motor abnormalities. Another main conclusion of our study is that both D1-immunoreactive and D1-negative spines are lost following MPTP-induced dopamine depletion in the monkey striatum, suggesting that both direct and indirect pathway neurons are affected in this animal model. Finally, a close correlation between striatal enrichment in CB immunoreactivity and decreased sensitivity to MPTP-induced DA denervation and striatal spine loss was shown, suggesting an involvement of calcium in the mechanism that underlies dopamine denervation-induced spine pathogenesis in parkinsonism.

Striatal spine loss has been reported in various rodent models of parkinsonism and humans with PD (Ingham et al. 1993; Cheng et al. 1997; Stephens et al. 2005; Zaja-Milatovic et al. 2005). Previous studies have reported an increased sensitivity for spine pathology in both human PD and MPTP-treated monkeys (Moratalla et al. 1992; Hornykiewicz 2001; Stephens et al. 2005; Zaja-Milatovic et al. 2005). The progressive and regional sensitivity of different striatal territories to degeneration of the dopaminergic nigrostriatal system is well established either in PD or animal models of parkinsonism (Moratalla et al. 1992; Damier et al. 1999; Dauer and Przedborski 2003; Iravani et al. 2005). Our findings demonstrate that striatal spine loss also follows a specific pattern of degeneration that appears to be highly dependent on the degree of dopamine denervation, further suggesting the important role of dopamine in regulating striatal spine plasticity. This dependence was particularly well demonstrated in the partially dopamine-depleted asymptomatic monkeys,

Fig. 5 (**a**, **b**) Light micrographs comparing the distribution of tyrosine hydroxylase (TH) immunostaining in a commissural striatal section of a partially DA-denervated MPTP-treated monkey with the distribution of calbindin (CB) immunostaining in the commissural striatum of a normal monkey (**b**). Note the striking correspondence between striatal areas poor in CB immunoreactivity in control monkeys and the regional pattern of DA denervation in the striatum of the MPTP-treated monkey. (**c**, **d**) Electron micrographs of CB-immunostained spines in commissural putamen (**c**) and caudate (**d**) from a control (**c**) and a severely dopamine-depleted MPTP monkey (**d**). (**e**, **f**) Quantitative comparative analysis of the ratio of CB-labeled vs. unlabeled spines in different regions of the commissural striatum of control and complete DA-denervated MPTP-treated animals. Significant, but opposite, differences (*) in ratios of labeled vs. unlabeled spines were found in the ventral caudate (Chi-square; $P = 0.005$) and ventral putamen (Chi-square; $P = 0.022$). Abbreviations: *Ac* anterior commissure, *CD* caudate, *GPe* globus pallidus, external segment, *IC* internal capsule, *LS* labeled spine, *Pu* putamen, *Te* terminal, *US* unlabeled spine. Scale bar in (**b**) = 1 mm [valid for (**a**)]. Scale bar in (**c**) and (**d**) = 0.5 μm

indicating that spine degeneration is an early pathogenic feature of PD that appears prior to the development of abnormal motor signs.

Interestingly, although considerable spine loss was found in the sensorimotor striatum of partially dopamine-depleted monkeys, this pathology did not result in significant motor impairment. Knowing the important role of spine plasticity in mediating proper glutamatergic transmission at corticostriatal and thalamostriatal synapses, one may wonder about the compensatory mechanisms that are put in place to mediate normal motor behavior despite considerable striatal spine pathology in the sensorimotor striatum. Another possibility is that striatal spine loss underlies non-motor, learning and cognitive deficits not thoroughly assessed in these animals, but known to occur in early MPTP-treated monkeys and humans prior to the development of parkinsonian motor deficits (Stern et al. 1990; Schneider and Pope-Coleman 1995).

Recent rodent data have provided convincing evidence that spine loss selectively affects striatal D2-containing neurons of the indirect pathway in reserpine-treated mice (Day et al. 2006). These findings were supported by quantitative immunoelectron microscopy data from 6-OHDA-treated adult rats showing a significant decrease in the number of striatal D1-negative spines. These observations are at odds with our data and findings of previous Golgi studies in both human parkinsonians and rodent models of parkinsonism showing a rather homogeneous loss of spines across large populations of Golgi-impregnated MSNs (Ingham et al. 1989, 1998; Stephens et al. 2005; Zaja-Milatovic et al. 2005). In each of these studies, there was a minimal degree of variability in the density of spines on randomly selected dendrites of Golgi-impregnated striatal neurons, suggesting a homogeneous spine loss across both populations of striatal MSNs. Our findings further support these data and also provide electron microscopic evidence for a reduced density of both D1-immunoreactive and D1-negative (presumably D2-containing) spines in the putamen of MPTP-treated monkeys. Whether these apparent discrepancies rely on species differences, chronic vs. acute dopamine toxin exposure, or other technical/sampling issues remains to be established.

Until recently, the mechanism(s) underlying striatal spine loss in PD remained unknown. However, recent compelling evidence suggests that the dysregulation of calcium concentrations in striatopallidal neurons ultimately leads to specific spine loss in reserpine-treated mice (Day et al. 2006). There is good evidence that L-type voltage-gated calcium channels (LVGCC) may be involved because the chronic administration of an LVGCC antagonist completely blocks spine loss in mice (Day et al. 2006). Our data showing a close correlation between the increased incidence of striatal spine loss and reduced CB immunoreactivity in MPTP-treated monkeys is in line with these observations and strongly support an important role of intracellular calcium buffering capabilities in protecting striatal neurons from spine pathology in PD. However, our electron microscopic results showing differences in ratios of CB-labeled vs. unlabeled spines between the caudate nucleus and putamen suggest that calbindin expression is not the only factor that contributes to the regional striatal heterogeneity in striatal spine loss in Parkinson's disease.

Although the broad implications of striatal spine loss in PD remain to be established, the fact that spines are the main targets of glutamatergic inputs from the cerebral cortex and thalamus (Smith and Bolam 1990; Raju et al. 2006) combined with functional evidence for highly specific interactions between convergent axo-spinous glutamatergic and dopaminergic afferents in the striatum (Bamford et al. 2004) suggests that this change in synaptic connectivity may contribute to ineffectively timed and patterned striatofugal activity, thereby leading to pathophysiological basal ganglia discharges in PD (Wichmann et al. 2007).

Acknowledgments The authors thank Dr. T. Wichmann for some of the MPTP-treated animals used in this study. This research was supported by NIH grant R01 NS 037948 to Y. Smith and the NIH base grant to the Yerkes National Primate Research Center (RR 00165).

References

Arbuthnott GW, Ingham CA and Wickens JR (2000) Dopamine and synaptic plasticity in the neostriatum. J Anat 196: 587–596.

Bamford NS, Robinson S, Palmiter RD, Joyce JA, Moore C and Meshul CK (2004) Dopamine modulates release from corticostriatal terminals. J Neurosci 24: 9541–9552.

Bergman H, Wichmann T and DeLong MR (1990) Reversal of experimental parkinsonism by lesions of the subthalamic nucleus. Science 249: 1436–1438.

Bernheimer H, Birkmayer W, Hornykiewicz O, Jellinger K and Seitelberger F (1973) Brain dopamine and the syndromes of Parkinson and Huntington. Clinical, morphological and neurochemical correlations. J Neurol Sci 20: 415–455.

Bogenpohl J, Pare J-P and Smith Y (2007) Subcellular localization of adenosine A2A receptors in the striatum and globus pallidus of monkey and rat. Ninth Triennial Meeting of the International Basal Ganglia Society, The Netherlands, September 2–6, 2007.

Cheng HW, Rafols JA, Goshgarian HG, Anavi Y, Tong J and McNeill TH (1997) Differential spine loss and regrowth of striatal neurons following multiple forms of deafferentation: a Golgi study. Exp Neurol 147: 287–298.

Damier P, Hirsch EC, Agid Y and Graybiel AM (1999) The substantia nigra of the human brain. II. Patterns of loss of dopamine-containing neurons in Parkinson's disease. Brain 122: 1437–1448.

Dauer W and Przedborski S (2003) Parkinson's disease: mechanisms and models. Neuron 39: 889–909.

Day M, Wang Z, Ding J, An X, Ingham CA, Shering AF, Wokosin D, Ilijic E, Sun Z, Sampson AR, Mugnaini E, Deutch AY, Sesack SR, Arbuthnott GW and Surmeier DJ (2006) Selective elimination of glutamatergic synapses on striatopallidal neurons in Parkinson disease models. Nat Neurosci 9: 251–259.

Herkenham M, Little MD, Bankiewicz K, Yang SC, Markey SP and Johannessen JN (1991) Selective retention of MPP+ within the monoaminergic systems of the primate brain following MPTP administration: an in vivo autoradiographic study. Neuroscience 40: 133–158.

Hersch SM, Ciliax BJ, Gutekunst CA, Rees HD, Heilman CJ, Yung KK, Bolam JP, Ince E, Yi H and Levey AI (1995) Electron microscopic analysis of D1 and D2 dopamine receptor proteins in the dorsal striatum and their synaptic relationships with motor corticostriatal afferents. J Neurosci 15: 5222–5237.

Hornykiewicz O (2001) Chemical neuroanatomy of the basal ganglia - normal and in Parkinson's disease. J Chem Neuroanat 22: 3–12.

Ingham CA, Hood SH and Arbuthnott GW (1989) Spine density on neostriatal neurones changes with 6-hydroxydopamine lesions and with age. Brain Res 503: 334–338.

Ingham CA, Hood SH, van Maldegem B, Weenink A and Arbuthnott GW (1993) Morphological changes in the rat neostriatum after unilateral 6-hydroxydopamine injections into the nigrostriatal pathway. Exp Brain Res 93: 17–27.

Ingham CA, Hood SH, Taggart P and Arbuthnott GW (1998) Plasticity of synapses in the rat neostriatum after unilateral lesion of the nigrostriatal dopaminergic pathway. J Neurosci 18: 4732–4743.

Iravani MM, Syed E, Jackson MJ, Johnston LC, Smith LA and Jenner P (2005) A modified MPTP treatment regime produces reproducible partial nigrostriatal lesions in common marmosets. Eur J Neurosci 21: 841–854.

Kolb B, Gorny G, Li Y, Samaha AN and Robinson TE (2003) Amphetamine or cocaine limits the ability of later experience to promote structural plasticity in the neocortex and nucleus accumbens. Proc Natl Acad Sci USA 100: 10523–10528.

Lei W, Jiao Y, Del Mar N and Reiner A (2004) Evidence for differential cortical input to direct pathway versus indirect pathway striatal projection neurons in rats. J Neurosci 24: 8289–8299.

Li Y, Kolb B and Robinson TE (2003) The location of persistent amphetamine-induced changes in the density of dendritic spines on medium spiny neurons in the nucleus accumbens and caudate-putamen. Neuropsychopharmacology 28: 1082–1085.

Moratalla R, Quinn B, DeLanney LE, Irwin I, Langston JW and Graybiel AM (1992) Differential vulnerability of primate caudate-putamen and striosome-matrix dopamine systems to the neurotoxic effects of 1-methyl-4-phenyl-1,2,3,6-tetrahydropyridine. Proc Natl Acad Sci USA 89: 3859–3863.

Norrholm SD, Bibb JA, Nestler EJ, Ouimet CC, Taylor JR and Greengard P (2003) Cocaine-induced proliferation of dendritic spines in nucleus accumbens is dependent on the activity of cyclin-dependent kinase-5. Neuroscience 116: 19–22.

Raju DV, Shah DJ, Wright TM, Hall RA and Smith Y (2006) Differential synaptology of vGluT2-containing thalamostriatal afferents between the patch and matrix compartments in rats. J Comp Neurol 499: 231–243.

Robinson TE and Kolb B (1999) Alterations in the morphology of dendrites and dendritic spines in the nucleus accumbens and prefrontal cortex following repeated treatment with amphetamine or cocaine. Eur J Neurosci 11: 1598–1604.

Schneider JS and Pope-Coleman A (1995) Cognitive deficits precede motor deficits in a slowly progressing model of parkinsonism in the monkey. Neurodegeneration 4: 245–255.

Smith Y and Bolam JP (1990) The output neurones and the dopaminergic neurones of the substantia nigra receive a GABA-containing input from the globus pallidus in the rat. J Comp Neurol 296: 47–64.

Soares J, Kliem MA, Betarbet R, Greenamyre JT, Yamamoto B and Wichmann T (2004) Role of external pallidal segment in primate parkinsonism: comparison of the effects of 1-methyl-4-phenyl-1,2,3,6-tetrahydropyridine-induced parkinsonism and lesions of the external pallidal segment. J Neurosci 24: 6417–6426.

Stephens B, Mueller AJ, Shering AF, Hood SH, Taggart P, Arbuthnott GW, Bell JE, Kilford L, Kingsbury AE, Daniel SE and Ingham CA (2005) Evidence of a breakdown of corticostriatal connections in Parkinson's disease. Neuroscience 132: 741–754.

Stern Y, Tetrud JW, Martin WR, Kutner SJ and Langston JW (1990) Cognitive change following MPTP exposure. Neurology 40: 261–264.

Villalba R, Verrault M and Smith Y (2006) Spine loss in the striatum of MPTP-treated monkeys: a correlation with the degree of striatal dopaminergic denervation. Society for Neuroscience, Atlanta, GA, October 14–18, 2006.

Villalba R, Lee H, Raju D and Smith Y (2007) Dopaminergic denervation and spine loss in the striatum of MPTP-treated monkeys. Ninth Triennial Meeting of the International Basal Ganglia Society, The Netherlands, September 2–6, 2007.

Wichmann T and Soares J (2006) Neuronal firing before and after burst discharges in the monkey basal ganglia is predictably patterned in the normal state and altered in parkinsonism. J Neurophysiol 95: 2120–2133.

Wichmann T, Kliem MA and DeLong MR (2001) Antiparkinsonian and behavioral effects of inactivation of the substantia nigra pars reticulata in hemiparkinsonian primates. Exp Neurol 167: 410–424.

Wichmann T, Smith Y and Vitek J (2007) Basal ganglia: anatomy and physiology. In: Factor S and Weiner W (eds) Parkinson's Disease: Diagnosis and Clinical Management. New York: Demos.

Zaja-Milatovic S, Milatovic D, Schantz AM, Zhang J, Montine K, Sami A, Deutch AY and Montine TJ (2005) Dendritic degeneration in neostriatal medium spiny neurons in Parkinson disease. Neurology 64: 545–547.

Prevention of Calbindin Recruitment into Nigral Dopamine Neurons from MPTP-Induced Degeneration in *Macaca fascicularis*

Masahiko Takada, Ken-ichi Inoue, Shigehiro Miyachi, Haruo Okado, and Atsushi Nambu

Abstract Dopaminergic neurons in the substantia nigra pars compacta that express the calcium-binding protein calbindin selectively survive the cell death period in Parkinson's disease. On the basis of this finding, we examined the preventive effect of calbindin recruitment into nigral dopamine neurons on toxic insults induced by 1-methyl-4-phenyl-1,2,3,6-tetrahydropyridine (MPTP). A recombinant adenoviral vector encoding the calbindin gene was injected unilaterally into the striatum of macaque monkeys. One to two weeks later, expression of calbindin through retrograde transduction was observed in cell bodies of nigral dopamine neurons on the side ipsilateral to vector treatment. In these monkeys, MPTP was administered systemically by repeated intravenous injections. Parkinsonian motor signs, such as akinesia, rigidity, and flexed posture, appeared less severely in the limbs contralateral to vector treatment. Histological analysis revealed that tyrosine hydroxylase immunoreactivity in the striatum was preserved better on the calbindin-recruited side, whereas α-synuclein was expressed in nigral dopamine neurons much more strongly on the nonrecruited side. These results indicate that gene delivery of calbindin into nigral dopamine neurons protects against MPTP-induced parkinsonian symptoms in monkeys.

1 Introduction

It is well known that Parkinson's disease is caused by striatal dopamine deficiency due to a loss of dopaminergic neurons from the substantia nigra pars compacta (SNc). There is accumulating evidence to indicate that neuroactive substances contained in nigral dopamine neurons are involved in the onset of their degeneration. Previous studies have shown that a subpopulation of the dopaminergic neurons that

M. Takada (✉), K.-i. Inoue, S. Miyachi, H. Okado, and A. Nambu
System Neuroscience Section, Primate Research Institute, Kyoto University Inuyama, Aichi, Tokyo, 484-8506, Japan
e-mail: takada-ms@igakuken.or.jp

H.J. Groenewegen et al. (eds.), *The Basal Ganglia IX*, Advances in Behavioral Biology 58, DOI 10.1007/978-1-4419-0340-2_29, © Springer Science+Business Media, LLC 2009

expresses the calcium-binding protein, calbindin, is invulnerable to parkinsonian insults (Yamada et al. 1990; Lavoie and Parent 1991; Gibb 1992; Damier et al. 1999). Approximately 20% of the total dopaminergic neurons contains calbindin, and these neurons are localized in the dorsal tier of the SNc (Yamada et al. 1990; Lavoie and Parent 1991). According to clinical reports with postmortem brains of parkinsonian patients, the calbindin-containing dopaminergic neurons selectively survive the cell death period (Gibb 1992; Damier et al. 1999). A similar finding has been reported in a primate model for Parkinson's disease, which is produced by 1-methyl-4-phenyl-1,2,3,6-tetrahydropyridine (MPTP) (German et al. 1992; Parent and Lavoie 1993). These data prompted us to test the hypothesis that the progress in nigral dopaminergic degeneration might be prevented by recruitment of calbindin into dopamine neurons in the ventral tier of the SNc that do not normally contain this calcium-binding protein. In the present study, a recombinant adenoviral vector encoding the calbindin gene was used to examine the preventive effect of calbindin recruitment into the dopaminergic neurons on MPTP-induced nigral dopaminergic degeneration. The adenoviral vector was injected into the striatum of macaque monkeys to be delivered to cell bodies of the dopaminergic neurons in the SNc via retrograde axonal transport (Fig. 1). In these monkeys, MPTP was systemically administered, and the differences between the calbindin-recruited and the nonrecruited side were analyzed both behaviorally and histologically.

2 Methods

2.1 Animals

Six adult macaque (Japanese and crab-eating) monkeys of either sex (*Macaca fascicularis*) weighing 3.0–5.0 kg were used for this study. The experiments were carried out in a special laboratory (biosafety level 2) designated for in vivo infectious experiments. The experimental protocol was approved by the Animal Care and Use Committee of the Tokyo Metropolitan Institute for Neuroscience, and all experiments were conducted according to the Guideline for the Care and Use of Animals at the Tokyo Metropolitan Institute for Neuroscience.

2.2 Surgical Procedures

For intrastriatal injections of the calbindin-expressing adenoviral vector, the monkeys were first sedated with ketamine hydrochloride (5 mg/kg, i.m.; Daiichi Sankyo Company, Tokyo, Japan) and xylazine hydrochloride (0.5 mg/kg, i.m.; Bayer, Osaka, Japan) and then anesthetized with sodium pentobarbital (20 mg/kg, i.v.; Dainippon Sumitomo Pharma, Osaka, Japan). They were seated in a primate

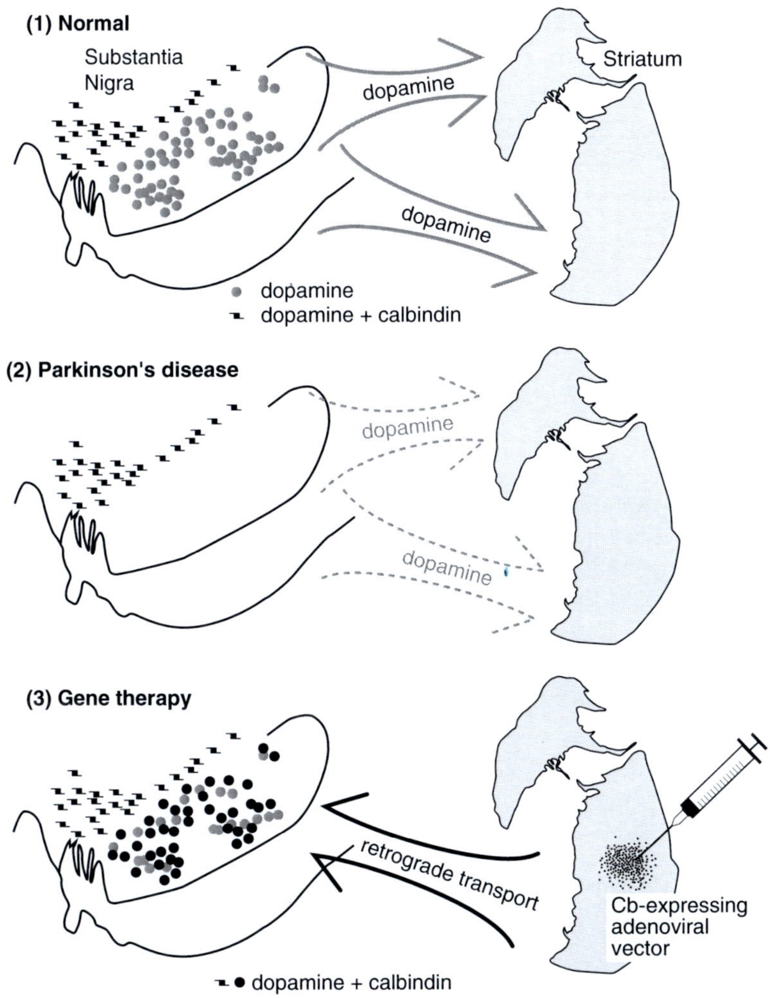

Fig. 1 Schematic diagram showing the present strategy of gene therapy for Parkinson's disease. A recombinant adenoviral vector encoding the calbindin gene is stereotaxically injected into the striatum to be delivered to cell bodies of dopaminergic neurons in the substantia nigra pars compacta (SNc) via retrograde axonal transport. This surgical treatment allows the expression of calbindin in dopaminergic neurons located in the ventral tier of the SNc that do not normally contain it and are vulnerable to parkinsonian insults. Thus, calbindin recruitment into nigral dopamine neurons may protect against the development of Parkinson's disease

chair with their head fixed in a stereotaxic frame attached to the chair. After partial removal of the skull, multiple injections of the vector were made unilaterally into the striatum, usually on the right side (i.e., the caudate nucleus and the putamen). A total of 50–100 μl of the vector preparation ($1.5 \times 1\ 0^{10}$ TU/ml) was injected at rostrocaudally different levels through a 50-μl Hamilton microsyringe.

2.3 MPTP Treatment

MPTP was administrated systemically by repeated intravenous injections through the major saphenous vein. A 0.1% solution of MPTP (hydrochloride salt; Sigma, MI, USA) in physiological saline was injected two to three times a week before the onset of parkinsonian motor signs. For one shot, the dose and volume were 0.5 mg and 0.5 ml per kg body weight, respectively. As parkinsonian symptoms developed, the number of MPTP injections gradually decreased once a week or more infrequently to make the motor signs stable.

2.4 Behavioral Analysis

Parkinsonian motor signs were evaluated according to a modified scale for rating monkey hemiparkinsonism. The modified scale was based on the scale by Smith et al. (1993) and improved for detecting the difference between the calbindin-recruited and the nonrecruited side in single monkeys. The monkeys underwent a battery of tests, which consist of gross motor skill (score 0–3), arm bent (score 0–3), and rigidity (score 0–3); higher scores represent severer impairments. This behavioral analysis was performed every third day before the onset of parkinsonian symptoms and then everyday after the onset.

2.5 Immunohistochemistry

The monkeys were anesthetized deeply with an overdose of sodium pentobarbital (50 mg/kg b.wt., i.v.) after induction with ketamine hydrochloride (10 mg/kg b.wt., i.m.) and perfused transcardially with 0.1 M phosphate buffered saline (PBS; pH 7.3), followed by 10% formalin dissolved in 0.1 M PBS. The brains were saturated with 30% sucrose at 4°C and cut serially into coronal sections 60 μm thick on a freezing microtome. The sections were divided into six groups. Three groups of sections were processed for immunohistochemistry for calbindin, tyrosine hydroxylase (TH), and α-synuclein. These sections were washed three times in 0.05 M PBS (pH 7.3), soaked in 1% skim milk for 1.5 h, and then incubated for 2 days at 4°C with mouse monoclonal anti-calbindin antibody (1:2,000; Swant, Switzerland), mouse monoclonal anti-TH antibody (1:2,000; Chemicon, France), and mouse monoclonal anti-α-synuclein antibody (1:800; BD Bioscience, CA, USA) in PBS containing 2% normal horse serum, 0.5% Triton X-100, and 0.05% NaN_3. Subsequently, the sections were incubated with biotinylated goat anti-mouse IgG antibody (1:200; Vector laboratories, CA, USA) in the same fresh incubation medium for 2 h, followed by ABC Elite (1:100; Vector) in 0.05 M PBS for 2 h. Finally, the sections were reacted for 10–20 min in 0.05 M Tris-HCl buffer (pH 7.6) containing 0.04% diaminobenzine tetrahydrochloride (Dojindo, Kumamoto, Japan), 0.04% $NiCl_2$, and 0.001% H_2O_2. The reaction time was adjusted to make the density of background immunostaining almost identical throughout the cases. After immunostaining, the sections were mounted onto gelatin-coated glass slides. One of the remaining groups of sections was mounted onto gelatin-coated glass slides and Nissl-stained with 1% Cresyl violet.

2.6 Histological Analysis

The density of TH-immunoreactive striatal terminals was measured from ten square areas of 250 μm^2 in the putamen per monkey brain by the aid of a computer-assisted imaging program (BZ Analyzer, Keyence, Tokyo, Japan). The number of nigral neurons immunoreactive for calbindin or α-synuclein was counted in every sixth section (360 μm apart). For cell counts, the SNc was delineated by reference to adjacent TH-immunostained sections.

3 Results

After injections of the calbindin-expressing adenoviral vector into the striatum on one side of the brain, a large number of nigral neurons on the same side exhibited calbindin immunoreactivity. These calbindin-positive neurons were located in both the dorsal and the ventral tiers of the SNc, while most of calbindin-positive neurons on the control side were restricted to the dorsal tier. The number of calbindin-positive neurons on the vector-injected side was about twice as many as those on the control side (Fig. 2c).

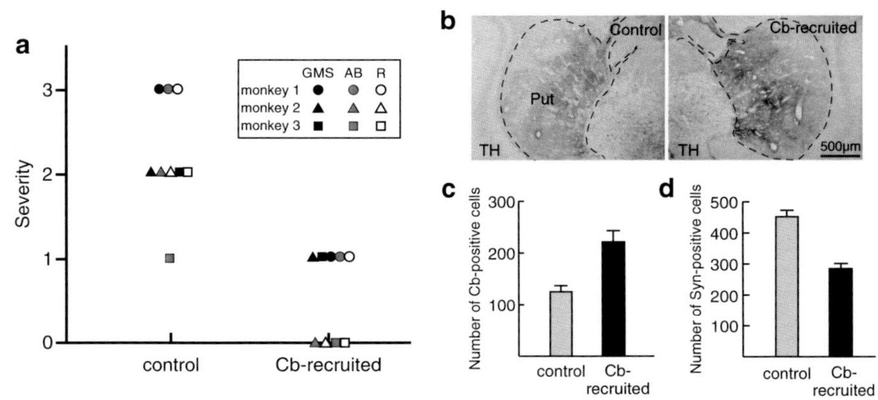

Fig. 2 (**a**) Behavioral analysis of parkinsonian motor signs. Three MPTP-treated monkeys underwent a battery of tests that consist of gross motor skill (GMS; score 0–3), arm bend (AB; score 0–3), and rigidity (R; score 0–3). Higher scores indicate severer impairments. In individual monkeys, the differences in parkinsonian motor signs were evaluated between the calbindin (Cb)-recruited and the nonrecruited (control) side. (**b**) TH immunoreactivity in the putamen on the Cb-recruited and nonrecruited (control) sides. (**c**) Number of Cb-positive cells in the SNc on the Cb-recruited and nonrecruited (control) sides. (**d**) Number of α-synuclein (Syn)-positive cells in the SNc on the Cb-recruited and nonrecruited (control) sides. For (**c**) and (**d**), data were obtained in the same monkeys as for (**a**)

The monkeys that were injected with the calbindin-expressing adenoviral vector did not display any parkinsonian motor signs. After MPTP administration, motor abnormalities were evaluated separately in the limbs on each side and scored in terms of gross motor skill, arm bent, and rigidity. In all of the monkeys examined, behavioral alterations occurred more severely on the side ipsilateral to calbindin recruitment (Fig. 2a). Furthermore, the monkeys preferentially used the arm contralateral to calbindin recruitment while they performed a food retrieval task.

Histological analysis was carried out to investigate the differences in immunoreactivity for TH and α-synuclein between the calbindin-recruited and the nonrecruited side. In the striatum, TH-immunoreactive axon terminals were denser on the calbindin-recruited side than on the nonrecruited side (Fig. 2b). Conversely, nigral neurons displaying intense α-synuclein immunoreactivity were seen much more frequently on the nonrecruited side than on the calbindin-recruited side (Fig. 2d).

4 Discussion

Nigral dopamine neurons that express calbindin selectively survive in human Parkinson patients and primate models of Parkinson's disease. These neurons are located in the dorsal tier of the SNc, whereas the dopaminergic neurons in the ventral tier do not normally contain calbindin. In the present study, we tested the hypothesis that the recruitment of calbindin into nigral dopamine neurons might prevent degeneration following MPTP treatment and the development of parkinsonian motor signs. Our results provide the first evidence that calbindin recruitment indeed protects the dopaminergic neurons against MPTP-induced degeneration. From a behavioral viewpoint, we confirmed that MPTP-treated monkeys exhibited parkinsonian motor signs (such as muscular rigidity and flexed posture) more severely in the limbs ipsilateral to the side injected with the calbindin-expressing adenoviral vector. Moreover, our histological analysis revealed that TH expression in the striatum was preserved better on the calbindin-recruited side. Conversely, many more neurons in the SNc displayed intense immunoreactivity for α-synuclein.

There is a consensus that intracellular Ca^{2+} overload triggers either necrotic or apoptotic cell death (Orrenius et al. 2003). In fact, a number of studies have shown that suppression of Ca^{2+} overload by Ca^{2+} chelators or Ca^{2+} channel blockers can rescue neurons that may otherwise die (Orrenius and Nicotera 1994; Murphy et al. 1996). It has been considered that calcium-binding proteins, including calbindin, parvalbumin, and calretinin, reduce intracellular free calcium (McMhon et al. 1998; Schwaller et al. 2002). Moreover, calbindin expression has been reported to promote survival of hippocampal neurons and some motoneurons (Mattson et al. 1991; Ho et al. 1996; Gary et al. 2000). Since overexcitation of nigral dopamine neurons has been implicated in the pathophysiology of Parkinson's disease, an increase in intracellular Ca^{2+} concentration driven by enhanced glutamatergic input may cause their degeneration (Coyle and Puttfarcken 1993; Piallat et al. 1996, 1999; Takada et al. 2000). Recently, it has been demonstrated that the unusual reliance on the voltage-gated L-type calcium channel, $Ca_v1.3$, for maintenance of rhythmic pacemaking activity of nigral dopamine neurons renders them vulnerable to stressors related to the onset of Parkinson's disease (Chan et al. 2007). The reliance on $Ca_v1.3$ increases with age, resulting in sustained elevations of cytosolic Ca^{2+} concentration and activity-dependent Ca^{2+} influx (Wilson and Callaway 2000). Many previous studies have indicated that the intracellular Ca^{2+} concentration is deeply involved in a series of events leading to cell death, for instance, mitochondrial stress and reactive oxygen species production (Krieger and Duchen 2002; Dauer and Przedborski 2003; Verkhratsky 2005). These suggest that intracellular Ca^{2+} level and activity-dependent Ca^{2+} influx are major determinants of degeneration in the dopaminergic neurons. Thus, there may be a possibility of protection against the progress in Parkinson's disease by introducing a mobile calcium-binding protein, such as calbindin, although no precise mechanism underlying a protective action of these proteins has so far been elucidated. Likewise, it has been reported that dopaminergic neurons containing calretinin survive in the brains of parkinsonian patients (Mouatt-Prigent et al. 1994).

Emphasis has been placed on the potential significance of gene therapy for Parkinson's disease. Several reports are so far available on therapeutic approaches with neurotrophic factors such as BDNF, enzymes related to dopamine synthesis such as TH and aromatic acid decarboxylase, and other key substances such as glutamate decarboxylase (Gash et al. 1996; During et al. 1998, 2001; Muramatsu et al. 2002; Fjord-Larsen et al. 2005). Intrastriatal transplantation of stem cells that differentiate into the dopaminergic neuron phenotype has been shown to improve behavioral aspects in parkinsonian rodents and monkeys induced by 6-hydroxydopamine and MPTP, respectively (Björklund et al. 2002; Takagi et al. 2005). The present results suggest that calbindin may be considered a novel potential target for gene therapy for Parkinson's disease by potentially protecting nigral dopamine neurons against degeneration.

References

Björklund LM, Sanchez-Pernaute R, Chung S, Andersson T, Chen IY, McNaught KS, Brownell AL, Jenkins BG, Wahlestedt C, Kim KS and Isacson O (2002) Embryonic stem cells develop into functional dopaminergic neurons after transplantation in a Parkinson rat model. Proc Natl Acad Sci USA 99: 2344–2349.

Chan CS, Guzman JN, Ilijic E, Mercer JN, Rick J, Tkatch T, Meredith GE and Surmeier DJ (2007) 'Rejuvention' protects neurons in mouse models of Parkinson's disease. Nature 447: 1081–1086.

Coyle JT and Puttfarcken P (1993) Oxidative stress, glutamate, and neurodegenerative disorders. Science 262: 689–695.

Damier P, Hirsch EC, Agid Y and Graybiel AM (1999) The substantia nigra of the human brain. I. Nigrosomes and the nigral matrix, a compartmental organization based on calbindin D(28K) immunohistochemistry. Brain 122: 1421–1436.

Dauer W and Przedborski S (2003) Parkinson's disease: mechanisms and models. Neuron 39: 889–909.

During MJ, Samulski RJ, Elsworth JD, Kaplitt MG, Leone P, Xiao X, Li J, Freese A, Taylor JR, Roth RH, Sladek JR Jr, O'Malley KL and Redmond DE Jr. (1998) In vivo expression of therapeutic human genes for dopamine production in the caudates of MPTP-treated monkeys using an AAV vector. Gene Ther 5: 820–827.

During MJ, Kaplitt MG, Stern MB and Eidelberg D (2001) Subthalamic GAD gene transfer in Parkinson disease patients who are candidates for deep brain stimulation. Hum Gene Ther 12: 1589–1591.

Fjord-Larsen L, Johansen JL, Kusk P, Tornoe J, Gronborg M, Rosenblad C and Wahlberg LU (2005) Efficient in vivo protection of nigral dopaminergic neurons by lentiviral gene transfer of a modified neurturin construct. Exp Neurol 195: 49–60.

Gary DS, Sooy K, Chan SL, Christakos S and Mattson MP (2000) Concentration- and cell type-specific effects of calbindin D28k on vulnerability of hippocampal neurons to seizure-induced injury. Mol Brain Res 75: 89–95.

Gash DM, Zhang Z, Ovadia A, Cass WA, Yi A, Simmerman L, Russell D, Martin D, Lapchak PA, Collins F, Hoffer BJ and Gerhardt GA (1996) Functional recovery in parkinsonian monkeys treated with GDNF. Nature 380: 252–255.

German DC, Manaye KF, Sonsalla PK and Brooks BA (1992) Midbrain dopaminergic cell loss in Parkinson's disease and MPTP-induced parkinsonism: sparing of calbindin-D28k-containing cells. Ann NY Acad Sci 648: 42–62.

Gibb WR (1992) Melanin, tyrosine hydroxylase, calbindin and substance P in the human midbrain and substantia nigra in relation to nigrostriatal projections and differential neuronal susceptibility in Parkinson's disease. Brain Res 581: 283–291.

Ho BK, Alexianu ME, Colom LV, Mohamed AH, Serrano F and Appel SH (1996) Expression of calbindin-D28K in motoneuron hybrid cells after retroviral infection with calbindin-D28K cDNA prevents amyotrophic lateral sclerosis IgG-mediated cytotoxicity. Proc Natl Acad Sci USA 93: 6796–6801.

Krieger C and Duchen MR (2002) Mitochondria, Ca21 and neurodegenerative disease. Eur J Pharmacol 447: 177–188.

Lavoie B and Parent A (1991) Dopaminergic neurons expressing calbindin in normal and parkinsonian monkeys. Neuroreport 2: 601–604.

Mattson MP, Rychlik B, Chu C and Christakos S (1991) Evidence for calcium-reducing and excito-protective roles for the calcium-binding protein calbindin-D28k in cultured hippocampal neurons. Neuron 6: 41–51.

McMahon A, Wong BS, Iacopino AM, Ng MC, Chi S and German DC (1998) Calbindin-D28k buffers intracellular calcium and promotes resistance to degeneration in PC12 cells. Mol Brain Res 54: 56–63.

Mouatt-Prigent A, Agid Y and Hirsch EC (1994) Does the calcium binding protein calretinin protect dopaminergic neurons against degeneration in Parkinson's disease? Brain Res 668: 62–70.

Muramatsu S, Fujimoto K, Ikeguchi K, Shizuma N, Kawasaki K, Ono F, Shen Y, Wang L, Mizukami H, Kume A, Matsumura M, Nagatsu I, Urano F, Ichinose H, Nagatsu T, Terao K, Nakano I and Ozawa K (2002) Behavioral recovery in a primate model of Parkinson's disease by triple transduction of striatal cells with adeno-associated viral vectors expressing dopamine-synthesizing enzymes. Hum Gene Ther 13: 345–354.

Murphy AN, Bredesen DE, Cortopassi G, Wang E and Fiskum G (1996) Bcl-2 potentiates the maximal calcium uptake capacity of neural cell mitochondria. Proc Natl Acad Sci USA 93: 9893–9898.

Orrenius S and Nicotera P (1994) The calcium ion and cell death. J Neural Transm Suppl 43: 1–11.

Orrenius S, Zhivotovsky B and Nicotera P (2003) Regulation of cell death: the calcium-apoptosis link. Nat Rev Mol Cell Biol 4: 552–565.

Parent A and Lavoie B (1993) The heterogeneity of the mesostriatal dopaminergic system as revealed in normal and parkinsonian monkeys. Adv Neurol 60: 25–24.

Piallat B, Benazzouz A and Benabid AL (1996) Subthalamic nucleus lesion in rats prevents dopaminergic nigral neuron degeneration after striatal 6-OHDA injection: behavioural and immunohistochemical studies. Eur J Neurosci 8: 1408–1414.

Piallat B, Benazzouz A and Benabid AL (1999) Neuroprotective effect of chronic inactivation of the subthalamic nucleus in a rat model of Parkinson's disease. J Neural Transm Suppl 55: 71–77.

Schwaller B, Meyer M and Schiffmann S (2002) 'New' functions for 'old' proteins: the role of the calcium-binding proteins calbindin D-28k, calretinin and parvalbumin, in cerebellar physiology. Studies with knockout mice. Cerebellum 1: 241–258.

Smith RD, Zhang Z, Kurlan R, McDermott M and Gash DM (1993) Developing a stable bilateral model of parkinsonism in rhesus monkeys. Neuroscience 52: 7–16.

Takada M, Matsumura M, Kojima J, Yamaji Y, Inase M, Tokuno H, Nambu A and Imai H (2000) Protection against dopaminergic nigrostriatal cell death by excitatory input ablation. Eur J Neurosci 12: 1771–1780.

Takagi Y, Takahashi J, Saiki H, Morizane A, Hayashi T, Kishi Y, Fukuda H, Okamoto Y, Koyanagi M, Ideguchi M, Hayashi H, Imazato T, Kawasaki H, Suemori H, Omachi S, Iida H, Itoh N, Nakatsuji N, Sasai Y and Hashimoto N (2005) Dopaminergic neurons generated from monkey embryonic stem cells function in a Parkinson primate model. J Clin Invest 115: 102–109.

Verkhratsky A (2005) Physiology and pathophysiology of the calcium store in the endoplasmic reticulum of neurons. Physiol Rev 85: 201–279.

Wilson CJ and Callaway JC (2000) Coupled oscillator model of the dopaminergic neuron of the substantia nigra. J Neurophysiol 83: 3084–3100.

Yamada T, McGeer PL, Baimbridge KG and McGeer EG (1990) Relative sparing in Parkinson's disease of substantia nigra dopamine neurons containing calbindin-D28K. Brain Res 526: 303–307.

Changes in the Subcellular Localization and Functions of GABA-B Receptors in the Globus Pallidus of MPTP-Treated Monkeys

Adriana Galvan, Bijli Nanda, Xing Hu, Yoland Smith, and Thomas Wichmann

Abstract We have analyzed whether the subcellular location of GABA-B receptors and the electrophysiologic effects of local administration of GABA-B receptor ligands in the external and internal segments of the globus pallidus (GPe and GPi, respectively) in monkeys change with the induction of parkinsonism. Rhesus monkeys were rendered parkinsonian by treatment with MPTP. Electron microscopy immunoperoxidase for GABA-B R1 was performed in GPe and GPi. Compared with normal monkeys, the overall density of GABA-B R1-immunoreactive dendrites remained the same, but the density of immunolabeled axons increased significantly in both pallidal segments of MPTP-treated monkeys. Single unit recordings in awake monkeys showed that blockade of GABA-B receptors did not consistently affect the activity of GPe or GPi. However, in all neurons tested, application of a GABA-B agonist inhibited the firing rate. These results suggest that dopamine depletion induces an increased presynaptic expression of GABA-B receptors in both pallidal segments, perhaps facilitating a more prominent role of GABA-B-mediated presynaptic regulation of neurotransmitter release in the pallidum.

1 Introduction

The selective loss of dopaminergic neurons in the substantia nigra pars compacta (SNc) in Parkinson's disease (PD) leads to complex changes in the activity within the basal ganglia–thalamocortical loops. According to current circuit models of

---

A. Galvan (✉), X. Hu, Y. Smith, and T. Wichmann
Yerkes Research Center, 954 Gatewood Road NE, Atlanta, GA 30329, USA
e-mail: agalvan@emory.edu

B. Nanda
Hindustan Institute of Medical Sciences and Research
Knowledge Park-III, Greater Noida, Up-201306

H.J. Groenewegen et al. (eds.), *The Basal Ganglia IX*, Advances in Behavioral Biology 58,
DOI 10.1007/978-1-4419-0340-2_30, © Springer Science+Business Media, LLC 2009

basal ganglia dysfunction in parkinsonism, decreased dopamine input to the striatum from the SNc leads to increased GABAergic inhibition of neurons in the external segment of the globus pallidus (GPe) and decreased GABAergic inhibition of the internal segment of the globus pallidus (GPi) and substantia nigra pars reticulata (SNr), the output nuclei of the basal ganglia (Wichmann and DeLong 2003a).

The effects of GABA are mediated by ionotropic GABA-A receptors and metabotropic GABA-B receptors. In this report we focus on the distribution and functions of GABA-B receptors in parkinsonian monkeys. GABA-B receptor subtypes are formed by a GABA-B1 subunit (GABA-B1a or GABA-B1b), which combines with a GABA-B2 subunit to form heteromeric receptors (Ulrich and Bettler 2007). Localization studies in rats and monkeys have shown that GABA-B receptors are expressed widely in the basal ganglia (Bischoff et al. 1999; Fritschy et al. 1999; Lu et al. 1999; Margeta-Mitrovic et al. 1999; Yung et al. 1999; Charara et al. 2000; Liang et al. 2000; Ng and Yung, 2001a,b; Boyes and Bolam, 2003; Galvan et al. 2004; Lacey et al. 2005). In GPe and GPi, postsynaptic GABA-B receptors are found in abundance in dendrites and also presynaptically in GABAergic and glutamatergic axons and terminals (Charara et al. 2000, 2004; Chen et al. 2004). In accordance with these anatomical studies, pre and postsynaptic GABA-B-mediated functions have been identified in GP. Presynaptic GABA-B receptors modulate the release of glutamate and GABA (Singh 1990; Chen et al. 2002; Kaneda and Kita 2005), while postsynaptic GABA-B receptors induce slow inhibitory postsynaptic potentials (IPSPs, Kaneda and Kita 2005). In in vivo experiments in awake monkeys, pharmacological activation of GABA-B receptors inhibits the activity of GPe and GPi neurons (Galvan et al. 2005). Exposure to GABA-B receptor antagonists produces a tendency to increase firing rates in these nuclei, suggesting that these receptors are tonically activated (Galvan et al. 2005; Kita et al. 2006). It has been suggested that the GABA-B responses in GPe and GPi cells are mediated by GABA release from pallidal but not striatal afferents (Kita et al. 2006).

Dopamine depletion caused by degeneration of the SNc alters the expression of several neurotransmitter receptors in the basal ganglia, including GABA-B receptors. In patients with Parkinson's disease (PD) and in animal models of the disease, the levels of protein or mRNA for GABA-B receptors are downregulated in GPe and upregulated in GPi (Calon et al. 2000, 2003; Johnston and Duty 2003). It remains unknown, however, if the subcellular and subsynaptic distribution of GABA-B receptors also undergoes changes with the induction of dopamine depletion. Such changes may profoundly affect GABAergic transmission. To further characterize the anatomical and functional plasticity of GABA-B receptors in parkinsonian condition, this study compares the subcellular localization of GABA-B R1 immunoreactivity in GPe and GPi between normal and MPTP (1-methyl-4-phenyl-1,2,3,6-tetrahydropyridine)-treated monkeys and explores the effects of GABA-B receptor compounds on neuronal activity in GPe and GPi of parkinsonian monkeys.

2 Materials and Methods

2.1 Animals and MPTP Treatment

Five Rhesus monkeys were used: three for the immunolocalization studies and two for the electrophysiological experiments. All experiments were carried out in accordance with the National Institutes of Health's Guide for the Care and Use of Laboratory Animals (Anonymous 1996) and the United States Public Health Service policy on the humane care and use of laboratory animals (amended 2002) and were approved by the Institutional Animal Care and Use Committee at Emory University. All animals (except one control monkey) were rendered parkinsonian with intracarotid and systemic injections of MPTP (total dose 7–10 mg/kg). Parkinsonian signs were stable for at least 6 weeks before the start of the immunohistochemical or electrophysiological experiments.

2.2 Immunohistochemical Localization of GABA-B Receptors

The monkeys were transcardially perfused with 4% paraformaldehyde and 0.1% glutaraldehyde in phosphate buffer (PB; 0.1 M, pH 7.4). The brains were cut into 10-mm-thick blocks in the frontal plane, and tissue sections (60 μm) were obtained with a vibratome, collected in cold phosphate-buffered saline (PBS; 0.01 M, pH 7.4), and treated with sodium borohydride (1% in PBS). The loss of dopaminergic fibers was confirmed with tyrosine hydroxylase immunolabeling (Chemicon International, Temecula, CA), using the protocol described later, except that 0.1% Triton X-100 was added to all immunoreagents.

To visualize GABA-B R1 immunolabeling in GPe and GPi of one control and two MPTP-treated monkeys, the tissue was processed as follows. The sections were cryoprotected, frozen at −80°C, thawed, and washed in PBS. After blocking of nonspecific sites with 10% normal goat serum (NGS) and 1% bovine serum albumin (BSA), the sections were incubated for 2 days at 4°C in the primary antibody solution (1:500 dilution; Chemicon, AB 1531). The sections were then incubated in biotinylated goat anti-guinea pig IgG (1:200; Vector labs, Burlingame, CA), followed by incubation in avidin-biotin complex (ABC) solution (1:100; Vectastain Standard kit, Vector labs). After rinsing in PBS and Tris buffer (0.05 M, pH 7.6), the sections were exposed to a solution containing 0.025% 3–3′-diaminobenzidine tetrahydrochloride (Sigma Chemicals, St Louis, MI), 0.01 M imidazole (Fisher Scientific, Norcross, GA), and 0.006% H_2O_2. The reaction was stopped by repeated washes in PBS and then in PB (0.1 M, pH 7.4). To prepare the sections for electron microscope (EM) observations, they were postfixed in osmium tetroxide (1% in PB) for 20 min and dehydrated in a graded series of ethanol and propylene oxide, adding uranyl acetate (1%) to 70% ethanol for 35 min. The sections were embedded in resin (Durcupan ACM, Fluka, Ft. Washington, PA) and placed in the oven for 48 h at 60°C. Blocks of GPe and GPi were cut and glued on the top of resin blocks. Serial ultrathin sections were obtained on an ultramicrotome (Leica Ultracut T2), collected onto Pioloform-coated single copper grids, stained with lead citrate, and analyzed with the electron microscope (Zeiss EM 10C).

The ultrastructural analysis was carried out on ultrathin sections collected from the surface of each block where the staining was optimal. The sections were scanned at 16,000–25,000×, and randomly selected fields containing immunoreactive elements were photographed. The labeled elements were identified based on standard ultrastructural features (Peters et al. 1991), and their density was expressed as number of elements per total tissue area analyzed.

2.3 *Local Administration of GABA-B Compounds and Extracellular Recording of Pallidal Units*

2.3.1 Surgery

Surgeries were performed using aseptic conditions under isoflurane anesthesia to position stainless steel recording chambers on the skull of the two monkeys used in these electrophysiological studies. The chambers were stereotactically aimed at the globus pallidus at an angle of 40° in the coronal plane. The recording chambers, together with standard metal head holders (Crist Instruments, Hagerstown, MD), were embedded into a dental acrylic "cap," which was affixed to the animal's skull. The recording and injection experiments began 1 week after surgery. The animals were awake during the recording sessions.

2.3.2 Recording and Injection Sessions

A combined recording-injection device was used for all experiments. This device permits local injections of small quantities (<1 μl) of drug solutions and extracellular electrophysiologic recording in the immediate vicinity (within 100 μm) of the injection site. As previously described (Kliem and Wichmann 2004), the device consists of a tungsten microelectrode (Frederick Haer Co., Bowdoinham, ME; Z = 0.7–2.5 MΩ) and a fused silica tube (Polymicro Technologies, Phoenix, AZ), which are both protected by a polyimide sleeve (O.D. = 0.5 mm; MicroLumen, Tampa, FL). The tip of the electrode extends 50–100 μm beyond the silica tubing. The silica tubing is attached to a gas-tight 1-ml syringe, operated by a pump for delivery of drugs or artificial cerebrospinal fluid [(aCSF) comprising (in mM concentrations) 143 NaCl, 2.8 KCl, 1.2 $CaCl_2$, 1.2 $MgCl_2$, 1 Na_2HPO_4, pH 7.2–7.4] via a micro-T connector (CMA, Solna, Sweden). A microdrive (MO-95B, Narishige; Tokyo, Japan) was used to lower the system into the brain.

We carried out initial electrophysiological mapping with standard electrophysiologic methods, using tungsten electrodes to identify GPe and GPi. For all recording sessions, the neuronal activity was amplified (DAM-80 pre-amplifier; WPI, Sarasota, FL and MDA-2 amplifier, BAK Mount Airy, MD), filtered (0.4–6.0 kHz, Krohn-Hite, Brockton, MA), displayed on an oscilloscope (DL1540; Yokogawa, Tokyo, Japan), and audioamplified. GPe and GPi were identified by their characteristic high-frequency discharge (interspersed with pauses, in the case of GPe) and their spatial relationship to surrounding nuclei. The electrophysiologic mapping information was used to place the recording-injection system in the GPe/GPi to record single neuronal activity. Neurons were recorded throughout a control period of at least 1 min prior to drug infusion, followed by the infusion period, and a postinfusion period lasting at least 10 min. The recorded activity was stored to a computer disk with a data acquisition system (Power 1401 with Spike2 interface, CED, Cambridge, UK) for off-line analysis.

2.3.3 Drugs

The drugs used in this study were the GABA-B receptor agonist baclofen (0.213 mg/ml; Sigma-Aldrich) and the GABA-B receptor antagonist CGP-55845 (0.402 mg/ml; Tocris). All solutions were filtered using a 0.2-μm pore size filter (Fisher Scientific, Hampton, NH) and infused at a rate of 0.3 μl/min for a total volume of 1 μl.

2.3.4 Data Analysis

Data collected during the preinfusion control period were compared to the "effect" period (considered from the beginning of the injection to 2 min after the end of the injection). Template-matching spike sorting (Spike 2, CED) was used to isolate action potential waveforms for individual units and to compute interspike intervals (ISIs). We then used the ISI information to calculate median discharge rates and construct ISI distribution histograms for each neuron, separately for the preinjection and effect periods, using algorithms in the Matlab programming environment (Mathworks, Natick, MA). The neuron's discharge rate (binned in 20-s intervals) during the "effect" period was considered different from baseline if at least three consecutive bins were outside of the range defined by the 10th or 90th percentile of baseline firing.

3 Results

In agreement with our previous data (Charara et al. 2000, 2004), GABA-B receptors were expressed mainly in dendrites and less frequently in unmyelinated axons in both GPe and GPi of untreated monkeys (Fig. 1a, c). In the MPTP-treated

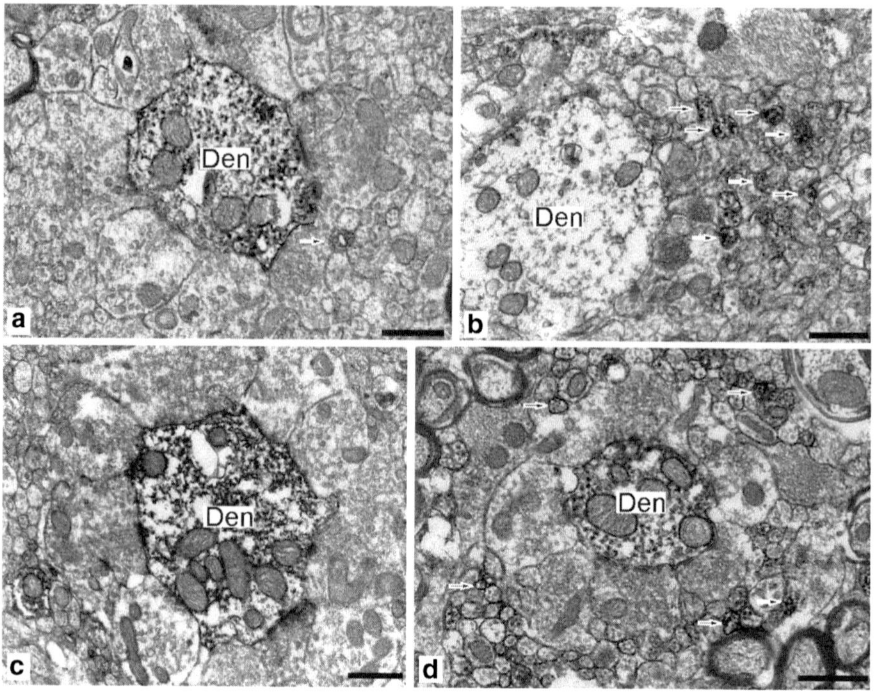

Fig. 1 GABA-B receptors are observed in small-diameter unmyelinated axons (*arrows*), and dendrites (Den) are observed in the GPe (**a**, **b**) and GPi (**c**, **d**) of a control (**a**, **c**) and a MPTP-treated monkey (**b**, **d**). Note the increased density of GABA-B-R1-receptor-positive axons in the GPe and GPi of MPTP-treated monkeys. Scale bars represent 0.5 μm in all cases

monkeys the overall density of GABA-B R1-immunoreactive dendrites was similar to that seen in the normal monkey, but there was a marked increase in the number of GABA-B R1 positive unmyelinated axons (Fig. 1b, d) in both GP segments. The density of dendrites in normal monkeys was 0.06 and 0.05 elements/μm^2 in the GPe and GPi, respectively. In MPTP-treated monkeys, there were 0.04 dendrites/μm^2, both in GPe and GPi. In the normal monkey, we found a density of 0.14 and 0.04 unmyelinated axons/μm^2 (GPe and GPi, respectively). In contrast, in the MPTP-treated monkeys, there were 0.27 unmyelinated axons/μm^2 in GPe and 0.4 unmyelinated axons/μm^2 in GPi.

Local injections of specific GABA-B agonists and antagonists were performed in GPe and GPi of MPTP-treated monkeys. In GPe, the effects of the GABA-B receptor antagonist CGP-55845 were variable: 3 of 13 cells showed increased firing, 6 showed a decreased activity, and 4 did not show any significant effect in response to drug application. As a group, GPe cells showed a median firing rate that was 28% below the baseline after CGP-55845 application (Fig. 2).

Injection of the GABA-B agonist baclofen into GPe induced a complete silencing in three cells tested (100% reduction in firing from baseline, Fig. 2). This effect was

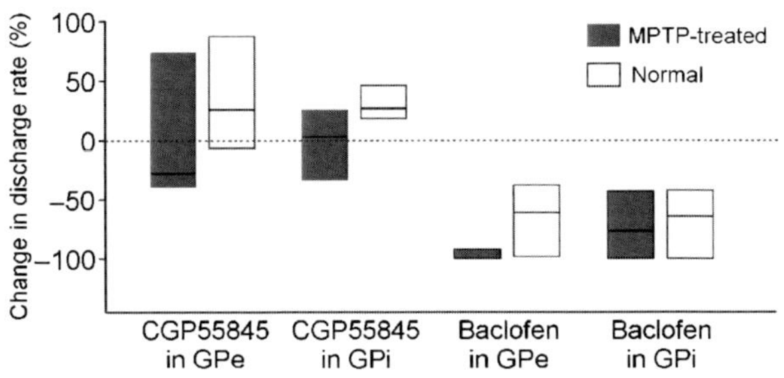

Fig. 2 Summary of changes in discharge rates of GPe and GPi neurons between MPTP-treated and normal monkeys after infusions of GABA-B receptor ligands. For each cell, changes in discharge rates were expressed as a percent change from baseline for each experiment. Medians of these percentages are reported (*line inside the boxes*), along with the 25th/75th percentiles (*top and bottom of boxes*). Data for normal monkeys are from Galvan et al. (2005)

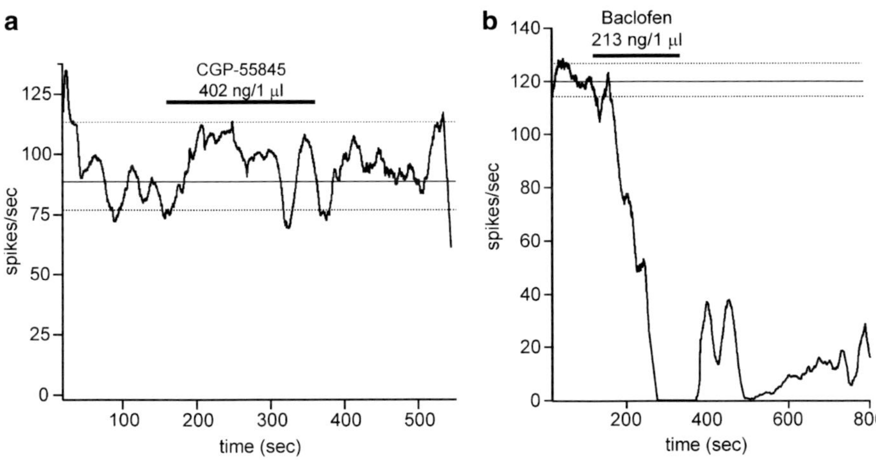

Fig. 3 Responses of GPi neurons to infusions of the GABA-B receptor antagonist CGP-55845 (**a**) and the GABA-B receptor agonist baclofen (**b**). *Thin solid lines* represent the median discharge rate during the baseline period, while *dashed lines* indicate 10th and 90th percentiles. The *thick solid line* represents the time and duration of the drug infusion (0.3 μl/min)

long-lasting; in one case the firing rate remained profoundly decreased for more than 60 min after the end of baclofen administration.

Similar injections were done in GPi of MPTP-treated monkeys. Among ten GPi neurons four responded to application of CGP-55845 with an increase in firing, three with a decrease in firing, and three showed no effect (e.g. in Fig. 3a). As a group, the median firing rate of GPi neurons was 3% above the basal firing rate after CGP-55845 administration (Fig. 2).

As shown in GPe, baclofen injections into GPi evoked a decrease in the firing rate of 5 out of 6 neurons (e.g. in Fig. 3b). The median firing rate of GPi neurons after baclofen administration was 64% below baseline.

In Fig. 2, the results obtained in MPTP-treated monkeys were compared against our data from an earlier study done in normal monkeys (Galvan et al. 2005). In normal monkeys, CGP-55845 induced an overall 25% and 26% increase in the firing rate of GPe (n = 5) and GPi (n = 5) neurons, respectively. Injections of baclofen, on the other hand, induced a significant decrease in firing of pallidal neurons in normal monkeys (62% in GPe, n = 7; 64% in GPi, n = 3).

4 Summary and Conclusions

In both pallidal segments of the MPTP-treated monkey, the density of GABA-B R1-labeled unmyelinated axons is markedly increased in comparison to normal monkeys. In GPe and GPi of MPTP-treated monkeys, activation of GABA-B receptors induced effects similar to those observed in the normal animal (Galvan et al. 2005), that is, a reduction of the firing rate. In contrast to the findings in the normal state, where GABA-B receptor blockade tends to increase the firing in GPe or GPi (Galvan et al. 2005; Kita et al. 2006), the effects of this intervention were more variable in the MPTP-treated animals, with a greater proportion of cells responding with decreases in discharge than in the normal state.

4.1 Effects of Dopaminergic Depletion on GABA-B Receptor Expression in the GP

In PD and dopamine-depleted animal models of parkinsonism, the loss of dopamine induces a cascade of alterations in the basal ganglia structures, including changes in the expression of different neurotransmitters. GABA-B receptor levels are decreased in GPe, but increased in GPi (Calon et al. 2000, 2003). In the circuit model of PD, it is proposed that striato-GPe afferents may be overactive, while the striato-GPi projection decreases its activity (Wichmann and DeLong 2003a, b). Thus, the reported changes in overall GABA-B receptor density are likely compensatory in nature counteracting the altered GABAergic transmission. Our study provides additional and complementary information showing changes in the subcellular localization of these receptors after dopaminergic depletion.

The increased density of GABA-B receptors in unmyelinated axons suggests an increased GABA-B-mediated presynaptic regulation of neurotransmitter release in the parkinsonian state than in the controls. The available data are not sufficient to determine the chemical phenotype and source(s) of the GABA-B receptor-positive unmyelinated axons. However, the fact that the majority of synaptic inputs to GPe cells originate from GABAergic striatal or intrapallidal axons (Smith et al. 1998) strongly suggests that a large proportion of these GABA-B R1-containing axons are, in fact, GABAergic.

4.2 Pharmacological Activation and Blockade of GABA-B Receptors in GPe and GPi of Parkinsonian Monkeys

The inhibitory effects of exogenous activation of pallidal GABA-B receptors with baclofen in MPTP-treated monkeys were not different from those previously recorded in controls (Galvan et al. 2005), suggesting that dopaminergic depletion does not affect the plasma membrane availability or pharmacological features of GABA-B receptors in GPe and GPi.

Local application of the GABA-B receptor antagonist CGP-55845 had variable effects on neuronal activity in GPe and GPi of MPTP-treated monkeys, while the majority of neurons showed a significant decrease in firing, a few cells displayed a dramatic increased activity, and others showed no response to the drug. As a group, GPe cells decreased their firing in response to CGP-55845, whereas GPi cells were not significantly affected. The observations from GPe neurons contrast with the results obtained in normal monkeys, in which either no effect or increased firing was recorded in response to CGP-55845 application (Galvan et al. 2005; Kita et al. 2006). At first glance, a decrease in firing after blockade of GABAergic receptors seems paradoxical. One possible explanation is that the antagonist application abolishes tonic activation of presynaptic GABA-B autoreceptors located in GABAergic terminals, thereby facilitating GABA release from these terminals, which would result in a decreased pallidal activity. The fact that our electron microscopic studies showed a significant increase in presynaptic GABA-B receptors in the pallidum of parkinsonian monkeys is consistent with these functional data. Our results with baclofen, however, are difficult to reconcile with those obtained after blockade of GABA-B receptors with CGP-55845.

In our experiments we cannot discern between pre- and postsynaptic GABA-B-mediated effects. It is likely that in the GP pre- and postsynaptic GABA-B receptors differ in subunit composition and pharmacology, as has been shown in other brain areas (Ong et al. 2001; Ulrich and Bettler 2007). The development of highly selective pharmacological agents for pre- or postsynaptic GABA-B receptors would be an essential tool to further analyze the functions of these receptors.

References

Anonymous (1996) Guide for the Care and Use of Laboratory Animals. Washington, DC: National Academy Press.

Bischoff S, Leonhard S, Reymann N, Schuler V, Shigemoto R, Kaupmann K and Bettler B (1999) Spatial distribution of GABA(B)R1 receptor mRNA and binding sites in the rat brain. J Comp Neurol 412: 1–16.

Boyes J and Bolam JP (2003) The subcellular localization of GABA(B) receptor subunits in the rat substantia nigra. Eur J Neurosci 18: 3279–3293.

Calon F, Morissette M, Goulet M, Grondin R, Blanchet PJ, Bedard PJ and Di Paolo T (2000) 125I-CGP 64213 binding to GABA(B) receptors in the brain of monkeys: effect of MPTP and dopaminomimetic treatments. Exp Neurol 163: 191–199.

Calon F, Morissette M, Rajput AH, Hornykiewicz O, Bedard PJ and Di Paolo T (2003) Changes of GABA receptors and dopamine turnover in the postmortem brains of parkinsonians with levodopa-induced motor complications. Mov Disord 18: 241–253.

Charara A, Heilman TC, Levey AI and Smith Y (2000) Pre- and postsynaptic localization of GABA(B) receptors in the basal ganglia in monkeys. Neuroscience 95: 127–140.

Charara A, Galvan A, Kuwajima M, Hall RA and Smith Y (2004) An electron microscope immunocytochemical study of GABAB R2 receptors in the monkey basal ganglia: a comparative analysis with GABAB R1 receptor distribution. J Comp Neurol 476: 65–79.

Chen L, Chan SC and Yung WH (2002) Rotational behavior and electrophysiological effects induced by GABA(B) receptor activation in rat globus pallidus. Neuroscience 114: 417–425.

Chen L, Boyes J, Yung WH and Bolam JP (2004) Subcellular localization of GABAB receptor subunits in rat globus pallidus. J Comp Neurol 474: 340–352.

Fritschy JM, Meskenaite V, Weinmann O, Honer M, Benke D and Mohler H (1999) GABAB-receptor splice variants GB1a and GB1b in rat brain: developmental regulation, cellular distribution and extrasynaptic localization. Eur J Neurosci 11: 761–768.

Galvan A, Charara A, Pare JF, Levey AI and Smith Y (2004) Differential subcellular and subsynaptic distribution of GABA(A) and GABA(B) receptors in the monkey subthalamic nucleus. Neuroscience 127: 709–721.

Galvan A, Villalba RM, West SM, Maidment NT, Ackerson LC, Smith Y and Wichmann T (2005) GABAergic modulation of the activity of globus pallidus neurons in primates: in vivo analysis of the functions of GABA receptors and GABA transporters. J Neurophysiol 94: 990–1000.

Johnston T and Duty S (2003) Changes in GABA(B) receptor mRNA expression in the rodent basal ganglia and thalamus following lesion of the nigrostriatal pathway. Neuroscience 120: 1027–1035.

Kaneda K and Kita H (2005) Synaptically released GABA activates both pre- and postsynaptic GABAB receptors in the rat globus pallidus. J Neurophysiol 94: 1104–1114.

Kita H, Chiken S, Tachibana Y and Nambu A (2006) Origins of GABA(A) and GABA(B) receptor-mediated responses of globus pallidus induced after stimulation of the putamen in the monkey. J Neurosci 26: 6554–6562.

Kliem MA and Wichmann T (2004) A method to record changes in local neuronal discharge in response to infusion of small drug quantities in awake monkeys. J Neurosci Methods 138: 45–49.

Lacey CJ, Boyes J, Gerlach O, Chen L, Magill PJ and Bolam JP (2005) GABA(B) receptors at glutamatergic synapses in the rat striatum. Neuroscience 136: 1083–1095.

Liang F, Hatanaka Y, Saito H, Yamamori T and Hashikawa T (2000) Differential expression of γ-aminobutyric acid type B receptor-1a and -1b mRNA variants in GABA and non-GABAergic neurons of the rat brain. J Comp Neurol 416: 475–495.

Lu XY, Ghasemzadeh MB and Kalivas PW (1999) Regional distribution and cellular localization of γ-aminobutyric acid subtype 1 receptor mRNA in the rat brain. J Comp Neurol 407: 166–182.

Margeta-Mitrovic M, Mitrovic I, Riley RC, Jan LY and Basbaum AI (1999) Immunohistochemical localization of GABA(B) receptors in the rat central nervous system. J Comp Neurol 405: 299–321.

Ng TK and Yung KK (2001a) Subpopulations of neurons in rat substantia nigra display GABA(B) R2 receptor immunoreactivity. Brain Res 920: 210–216.

Ng TK and Yung KK (2001b) Differential expression of GABA(B)R1 and GABA(B)R2 receptor immunoreactivity in neurochemically identified neurons of the rat neostriatum. J Comp Neurol 433: 458–470.

Ong J, Bexis S, Marino V, Parker DAS, Kerr DIB and Froestl W (2001) CGP 36216 is a selective antagonist at GABAB presynaptic receptors in rat brain. Eur J Pharmacol 415: 191–195.

Peters A, Palay S and Webster HD (1991) The Fine Structure of the Nervous System, Third Edition. New York: Oxford University Press.
Singh R (1990) GABA-B receptors modulate glutamate release in the rat caudate and globus pallidus. Soc Neurosci Abstr 16: 429–428.
Smith Y, Shink E and Sidibe M (1998) Neuronal circuitry and synaptic connectivity of the basal ganglia. Neurosurg Clin North Am 9: 203–222.
Ulrich D and Bettler B (2007) GABAB receptors: synaptic functions and mechanisms of diversity. Curr Opin Neurobiol 17: 298–303.
Wichmann T and DeLong MR (2003a) Pathophysiology of Parkinson's disease: the MPTP primate model of the human disorder. Ann NY Acad Sci 991: 199–213.
Wichmann T and DeLong MR (2003b) Functional neuroanatomy of the basal ganglia in Parkinson's disease. Adv Neurol 91: 9–18.
Yung KK, Ng TK and Wong CK (1999) Subpopulations of neurons in the rat neostriatum display GABABR1 receptor immunoreactivity. Brain Res 830: 345–352.

Morphogenesis of Rodent Neostriatum Following Early Developmental Dopamine Depletion

Pepijn van den Munckhof, Vladimir V. Rymar, Kelvin C. Luk, Lifeng Gu, Nienke S. Weiss, Pieter Voorn, and Abbas F. Sadikot

Abstract Neurons of the neostriatum are organized in anatomically and chemically distinct patch and matrix compartments. Dopaminergic neurons of the substantia nigra and ventral tegmental area provide early input to the embryonic neostriatum and may play an important role in neostriatal morphogenesis. Our group and others have demonstrated early loss of dopaminergic neurons in the Pitx3-deficient *aphakia* mouse, associated with a >90% reduction in neostriatal dopamine levels and a hypokinetic movement disorder. Here, we show scant or absent dopaminergic fibers in the adult *aphakia* dorsolateral and medial neostriatum, altering the chemical anatomy of mu-opioid receptor stained patches. Furthermore, neostriatal neuron numbers are comparable in *aphakia* and wild-type animals, but the volume of the *aphakia* neostriatum and its neurons is significantly reduced. We propose that early nigrostriatal dopaminergic loss in *aphakia* mice results in loss of trophic support to the developing neostriatum.

1 Introduction

The neostriatum, which is the major component of the basal ganglia, displays a complex organization of neurochemical systems that relates to its neuroanatomical connections. Two neurochemically distinct compartments, the patches and matrix, are arranged in a mosaic fashion. They receive different afferents from the cortex

P. van den Munckhof (✉), N.S. Weiss, and P. Voorn
Department of Anatomy and Neurosciences, Neurosciences Campus Amsterdam,
VU University Medical Center, PO Box 7057, 1007 MB Amsterdam, The Netherlands
e-mail: p.vandenmunckhof@amc.nl

V.V. Rymar, K.C. Luk, L. Gu, and A.F. Sadikot
Department of Neurology and Neurosurgery, Montreal Neurological Institute,
McGill University, 3801 University Street, Montreal, Quebec H3A 2B4, Canada

H.J. Groenewegen et al. (eds.), *The Basal Ganglia IX*, Advances in Behavioral Biology 58, 399
DOI 10.1007/978-1-4419-0340-2_31, © Springer Science+Business Media, LLC 2009

and midbrain dopaminergic (DA) neurons and give rise to separate projection systems to the midbrain. In rodents, DA inputs to the patches arise from a ventral set of neurons in the substantia nigra pars compacta (vSNc) and from islands of DA neurons in the SN pars reticulata (SNr), whereas DA inputs to the matrix arise from a dorsal set of neurons in the ventral tegmental area (VTA) and the dorsal tier of the SNc (dSNc) (Gerfen et al. 1987a). The patches project to SNc neurons and the matrix projects to SNr neurons (Gerfen 1985). Neurons of the neostriatum are born in the lateral ganglionic eminence (LGE), a transient embryonic neuroepithelial structure bordering the lateral ventricles in the basal telencephalon (Fernández et al. 1979), with patch neurons born earlier than matrix neurons (Van der Kooy and Fishell 1987). Postnatally, the neostriatum undergoes extensive naturally occurring cell death that eliminates approximately 25–30% of its neurons, sparing those neurons with early projections to the midbrain (Fentress et al. 1981; Fishell and Van der Kooy 1991).

Dopamine and its receptor binding sites are present early in embryonic brain development, prior to the onset of synaptogenesis, suggesting a role in development independent of its role at the mature synapse (Lauder 1988). In mice, DA neurons differentiate in the SN and VTA at embryonic day (E) 10 (Di Porzio et al. 1990; Kawano et al. 1995). By E13 their extended axons reach the LGE, and the dopamine content of the forebrain remains high from E13 onward throughout the period of neostriatal neurogenesis (Ohtani et al. 2003). Indeed, early embryonic dopamine has been shown to modulate cell cycle regulation in the murine LGE, resulting in an increase in cell output (Ohtani et al. 2003). Mechanical and pharmacological lesions of DA inputs to the rodent neostriatum throughout development alter the neurochemical properties of the patches (Lança et al. 1986; Gerfen et al. 1987b; Snyder-Keller 1991; Van der Kooy and Fishell 1992), but only mechanical lesions during the first postnatal week seem to cause substantial death of neostriatal neurons (Van der Kooy 1996).

We and other groups recently discovered that the naturally occurring *aphakia* (*ak*) mouse, bearing deletions in the promoter and exon-1 of the homeobox gene *Pitx3* (Semina et al. 2000; Rieger et al. 2001), shows marked developmental loss of DA neurons in the SN and VTA (Nunes et al. 2003; Van den Munckhof et al. 2003; Hwang et al. 2003; Smidt et al. 2004). The DA cell loss in the SN starts as early as E12.5 (Smidt et al. 2004), whereas those of the VTA disappear in the first postnatal weeks (Van den Munckhof et al. 2003). In adult *ak* mice, DA neurons are reduced by 71% in the SN, with relative sparing of dSNc, and by 52% in the VTA (Van den Munckhof et al. 2003; Sadikot et al. 2005). In the adult *ak* neostriatum, dopamine levels are reduced by 93% (Van den Munckhof et al. 2003), together with significantly decreased dopamine D_3 receptor levels and increased levels of dopamine D_2 receptor and enkephalin (Van den Munckhof et al. 2006). *Ak* mice exhibit reduced spontaneous locomotor behavior that can be rescued by injection of L-dopa (Hwang et al. 2005; Van den Munckhof et al. 2006).

We use the *ak* mouse to study the effects of early embryonic dopamine depletion on the development of the neostriatum since the specific pattern of developmental DA cell loss in this mouse potentially offers new insights in the role of dopamine in the development of the neostriatum and its patch/matrix compartmentalization. First, dopamine depletion in *ak* mice occurs throughout the period of embryonic neostriatal neurogenesis and postnatal cell death. Second, the *ak* cell loss predominantly affects DA neurons in the vSNc and those occupying the SNr that project to the patches, whereas DA neurons in the dSNc and VTA that innervate the matrix are relatively spared. Third, these lost vSNc neurons are normally the main projection neurons of the patches. Thus, in *ak* mice the neostriatum develops in hypodopaminergic conditions, and the developing patches lack their proper afferent and efferent vSNc neurons.

Here, immunohistochemical studies demonstrate scant or absent DA fibers in the dorsolateral and medial *ak* neostriatum, altering the chemical anatomy of mu-opioid receptor (MOR)-stained patches. Using unbiased stereology, we show comparable neostriatal neuron numbers in adult *ak* and control animals. However, the neuronal soma size is significantly reduced in *ak* mutants, as estimated using the nuclear probe. We, therefore, propose that early nigrostriatal DA loss in *ak* mice results in loss of trophic support to the developing neostriatum, which is reflected in changes in neuronal size rather than in reduced cell number.

2 Methods

All animal procedures were performed in accordance with the Canadian Council on Animal Care guidelines for the use of animals in research and were approved by the McGill University Animal Care Committee. *Ak* mice homozygous for the Pitx3 mutation were purchased from Jackson Laboratories (Maine, USA) and backcrossed to C57BL/6 mice. A total of five adult male mutant and five adult male wild-type C57BL/6 mice were used. The animals were transcardially perfused with 4% paraformaldehyde [(PFA); in phosphate buffer, pH 7.4] at postnatal day 100–170. Brains were then removed, postfixed for 12 h, and cryoprotected by immersion in buffered sucrose (30%, pH 7.4) for an additional 48 h before cutting. Sectioning was performed using a freezing microtome (Micron, Germany) at 50 μm. Floating coronal sections were collected in phosphate buffered saline (PBS) as six separate sets so that each set contained every sixth serial section.

One series of sections was immunostained for tyrosine hydroxylase (TH) and another for the MOR, using an avidin–biotin-peroxidase complex (ABC) method as previously described (Van den Munckhof et al. 2003). Antibodies used were mouse anti-TH (1:1,000, Immunostar, USA) and rabbit anti-MOR (1:750, Immunostar, USA). Floating sections were incubated in anti-TH or anti-MOR antibody overnight, washed in PBS three times, and incubated for 1 h with biotinylated anti-mouse or anti-rabbit secondary antibody (1:200, Vector, USA). Following another set of washes, ABC (Vector, USA) was used as per manufacturer's instructions and the final reaction was revealed by 3′3′-diaminobenzidene (DAB) in the presence of 0.006% hydrogen peroxide. A third series of sections was Nissl-stained in 0.1%

Cresyl violet. The sections were dehydrated and coverslipped. Photographs were captured using a digital camera (Microfire, Optronics, USA).

Unbiased estimates of neuron numbers in the neostriatum were obtained using the optical dissector method (Gundersen et al. 1988) as previously described (Van den Munckhof et al. 2003). The entire rostro-caudal extent of the neostriatum was examined in a 1:6 series of Nissl-stained coronal sections using an Olympus BX40 microscope equipped with an XYZ movement-sensitive stage and Stereo Investigator software (Microbrightfield, USA). The first and last coronal sections with visible caudate-putamen were used as rostral and caudal limits of the reference volume. The corpus callosum, globus pallidus, external capsule, lateral ventricle, and anterior commissure were used as boundaries for the area of analysis. In the more caudal sections, the borders of the caudate-putamen were the external capsule, globus pallidus, bed nucleus of the stria terminalis, substantia innominata, and dorsal amygdala. The boundary of the neostriatum and ventral striatum in precommissural sections was delineated by a line drawn from the ventral tip of the lateral ventricle to the dorsal border of the piriform cortex, as described in detail in previous work (Sadikot and Sasseville 1997). In each section, the area of analysis was traced at 10 × magnification to estimate the surface area. Nissl cell counts were performed at 100× magnification (oil, NA 1.25) using a 60 × 60 μm counting frame. A 10-μm dissector was placed 1 μm below the surface of the section at counting sites located at 500-μm intervals. Glia were excluded on the basis of size (<7.0 μm diameter) and morphology. Cavalieri's method was used to estimate the volume of the reference region (Oorschot 1996; Luk and Sadikot 2001).

For analysis of cell size, a four-ray isotropic nucleator probe (Gundersen et al. 1988) was applied to estimate the cross-sectional area and volume of neostriatal neurons. The longest axis of each cell was determined during application of the nucleator, and the cells were divided into three subgroups based on size. Again, glia were excluded on the basis of size (<7.0 μm diameter) and morphology.

Statistical analysis was performed by one-way ANOVA with the Student–Newman–Keuls post hoc test for comparison between groups.

3 Results

To study the effects of developmental midbrain DA cell loss on the neurochemistry of the neostriatum, the brains of adult wild-type and *ak* mice were immunostained and examined at multiple coronal levels for TH, the key enzyme for dopamine synthesis (Nagatsu et al. 1964), and MOR, a postsynaptic neurochemical marker of the patches (Herkenham and Pert 1981; Wang et al. 1996). As previously reported, dopamine levels are reduced by 93% in the *ak* neostriatum and by 69% in the *ak* nucleus accumbens (Van den Munckhof et al. 2003). TH immunostaining shows significant loss of neostriatal TH-positive fibers in *ak* mice, with scant or absent input to the dorsolateral and medial neostriatum and relative preservation of inputs to the ventrolateral neostriatum and the nucleus accumbens (Fig. 1). Within the relatively spared ventrolateral TH immunostaining, which by itself also seems to

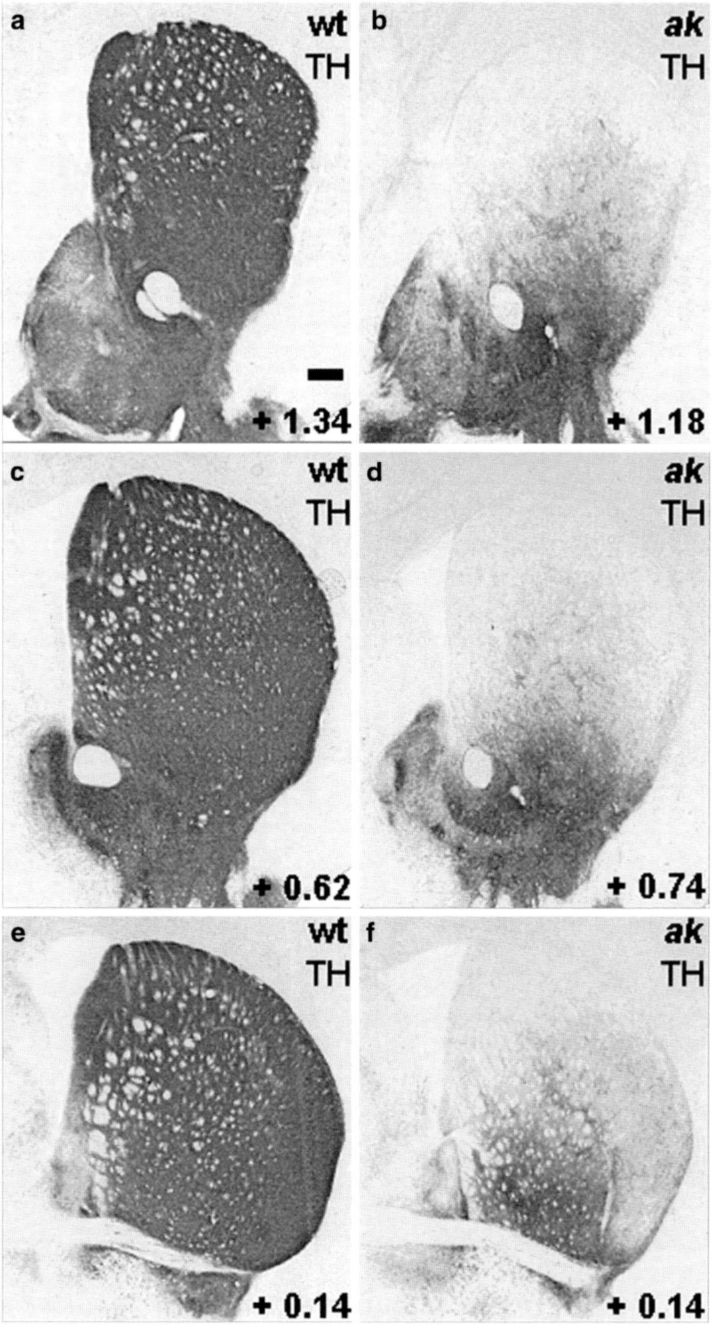

Fig. 1 Loss of neostriatal dopaminergic fibers in *aphakia* (*ak*) mice. Tyrosine hydroxylase (TH) immunohistochemistry in adult *ak* (**b**, **d**, **f**) and wild-type (wt) (**a**, **c**, **e**) mice demonstrates marked loss of DA innervation of the neostriatum. DA terminals are largely absent in the dorsolateral and medial neostriatum and relatively preserved in the ventrolateral neostriatum and the nucleus accumbens. Note the patches of DA fibers visible on a background of lower innervation density in the ventrolateral *ak* neostriatum (**b**, **d**, **f**). Scale bar: 250 μm

contain a dorsal to ventral density gradient of DA fibers, patches of DA fibers are visible on a background of lower innervation density (Fig. 1b, d, f).

Immunostaining for MOR shows a similar dorsoventral gradient in the neostriatum. MOR patches are absent in the dorsolateral and medial part, lightly stained in the upper part of the ventrolateral neostriatum, and relatively preserved in its lower parts (Fig. 2).

To determine whether developmental dopamine depletion influences the number of neurons in the adult *ak* neostriatum, the total number of neurons in the neostriatum was determined on Nissl-stained sections using the optical fractionator technique. Unbiased stereology shows similar neuronal number in the neostriatum of *ak* mice compared with that in wild-type controls (Fig. 3a). The average number of neostriatal neurons was 1.35 ± 0.13 million for *ak* mice and 1.39 ± 0.09 million for wild-type controls (±SD, $n = 5$). These numbers are similar to results obtained in previous studies of the neostriatum of adult C57BL/6 mice (Fentress et al. 1981).

Morphological analysis of the *ak* neostriatum shows a significant reduction in neostriatal volume (±18%) compared with that of wild-type controls – 5.3 vs. 6.5 mm^3 (Fig. 3b, $p < 0.05$, $n = 5$). Neurons in *ak* mice show a shift in frequency distribution to neurons with a smaller long axis compared with wild-type controls (Fig. 3c), with a significant reduction in neuronal volume (±29%) - 295 vs. 413 μm^3 (Fig. 3d, $p < 0.05$, $n = 5$).

4 Discussion and Perspectives

Our understanding on the role of dopamine in neostriatal development stems from animal studies involving pharmacological or mechanical lesions of DA neurons and transgenic mice. Embryonic and perinatal 6-hydroxydopamine (6-OHDA) lesions and late embryonic and neonatal mechanical lesions of the SN in rats result in a reduction of MOR binding but do not cause substantial disruption of patch-matrix compartmentalization (Lança et al. 1986; Gerfen et al. 1987b; Snyder-Keller 1991; Van der Kooy and Fishell 1992). Late embryonic mechanical lesions of the SN do not cause obvious neostriatal shrinkage or cell death (Van der Kooy and Fishell 1992), whereas first postnatal week mechanical lesions cause a greater than 50% shrinkage of the ipsilateral neostriatum due to death of neostriatal neurons and disruption of patch/matrix compartmentalization (Van der Kooy 1996). On the other hand, first postnatal week 6-OHDA lesions of the SN do not produce substantial neostriatal shrinkage (Caboche et al. 1991). Transgenic dopamine-deficient mice seem to have a smaller neostriatum with more compact neurons, although no volumes were reported (Zhou and Palmiter 1995). The size of the neostriatum in D_1 receptor knockout mice is reduced by 22%, but no quantative data on neostriatal cell numbers were reported (Drago et al. 1998).

Thus, although dopamine clearly regulates MOR expression, a uniform conclusion on the developmental role of dopamine in neostriatal neurogenesis and patch/matrix compartmentalization cannot easily been drawn from these results.

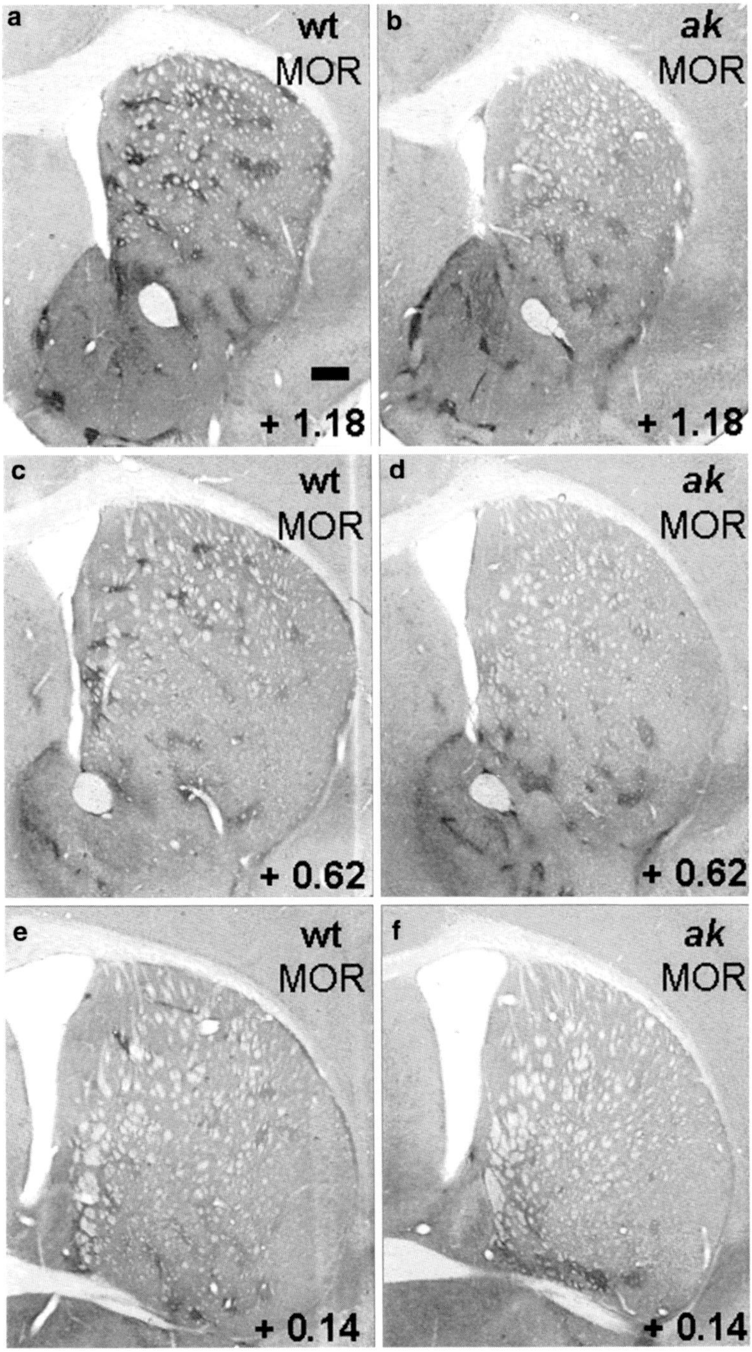

Fig. 2 Loss of neostriatal mu opioid receptor (MOR) expressing patches in *aphakia* (*ak*) mice. MOR immunohistochemistry in adult *ak* (**b**, **d**, **f**) and wild-type (wt) (**a**, **c**, **e**) mice demonstrates marked loss of MOR expression in the neostriatum. MOR patches are absent in the dorsolateral and medial neostriatum, lightly stained in the upper parts of the ventrolateral neostriatum, and relatively preserved in its lower parts (**b**, **d**, **f**). Scale bar: 250 μm

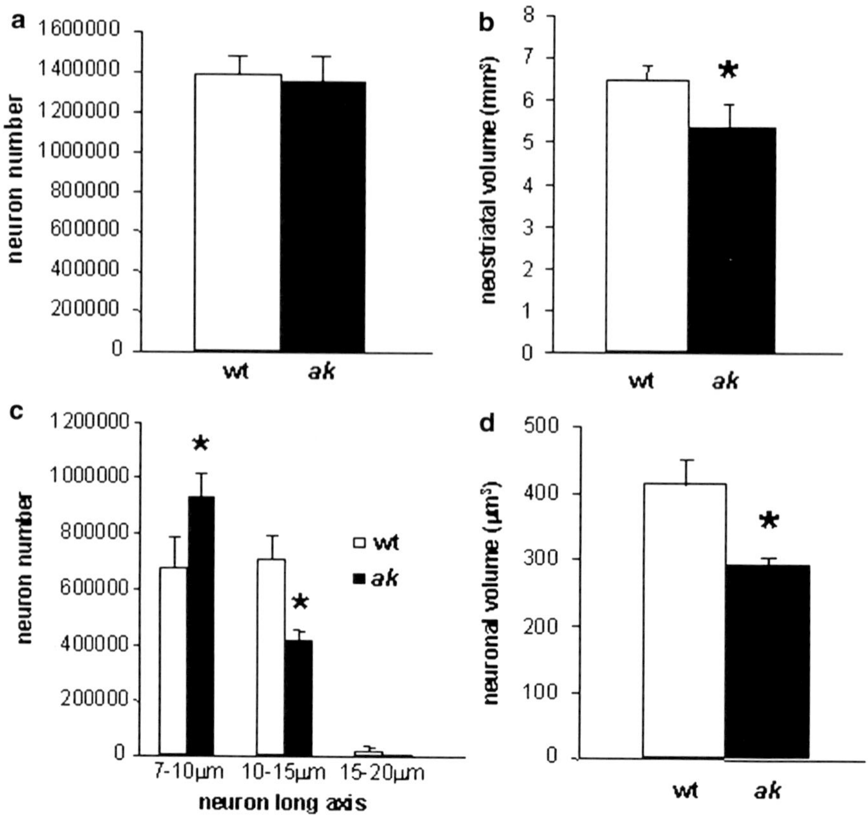

Fig. 3 Developmental dopaminergic depletion in *aphakia* (*ak*) mice does not alter neostriatal neuron numbers but reduces neostriatal volume and neuronal cell size. (**a**) Stereological estimates of the total number of neostriatal neurons using the optical fractionator technique are similar for adult *ak* mice compared with wild-type (wt) controls. The average number of neostriatal neurons is 1.4 million ($n = 5$). (**b**) Stereological estimates of neostriatal volume using Cavalieri's method show a significantly reduced neostriatal volume (±18%) in adult *ak* mice compared with wt controls – 5.3 vs. 6.5 mm^3 ($p < 0.05$, $n = 5$). (**c**) Stereological estimates of frequency distribution of neostriatal neuronal long axis, determined with a four-ray isotropic nucleator probe, show a shift in frequency distribution to neurons with a smaller long axis in *ak* mice compared with wt controls. (**d**) *Ak* mice have significantly smaller neostriatal neurons (±29%) compared with wt controls – 295 vs. 413 µm^3 ($p < 0.05$, $n = 5$)

4.1 *Neostriatal DA Afferents in* ak *Mice Are Reduced in a Dorsal to Ventral Manner*

DA inputs to the patches arise from vSNc neurons and islands of SNr neurons, whereas DA inputs to the matrix arise from dSNc and VTA neurons (Gerfen et al. 1987a). Since vSNc/SNr DA neurons are lost and dSNc/VTA DA neurons relatively spared in *ak* mice (Van den Munckhof et al. 2003; Sadikot et al. 2005), TH immunostaining of the *ak* neostriatum in the present study was expected to show absence of DA inputs to the patches and relative sparing of DA inputs to the matrix. Interestingly, however, TH immunostaining showed a dorsoventral gradient in the neostriatum, with scant or absent input to the dorsolateral and medial neostriatum, without any residual input to the matrix and relative preservation of inputs to the ventrolateral striatum. Within the relatively spared ventrolateral TH immunostaining, patches of DA fibers were preserved. So apparently, compensatory innervation from spared dSNc/VTA DA neurons is responsible for these preserved patches of DA fibers. And although *ak* dSNc/VTA DA neurons are relatively spared, those responsible for DA input to the matrix of the dorsolateral/medial neostriatum seem to be gone.

Alternatively, the supposed reciprocal connections between vSNc/SNr and patches and dSNc/VTA and matrix are not strictly separated but more intermingled in *ak* mice. Or, more fundamentally, these mice exhibit reciprocal connections between vSNc/SNr DA neurons and dorsolateral/medial neostriatum and dSNc/VTA DA neurons and ventrolateral neostriatum. Future experiments with anterograde and retrograde axonal tracers in *ak* neostriatum and SNc/VTA will explain the observed neostriatal TH immunostaining.

4.2 Reduced Neostriatal DA Afferents in ak *Mice Reduce Patchy MOR Expression*

Much evidence has suggested that the patchy postsynaptic MOR expression in the neostriatum is associated with nigrostriatal DA afferent terminals. Indeed, mechanical, 6-OHDA, and *N*-methyl-4-phenyl-1,2,3,6-tetrahydropyridine (MPTP) lesions of the rodent SN produce significant depletion of MOR-expressing or MOR-ligand binding patches (Pollard et al. 1978; Gardner et al. 1980; Bowen et al. 1982; Van der Kooy 1984; Gerfen et al. 1987b; Sirinathsinghji and Dunnett 1989; Caboche et al. 1991; Van der Kooy and Fishell 1992). Our current results of TH and MOR immunostaining in the *ak* neostriatum further support this strong association. The density of the patches of MOR expression in the *ak* neostriatum consequently follows the dorsoventral density gradient of TH immunostaining. MOR patches are absent in those neostriatal regions devoid of DA fibers and get denser in areas with relatively preserved DA innervation. Thus, the presence of DA nigrostriatal projections is critical for patchy neostriatal MOR expression. Whether dopamine itself or another factor released by DA neurons is the critical inducer for MOR expression cannot be concluded from the current data. Neither can it be determined whether the absence of dopamine in the dorsolateral and medial neostriatum of *ak* mice affects the embryonic aggregation of early born patch neurons into discrete patches embedded in the later born matrix neurons in these neostriatal subregions. This remains to be determined with intraperitoneal bromodeoxyuridine (BrdU) injections in timed pregnant *ak* mice during the period of patch/matrix neurogenesis and these experiments are currently performed.

4.3 Developmental Dopamine Depletion in ak *Mice Does Not Alter Neostriatal Neuron Numbers but Reduces Neostriatal Volume and Neuronal Cell Size*

Early embryonic dopamine has been shown to modulate cell cycle regulation in the murine LGE, resulting in an increase in cell output (Ohtani et al. 2003). Experimental quantitative data elucidating the developmental role of dopamine in neostriatal neurogenesis are, however, scarce. Van der Kooy (1996) reported massive neostriatal shrinkage and more than 40% loss of neostriatal neurons following first postnatal week mechanical lesions of the SN but suggested that concomitant lesioning of striatonigral projections could be crucial for this observation, since first postnatal week pharmacological lesions of the SN and transgenic dopamine deficiency did not cause obvious shrinkage (Caboche et al. 1991; Zhou and Palmiter 1995).

Our current stereology results do not attribute a clear role to dopamine in neostriatal neurogenesis or neuronal survival in *ak* mice. Total neostriatal neuron numbers in adult *ak* mice are no different from wild type mice. Since the *ak* neostriatum contains a dorsal to ventral density gradient of DA fiber input, an analysis comparing neuron numbers in the different subregions of the *ak* neostriatum might provide further “proof” for the dopamine independency of neostriatal neurogenesis in vivo. In addition, quantification of *ak* neostriatal neurons during embryonic and postnatal development could determine whether compensatory plasticity from other afferent systems, e.g., corticostriatal inputs, may allow for survival of the appropriate number of neostriatal neurons.

The current volumetric results point to a trophic role of dopamine in neostriatal development. Adult *ak* mice have a 18% smaller neostriatum and 29% smaller neuronal soma size. Similar reductions in neostriatal volume were reported in dopamine-deficient mice (Zhou and Palmiter 1995) and D_1 receptor knockout mice (Drago et al. 1998). Thus, dopamine might play a specific role in neostriatal cell growth and differentiation. Again, an analysis comparing neuronal volumes in the neostriatal subregions with different density of DA fibers input might further unravel this trophic role of dopamine in neostriatal development.

References

Bowen WD, Pert CB and Pert A (1982) Nigral 6-hydroxydopamine lesions equally decrease mu and delta opiate binding to striatal patches: further evidence for a conformationally malleable type 1 opiate receptor. Life Sci 31: 1679–1682.

Caboche J, Rogard M and Besson MJ (1991) Comparative development of D1-dopamine and mu opiate receptors in normal and in 6-hydroxydopamine-lesioned neonatal rat striatum: dopaminergic fibers regulate mu but not D1 receptor distribution. Brain Res Dev Brain Res 58: 111–122.

Di Porzio U, Zuddas A, Cosenza-Murphy DB and Barker JL (1990) Early appearance of tyrosine hydroxylase immunoreactive cells in the mesencephalon of mouse embryos. Int J Dev Neurosci 8: 523–532.

Drago J, Padungchaichot P, Accili D and Fuch S (1998) Dopamine receptors and dopamine transporter in brain function and addictive behaviors: insights from targeted mouse mutants. Dev Neurosci 20: 188–203.

Fentress JC, Stanfield BB and Cowan WM (1981) Observations on the development of the striatum in mice and rats. Anat Embryol 163: 275–298.

Fernández V, Bravo H, Kuljis R and Fuentes I (1979) Autoradiographic study of the development of the neostriatum in the rabbit. Brain Behav Evol 16: 113–128.

Fishell G and Van der Kooy D (1991) Pattern formation in the striatum: neurons with early projections to the substantia nigra survive the cell death period. J Comp Neurol 312: 33–42.

Gardner EL, Zukin RS and Makman MH (1980) Modulation of opiate receptor binding in striatum and amygdala by selective mesencephalic lesions. Brain Res 194: 232–239.

Gerfen CR (1985) The neostriatal mosaic. I. Compartmental organization of projections from the striatum to the substantia nigra in the rat. J Comp Neurol 236: 454–476.

Gerfen CR, Herkenham M and Thibault J (1987a) The neostriatal mosaic. II. Patch and matrix-directed mesiostriatal dopaminergic and non-dopaminergic systems. J Neurosci 7: 3915–3934.

Gerfen CR, Baimbridge KG and Thibault J (1987b) The neostriatal mosaic. III. Biochemical and developmental dissociation of patch-matrix mesostriatal systems. J Neurosci 7: 3935–3944.

Gundersen HJ, Bagger P, Bendtsen TF, Evans SM, Korbo L, Marcussen N, Möller A, Nielsen K, Nyengard JR, Pakkenberg B et al (1988) The new stereological tools: disector, fractionator, nucleator and point sampled intercepts and their use in pathological research and diagnosis. APMIS 96: 857–881.

Herkenham M and Pert CB (1981) Mosaic distribution of opiate receptors, parafascicular projections and acetylcholinesterase in rat striatum. Nature 291: 415–418.

Hwang DY, Ardayfio P, Kang UJ, Semina EV and Kim KS (2003) Selective loss of dopaminergic neurons in the substantia nigra of Pitx3-deficient aphakia mice. Brain Res Mol Brain Res 114: 123–131.

Hwang DY, Fleming SM, Ardayfio P, Moran-Gates T, Kim H, Tarazi FI, Chesselet MF and Kim KS (2005) 3,4-Dihydroxyphenylalanine reverses the motor deficits in Pitx3-deficient aphakia mice: behavioral characterization of a novel genetic model of Parkinson's disease. J Neurosci 25: 2132–2137.

Kawano H, Ohyama K, kawamura K and Nagatsu I (1995) Migration of dopaminergic neurons in the embryonic mesencephalon of mice. Brain Res Dev Brain Res 86: 101–113.

Lança AJ, Boyd S, Kolb BE and Van der Kooy D (1986) The development of a patchy organization of the rat striatum. Brain Res Dev Brain Res 27: 1–10.

Lauder JM (1988) Neurotransmitters as morphogens. Prog Brain Res 73: 365–387.

Luk KC and Sadikot AF (2001) GABA promotes survival but not proliferation of parvalbumin-immunoreactive interneurons in rodent neostriatum: an in vivo study with stereology. Neuroscience 104: 93–103.

Nagatsu T, Levitt M and Udenfriend S (1964) Tyrosine hydroxylase. The initial step in norepinephrine biosynthesis. J Biol Chem 239: 2910–2917.

Nunes I, Tovmasian LT, Silva RM, Burke RE and Goff SP (2003) Pitx3 is required for development of substantia nigra dopaminergic neurons. Proc Natl Acad Sci USA 100: 4245–4250.

Ohtani N, Goto T, Waeber C and Bhide PG (2003) Dopamine modulates cell cycle in the lateral ganglionic eminence. J Neurosci 23: 2840–2850.

Oorschot DE (1996) Total number of neurons in the neostriatal, pallidal, subthalamic, and substantia nigra nuclei of the rat basal ganglia: a stereological study using the cavalieri and optical dissector methods. J Comp Neurol 366: 580–599.

Pollard H, Llorens C, Schwartz JC, Gros C and Dray F (1978) Localization of opiate receptors and enkephalins in the rat striatum in relationship with the nigrostriatal dopaminergic system: lesion studies. Brain Res 151: 392–398.

Rieger DK, Reichenberger E, McLean W, Sidow A and Olsen BR (2001) A double-deletion mutation in the *Pitx3* gene causes arrested lens development in aphakia mice. Genomics 72: 61–72.

Sadikot AF and Sasseville R (1997) Neurogenesis in the mammalian neostriatum and nucleus accumbens: parvalbumin-immunoreactive GABAergic interneurons. J Comp Neurol 389: 193–211.

Sadikot AF, Luk KC, Van den Munckhof P, Rymar VV, Leung K, Gandhi R and Drouin J (2005) Pitx3 is necessary for survival of midbrain dopaminergic neuron subsets relevant to Parkinson's disease. In: Bolam JP, Ingham CA and Magill PJ (eds) The Basal Ganglia. VIII. Advances in Behavioral Biology, Springer, New York, pp 265–274.

Semina EV, Murray JC, Reiter R, Hrstka RF and Graw J (2000) Deletion in the promoter region and altered expression of *Pitx3* homeobox gene in aphakia mice. Hum Mol Genet 9: 1575–1585.

Sirinathsinghji DJS and Dunnett SB (1989) Disappearance of the mu-opiate receptor patches in the rat neostriatum following lesioning of the ipsilateral nigrostriatal dopamine pathway with 1-methyl-4-phenylpyridinium ion (MPP^+): restoration by embryonic nigral dopamine grafts. Brain Res 504: 115–120.

Smidt MP, Smits SM, Bouwmeester H, Hamers FP, Van der Linden AJ, Hellemons AJ, Graw J and Burbach JP (2004) Early developmental failure of substantia nigra dopamine neurons in mice lacking the homeodomain gene *Pitx3*. Development 131: 1145–1155.

Snyder-Keller AM (1991) Development of striatal compartmentalization following pre- or postnatal dopamine depletion. J Neurosci 11: 810–821.

Van den Munckhof P, Luk KC, Ste-Marie L, Montgomery J, Blanchet PJ, Sadikot AF and Drouin J (2003) Pitx3 is required for motor activity and for survival of a subset of midbrain dopaminergic neurons. Development 130: 2535–2542.

Van den Munckhof P, Gilbert F, Chamberland M, Lévesque D and Drouin J (2006) Striatal neuroadaption and rescue of locomotor deficit by L-dopa in aphakia mice, a model of Parkinson's disease. J Neurochem 96: 160–170.

Van der Kooy D (1984) Developmental relationships between opiate receptors and dopamine in the formation of caudate-putamen patches. Brain Res Dev Brain Res 14: 300–303.

Van der Kooy D (1996) Early postnatal lesions of the substantia nigra produce massive shrinkage of the rat striatum, disruption of patch neuron distribution, but no loss of patch neurons. Brain Res Dev Brain Res 94: 242–245.

Van der Kooy D and Fishell G (1987) Neuronal birthdate underlies the development of striatal compartments. Brain Res 401: 155–161.

Van der Kooy D and Fishell G (1992) Embryonic lesions of the substantia nigra prevent the patchy expression of opiate receptors, but not the segregation of patch and matrix compartment neurons, in the developing striatum. Brain Res Dev Brain Res 66: 141–145.

Wang H, Moriwaki A, Wang JB, Uhl GR and Pickel VM (1996) Ultrastructural immunocytochemical localization of mu opioid receptors and Leu5-enkephalin in the patch compartment of the rat caudate-putamen nucleus. J Comp Neurol 375: 659–674.

Zhou QY and Palmiter RD (1995) Dopamine-deficient mice are hypoactive, adipsic, and aphagic. Cell 83: 1197–1209.

Upregulation of NAD(P)H:Quinone Oxidoreductase (NQO1) in Glial Cells of 6-Hydroxydopamine-Lesioned Substantia Nigra in the Rat

Andrea C. Kil, Benjamin Drukarch, Allert J. Jonker, Henk J. Groenewegen, and Pieter Voorn

Abstract Dopamine quinone toxicity has been implicated in the degeneration of nigral dopaminergic (DA) neurons in Parkinson's disease (PD). NAD(P)H:quinone oxidoreductase (NQO1) may protect against this quinone toxicity. In Parkinsonian brains, levels of NQO1 are increased in reactive glia cells that are located around the remaining DA neurons in the substantia nigra pars compacta (SNc), suggesting a neuroprotective role of NQO1.

It is not known at which stage of the disease process the upregulation of glial NQO1 starts. Furthermore, it is at present not clear whether NQO1 indeed plays a neuroprotective role in the disease process. As a first step to experimentally study a potential neuroprotective role of NQO1, it was examined whether activation of glia cells and changes in the distribution of NQO1-positive glia cells and the expression levels of glial NQO1, as seen in PD brains, also occurred in a 6-hydroxy-dopamine (6-OHDA) rat model of PD.

Our results show that astroglia cells and microglia cells were activated. Furthermore, NQO1 was upregulated in astroglia cells in the SNc in those areas in which DA neurons degenerated. The time course and pattern of upregulation of NQO1 paralleled those of the degeneration of DA neurons. Activated microglia were seen at a later stage during the course of degeneration of DA neurons.

In conclusion, in the present model, astroglia cells and microglia cells are activated in response to 6-OHDA-induced oxidative stress. Furthermore, levels of NQO1 are increased in astroglia cells. The findings in the present model are in line with the findings as seen in parkinsonian brains. The 6-OHDA rat model of PD is, therefore, suitable for further research to examine a potential neuroprotective role of NQO1.

A.C. Kil, B. Drukarch, A.J. Jonker, H.J. Groenewegen, and P. Voorn (✉)
Department of Anatomy and Neurosciences, Neuroscience Campus Amsterdam, VU University Medical Center, PO Box 7057, 1007, MB Amsterdam, The Netherlands
e-mail: p.voorn@vumc.nl

H.J. Groenewegen et al. (eds.), *The Basal Ganglia IX*, Advances in Behavioral Biology 58, DOI 10.1007/978-1-4419-0340-2_32,

1 Introduction

An important pathological hallmark of Parkinson's disease (PD) is the degeneration of dopaminergic (DA) neurons in the substantia nigra pars compacta (SNc), which eventually leads to the dopamine deficiency characteristic for PD. There are strong indications that oxidative stress due to dopamine auto-oxidation plays a major role in the degeneration of DA neurons in PD (Spencer et al. 1998; Hirsch et al. 1988; Drukarch and van Muiswinkel 2000).

One type of highly reactive products induced by dopamine auto-oxidation is the dopamine quinone, which is toxic to cells in several ways. Quinones can react with cysteinyl groups, thereby forming cysteinyl-catechol adducts. This inactivates cysteinyl-rich proteins, such as glutathione. Quinones can also participate in redox cycling, a process that generates reactive oxygen species such as superoxide radicals, hydrogen peroxide, and hydroxyl radicals (Smythies and Galzigna 1998; Stokes et al. 1999; Bindoli et al. 1992).

DA neurons contain many compounds that can counteract the effects of (quinone-related) oxidative stress products (Zhang et al. 1994; Damier et al. 1993). In addition, surrounding glial cells, which are abundantly present in the substantia nigra, are specialized in protecting neurons against the harmful effects of oxidative stress (reviewed by Hirsch 2000; Damier et al. 1993). We hypothesized that in PD degeneration of DA neurons may involve inadequate protection by glial cells from or exacerbation of quinone-related toxicity and subsequent oxidative stress in the SNc.

One of the defense mechanisms of glial cells against oxidative stress is the increased production of antioxidant enzymes. An example of this is the observed increase in the density of glutathione peroxidase immunopositive glia cells in the SNc around remaining DA neurons in PD brains, compared with glial cells surrounding degenerated neurons (Damier et al. 1993).

We previously described another antioxidant enzyme in astroglia cells in the SNc, NAD(P)H:quinone oxidoreductase (NQO1), which might contribute to protection of neurons against quinone-related toxicity (van Muiswinkel et al. 2004). NQO1 (EC 1.6.99.2), formerly known as DT-diaphorase, is a flavoprotein that can reduce quinones via a two-step reduction. The principal function of NQO1 is thought to be the detoxification of quinones, thereby protecting the cells against quinone-related toxicity.

In our previous study we reported increased levels of NQO1 in astroglia cells in the SNc of PD brains than in normal brains (van Muiswinkel et al. 2004). Furthermore, these astroglia cells were found to be reactive, and also activated microglia cells were present during the course of the disease. Interestingly, the observed increase in expression of NQO1 was most prominent during ongoing degeneration of DA neurons and had disappeared when the loss of DA neurons was almost complete (van Muiswinkel et al. 2004). These data suggest that NQO1 is induced in an attempt of glial cells to protect DA neurons from quinone-related toxicity and oxidative stress.

As a first step to experimentally study such a role, we investigated in a rat model of PD whether 6-hydroxydopamine (6-OHDA)-induced quinone-related toxicity would be accompanied by changes in glial expression levels of NQO1 and glial reactivity, similar to those seen in PD brains.

2 Materials and Methods

2.1 *Animals*

Female Wistar rats (200 ± 20 g, Harlan, Zeist, The Netherlands) were used in this study. All experiments were approved by the VU University Medical Center animal ethics committee. This was according to EU regulations and guidelines established by the Royal Dutch Academy of Sciences.

2.2 *Intracerebral Injections of 6-Hydroxydopamine*

Ninety rats were anesthetized with a mixture of Ketamine and Rompun® (15 mg/kg Ketamine and 50 mg/kg Rompun, im) and received a stereotactic injection of 6-OHDA (Sigma, St. Louis, MO, USA) 2 μg in 2 μl 0.9% NaCl containing 0.05% ascorbic acid (Sigma) in the substantia nigra pars reticulata (SNr) of the right hemisphere at coordinates 9.0 mm anterior, 2.0 mm lateral, and 8.5 mm below the cortical surface, according to the atlas of Paxinos and Watson (1998) and on the basis of a previous study. Lidocaine spray (Astra Pharmaceutica, Zoetermeer, The Netherlands), 100 mg/ml, was used as a local anesthetic and applied to the wound surface. Animals were sacrificed at nine different time points, i.e., 3, 6, 24, 72, 96 h, 1, 3 weeks, or 3 months postlesion.

Ten rats received a sham lesion by a stereotactic injection of 2 μl 0.9% NaCl containing 0.05% ascorbic acid (Sigma). These animals were sacrificed at time point 1 week postlesion. Ten untreated rats were used to establish the normal distribution patterns of the different stainings used in this study.

The animals were anesthetized with sodium pentobarbital (Sanofi Sante BV, The Netherlands; 60 mg/kg, ip) and perfused transcardially with 4% paraformaldehyde (Merck, Germany) fixative, containing 0.25% picric acid (Merck, Germany) in 0.1 M phosphate buffer, pH 7.4. Brains were taken out of the skull, immersion-fixed for 60 min in the same fixative, and then immersed in 30% sucrose in phosphate buffer, overnight at 4°C. The brains were sectioned on a freezing microtome at a thickness of 30 μm and the sections were (immuno)stained.

2.3 *Immunohistochemistry*

As primary antibody against NQO1 rabbit-anti-QR [anti-quinone-reductase; Kelly et al. (2000) – kindly provided by Dr. Philip Sherratt, Schering-Plough Research Institute, NJ, USA – a polyclonal antibody, raised against full-length recombinant NQO1 from rat] was used.

Immunohistochemistry for NQO1 (rabbit-anti-QR, 1:400), tyrosine hydroxylase (mouse-anti-rat TH; Incstar, Stillwater, MN, USA; 1:2,000), GFAP (mouse-anti-GFAP

Sigma, St. Louis, MO, USA; 1:2,000), or OX-6 (mouse-anti-OX-6, kindly provided by E. Döpp, Department of Molecular Cell Biology and Immunology, VU University Medical Center, Amsterdam; 1:300) was performed on free-floating sections.

NQO1 and tyrosine hydroxylase antibody incubations were in TBS-Tx (0.5% Triton X-100 in TRIS-buffered saline, pH 8.0), and rinses between incubation steps were in the same buffer without Tx. As secondary antibody biotinylated-goat-anti-rabbit IgG (DAKO diagnostics BV, Copenhagen, Denmark, 1:100) or biotinylated-horse-anti-mouse IgG (Vector Laboratories, Burlingame, CA, USA; 1:100) was used. Peroxidase was visualized using an ABC immunoperoxidase kit (Vector Laboratories; 1:200) and 3,3 diaminobenzidine tetrahydrochloride dehydrate (Sigma, St. Louis, MO, USA; 0.5 mg/ml) as chromogen.

Immunofluorescence for GFAP or OX-6 was visualized with an Alexa 594-conjugated-horse-anti-mouse antibody (Molecular Probes; 1:200). For immunofluorescence with rabbit anti-QR (1:600), biotinylated-goat-anti-rabbit IgG was used as secondary antibody. Peroxidase was visualized using an ABC immunoperoxidase kit (Vector laboratories; 1:200) and Alexa 488-tyramide (TSA#22 kit, Molecular Probes, 1:100) as chromogen.

2.4 LY 83583-Mediated Enzyme Histochemistry

LY83583-mediated enzyme histochemistry was performed on free-floating microtome sections, according to the method of Murphy et al. (1998). LY83583 (Sigma, St. Louis, MO, USA) was used in a concentration of 1 μM.

2.5 Fluorojade Histochemistry

Fluorojade-B (Chemicon, Temecula, CA, USA) histochemistry was performed on slide-mounted microtome sections according to Zuch et al. (2000). The Fluorojade-B staining was examined using epifluorescence microscopy.

2.6 Digital Images

The sections were digitized with a Xillix MicroImager camera attached to a MCID Elite image analysis system (Imaging Research, Ontario, Canada). Image compilations were made of the entire substantia nigra with objective 10× using the MCID tiling tool. Images of autofluorescence (see later) were made on water-mounted sections prior to NQO1 immunohistochemistry. Adobe Photoshop (version 5.5) was used for image processing (adjustment of contrast and brightness and compilation of figures).

3 Results

To study whether 6-OHDA-induced quinone-related toxicity would be accompanied by changes in glial expression levels of NQO1 and glial reactivity in a rat model of PD, degeneration of dopaminergic neurons was induced by injecting a small amount of 6-OHDA in the substantia nigra pars reticulata (SNr), close to the substantia nigra pars compacta (SNc). Then, at the indicated time points, samples were taken and stained for several brain cell markers.

First, the temporal degeneration pattern of the DA neurons was determined using a staining for tyrosine hydroxylase (TH), a marker for DA neurons. Fluorojade was used as a marker for degenerating neurons, and autofluorescence was used as a possible marker for neuronal damage. Second, changes in glial expression levels of NQO1 were examined using NQO1 immunohistochemistry and NQO1 enzyme histochemistry. Glial reactivity was investigated by staining for GFAP, marking reactive glial cells, and for OX-6, which is present in microglial cells that are activated.

3.1 Degeneration Pattern of DA Neurons After 6-OHDA Administration

Loss of TH-immunopositive neurons became apparent as of 24 h postlesion, at first in the ventromedial part of the SNc (Fig. 1b) and the lateral part of the ventral tegmental area (VTA). After 96 h, cell loss became apparent in the medial part of the SNc and after a week almost all DA neurons in the ventromedial and medial part of the SNc had disappeared. At this point in time, cell loss was also seen in the lateral part of the SNc (Fig. 1c, d). The lesion size had not changed at time point 3 weeks and 3 months postlesion.

3.2 Fluorojade-B

Fluorojade-B staining of the SN at the various time points showed a pattern opposite to that described earlier for the changes in TH immunoreactivity. Fluorojade-B staining became visible in regions where TH immunoreactivity disappeared. Fluorojade-B-positive neurons first became visible at 24 h postlesion in the ventromedial part of the ipsilateral SNc (Fig. 2c). The lateral SNc contained Fluorojade-B-positive neurons after 1-week survival time (Fig. 2e). No Fluorojade-positive cells were present in the SNc contralateral to the side of 6-OHDA injection (not shown).

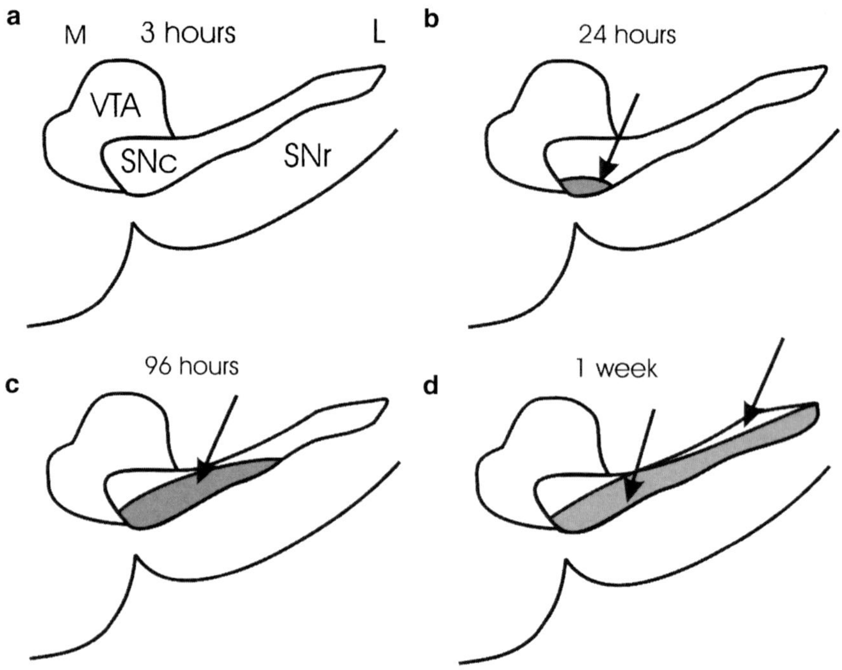

Fig. 1 Temporal degeneration pattern of TH immunoreactivity in neurons in the substantia nigra pars compacta. The drawings show loss of TH immunoreactivity in the SNc at 3 h (**a**), 24 h (**b**), 96 h (**c**), and 1 week (**d**) after 6-OHDA administration. At 3 h postlesion, no loss of TH immunoreactivity is visible (**a**). (**b**) shows loss of TH immunoreactivity in the ventromedial part of the pars compacta [*arrow in* (**b**)] at 24 h postlesion. The *arrow* in (**c**) points to loss of TH immunoreactivity in the medial part of the pars compacta at 96 h postlesion. One week postlesion loss of TH immunoreactivity is visible throughout the ventral past of the pars compacta [arrows in (**d**)]. Abbreviations: *SNc* substantia nigra pars compacta, *SNr* substantia nigra pars reticulata, *VTA* ventral tegmental area, *M* medial, *L* lateral

3.3 Autofluorescence

As of 3 h postlesion until 3 weeks after 6-ODA administration, autofluorescent cells were visible in the SNc, on the lesioned side (Fig. 2). At the time points 3 until 24 h postlesion, the autofluorescence was visible in the ventromedial part of the SNc (Fig. 2b, d). The autofluorescence was present in neurons that looked normal, according to shape and normal distribution of TH immunoreactivity in the cells, as seen in adjacent sections (Fig. 3a, b). At the time point 1 week, the autofluorescence was also visible in the middle and lateral parts of the SNc (Fig. 2f). Autofluorescence was visible until 3 weeks after 6-OHDA administration (Fig. 3c), at which point autofluorescent cell fragments of degenerated neurons were found in the SNc (Fig. 3c, d).

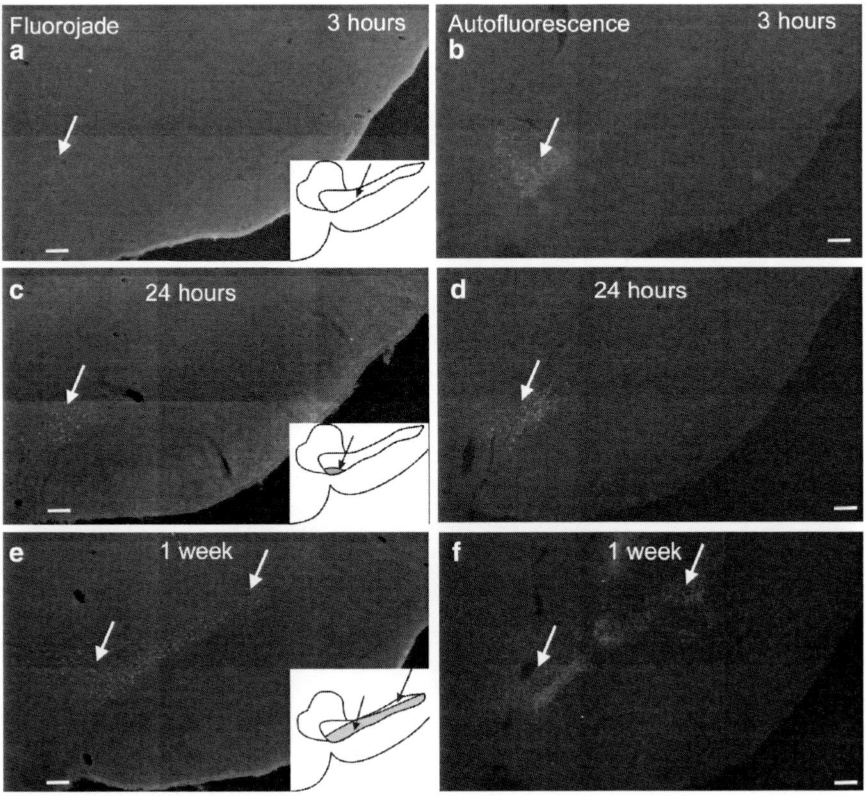

Fig. 2 Patterns of Fluorojade-B staining (**a**, **c**, **e**) and autofluorescence (**b**, **d**, **e**) at 3 h (**a**, **b**), 24 h (**c**, **d**), and 1 week (**e**, **f**) postlesion. At 3 h postlesion, Fluorojade-B staining is absent. Autofluorescence is present in neurons in the ventromedial pars compacta (**a**, **b**). At 3, 24, and 48 h postlesion, autofluorescence is present in the ventromedial pars compacta (**a–c**). At 96 h after 6-OHDA administration, autofluorescence becomes visible in the medial part of pars compacta (**d**). At 1 and 3 weeks postlesion, autofluorescence also appears laterally in the ventral part of pars compacta (**e**, **f**). Bars indicate 100 μm

The neurons that had survived after 3 weeks did not contain autofluorescence. (Fig. 3c, d). Three months after 6-OHDA administration no autofluorescent cells were visible in the SNc (not shown).

3.4 Activation of Microglia

In sham-lesioned animals, activated OX-6-immunopositive microglial cells were visible along the needle track (Fig. 4a). There was no OX-6 immunoreactivity present in the SNc. There was a strong activation of microglia, as seen by the OX-6 immunoreactivity in these cells, in the ventral part of the SNc at 1 week after 6-OHDA

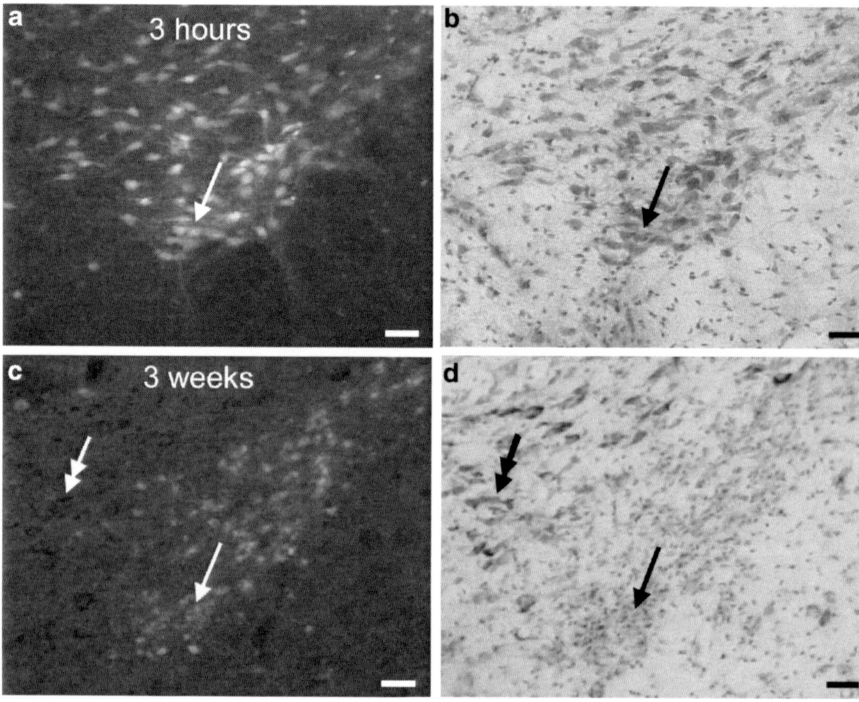

Fig. 3 Detailed pictures of autofluorescence (**a**, **c**) and Nissl (**b**, **d**) at 3 h and 3 weeks postlesion. (**a**) shows a high magnification of autofluorescent cells at 3 h postlesion. (**b**) shows a Nissl staining of the same cells. The cells have a neuronal morphology and appear healthy. The *arrows* in (**a**) and (**b**) point to the same cells. (**c**) shows a high magnification of autofluorescence at 3 weeks postlesion and (**d**) shows a Nissl stain of the same region. *Arrow* in (**c**) points to the autofluorescence, which at this time point is visible in cell fragments, as confirmed by the Nissl stain [arrow in (**d**)]. The *double-headed arrows* in (**c**) and (**d**) point to cells that survived the 6-OHDA-induced oxidative stress. These cells appear normal and do not contain autofluorescence. Bars indicate 25 μm

administration (Fig. 4b). At 3 weeks postlesion, OX-6-positive activated microglia were visible throughout the SNc (Fig. 4c). The OX-6-positive microglia were located around neurons containing autofluorescence (Fig. 4d, e).

No OX-6-immunopositive microglia were seen in the SNc at earlier time points (not shown).

3.5 *Upregulation of NQO1 Immunoreactivity and NQO1 Enzyme Activity in Glial Cells*

In nontreated animals, NQO1 immunoreactivity and NQO1 enzyme activity were present in glial cells in the SNc and the SNr (Fig. 5a, b, e, f). The NQO1-immunopositive glial cells in the SNc were GFAP immunonegative, whereas the

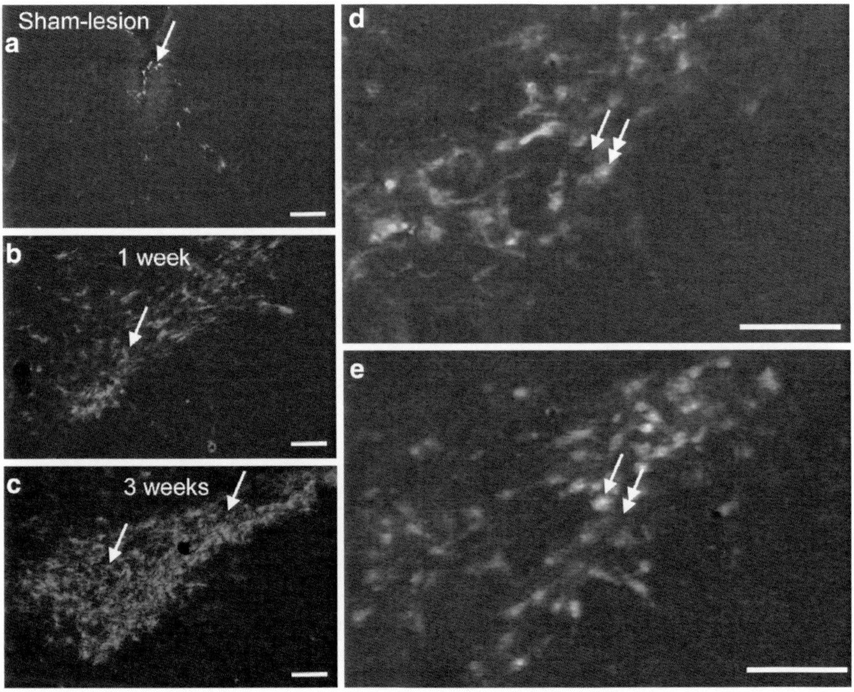

Fig. 4 OX 6 immunoreactivity in SN microglial cells: Activated OX-6 immunoreactive microglial cells in a sham-lesioned animal (**a**), at 1 week (**b**), and 3 weeks (**c**) after 6-OHDA-administration. Detailed pictures of OX-6 immunoreactive microglial cells (**d**) and autofluorescence (**e**) at 1 week postlesion. In the sham-lesioned animal, activated microglial cells are visible around the needle track [*arrow* in (**a**)]. The *arrow* in (**b**) points to activated microglial cells in the ventral part of the substantia nigra pars compacta. In (**c**), activated microglial cells are present throughout the SNc. (**d**) shows a detailed picture of OX-6 immunoreactive microglial cells in the SNc. (**e**) shows a picture of the same area in which autofluorescent cells are visible. The *arrows* in (**d**) and (**e**) point to an autofluorescent cell that is OX-6 immunonegative, whereas the *double-headed arrow* points to an OX-6-immunopositive microglia cell that does not contain autofluorescence. The OX-6-immunopositive microglia are situated close to the autofluorescent cells. Bars indicate 100 μm (**a-c**) or 50 μm (**d**, **e**)

NQO1-immunopositive glial cells in the SNr were GFAP immunopositive (Fig. 5d, h). On the basis of the shape and size of the cells as well as the similarities between the NQO1-immunopositive cells in the SNc and the NQO1-immunopositive and GFAP-immunopositive glial cells in the SNr, we concluded that the NQO1-immunopositive cells in the SNc are glial cells.

After 6-OHDA administration in the substantia nigra, an upregulation of NQO1 immunoreactivity and NQO1 enzyme activity was visible in glial cells in the SNc, compared with nontreated and sham-lesioned animals. In the SNc this upregulation was at first visible in the ventromedial part at 24 h postlesion (Figs. 6a and 7a). At 96 h postlesion, increased glial NQO1 was present in the ventromedial and in the

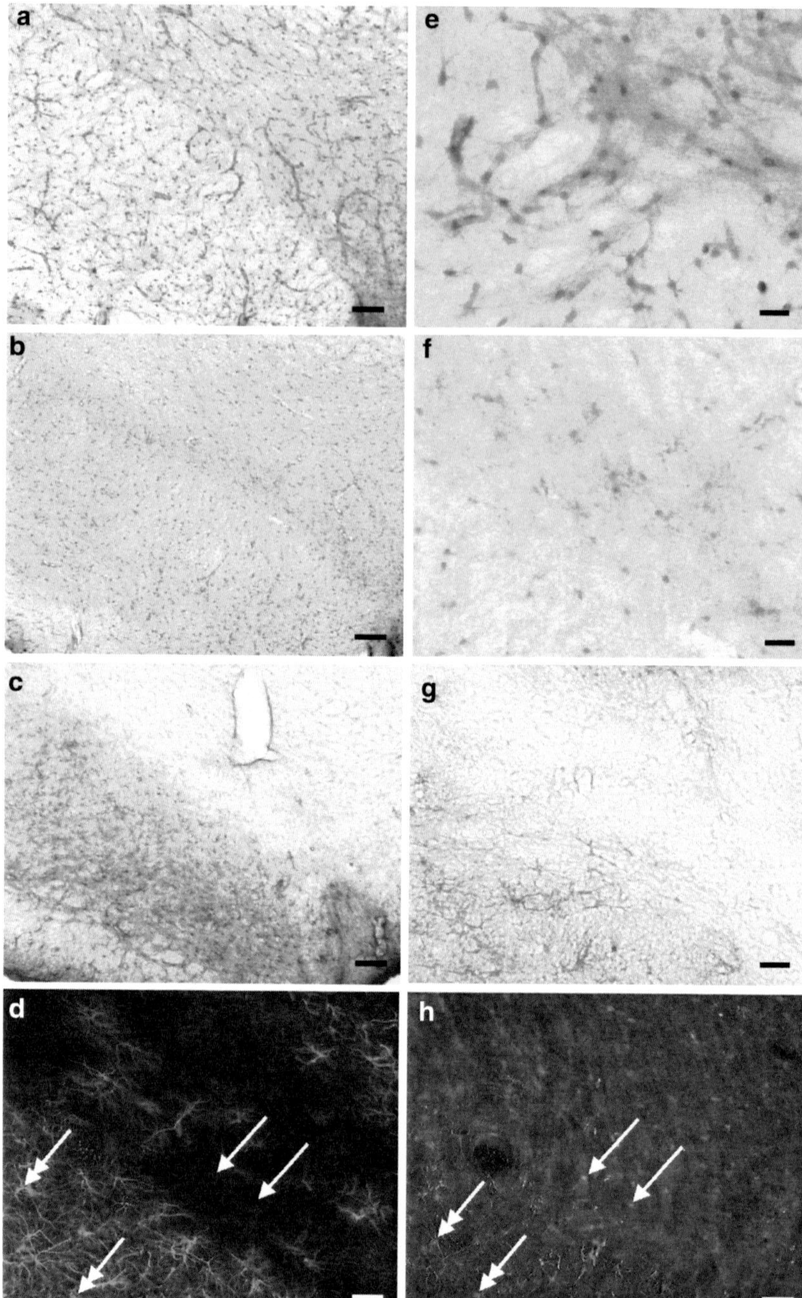

Fig. 5 Distribution of NQO1 and GFAP in nontreated animals: Pictures show NQO1 immunoreactivity (**a**, **e**), NQO1 enzyme activity (**b**, **f**), GFAP immunoreactivity (**c**, **g**), and double-labeling of NQO1 and GFAP (**d**, **h**) in SN glial cells of nontreated animals. (**a**, **e**) (at higher magnification) show NQO1 enzyme activity in glial cells in the substantia nigra pars compacta and pars reticulata. (**b**, **f**) (at higher magnification) show NQO1 immunoreactivity in glial cells in the substantia nigra pars compacta and pars reticulata. (**c**, **g**) (at higher magnification) show GFAP immunoreactivity

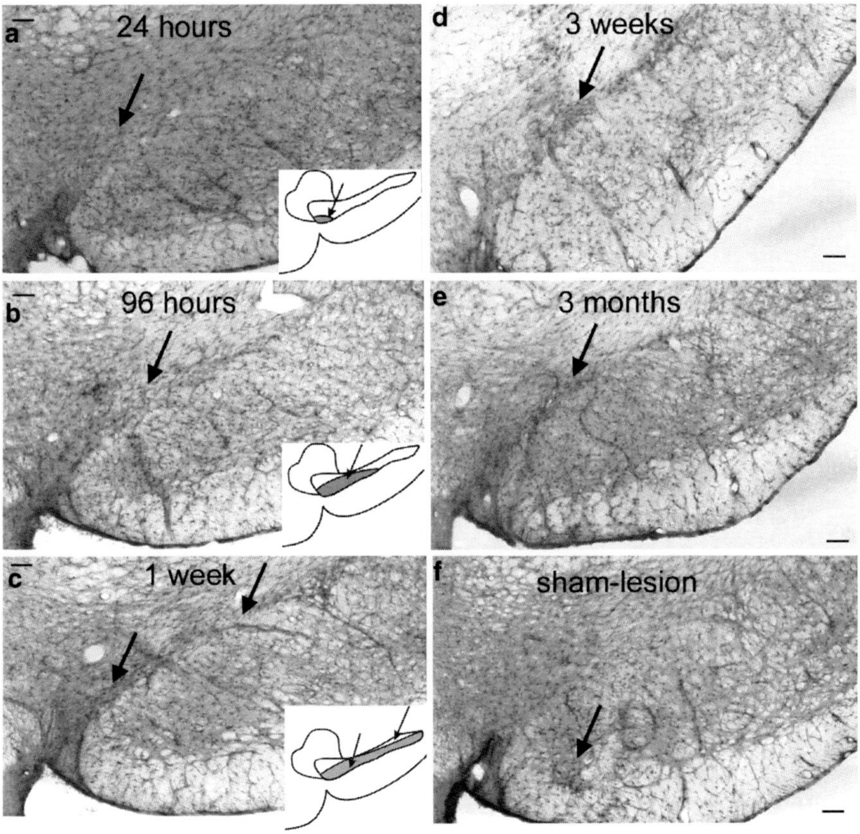

Fig. 6 NQO1 enzyme activity in SN glial cells. The pictures show NQO1 enzyme activity in the SN at 24 h (**a**), 96 h (**b**), 1 week (**c**), 3 weeks (**d**), and 3 months (**e**) after 6-OHDA administration. (**f**) shows NQO1 enzyme activity in a sham-lesioned animal. The insets in (**a–c**) show the degeneration pattern of DA neurons at the same time points postlesion. At time point 24 h (**a**) postlesion, cells in the ventromedial part of the SNc show a small increase in NQO1 enzyme activity [*arrow* in (A)]. (**b**) shows upregulation of NQO1 enzyme activity in the medial part of the SNc at 96 h postlesion. *Arrows* in (**c**) point at an increase of NQO1 enzyme activity in the ventral part of the SNc. Three weeks (**d**) and 3 months (**e**) postlesion NQO1 enzyme activity is still increased in the ventral part of the SNc, but to a lesser extent than at time point 1 week postlesion. (**f**) shows NQO1 enzyme activity in a sham-lesioned animal. One week after saline administration no changes are visible in the NQO1 enzyme activity in the SNc. *Arrow* in (**f**) points to an increase of NQO1 enzyme activity around the site of injection. Bars indicate 100 μm

in glial cells in the substantia nigra pars reticulata. Note the absence of GFAP immunoreactivity in glial cells in the pars compacta. (**d**, **h**) show a double-labeling of GFAP immunohistochemistry (**d**) and NQO1 immunohistochemistry (**h**). *Double-headed arrows* point to glial cells in the substantia nigra pars reticulata that contain GFAP immunoreactivity and NQO1 immunoreactivity, whereas the *arrows* point to NQO1-immunopositive glial cells in the pars compacta that do not contain GFAP immunoreactivity. Bars indicate 50 μm (**a–d**) and 25 μm (**e–h**)

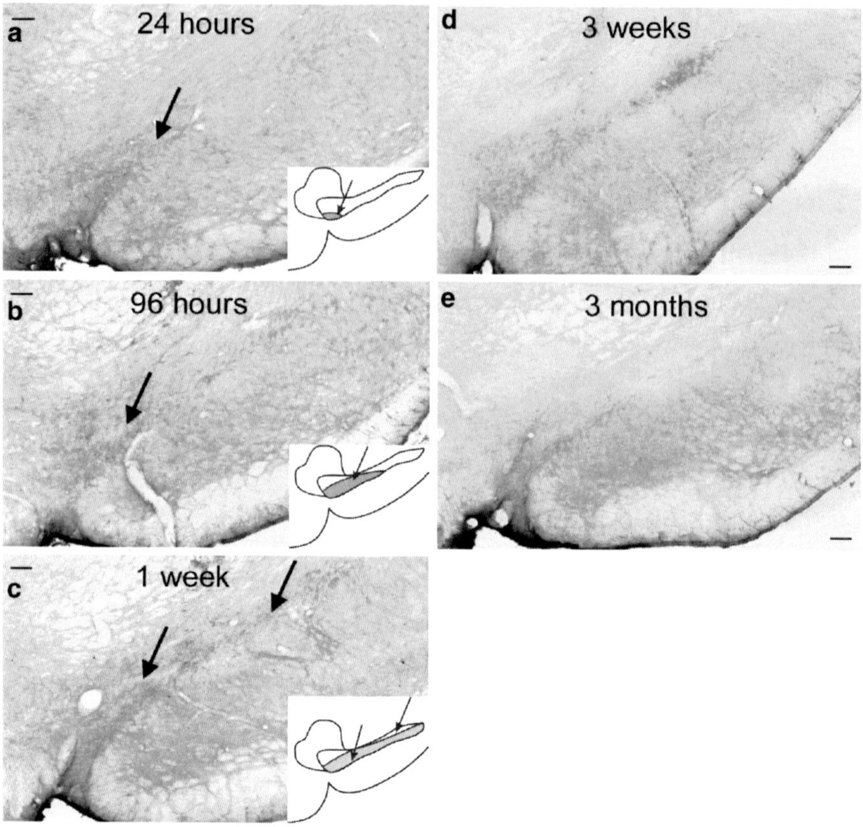

Fig. 7 NQO1 immunoreactivity in SN glial cells. NQO1 immunoreactivity at 24 h (**a**), 96 h (**b**), 1 week (**c**), 3 weeks (**d**), and 3 months (**e**) after 6-OHDA administration. The insets in (**a–c**) show the degeneration pattern of DA neurons at the same time points postlesion. A small increase of NQO1 is visible in the ventromedial part of the SNc at 24 h [*arrow* in (**a**)]. The *arrow* in (**b**) points at the presence of increased NQO1 immunoreactivity in the ventromedial SNc at 96 h postlesion. (**c**) shows increased NQO1 immunoreactivity in the ventral part of the SNc [*arrows* in (**c**)]. In (**d**) increased NQO1 immunoreactivity is visible in the ventral part of the SNc. (**e**) shows faint NQO1 immunoreactivity in the SNc at 3 months postlesion. Bars indicate 100 μm

medial part of the SNc (Figs. 6b and 7b). After 1 week, the upregulation of glial NQO1 was also seen laterally in the SNc (Figs. 6c and 7c). The highest levels of NQO1 immunoreactivity and enzyme activity were visible at the time points 1 and 3 weeks (Figs. 6d and 7d). Three months after 6-OHDA administration, levels of NQO1 enzyme activity were still increased to some extent in the ventral part of the SNc, but to a lesser degree than in the preceding time points (Figs. 6e and 7e). In the SNr, the upregulation was visible at the site of injection.

In the sham-lesioned animals at 1 week postlesion, an upregulation of NQO1 immunoreactivity and enzyme activity was visible in the SNr at the site of the injection (Fig. 6f). There was no upregulation of NQO1 visible in the SNc (Fig. 6f).

3.6 Induction of Reactive Glial Cells and Upregulation of GFAP Immunoreactivity in Reactive Glial Cells

In nontreated animals, glial cells in the pars reticulata were GFAP immunopositive, whereas in the SNc no GFAP-immunopositive glial cells were present, except for a few scattered GFAP-immunopositive glial cells (Fig. 5c, d, g).

After 6-OHDA administration, glial cells in SNc became GFAP positive and the number of GFAP-positive glial cells as well as the amount of GFAP immunoreactivity per glia cell in the SNr increased, compared with nontreated animals (Fig. 8). In the SNc, glial cells showed GFAP immunoreactivity as of 24 h postlesion. At this time point, a small increase of GFAP immunoreactivity was visible in the SNc, in the VTA and dorsally to the SNc (Fig. 8a). This induction of reactive glial cells spread in a dorsolateral direction. After 96 h, GFAP immunoreactivity was also visible in the medial part of the SNc, and at time point 1 week postlesion, GFAP-positive glial cells were present throughout the ventral part of the SNc (Fig. 8b, c). Three weeks after 6-OHDA administration, an induction of GFAP immunoreactivity was visible in the SNc, but to a lesser extent than at earlier time points (Fig. 8d). At 3 months postlesion, faint GFAP immunoreactivity was present in the SNc (Fig. 8e). In the sham-lesioned animals, an induction and upregulation of GFAP immunoreactivity was visible in the SNr at the site of the injection, but not in the SNc (Fig. 8f). On the nonlesioned side of animals that received a 6-OHDA injection, no induction or upregulation of GFAP immunoreactivity was visible in the substantia nigra.

3.7 Comparison of the Temporal Patterns of the Stainings Used

When the temporal patterns of the different stainings were compared, it appeared that the patterns of NQO1 immunoreactivity upregulation, NQO1 enzyme activity upregulation, and induction of GFAP immunoreactivity were identical (Fig. 9). These upregulations and induction became apparent at the same time point when a decrease of TH immunoreactivity was visible.

Fluorojade, a marker of neuronal degeneration, became visible earlier during the time course of degeneration and disappeared as soon as the neurons had degenerated. Autofluorescence, as a marker for oxidative stress, became apparent very early during the course of degeneration.

At the point where almost all TH immunoreactivity had disappeared, OX-6 immunoreactivity, i.e., activation of microglia, was visible.

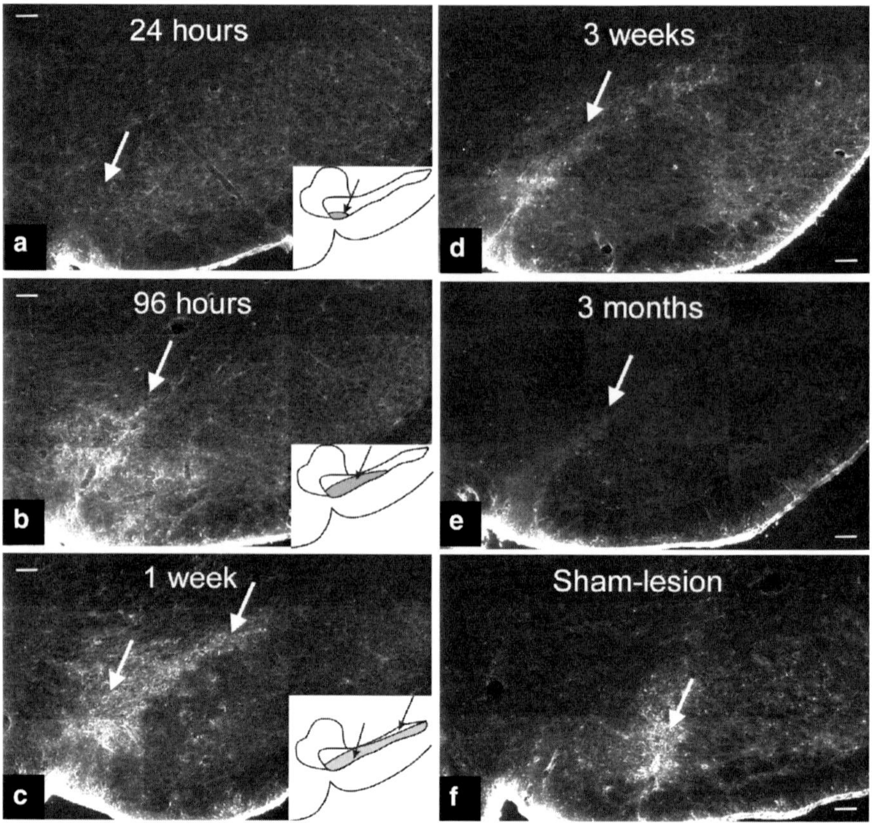

Fig. 8 GFAP immunoreactivity in SN glial cells The pictures show GFAP immunoreactivity in the SN at 24 h (**a**), 96 h (**b**), 1 week (**c**), 3 weeks (**d**), and 3 months (**e**) after 6-OHDA administration. (**f**) shows GFAP immunoreactivity in a sham-lesioned animal. The insets in (**a–c**) show the degeneration pattern of DA neurons at the same time points postlesion. At time point 24 h postlesion (**a**), cells in the ventromedial part of the SNc show a small increase in GFAP immunoreactivity [*arrow* in (**a**)]. The *arrow* in (**b**) points at an upregulation of GFAP immunoreactivity in the medial part of the SNc at 96 h postlesion. In (**c**), an upregulation of GFAP immunoreactivity is visible in the ventral part of the SNc at 1 week after 6-OHDA administration [*arrows* in (**c**)]. At 3 weeks (**d**) GFAP immunoreactivity is still increased in the ventral part of the SNc, but to a lesser extent than at time point 1 week postlesion [*arrow* in (**d**)]. Figure (**e**) shows faint GFAP immunoreactivity at 3 months postlesion. (**f**) shows GFAP immunoreactivity in a sham-lesioned animal. The *arrow* in (**f**) points at GFAP immunoreactive cells present along the needle tract. GFAP immunoreactivity is absent from the SNc. Bars indicate 100 μm

4 Discussion

In the present study, we found an increase in reactive astroglia cells, activation of microglia cells, and increased levels of NQO1 in astroglia cells in the substantia nigra pars compacta of 6-OHDA-lesioned rats. The increase in reactive astroglia as

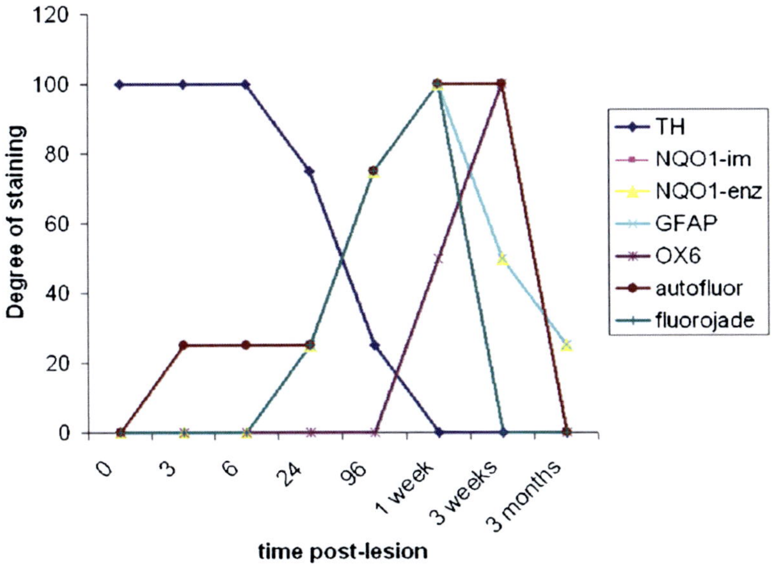

Fig. 9 Temporal patterns of (immuno)markers for degeneration and glial reactivity. The graphs show the intensity of (immuno)staining in the SNc at 3 h, 24 h, 96 h, 1 week, 3 weeks, and 3 months after 6-OHDA administration. Values are normalized in relation to the maximum staining intensity for each individual marker in the SNc. First, autofluorescence, a possible marker for oxidative stress in neurons, becomes apparent. This is followed by Fluorojade, a degeneration marker. At the same time point that Fluorojade becomes visible in DA neurons, loss of TH immunoreactivity in these neurons becomes apparent. Astroglia cells become reactive, as shown by an increase in GFAP immunoreactivity, at the time degeneration and loss of TH immunoreactivity start to develop. At the same time, NQO1 immunoreactivity and NQO1 enzyme activity are upregulated in glial cells. At 1 week postlesion, degeneration is complete and Fluorojade staining disappears. At this time point, GFAP immunostaining, NQO1 immunoreactivity, and enzyme activity decrease to low levels. Microglia become activated, as shown by OX6 immunoreactivity, at the time when degeneration of DA neurons is complete. At 3 months postlesion, OX6 immunoreactivity and autofluorescence have disappeared

well as the upregulation of NQO1 developed in a pattern that was identical to that of the degeneration of DA neurons. Activated microglia cells were seen at a later stage during the course of neuronal degeneration.

Astroglia cells and microglia cells are known to become reactive or activated in response to oxidative stress and tissue damage (Depino et al. 2003; Kohutnicka et al. 1998; Teismann and Schulz 2004). 6-OHDA induces oxidative stress and quinone-related toxicity via the formation of quinines (Blum et al. 2001; Napolitano et al. 1999). A new phenomenon found in the present study, the presence of autofluorescence in neurons, may be an indicator of the oxidative stress and quinone-related toxicity, induced in the DA neurons by 6-OHDA. It is known that advanced lipid peroxidation products and glycoxidation products that are formed when cells are under oxidative stress are autofluorescent (Tsuchida et al. 1987). The autofluorescence became apparent as of 3 h after administration of the 6-OHDA and was seen in neurons in the regions where loss of TH immunoreactivity and Fluorojade-B-positive neurons was seen at later time points. From the loss of TH immunoreactivity in combination with the presence of Fluorojade-B, a marker that stains degenerating neurons (Schmued and Hopkins 2000), we may conclude that degeneration of neurons took place in these regions. The autofluorescence may indicate the presence of oxidative stress that preceded and, most likely, induced this degeneration of the DA neurons. At a later time point, 3 weeks postlesion, autofluorescence was still present in fragments of degenerated neurons, but was absent from neurons that survived the 6-OHDA-induced oxidative stress. These fragments of degenerated neurons were also Fluorojade-B-positive. At 3 months postlesion, autofluorescence as well as Fluorojade-B were absent, indicating that levels of oxidative stress and degeneration of dopaminergic neurons were low at this time point. This is according to the expectation that at 3 months postlesion, the lesion is complete, and the acute 6-OHDA-induced oxidative stress is at a low level.

Activated microglia were seen mainly 1 and 3 weeks postlesion and were located around the cell fragments that contained autofluorescence. It is not known what the exact role of the microglia is in the degeneration process. On the one hand, microglia cells can protect neurons by removing potentially deleterious cellular debris and by promoting tissue repair by secreting growth factors (Orr et al. 2002; Kreutzberg 1996). On the other hand, microglia cells start to excrete proinflammatory cytokines and chemokines, in answer to neuronal damage. These cytokines and chemokines can cause neuronal damage, thereby perpetuating the neurodegenerative process, rather than protecting the neurons against degeneration (Orr et al. 2002). In the present study, the presence of activated microglia around the autofluorescent cell fragments suggests a role in protecting the neurons by clearing the tissue from cellular debris that remained after the degeneration of the DA neurons. A contribution of the microglia to the neurodegeneration in the present study is not likely, since the activated microglia first appeared at 1 week after 6-OHDA administration, a time point at which most DA neurons had already degenerated.

Increased levels of NQO1 were found in reactive astroglia in the SNc. One of the defense mechanisms that is activated by reactive astroglia after 6-OHDA-induced oxidative stress is the antioxidant responsive element (ARE), which is an enhancer element that is upstream of many phase-II detoxification and antioxidant enzymes such as NQO1, heme-oxygenase-I, and glutathione-*S*-transferase (Jakel et al. 2005; Friling et al. 1990; Rushmore and Pickett 1990a, 1991b; Rushmore et al. 1990b, 1991a). Interestingly, besides oxidative stress, the presence of quinolinic compounds is also a trigger to activate ARE (Nguyen et al. 2004). Since 6-OHDA induces oxidative stress via the formation of quinones, this may explain the upregulation of NQO1 in reactive astroglia, as seen in the present study.

The reactivity of astroglia cells and the upregulation of NQO1 in these cells were restricted to areas in the SN pars compacta in which DA neurons degenerated, as seen by the loss of TH immunoreactivity and the presence of Fluorojade in these DA neurons. These astroglial features were most likely in response to the damage occurring in neurons rather than in glia cells. The autofluorescence was present in neurons only and not in glia cells, suggesting that the oxidative stress, induced by 6-OHDA, occurred in neurons and not in glia cells. The upregulation of NQO1 and reactivity of the astroglia cells suggest a role in protecting the neurons against the induced oxidative stress and quinone-related toxicity. A more general and widespread upregulation of NQO1 in the pars compacta and the pars reticulata would have been expected if the astroglia cells were protecting themselves against oxidative stress.

At time point 3 months postlesion, levels of NQO1 and reactivity of astroglia cells decreased, but were still at a higher level than in the control situation. At that time point, most DA neurons had degenerated and a large part of the pars compacta was void of DA neurons. Autofluorescence and Fluorojade-B were absent, suggesting that the acute 6-OHDA-induced oxidative stress levels and levels of degeneration were low. Since astroglia respond to increased oxidative stress levels, a decrease in oxidative stress levels may explain the decrease in reactivity of astroglia and decrease of NQO1 enzyme activity.

The patterns of activation of glia cells and changes in NQO1 expression levels as well as the time course of these features during the course of degeneration of DA neurons in the 6-OHDA-lesioned rats are in line with the findings in the parkinsonian brains. In humans, the activation of astroglia and the increase in levels of NQO1 were most prominent during the active stage of disease, when degeneration of DA neurons took place and decreased at the end stage of the disease, when virtually all DA neurons had disappeared (van Muiswinkel et al. 2004). This was also seen in the 6-OHDA rat model, where the highest levels of reactive astroglia and NQO1 enzyme activity were seen at 1 and 3 weeks postlesion, when degeneration of DA neurons occurred, whereas the levels decreased at 3 months postlesion, a time point at which most DA neurons had degenerated. We may, therefore, conclude that this 6-OHDA-rat model of Parkinson's disease is suitable for examining a potential neuroprotective role of NQO1.

To summarize, activation of glia cells and changes in levels of NQO1 take place in the substantia nigra pars compacta of 6-OHDA-lesioned rats, as seen in parkinsonian brains. It is, therefore, a suitable model to examine a potential neuroprotective role of NQO1. Since quinone-related toxicity is implicated in the pathogenesis of PD and changes of NQO1 expression are found in parkinsonian brains, NQO1 may be a potential target for the development of neuroprotective strategies for PD (Drukarch and van Muiswinkel 2000; van Muiswinkel et al. 2004). Further research is focused on the examination of a potential neuroprotective role of NQO1 in the 6-OHDA rat model of PD. Within this context, our finding of neuronal autofluorescence as an early marker for cellular distress may provide for an excellent readout to evaluate the influence of manipulation of NQO1 activity on neuronal survival.

Acknowledgments The excellent drawing skills of Mr. Dirk de Jong in preparing the figures are gratefully acknowledged.

References

Bindoli A, Rigobello MP and Deeble DJ (1992) Biochemical and toxicological properties of the oxidation products of catecholamines. Free Radic Biol Med 13: 391–405.

Blum D, Torch S, Lambeng N, Nissou M, Benabid AL, Sadoul R and Verna JM (2001) Molecular pathways involved in the neurotoxicity of 6-OHDA, dopamine and MPTP: contribution to the apoptotic theory in Parkinson's disease. Prog Neurobiol 65: 135–172.

Damier P, Hirsch EC, Zhang P, Agid Y and Javoy-Agid F (1993) Glutathione peroxidase, glial cells and Parkinson's disease. Neuroscience 52: 1–6.

Depino AM, Earl C, Kaczmarczyk E, Ferrari C, Besedovsky H, del Rey A, Pitossi FJ and Oertel WH (2003) Microglial activation with atypical proinflammatory cytokine expression in a rat model of Parkinson's disease. Eur J Neurosci 18: 2731–2742.

Drukarch B and van Muiswinkel FL (2000) Drug treatment of Parkinson's disease. Time for phase II. Biochem Pharmacol 59: 1023–1031.

Friling RS, Bensimon A, Tichauer Y and Daniel V (1990) Xenobiotic-inducible expression of murine glutathione-*S*-transferase Ya subunit gene is controlled by an electrophile-responsive element. Proc Natl Acad Sci USA 87: 6258–6262.

Hirsch E, Graybiel AM and Agid YA (1988) Melanized dopaminergic neurons are differentially susceptible to degeneration in Parkinson's disease. Nature 334: 345–348.

Hirsch EC (2000) Glial cells and Parkinson's disease. J Neurol 247 Suppl 2: 1158–1162.

Jakel RJ, Kern JT, Johnson DA and Johnson JA (2005) Induction of the protective antioxidant response element pathway by 6-hydroxydopamine in vivo and in vitro. Toxicol Sci 87: 176–186.

Kelly VP, Ellis EM, Manson MM, Chanas SA, Moffat GJ, McLeod R, Judah DJ, Neal GE and Hayes JD (2000) Chemoprevention of aflatoxin B1 hepatocarcinogenesis by coumarin, a natural benzopyrone that is a potent inducer of aflatoxin B1-aldehyde reductase, the glutathione *S*-transferase A5 and P1 subunits, and NAD(P)H:quinone oxidoreductase in rat liver. Cancer Res 60: 957–969.

Kohutnicka M, Lewandowska E, Kurkowska-Jastrzebska I, Czlonkowski A and Czlonkowska A (1998) Microglial and astrocytic involvement in a murine model of Parkinson's disease induced by 1-methyl-4-phenyl-1,2,3,6-tetrahydropyridine (MPTP). Immunopharmacology 39: 167–180.

Kreutzberg GW (1996) Microglia: a sensor for pathological events in the CNS. Trends Neurosci 19: 312–318.

Murphy TH, So AP and Vincent SR (1998) Histochemical detection of quinone reductase activity in situ using LY 83583 reduction and oxidation. J Neurochem 70: 2156–2164.

Napolitano A, Pezzella A and Prota G (1999) New reaction pathways of dopamine under oxidative stress conditions: nonenzymatic iron-assisted conversion to norepinephrine and the neurotoxins 6-hydroxydopamine and 6, 7-dihydroxytetrahydroisoquinoline. Chem Res Toxicol 12: 1090–1097.

Nguyen T, Yang CS and Pickett CB (2004) The pathways and molecular mechanisms regulating Nrf2 activation in response to chemical stress. Free Radic Biol Med 37: 433–441.

Orr CF, Rowe DB and Halliday GM (2002) An inflammatory review of Parkinson's disease. Prog Neurobiol 68: 325–340.

Paxinos G and Watson C (1998) The Rat Brain in Stereotaxic Coordinates, New York: Academic Press

Rushmore TH and Pickett CB (1990a) Transcriptional regulation of the rat glutathione *S*-transferase Ya subunit gene. Characterization of a xenobiotic-responsive element controlling inducible expression by phenolic antioxidants. J Biol Chem 265: 14648–14653.

Rushmore TH, King RG, Paulson KE and Pickett CB (1990b) Regulation of glutathione *S*-transferase Ya subunit gene expression: identification of a unique xenobiotic-responsive element controlling inducible expression by planar aromatic compounds. Proc Natl Acad Sci USA 87: 3826–3830.

Rushmore TH, Morton MR and Pickett CB (1991a) The antioxidant responsive element. Activation by oxidative stress and identification of the DNA consensus sequence required for functional activity. J Biol Chem 266: 11632–11639.
Rushmore TH and Pickett CB (1991b) Xenobiotic responsive elements controlling inducible expression by planar aromatic compounds and phenolic antioxidants. Methods Enzymol 206: 409–420.
Schmued LC and Hopkins KJ (2000) Fluoro-Jade B: a high affinity fluorescent marker for the localization of neuronal degeneration. Brain Res 874: 123–130.
Smythies J and Galzigna L (1998) The oxidative metabolism of catecholamines in the brain: a review. Biochim Biophys Acta 1380: 159–162.
Spencer JP, Jenner P, Daniel SE, Lees AJ, Marsden DC and Halliwell B (1998) Conjugates of catecholamines with cysteine and GSH in Parkinson's disease: possible mechanisms of formation involving reactive oxygen species. J Neurochem 71: 2112–2122.
Stokes AH, Hastings TG and Vrana KE (1999) Cytotoxic and genotoxic potential of dopamine. J Neurosci Res 55: 659–665.
Teismann P and Schulz JB (2004) Cellular pathology of Parkinson's disease: astrocytes, microglia and inflammation. Cell Tissue Res 318: 149–161.
Tsuchida M, Miura T and Aibara K (1987) Lipofuscin and lipofuscin-like substances. Chem Phys Lipids 44: 297–325.
van Muiswinkel FL, de Vos RA, Bol JG, Andringa G, Jansen Steur EN, Ross D, Siegel D and Drukarch B (2004) Expression of NAD(P)H:quinone oxidoreductase in the normal and Parkinsonian substantia nigra. Neurobiol Aging 25: 1253–1262.
Zhang P, Anglade P, Hirsch EC, Javoy-Agid F and Agid Y (1994) Distribution of manganese-dependent superoxide dismutase in the human brain. Neuroscience 61: 317–330.
Zuch CL, Nordstroem VK, Briedrick LA, Hoernig GR, Granholm AC and Bickford PC (2000) Time course of degenerative alterations in nigral dopaminergic neurons following a 6-hydroxy-dopamine lesion. J Comp Neurol 427: 440–454.

Clioquinol Protects Against Cell Death in Parkinson's Disease Models In Vivo and In Vitro

Simon Wilkins, Colin L. Masters, Ashley I. Bush, Robert A. Cherny, and David I. Finkelstein

Abstract Parkinson's disease (PD) is characterized by the loss of dopaminergic neurons located in the substantia nigra (SN). Data from our group and others indicate that metals, oxidative stress, and bioavailable reductants provide a possible mechanism for the neurodegeneration observed in PD. 6-Hydroxydopamine (6-OHDA) injection into the nigra of mice resulted in quantified loss of dopaminergic neurons. Oral administration of the metal–protein binding compound clioquinol (CQ) commencing on the day of lesion led to a significant reduction in lesion size. This finding elaborates upon our previous study that long-term pre-treatment with CQ reduced the susceptibility of SN neurons to another neurotoxin, 1-methyl-4-phenyl-1,2,3,6-tetrapyridine (MPTP) (Kaur et al. 2003), suggesting metals as a common pathway for propagation of these lesions. Human neuroblastoma M17 cells were susceptible to metal, 6-OHDA and dopamine-induced cell death that was partially recoverable by co-incubation with CQ or catalase. These results support the concept that CQ or a similar moderate-affinity transition metal ligand could modulate neuronal oxidative stress and therefore may be a novel class of drug that may be useful for the treatment of PD.

1 Introduction

Parkinson's disease (PD) is characterized by the loss of greater than 70% of dopaminergic neurons in the substantia nigra (SN) region of the brain. The mechanisms leading to SN cell loss are not completely understood (Lang and Lozano 1998), although oxidative stress and a deficiency in energy metabolism are thought to be involved (Fahn and Cohen 1992; Hirsch 1993; Jenner and Olanow 1996). There is widespread evidence that oxidative stress occurs in the SN and that this is increased in PD (Jenner 2003). Reactive oxygen species are produced during normal cellular

S. Wilkins, C.L. Masters, A.I. Bush, R.A. Cherny, and D.I. Finkelstein (✉)
The University of Melbourne, Melbourne, VIC 3010, Australia
e-mail: dfinkelstein@mhri.edu.au

H.J. Groenewegen et al. (eds.), *The Basal Ganglia IX*, Advances in Behavioral Biology 58, DOI 10.1007/978-1-4419-0340-2_33,

metabolism e.g. hydrogen peroxide (H_2O_2), superoxide and hydroxyl radicals (Halliwell et al. 1992). These molecules can damage DNA, lipids and proteins leading to neuronal cell death from oxidative stress.

Many *in vivo* experimental models of PD exist involving neurotoxins e.g. 1-methyl-4-phenyl-1,2,3,6-tetrahydropyridine (MPTP), 6-hydroxy-dopamine (6-OHDA), rotenone or paraquat. One of the most commonly used, 6-OHDA, is a neurotoxin that rapidly kills catecholaminergic neurons and is used both *in vitro* and *in vivo* (Blum et al. 2001; Schober 2004). 6-OHDA is a dopamine (DA) analogue formed by the interaction of DA and oxygen creating a reactive molecule that does not cross the blood-brain barrier (Blum et al. 2001). The use of this model is highly relevant in PD research as there is evidence that 6-OHDA accumulates in PD patients (Andrew et al. 1993) and is involved in the pathogenesis of PD (Jellinger et al. 1995). The animal model involves injecting 6-OHDA into the SN of the brain, causing a biochemical cascade and subsequent cell loss. 6-OHDA is thought to act on mitochondria (Wu et al. 1996) although the endoplasmic reticulum (ER) may also be affected (Ryu et al. 2002). The neurotoxic mechanism of 6-OHDA involves oxygen radical formation (Cohen and Heikkila 1974), and the oxidative stress effects of 6-OHDA can be alleviated by elevation of superoxide dismutase (Bensadoun et al. 1998) or glutathione peroxidase (Callio et al. 2005).

Clioquinol (CQ) is an orally bioavailable anti-oxidant that crosses the blood-brain barrier and potently inhibits metal-mediated hydrogen peroxide production (Barnham et al. 2004). Unlike traditional high-affinity chelators such as EDTA, CQ does not cause bulk excretion of metals but instead acts as an ionophore to redistribute metals from areas of superabundance to those that may be deficient (Ding et al. 2005). We have previously shown that long-term pre-treatment with CQ reduced the susceptibility of SN neurons to the neurotoxin MPTP (Kaur et al. 2003). In this study we have devised a new experimental paradigm using 6-OHDA, a neurotoxin involved in PD, combined with a model of no prior drug pre-treatment in order to investigate the acute phase of PD. We therefore investigated the effect of CQ modulating the physiological metal milieu and any resultant effects on cell death in both an *in vivo* 6-OHDA animal model and an *in vitro* cell model.

2 Methods

2.1 Mice

All methods conformed to the Australian National Health and Medical Research Council published code of practice for animal research and were approved by the University of Melbourne Animal Ethics Committee. Balbc and C57Bl/6 mice were purchased from the Animal Resources Centre (Western Australia). The animals were killed by an overdose of sodium pentobarbitone (Lethobarb; 100 mg/kg) and perfused with 30 ml of warmed (37°C) 0.1 M PBS, pH 7.4, with heparin (1 U/ml), followed by 30 ml of chilled 4% w/v paraformaldehyde (Sigma, St. Louis, MO) and 0.1 M phosphate buffer (4°C), pH 7.4. The brains were then removed and left at 4°C overnight in 30% w/v sucrose in PBS before being frozen and sectioned on a cryostat.

2.2 6-OHDA Toxin Lesioning

A partial lesion of the SN was produced in the mice by injecting the neurotoxin 6-OHDA into the right SN. The mouse was injected with atropine (0.5 mg/kg) to reduce respiratory tract secretions together with xylazine (10 mg/kg) to produce sedation (intramuscular injection with a 27-gauge needle). After 5 min mice were induced into anaesthesia by inhalation of 3–4% isoflurane carried by oxygen and then maintained at 1.5–2.5% isoflurane. Heads were secured in a stereotaxic frame with the bite bar 3 mm above horizontal. A solution of 6-OHDA (1.65 mg/ml, Sigma) was prepared with ascorbic acid (0.2 mg/ml) and kept on ice until the time of injection. A 26-gauge needle attached via tubing to a 500-µl Hamilton syringe mounted in a syringe pump (Bioanalytical Systems Inc., West Lafayette, IN, USA) was inserted into the right SN through a small hole drilled through the top of the skull. The needle was left to settle for 2 min before 2 µl (2.5–3.3 µg, 1.5–1.65 mg/ml) of 6-OHDA was injected slowly (0.5 µl/min) into the right SN (anteroposterior, 3.0 mm; lateral, 1.1 mm; dorsoventral, 4.7 mm, with respect to Bregma) (Franklin and Paxinos 1997; Parish et al. 2001). On completion of the injection, the needle was left in place for 2 min and then slowly withdrawn at a rate of 1 mm/min. After surgery, the skin was sutured, antiseptic (1% w/w iodine, Betadine; Faulding and Company, South Australia) was applied to the wound, and the mice were left in a warmed cage to recover. Paracetamol (100 mg/kg) was administered in drinking water as an analgesic after surgery.

2.3 Clioquinol Feeding

CQ was suspended in a standard suspension vehicle (SSV; NaCl 0.9% w/v, carboxy methyl cellulose 0.05% w/v, benzyl alcohol 0.05% v/v, Tween 80 0.04% v/v) and was delivered by oral gavage at a daily dosage of 5 or 30 mg/kg for a period of 14 consecutive days post lesion as described previously (Cherny et al. 2001); controls received SSV alone.

2.4 Histology, Estimation of Lesion Size and Stereological Cell Counts

C57Bl/6 brains were cut into 30-µm sections in a 1:3 series and then stained with Neutral Red. The total number of DA neurons in the SN was estimated using a fractionator sampling design (Finkelstein et al. 2000; Stanic et al. 2004; West and Gundersen 1990). Counts were made at regular predetermined intervals (*x* 140 μm, *y* 140 μm). Systematic samples of the area occupied by the nuclei were made from a random starting point. An unbiased counting frame of known area (45 μm × 35 μm) was superimposed on the image of the tissue sections using MBF Bioscience Stereology Investigator 7 software using a Leica DMLB microscope.

2.5 MTT Assay for Determination of Human Neuroblastoma M17 Cell Viability

The human dopaminergic neuroblastoma cell line M17 was maintained in OPTI-MEM media (Gibco) supplemented with 10% fetal calf serum, nonessential amino acids, sodium pyruvate, and antibiotics. Cells were incubated at 37°C in a humidified atmosphere of 95% air and 5% CO_2. The cells were plated out into 48-well culture plates at 4×10^4 cells per well. The cells were left to settle overnight and then incubated with drugs for 24 h prior to being subjected to MTT [3-(4,5-dimethylthiazol-2-yl)-2,5-diphenyl tetrazolium bromide] assays for cell viability. Active mitochondria convert the yellow MTT tetrazolium salt to purple formazan, which can be measured by a spectrophotometer (Mosmann 1983). The mixture of media and reagents e.g. DA was aspirated off and media containing 5 mg/ml MTT was added and the plate incubated for 1 h at 37°C. The remaining media/MTT solution was aspirated off and the cells and MTT were then solubilized with DMSO, and absorbance was read at 595 nm on a Wallac Victor 2 plate reader.

3 Results

3.1 Effects of CQ on Mice Lesioned with 6-OHDA

The effects of CQ were tested on 30 Balbc mice lesioned with 6-OHDA. The mice were divided into three equal groups and for 14 consecutive days post-lesion mice were fed via oral gavage either SSV or CQ at 5 or 30 mg/kg/day. Figure 1 shows neutral red staining of typical lesions for SSV control mice and mice treated with CQ at 30 mg/kg/day. After 14 days SSV vehicle alone, mice had lesions of about 25% whereas mice treated with 30 mg/kg/d CQ had average lesion sizes of 7% (Fig. 2a). Mice treated with CQ at 30 mg/kg/day had significantly smaller lesions than control mice ($p < 0.05$, one-way ANOVA and post hoc Dunnett's multiple comparison test). In a second trial, sixteen C57Bl/6 mice were lesioned with 6-OHDA and were divided into two equal groups. One group was treated with SSV vehicle alone for 14 days post-lesion and the other with CQ 30 mg/kg/day. Lesion sizes reduced from 49% in SSV control mice to 20.4% in CQ-treated mice [Fig. 2b; $n = 6$ (Normal), 7 (SSV), 9 (CQ); $^{***}p < 0.001$ one-way ANOVA with post hoc Bonferroni's multiple comparison test].

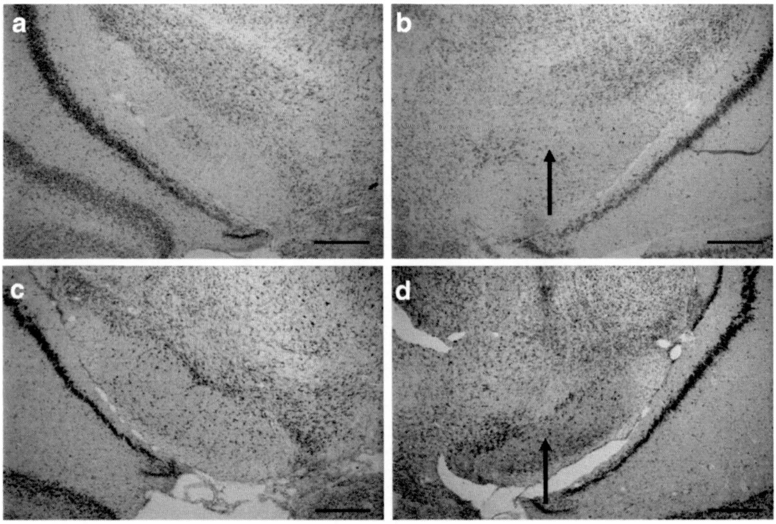

Fig. 1 Representative micrographs showing neutral red staining of cells and lesions in the SN following 6-OHDA treatment with and without subsequent treatment with clioquinol (CQ). (**a**) SSV vehicle alone (control) left non-lesioned side, (**b**) SSV vehicle alone (control) left lesioned side, (**c**) CQ-untreated left side, (**d**) CQ-treated lesioned side. *Arrows* indicate the site of lesion. SSV-fed control mice had an average 25% sized lesions whereas mice receiving 30 mg/kg CQ had an average lesion size of 7%. Scale Bar is 250 μm

3.2 Effects of DA, Metals and CQ in an In Vitro Cellular Model

Human neuroblastoma cells (M17) were exposed to various concentrations of 6-OHDA for 24 h and cell viability was assessed via MTT assay. 6-OHDA caused a dose-dependent decrease in cell viability as concentrations increased to 200 μM at which almost all cells had died (data not shown). 6-OHDA was very potent, but to better model the DA-rich environment in the brain, DA was used for further cell assays. Dopamine caused a dose-dependent decrease in cell viability as concentrations increased to 2 mM (Fig. 3). We calculated the IC_{50} of DA to be 500 μM and used this as starting point to examine the effects of coincubation of other reagents on DA-induced cytotoxicity. We investigated the effects on the cells of co-incubation of the cells with DA and metals. Copper (II) chloride significantly reduced M17 cell viability at concentrations of 200 and 400 μM (Fig. 4a). The addition of 50–400 μM Copper (II) chloride with DA increased cell mortality (Fig. 4a). Iron (II) chloride and Iron (III) chloride were not toxic to M17 cells at concentrations up to 400 μM. Incubation of 400 μM Fe^{2+} or Fe^{3+} with DA showed some protection against DA-induced cell death (Fig. 4b, c). Coincubation of CQ (1 μM) and catalase (2,000 U) was completely non-toxic to cells (Fig. 4d). When cells were incubated with CQ in the presence of DA there was a significant reduction of DA-induced cell

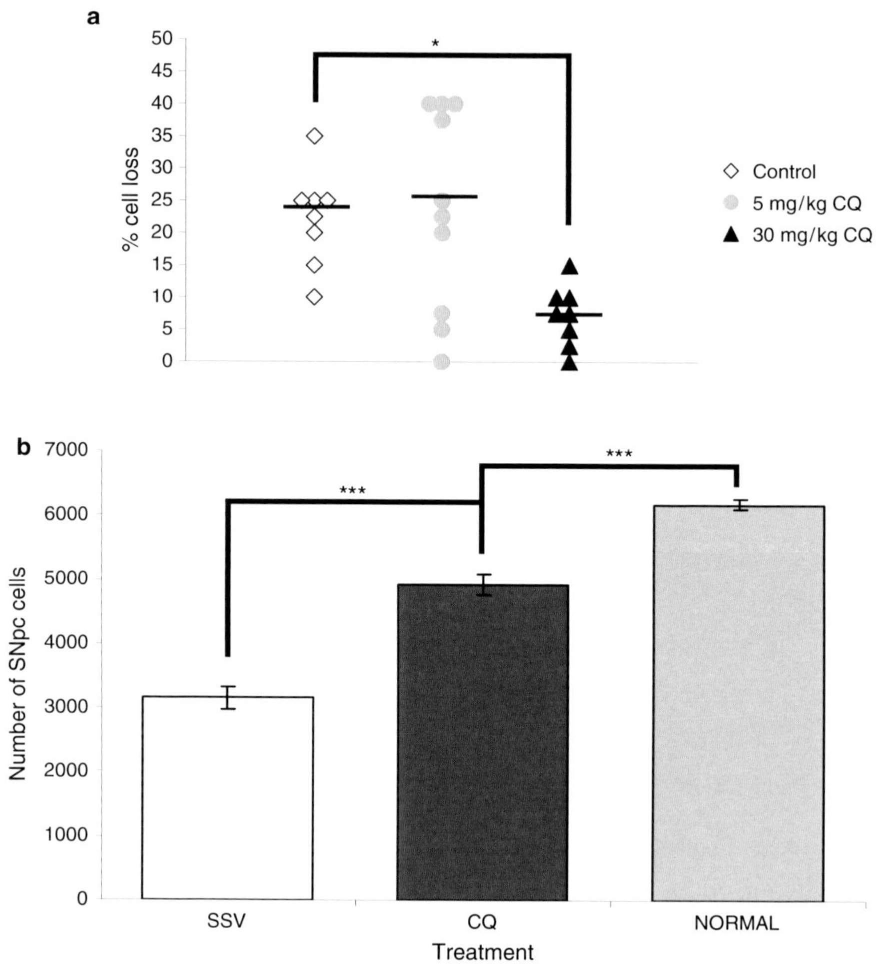

Fig. 2 (**a**) Graph showing the relationship between cell loss in the SN of Balbc mice and the drug administered following 6-OHDA treatment. Mice were lesioned with 6-OHDA and then dosed with SSV (control) or CQ on 14 consecutive days after surgery. Each point indicates an individual mouse whether control mice (*open diamonds*), 5 mg/kg/day CQ-treated mice (*grey circles*) or 30 mg/kg/day CQ-treated mice (*black triangles*). *Bars* indicate mean average lesion size for each group. CQ at 30 mg/kg/day significantly reduced lesion sizes in treated mice [n = 8 (SSV), 10 (5 mg/kg/day), 8 (30 mg/kg/day); $*p < 0.05$ one-way ANOVA with post hoc Dunnett's multiple comparison test]. (**b**) Graph showing the relationship between cell loss in the SN of C57Bl/6 mice and the CQ administered following 6-OHDA treatment. Mice were lesioned with 6-OHDA and then dosed with SSV (control vehicle) or CQ on 14 consecutive days after surgery. CQ at 30 mg/kg/day significantly reduced lesion sizes in treated mice from 49% of control values to 20.4% [Data ± SEM, n = 6 (normal), 7 (SSV), 9 (CQ); $***p < 0.001$ one-way ANOVA with post hoc Bonferroni's multiple comparison test]

death by 56% (Fig. 4d). Addition of catalase also reduced DA-induced cell death by 37%. Neither CQ nor catalase reduced cell death when 100 μM Cu^{2+} or Fe^{2+} and DA were present (Fig. 4d).

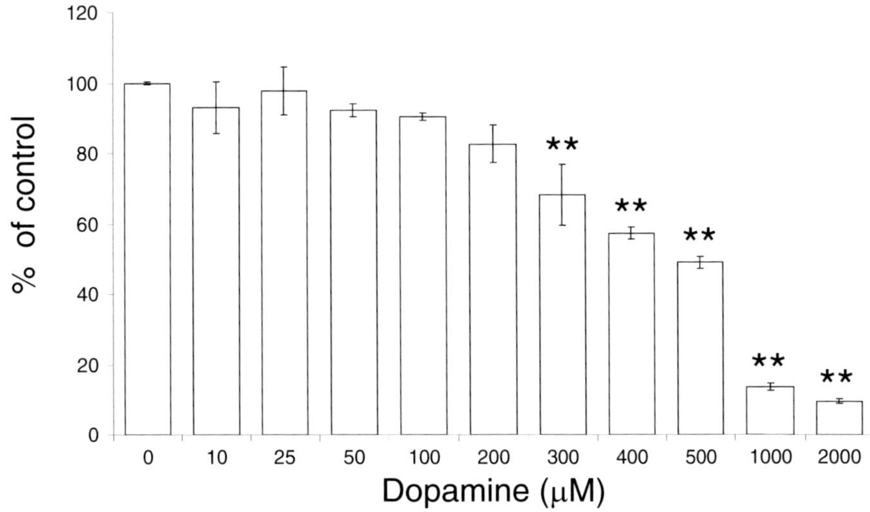

Fig. 3 Graph showing the dose-dependent effect of increasing concentrations of dopamine on M17 neuroblastoma cell viability. Cells were plated out and incubated with different concentrations of dopamine diluted in media for 24 h. Cell viability was detected with a MTT assay. The graph shows an IC_{50} of dopamine to be approximately 500 μM. Each value is the mean ± standard error ($n = 3$ experiments carried out in duplicate) expressed as a percentage of the control. $^{**}p < 0.01$ comparison to control (0 μM DA) (one-way ANOVA with post hoc Dunnett's multiple comparison test)

4 Discussion

Clioquinol (CQ; 5-chloro-7-iodo-8-hydroxyquinoline), an orally bioavailable ligand with moderate affinity for copper, zinc and iron, has shown encouraging neuroprotective results in the MPTP mouse model (Kaur et al. 2003). MPTP and 6-OHDA are two of the most widely used models for PD (Schober 2004). In this study CQ reduced 6-OHDA lesion sizes in Balbc and C57Bl/6 mice without prior drug pretreatment. These data suggest that nigral toxicity in both MPTP and 6-OHDA models has common pathways that involve metals and oxidative stress. Although some of the biochemistry of the MPTP and 6-OHDA models is different there are similarities in the way cells are killed by these toxins (Blum et al. 2001). MPTP is converted into 1-methyl-4-phenylpyridinium (MPP+) by monoamine oxidase in glial cells, which is then in turn taken up by neurons through a DA transporter whereas 6-OHDA enters neuronal cells directly through a DA transporter. Inside the cell both 6-OHDA and MPP+ interact with Fe^{2+} resulting in hydrogen peroxide production leading to oxidative stress and cell death. Both toxins also directly inhibit mitochondrial function leading to cell death (Blum et al. 2001). The pathology of both models shows the reduction of striatal DA levels and a loss of immunoreactive striatal tyrosine hydroxylase fibres and neurons (Schober 2004). One of the key similarities between the models is the role of both iron and hydrogen peroxide production. We investigated this hypothesis in a M17 neuroblastoma cell line model.

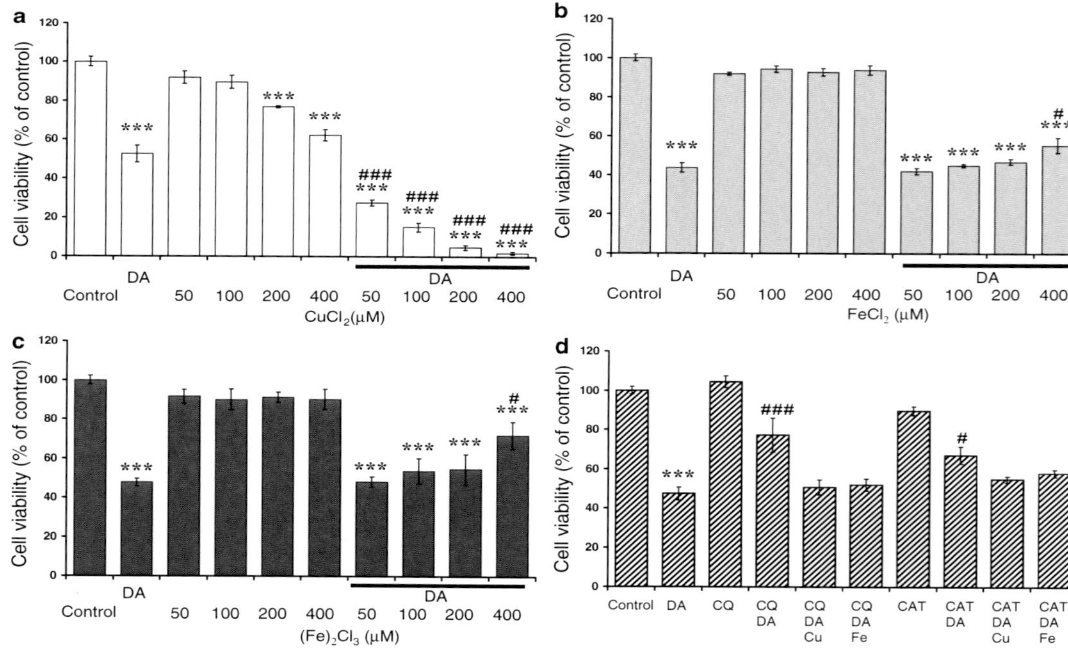

Fig. 4 Cells were plated out and incubated with 500 μM dopamine diluted in media for 24 h in the absence or presence of different concentrations of metal ions. Cell viability was detected via MTT assay. Each value is the mean ± standard error ($n = 4$) expressed as a percentage of the control. (**a**) Effects of copper with and without dopamine, (**b**) effects of Fe^{2+} with and without dopamine, (**c**) effects of Fe^{3+} with and without dopamine. $^{***}p < 0.001$ comparison to control, $\#p < 0.05$ comparison to DA, $\#\#\#p < 0.001$ comparison to DA, (**d**) effects of CQ and catalase with and without dopamine and metals. DA, Dopamine 500 μM; CQ 1 μM; CAT, catalase 2,000 units; Fe, Fe^{2+} 100 μM; and Cu, Cu^{2+} 100 μM. $^{***}p < 0.001$ comparison to control, $\#p < 0.05$ comparison to DA, $\#\#\#p < 0.001$ comparison to DA (one-way ANOVA with post hoc Bonferroni's multiple comparison test). CQ and catalase reduced DA-induced cell death by 56% and 37%, respectively

Neuroblastoma cells containing elevated levels of DA were used as a model of SN cells in the brain. After cell loss in the SN-lesioned animal, the remaining neurons compensate for the loss of cells by increasing both the amount of DA produced and the rate of turnover (Stanic et al. 2003). Dopamine kills neurons *in vivo* by interacting with cellular iron and producing hydrogen peroxide (Blum et al. 2001). 6-OHDA and DA were toxic to M17 cells in a dose-dependent manner. However, CQ protected M17 cells against DA-induced cell death. Copper that was toxic at certain concentrations increased the toxic effects of DA on cells. Copper converts DA to toxic byproducts e.g. *o*-quinone and hydrogen peroxide resulting in increased cell death (Izumi et al. 2005; Paris et al. 2005). Adding Fe^{2+} or Fe^{3+} did not result in increased cell death due to iron interacting with DA and forming the non-toxic aminochrome and little or no hydrogen peroxide (Izumi et al. 2005; Paris et al. 2005). Addition of high concentrations of Fe^{2+} and Fe^{3+} offered some protection against DA-induced cell death. However, caution must be taken with the interpretation of cell culture results (Halliwell 2003), and any cellular protection seen against DA with high concentrations of Fe^{2+} and Fe^{3+} may be specific to cell culture rather than the *in vivo* condition due to the wide body of evidence implicating iron in the pathogenesis of PD. Catalase partially protected cells in the presence of DA, indicating that hydrogen peroxide was produced in the culture in the presence of background levels of metals in the media. Trace metal levels in cell culture media were detected as 6.5 μM Fe and 0.2 μM Cu (unpublished observations). It is therefore clear that CQ and to a lesser extent catalase protected cells in the media with these background levels of metal, but not when much higher concentrations of copper or iron were added in the presence of DA. The protective effect of CQ in the cell model again highlights the role of metal in catecholamine-induced cell death.

The association between iron and PD is complex and finely balanced; insufficient iron can lead to neurological deficits, but excess iron will cause neuronal loss (Fredriksson et al. 1999). Iron plays a key role in the pathogenesis of PD and there is much evidence showing a selective increase in iron in the SN of PD patients (Berg et al. 2001; Faucheux et al. 2002), in rats following 6-OHDA lesions (Ben-Shachar et al. 1991), and in primates following MPTP treatment (Mochizuki et al. 1994; Temlett et al. 1994). Direct injection of iron into the brain produces neurotoxic effects comparable to those induced by 6-OHDA, which suggests involvement of iron in 6-OHDA-induced neuronal degeneration (Ben-Shachar and Youdim 1991). Intranigral injection of iron has also been put forward as a potential model for PD (Sziraki et al. 1998). Whether the increase in iron in the SN is a causative factor in PD or a secondary phenomenon is yet to be resolved (Berg et al. 2001; Thompson et al. 2001). However, prevention of the toxic effects of 6-OHDA with iron-chelating compounds or vitamin E provides further evidence for the involvement of iron and oxidative stress in the mechanism of toxicity of 6-OHDA (Ben-Shachar et al. 1991; Cadet et al. 1989; Knoll 1986; Perumal et al. 1992; Takano et al. 2007).

In the current study it is likely that the reduction in metals available for oxidative stress mechanisms within the brain reduced the deleterious effects of 6-OHDA by leading to a reduction in oxidative stress resulting in better dopaminergic SN neuron survival in CQ-treated animals. The introduction of CQ breaks the chain of DA–metal–oxidative stress–cell death and so protects cells in the *in vivo* MPTP model (Kaur et al. 2003), in the 6-OHDA model and in the *in vitro* cellular model in this study. These results suggest that CQ and similar agents may be useful for the prevention and treatment of Parkinson's disease in the future.

Acknowledgements This study was funded by Prana Biotechnology Ltd., Australian Brain Foundation as well as funds from the National Health and Medical Research Council.

References

Andrew R, Watson DG, Best SA, Midgley JM, Wenlong H and Petty RK (1993) The determination of hydroxydopamines and other trace amines in the urine of parkinsonian patients and normal controls. Neurochem Res 18: 1175–1177.

Barnham KJ, Haeffner F, Ciccotosto GD, Curtain CC, Tew D, Mavros C, Beyreuther K, Carrington D, Masters CL, Cherny RA, Cappai R and Bush AI (2004) Tyrosine gated electron transfer is key to the toxic mechanism of Alzheimer's disease beta-amyloid. FASEB J 18: 1427–1429.

Bensadoun JC, Mirochnitchenko O, Inouye M, Aebischer P and Zurn AD (1998) Attenuation of 6-OHDA-induced neurotoxicity in glutathione peroxidase transgenic mice. Eur J Neurosci 10: 3231–3236.

Ben-Shachar D and Youdim MB (1991) Intranigral iron injection induces behavioral and biochemical "parkinsonism" in rats. J Neurochem 57: 2133–2135.

Ben-Shachar D, Eshel G, Finberg JP and Youdim MB (1991) The iron chelator desferrioxamine (Desferal) retards 6-hydroxydopamine-induced degeneration of nigrostriatal dopamine neurons. J Neurochem 56: 1441–1444.

Berg D, Gerlach M, Youdim MB, Double KL, Zecca L, Riederer P and Becker G (2001) Brain iron pathways and their relevance to Parkinson's disease. J Neurochem 79: 225–236.

Blum D, Torch S, Lambeng N, Nissou M, Benabid AL, Sadoul R and Verna JM (2001) Molecular pathways involved in the neurotoxicity of 6-OHDA, dopamine and MPTP: contribution to the apoptotic theory in Parkinson's disease. Prog Neurobiol 65: 135–172.

Cadet JL, Katz M, Jackson-Lewis V and Fahn S (1989) Vitamin E attenuates the toxic effects of intrastriatal injection of 6-hydroxydopamine (6-OHDA) in rats: behavioral and biochemical evidence. Brain Res 476: 10–15.

Callio J, Oury TD and Chu CT (2005) Manganese superoxide dismutase protects against 6-hydroxydopamine injury in mouse brains. J Biol Chem 280: 18536–18542.

Cherny RA, Atwood CS, Xilinas ME, Gray DN, Jones WD, McLean CA, Barnham KJ, Volitakis I, Fraser FW, Kim Y, Huang X, Goldstein LE, Moir RD, Lim JT, Beyreuther K, Zheng H, Tanzi RE, Masters CL and Bush AI (2001) Treatment with a copper–zinc chelator markedly and rapidly inhibits beta-amyloid accumulation in Alzheimer's disease transgenic mice. Neuron 30: 665–676.

Cohen G and Heikkila RE (1974) The generation of hydrogen peroxide, superoxide radical, and hydroxyl radical by 6-hydroxydopamine, dialuric acid, and related cytotoxic agents. J Biol Chem 249: 2447–2452.

Ding WQ, Liu B, Vaught JL, Yamauchi H and Lind SE (2005) Anticancer activity of the antibiotic clioquinol. Cancer Res 65: 3389–3395.

Fahn S and Cohen G (1992) The oxidant stress hypothesis in Parkinson's disease: evidence supporting it. Ann Neurol 32: 804–812.

Faucheux BA, Martin ME, Beaumont C, Hunot S, Hauw JJ, Agid Y and Hirsch EC (2002) Lack of up-regulation of ferritin is associated with sustained iron regulatory protein-1 binding activity in the substantia nigra of patients with Parkinson's disease. J Neurochem 83: 320–330.

Finkelstein DI, Stanic D, Parish CL, Tomas D, Dickson K and Horne MK (2000) Axonal sprouting following lesions of the rat substantia nigra. Neuroscience 97: 99–112.

Franklin KBJ and Paxinos G (1997) The mouse brain in stereotaxic coordinates. San Diego: Academic Press.

Fredriksson A, Schroder N, Eriksson P, Izquierdo I and Archer T (1999) Neonatal iron exposure induces neurobehavioural dysfunctions in adult mice. Toxicol Appl Pharmacol 159: 25–30.

Halliwell B (2003) Oxidative stress in cell culture: an under-appreciated problem? FEBS Lett 540: 3–6.

Halliwell B, Gutteridge JM and Cross CE (1992) Free radicals, antioxidants, and human disease: where are we now? J Lab Clin Med 119: 598–620.

Hirsch EC (1993) Does oxidative stress participate in nerve cell death in Parkinson's disease? Eur Neurol 33 Suppl 1: 52–59.

Izumi Y, Sawada H, Yamamoto N, Kume T, Katsuki H, Shimohama S and Akaike A (2005) Iron accelerates the conversion of dopamine-oxidized intermediates into melanin and provides protection in SH-SY5Y cells. J Neurosci Res 82: 126–137.

Jellinger K, Linert L, Kienzl E, Herlinger E and Youdim MB (1995) Chemical evidence for 6-hydroxydopamine to be an endogenous toxic factor in the pathogenesis of Parkinson's disease. J Neural Transm Suppl 46: 297–314.

Jenner P (2003) Oxidative stress in Parkinson's disease. Ann Neurol 53 Suppl 3: S26–S36; discussion S36–S28.

Jenner P and Olanow CW (1996) Oxidative stress and the pathogenesis of Parkinson's disease. Neurology 47: S161–S170.

Kaur D, Yantiri F, Rajagopalan S, Kumar J, Mo JQ, Boonplueang R, Viswanath V, Jacobs R, Yang L, Beal MF, DiMonte D, Volitaskis I, Ellerby L, Cherny RA, Bush AI and Andersen JK (2003) Genetic or pharmacological iron chelation prevents MPTP-induced neurotoxicity in vivo: a novel therapy for Parkinson's disease. Neuron 37: 899–909.

Knoll J (1986) The pharmacology of (-)deprenyl. J Neural Transm Suppl 22: 75–89.

Lang AE and Lozano AM (1998) Parkinson's disease. First of two parts. N Engl J Med 339: 1044–1053.

Mochizuki H, Imai H, Endo K, Yokomizo K, Murata Y, Hattori N and Mizuno Y (1994) Iron accumulation in the substantia nigra of 1-methyl-4-phenyl-1,2,3,6-tetrahydropyridine (MPTP)-induced hemiparkinsonian monkeys. Neurosci Lett 168: 251–253.

Mosmann T (1983) Rapid colorimetric assay for cellular growth and survival: application to proliferation and cytotoxicity assays. J Immunol Methods 65: 55–63.

Paris I, Martinez-Alvarado P, Cardenas S, Perez-Pastene C, Graumann R, Fuentes P, Olea-Azar C, Caviedes P and Segura-Aguilar J (2005) Dopamine-dependent iron toxicity in cells derived from rat hypothalamus. Chem Res Toxicol 18: 415–419.

Parish CL, Finkelstein DI, Drago J, Borrelli E and Horne MK (2001) The role of dopamine receptors in regulating the size of axonal arbors. J Neurosci 21: 5147–5157.

Perumal AS, Gopal VB, Tordzro WK, Cooper TB and Cadet JL (1992) Vitamin E attenuates the toxic effects of 6-hydroxydopamine on free radical scavenging systems in rat brain. Brain Res Bull 29: 699–701.

Ryu EJ, Harding HP, Angelastro JM, Vitolo OV, Ron D and Greene LA (2002) Endoplasmic reticulum stress and the unfolded protein response in cellular models of Parkinson's disease. J Neurosci 22: 10690–10698.

Schober A (2004) Classic toxin-induced animal models of Parkinson's disease: 6-OHDA and MPTP. Cell Tissue Res 318: 215–224.

Stanic D, Finkelstein DI, Bourke DW, Drago J and Horne MK (2003) Timecourse of striatal re-innervation following lesions of dopaminergic SNpc neurons of the rat. Eur J Neurosci 18: 1175–1188.

Stanic D, Tripanichkul W, Drago J, Finkelstein DI and Horne MK (2004) Glial responses associated with dopaminergic striatal reinnervation following lesions of the rat substantia nigra. Brain Res 1023: 83–91.

Sziraki I, Mohanakumar KP, Rauhala P, Kim HG, Yeh KJ and Chiueh CC (1998) Manganese: a transition metal protects nigrostriatal neurons from oxidative stress in the iron-induced animal model of parkinsonism. Neuroscience 85: 1101–1111.

Takano K, Kitao Y, Tabata Y, Miura H, Sato K, Takuma K, Yamada K, Hibino S, Choshi T, Iinuma M, Suzuki H, Murakami R, Yamada M, Ogawa S and Hori O (2007) A dibenzoylmethane derivative

protects dopaminergic neurons against both oxidative stress and endoplasmic reticulum stress. Am J Physiol Cell Physiol 293: C1884–C1894.
Temlett JA, Landsberg JP, Watt F and Grime GW (1994) Increased iron in the substantia nigra compacta of the MPTP-lesioned hemiparkinsonian African green monkey: evidence from proton microprobe elemental microanalysis. J Neurochem 62: 134–146.
Thompson KJ, Shoham S and Connor JR (2001) Iron and neurodegenerative disorders. Brain Res Bull 55: 155–164.
West MJ and Gundersen HJ (1990) Unbiased stereological estimation of the number of neurons in the human hippocampus. J Comp Neurol 296: 1–22.
Wu Y, Blum D, Nissou MF, Benabid AL and Verna JM (1996) Unlike MPP+, apoptosis induced by 6-OHDA in PC12 cells is independent of mitochondrial inhibition. Neurosci Lett 221: 69–71.

Oscillatory Activity and Synchronization in the Basal Ganglia Network in Rodent Models of Parkinson's Disease

Judith R. Walters, Patrick L. Tierney, and Debra A. Bergstrom

Abstract The efficacy of deep brain stimulation (DBS) in the subthalamic nucleus (STN) in Parkinson's disease (PD) has focused attention on the role of STN firing patterns in PD symptomatology. Oscillatory activity in the beta frequency range is of special interest as local field potential (LFP) recordings in the STN in bradykinetic PD patients during DBS electrode placement show prominent activity in this frequency range. As increased synchronization between the globus pallidus pars externus (GP) and STN has been implicated in the emergence of oscillatory activity in the STN, the effect of dopamine loss and increased dopamine stimulation on beta range activity in paired GP-STN single unit and LFP recordings was examined in intact or nigrostriatally lesioned, locally anesthetized, and immobilized rats. Spike-triggered waveform averages (STWAs) show that GP spiking becomes significantly more synchronized with STN LFP oscillations in the beta range after dopamine cell lesion. Coherences between GP spiking and STN LFP and between STN spiking and STN LFP are increased after dopamine cell lesion. A desynchronizing effect of increased dopamine receptor stimulation with apomorphine on GP-STN relationships in both intact and lesioned rats is observed. Results support a role for increased synchronization between STN and GP in the emergence of beta range activity in the STN and a greater impact of GP on STN activity after decreased dopamine receptor stimulation.

1 Introduction

The efficacy of deep brain stimulation (DBS) in Parkinson's disease (PD) and the availability of data from PD patients during electrode implantation have focused attention on the pathological role of synchronized and oscillatory activity

J.R. Walters (✉), P.L. Tierney, and D.A. Bergstrom
Neurophysiological Pharmacology Section, National Institutes of Health, National Institute of Neurological Disorders and Stroke, Building 35, Room 1C905, 35 Convent Drive, Bethesda, MD, 20892, USA
e-mail: waltersj@ninds.nih.gov

H.J. Groenewegen et al. (eds.), *The Basal Ganglia IX*, Advances in Behavioral Biology 58, DOI 10.1007/978-1-4419-0340-2_34,

in basal ganglia networks after loss of dopamine. Oscillatory activity in the beta frequency range is of particular interest in this regard. Broadly defined as 8–35 Hz, or more narrowly defined as 12- or 15–30 Hz, depending on the study, increased beta power is observed in subthalamic nucleus (STN) LFP recordings in PD patients at rest and hypothesized to be relevant to the emergence of bradykinesia in these individuals (Brown et al. 2001; Marsden et al. 2001; Cassidy et al. 2002; Levy et al. 2002; Priori et al. 2002, 2004; Williams et al. 2002; Brown 2003, 2007; Kuhn et al. 2004; Alegre et al. 2005; Doyle et al. 2005; Foffani et al. 2005; Alonso-Frech et al. 2006; Weinberger et al. 2006). Beta frequency activity in the STN is reduced following L-dopa treatment and during movement in PD patients off medication, while stimulation of the STN at beta frequencies promotes bradykinesia in PD patients (Chen et al. 2007). In addition to bradykinesia, rigidity and tremor have also been linked to the presence of oscillatory activity in basal ganglia circuits (Krack et al. 1998; Hurtado et al. 1999; Lemstra et al. 1999; Deuschl et al. 2000; Levy et al. 2000; Kuhn et al. 2006; Shapiro et al. 2007; Tabbal et al. 2008).

These observations have lead to interest in how pathological firing patterns emerge in the basal ganglia in PD. The present chapter reviews results from a study exploring one aspect of emergent oscillatory activity in the basal ganglia network in a rodent model of PD: specifically, the relationship between increases in beta frequency oscillations in the STN and expression of beta range activity in the globus pallidus pars externus (GP), a nucleus that both projects to and receives from the STN.

It is a challenge, in league with the chicken or the egg question, to determine where and how oscillations start in a network. One approach to this problem is to identify the conditions under which the oscillatory firing patterns emerge and then define the relationships between the components of the network that become engaged. Clearly, decreased dopamine receptor stimulation is a condition that strongly facilitates the emergence of oscillatory activity in the basal ganglia. Increases in basal ganglia oscillatory activity have been reported in nonhuman primates and rodents after dopamine cell lesion via MPTP treatment, 6-hydroxydopamine infusion, or dopamine receptor antagonist administration; this has allowed investigators to use the dopamine depleted animal models of PD to explore relationships between oscillatory activity in different components of the basal ganglia network. A number of studies have focused on the effect of dopamine cell lesion on firing patterns that emerge in the basal ganglia when cortical input is synchronized, as occurs during anesthesia where the dominant activity in the cortex has a "slow wave" ~1-Hz periodicity (Magill et al. 2001; Tseng et al. 2001; Kasanetz et al. 2002; Murer et al. 2002; Belluscio et al. 2003, 2007; Walters et al. 2005, 2007; Mallet et al. 2006; Parr-Brownlie et al. 2007; Zold et al. 2007a, b; Aravamuthan et al. 2008), or during high-voltage spindles when synchronized activity emerges in the 5–13-Hz range (Dejean et al. 2008). Results have provided evidence for dopamine cell lesion-induced alterations in striatal processing of cortical input to the indirect pathway and reduction in transmission of cortical activity through the direct

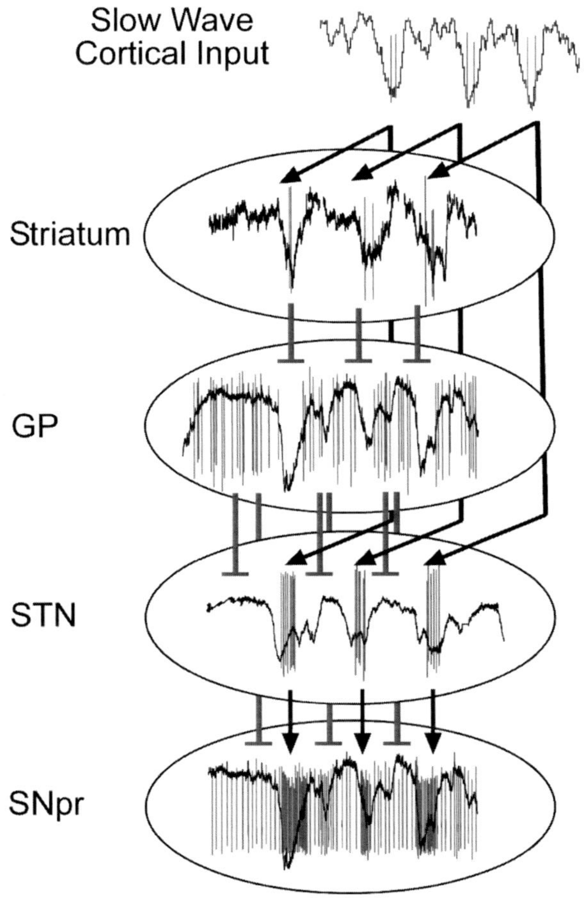

Fig. 1 Phase model illustrating a hypothesized scheme for passage of oscillatory signals in the slow ~1-Hz range through the corticobasal ganglia pathway after dopamine loss in anesthetized rats. *Black lines with arrows* indicate excitatory connections between nuclei and *gray lines with bars* indicate inhibitory connections. Examples of striatal, GP, STN, and SNpr spiking activity recorded simultaneously with SNpr LFP are shown with SNpr LFP as a common temporal reference (indicated in *black*). Cortical oscillatory inputs to the striatum and STN are indicated. Results are consistent with loss of dopamine enhancing transmission of oscillatory activity primarily through the indirect pathway, resulting in robust oscillatory activity in the SNpr. Implied in this model are (1) D2 receptor-bearing striatopallidal neurons transmitting patterned activity from the cortex to the GP, (2) striatally mediated pauses in inhibitory GP activity contributing to the timing of bursts in STN neuronal activity, and (3) pauses in inhibitory GP output coinciding with bursts in excitatory STN output supporting enhanced oscillatory activity in SNpr/GPi spike trains. Abbreviations: *GP* globus pallidus pars externus, *STN* subthalamic nucleus, *SNpr* substantia nigra pars reticulata, *arrows* excitatory transmission, *bars* inhibitory transmission

pathway (Tseng et al. 2001; Murer et al. 2002; Mallet et al. 2006; Belluscio et al. 2007; Zold et al. 2007b; Dejean et al. 2008). Phase relationships assessed through simultaneous recordings in different basal ganglia nuclei in the urethane-anesthetized rat model of PD support the view that after loss of dopamine, increases in oscillatory activity in the indirect pathway are facilitated by changes in striatal input to the GP, and increases in oscillatory activity in the STN appear to be supported by convergent antiphasic inhibitory input from the GP and excitatory oscillatory input from the cortex via the hyperdirect pathway (Fig. 1). Furthermore, increased oscillatory activity in substantia nigra pars reticulata (SNpr) spike trains in this PD model appears organized by convergent antiphasic inhibitory and excitatory oscillatory inputs from the GP and STN, respectively (Walters et al. 2007).

While these observations provide insight into how loss of dopamine facilitates the entrainment of basal ganglia activity to cortical slow wave activity in the anesthetized rat, it is unclear whether the same principles apply to the emergence of faster frequency beta range oscillatory activity observed in the STN in PD patients. Nevertheless, investigations utilizing in vitro preparations have called attention to the potential importance of changes in GP input to the STN in generating increased beta activity in STN firing pattern after loss of dopamine (Bevan et al. 2000, 2002a, b, 2006, 2007; Shen et al. 2003; Floran et al. 2004; Baufreton et al. 2005; Hallworth and Bevan 2005; Shen and Johnson 2005; Baufreton and Bevan 2008). This chapter reexamines data from an earlier study conducted in locally anesthetized, artificially respired, immobilized rats involving simultaneous recordings in the STN and GP after dopamine cell lesion (Tierney et al. 2003). The goals were to determine whether beta frequency range (15–30 Hz) activity is increased in the rodent STN and GP after unilateral dopamine cell lesion, whether GP and STN activity becomes more synchronized in this frequency range, and whether dopamine agonist administration reduces this synchronization.

2 Results

2.1 *Beta Frequency Activity in Paired GP-STN Recordings: Firing Rate and Pattern*

Relationships between spike train firing patterns and LFPs in GP and STN were evaluated in a rat model of PD using extracellular single unit recording techniques as described previously (Kreiss et al. 1997; Allers et al. 2000; Ruskin et al. 2003). Four to six weeks after dopamine cell lesion, firing rates in the STN were significantly increased and firing rates in the GP were significantly decreased (Fig. 2a). In addition, analysis of spike trains showed that the coefficient of variation of the interspike intervals (ISI CV) in the GP significantly increased after dopamine cell lesion (intact: 0.65 ± 0.12, $n = 15$; lesioned: 1.00 ± 0.11, $n = 19$), indicating that GP firing rate becomes more variable and potentially more patterned. However, no significant change was observed in STN ISI CV following dopamine cell lesion (intact: 1.17 ± 0.11, $n = 16$; lesioned: 1.20 ± 0.10, $n = 24$), consistent with results

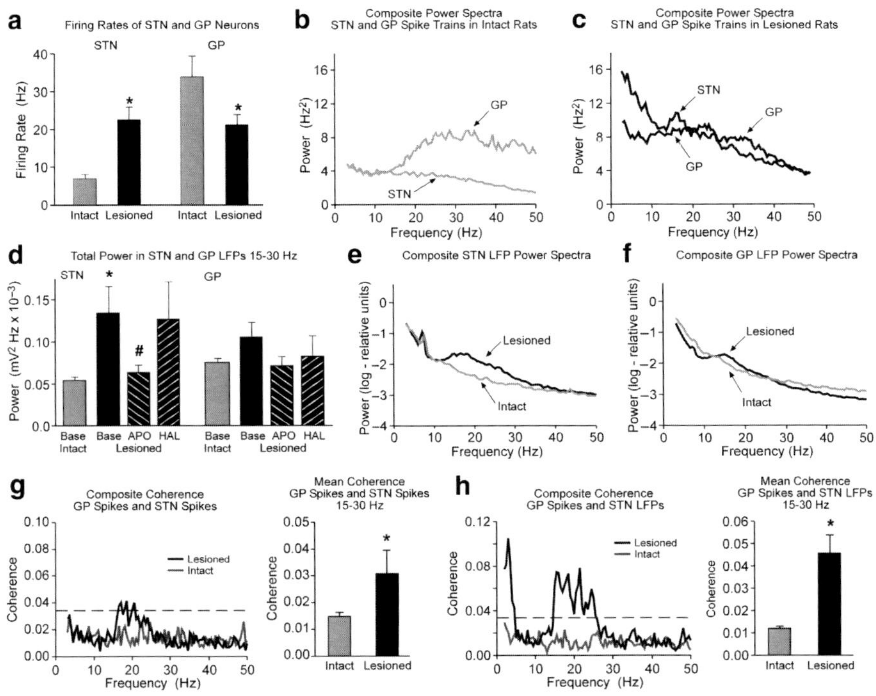

Fig. 2 Rate, beta power, and spike-LFP coherence in paired GP-STN recordings in intact and dopamine cell-lesioned rats. Recordings were conducted as described in Ruskin et al. (2003) with LFP recordings and data analyses conducted as described in Walters et al. (2007). For data analyses, 180 s of spike train and LFP recordings that were representative and free from artifacts were selected. (**a**) Firing rates significantly increased in the STN and significantly decreased in the GP following dopamine cell lesion (n = 15–24 spike trains from 7–9 rats). (**b**, **c**). Following dopamine cell lesion, spectral analyses of STN and GP spike trains showed increased overlap in the beta frequency range. (**d–f**) STN LFPs from lesioned rats showed significant increases in power in the 15–30-Hz beta range. GP LFPs from lesioned rats also showed increases in beta power, with significant increases observed in the 15–18-Hz range but not over the entire 15–30-Hz range. Lesion-induced increase in STN beta power was significantly reduced by systemic apomorphine administration (0.05 mg/kg, i.v.); haloperidol (0.2 mg/kg, i.v.) reversed the dopamine agonist-induced decrease in LFP power. This agonist-induced decrease in STN LFP beta power was not observed in intact rats (data not shown; apomorphine 0.3 mg/kg) or in GP LFP beta power of lesioned rats. Aforementioned graphs represent averages of individual power spectra from 7–9 LFP or spike/LFP recordings with 1 recording per rat. (**g**, **h**) Coherence analyses demonstrated that GP spiking activity became more time locked to the beta frequency oscillations in STN spiking and LFPs after dopamine cell lesion. *Dashed line* indicates statistical significance of p = 0.05. *Significantly different from intact; # significantly different from baseline and haloperidol

from earlier studies in locally anesthetized rats (Kreiss et al., 1997; Allers et al. 2005) and in contrast to results observed in urethane, ketamine, or chloral hydrate anesthetized rats (Hollerman and Grace 1992; Hassani et al. 1996; Perier et al. 2000; Vila et al. 2000; Magill et al. 2001; Ni et al. 2001; Breit et al. 2005, 2006; Walters et al. 2005, 2007; Parr-Brownlie et al. 2007).

To examine relationships between oscillatory activity in the STN and GP, four different approaches were taken. The first approach involved comparisons of oscillatory power in spike trains recorded from the two nuclei in intact and lesioned rats. Spike trains were converted to waveforms (CED Spike2 scripts, sampling rate of 1 KHz); power spectra were constructed over the range of 3–50 Hz with a 0.5-Hz resolution and averaged power spectra were compared for intact and lesioned rats. Results (Fig. 2b, c) show that in intact rats spike train power in the beta range was greater in the GP than in the STN. This was consistent with the GP having a faster firing rate than the STN in the intact state and, therefore, having greater potential for power in beta frequencies. After dopamine cell lesion, however, consistent with the increase in STN firing rates and the decrease in GP firing rates (Fig. 2a), spike train power in the beta frequency range was more comparable in the two nuclei. There was more overlap between the GP and STN in the distribution of spike train power in the beta frequency range in the lesioned rat than in the intact rat. These changes in firing rates and pattern in the STN and GP spike trains suggested that there was potential for increased synchronization of spiking activity between the two nuclei in the beta frequency range after dopamine cell lesion.

2.2 Beta Frequency Activity in Paired GP–STN Recordings: LFP Power

A second approach to examining relationships between oscillatory activity in the STN and GP involved comparison of local field potentials (LFPs), recorded simultaneously with single unit activity as described previously (Walters et al. 2007). Spectral analysis of LFPs from recordings in the STN of awake rats showed a significant increase in total power in the 15–30-Hz frequency range in STN LFPs of lesioned rats compared with intact rats (Fig. 2d, e). A significant increase in beta power was also observed in the simultaneously recorded GP LFPs, although in the narrower frequency range of 15–18 Hz (Fig. 2d, f).

Notably, in lesioned rats STN LFP beta power (15–30 Hz) was markedly decreased by i.v. administration of the D1/D2 dopamine receptor agonist apomorphine (Fig. 2d). Apomorphine's effect in lesioned rats was effectively reversed by the dopamine antagonist haloperidol. The agonist-induced decrease in LFP beta power observed in the STN in the lesioned rats was not observed in intact rats (data not shown) or in the GP of lesioned rats (Fig. 2d).

2.3 Beta Frequency Activity in Paired GP–STN Recordings: Coherence and Spike–Triggered LFP Waveforms

A third approach to examining relationships between oscillatory activity in the STN and GP involved assessment of coherence between spike trains converted to waveforms and LFPs in STN and GP in the beta frequency range. Coherence was calculated with Spike2 scripts with 88 nonoverlapping windows, a 0.5-Hz resolution on 180-s recording epochs, and confidence levels determined as described by Rosenberg et al. (1989). Coherences between GP and STN spike trains (Fig. 2g), GP spike trains and STN LFP (Fig. 2h), and STN spike trains and STN LFP (data not shown) were all significantly increased in the lesioned preparation, supporting the view that increases in beta activity in the GP and STN in the rodent model of PD involve increased synchronization between the two nuclei in this frequency range.

A fourth approach to examine relationships between GP and STN involved analyzing spike-triggered waveform average (STWAs) peak to trough amplitudes (Fig. 3). STN spike-triggered STN LFP waveform averages in lesioned rats had mean peak to trough amplitudes 3.5 times those observed in intact rats (Fig. 3b), and GP spike-triggered GP LFP waveform average peak to trough amplitudes were 2.5 times those observed in intact rats (Fig. 3e). These data show that spiking activity becomes significantly more synchronized to LFP fluctuations in the beta frequency range in each of these nuclei after dopamine cell lesion. As LFP fluctuations reflect net potential changes of the neuronal population surrounding the recording electrode, increased LFP peak to trough amplitude within a given frequency range implies increased synchronization of activity in that frequency range within the GP and the STN.

Moreover, significant increases in peak to trough amplitudes were also observed when relationships between STN spikes and GP LFPs or GP spikes and STN LFPs were compared in intact and lesioned rats. Mean peak to trough amplitudes of STN spike-triggered GP LFP waveforms were 2.4 times greater after dopamine cell lesion (Fig. 3d). Interestingly, the most dramatic increase noted in STWAs in the lesioned rats was in GP spike-triggered STN LFP waveform peak to trough amplitudes (Fig. 3a). Lesioned rats had mean GP spike-triggered STN LFP amplitudes 27 times those observed in the intact rats, indicating a notably greater change in the tendency for the GP spikes to be synchronized with the LFPs in the STN than for STN spikes to be synchronized with GP LFPs following loss of dopamine. To further explore these relationships, STWAs for intact and lesioned STN spike-triggered GP LFP waveforms were normalized by the paired GP spike-triggered GP LFP, and GP spike-triggered STN LFP waveforms were normalized by STN spike-triggered STN LFP waveforms (Fig. 3c, f). The pronounced effect of dopamine cell lesion on GP spike-STN LFP relationships is evident in these bar graphs.

These data support the view that that loss of dopamine is associated with a substantial increase in synchronization of firing patterns in the beta frequency range within the STN and GP, and between the STN and GP, with the most notable change in the lesioned animals occurring in measures of synchronization of GP spiking with respect to STN firing patterns. Moreover, the phase locking between GP spikes and STN LFPs was attenuated by apomorphine administration in both intact and lesioned rats; apomorphine's effects were effectively reversed in intact rats and partially reversed in lesioned rats by subsequent administration of haloperidol (Fig. 3g).

Examination of the timing of GP spikes relative to the peaks and troughs in the STN LFP oscillations provided further insight into alterations in the relationship between GP activity and firing pattern in the STN induced by dopamine cell lesion.

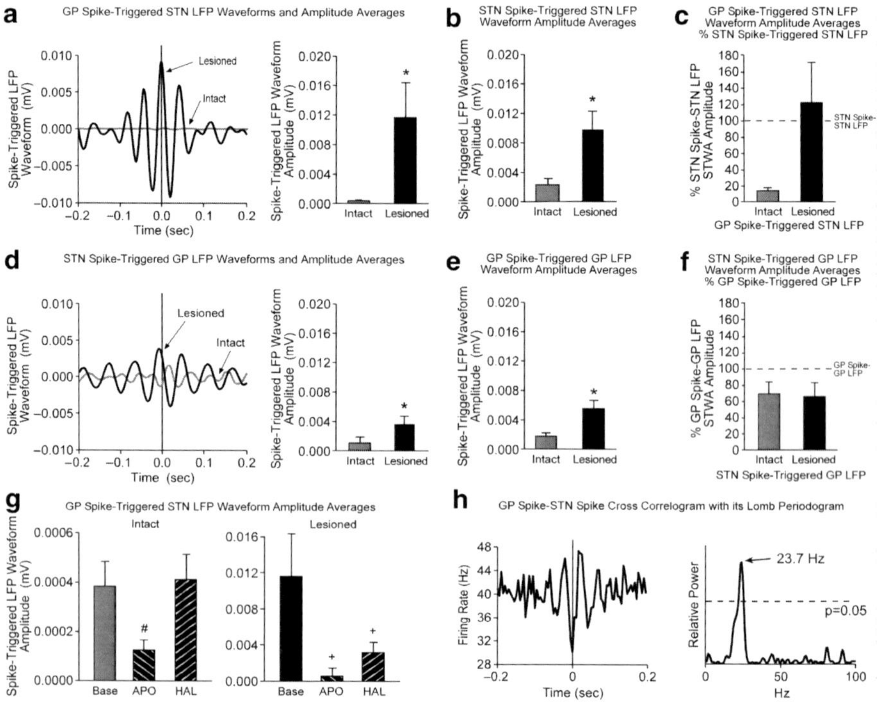

Fig. 3 Spike-triggered waveform averages (SWTA) of GP and STN spikes and LFPs in intact and dopamine cell-lesioned rats. Recordings were conducted as described in Ruskin et al. (2003) with LFP recordings and data analyses conducted as described in Walters et al. (2007). (**a**) GP spike-triggered STN LFP waveform averages were significantly increased following dopamine cell lesion. Timing of the GP spikes was consistently aligned with the peak of the STN LFP oscillations. Graph at left shows a GP spike-triggered STN LFP waveform from a single GP spike/STN LFP paired recording. (**b**) Increased synchronization between STN spike and STN LFPs in the beta frequency range was also observed; STWAs in lesioned animals were 3.6 times greater than STWAs from intact animals. (**c**) STWAs for intact and lesioned GP spike-triggered STN LFP waveforms were normalized by the means for the relevant STN-triggered STN LFP waveforms. There was a large difference in the relationships between GP spikes to STN LFPs in intact vs. lesioned rats, when data were expressed as a % of STN spike to STN LFP, suggesting increased synchronization and impact of GP spikes on STN LFP after dopamine cell lesion. (**d**) STN spike-triggered GP LFP waveform averages were significantly increased following dopamine cell lesion. Graph at left shows a STN spike-triggered GP LFP waveform from a single STN spike-GP LFP paired recording. STN neurons tended to spike shortly after the peaks of the GP LFPs. (**e**) GP spike-triggered GP LFP waveform averages were 2.5 times greater in lesioned animals. (**f**) STWAs for intact and lesioned STN spike-triggered GP LFP STWAs for intact and lesioned were normalized by the means for the relevant GP-triggered GP LFP waveforms. There was no difference in the relationships between STN spikes to GP LFPs in intact vs. lesioned rats, when data were expressed as a % of GP spike to GP LFP. (**g**) Synchronization between GP spikes and STN LFPs was significantly decreased by dopamine agonist administration in both intact and lesioned rats. This desynchronizing effect was antagonized by haloperidol effectively in intact rats and partially in lesioned rats. Similar results were obtained with coherence measures in lesioned animals (data not shown). (**h**) GP spike-triggered STN spike cross-correlogram showed that spiking activity was significantly correlated and with an antiphase relationship (~180°). The Lomb periodogram power spectrum shows that the dominant frequency is in the beta frequency range. *Bar graphs* represent averages of results from 7–9 spike/LFP paired recordings with 1 recording per rat. *Significantly different from baseline and haloperidol; #significantly different from baseline

STWA data showed that GP spikes, on average, were aligned with the peaks of the STN LFP oscillations, indicating that GP input to the STN most frequently correlated with times when STN neurons were most hyperpolarized, i.e., inhibited, consistent with an impact of inhibitory GP input on STN firing pattern. In contrast, STN neurons tended to spike shortly after the peaks of the GP LFP oscillations, as opposed to spiking at the troughs of the GP oscillations as might be expected from the fact that STN input to the GP is excitatory. The evidence that the excitatory STN input more frequently occurred when GP neurons were less likely to be fired, i.e., not, on average, directly triggering GP firing, is consistent with the view that alterations in inhibitory input from the striatum have an enhanced impact on GP's spike timing after dopamine cell lesion. Local axon collaterals within the GP may also play a role in organizing the expression of beta frequency synchronization observed in this preparation (Kita 2007).

The cross-correlogram in Fig. 3h illustrates data from a pair of simultaneously recorded GP-STN neurons that display synchronized antiphasic oscillatory activity at beta frequency. Data from phase plots of STWAs of 8 pairs of GP-STN neurons from lesioned rats showed that during baseline GP spikes occurred from 11–33 ms after the STN spike trigger with a mean of 21 ms. Similarly, mean STN spiking occurred 17–41 ms after the GP spike trigger with a mean of 30 ms. Adding of the mean delays yields a period of 51 ms and supports the view that after dopamine cell lesion, GP and STN activity becomes more synchronized with oscillatory periods in the range of 20 Hz.

3 Discussion

3.1 Loss of Dopamine and Emergence of Synchronized Activity in Basal Ganglia

Data reviewed earlier show that the level of dopamine receptor stimulation regulates expression of beta frequency activity in two interconnected basal ganglia nuclei, the STN and the GP, in a rodent model of PD. Removal of dopaminergic input or administration of dopamine agonists affected not only firing rates in the GP and STN, but also levels of synchronized and oscillatory activity within this network. These observations support the hypothesis that loss of dopamine is associated with increases in beta activity in the STN in PD patients. They also underscore the potential of rodent models of PD for gaining further insight into the relationship between processes generating these oscillations and dopamine's ability to modulate their expression in basal ganglia circuits.

As discussed earlier, Murer et al. have provided perspective on processes contributing to increases in ~1-Hz synchronization in basal ganglia network in the anesthetized rat model of PD, demonstrating that striatal neurons are more depolarized after dopamine cell lesion and fire more frequently in conjunction with the synchronized cortical input present in the anesthetized rat (Tseng et al. 2001; Murer et al. 2002). These observations, together with other studies (Mallet et al. 2006; Walters et al. 2007), argue that after loss of dopamine, changes in striatal processing of synchronized 1-Hz cortical input contribute to the emergence of the 1-Hz oscillatory activity throughout the indirect pathway in the basal ganglia of anesthetized rats. The hypothesis that loss of dopamine facilitates passage of synchronized activity through the basal ganglia network by disrupting striatal "filtering" of oscillatory components of cortical input (Murer et al. 2002) provides a useful framework for envisioning how loss of dopamine could alter basal ganglia dynamics. This model incorporates aspects of the older rate model (Albin et al. 1989; DeLong 1990), predicting that decreases in dopamine receptor stimulation would reduce activity in the direct pathway and facilitate activity in the indirect pathway, and adds input from the hyperdirect pathway and consideration of phase relationships to the overall dynamics (Fig. 1). Nevertheless, other processes may also contribute to the emergence of oscillatory and bursty activity in the basal ganglia in PD, such as effects of dopamine loss on activity in thalamocortical projections and cortical networks that may further contribute to changes in basal ganglia activity (Leblois et al. 2006; Mallet et al. 2006; Li et al. 2007; Gatev and Wichman 2008; Zaidel et al. 2008). Data also support a role for cortical input to the STN (Mallet et al. 2008) and direct effects of dopaminergic denervation on local processes within the STN facilitating resonance with the GP in the beta range (Bevan et al. 2000, 2002a, b, 2006, 2007; Shen et al. 2003; Burton et al. 2005; Shen and Johnson 2005; Baufreton and Bevan 2008).

The potential contribution of a pacemaker resonance between GP and STN to the emergence of oscillatory activity in these nuclei was initially highlighted by studies in organotypic cultures demonstrating that GP and STN nuclei have the potential to engage in reciprocal oscillatory activity in the range of 0.4–2 Hz (Plenz and Kitai 1999). In vitro studies in brain slices support the view that membrane properties of STN neurons are conducive to entrainment of organized oscillatory activity in the beta frequency range by GABAergic input and that loss of dopaminergic innervation in the STN may contribute to changes in STN response that promote STN-GP beta frequency resonance (Bevan et al. 2000, 2002a, b, 2006, 2007; Shen et al. 2003; Baufreton et al. 2005; Shen and Johnson 2005; Baufreton and Bevan 2008). In addition, the specifics of interconnections between the two nuclei and evidence for collateral inhibition within the GP may further facilitate potential for resonance (Kita 2007; Sadek et al. 2007; Zold et al. 2007a). Relevant to these considerations is evidence for a subpopulation of GP neurons that fires in an antiphase relationship to the majority of the GP neurons over frequencies ranging from ~1-Hz slow wave (Walters et al. 2007; Zold et al. 2007a) to multisecond periodicities (Tierney et al. 2002; unpublished observations). Divergent subpopulations in GP have also emerged in in vitro studies (Kita 2007) and in vivo in response to systemic administration of apomorphine, where some neurons show rate decreases (Kelland et al. 1995) while the majority show rate increases (Bergstrom et al. 1982). Thus, dynamic interactions of neuronal subpopulations within the GP in concert with changes induced by loss of dopamine in the STN and striatum are all potential contributors to the emergence of increased oscillatory activity and synchronization in the GP–STN network.

The present results do not resolve the question of which of these processes are most critical to the emergence of beta oscillations in the STN in the rat model of PD, but the data clearly show that increased coherence between GP output and STN activity occurs after dopamine cell lesion, leaving open a role for GP–STN resonance, as well as for alterations in striatal input to the GP in modulating GP–STN relationships as has been hypothesized for the ~1-Hz oscillations in the urethane-anesthetized rat model. A role for changes in striatal processing of faster frequency oscillations after loss of dopamine is supported by a report from Dejean et al. (2008) of increased coherence between cortex and striatum in the beta frequency range in the awake rat after dopamine cell lesion. Increased synchronization of oscillatory activity in cortex and striatum has also been associated with dopaminergic depletion in awake behaving mice (Costa et al. 2006). These observations are consistent with a role of the striatum, via the GP, in entraining synchronized activity in the STN after loss of dopamine, in conjunction with input to STN from cortex via the hyperdirect pathway. Also consistent with alterations in striatal input to the GP after dopamine loss (Liang et al. 2008) is the observation that mean firing rates are reduced in GP and increased in the STN with the result that mean firing rates of neurons in both nuclei are in the beta frequency range. Convergence of firing rates may enhance the potential for increased synchronization of spiking in this frequency. These results are in agreement with previous observation from this preparation showing that under normal conditions mean firing rate is increased in both STN and GP in association with dopamine agonist administration, while in lesioned animals dopamine agonists tend to induce a decrease in activity in STN, while further stimulating activity of many GP neurons (Bergstrom and Walters 1981; Bergstrom et al. 1982; Walters et al. 1987; Carlson et al. 1990; Kreiss et al. 1996, 1997; Allers et al. 2000, 2005).

The present data also show a desynchronizing effect of dopamine agonist administration on GP–STN relationships in the beta frequency range in both intact and lesioned rats, which is reversed by dopamine antagonist administration. Increased stimulation of dopamine receptors attenuates GP–STN synchronization at beta frequencies. As these results are obtained in immobilized rats, they argue that the modulatory effects of dopamine agonists on aspects of neuronal synchronization within the GP and STN network in the lesioned rats are not due to an indirect/or direct effect of altered motor activity levels. Sharott et al. (2005) have also observed reductions in beta frequency power in association with and just preceding apomorphine-induced rotations in LFP recordings from the STN and electrocorticograms (ECoG) over the frontal cortex in behaving hemiparkinsonian rats.

3.2 Beta Activity in Additional Awake Rodent Models of PD

Beta frequency activity has also been observed in basal ganglia nuclei in other awake rodent models of PD studied in our laboratory: head-restrained rats treated with neuroleptics and awake behaving rats with chronically implanted electrodes and unilateral 6-OHDA-induced dopamine cell lesions. In studies in head-restrained rats, parkinsonism was induced by administration of the D2 dopamine receptor antagonist eticlopride after rats were trained to rest quietly in a full body hammock and equipped with a cranial plate stereotaxically positioned to guide electrode placement. Single unit and LFP recordings conducted in the GP with glass micropipette electrodes showed significant increases in GP LFP power and synchronization between GP spiking and GP LFP in the beta frequency range (14–30 Hz) (Ghiglieri et al. 2003) following dopamine receptor blockade. In experiments in awake behaving rats with unilateral dopamine cell lesions, electrodes were chronically implanted bilaterally in the SNpr. Recordings showed that loss of dopamine was associated with increased expression of beta frequency activity in the 12–25-Hz range in the SNpr LFP and increased synchronization of SNpr spikes and SNpr LFPs in the lesioned hemisphere during rest. When rats walked on a circular treadmill, a significant reduction in SNpr LFP beta power in both lesioned and nonlesioned hemispheres was observed (Avila et al. 2007). Moreover, SNpr LFP power and SNpr spike/LFP synchronization in a 25–40-Hz high-beta/low-gamma frequency band were selectively increased in the lesioned hemisphere during effort to walk in these lesioned rats with impaired motor function (Avila et al. 2008).

3.3 Observations in Rodent Models Consistent with Changes Observed in PD Patients: Conclusions

In sum, data support a critical role for dopamine in modulating organization of synchronized and oscillatory activity in basal ganglia networks. Observations in rodent models of PD are consistent with data from MPTP-lesioned nonhuman primates showing increased synchronization of activity within the GP after dopamine cell lesion (Nini et al. 1995; Bergman et al. 1998; Raz et al. 2000, 2001; Heimer et al. 2006). Results in the rodent model are also consistent with findings that oscillatory activity in the beta range is observed in STN spike trains and STN LFPs from PD patients off dopaminergic medications and is reduced by voluntary movement and by L-dopa or dopamine agonist treatment (Brown et al. 2001; Cassidy et al. 2002; Levy et al. 2002; Priori et al. 2002; Williams et al. 2002; Kuhn et al. 2004, 2006; Alegre et al. 2005; Doyle et al. 2005; Alonso-Frech et al. 2006; Weinberger et al. 2006).

Questions remain, however (for review, Brown and Williams 2005; Brown 2007; Galvan and Wichmann 2008), including the extent to which synchronized and oscillatory activity in the basal ganglia results primarily from alterations in striatal filtering of cortical input or is generated by local changes in dopamine input engendering reverberations in subcomponents of the basal ganglia network. Finally, critical and unresolved questions include how alterations in basal ganglia output affect downstream circuits and how these changes in turn impact on basal ganglia function (Gatev and Wichmann 2008) and ultimately contribute to the symptoms of Parkinson's disease.

Acknowledgments The Intramural Research Program of the NINDS, NIH supported this research.

References

Albin RL, Young AB and Penney JB (1989) The functional anatomy of basal ganglia disorders. Trends Neurosci 12: 366–375.

Alegre M, Alonso-Frech F, Rodriguez-Oroz M, Guridi J, Zamarbide I, Valencia M, Manrique M, Obeso J and Artieda J (2005) Movement-related changes in oscillatory activity in the human nucleus: ipsilateral vs. contralateral movements. Eur J Neurosci 22: 2315–2324.

Allers KA, Kreiss DS and Walters JR (2000) Multisecond oscillations in the subthalamic nucleus: effects of apomorphine and dopamine cell lesion. Synapse 38: 38–50.

Allers KA, Bergstrom DA, Ghazi LJ, Kreiss DS and Walters JR (2005) MK801 and amantadine exert different effects on subthalamic neuronal activity in a rodent model of Parkinson's disease. Exp Neurol 191: 104–118.

Alonso-Frech F, Zamarbide I, Alegre M, Rodriguez-Oroz M, Guridi J, Manrique M, Valencia M, Artieda J and Obeso J (2006) Slow oscillatory activity and levodopa-induced dyskinesias in Parkinson's disease. Brain 129: 1748–1757.

Aravamuthan BR, Bergstrom DA, French RA, Taylor JJ, Parr-Brownlie LC and Walters JR (2008) Altered neuronal activity relationships between the pedunculopontine nucleus and motor cortex in a rodent model of Parkinson's disease. Exp Neurol 213: 268–280.

Avila I, Parr-Brownlie LC, Bergstrom DA, Gonzales KK, Poloskey SL, Castaneda E and Walters JR (2007) Firing rate and local field potential changes in a basal ganglia output nucleus during walking and rest in the hemiparkinsonian rat. Soc Neurosci Abstr 2007 Online Program No. 622.8.

Avila I, Parr-Brownlie LC, Bergstrom DA, Castaneda E and Walters JR (2008) Beta and low gamma frequency synchronization in basal ganglia output during rest and walk in the hemiparkinsonian rat. Mov Disord 23: S299.

Baufreton J and Bevan MD (2008) D-2-like dopamine receptor-mediated modulation of activity-dependent plasticity at GABAergic synapses in the subthalamic nucleus. J Physiol 586: 2121–2142.

Baufreton J, Atherton JF, Surmeier DJ and Bevan MD (2005) Enhancement of excitatory synaptic integration by GABAergic inhibition in the subthalamus nucleus. J Neurosci 25: 8505–8517.

Belluscio MA, Kasanetz F, Riquelme LA and Murer MG (2003) Spreading of slow cortical rhythms to the basal ganglia output nuclei in rats with nigrostriatal lesions. Eur J Neurosci 17: 1046–1052.

Belluscio MA, Riquelme LA and Murer MG (2007) Striatal dysfunction increases basal ganglia output during motor cortex activation in parkinsonian rats. Eur J Neurosci 25: 2791–2804.

Bergman H, Feingold A, Nini A, Raz A, Slovin H, Abeles M and Vaadia E (1998) Physiological aspects of information processing in the basal ganglia of normal and parkinsonian primates. Trends Neurosci 21: 32–38.

Bergstrom DA and Walters JR (1981) Neuronal responses of the globus pallidus to systemic administration of D-amphetamine: investigation of the involvement of dopamine, norepinephrine, and serotonin. J Neurosci 1: 292–299.

Bergstrom DA, Bromley SD and Walters JR (1982) Apomorphine increases the activity of rat globus pallidus neurons. Brain Res 238: 266–271.

Bevan MD, Wilson CJ, Bolam JP and Magill PJ (2000) Equilibrium potential of $GABA_A$ current and implications for rebound burst firing in rat subthalamic neurons in vitro. J Neurophysiol 83: 3169–3172.

Bevan MD, Magill PJ, Hallworth NE, Bolam JP and Wilson CJ (2002a) Regulation of the timing and pattern of action potential generation in rat subthalamic neurons in vitro by GABA-A IPSPs. J Neurophysiol 87: 1348–1362.

Bevan MD, Magill PJ, Terman D, Bolam JP and Wilson CJ (2002b) Move to the rhythm: oscillations in the subthalamic nucleus-external globus pallidus network. Trends Neurosci 25: 525–531.

Bevan MD, Atherton JF and Baufreton J (2006) Cellular principles underlying normal and pathological activity in the subthalamic nucleus. Curr Opin Neurobiol 16: 621–628.

Bevan MD, Hallworth NE and Baufreton J (2007) GABAergic control of the subthalamic nucleus. Prog Brain Res 160: 173–188.

Breit S, Lessmann L, Benazzouz A and Schulz JB (2005) Unilateral lesion of the pedunculopontine nucleus induces hyperactivity in the subthalamic nucleus and substantia nigra in the rat. Eur J Neurosci 22: 2283–2294.

Breit S, Lessmann L, Unterbrink D, Popa RC, Gasser T and Schulz JB (2006) Lesion of the pedunculopontine nucleus reverses hyperactivity of the subthalamic nucleus and substantia nigra pars reticulata in a 6-hydroxydopamine rat model. Eur J Neurosci 24: 2275–2282.

Brown P (2003) Oscillatory nature of human basal ganglia activity: relationship to the pathophysiology of Parkinson's disease. Mov Disord 18: 357–363.

Brown P (2007) Abnormal oscillatory synchronisation in the motor system leads to impaired movement. Curr Opin Neurobiol 17: 656–664.

Brown P and Williams D (2005) Basal ganglia local field potential activity: character and functional significance in the human. Clin Neurophysiol 116: 2510–2519.

Brown P, Oliviero A, Mazzone P, Insola A, Tonali P and DiLazzaro V (2001) Dopamine dependency of oscillations between subthalamic nucleus and pallidum in Parkinson's disease. J Neurosci 21: 1033–1038.

Carlson JH, Bergstrom DA, Demo SD and Walters JR (1990) Nigrostriatal lesion alters neurophysiological responses to selective and nonselective D-1 dopamine and D-2 dopamine agonists in rat globus pallidus. Synapse 5: 83–93.

Cassidy M, Mazzone P, Oliviero A, Insola A, Tonali P, Di Lazzaro V and Brown P (2002) Movement-related changes in synchronization in the human basal ganglia. Brain 125: 1235–1246.

Chen CC, Litvak V, Gilbertson T, Kuhn A, Lu CS, Lee ST, Tsai CH, Tisch S, Limousin P, Hariz M and Brown P (2007) Excessive synchronization of basal ganglia neurons at 20 Hz slows movement in Parkinson's disease. Exp Neurol 205: 214–221.

Costa RM, Lin SC, Sotnikova TD, Cyr M, Gainetdinov RR, Caron MG and Nicolelis MAL (2006) Rapid alterations in corticostriatal ensemble coordination during acute dopamine-dependent motor dysfunction. Neuron 52: 359–369.

Dejean C, Gross CE, Bioulac B and Boraud T (2008) Dynamic changes in the cortex-basal ganglia network after dopamine depletion in the rat. J Neurophysiol 100: 385–396.

DeLong MR (1990) Primate models of movement disorders of basal ganglia origin. Trends Neurosci 13: 281–285.

Deuschl G, Raethjen J, Baron R, Lindemann M, Wilms H and Krack P (2000) The pathophysiology of parkinsonian tremor: a review. J Neurol 247: V/33-V/48.

Doyle LMF, Kuhn AA, Hariz M, Kupsch A, Schneider GH and Brown P (2005) Levodopa-induced modulation of subthalamic beta oscillations during self-paced movements in patients with Parkinson's disease. Eur J Neurosci 21: 1403–1412.

Floran B, Floran L, Erlij D and Aceves J (2004) Dopamine D4 receptors inhibit depolarization-induced [3H]GABA release in the rat subthalamic nucleus. Eur J Pharmacol 498: 97–102.

Foffani G, Bianchi A, Baselli G and Priori A (2005) Movement-related frequency modulation of beta oscillatory activity in the human subthalamic nucleus. J Physiol 568: 699–711.

Galvan A and Wichmann T (2008) Pathophysiology of parkinsonism. Clin Neurophysiol 119: 1459–1474.

Gatev P and Wichmann T (2009) Interactions between cortical rhythms and spiking activity of single basal ganglia neurons in the normal and parkinsonian state. Cereb Cortex 19: 1330–1344.

Ghiglieri V, Soucy AL, Itoga CA, Bergstrom DA and Walters JR (2003) Dopamine D2 receptor blockade affects relationships between single unit neuronal activity and local field potential power in the globus pallidus of head-restrained and freely moving rats. Soc Neurosci Abstr 2003 Online Program No. 273.4.

Hallworth NE and Bevan MD (2005) Globus pallidus neurons dynamically regulate the activity pattern of subthalamic nucleus neurons through the frequency-dependent activation of postsynaptic $GABA_A$ and $GABA_B$ receptors. J Neurosci 25: 6304–6315.

Hassani OK, Mouroux M and Féger J (1996) Increased subthalamic neuronal activity after nigral dopaminergic lesion independent of disinhibition via the globus pallidus. Neuroscience 72: 105–115.

Heimer G, Rivlin-Etzion M, Bar-Gad I, Goldberg JA, Haber SN and Bergman H (2006) Dopamine replacement therapy does not restore the full spectrum of normal pallidal activity in the 1-methyl-4-phenyl-1,2,3,6-tetra-hydropyridine primate model of parkinsonism. J Neurosci 26: 8101–8114.

Hollerman JR and Grace AA (1992) Subthalamic nucleus cell firing in the 6-OHDA-treated rat: basal activity and response to haloperidol. Brain Res 590: 291–299.

Hurtado JM, Gray CM, Tamas LB and Sigvardt KA (1999) Dynamics of tremor-related oscillations in the human globus pallidus: a single case study. Proc Natl Acad Sci USA 96: 1674–1679.

Kasanetz F, Riquelme LA and Murer MG (2002) Disruption of the two-state membrane potential of striatal neurones during cortical desynchronisation in anaesthetised rats. J Physiol 543: 577–589.

Kelland MD, Soltis RP, Anderson LA, Bergstrom DA and Walters JR (1995) In vivo characterization of two cell types in the rat globus pallidus which have opposite responses to dopamine receptor stimulation: comparison of electrophysiological properties and responses to apomorphine, dizocilpine, and ketamine anesthesia. Synapse 20: 338–350.

Kita H (2007) Globus pallidus external segment. Prog Brain Res 160: 111–133.

Krack P, Benazzouz A, Pollak P, Limousin P, Piallat B, Hoffmann D, Xie J and Benabid AL (1998) Treatment of tremor in Parkinson's disease by subthalamic nucleus stimulation. Mov Disord 13: 907–914.

Kreiss DS, Anderson LA and Walters JR (1996) Apomorphine and dopamine D-1 receptor agonists increase the firing rates of subthalamic nucleus neurons. Neuroscience 72: 863–876.

Kreiss DS, Mastropietro CW, Rawji SS and Walters JR (1997) The response of subthalamic nucleus neurons to dopamine receptor stimulation in a rodent model of Parkinson's disease. J Neurosci 17: 6807–6819.

Kuhn AA, Williams D, Kupsch A, Limousin P, Hariz M, Schneider GH, Yarrow K and Brown P (2004) Event-related beta desynchronization in human subthalamic nucleus correlates with motor performance. Brain 127: 735–746.

Kuhn AA, Kupsch A, Schneider GH and Brown P (2006) Reduction in subthalamic 8–35 Hz oscillatory activity correlates with clinical improvement in Parkinson's disease. Eur J Neurosci 23: 1956–1960.

Leblois A, Boraud T, Meissner W, Bergman H and Hansel D (2006) Competition between feedback loops underlies normal and pathological dynamics in the basal ganglia. J Neurosci 26: 3567–3583.

Lemstra AW, Verhagen Metman L, Lee JI, Dougherty PM and Lenz FA (1999) Tremor-frequency (3–6 Hz) activity in the sensorimotor arm representation of the internal segment of the globus pallidus in patients with Parkinson's disease. Neurosci Lett 267: 129–132.

Levy R, Hutchison WD, Lozano AM and Dostrovsky JO (2000) High-frequency synchronization of neuronal activity in the subthalamic nucleus of parkinsonian patients with limb tremor. J Neurosci 20: 7766–7775.

Levy R, Ashby P, Hutchison WD, Lang AE, Lozano AM and Dostrovsky JO (2002) Dependence of subthalamic nucleus oscillations on movement and dopamine in Parkinson's disease. Brain 125: 1196–1209.

Li S, Arbuthnott GW, Jutras MJ, Goldberg JA and Jaeger D (2007) Resonant antidromic cortical circuit activation as a consequence of high-frequency subthalamic deep-brain stimulation. J Neurophysiol 98: 3525–3537.

Liang L, DeLong MR and Papa SM (2008) Inversion of dopamine responses in striatal medium spiny neurons and involuntary movements. J Neurosci 28: 7537–7547.

Magill PJ, Bolam JP and Bevan MD (2001) Dopamine regulates the impact of the cerebral cortex on the subthalamic nucleus-globus pallidus network. Neuroscience 106: 313–330.

Mallet N, Ballion B, Le Moine C and Gonon F (2006) Cortical inputs and GABA interneurons imbalance projection neurons in the striatum of parkinsonian rats. J Neurosci 26: 3875–3884.

Mallet N, Pogosyan A, Sharott A, Csicsvari J, Bolam JP, Brown P and Magill PJ (2008) Disrupted dopamine transmission and the emergence of exaggerated beta oscillations in subthalamic nucleus and cerebral cortex. J Neurosci 28: 4795–4806.

Marsden JF, Limousin-Dowsey P, Ashby P, Pollak P and Brown P (2001) Subthalamic nucleus, sensorimotor cortex and muscle interrelationships in Parkinson's disease. Brain 124: 378–388.

Murer MG, Tseng KY, Kasanetz F, Belluscio M and Riquelme LA (2002) Brain oscillations, medium spiny neurons, and dopamine. Cell Mol Neurobiol 22: 611–632.

Ni ZG, Bouali-Benazzouz R, Gao DM, Benabid AL and Benazzouz A (2001) Time-course of changes in firing rates and firing patterns of subthalamic nucleus neuronal activity after 6-OHDA-induced dopamine depletion in rats. Brain Res 899: 142–147.

Nini A, Feingold A, Slovin H and Bergman H (1995) Neurons in the globus pallidus do not show correlated activity in the normal monkey, but phase-locked oscillations appear in the MPTP model of parkinsonism. J Neurophysiol 74: 1800–1805.

Parr-Brownlie LC, Poloskey SL, Flanagan KK, Eisenhofer G, Bergstrom DA and Walters JR (2007) Dopamine lesion-induced changes in subthalamic nucleus activity are not associated with alterations in firing rate or pattern in layer V neurons of the anterior cingulate cortex in anesthetized rats. Eur J Neurosci 26: 1925–1939.

Perier C, Agid Y, Hirsch EC and Feger J (2000) Ipsilateral and contralateral subthalamic activity after unilateral dopaminergic lesion. NeuroReport 11: 3275–3278.

Plenz D and Kitai ST (1999) A basal ganglia pacemaker formed by the subthalamic nucleus and external globus pallidus. Nature 400: 677–682.

Priori A, Foffani G, Pesenti A, Bianchi A, Chiesa V, Baselli G, Caputo E, Tamma F, Rampini P, Egidi M, Locatelli M, Barbieri S and Scarlato G (2002) Movement-related modulation of neural activity in human basal ganglia and its L-DOPA dependency: recordings from deep brain stimulation electrodes in patients with Parkinson's disease. Neurol Sci 23: S101-S102.

Priori A, Foffani G, Pesenti A, Tamma F, Bianchi A, Pellegrini M, Locatelli M, Moxon K and Villani R (2004) Rhythm-specific pharmacological modulation of subthalamic activity in Parkinson's disease. Exp Neurol 189: 369–379.

Raz A, Vaadia E and Bergman H (2000) Firing patterns and correlations of spontaneous discharge of pallidal neurons in the normal and the tremulous 1-methyl-4-phenyl-1,2,3,6-tetrahydropyridine vervet model of parkinsonism. J Neurosci 20: 8559–8571.

Raz A, Frechter-Mazar V, Feingold A, Abeles M, Vaadia E and Bergman H (2001) Activity of pallidal and striatal tonically active neurons is correlated in MPTP-treated monkeys but not in normal monkeys. J Neurosci 21 (RC128): 1–5.

Rosenberg JR, Amjad AM, Breeze P, Brillinger DR and Halliday DM (1989) The Fourier approach to the identification of functional coupling between neuronal spike trains. Prog Biophys Mol Biol 53: 1–31.

Ruskin DN, Bergstrom DA, Tierney PL and Walters JR (2003) Correlated multisecond oscillations in firing rate in the basal ganglia: modulation by dopamine and the subthalamic nucleus. Neuroscience 117: 427–438.

Sadek AR, Magill PJ and Bolam JP (2007) A single-cell analysis of intrinsic connectivity in the rat globus pallidus. J Neurosci 27: 6352–6362.

Shapiro MB, Vaillancourt DE, Sturman MM, Verhagen Metman L, Bakay RAE and Corcos DM (2007) Effects of STN DBS on rigidity in Parkinson's disease. IEEE Trans Neural Syst Rehabil Eng 15: 173–181.

Sharott A, Magill PJ, Harnack D, Kupsch A, Meissner W and Brown P (2005) Dopamine depletion increases the power and coherence of beta-oscillations in the cerebral cortex and subthalamic nucleus of the awake rat. Eur J Neurosci 21: 1413–1422.

Shen KZ and Johnson SW (2005) Dopamine depletion alters responses to glutamate and GABA in the rat subthalamic nucleus. NeuroReport 16: 171–174.

Shen KZ, Zhu ZT, Munhall A and Johnson SW (2003) Dopamine receptor supersensitivity in rat subthalamus after 6-hydroxydopamine lesions. Eur J Neurosci 18: 2967–2974.

Tabbal SD, Ushe M, Mink JW, Revilla FJ, Wernle AR, Hong M, Karimi M and Perlmutter JS (2008) Unilateral subthalamic nucleus stimulation has a measurable ipsilateral effect on rigidity and bradykinesia in Parkinson disease. Exp Neurol 211: 234–242.

Tierney PL, Bergstrom DA, Ruskin DN and Walters JR (2002) Correlation between multisecond oscillations in firing rate in basal ganglia neurons: modulation by dopamine receptor stimulation and association with hippocampal theta rhythm. Soc Neurosci Abstr 2002 Online Program No. 765.3.

Tierney PL, Bergstrom DA, Soucy AL, Itoga CA, Allers KA, Ruskin DN and Walters JR (2003) Single unit activity and local field potentials in paired recordings from subthalamic nucleus and globus pallidus neurons in awake rats: effect of alterations in dopamine receptor stimulation. Soc Neurosci Abstr 2003 Online Program No. 273.5.

Tseng KY, Kasanetz F, Kargieman L, Riquelme LA and Murer MG (2001) Cortical slow oscillatory activity is reflected in the membrane potential and spike trains of striatal neurons in rats with chronic nigrostriatal lesions. J Neurosci 21: 6430–6439.

Vila M, Perier C, Feger J, Yelnik J, Faucheux B, Ruberg M, Raisman-Vozari R, Agid Y and Hirsch EC (2000) Evolution of changes in neuronal activity in the subthalamic nucleus of rats with unilateral lesion of the substantia nigra assessed by metabolic and electrophysiological measurements. Eur J Neurosci 12: 337–344.

Walters JR, Bergstrom DA, Carlson JH, Chase TN and Braun AR (1987) D1 dopamine receptor activation required for postsynaptic expression of D2 agonist effects. Science 236: 719–722.

Walters JR, Hu D, Itoga CA, Parr-Brownlie LC and Bergstrom DA (2005) Do local field potentials reflect synchronized spiking activity of neuronal populations in the basal ganglia? Studies in a rodent model of Parkinson's disease. In: Bolam JP, Magill PJ, Ingham C (eds) The Basal Ganglia VIII. Springer, New York, pp 37–46.

Walters JR, Hu D, Itoga CA, Parr-Brownlie LC and Bergstrom DA (2007) Phase relationships support a role for coordinated activity in the indirect pathway in organizing slow oscillations in basal ganglia output after loss of dopamine. Neuroscience 144: 762–776.

Weinberger M, Mahant N, Hutchison W, Lozano A, Moro E, Hodaie M, Lang A and Dostrovsky J (2006) Beta oscillatory activity in the subthalamic nucleus and its relation to dopaminergic response in Parkinson's disease. J Neurophysiol 96: 3248–3256.

Williams D, Tijssen M, van Bruggen G, Bosch A, Insola A, Di Lazzaro V, Mazzone P, Oliviero A, Quartarone A, Speelman H and Brown P (2002) Dopamine-dependent changes in the functional connectivity between basal ganglia and cerebral cortex in humans. Brain 125: 1558–1569.

Zaidel A, Moran A, Marjan G, Bergman H and Israel Z (2008) Prior pallidotomy reduces and modifies neuronal activity in the subthalamic nucleus of Parkinson's disease patients. Eur J Neurosci 27: 1308–1310.

Zold CL, Ballion B, Riquelme LA, Gonon F and Murer MG (2007a) Nigrostriatal lesion induces D2-modulated phase-locked activity in the basal ganglia of rats. Eur J Neurosci 25: 2131–2144.

Zold CL, Larramendy C, Riquelme LA and Murer MG (2007b) Distinct changes in evoked and resting globus pallidus activity in early and late Parkinson's disease experimental models. Eur J Neurosci 26: 1267–1279.

Behavioural Correlates of Dopaminergic Agonists' Dyskinetic Potential in the 6-OHDA-Lesioned Rat

Anna R. Carta, Lucia Frau, Annalisa Pinna, and Micaela Morelli

Abstract Prolonged treatment with L-DOPA induces highly disabling dyskinesia in patients affected by Parkinson's disease (PD). In contrast, dopamine receptor agonist treatments display dyskinetic outcome variably, depending on pharmacokinetic/pharmacodynamic drug profile. The present study was aimed at assessing behavioural correlates of intense or mild dyskinesia displayed by the different dopamine receptor stimulation in the 6-hydroxydopamine rat model of PD. Sensitization of contralateral turning behaviour (SCT) and abnormal involuntary movements (AIMs) were assessed as behavioural correlates of dyskinetic responses during subchronic stimulation of the D_1 receptor by SKF38393 and the D_2/D_3 receptors by ropinirole. Similarly to what already has been described for L-DOPA, subchronic SKF38393 caused AIMs and SCT whereas ropinirole elicited SCT only, indicating that both drugs induced some dyskinetic response, albeit of different type. Results suggest that presence of SCT alone or SCT plus AIMs might represent correlates of the differential severity of dyskinetic movements induced by treatment with low (ropinirole) or high (SKF38393) dyskinetic potential. Evaluation of both behavioural responses represents a useful test to predict the dyskinetic potential of drugs in preclinical screening.

1 Introduction

Motor complications comprising fluctuations of drug response and dyskinesia represent a major limitation in the long-term use of most dopaminergic therapies in Parkinson's disease (PD). Prolonged treatment with the dopamine precursor L-DOPA induces the most disabling dyskinesia in PD patients and in parkinsonian nonhuman primates. Dopamine agonists used in clinic or tested in primate models of PD display different dyskinetic outcome, depending upon a complex interaction

A. Carta (✉), L. Frau, A. Pinna, and M. Morelli
Department of Toxicology, University of Cagliari, Via Ospedale 72, 09124 Cagliari, Italy
e-mail: acarta@unica.it

of pharmacodynamic and pharmacokinetic characteristics (Jenner 2003). Agonists selectively stimulating the D_1 receptor have produced contrasting results on their liability to induce dyskinesia in primates (Andringa et al. 1999a; Blanchet et al. 1995, 1996; Calon et al. 1999; Goulet et al. 1996; Hurley and Jenner 2006). Moreover, no D_1 agonists are available for clinical use, although two reports have described a dyskinetic response to D_1 agonists similar to L-DOPA, in PD patients previously primed with L-DOPA (Giardina and Williams 2001; Rascol et al. 2001). D_2 dopamine receptor agonists given de novo are less likely to induce dyskinesia, since mild dyskinetic responses have been described after prolonged treatment with several D_2 or D_2/D_3 agonists in humans and primates (Adler et al. 1997; Pearce et al. 1998; Rascol et al. 2000; Reichmann 2000; Sethi et al. 1998; Smith et al. 2006). However, other studies have reported intense dyskinesia in primates treated with D_2 agonists, suggesting that drug pharmacokinetics plays a main role, together with the pharmacodynamic profile, in determining dyskinetic potential. Hence, long-acting D_2 agonists, as ropinirole and cabergoline, own the lowest dyskinetic potential, whereas short-acting compounds are able to induce more intense dyskinesia (Andringa et al. 1999b; Blanchet et al. 1995; Calon et al. 1995). Therefore, the only assessment of drug pharmacodynamic properties leaves the evaluation of their dyskinetic potential largely unpredictable.

In 6-hydroxydopamine (6-OHDA)-lesioned rats, repeated treatment with L-DOPA, or the D_1/D_2 agonist apomorphine, has been extensively characterized in their behavioural outcome, leading to the assessment of a rat model of L-DOPA-induced dyskinesia (LID). Chronic treatment with these drugs induces a range of abnormal motor responses, including sensitization of contralateral turning behaviour (SCT), and abnormal involuntary movements (AIMs), taken as a rat correlate of LID (Carta et al. 2006; Henry et al. 1998; Lundblad et al. 2002; Pinna et al. 2006). Conversely, behavioural correlates in the rat of the different dyskinetic potential displayed by D_1 and D_2 dopaminergic agonists have been poorly investigated. Whereas D_1 agonists have not been characterized to date, more data are available on D_2 receptor agonists, albeit incomplete (Delfino et al. 2004; Lindgren et al. 2007; Carta et al. 2006; Ravenscroft et al. 2004).

To characterize behavioural correlates of dopaminergic agonist-induced dyskinesia in the rat, we evaluated a subchronic treatment with the D_1 agonist SKF38393 and the D_2/D_3 agonist ropinirole, in the rat 6-OHDA model of PD. Since several reports have shown development of tolerance to the pharmacological effects of long-acting D_1 agonists, the short-acting D_1 agonist SKF38393 was chosen for the study (Andringa et al. 1999a; Smith et al. 2002; Blanchet et al. 1996; Setler et al. 1978). Moreover, we evaluated a subchronic treatment with the D_2/D_3 agonist ropinirole, which is known from clinical reports to induce dyskinesia of mild intensity in PD patients (Rascol et al. 2000).

Two doses of SKF38393 and two doses of ropinirole were selected, on the basis of previous studies and of preliminary dose–response experiments (Fukuzaki et al. 2000; Matsuda et al. 1992; Morelli et al. 1989; Van de Witte et al. 1998). The lowest doses eliciting a contralateral turning response were selected for both drugs. In addition, a dose able to elicit a robust contralateral turning was chosen, in order to evaluate the induction of AIMs in response to a full dosage.

2 Experimental Procedures

2.1 Subjects and 6-OHDA Lesion

Male Sprague–Dawley rats (Charles River, Italy) were used in all experiments. The rats were maintained on a 12-h light–dark cycle (lights on 08:00–20:00 h) with food and water available ad libitum. All experiments were carried out in accordance with the European Communities Council Directive (86/609/EEC).

6-OHDA-HCl (8 μg in 4 μl of 0.05% ascorbic acid in saline) was injected unilaterally in the left medial forebrain bundle of chloral hydrate-anaesthetized rats (275–300 g), at coordinates $A = -2.2$, $L = +1.5$, and $V = -7.9$, according to the atlas of Pellegrino et al. (1979). The rats were pre-treated with 10 mg/kg (i.p.) of desipramine to prevent damage to noradrenergic neurons.

2.2 Drugs

6-OHDA–HCl, desipramine and SKF38393 were purchased from Sigma-Aldrich (St Louis, MO, USA); ropinirole was kindly provided by GlaxoSmithKline. Desipramine was injected in a volume of 0.3 ml i.p./100 g body weight. SKF38393 and ropinirole were dissolved in distillate water and injected s.c. (SKF38393) or i.p. (ropinirole) in a volume of 0.1 ml/100 g body weight.

2.3 Assessment of Nigrostriatal Lesion: Adjusting Steps and Cylinder Test

Adjusting step test and cylinder test were used to assess nigrostriatal lesion (Olsson et al. 1995; Schallert et al. 2000). Two weeks after 6-OHDA lesion the adjusting step test was repeated twice a day for 3 days. Only rats displaying no more than two steps with the right forepaw (contralateral to the lesion) in the forehand direction were included in the experiment. For the cylinder test, the rats were placed in a plexiglass cylinder and the forepaw used during exploratory behaviour of the cylinder was evaluated for 3 min. The test was performed under red light and forelimb preference displayed in placements made was calculated as a percentage of the total number of placements. Rats using the right forepaw more than 20% of total placements were eliminated from the study. These criteria led to selection of rats with a depletion of dopamine in the striatum exceeding 80% (Chang et al. 1999). Behavioural tests were confirmed by in situ hybridization of tyrosine hydroxylase mRNA in the substantia nigra pars compacta (data not shown).

2.4 Drug Treatment and Behavioural Tests

Six weeks after 6-OHDA infusion, the rats were treated for 19 days as follows: vehicle ($N = 10$), SKF38393 (1 mg/kg s.c. once/day; $N = 5$), SKF38393 (3 mg/kg s.c. once/day; $N = 10$), ropinirole (2 mg/kg i.p. once/day; $N = 5$) and ropinirole (5 mg/kg i.p. once/day; $N = 10$).

SCT and AIMs were measured throughout treatments in hemispherical plexiglass bowls (50-cm diameter) on alternate days. Rats were placed in each apparatus 30 min before drug administration in order to acclimatise and to extinguish any spontaneous turning behaviour. SCT was evaluated with an automated rotameter system (Carnegie Medicine, Sweden). The total number of contralateral turns performed for 180 min after drug injections was measured. AIMs (axial, limb and orolingual according to their topographic distribution) were observed individually from the first minute, every 20 min (starting from the 20th minute after drug administration), for 180 min [For details, see Pinna et al. (2006)].

2.5 Statistics

Mean ± S.E.M. of contralateral turns in 180-min testing period (total number of turns) and mean ± S.E.M. of seconds spent by rats in axial, limb and orolingual AIMs during 1-min testing-period (average of eight evaluations) were calculated. The development of SCT and AIMs sensitization during subchronic treatment with SKF38393 or ropinirole was assessed by means of linear regression, according to which a positive slope coefficient was considered as an index of sensitized response. Significant differences between groups were evaluated by two-way analysis of variance (ANOVA) followed by LSD post hoc test ($P < 0.05$).

3 Results

Subchronic treatment with SKF38393 elicited a contralateral turning response, which sensitized reaching a plateau from test day 6, as indicated by linear regression analysis (SKF38393 1 mg/kg: slope coefficient = 348.7 ± 49.71, $r^2 = 0.89$, $P = 0.0004$; SKF 3 mg/kg: slope coefficient = 261.2 ± 52.36, $r^2 = 0.7805$, $P = 0.016$) (Fig. 1a).

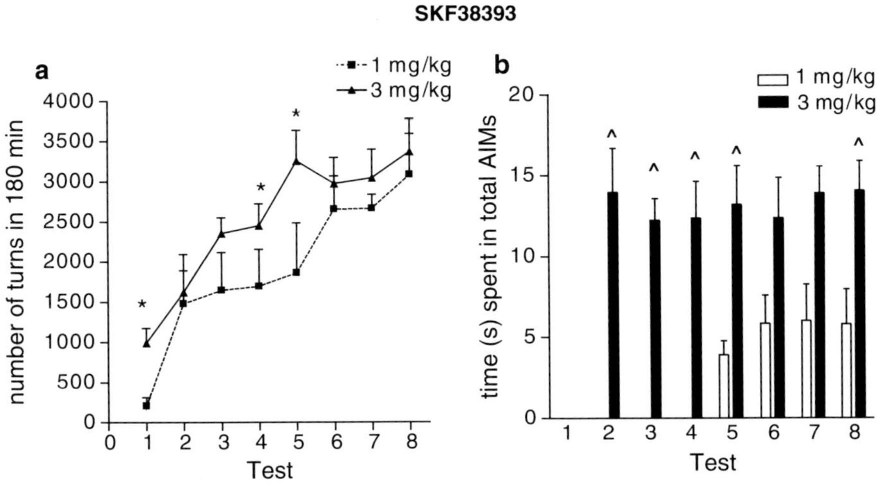

Fig. 1 Mean ± S.E.M. of total turns in 180-min testing period (**a**) and mean ± S.E.M of total seconds spent by unilaterally 6-OHDA-lesioned rats in AIMs (**b**) in the first minute of each 20-min testing period, during intermittent subchronic administration of SKF38393 (1 or 3 mg/kg s.c.). $*P < 0.05$, contralateral turning induced by SKF38393 (3 mg/kg) vs. SKF38393 (1 mg/kg); $\wedge P < 0.05$, AIMs induced by SKF38393 (3 mg/kg) vs. SKF38393 (1 mg/kg)

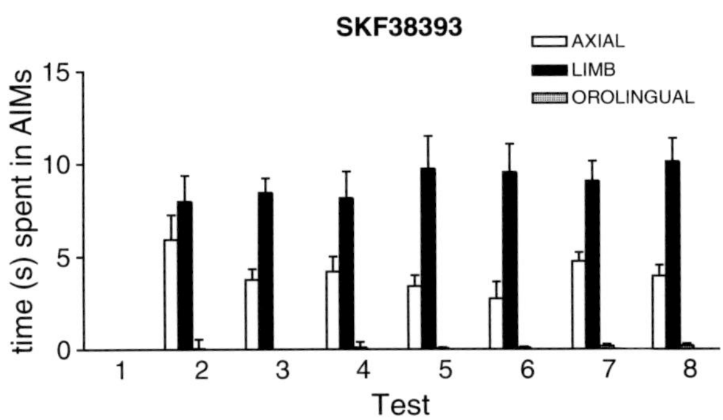

Fig. 2 Mean ± S.E.M of total seconds spent by unilaterally 6-OHDA-lesioned rats in limb, axial or orolingual AIMs, in the first minute of each 20-min testing period, during intermittent subchronic administration of SKF38393 (3 mg/kg s.c.)

SKF38393-induced (1 or 3 mg/kg) contralateral turning was dose dependent. Two-way ANOVA followed by LSD post hoc test revealed a significant difference on days 1, 4 and 5 ($P < 0.05$). Limb and axial AIMs were observed during treatment with both SKF38393 doses, whereas orolingual AIMs were never observed, as detailed in Fig. 2 for the higher dose. Neither doses displayed sensitization of AIMs, which remained stable till the last day of treatment (Figs. 1b and 2). Whereas SKF38393 at 1 mg/kg elicited AIMs of low intensity from the fifth test day, the

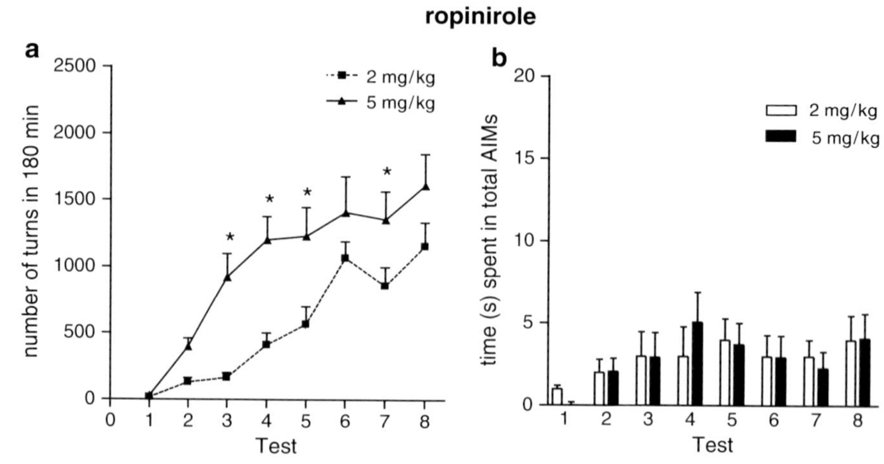

Fig. 3 Mean ± S.E.M. of total turns in 180-min testing period (**a**) and mean ± S.E.M of total seconds spent by unilaterally 6-OHDA-lesioned rats in AIMs (**b**) in the first minute of each 20-min testing period, during intermittent subchronic administration of ropinirole (2 or 5 mg/kg i.p.). $^{*}P < 0.05$, contralateral turning induced by ropinirole (2 mg/kg) vs. ropinirole (5 mg/kg)

higher dose induced more intense AIMs from the second test day. Two-way ANOVA followed by LSD post hoc test revealed a significant difference in AIM induction between 3 and 1 mg/kg SKF38393 on days 2, 3, 4, 5 and 8 ($P < 0.0\ 5$) (Fig. 1b).

Subchronic ropinirole induced a contralateral turning response, which sensitized throughout the treatment for both doses (ropinirole 2 mg: slope coefficient = 172.3 ± 21.02, $r^2 = 0.918$, $P < 0.001$; ropinirole 5 mg: slope coefficient = 206.6 ± 33.04, $r^2 = 0.87$, $P < 0.001$). Contralateral turning was dose dependent; two-way ANOVA followed by LSD post hoc test revealed a significant difference in turning induction between 5 and 2 mg/kg ropinirole on days 3, 4, 5 and 7 ($P < 0.05$) (Fig. 3a). In contrast, AIMs were almost absent and characterized by scarce episodes of limb AIMs (Fig. 3b).

The higher dose of ropinirole (5 mg/kg) produced, after 10 days of treatment, strong aggressiveness. Even higher aggressiveness and irregular locomotion and turning behaviour were observed after higher doses (data not shown).

4 Discussion

Prolonged treatment of PD patients with L-DOPA is well known to produce disabling dyskinesia. In contrast, the dyskinetic profile of dopaminergic agonists is more controversial and is hard to predict, depending on pharmacodynamic/pharmacokinetic factors.

By characterising the different behavioural outcomes of a subchronic treatment with SKF38393 or ropinirole the present study shows that subchronic SKF38393 and ropinirole induce a different behavioural dyskinetic response in the 6-OHDA-lesioned rat.

SKF38393, similarly to L-DOPA, was associated with the development of AIMs and SCT, which may represent a correlate of high dyskinetic potential in the preclinical evaluation of dopaminergic drugs. Conversely, ropinirole was associated with the development of SCT only, which may represent a model of mild dyskinetic responses induced by this drug in clinic.

4.1 High Dyskinetic Response After D_1 Receptor Stimulation by SKF38393

In the 6-OHDA-lesioned rat, SCT has been described upon treatment with highly dyskinetic drugs such as L-DOPA and apomorphine, as well as upon administration of D_2 receptor agonists with a lower clinical dyskinetic outcome (Delfino et al. 2004; Henry et al. 1998; Marin et al. 2006; Pinna et al. 2006; Lane et al. 2006). In contrast, AIMs were associated with treatments with a high dyskinetic potential, as L-DOPA, whereas they were absent upon treatments with a mild dyskinetic potential, as ropinirole (Carta et al. 2006; Delfino et al. 2004; Lundblad et al. 2002). Interestingly, in previous studies it was observed that a treatment void of dyskinetic potential, as A_{2A} receptor antagonists, failed to elicit both SCT and AIMs (Pinna et al. 2001; Xiao et al. 2006).

In the present study, we observed that abnormal behaviours induced by repeated SKF38393 (SCT and AIMs) were similar to those described upon repeated L-DOPA, therefore suggesting that repeated treatment with this D_1 agonist was characterized by an intense dyskinetic response. Accordingly, short-acting D_1 agonists, as the drugs used here, have been associated with the development of dyskinesia in primates (Andringa et al. 1999a; Blanchet et al. 1995; Calon et al. 1999; Goulet et al. 1996; Hurley and Jenner 2006; Lundblad et al. 2002; Pinna et al. 2006).

Interestingly, SKF38393-induced AIMs were of lower intensity than AIMs previously described upon subchronic L-DOPA administration, even when compared with a low dose of this drug (Lundblad et al. 2002; Pinna et al. 2006). Moreover, in contrast to L-DOPA, orolingual AIMs were never observed upon SKF38393 administration. Therefore, D_1 receptor's repeated stimulation, although causing a range of abnormal behaviours similar to L-DOPA, was not able to elicit a response of similar intensity, suggesting that concurrent D_1/D_2 receptor stimulation might be required to induce the most invalidating dyskinesia.

4.2 Low Dyskinetic Response After D_2 Receptor Stimulation by Ropinirole

Repeated D_2/D_3 receptor stimulation elicited SCT but failed to cause AIMs at either doses tested. This result indicates that ropinirole induced a dyskinetic response of lower intensity than that induced by the D_1 agonist and previously described L-DOPA treatment. Our results are in line with a previous study reporting SCTs not associated with abnormal movements, in response to a lower dose of ropinirole

(Ravenscroft et al. 2004). The lower intensity of SCT induced by ropinirole than that induced by SKF38393 could lead to the conclusion that a different drug potency might account for lack of AIMs upon ropinirole treatment. However, it is interesting to notice that turning intensity reached here by ropinirole treatment was similar to that reported upon a L-DOPA treatment also causing development of AIMs (see Pinna et al. 2006). Therefore, results argue against a role of drug potency in the differential induction of abnormal behaviours. Other factors, such as the selective D_2/D_3 receptor stimulation, might play a role. Accordingly, preferential D_3 receptor agonists have been proven to elicit lower dyskinesia when used in PD models (Guigoni et al. 2005; Guttman and Jaskolka 2001; Silverdale et al. 2004). Moreover, the pharmacokinetic profile of ropinirole, which is considered a long-acting compound, might account for different appearance of dyskinetic behaviours in response to this drug as compared with the higher dyskinetic D_2 agonists and the D_1 agonist used here (Blanchet et al. 1995; Calon et al.1995; Delfino et al. 2004).

5 Conclusions

AIMs have been previously described upon D_1/D_2 receptor stimulation with L-DOPA or apomorphine or upon D_2 receptor stimulation with quinpirole (Delfino et al. 2004; Lundblad et al. 2002). Here, we show that AIMs can be also elicited by D_1 agonist treatment. Therefore, the present results, together with previous studies, suggest that appearance of AIMs and SCT is not related to stimulation of dopamine receptors of a specific type, rather their appearance seems to reflect the dyskinetic potential of drugs. In this view, the presence of SCT only could be undertaken as a useful model for the mild dyskinesia observed in primates and humans upon administration of low dyskinetic dopaminergic agonists, whereas AIMs might represent a rat model of intense dyskinesia induced by highly dyskinetic drugs. Importantly, the different dyskinetic potential of drugs from low to high would be underestimated solely by evaluation of AIMs.

Acknowledgements The authors thank GlaxoSmithKline (Harlow, Essex, UK) for providing ropinirole. This work was supported by a grant from MURST – project FIRB (RBNE03YA3L-2005).

References

Adler CH, Sethi KD, Hauser RA, Davis TL, Hammerstad JP, Bertoni J, Taylor RL, Sanchez-Ramos J and O'Brien CF (The Ropinirole Study Group Neurology) (1997) Ropinirole for the treatment of early Parkinson's disease. Eur J Neurol 49: 393–399.

Andringa G, Lubbers L, Drukarch B, Stoof JC and Cools AR (1999a) The predictive validity of the drug-naive bilaterally MPTP-treated monkey as a model of Parkinson's disease: effects of L-DOPA and the D1 agonist SKF 82958. Behav Pharmacol 10: 175–182.

Andringa G, Vermeulen RJ, Drukarch B, Renier WO, Stoof JC and Cools AR (1999b) The validity of the pretreated, unilaterally MPTP-treated monkeys as a model of Parkinson's disease: a detailed behavioural analysis of the therapeutic and undesired effects of the D2 agonist quinpirole and the D1 agonist SKF 81297. Behav Pharmacol 10: 163–173.

Blanchet PJ, Gomez-Mancilla B and Bedard PJ (1995) DOPA-induced "peak dose" dyskinesia: clues implicating D2 receptor-mediated mechanisms using dopaminergic agonists in MPTP monkeys. J Neural Transm 45: S103-S112.

Blanchet PJ, Grondin R and Bédard PJ (1996) Dyskinesia and wearing-off following dopamine D1 agonist treatment in drug-naive 1-methyl-4-phenyl-1,2,3,6-tetrahydropyridine-lesioned primates. Mov Disord 11: 91–94.

Calon F, Goulet M, Blanchet PJ, Martel JC, Piercey MF, Bédard PJ and Di Paolo T (1995) Levodopa or D2 agonist induced dyskinesia in MPTP monkeys: correlation with changes in dopamine and GABAA receptors in the striatopallidal complex. Brain Res 680: 43–52.

Calon F, Morissette M, Goulet M, Grondin R, Blanchet PJ, Bedard PJ and Di Paolo T (1999) Chronic D1 and D2 dopaminomimetic treatment of MPTP-denervated monkeys: effects on basal ganglia GABA(A)/benzodiazepine receptor complex and GABA content. Neurochem Int 35: 81–91.

Carta AR, Pinna A and Morelli M (2006) How reliable is the behavioural evaluation of dyskinesia in animal models of Parkinson's disease? Behav Pharmacol 17: 393–402.

Chang JW, Wachtel SR, Young D and Kang UJ (1999) Biochemical and anatomical characterization of forepaw adjusting steps in rat models of Parkinson's disease: studies on medial forebrain bundle and striatal lesions. Neuroscience 88: 617–628.

Delfino MA, Stefano AV, Ferrario JE, Taravini IR, Murer MG and Gershanik OS (2004) Behavioral sensitization to different dopamine agonists in a parkinsonian rodent model of drug-induced dyskinesias. Behav Brain Res 152: 297–306.

Fukuzaki K, Kamenosono T and Nagata R (2000) Effects of ropinirole on various parkinsonian models in mice, rats, and cynomologus monkeys. Pharmacol Biochem Behav 65: 503–508.

Giardina WJ and Williams M (2001) Adrogolide HCl (ABT-431; DAS-431), a prodrug of the dopamine D1 receptor agonist, A-86929: preclinical pharmacology and clinical data. CNS Drug Rev Fall 7: 305–316.

Goulet M, Grondin R, Blanchet PJ, Bédard PJ and Di Paolo T (1996) Dyskinesias and tolerance induced by chronic treatment with a D1 agonist administered in pulsatile or continuous mode do not correlate with changes of putaminal D1 receptors in drug-naive MPTP monkeys. Brain Res 719: 129–137.

Guigoni C, Aubert I, Li Q, Gurevich VV, Benovic JL, Ferry S, Mach U, Stark H, Leriche L, Håkansson K, Bioulac BH, Gross CE, Sokoloff P, Fisone G, Gurevich EV, Bloch B and Bezard E (2005) Pathogenesis of levodopa-induced dyskinesia: focus on D1 and D3 dopamine receptors. Parkinsonism Relat Disord 11: S25–S29.

Guttman M and Jaskolka J (2001) The use of pramipexole in Parkinson's disease: are its actions D(3) mediated? Parkinsonism Relat Disord 7: 231–234.

Henry B, Crossman AR and Brotchie JM (1998) Characterization of enhanced behavioural responses to L-DOPA following repeated administration in the 6-hydroxydopamine-lesioned rat model of Parkinson disease. Exp Neurol 151: 334–342.

Hurley MJ and Jenner P (2006) What has been learnt from study of dopamine receptors in Parkinson's disease? Pharmacol Ther 111: 715–728.

Jenner P (2003) Dopamine agonists, receptor selectivity and dyskinesia induction in Parkinson's disease. Curr Opin Neurol 16: S3–S7.

Lane EL, Cheetham SC and Jenner P (2006) Does contraversive circling in the 6-OHDA-lesioned rat indicate an ability to induce motor complications as well as therapeutic effects in Parkinson's disease? Exp Neurol 97: 284–290.

Lindgren HS, Rylander D, Ohlin KE, Lundblad M and Cenci MA (2007) The "motor complication syndrome" in rats with 6-OHDA lesions treated chronically with L-DOPA: relation to dose and route of administration. Behav Brain Res 177: 150–159.

Lundblad M, Andersson M, Winkler C, Kirik D, Wierup N and Cenci MA (2002) Pharmacological validation of behavioural measures of akinesia and dyskinesia in a rat model of Parkinson's disease. Eur J Neurosci 15: 120–132.

Marin C, Rodriguez-Oroz MC and Obeso JA (2006) Motor complications in Parkinson's disease and the clinical significance of rotational behavior in the rat: have we wasted our time? Exp Neurol 197: 269–274.

Matsuda H, Hiyama Y, Terasawa K, Watanabe H and Matsumoto K (1992) Enhancement of rotational behavior induced by repeated administration of SKF38393 in rats with unilateral nigrostriatal 6-OHDA lesions. Pharmacol Biochem Behav 42: 213–218.

Morelli M, Fenu S, Garau L and Di Chiara G (1989) Time and dose dependence of the 'priming' of the expression of dopamine receptor supersensitivity. Eur J Pharmacol 162: 329–335.

Olsson M, Nikkhah G, Bentlage C and Bjorklund A (1995) Forelimb akinesia in the rat Parkinson model: differential effects of dopamine agonists and nigral transplants as assessed by a new stepping test. J Neurosci 15: 3863–3875.

Pearce RK, Banerji T, Jenner P and Marsden CD (1998) De novo administration of ropinirole and bromocriptine induces less dyskinesia than L-dopa in the MPTP-treated marmoset. Mov Disord 13: 234–241.

Pellegrino A, Pellegrino AS and Cushman AJ (1979) Stereotaxic Atlas of the Rat Brain. Plenum Press, New York.

Pinna A, Fenu S and Morelli M (2001) Motor stimulant effects of the adenosine A2A receptor antagonist SCH 58261 do not develop tolerance after repeated treatments in 6-hydroxydopamine-lesioned rats. Synapse 39: 233–238.

Pinna A, Pontis S and Morelli M (2006) Expression of dyskinetic movements and turning behaviour in subchronic L-DOPA 6-hydroxydopamine-treated rats is influenced by the testing environment. Behav Brain Res 171: 175–178.

Rascol O, Brooks DJ, Korczyn AD, De Deyn PP, Clarke CE and Lang AE (2000) A five-year study of the incidence of dyskinesia in patients with early Parkinson's disease who were treated with ropinirole or levodopa. N Engl J Med 342: 1484–1491.

Rascol O, Nutt JG, Blin O, Goetz CG, Trugman JM, Soubrouillard C, Carter JH, Currie LJ, Fabre N, Thalamas C, Giardina WW and Wright S (2001) Induction by dopamine D1 receptor agonist ABT-431 of dyskinesia similar to levodopa in patients with Parkinson disease. Arch Neurol 58: 249–254.

Ravenscroft P, Chalon S, Brotchie JM and Crossman AR (2004) Ropinirole versus L-DOPA effects on striatal opioid peptide precursors in a rodent model of Parkinson's disease: implications for dyskinesia. Exp Neurol 185: 36–46.

Reichmann H (2000) Long-term treatment with dopamine agonists in idiopathic Parkinson's disease. J Neurol 247: S17–S19.

Sethi KD, O'Brien CF, Hammerstad JP, Adler CH, Davis TL, Taylor RL, Sanchez-Ramos J, Bertoni JM and Hauser RA (1998) Ropinirole for the treatment of early Parkinson disease: a 12-month experience. Arch Neurol 55: 1211–1216.

Setler PE, Sarau HM, Zirkle CL and Saunders HL (1978) The central effects of a novel dopamine agonist. Eur J Pharmacol 50: 419–430.

Schallert T, Fleming SM, Leasure JL, Tillerson JL, Bland ST (2000) CNS plasticity and assessment of forelimb sensorimotor outcome in unilateral rat models of stroke, cortical ablation, parkinsonism and spinal cord injury. Neuropharmacology 39: 777–787.

Silverdale MA, Nicholson SL, Ravenscroft P, Crossman AR, Millan MJ and Brotchie JM (2004) Selective blockade of D(3) dopamine receptors enhances the anti-parkinsonian properties of ropinirole and levodopa in the MPTP-lesioned primate. Exp Neurol 188: 128–138.

Smith LA, Jackson MJ, Al-Barghouthy G and Jenner P (2002) The actions of a D-1 agonist in MPTP treated primates show dependence on both D-1 and D-2 receptor function and tolerance on repeated administration. J Neural Transm 109: 123–140.

Smith LA, Jackson MJ, Johnston L, Kuoppamaki M, Rose S, Al-Barghouthy G, Del Signore S and Jenner P (2006) Switching from levodopa to the long-acting dopamine D2/D3 agonist piribedil reduces the expression of dyskinesia while maintaining effective motor activity in MPTP-treated primates. Clin Neuropharmacol 29: 112–125.

Van de Witte SV, Drukarch B, Stoof JC and Voorn P (1998) Priming with L-DOPA differently affects dynorphin and substance P mRNA levels in the striatum of 6-hydroxydopamine-lesioned rats after challenge with dopamine D1-receptor agonist. Brain Res Mol Brain Res 61: 219–223.

Xiao D, Bastia E, Xu YH, Benn CL, Cha JH, Peterson TS, Chen JF and Schwarzschild MA (2006) Forebrain adenosine A2A receptors contribute to L-3,4-dihydroxyphenylalanine-induced dyskinesia in hemiparkinsonian mice. J Neurosci 26: 13548–13555.

Basal Ganglia and Behaviour: Behavioural Effects of Deep Brain Stimulation in Experimental Neurological and Psychiatric Disorders

Thibaut Sesia, Sonny Tan, Rinske Vlamings, Lee Wei Lim, Veerle Visser-Vandewalle, and Yasin Temel

Abstract The use of deep brain stimulation (DBS) to control severely disabling neurological and psychiatric conditions is an exciting and fast emerging area of neuroscience. Deep brain stimulation has generally the same clinical effects as a lesion with respect to the improvement of clinical disability, but has more advantages such as its adjustability and reversibility. To this day, fundamental knowledge regarding the application of electrical currents to deep brain structures is far from complete. Despite improving key symptoms in movement disorders, DBS can be associated with the occurrence of a variety of changes in cognitive and limbic functions both in humans and animals. Furthermore, in psychiatric disorders, DBS is primarily used to evoke cognitive and limbic changes to reduce the psychiatric disability. Preclinical DBS experiments have been carried out to investigate the mechanisms underlying the clinical effects of DBS for at least three (interrelated) reasons: to increase our scientific knowledge, to optimize/refine the technology, or to prevent/reduce side effects. In this review, we will discuss the behavioural effects of DBS in experimental neurological and psychiatric disorders.

1 Introduction

The use of stimulation electrodes implanted in the brain to control severely disabling neurological and psychiatric conditions is an exciting and rapidly emerging area of neuroscience (Wichmann and DeLong 2006). To date, several clinical studies have been performed to treat intractable neurological and psychiatric disorders such as Parkinson's disease (PD), obsessive compulsive disorder (OCD), depression, and Tourette's syndrome, using deep brain stimulation (DBS) of basal ganglia nuclei (Wichmann and DeLong 2006; Hardesty and Sackeim 2007). Significant therapeutic effects were achieved with DBS, but at the same time in some patients behavioural

T. Sesia, S. Tan, R. Vlamings, L.W. Lim, V. Visser-Vandewalle, and Y. Temel (✉)
Department of Neurosurgery, University Hospital Maastricht, P. Debyelaan 25, 6202 AZ, Maastricht, The Netherlands
e-mail: y.temel@np.unimaas.nl

H.J. Groenewegen et al. (eds.), *The Basal Ganglia IX*, Advances in Behavioral Biology 58,
DOI 10.1007/978-1-4419-0340-2_36,

changes, both positive and negative, were observed. For instance, prominent antidepressant effects were found during stimulation of the ventral capsule/ventral striatum for OCD (Greenberg et al. 2006). These unexpected findings motivate preclinical researchers to evaluate the applicability of DBS in other psychiatric disorders.

The underlying mechanisms of electrical brain stimulation are currently being extensively investigated. Some authors assume a local inhibitory effect (Filali et al. 2004) and others a more complex effect of stimulation. The latter suggests a local inhibition of the soma and the generation of action potentials in the surrounding axons (McIntyre et al. 2004a, b, 2006), thereby superimposing a rhythmic output of the stimulated nucleus (Meissner et al. 2005). The combination of orthodromic and antidromic actions of electrical stimulation is more and more emphasized. Recently, it was shown that the electrophysiological changes in the prefrontal cortex during nucleus accumbens stimulation are likely to be an antidromic effect (McCracken and Grace 2007). In parallel to the efforts to understand neuronal mechanisms, investigations also focus on the network effect of DBS on the directly connected target nuclei (i.e. monosynaptic effects) (Tai et al. 2003) and the indirectly connected target nuclei (i.e. polysynaptic effects) (Temel et al. 2007).

Currently, DBS is mainly used for the treatment of patients suffering from intractable movement disorders. This is particularly true for PD patients who receive stimulation of the subthalamic nucleus (STN). Despite improving key motor symptoms, DBS can be associated with the occurrence of a variety of changes in cognitive and limbic functions both in humans and animals (Bejjani et al. 1999; Piasecki and Jefferson 2004; Temel et al. 2005a, 2006b; Baunez et al. 2007). Although primarily considered a motor nucleus, the STN is also involved in cognitive and limbic functions (Temel et al. 2005a). Furthermore, the indications for DBS are expanding to psychiatric disorders with the aim of evoking cognitive and limbic changes to reduce psychiatric disability. In this chapter, we will discuss the behavioural effects of DBS in experimental neurological and psychiatric conditions.

2 Deep Brain Stimulation for Non-motor Symptoms of Movement Disorders

PD and Huntington's disease (HD) are both *neurological and psychiatric* disorders and have key motor symptoms such as rigidity and chorea, respectively. Most of the current treatments are targeting principally these symptoms. Interestingly both PD and HD include cognitive or mood-related changes in their core symptomatology (Craufurd et al. 2001; Temel et al. 2005a; Paulsen and Conybeare 2005).

Reaction time (RT) performance is frequently used in animal models to evaluate both cognitive and motor performance (Blokland 1998). This task is based on operant conditioning and is very suitable to use in DBS experiments, since the subject can be tested repeatedly while cognitive and motor performance are evaluated simultaneously. There are various forms of RT tasks generally divided into simple RT (SRT) and choice/complex RT (CRT) tasks.

Darbaky and co-workers (2003) tested the effects of STN DBS in a rat model of PD, the unilateral 6-hydroxydopamine (6-OHDA) model. Rats were trained to perform a CRT task, in which they were required to sustain a nose-poke in a central hole and wait for a light presented either at the left- or right-side hole before responding by making a nose-poke in the illuminated hole. The motor effects of STN DBS on CRT task performance were not as strong as on basic motor tests. STN DBS had beneficial effects only in animals that were able to perform the task after the unilateral 6-OHDA lesion, by slightly improving their neglect towards the contralateral side of the lesion. In animals not able to perform the task after the dopaminergic lesion, STN DBS had no "awakening" effect and was unable to help them perform the task. The authors suggested that STN DBS was beneficial in alleviating the motor deficit, but unable to improve performance when a cognitive load was required (Darbaky et al. 2003).

Temel and associates reported on the effects of *bilateral* STN DBS in rats (Desbonnet et al. 2004). The stimulation parameters were adjusted to prevent histological damage in their rat models. DBS was applied to the STN of non-depleted rats with various stimulation parameters during a CRT task. Results showed a significant linear decrease in premature responding with decreasing amplitudes and high frequencies only. It was suggested that a decrease in premature responding could be attributed to a stimulation-dependent increased activation of pre-motor cortical areas associated with the "motor readiness potential" resulting in modulated cognitive performance (Desbonnet et al. 2004). The same group conducted a second study in which bilateral STN DBS was applied to parkinsonian rats, again during a CRT task (Temel et al. 2005b) (Fig. 1). Stimulations were performed at 130 Hz (frequency) and 60 μs (pulse width) because these parameters had been shown to be effective in the previous study (Desbonnet et al. 2004). In addition, varying amplitudes of 1, 3, 30, and 150 μA were applied. Bilateral STN DBS with an amplitude of 3 μA significantly decreased 6-OHDA-induced deficits in premature responding. Stimulation with an amplitude of 30 μA reversed the lesion-induced motor deficits (Fig. 2). These data showed that bilateral STN DBS could acutely and separately influence the 6-OHDA-induced motor- and cognitive deficits. Temel et al. (2005b) attributed these findings to unique physiological properties of the basal ganglia-thalamocortical motor and associative circuits, responsible for specific motor and cognitive performances.

Baunez and Robbins (1997, 1999) reported the effects of bilateral STN lesions in the five-choice serial reaction time task. This task is particularly adapted to assess divided and sustained attention (Robbins 2002). Bilateral STN lesion effects had been characterized in this task and they revealed multiple and independent deficits, suggestive of attention deficits, disinhibition, and perseverative problems as well as a motivational excess (Baunez and Robbins 1997). A further study characterized the effects of bilateral 6-OHDA striatal lesions mainly as a slight attention deficit and perseverative behaviour in addition to motor impairment. Most of the cognitive impairments induced by this dopaminergic depletion were not alleviated by a bilateral STN lesion (Baunez and Robbins 1999). Bilateral STN DBS applied at similar parameters to those used by Darbaky et al. (2003) induced comparable

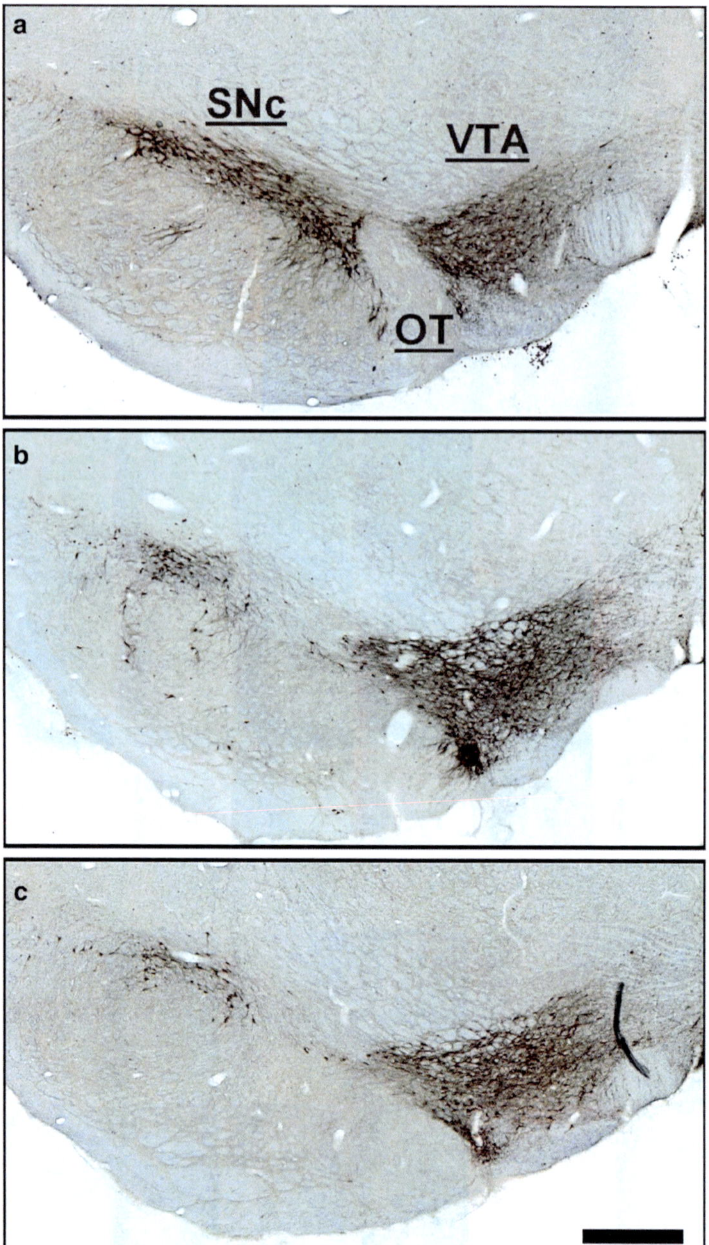

Fig. 1 Representative low-power photomicrographs of 30-μm frontal sections immunoprocessed for TH, showing the ventral tegmental area (VTA) and the SNc in the brain of a rat subjected to stereotactic bilateral vehicle injection into the striatum (**a**), a rat subjected to stereotactic bilateral 6-OHDA injection into the striatum (**b**), and a rat subjected to both stereotactic bilateral 6-OHDA injection into the striatum and stereotactic implantation of electrodes to stimulate the STN (**c**). The photomicrographs show similar frontal levels in the midbrain in which the optic tract (OT) divides the TH-immunoreactive cells into the VTA (medial) and SNc (lateral) regions. Bilateral 6-OHDA lesions resulted in a substantial reduction in the number of TH-immunoreactive cells in the SNc (**b**, **c**) (scale bar = 500 μm). Adapted from Vlamings et al. (2007)

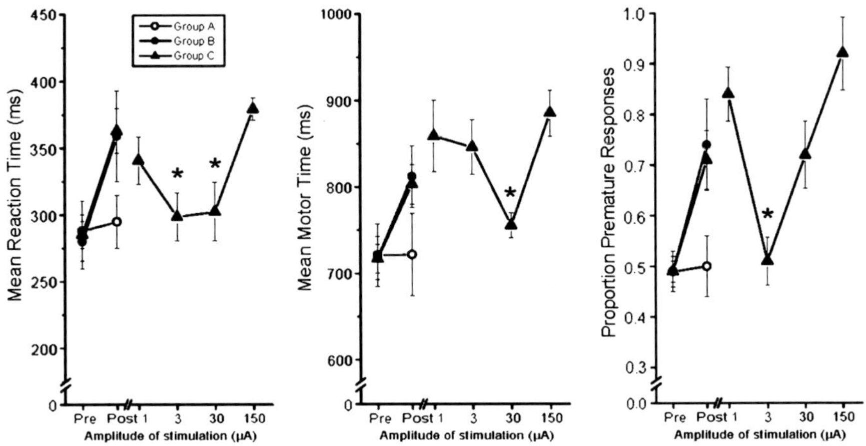

Fig. 2 Results of the behavioural testing of rats subjected to vehicle injection into the striatum (group A), rats subjected to 6-OHDA injection into the striatum (group B), and rats subjected to both 6-OHDA injection into the striatum and implantation of electrodes to stimulate the STN (group C). Data are shown as mean ± SEM, and are given as mean reaction time (RT; on the *left*), mean motor time (MT; *middle*) and proportion of premature responses (PR; on the *right*) in a specific choice reaction time task (36). On the *x*-axis, data are organized in the following order: preoperative data from all rats (Pre), postoperative data from all rats (Post; note that these data also represent the stimulation off data from rats in group C), and stimulation-dependent data from rats in group C as a function of the stimulation amplitude. Repeated measures ANOVA revealed significant differences in the postoperative data between the groups ($p < 0.05$) as well as significant differences between the postoperative data and the stimulation-dependent data of group C. The p values from the corresponding post hoc tests are provided as *asterisk* ($p < 0.0125$). Note that bilateral STN DBS significantly improved RT at both 3 μm and 30 μA, MT only at 30 μA, and PR only at 3 μA. Adapted from Temel et al. (2005b)

deficits to those observed after STN lesions in intact rats (attention deficit, perseverative behaviour, as well as increased perseverations towards the food container; the latter observation being suggestive of a motivational excess), although some of these deficits were only transient under DBS (Baunez et al. 2007). Interestingly, when DBS was applied in parkinsonian rats, it did not impair their performance further, but did not alleviate the deficits induced by the dopaminergic lesion either. The most relevant effect observed after STN DBS in both intact and parkinsonian rats was the increased perseveration towards the food container, suggestive of increased motivation for the food reward (Baunez et al. 2007).

A model of a different movement disorder is the transgenic rat model (tgHD) of HD. Temel and associates studied in rats the effects of globus pallidus (GP, equivalent of the external segment of the globus pallidus in primates) DBS on cognitive and motor symptoms in this tgHD model, which recently has been generated and carries a truncated huntingtin cDNA fragment with 51 CAG repeats (van Hörsten et al. 2003; Cao et al. 2006). After these two model-validation studies, the authors implanted tgHD rats with stimulating electrodes at the level of the GP (Temel et al. 2006a). Rats were evaluated in a CRT and an open field task. Stimulation of the GP

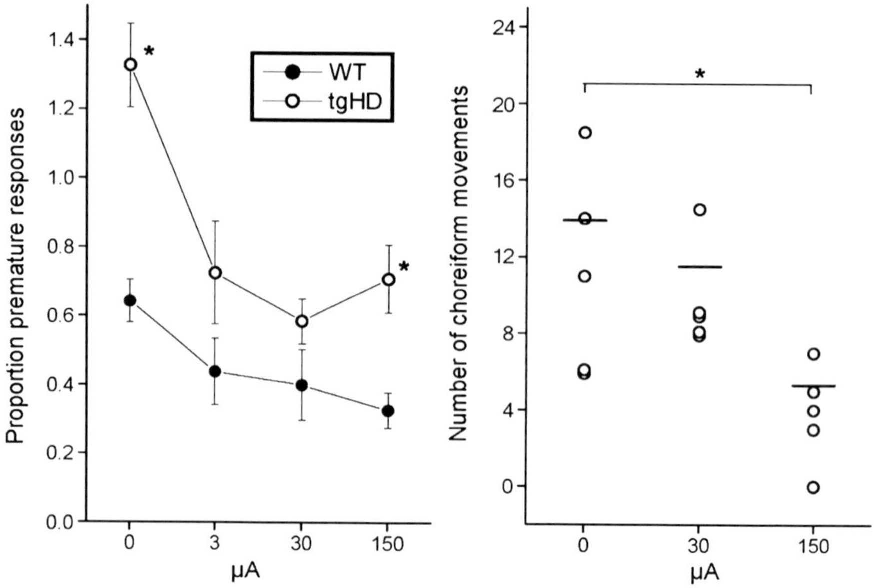

Fig. 3 Means and standard errors of the proportion of premature responses (PR) (*left figure*) and the number of choreiform movements (*right figure*) in both wild-type (WT) and transgenic Huntington's disease rats (tgHD) in relation to different stimulation amplitudes. Deep brain stimulation of the globus pallidus (GP) significantly improved the cognitive deficit (PR) and the motor disorder (choreiform movements) in tgHD rats. *$p < 0.05$. Adapted from Temel et al. (2006a)

clearly reduced the number of premature responses. Premature responding is thought to reflect the inability to inhibit unwanted responses (Temel et al. 2005a) and is a key feature of cognitive dysfunction in HD (Aron et al. 2003). DBS with medium amplitude (30 μA) decreased premature responses most effectively, while DBS with the higher amplitude (150 μA) significantly reduced the choreiform movements in the tgHD rats (Fig. 3).

3 Deep Brain Stimulation in Psychiatric Disorders

3.1 Obsessive-Compulsive Disorder

Van Kuyck et al. (2003) used a rodent model for obsessive-compulsive disorder (OCD) and evaluated the effects of DBS of the nucleus accumbens (NAc) in a freely moving animal, in particular the more medial shell part. Rats that received 8-OH-DPAT (8-hydroxy-2 di *n* propylamino-tetralin hydrobromide), a 5-hydroxytryptamine (i.e. serotonin) 1A agonist, showed a reduction of spontaneous alternation behaviour in the T-maze. According to the authors, this behaviour models the compulsive and repetitive behaviour of patients suffering from OCD. Both 5-HT lesions and DBS

(5 Hz and 100 μA) significantly reduced spontaneous alternation behaviour, which reflects an increase in compulsive behaviour. The authors argued a possible lack of power of this model. Additionally, this study only used 5-Hz stimulation; whereas clinical studies demonstrated that high-frequency stimulation of the NAc was effective in reducing OCD symptoms (Okun et al. 2007). Interestingly, the same authors observed an amplitude-dependent reduction of compulsion, effective only with high-frequency stimulation using a schedule-induced polydipsia model in a different study (Woods et al. 1993; van Kuyck et al. 2007).

Others have also evaluated the effect of high-frequency stimulation of the NAc, the STN and the amygdala on an alternative OCD model with quinpirole-induced compulsive checking behaviour (Winter et al. 2004; Jalali et al. 2004). The more lateral core part of the NAc and the STN differentially affected compulsive checking behaviour; although both reduced the compulsive behaviour, stimulation of the STN did not prevent the induction of the OCD-like behaviour.

These results clearly show that the basal ganglia (e.g. NAc and STN) are involved in compulsive behaviour. Further investigations are necessary to evaluate the therapeutic value of the NAc as a potential target for DBS in animal models of OCD. More emphasis on the effects of different stimulation paradigms is needed.

3.2 Animal Models of Depression

The forced swim task (FST) is a well-validated animal model for depression-like behaviour sensitive to changes in 5-HT (Cryan et al. 2002). The FST task has originally been developed by Porsolt and colleagues (1977) and was later modified by Detke and Lucki (1996). This test is based on exposure to a learned inescapable stressor, and the measurement of immobility that is thought to reflect a failure of persistence in escape-directed behaviour, also referred as behavioural despair (Cryan et al. 2002, 2004; Porsolt et al. 1977). The duration of immobility has been shown to correlate significantly with the antidepressant effect of antidepressant drugs: the lower the duration of immobility, the greater the antidepressant effect of the drug.

In this rat model of depression, Temel and co-workers (2007) showed that DBS of the STN worsened depressive symptoms. In more detail, bilateral DBS of the STN (130 Hz, 60 μs, 150 μA) caused a striking increase in immobility and a decrease in climbing time of stimulated rats compared to non-stimulated controls, thereby indicative of the induction of behavioural "despair". This effect was present both in normal and parkinsonian animals. Importantly, the effects of DBS of the STN stimulation in the FST were completely prevented by a prior course of treatment with the selective 5-HT reuptake inhibitor, citalopram, at a dose (10 mg/kg s.c., once daily for 14 days) that by itself had no significant effect in non-stimulated controls. To exclude any motor effect of STN DBS that may have confounded the results in the FST, it was demonstrated that in an open field paradigm, DBS of the STN had no effect on normal motor activity (Temel et al. 2007). In contrast, DBS of the STN (using the same stimulation parameters) reversed motor-time deficits in parkinsonian rats in agreement with previous results (Temel et al. 2005b). These data show that STN DBS in rats causes an acute induction of depression-like behaviour. This link between the STN and the 5-HT system supports the existence of a novel "motor-limbic interface" that may contribute to mood disturbances in basal ganglia disease.

3.3 *Animal Models of Panic Disorder*

While investigating the effect of striatal DBS on the dopaminergic and serotonergic system, Hano et al. (1978) reported violent running, rearing, backing, and shaking during NAc stimulation; this was interpreted as aggressive behaviour. Additionally they noted a decrease in serotonin levels within the ventral and dorsal raphe nuclei. Nonetheless, the effects of NAc stimulation are highly diverse in the literature; ranging from fear alleviation, arrest, escape behaviour to no behavioural effects at all (Wright et al. 1977; Wilson 1983; Murer and Pazo 1993). These behavioural changes may be mediated through the NAc connections with structures such as the amygdala, the prefrontal cortex, and the periaqueductal gray (PAG). These structures and their interconnections are thought to be neuronal substrates for neuropsychiatric disorders, e.g. anxiety and panic disorders.

Stimulation of the dorsal PAG is a well-known animal model for panic disorders (Mongeau and Marsden 1997). It has been demonstrated that DBS of this area in rats leads to "escape behaviour" characterized by running, jumping, and galloping (Bandler et al. 1985). To date, most experiments involving PAG stimulation were designed to evoke rather than to inhibit the panic-like behaviour, often using a simple paradigm of stimulation (i.e. a limited range of amplitudes or frequencies). In a recent study, Lim and co-workers addressed the question whether DBS (now using a brain stimulation paradigm using state-of-art technology) at high frequencies of the PAG could *inhibit* panic-like behaviour (Lim et al. 2007). In addition, in the same study they investigated the effect of different stimulation frequencies on escape behaviour. All animals were subjected to the stimulation parameters at pulse width (100 μs), frequency (1–300 Hz), and amplitude (1–650 μA). Their results showed that stimulation with any stimulation frequency between 1 and 300 Hz induced escape behaviour. Furthermore, they also noted a post-stimulation effect in which stimulated animals displayed intense fear or immobility after the animals were placed back to the same open field arena of stimulation. It is noteworthy, that this unconditioned fear or anxiety generated by the dorsal lateral PAG stimulation has been considered for future research aimed at the development of anti-panic and anti-anxiety drugs.

4 Discussion

In this review, we have summarized the behavioural effects of DBS in experimental models for some neurological and psychiatric diseases. The reports can be divided into studies focusing on models of movement disorders and models of psychiatric disorders.

With respect to movement disorders, experiments have shown that STN DBS in parkinsonian and in intact rats have profound effects on cognitive and motivational parameters, while improving parkinsonian motor deficits (Darbaky et al. 2003; Desbonnet et al. 2004; Temel et al. 2005b; Baunez et al. 2007). The involvement of the STN in non-motor functions was already demonstrated by Baunez and associates (Baunez et al. 1995; Baunez and Robbins 1997) before STN DBS was performed on such a large scale in PD patients. Due to the profound effects of STN DBS on cognitive and limbic functions, clinicians should evaluate each individual PD patient carefully before considering them for STN DBS. At the same time, these results suggest that STN might also be a target for DBS in cognitive and limbic disorders. For instance, it has been shown that STN DBS in rats increased perseverative visits to the food container. This effect, interpreted as an increased motivation for the food reward, is in line with the effects described after STN lesions on motivation for food (Kirkwood et al. 2000; Baunez et al. 2002). Interestingly, STN lesions increase motivation for food, but decrease motivation for cocaine (Baunez et al. 2005). Taking into account the fact that STN DBS mostly mimics the behavioural effects of STN lesions, it is thus tempting to suggest that STN DBS could be an interesting surgical therapeutic tool for the treatment of cocaine addiction.

In the transgenic rat model of HD, DBS is currently being evaluated as a treatment option. HD is characterized by progressive deterioration in cognitive and motor functions (Gardian and Vecsei 2004). Cognitive symptoms often precede the choreiform movement disorder (Kirkwood et al. 2000). Although the movement disorder can be treated pharmacologically to some extent, no effective therapies are available to treat the cognitive symptoms (Gardian and Vecsei 2004). Therefore, it is usually the cognitive dysfunction that determines the quality of life of the HD patient. The results of the pilot studies presented in this review show that DBS of the GP has the potential to improve cognitive and motor symptoms of transgenic HD rats. Currently, experiments are carried out to further evaluate the potential of DBS in animal models of HD.

In the field of psychiatric disorders such as OCD, depression, and anxiety, DBS is being evaluated as a possible tool to treat refractory patients. However, before DBS can be applied to new indications, preclinical studies are required to test its potential effects. There are only few preclinical studies published in this area, but the expectation is that the number of studies will increase substantially over time. The few published reports are promising and support the usefulness of DBS in psychiatric disorders. Since currently available models failed to face validity, an alternative approach may be to study core symptoms common to multiple disorders. Experiments are carried out in our group to map a number of impulsive behaviours (e.g. impulsive action, impulsive choice) within the basal ganglia nuclei.

Acknowledgements The deep brain stimulation program of Yasin Temel and colleagues received support from the Dutch Medical Research Council (ZonMw and NWO), the Dutch Brain Foundation (Hersenstichting Nederland), the Prinses Beatrix Foundation, and the FP6 Marie Curie Fellowship (MEST-CT-2005-020589).

References

Aron AR, Watkins L, Sahakian BJ, Monsell S, Barker RA and Robbins TW (2003) Task-set switching deficits in early-stage Huntington's disease: Implications for basal ganglia function. J Cogn Neurosci 2003: 629–642.

Bandler R, Depaulis A and Vergnes M (1985) Identification of midbrain neurones mediating defensive behaviour in the rat by microinjections of excitatory amino acids. Behav Brain Res 15: 107–119.

Baunez C and Robbins TW (1997) Bilateral lesions of the subthalamic nucleus induce multiple deficits in an attentional task in rats. Eur J Neurosci 9: 2086–2099.

Baunez C and Robbins TW (1999) Effects of dopamine depletion of the dorsal striatum and further interaction with subthalamic nucleus lesions in an attentional task in the rat. Neuroscience 92: 1343–1356.

Baunez C, Nieoullon A and Amalric M (1995) In a rat model of parkinsonism, lesions of the subthalamic nucleus reverse increases of reaction time but induce a dramatic premature responding deficit. J Neurosci 15: 6531–6541.

Baunez C, Amalric M and Robbins TW (2002) Enhanced food-related motivation after bilateral lesions of the subthalamic nucleus. J Neurosci 22: 562–568.

Baunez C, Dias C, Cador M and Amalric M (2005) The subthalamic nucleus exerts opposite control on cocaine and 'natural' rewards. Nat Neurosci 8: 484–489.

Baunez C, Christakou A, Chudasama Y, Forni C and Robbins TW (2007) Bilateral high-frequency stimulation of the subthalamic nucleus on attentional performance: Transient deleterious effects and enhanced motivation in both intact and parkinsonian rats. Eur J Neurosci 25: 1187–1194.

Bejjani BP, Damier P, Arnulf I, Thivard L, Bonnet AM, Dormont D, Cornu P, Pidoux B, Samson Y and Agid Y (1999) Transient acute depression induced by high-frequency deep-brain stimulation. N Engl J Med 340: 1476–1480.

Blokland A (1998) Reaction time responding in rats. Neurosci Biobehav Rev 22: 847–864.

Cao C, Temel Y, Blokland A, Ozen H, Steinbusch HW, Vlamings R, Nguyen HP, von Hörsten S, Schmitz C and Visser-Vandewalle V (2006) Progressive deterioration of reaction time performance and choreiform symptoms in a new Huntington's disease transgenic rat model. Behav Brain Res 170: 257–261.

Craufurd D, Thompson JC and Snowden JS (2001) Behavioral changes in Huntington disease. Neuropsychiatry Neuropsychol Behav Neurol 14: 219–226.

Cryan JF, Markou A and Lucki I (2002) Assessing antidepressant activity in rodents: Recent developments and future needs. Trends Pharmacol Sci 23: 238–245.

Cryan JF, O'Leary OF, Jin SH, Friedland JC, Ouyang M, Hirsch BR, Page ME, Dalvi A, Thomas SA and Lucki I (2004) Norepinephrine-deficient mice lack responses to antidepressant drugs, including selective serotonin reuptake inhibitors. Proc Natl Acad Sci USA 101: 8186–8191.

Darbaky Y, Forni C, Amalric M and Baunez C (2003) High frequency stimulation of the subthalamic nucleus has beneficial antiparkinsonian effects on motor functions in rats, but less efficiency in a choice reaction time task. Eur J Neurosci 18: 951–956.

Desbonnet L, Temel Y, Visser-Vandewalle V, Blokland A, Hornikx V and Steinbusch HW (2004) Premature responding following bilateral stimulation of the rat subthalamic nucleus is amplitude and frequency dependent. Brain Res 1008: 198–204.

Detke MJ and Lucki I (1996) Detection of serotonergic and noradrenergic antidepressants in the rat forced swimming test: the effects of water depth. Behav Brain Res 73: 43–46.

Filali M, Hutchison WD, Palter VN, Lozano AM and Dostrovsky JO (2004) Stimulation-induced inhibition of neuronal firing in human subthalamic nucleus. Exp Brain Res 156: 274–281.

Gardian G and Vecsei L (2004) Huntington's disease: Pathomechanism and therapeutic perspectives. J Neural Transm 111: 1485–1494.

Greenberg BD, Malone DA, Friehs GM, Rezai AR, Kubu CS, Malloy PF, Salloway SP, Okun MS, Goodman WK and Rasmussen SA (2006) Three-year outcomes in deep brain stimulation for highly resistant obsessive-compulsive disorder. Neuropsychopharmacology 31: 2384–2393.

Hano J, Przewlocki R, Smialowska M, Chłapowska M and Rokosz-Pelc A (1978) The effect of electric stimulation of caudate nucleus and nucleus accumbens septi on serotonergic neurons in the rat brain. Pol J Pharmacol Pharm 30: 475–481.

Hardesty DE and Sackeim HA (2007) Deep brain stimulation in movement and psychiatric disorders. Biol Psychiatry 61: 831–835.

Jalali R, Bergmann O, Hosmann K, Kupsch A, Juckel D, Morgenstern R and Winter C (2004) High frequency stimulation of the subthalamic nucleus reduces quinpirole induced compulsive checking behavior in rats. Program No.115.8. 2004 Abstract Viewer/Itinerary Planner. Washington, DC: Society for Neuroscience, p. 118.

Kirkwood SC, Siemers E, Hodes ME, Conneally PM, Christian JC and Foroud T (2000) Subtle changes among presymptomatic carriers of the Huntington's disease gene. J Neurol Neurosurg Psychiatry 69: 773–779.

Lim LW, Temel Y, Hameleers R, Vlamings R, Sesia T, Steinbusch HW, Visser-Vandewalle V, Griez E and Temel Y (2007) Escape behaviour induced by electrical deep brain stimulation of the periaqueductal gray and ventromedial hypothalamus. Br Neurosci Assoc Abstr P64.02 (abstract).

McCracken CB and Grace AA (2007) High-frequency deep brain stimulation of the nucleus accumbens region suppresses neuronal activity and selectively modulates afferent drive in rat orbitofrontal cortex in vivo. J Neurosci 27: 12601–12610.

McIntyre CC, Grill WM, Sherman DL and Thakor NV (2004a) Cellular effects of deep brain stimulation: Model-based analysis of activation and inhibition. J Neurophysiol 91: 1457–1469.

McIntyre CC, Savasta M, Kerkerian-Le Goff L and Vitek JL (2004b) Uncovering the mechanism(s) of action of deep brain stimulation: Activation, inhibition, or both. Clin Neurophysiol 115: 1239–1248.

McIntyre CC, Butson CR, Maks CB and Noecker AM (2006) Optimizing deep brain stimulation parameter selection with detailed models of the electrode-tissue interface. Conf Proc IEEE Eng Med Biol Soc 1: 893–895.

Meissner W, Leblois A, Hansel D, Bioulac B, Gross CE, Benazzouz A and Boraud T (2005) Subthalamic high frequency stimulation resets subthalamic firing and reduces abnormal oscillations. Brain 128: 2372–2382.

Mongeau R and Marsden CA (1997) Effect of central and peripheral administrations of cholecystokinin-tetrapeptide on panic-like reactions induced by stimulation of the dorsal periaqueductal grey area in the rat. Biol Psychiatry 42: 335–344.

Murer MG and Pazo JH (1993) Behavioral responses induced by electrical stimulation of the caudate nucleus in freely moving cats. Behav Brain Res 57: 9–19.

Okun MS, Mann G, Foote KD, Shapira NA, Bowers D, Springer U, Knight W, Martin P and Goodman WK (2007) Deep brain stimulation in the internal capsule and nucleus accumbens region: Responses observed during active and sham programming. J Neurol Neurosurg Psychiatry 78: 310–314.

Paulsen JS and Conybeare RA (2005) Cognitive changes in Huntington's disease. Adv Neurol 96: 209–225.

Piasecki SD and Jefferson JW (2004) Psychiatric complications of deep brain stimulation for Parkinson's disease. J Clin Psychiatry 65: 845–849.

Porsolt RD, Le Pichon M and Jalfre M (1977) Depression: A new animal model sensitive to antidepressant treatments. Nature 266: 730–732.

Robbins TW (2002) The 5-choice serial reaction time task: Behavioural pharmacology and functional neurochemistry. Psychopharmacology (Berl) 163: 362–380.

Tai CH, Boraud T, Bezard E, Bioulac B, Gross C and Benazzouz A (2003) Electrophysiological and metabolic evidence that high-frequency stimulation of the subthalamic nucleus bridles neuronal activity in the subthalamic nucleus and the substantia nigra reticulata. FASEB J 17: 1820–1830.

Temel Y, Blokland A, Steinbusch HW and Visser-Vandewalle V (2005a) The functional role of the subthalamic nucleus in cognitive and limbic circuits. Prog Neurobiol 76: 393–413.

Temel Y, Visser-Vandewalle V, Aendekerk B, Rutten B, Tan S, Scholtissen B, Schmitz C, Blokland A and Steinbusch HW (2005b) Acute and separate modulation of motor and cognitive performance in parkinsonian rats by bilateral stimulation of the subthalamic nucleus. Exp Neurol 193: 43–52.

Temel Y, Cao C, Vlamings R, Blokland A, Ozen H, Steinbusch HW, Michelsen KA, von Hörsten S, Schmitz C and Visser-Vandewalle V (2006a) Motor and cognitive improvement by deep brain stimulation in a transgenic rat model of Huntington's disease. Neurosci Lett 406: 138–141.

Temel Y, Kessels A, Tan S, Topdag A, Boon P and Visser-Vandewalle V (2006b) Behavioural changes after bilateral subthalamic stimulation in advanced Parkinson disease: A systematic review. Parkinsonism Relat Disord 12: 265–272.

Temel Y, Boothman LJ, Blokland A, Magill PJ, Steinbusch HW, Visser-Vandewalle V and Sharp T (2007) Inhibition of 5-HT neuron activity and induction of depressive-like behavior by high-frequency stimulation of the subthalamic nucleus. Proc Natl Acad Sci USA 104: 17087–17092.

van Kuyck K, Demeulemeester H, Feys H, De Weerdt W, Dewil M, Tousseyn T, De Sutter P, Gybels J, Bogaerts K, Dom R and Nuttin B (2003) Effects of electrical stimulation or lesion in nucleus accumbens on the behaviour of rats in a T-maze after administration of 8-OH-DPAT or vehicle. Behav Brain Res 140: 165–173.

van Kuyck K, Gabriëls L, Cosyns P, Arckens L, Sturm V, Rasmussen S and Nuttin B (2007) Behavioural and physiological effects of electrical stimulation in the nucleus accumbens: A review. Acta Neurochir Suppl 97: 375–391.

Vlamings R, Visser-Vandewalle V, Koopmans G, Joosten EA, Kozan R, Kaplan S, Steinbusch HW and Temel Y (2007) High frequency stimulation of the subthalamic nucleus improves speed of locomotion but impairs forelimb movement in Parkinsonian rats. Neuroscience 148: 815–823.

von Hörsten S, Schmitt I, Nguyen HP, Holzmann C, Schmidt T, Walther T, Bader M, Pabst R, Kobbe P, Krotova J, Stiller D, Kask A, Vaarmann A, Rathke-Hartlieb S, Schulz JB, Grasshoff U, Bauer I, Vieira-Saecker AM, Paul M, Jones L, Lindenberg KS, Landwehrmeyer B, Bauer A, Li XJ and Riess O (2003) Transgenic rat model of Huntington's disease. Hum Mol Genet 12: 617–624.

Wichmann T and DeLong MR (2006) Deep brain stimulation for neurologic and neuropsychiatric disorders. Neuron 52: 197–204.

Wilson WJ (1983) Nucleus accumbens inhibits specific motor but not nonspecific classically conditioned responses. Brain Res Bull 10: 505–515.

Winter C, Mundt A, Jalali R, Joel D, Harnack D, Morgenstern R, Juckel G, Kupsch A (2008) High frequency stimulation and temporary inactivation of the subthalamic nucleus reduce quinpirole-induced compulsive checking behavior in rats. Exp Neurol 210(1): 217–228.

Woods A, Smith C, Szewczak M, Dunn RW, Cornfeldt M and Corbett R (1993) Selective serotonin re-uptake inhibitors decrease schedule-induced polydipsia in rats: A potential model for obsessive compulsive disorder. Psychopharmacology (Berl) 112: 195–198.

Wright J, Kelly D, Mitchell-Heggs N and Frankel R (1977) Respiratory changes induced by intracranial stimulation: Anatomical localizing value and related functional effects in rhesus monkeys. In: Sweet WH, Obrador S, Martin-Rodriguez JG (eds) Neurosurgical Treatment in Psychiatry, Pain, and Epilepsy. University Park Press, Baltimore, pp. 751–756.

Modeling Nonmotor Symptoms of Parkinson's Disease in Genetic Mouse Models

Sheila M. Fleming and Marie-Francoise Chesselet

Abstract In addition to the cardinal motor symptoms associated with Parkinson's disease (PD), there are also several nonmotor symptoms that can occur, many of which can significantly affect the quality of life of patients. These nonmotor symptoms are currently of interest because they may provide new insight into the pathophysiology of PD as well as provide important targets for potential treatments. Mice overexpressing human wild-type alpha-synuclein display many nonmotor impairments including deficits in olfaction, autonomic function, cognitive function, and anomalies in measures of anxiety, making these mice a good model to study the mechanisms underlying nonmotor impairments associated with PD.

1 Introduction

The clinical presentation of Parkinson's disease (PD) is dominated by the classical triad of motor symptoms including akinesia/bradykinesia, rigidity, and resting tremor. However, patients present not only with other motor symptoms such as balance problems and dystonia but also with a host of nonmotor symptoms. Most of these symptoms are not improved by standard dopaminergic therapies (L-dopa or dopamine agonists) suggesting they are not due to the loss of nigrostriatal dopaminergic neurons. Importantly, several recent studies (see below) have confirmed that some of these nonmotor, nondopaminergic symptoms can precede, sometimes by many years, the cardinal neurological symptoms. One possible reason is that they are caused by pathology in extranigral brain regions, which may occur before nigrostriatal cell loss. The mechanisms of nonmotor symptoms of PD, however, are poorly understood.

Research on the mechanism of nonmotor symptoms has been hindered, in part, by the lack of an appropriate animal model that can reproduce these nonmotor symptoms.

S.M. Fleming and M.-F. Chesselet (✉)
Departments of Neurology and Neurobiology, The David Geffen School of Medicine at UCLA, Los Angeles, CA 90095-1769, USA
e-mail: mchesselet@mednet.ucla.edu

H.J. Groenewegen et al. (eds.), *The Basal Ganglia IX*, Advances in Behavioral Biology 58, DOI 10.1007/978-1-4419-0340-2_37,

This is not surprising considering that the most widely studied models of PD are based on the selective destruction of nigrostriatal dopaminergic neurons with neurotoxins such as 6-hyroxydopamine, 1-methyl 4-phenyl 1, 2, 3, 6-tetrahydropyridine (MPTP), paraquat, or rotenone. The identification of mutations causing PD in humans has led to the generation of genetic mouse models expressing PD-causing mutations. One major advantage of these models is that when mutations are broadly expressed through a knock-out strategy or by driving the transgene with an appropriate promoter, one may expect to reproduce extranigral pathology and related nonmotor symptoms. To date, however, few models have been tested for the presence of nonmotor symptoms with the exception of a number of cognitive tasks (Freichel et al. 2007). We have examined in detail a variety of behavioral endpoints in mice overexpressing alpha-synuclein under the Thy1 promoter (Thy1-aSyn). Between 3 and 9 months of age, these mice have no loss of dopaminergic (DA) nigrostriatal neurons but exhibit a wide range of motor and nonmotor symptoms and may therefore provide a useful model of premanifest PD at this age.

2 Thy1-aSyn Mice

2.1 Neuropathological Alterations in Thy1-aSyn Mice

The Thy1 promoter drives overexpression of alpha-synuclein in many neurons, including nigrostriatal dopaminergic neurons (Rockenstein et al. 2002; Mortazavi et al., unpublished observations). Western blot analysis suggests that the level of overexpression of alpha-synuclein is approximately tenfold the level of the endogenous protein (Rockenstein et al. 2002). Thy1-aSyn mice exhibit proteinase K-resistant alpha-synuclein aggregates in many brain regions (Fernagut et al. 2007). At 4 months of age, these protein aggregates are present in the substantia nigra, the olfactory bulb, the locus coeruleus, and other brain areas but not in the cerebral cortex despite high levels of overexpression of alpha-synuclein in this region (Fernagut et al. 2007; Hutson and Chesselet, unpublished observations). With age, the number and size of these proteinase K-resistant alpha-synuclein aggregates increases and they are found in more brain regions (Hutson and Chesselet, unpublished observations). Despite the presence of clear pathological accumulation of alpha-synuclein, the mice do not lose nigrostriatal DA neurons up to 8 months of age (Fernagut et al. 2007). Preliminary data, however, indicate a significant loss of TH-positive terminals in the lateral striatum at 14 months of age, suggesting progressive nigrostriatal pathology in these mice (Mortazavi and Chesselet, unpublished observations).

The widespread alpha-synuclein pathology without DA cell loss observed in young Thy1-aSyn mice is reminiscent of the distribution of alpha-synuclein aggregates described by Braak and collaborators for "incidental Lewy Body disease" (Braak et al. 2003). Braak hypothesized that the presence of alpha-synuclein pathology in the brains of individuals who died without frank neurological symptoms may correspond to an early, premanifest, stage of PD (Braak et al. 2003).

Accordingly, our data suggest that the Thy1-aSyn mice could represent a model of premanifest PD. To test this hypothesis, we examined a number of functions that are often altered prior to the occurrence of typical neurological symptoms in patients with PD (Langston 2006).

2.2 *Nonmotor Deficits in Thy1-aSyn Mice*

2.2.1 Olfactory Deficits

Olfactory dysfunction is frequent in PD patients and can precede the onset of motor symptoms by several years (Langston 2006; Ponsen et al. 2004; Ross et al. 2007). Furthermore, PD occurs more frequently in relatives of PD patients with olfactory dysfunction than in relatives who do not show olfactory deficits (Berendse et al. 2001). Importantly, carriers of the LRRK2 mutation, a cause of familial PD, exhibit deficits in olfactory function before expressing motor symptoms characteristic of PD (Ferreira et al. 2007). Therefore, we examined olfactory function with a variety of tests in Thy1-aSyn mice (Fleming et al. 2008). For the "block test" (Tillerson et al. 2006), the mouse is exposed to wooden blocks previously placed in the mouse bedding for 7 days to acquire its own "self-odor." After repeated exposure to four blocks with self-odor, the mouse is then exposed to three self-odor blocks and one block carrying the odor of another mouse. This elicits a marked increase in block sniffing in wild-type mice, but this response, although present, is strongly attenuated in Thy1-aSyn mice. To control for the interference of a possible deficit in social recognition, a similar test was conducted by repeatedly exposing mice to a neutral odor such as lemon or coconut (Bielsky et al. 2004). After repeated presentation of one type of odor, both mice habituate to the scent, as indicated by a marked decrease in sniffing the odor. Upon presentation of a novel scent, wild-type mice resume intense sniffing, an effect again attenuated in transgenic mice. This effect was observed with several pairs of odors, including closely related citrus scents. This is of particular interest because studies in carriers of the LRRK2 mutation show particularly strong deficits in detection and discrimination of citrus scents (Hentschel et al. 2005). Finally, Thy1-aSyn mice also take much longer than wild-type mice to find a scented food pellet buried in their bedding. Because both Thy1-aSyn and wild-type mice find the pellet with the same latency when it is put on top of the bedding, the difference in finding the buried pellet cannot be attributed to differences in interest for food or locomotor activity. Together with the other tests, these data suggest that olfactory function is impaired (although not abolished) in the transgenic mice (Fleming et al. 2008). This is reminiscent of observations in PD patients (Doty et al. 1988).

The mechanisms of olfactory deficits in PD patients are not known. An increased number of dopaminergic neurons and/or the presence of Lewy Bodies in olfactory brain regions could play a role (Huisman et al. 2004; Braak et al. 2003). In addition, studies in another line of mice overexpressing alpha-synuclein revealed a decrease in neurogenesis, which could also contribute to the olfactory deficits (Winner et al. 2004, 2007). The Thy1-aSyn mice now provide a useful model to decipher the mechanisms of olfactory dysfunction in a model that expresses PD-like pathology, i.e., insoluble alpha-synuclein aggregates in the olfactory bulb.

2.2.2 Anomalies in Circadian Rhythm

Patients with PD suffer from alterations in their sleep pattern, including insomnia and daytime sleepiness (Commella 2007). The sleep pattern of Thy1-aSyn mice is not known. However, preliminary data reveal a marked increase in spontaneous running on a running wheel during the beginning of the light phase in Thy1-aSyn compared to wild-type mice, suggesting anomalies in the mouse diurnal cycle.

2.2.3 Anxiety and Depression

Over 80% of patients with PD suffer from some form of mood and/or affective disorder, including clinical depression, apathy, and anxiety (Shiba et al. 2000). The use of classical tests to detect depressive states, such as the Porsolt test, is limited in the Thy1-aSyn mice by their inability to swim. When put in water, the mice would rapidly drown if not actively rescued by the investigator. However, Thy1-aSyn mice show prominent hindlimb clasping, which has been interpreted in some studies as evidence of depression. We have examined the behavior of Thy1-aSyn mice in a variety of tests known to detect increased anxiety. In the light-dark box test, latency to enter the light chamber significantly increases with the age in transgenic mice whereas it does not significantly change in wild-type mice, suggesting that the mice may develop anxious behavior as they age. Curiously, young Thy1-aSyn mice spend significantly more time in the light chamber during their first visit than do wild-type mice. This would suggest a decrease in anxiety, which is compatible with their performance in a test of social anxiety, the mirrored chamber test. The reason for these paradoxical behaviors is not known but they could reflect increased impulsivity (Baunez et al. 2007; Amalric et al. 1995).

2.2.4 Cognitive Deficits

Cognitive deficits in PD are present in a considerable subset of patients and affect implicit memory, a striatal function, more than declarative or working memory (Knowlton et al. 1996; Dubois and Pillon 1997). Again, the profound motor deficits shown in water by the Thy1-aSyn mice at an early age prevent the use of classical tests such as the water maze. Instead, we tested Thy1-aSyn mice in a reversal learning task where mice first learn to nose-poke an illuminated bin to receive a reward. Following nose-poke training, the animals are presented with two illuminated bins and then have to learn to poke one of the bins to receive a reward. Once this is learned, the reversal session begins where the opposite light is now linked to the reward (Jentsch and Taylor 2001). We found that Thy1-aSyn eventually learnt the task but were less accurate at learning the reversal task compared to wild-type mice, indicating the presence of cognitive impairments in the Thy1-aSyn mice.

2.2.5 Gastrointestinal Dysfunction

A variety of digestive symptoms have been observed in PD patients, who are known to have widespread Lewy bodies and alpha-synuclein pathology in peripheral neurons innervating the digestive tract, as well as centrally in the motor nucleus of the vagal nerve (Braak et al. 2006; Pfeiffer 2005; Wakabayashi et al. 1988). Retrospective studies have identified chronic constipation as a predictor of PD (Abbott et al. 2001, 2007). This symptom is present in PD, as already described by James Parkinson in his original publication on PD (Parkinson 2002). Preliminary findings suggest that Thy1-aSyn mice have deficits in colonic motility compared to wild-type mice (Wang et al. 2008).

2.2.6 Cardiovascular Dysfunction

The most frequent cardiovascular symptom in PD patients is orthostatic hypotension, although it cannot be excluded that minor deficits in cardiovascular homeostasis are also present in many patients (Goldstein et al. 2002). Indeed, some studies suggest that an early loss of noradrenergic innervation of the heart exists in these patients (Goldstein et al. 2007). Since orthostatic hypotension is a classic symptom in human PD patients, we tested the response of the cardiac autonomic nervous system in the Thy1-aSyn mice to transient reductions in blood pressure induced by bolus infusions of nitroprusside. We found that Thy1-aSyn mice showed a blunted heart rate response to the transient decrease in blood pressure compared to wild-type mice, which indicates autonomic dysfunction and susceptibility to poor blood pressure regulation (Fleming et al. 2007).

2.3 Progressive Sensorimotor Dysfunction in Thy1-aSyn Mice

Although the Thy1 aSyn mice do not lose dopaminergic terminals until after 8 months of age (Fernagut et al. 2007; Mortazavi and Chesselet, unpublished observations), they display progressive anomalies in challenging tests of motor behavior starting at 2 months of age. We have used a battery of behavioral tests known, from other models, to be sensitive to nigrostriatal deficits (Goldberg et al. 2003; Hwang et al. 2005). For example, in Pitx3 mice, which lose nigrostriatal dopaminergic neurons during early postnatal development, these tests detected motor anomalies that were reversed by treatment with L-dopa (Hwang et al. 2005). Thy1-aSyn mice exhibit deficits in the pole test, on the challenging beam, in the cylinder, and in nest building (Fleming et al. 2004). However, in the Thy1-aSyn mice, these deficits are not reversed by L-dopa or apomorphine, as expected from the absence of dopaminergic cell loss in these mice (Fleming et al. 2006). This observation suggests that the motor deficits expressed by Thy1-aSyn mice do not correspond to "parkinsonism" but reveal early effects of the mutation on the motor system. The mechanisms leading to these deficits may include decreased cortical noradrenalin (Maidment, unpublished observations), synaptic dysfunction of corticostriatal synapses (Wu et al. 2005), and/or alterations in cellular properties of nigrostriatal dopaminergic neurons, as suggested by the existence of profound changes in mRNA encoding ion channels in these neurons (Mortazavi et al. 2007).

Table 1 Comparison of early stage nonmotor symptoms in Parkinson's disease (PD) and in Thy1-aSyn Mice

Domain	Manifestations in PD patients	Abnormalities observed in Thy1-aSyn mice
Olfaction	Impairments in odor detection, discrimination, and identification	Impairments in odor detection and discrimination
Sleep disorders	Excessive daytime sleepiness, REM sleep behavior disorder	Abnormal circadian rhythm: Increased wheel running during the light cycle
Neuropsychiatric symptoms	Anxiety and depression	Increased anxiety in the light-dark box in older mice. Decreased anxiety in younger mice
Cognitive dysfunction	Problems in planning, impulse control, and mental flexibility	Impaired reversal learning
Gastrointestinal dysfunction	Weight loss, abnormal gastric emptying, and constipation	Impaired colonic motility and reduced weight
Autonomic cardiovascular dysfunction	Orthostatic hypotension, sympathetic denervation, and reduced heart rate variability	Impaired baroreceptor function: decreased response to transient hypotension

3 Conclusion

The abnormal accumulation of the alpha-synuclein links sporadic PD, where all patients present with Lewy Bodies and Lewy neurites throughout the central and peripheral nervous system, to the high levels of alpha-synuclein due to gene multiplication that causes PD in familial forms of the disease (Singleton et al. 2003; Chartier-Harlin et al. 2004). We have shown that mice overexpressing human wild-type alpha-synuclein under the Thy1 promoter present a broad range of nonmotor symptoms in behavioral domains that are also affected at early stages of PD (Table 1). These mice may therefore provide a useful model to study the mechanisms of these nonmotor symptoms, which may constitute early manifestations of alpha-synuclein-induced pathology, both in mice and in humans. Furthermore, these observations provide new behavioral endpoints for testing disease-modifying therapies designed to interfere with the pathophysiological processes leading to PD in humans.

Acknowledgments We are grateful to our collaborators Drs. Eliezer Masliah, UCSD, for the gift of the Thy1-aSyn mice; Michael Levine, Director of the UCLA Parkinson disease mouse models repository for help with the mouse colony; Yvette Tache and Lixin Wang, UCLA, for studies of gastrointestinal dysfunction; Ken Roos and Maria Jordan, UCLA, for studies of cardiovascular function; David Jentsch for studies of cognition; Christopher Colwell for studies of circadian rhythm; Pierre-Olivier Fernagut, Bernd Meurers, and Farzad Mortazavi, UCLA, for additional data on the Thy1-aSyn mice; and Che Hutson, Nicole Tetreault, Caitlin Mulligan, and Eddie Garcia for help with unpublished experiments. This work was supported by PHS grants P50-NS038367 and U54-ES12078, the American Parkinson Disease Association and the Chen family.

References

Abbott RD, Petrovitch H, White LR, Masaki KH, Tanner CM, Curb JD, Grandinetti A, Blanchette PL, Popper JS and Ross GW (2001) Frequency of bowel movements and the future risk of Parkinson's disease. Neurology 57: 456–462.

Abbott RD, Ross GW, Petrovitch H, Tanner CM, Davis DG, Masaki KH, Launer LJ, Curb JD and White LR (2007) Bowel movement frequency in late-life and incidental Lewy bodies. Mov Disord 22: 1581–1586.

Amalric M, Moukhles H, Nieoullon A and Daszuta A (1995) Complex deficits on reaction time performance following bilateral intrastriatal 6-OHDA infusion in the rat. Eur J Neurosci 7: 972–980.

Baunez C, Christakou A, Chudasama Y, Forni C and Robbins TW (2007) Bilateral high-frequency stimulation of the subthalamic nucleus on attentional performance: Transient deleterious effects and enhanced motivation in both intact and parkinsonian rats. Eur J Neurosci 25: 1187–1194.

Berendse H W, Booij J, Francot CM, Bergmans PL, Hijman R, Stoof JC and Wolters EC (2001) Subclinical dopaminergic dysfunction in asymptomatic Parkinson's disease patients' relatives with a decreased sense of smell. Ann Neurol 50: 34–41.

Bielsky IF, Hu SB, Szegda KL, Westphal H and Young LJ (2004) Profound impairment in social recognition and reduction in anxiety-like behavior in vasopressin V1a receptor knockout mice. Neuropsychopharmacology 29: 483–493.

Braak H, Del Tredici K, Rub U, de Vos RA, Jansen Steur EN and Braak E (2003) Staging of brain pathology related to sporadic Parkinson's disease. Neurobiol Aging 24: 197–211.

Braak H, de Vos RA, Bohl J and Del Tredici K (2006) Gastric alpha-synuclein immunoreactive inclusions in Meissner's and Auerbach's plexuses in cases staged for Parkinson's disease-related brain pathology. Neurosci Lett 396: 67–72.

Chartier-Harlin MC, Kachergus J, Roumier C, Mouroux V, Douay X, Lincoln S, Levecque C, Larvor L, Andrieux J, Hulihan M, Waucquier N, Defebvre L, Amouyel P, Farrer M and Destée A (2004) Alpha-synuclein locus duplication as a cause of familial Parkinson's disease. Lancet 364: 1167–1169.

Commella CL (2007) Sleep disorders in Parkinson's disease: An overview. Mov Disord 22(S17): S367–S373.

Doty RL, Deems DA and Stellar S (1988) Olfactory dysfunction in parkinsonism: A general deficit unrelated to neurologic signs, disease stage, or disease duration. Neurology 38: 1237–1244.

Dubois B and Pillon B (1997) Cognitive deficits in Parkinson's disease. J Neurol 244: 2–8.

Fernagut PO, Hutson CB, Fleming SM, Tetreaut NA, Salcedo J, Masliah E and Chesselet MF (2007) Behavioral and histopathological consequences of paraquat intoxication in mice: Effects of alpha-synuclein overexpression. Synapse 61: 991–1001.

Ferreira JJ, Guedes LC and Rosa MM (2007) High prevalence of LRRK2 mutations in familial and sporadic Parkinson's disease in Portugal. Mov Disord 22: 1194–1201.

Fleming SM, Salcedo J, Fernagut PO, Rockenstein E, Masliah E, Levine MS and Chesselet M-F (2004) Early and progressive sensorimotor anomalies in mice overexpressing wild-type human alpha-synuclein. J Neurosci 24: 9434–9440.

Fleming SM, Salcedo J, Hutson CB, Rockenstein E, Masliah E, Levine MS and Chesselet M-F (2006) Behavioral effects of dopaminergic agonists in transgenic mice overexpressing human wildtype alpha-synuclein. Neuroscience 142: 1245–1253.

Fleming SM, Jordan, MC, Masliah E, Chesselet M-F and Roos KP (2007) Alterations in baroreceptor function in transgenic mice overexpressing human wildtype alpha synuclein. Program No. 50.9. 2007 Neuroscience Meeting Planner. San Diego, CA: Society for Neuroscience.

Fleming SM, Tetreault NA, Mulligan CK, Hutson CB, Masliah E and Chesselet MF (2008) Olfactory deficits in mice overexpressing human wildtype alpha-synuclein. Eur J Neurosci 28: 247–256.

Freichel C, Neumann M, Ballard T, Müller V, Woolley M, Ozmen L, Borroni E, Kretzschmar HA, Haass C, Spooren W and Kahle PJ (2007) Age-dependent cognitive decline and amygdala pathology in alpha-synuclein transgenic mice. Neurobiol Aging 28: 1421–1435.

Goldberg MS, Fleming SM, Palacino JJ, Cepeda C, Lam HA, Bhatnagar A, Meloni EG, Wu N, Ackerson LC, Klapstein GJ, Gajendiran M, Roth BL, Chesselet MF, Maidment NT, Levine MS and Shen J (2003) Parkin-deficient mice exhibit nigrostriatal deficits but not loss of dopaminergic neurons. J Biol Chem 278: 43628–43635.

Goldstein DS, Holmes CS and Dendi R (2002) Orthostatic hypotension from sympathetic denervation in Parkinson's disease. Neurology 58: 1247–1255.

Goldstein DS, Sharabi Y and Karp BI (2007) Cardiac sympathetic denervation preceding motor signs in Parkinson disease. Clin Auton Res 17: 118–1121.

Hentschel K, Furtado S, Markopoulou K, Doty RL, Uitti RJ and Wszolek ZK (2005) Differences in olfactory dysfunction between sporadic PD and familial Parkinson's disease kindreds by UPSIT subgroup analysis. Mov Disord 20(Suppl 10): S55.

Huisman E, Uylings HB and Hoogland PV (2004) A 100% increase of dopaminergic cells in the olfactory bulb may explain hyposmia in Parkinson's disease. Mov Disord 19: 687–692.

Hwang DY, Fleming SM, Ardayfio P, Moran-Gates T, Kim H, Tarazi FI, Chesselet MF and Kim KS (2005) 3,4-Dihydroxyphenylalanine reverses the motor deficits in Pitx3-deficient aphakia mice: Behavioral characterization of a novel genetic model of Parkinson's disease. J Neurosci 25: 2132–2137.

Jentsch JD and Taylor JR (2001) Impaired inhibition of conditioned responses produced by subchronic administration of phencyclidine to rats. Neuropsychopharmacology 24: 66–74.

Knowlton BJ, Mangels JA and Squire LR (1996) A neostriatal habit learning system in humans. Science 273: 1399–1402.

Langston JW (2006) The Parkinson's complex: Parkinsonism is just the tip of the iceberg. Ann Neurol 59: 591–596.

Mortazavi F, Meurers B, Oh MS, Elashoff D, Masliah E and Chesselet MF (2007) Transcriptome analysis of laser-captured dopaminergic neurons in mice overexpressing human wildtype alpha-synuclein. Program No. 151.21 2007 Abstract Viewer/Itinerary Planner. San Diego, CA: Society for Neuroscience.

Parkinson J (2002) An essay on the shaking palsy. 1817. J Neuropsychiatry Clin Neurosci 14: 223–236.

Pfeiffer RF (2005) Gastrointestinal dysfunction in Parkinson's disease. Lancet Neurol 10: 283–287.

Ponsen MM, Stoffers D, Booij J, van Eck-Smit BL, Wolters ECh and Berendse HW (2004) Idiopathic hyposmia as a preclinical sign of Parkinson's disease. Ann Neurol 56: 173–181.

Rockenstein E, Mallory M, Hashimoto M, Song D, Shults CW, Lang I and Masliah E (2002) Differential neuropathological alterations in transgenic mice expressing alpha-synuclein from the platelet-derived growth factor and Thy-1 promoters. J Neurosci Res 68: 568–578.

Ross GW, Petrovitch H, Abbott RD, Tanner CM, Popper J, Masaki K, Launer L and White LR (2007) Association of olfactory dysfunction with risk for future Parkinson's disease. Ann Neurol 63: 167–173.

Shiba M, Bower JH, Maraganore DM, McDonnell SK, Peterson BJ, Ahlskog JE, Schaid DJ and Rocca WA (2000) Anxiety disorders and depressive disorders preceding Parkinson's disease: A case-control study Mov Disord 15: 669–677.

Singleton AB, Farrer M, Johnson J, Singleton A, Hague S, Kachergus J, Hulihan M, Peuralinna T, Dutra A, Nussbaum R, Lincoln S, Crawley A, Hanson M, Maraganore D, Adler C, Cookson MR, Muenter M, Baptista M, Miller D, Blancato J, Hardy J and Gwinn-Hardy K (2003) alpha-Synuclein locus triplication causes Parkinson's disease. Science 302: 841.

Tillerson JL, Caudle WM, Parent JM, Gong C, Schallert T and Miller GW (2006) Olfactory discrimination deficits in mice lacking the dopamine transporter or the D2 dopamine receptor. Behav Brain Res 172: 97–105.

Wakabayashi K, Takahashi H, Takeda S, Ohama E and Ikuta F (1988) Parkinson's disease: The presence of Lewy bodies in Auerbach's and Meissner's plexuses. Acta Neuropathol 76: 217–221.

Wang L, Fleming SM, Chesselet MF and Taché Y (2008) Abnormal colonic motility in mice overexpressing human wild-type alpha-synuclein. Neuroreport 19: 873–876.

Winner B, Lie DC, Rockenstein E, Aigner R, Aigner L, Masliah E, Kuhn HG and Winkler J (2004) Human wild-type alpha-synuclein impairs neurogenesis. J Neuropathol Exp Neurol 63: 1155–1166.

Winner B, Rockenstein E, Lie DC, Aigner R, Mante M, Bogdahn U, Couillard-Depres S, Masliah E and Winkler J (2007) Mutant alpha-synuclein exacerbates age-related decrease of neurogenesis. Neurobiol Aging 29: 913–925.

Wu N, Cepeda C, Masliah E and Levine MS (2005) Abnormal glutamate and dopamine receptor function in the striatum of α-synuclein-overexpressing mice. Program No. 85.12 AbstractViewer/Itinerary Planner. Washington, DC: Society for Neuroscience.

Differential Expression of Doublecortin-Like Kinase Gene Products in the Striatum of Behaviorally Hyperresponsive Rats

Pieter Voorn, Tessa Hartog, Allert Jan Jonker, Louk J.M.J. Vanderschuren, and Erno Vreugdenhil

Abstract Behavioral hyperresponsiveness entails major changes in neuronal plasticity, such as augmented striatal dopaminergic neurotransmission and changes in neuronal morphology. These two processes may be related via mechanisms involving Ca2+/calmodulin-dependent protein kinases (CaMK). Dopaminergic regulation of a newly characterized CaMK-like gene, i.e., doublecortin-like kinase (DCLK), has been reported. However, it is not known if and how CaMK-like splice variants of the gene are responsive in behavioral hyperresponsiveness induced by enhanced dopaminergic neurotransmission. We, therefore, measured with quantitative in situ hybridization levels of mRNA encoding the CaMK domain-containing gene product, called DCLK or a CaMK domain-lacking variant called CaMK-related peptide (CARP) in the rat striatum. We have applied three forms of behavioral hyperresponsiveness, viz. administration of selective dopamine agonists to animals with a unilaterally dopamine-depleted striatum and psychomotor sensitization to amphetamine or morphine. DCLK mRNA only showed changes after dopamine depletion, but not after subsequent treatment with dopamine agonists during expression of amphetamine or morphine sensitization. In contrast, CARP mRNA was dramatically increased in the dopamine-depleted striatum after administration of a dopamine D1 agonist. Furthermore, CARP was upregulated and downregulated during expression of amphetamine and morphine sensitization, respectively.

CARP is suggested to play a role in morphological plasticity. We, therefore, propose that the observed changes are involved in dopamine-induced changes in striatal neuronal circuitry. A hypothetical model of action of CARP is presented.

P. Voorn (✉), T. Hartog, A.J. Jonker, L.J.M.J. Vanderschuren, and E. Vreugdenhil
Department of Anatomy and Neurosciences, Neuroscience Campus Amsterdam, VU University Medical Center, P.O. Box 7057, 1007 MB Amsterdam, The Netherlands
e-mail: p.voorn.@vumc.nl

H.J. Groenewegen et al. (eds.), *The Basal Ganglia IX*, Advances in Behavioral Biology 58, DOI 10.1007/978-1-4419-0340-2_38,

1 Introduction

Changes in dopaminergic neurotransmission in the striatum are associated with malfunctions in a variety of functions including motor control, cognition, emotion, and motivation. In the well-known dopamine-depletion model for Parkinson's disease, animals are unilaterally deprived of the ascending mesostriatal dopaminergic system which results in a dramatically increased sensitivity to dopaminomimetics, such as L-DOPA, in the striatum. Administration of L-DOPA leads to profound disturbances in motor control that have been compared to the dyskinesia symptoms found in Parkinson's disease patients after prolonged use of the drug.

Changes in cognition and motivation associated with striatal dopaminergic neurotransmission are also critical for drug abuse. Repeated exposure to amphetamine or morphine leads to increased dopaminergic neurotransmission and increased behavioral responsiveness (psychomotor sensitization), a phenomenon that has been suggested to involve enhanced sensitivity of the neural circuitry engaged in assessing incentive value of drugs and drug-associated stimuli (Vanderschuren and Kalivas 2000; Robinson and Kolb 2004). The mesoaccumbens dopamine system is considered to be a crucial part of this circuitry (Berridge and Robinson 1998; Cardinal et al. 2002; Schultz and Dickinson 2000). However, the involvement of striatal dopamine in sensitization and addiction is not restricted to the nucleus accumbens. In fact, dopaminergic hyperresponsivity in the caudate-putamen coincides with behavioral sensitization (Kolta et al. 1985; Paulson and Robinson 1995). Moreover, the well-established role of dorsal striatal dopamine in procedural learning (Packard and Knowlton 2002) suggests that altered striatal dopamine function may play a prominent role in late stages of addiction (Everitt and Wolf 2002). Indeed, dorsal striatal dopamine has been implicated in well-established, cue-controlled cocaine seeking (Ito et al. 2002; Vanderschuren et al. 2005). Furthermore, the dorsal striatum has also been shown to be involved in the reinstatement of drug seeking after abstinence (Fuchs et al. 2006). Altered striatal function has also been observed during cocaine craving in humans (Garavan et al. 2000; Volkow et al. 2006; Wong et al. 2006).

The various treatments that lead to increased behavioral responsiveness entail a wide range of molecular changes (Berke et al. 1998; Jacobs et al. 2003) some of which are responsible for the enhanced sensitivity to dopamine while others are related to other indices of plasticity. An intriguing possibility is that part of the observed behavioral symptoms and changes depend on morphological plasticity, i.e., changes in striatal neuronal morphology and, hence, circuitry. Such morphological changes have indeed been found. Robinson and Kolb (2004) saw changes in dendritic branching, length of dendrites, and density and number of dendritic spines of medium-sized spiny striatal neurons after treatment with different drugs of abuse including morphine and amphetamine. Furthermore, depletion of dopamine via intracerebral injection of 6-hydroxydopamine affects dendrite and synapse morphology of the same striatal cell population (Ingham et al. 1993).

The molecular background of these structural alterations is only beginning to be unraveled. Ca2+/calmodulin-dependent protein kinase (CaMK) enzymes play an important role in neural plasticity. CaMKII and CaMKIV enzyme activity can be regulated by dopamine (Brown et al. 2005; Kebabian 1997; Takeuchi et al. 2002) and CaMKII is associated with cocaine-, amphetamine- and haldol-induced effects on neural plasticity (Anderson et al. 2008; Iwata et al. 1997; Meshul and Tan 1994). Moreover, Ca2+/calmodulin-dependent protein kinases are implicated in psychostimulant-induced behavioral sensitization (Licata and Pierce 2003). Selective stimulation of striatal dopamine D1 receptors has been shown to strongly enhance expression levels of a splice variant (ANIA-4) of a CaMK-like gene, doublecortin-like kinase (DCLK) (Berke et al. 1998). The murine DCLK gene generates four different products, three of which contain a CaMK catalytic domain. In a fourth alternatively spliced variant the catalytic domain is absent. It may act as an inhibitor of CaMK activity since it shows great similarity to the autoinhibitory domain of CaMK-IV, hence its name: "CaMK-related peptide" – CARP (Vreugdenhil et al. 1999).

The induction of ANIA-4 in the striatum after cocaine treatment or administration of a dopamine D1 agonist to unilaterally dopamine-depleted animals as observed by Berke et al. (1998) indicates an important regulatory role for dopamine and suggests that dopamine may influence striatal neuronal plasticity via a signal transduction route involving one (or more) DCLK gene products. The question remains which of the transcripts is responsive to dopamine: the CaMK domain-containing DCLK transcripts or CARP? Therefore, in the present study we investigated the response of DCLK expression in several experimental paradigms used to induce enhanced (behavioral) responsiveness to dopamine and in which neuromorphological changes have been found: administration of dopamine D1 and D2 agonists to animals that previously sustained a unilateral lesion of the ascending dopaminergic system and psychomotor sensitization to amphetamine and morphine. DCLK expression was assessed in the striatum using quantitative in situ hybridization with cRNA probes specific for DCLK or for CARP transcripts.

2 Materials and Methods

2.1 Animals and Tissue Preparation

Male Wistar rats were obtained from Harlan (Horst, The Netherlands). The rats were housed two per cage in Macrolon cages under controlled conditions (lights on from 7.00 a.m. to 7.00 p.m.) and were allowed to acclimatize to the housing conditions for at least 1 week before use. Food and water were available ad libitum. All experiments were approved by the Animal Care Committee of the Vrije Universiteit of Amsterdam. Animals were terminated by decapitation. The brains were dissected out of the skull, quickly frozen in isopentane cooled in ethanol with pulverized dry ice and stored at −80°C. Frozen brains were horizontally or frontally sectioned in a Leica (2800) cryostat. Sections (thickness = 14 μm) were thaw-mounted on gelatin–chrome-alum-coated slides which were then stored desiccated at −80° C until used for in situ hybridization or immunohistochemistry.

2.2 Unilateral Depletion of Dopamine and Exposure to Dopamine Agonists

Twenty-four rats (300–350 g) were used. The animals were anesthetized with Nembutal (60 mg/kg, ip, Sanofi Sante BV, Maassluis, the Netherlands) and administered DMI (10 mg/kg ip, Sigma Chemical Co, St. Louis, MO) 30 min prior to intracerebral injection of 6-hydroxydopamine (6-OHDA, Sigma Chemical Co, St. Louis, MO) to protect noradrenergic neurons. Lidocaine (Astra Pharmaceutica BV, Zoetermeer, Netherlands) was applied as a spray-on local anesthetic to the periost. 6-OHDA (8 μg in 3 μl of 0.05% ascorbic acid) was injected unilaterally with 30 s time intervals over a period of 5 min into the ascending dopaminergic fiber bundle using a Hamilton needle (−3.4 mm caudal to Bregma, 1.7 mm lateral, and 8.6 mm below the cortical surface, Paxinos and Watson 1986). The needle was left in place for 10 more minutes before retracting.

Three weeks postoperation, the animals were administered a single dose of dopamine D1 agonist SKF 38393 (5 mg/kg sc, RBI, Natick, MA), D2 agonist quinpirole (0.2 mg/kg ip, Lilly) or saline (0.9% ip). Two hours later, animals were terminated and brain tissue was further processed as mentioned above.

2.3 Psychomotor Sensitization to Amphetamine or Morphine

Animals were briefly handled on the 2 days preceding the beginning of drug treatment and on the 2 days preceding drug challenges. The treatment regimens for amphetamine and morphine sensitization elicit robust and reproducible behavioral sensitization in our laboratory (Vanderschuren et al. 1997, 1999a, b). During pretreatment in the amphetamine sensitization experiment, the animals (n = 10 per group) were once daily injected with D-amphetamine sulphate (2.5 mg/kg, ip; OPG, Utrecht, The Netherlands) or saline for 5 consecutive days. Three weeks after the last pretreatment injection, half of the animals of both pretreatment groups received a challenge injection with 1 mg/kg amphetamine (ip) and the other half received a saline (ip) challenge. Pretreatment in the morphine experiment consisted of a daily injection with morphine hydrochloride (10 mg/kg, sc; OPG) or saline (n = 11 per group) for 14 days. Three weeks after cessation of pretreatment, six of the morphine- and five of the saline-pretreated rats received a challenge with morphine (5 mg/kg, sc), while the remaining rats were injected with saline (sc). One hour and 45 min after the last injection with amphetamine, morphine, or saline the animals were sacrificed.

2.4 Tyrosine Hydroxylase Immunostaining

Immunostaining for tyrosine hydroxylase (TH) was performed to determine the extent of the 6-OHDA lesions. Briefly, TH immunoreactivity was visualized in an immunoperoxidase staining protocol, using mouse-anti-TH (1:2,000, Incstar, Stillwater, MN) as a primary antibody, and 3,3′-diaminobenzidine hydrochloride (with H_2O_2; Sigma Chemical Co., St. Louis, MO) as peroxidase enzyme substrate. Slides were thawed, formaldehyde-fixed, immunostained, dehydrated, and cover-slipped in Entellan.

2.5 In Situ Hybridization

Tissue sections were thawed, fixed in 4% paraformaldehyde, acetylated in acetic anhydride, dehydrated, and delipidated. The sections were hybridized overnight with [^{35}S]-UTP labeled cRNA probes (1 × 10^6 cpm per section, see below for preparation) for DCLK, CARP, and preproenkephalin (ppE) at 45°C in hybridization buffer. Stringency washings were performed by incubating the slides two times 15 min in 2× SSC + 50% formamide at 40°C, 10 min in RNASe A (Boehringer Mannheim, Germany) in 2× SSC, another two times 15 min in 2× SSC + 50% formamide at 40°C, followed by a final washing in 2× SSC. Subsequently, the sections were air dried and exposed to Kodak Biomax X-ray films.

The cRNA probes were transcribed in vitro from the full length ppE (Yoshikawa et al. 1984) cDNA and DCLK-specific and CARP-specific cDNA (Vreugdenhil et al. 1999), which were, respectively, subcloned into pSP64 (pYSEA1) and pGEM-T. cDNA was polymerase chain reaction (PCR)-amplified with specific primers flanking the RNA polymerase promotor sites, gel-purified, and used as template for in vitro transcription. After in vitro transcription, the PCR product was degraded with RNAse-free DNAse I and cRNA probes were partially hydrolyzed to create 150–200 base fragments. The cRNA probes were synthesized in the presence of [^{35}S]-UTP (NEN, specific activity 1,100–1,500 Ci/mmol) according to the protocol provided by the manufacturer of the RNA polymerases (Promega, Madison, WI), with 500 ng of PCR fragments as template.

2.6 Measurements and Quantifications

Exposed X-ray films (DCLK 24 h, CARP 166 h, ppE 18½ h) were developed in Kodak D-19 and fixed in Kodak RA 3000. The autoradiograms were digitized and quantified using an MCID M5$^+$ image analysis system (Imaging Research, St. Catherines, ON, Canada). The amount of hybridized DCLK, CARP, and ppE probes in the brain sections was determined by conversion of pixel gray levels as determined with densitometry in the autoradiograms to radioactivity per mg tissue ("probe radioactivity") using co-exposed [^{35}S]-cross calibrated [^{14}C] standards (ARC, St. Louis, MO). For each animal, the mean hybridized probe radioactivity in the striatum of two hemispheres in 2–3 brain sections was measured. The striatum was manually delineated in these sections at rostrocaudal level Bregma 1.00 mm and (in the lesion experiments) dorsalventral level Bregma −4.10. Background measurements were taken in the corpus callosum in each section in both hemispheres and subtracted from the measured densities in the striata. Subsequently, data from the two hemispheres were averaged in the amphetamine and morphine experiments, and kept separate in the 6-OHDA lesion experiments. The changes in probe radioactivity as a consequence of lesion and treatment were expressed as the ratio of the lesioned and nonlesioned striatum.

Mean probe radioactivity and ratio of lesioned/nonlesioned striatum were analyzed using a one sample t test with test value 100 (set percentage value for nonlesioned striatum) for effect of the 6-OHDA lesion, a one-way ANOVA for drug treatment in the 6-OHDA lesion experiments and two-way ANOVA for pretreatment and challenge in the amphetamine and morphine sensitization experiments. Post hoc analysis was performed using a least significant difference (LSD) pair wise multiple comparison test.

3 Results

3.1 *Extent and Effect of the 6-OHDA Lesions*

Tyrosine hydroxylase – immunostaining and in situ hybridization for preproenkephalin were performed to determine the effectiveness of the 6-OHDA lesions. Only rats without visible immunostaining in the lesioned striatum were included in the study. This criterion resulted in the exclusion of five animals with an incomplete depletion of tyrosine hydroxylase immunostaining. The experimental groups were lesion followed by saline administration ($n = 6$), lesion followed by administration of SKF-38393 ($n = 6$), and lesion followed by administration of quinpirole ($n = 7$). All rats included in the study showed an upregulation of ppE mRNA on the lesioned side (saline: 151 ± 3%, quinpirole: 146 ± 9%, SKF-38393: 150 ± 5%), in agreement with previous findings and the literature (Angulo and McEwen 1994).

3.2 *Changes in Levels of DCLK mRNA*

In the 6-OHDA-lesion experiments, a small difference in amount of hybridized probe was detectable between the lesioned and the nonlesioned striatum (Fig. 1a), with the lesioned side showing higher intensity of labeling than the nonlesioned side. These differences were significant in all three treatment groups (Sal $t_{(5.59)}p <$ 0.01, Quin $t_{(9.57)}p < 0.001$, SKF $t_{(3.17)}p < 0.05$) (Fig. 2). Comparison of the ratios of the three treatment groups using a one-way ANOVA showed no significant differences between groups ($F_{(2,15)} = 1.88$ NS) (Fig. 2).

In the amphetamine- and morphine-induced psychomotor sensitization experiments, no significant changes whatsoever were induced by pretreatment or challenge

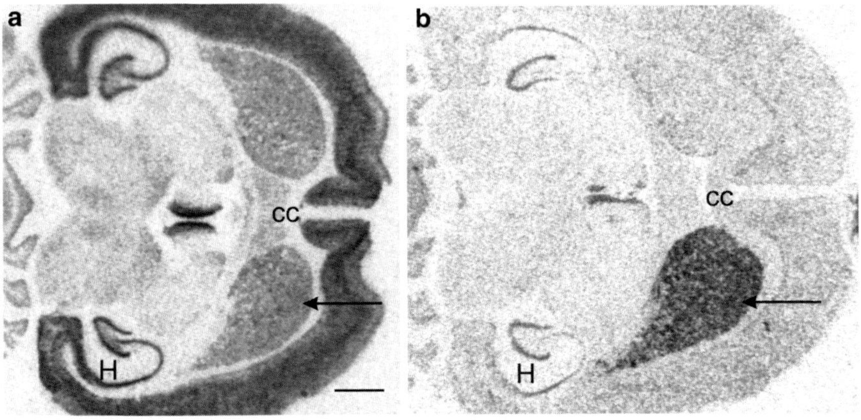

Fig. 1 Horizontal sections through the brain of 6-OHDA-lesioned rats exposed to saline (**a**) and SKF-38393 (**b**). Rostral is to the right. Section in (**a**) is hybridized for DCLK mRNA, whereas section in (**b**) is hybridized for CARP mRNA. *Arrows* point to the lesioned, i.e., dopamine-depleted, striatum. Note higher labeling intensity in the lesioned side for both DCLK and CARP. Labeling for CARP mRNA in the nonlesioned striatum is very weak. Abbreviatons: *cc* corpus callosum, *H* hippocampus. Bar = 2 mm

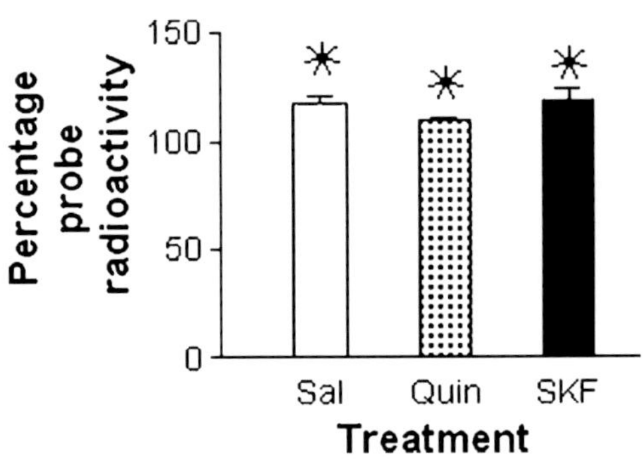

Fig. 2 DCLK mRNA levels in 6-OHDA lesioned striatum after acute exposure to saline (Sal), quinpirole 0.2 mg/kg (Quin), or SKF-38393 5 mg/kg (SKF). mRNA levels are expressed as mean percentage ± SEM DCLK hybridized probe radioactivity of the nonlesioned striatum. *Asterisks* indicate significant difference ($p < 0.05$) between lesioned and nonlesioned striatum (One sample t test)

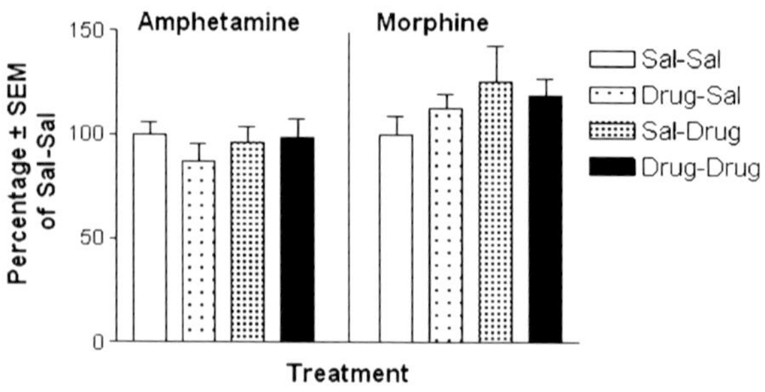

Fig. 3 DCLK mRNA levels in striatum after challenge exposure to amphetamine (*left panel*) or morphine (*right panel*) preceded by pretreatment with amphetamine or morphine, respectively, or with saline. Legends denote "pretreatment-challenge", e.g., "Sal-Sal" stands for "saline pretreatment – saline challenge." mRNA levels are expressed as percentage hybridized probe radioactivity of the saline pretreatment – saline challenge control. All values are mean ± SEM

with either drug (For amphetamine: $F_{(\text{pretreatment})(1,16)} = 0.43$ NS, $F_{(\text{challenge})(1,16)} = 0.21$ NS, $F_{(\text{pretreatment} \times \text{challenge})(1,16)} = 0.78$ NS; For morphine: $F_{(\text{pretreatment})(1,16)} = 0.08$ NS, $F_{(\text{challenge})(1,16)} = 0.18$ NS, $F_{(\text{pretreatment} \times \text{challenge})(1,16)} = 0.81$ NS) (Fig. 3).

3.3 Changes in Levels of CARP mRNA

Levels of CARP mRNA were extremely low in the striatum of saline-treated control animals or in the nonlesioned striatum (Fig. 1b). In the 6-OHDA lesion experiments, the experimental group treated with SKF-38393 stood out with high levels of CARP mRNA in the dopamine-depleted striatum that were almost six times higher than those in the nonlesioned striatum (575%, one sample *t* test $t_{(7.63)}p < 0.01$) (Figs. 1b and 4). The lesion–nonlesion differences in the quinpirole- and saline-treated groups were much smaller (saline: 122% and quinpirole: 119%), but significant (saline $t_{(5.99)}p < 0.01$, quinpirole $t_{(3.17)}p < 0.05$). Furthermore, group comparison showed that the lesioned/nonlesioned ratio of the SKF-38393 group was significantly higher than that of the saline and quinpirole groups (one-way ANOVA $F_{(\text{treatment})(2,15)} = 70.53$ $p < 0.001$, followed by LSD $p < 0.001$).

In the amphetamine and morphine psychomotor sensitization experiments, significant drug-induced changes were seen in CARP mRNA levels (Figs. 5 and 6). Amphetamine pretreatment had profoundly altered the response to amphetamine exposure on CARP expression. That is, in the amphetamine-sensitized group (amphetamine pretreatment followed by amphetamine challenge), CARP expression was twofold higher than in the other three groups (two-way ANOVA: $F_{(\text{pretreatment})(1,17)} = 20.08$ $p < 0.001$, $F_{(\text{challenge})(1,17)}$ 29.68 $p < 0.001$; $F_{(\text{pretreatment} \times \text{challenge})(1,17)} = 11.09$ $p < 0.01$) (Fig. 6).

In the morphine sensitization experiment, the drug-induced changes were opposite to those in the amphetamine sensitization experiment, as acute exposure to

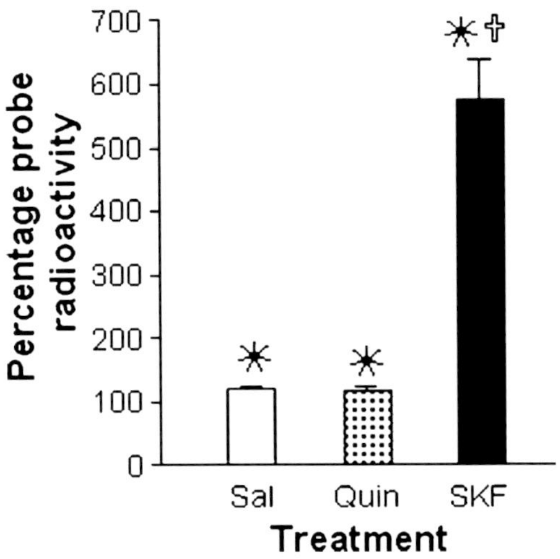

Fig. 4 CARP mRNA levels in 6-OHDA lesioned striatum after acute exposure to saline (Sal), quinpirole (Quin), or SKF-38393 (SKF). mRNA levels are expressed as mean percentage ± SEM CARP hybridized probe radioactivity of the nonlesioned striatum. *Asterisks* indicate significant difference ($p < 0.05$) between lesioned and nonlesioned striatum. *Cross* indicates significant difference ($p < 0.001$) from other groups in post hoc LSD test after one-way ANOVA

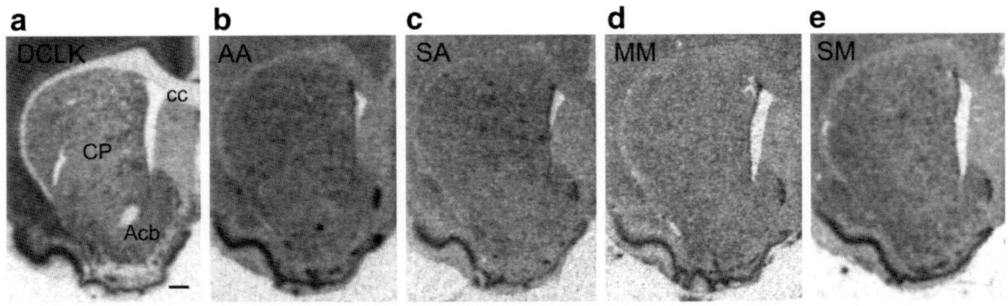

Fig. 5 Transversal sections through the striatum hybridized for DCLK (**a**) or CARP (**b–e**). The DCLK-hybridized section in (**a**) is from an animal pretreated and challenged with saline. Parts (**b**) and (**c**) show the effect of repeated and single treatment with amphetamine: (**b**) amphetamine-pretreated and amphetamine challenged (AA) and (**c**) saline-pretreated and amphetamine-challenged (SA). Parts (**d**) and (**e**) show the effect of repeated and single treatment with morphine: (**d**) morphine-pretreated and morphine challenged (MM) and (**e**) saline-pretreated and morphine-challenged (SM). Background labeling is higher than in the 6-OHDA experiment. This is due to greater section thickness (30 μm compared to 14 μm in the "6-OHDA sections" shown in Fig. 1). Abbreviations: *Acb* nucleus accumbens, *cc* corpus callosum, *CP* caudate-putamen. Bar = 600 μm

morphine seemed to decrease levels of CARP mRNA (two-way ANOVA: $F_{(\mathrm{pretreatment})(1,20)} = 0.89$ NS, $F_{(\mathrm{challenge})(1,20)} = 5.72$ $p < 0.05$, $F_{(\mathrm{pretreatment} \times \mathrm{challenge})(1,17)} = 1.94$ NS). This effect was especially apparent in the morphine-pretreated group challenged with morphine (Fig. 6). Mean CARP mRNA levels in this latter group were significantly

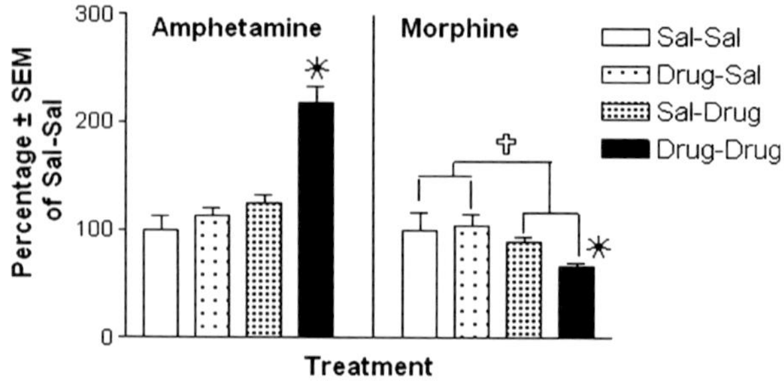

Fig. 6 CARP mRNA levels in striatum after challenge exposure to amphetamine (*left panel*) or morphine (*right panel*) which was preceded by pretreatment with amphetamine or morphine, respectively, or with saline. Legends denote "pretreatment-challenge." mRNA levels are expressed as percentage of the saline pretreatment – saline challenge control. All values are mean ± SEM. *Asterisk* in amphetamine panel indicates significant difference ($p < 0.001$ post hoc LSD test after two-way ANOVA) from other groups. *Cross* in morphine panel represents significant effect ($p < 0.05$) of challenge in two-way ANOVA. *Asterisk* in morphine panel indicates significant difference ($p < 0.05$ post hoc LSD test after two-way ANOVA) between Drug–Drug group and the Sal–Sal and Drug–Sal groups

lower than in the morphine- and saline-pretreated groups challenged with saline (LSD $p < 0.05$), but not significantly different from those in the saline-pretreated morphine-challenged group. There was no significant difference between the groups of animals pretreated with saline and challenged with either saline or morphine (LSD $p = 0.49$).

4 Discussion

The mRNA levels of the DCLK-gene splicing products DCLK and CARP in the striatum were differentially affected in the three paradigms that were used to induce behavioral hyperresponsiveness in the present study. Changes were most prominent in levels of CARP mRNA. A unilateral lesion with 6-hydroxydopamine (6-OHDA) increased both CARP and DCLK mRNA levels. Administration of a selective dopamine D1 agonist 2 weeks postlesion caused a huge rise in CARP mRNA, but had no effect on DCLK mRNA. The regulation of CARP mRNA showed a sensitized response to amphetamine after repeated administration of the drug, in the form of an increase in mRNA levels. In contrast, expression of morphine sensitization resulted in a decrease in CARP mRNA.

CARP has previously been identified as ANIA-4, a differentially regulated gene after 6-OHDA lesion combined with administration of D1 and D2 dopamine receptor agonists (Berke et al. 1998). Quantitative in situ hybridization showed that administration of the D1 dopamine receptor agonist SKF-38393 to 6-OHDA-lesioned animals or eticlopride and cocaine to naive rats led to an increase in ANIA-4 mRNA levels in the striatum (Berke et al. 1998). From these experiments it was not clear whether the cRNA probe that was used to detect ANIA-4 was selective for ANIA-4 or might also have hybridized to DCLK. The present results confirm that it is ANIA-4/CARP that is upregulated after the described treatments, i.e., after dopamine D1 receptor activation, and not DCLK.

4.1 Apoptosis

The DCLK gene is associated with different forms of neural plasticity, ranging from neurogenesis to cytoskeleton stability to apoptosis (Fitzsimons et al. 2008; Koizumi et al. 2006; Schenk et al. 2007; Shu et al. 2006). Schenk et al. (2007) have recently shown that CARP has a clear proapoptotic effect. If apoptosis is induced in NG108 cells, CARP accelerates cell death (Kruidering et al. 2001; Schenk et al. 2007). In the present experiments, it is doubtful whether the rise in CARP mRNA levels is correlated to apoptosis. Apoptosis in the striatum has, to our knowledge, not been reported in the experimental paradigm employing 6-OHDA lesions combined with the administration of dopamine agonists. After neurotoxic lesions with MPP^+, which are comparable to those in the present study, no neuronal cell damage or cell loss was found in the dopamine-depleted striatum (Canudas et al. 2000). Toxic damage to the striatum has been seen after intrastriatal injection of SKF-38393 (Kelley et al. 1990), but the observed increase in CARP mRNA is probably not a substance, i.e., SKF-38393, -dependent effect, since cocaine, eticlopride, and amphetamine gave the same response of CARP mRNA regulation (Berke et al. 1998; present results). It may be that increased levels of dopamine after amphetamine have an apoptotic effect via the D2 receptor (Shinkai et al. 1997). Cell death after amphetamine via a process resembling apoptosis has been seen in several cell types and brain regions including the striatum (Cadet et al. 2003; Kita et al. 2003; Mitchell et al. 1999). However, the doses at which amphetamine generates its neurotoxic effects range from 10 to 40 mg/kg, in some cases repeatedly administered with intervals of several hours (Davidson et al. 2001; Kita et al. 2003, for review). These doses are far higher than those used in the present study (challenge: 1 mg/kg, pretreatment: 2.5 mg/kg), making it unlikely that the currently applied treatment regimen led to appreciable levels of apoptosis in the striatum. In support of this conjecture, Bowyer et al. (1998) did not find toxic effects of amphetamine after a single or once repeated administration of 3.75 mg/kg. Nevertheless, it cannot be fully excluded that the rise in striatal CARP mRNA levels in the amphetamine-sensitized animals is not related to apoptosis.

A similar conclusion – that the rise in striatal CARP is probably not related to apoptosis – may hold for morphine, although it is based on different grounds. After morphine treatment, a decrease was observed in CARP mRNA levels, not an increase as seen in apoptotic neurons (Kruidering et al. 2001). Nevertheless, chronic morphine in doses comparable to those used in the present study may well lead to apoptosis of striatal neurons, as can be inferred from the morphine-induced increase in pro-apoptotic Fas receptor and concurrent down-regulation of Bcl-2 (Boronat et al. 2001). Other studies also support a positive correlation between apoptosis and morphine exposure (Hu et al. 2002; Kugawa et al. 2000; Mao et al. 2002). If the morphine treatment as applied in the present experiments does lead to apoptosis – which is at present not known – such a process is apparently not accelerated by an enhanced expression of CARP.

4.2 Neuronal Morphology

Besides apoptosis, the DCLK gene products may play a role in plastic changes in neuronal morphology. The expression of CARP, but not DCLK, is induced in the rat dentate gyrus by systemic administration of kainic acid, eliciting seizures (Vreugdenhil et al. 1999). In this model, major changes are induced in neuronal morphology and synaptic organization (Buckmaster and Dudek 1997; von Campe et al. 1997), whereas apoptosis is not observed. In the striatum, neuronal morphology is altered by denervation of dopaminergic inputs. Lesion-induced depletion of dopamine affects spine density, length and shape of dendrites, axon endings, and synaptic specializations of striatal (medium-sized spiny) projection neurons (Hattori and Fibiger 1982; Ingham et al. 1989, 1991, 1993, 1998; Meredith et al. 1995; Nitsch and Riesenberg 1995). These lesion-induced changes are apparently correlated with the presently observed increases in CARP and DCLK mRNA expression. A reason for the small magnitude of the changes in CARP and DCLK mRNA levels may be that at the time point of measuring, i.e., 3 weeks postlesion, mRNA levels are going back to steady-state levels. However, shortly (2 h) after exposing the lesioned animals - that by now had developed dopamine receptor supersensitivity as a consequence of the lesion - to a dopamine D1 agonist, a huge, selective upregulation of CARP mRNA was seen. At present, it is not known if administration of dopaminomimetics following dopamine depletion exacerbates or restores structural modifications of striatal circuitry. The fact that atrophy of striatal dendrites has been seen in Parkinson's disease stresses the relevance of the morphological findings and suggests that L-DOPA treatment does at least not prevent morphological changes (McNeill et al. 1988). On the contrary, Robinson and Kolb (1999) suggest that a rise in monoamine levels, and thus enhanced dopaminergic neurotransmission (in their paradigms induced by amphetamine or cocaine), is responsible for subsequent striatal structural changes. If dopamine receptor activation does indeed induce (further) changes in striatal morphology, the question is what the effect is of a concomitant D1-induced rise in CARP mRNA levels.

Repeated exposure to amphetamine results in alterations in dendritic morphology of striatal medium-sized spiny neurons. Increases have been seen in the density of dendritic spines in caudate-putamen and nucleus accumbens, and in the number of dendritic branches, the incidence of so-called branched spines and the length of dendrites in nucleus accumbens and cortical regions (Robinson and Kolb 2004). These morphological changes seem positively correlated to the rise in CARP mRNA levels seen in the present experiments. That is, our data clearly show that the response of CARP mRNA regulation to amphetamine depends on previous exposure to amphetamine, i.e., CARP mRNA expression is profoundly augmented during expression of amphetamine sensitization. In view of the coupling of CARP regulation to the D1 receptor (Berke et al. 1998; present study), this response is probably related to D1 receptor supersensitivity that accompanies amphetamine-induced behavioral sensitization (Vanderschuren and Kalivas 2000, for review). However, other mechanisms than the striatal D1 receptor pathway are responsible for the increase in behavioral responsiveness after amphetamine sensitization (Vanderschuren and Kalivas 2000). The striatal upregulation of CARP mRNA through the D1 receptor route may therefore be part of a mechanism involved in the regulation of structural modifications.

Supersensitivity of the dopamine D1 receptor pathway also develops in morphine-induced behavioral sensitization (Vanderschuren and Kalivas 2000). However, in the present paradigm employing morphine, the effect on CARP mRNA was opposite to that after amphetamine: CARP mRNA levels decreased during expression of morphine sensitization. As regards morphology, repeated, forced administration of morphine in doses identical to those used in the present paper induces a decrease in the number of dendritic spines and number of dendritic branches in nucleus accumbens and specific cortical regions (Robinson and Kolb 1999). Self-administration of morphine results in an even bigger decrease in number of spines in nucleus accumbens (Robinson et al. 2002). Thus, morphine and amphetamine exposure induce opposite changes in striatal neuronal morphology. Intriguingly, for both amphetamine and morphine, the direction of morphological changes correlates with that of CARP expression. The reason for the decrease in CARP mRNA levels after repeated morphine may be that morphine and not amphetamine or a D1 agonist was used as a challenge, as stimulation of mu and delta opioid vs. D1 receptors has opposite effects on striatal adenylate cyclase activity. Thus, a direct, inhibitory effect of opioid receptor activation may have overshadowed an indirect stimulatory effect of increased dopaminergic neurotransmission (Schoffelmeer et al. 1985). A challenge with a D1 agonist or amphetamine might well have induced an increase instead of a decrease in CARP mRNA levels in view of the fact that repeated morphine induces cross-sensitization to amphetamine (Vanderschuren et al. 1997).

4.3 Mechanisms

How can CARP affect morphological plasticity? Amphetamine and morphine exposure or administration of a dopamine D1 agonist to unilaterally dopamine-depleted animals are associated with changes in neurotrophic factors, growth factors, and TrkB receptors and their signal transduction pathways (Guillin et al. 2001; Iwakura et al. 2008; Meredith and Steiner 2006; Numan and Seroogy 1997; Numan et al. 1998). This may play a crucial role in the aforementioned morphological changes of striatal neuronal circuitry. CARP may affect these pathways in several ways. Its structure suggests that CARP is a substrate for mitogen-activated protein kinases (MAPK), enzymes that play a key role in signal transduction of neural growth and trophic factors. Moreover, CARP has recently been shown (Schenk et al. 2007) to be capable of binding to GRB2 (Growth factor receptor-bound protein 2; Lowenstein et al. 1992; Matuoka et al. 1992), an adaptor protein that links growth receptors and receptor tyrosine kinases to the Ras signaling pathway. GRB2-Ras signaling has multiple functions and seems to be required for cellular differentiation (Conti et al. 2001). Binding of CARP to GRB2 may thus interfere with activation of the Ras pathway by neurotrophic factor-signaling elicited by dopaminomimetics (see Fig. 7). In this way, CARP would counteract the induction of morphological alterations. It will be interesting to see if suppressing a putative, inhibitory feed-back regulation by CARP in the present three paradigms results in extreme morphological adaptations.

5 Conclusions

In the three paradigms used in the present study to increase behavioral responsiveness to dopamine and/or dopaminomimetics, major changes were found in mRNA levels of the DCLK gene splice variant CARP. We propose that the regulatory changes in CARP primarily reflect a dopamine-controlled modulatory effect on the

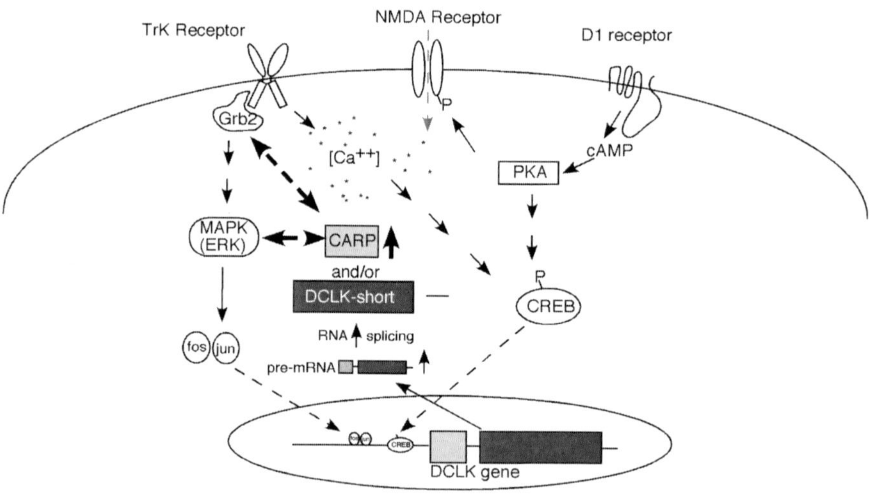

Fig. 7 Hypothetical mechanism of action via which dopamine can influence neurotrophic signaling through DCLK splice products. Dopamine D1 receptor stimulation controls DCLK gene transcription via CREB and induces formation of the splice variants DCLK-short (Engels et al. 2004) and CARP. CARP, in turn, may bind to GRB2 (Growth factor receptor-bound protein 2) which is associated with tyrosine kinase (Trk) receptors and MAPK in the Ras signaling pathway

reorganization of striatal neuronal circuitry that is induced in these paradigms. Since the structural modifications in psychomotor sensitization and in Parkinson's disease likely contribute to the persistent character of addictive behavior and motor dysfunction (McNeill et al. 1988; Robinson and Kolb 2004), respectively, it may be of therapeutic relevance to investigate how to exert control of CARP regulation.

References

Anderson SM, Famous KR, Sadri-Vakili G, Kumaresan V, Schmidt HD, Bass CE, Terwilliger EF, Cha JH and Pierce RC (2008) CaMKII: A biochemical bridge linking accumbens dopamine and glutamate systems in cocaine seeking. Nat Neurosci 11: 344–353.

Angulo JA and McEwen BS (1994) Molecular aspects of neuropeptide regulation and function in the corpus striatum and nucleus accumbens. Brain Res Brain Res Rev 19: 1–28.

Berke JD, Paletzki RF, Aronson GJ, Hyman SE and Gerfen CR (1998) A complex program of striatal gene expression induced by dopaminergic stimulation. J Neurosci 18: 5301–5310.

Berridge KC and Robinson TE (1998) What is the role of dopamine in reward: Hedonic impact, reward learning, or incentive salience? Brain Res Brain Res Rev 28: 309–369.

Boronat MA, Garcia-Fuster MJ and Garcia-Sevilla JA (2001) Chronic morphine induces up-regulation of the pro-apoptotic Fas receptor and down-regulation of the anti-apoptotic Bcl-2 oncoprotein in rat brain. Br J Pharmacol 134: 1263–1270.

Bowyer JF, Peterson SL, Rountree RL, Tor-Agbidye J and Wang GJ (1998) Neuronal degeneration in rat forebrain resulting from D-amphetamine-induced convulsions is dependent on seizure severity and age. Brain Res 809: 77–90.

Brown AM, Deutch AY and Colbran RJ (2005) Dopamine depletion alters phosphorylation of striatal proteins in a model of Parkinsonism. Eur J Neurosci 22: 247–256.

Buckmaster PS and Dudek FE (1997) Network properties of the dentate gyrus in epileptic rats with hilar neuron loss and granule cell axon reorganization. J Neurophysiol 77: 2685–2696.

Cadet JL, Jayanthi S and Deng X (2003) Speed kills: Cellular and molecular bases of methamphetamine-induced nerve terminal degeneration and neuronal apoptosis. FASEB J 17: 1775–1788.

Canudas AM, Friguls B, Planas AM, Gabriel C, Escubedo E, Camarasa J, Camins A and Pallas M (2000) MPP(+) injection into rat substantia nigra causes secondary glial activation but not cell death in the ipsilateral striatum. Neurobiol Dis 7: 343–361.

Cardinal RN, Parkinson JA, Hall J and Everitt BJ (2002) Emotion and motivation: The role of the amygdala, ventral striatum, and prefrontal cortex. Neurosci Biobehav Rev 26: 321–352.

Conti L, Sipione S, Magrassi L, Bonfanti L, Rigamonti D, Pettirossi V, Peschanski M, Haddad B, Pelicci P, Milanesi G, Pelicci G and Cattaneo E (2001) Shc signaling in differentiating neural progenitor cells. Nat Neurosci 4: 579–586.

Davidson C, Gow AJ, Lee TH and Ellinwood EH (2001) Methamphetamine neurotoxicity: Necrotic and apoptotic mechanisms and relevance to human abuse and treatment. Brain Res Brain Res Rev 36: 1–22.

Engels BM, Schouten TG, van DJ, Gosens I and Vreugdenhil E (2004) Functional differences between two DCLK splice variants. Brain Res Mol Brain Res 120: 103–114.

Everitt BJ and Wolf ME (2002) Psychomotor stimulant addiction: A neural systems perspective. J Neurosci 22: 3312–3320.

Fitzsimons CP, Ahmed S, Wittevrongel CF, Schouten TG, Dijkmans TF, Scheenen WJ, Schaaf MJ, de Kloet ER and Vreugdenhil E (2008) The microtubule-associated protein doublecortin-like regulates the transport of the glucocorticoid receptor in neuronal progenitor cells. Mol Endocrinol 22: 248–262.

Fuchs RA, Branham RK and See RE (2006) Different neural substrates mediate cocaine seeking after abstinence versus extinction training: A critical role for the dorsolateral caudate-putamen. J Neurosci 26: 3584–3588.
Garavan H, Pankiewicz J, Bloom A, Cho JK, Sperry L, Ross TJ, Salmeron BJ, Risinger R, Kelley D and Stein EA (2000) Cue-induced cocaine craving: Neuroanatomical specificity for drug users and drug stimuli. Am J Psychiatry 157: 1789–1798.
Guillin O, Diaz J, Carroll P, Griffon N, Schwartz JC and Sokoloff P (2001) BDNF controls dopamine D3 receptor expression and triggers behavioural sensitization. Nature 411: 86–89.
Hattori T and Fibiger HC (1982) On the use of lesions of afferents to localize neurotransmitter receptor sites in the striatum. Brain Res 238: 245–250.
Hu S, Sheng WS, Lokensgard JR and Peterson PK (2002) Morphine induces apoptosis of human microglia and neurons. Neuropharmacology 42: 829–836.
Ingham CA, Hood SH and Arbuthnott GW (1989) Spine density on neostriatal neurones changes with 6-hydroxydopamine lesions and with age. Brain Res 503: 334–338.
Ingham CA, Hood SH and Arbuthnott GW (1991) A light and electron microscopical study of enkephalin-immunoreactive structures in the rat neostriatum after removal of the nigrostriatal dopaminergic pathway. Neuroscience 42: 715–730.
Ingham CA, Hood SH, van MB, Weenink A and Arbuthnott GW (1993) Morphological changes in the rat neostriatum after unilateral 6-hydroxydopamine injections into the nigrostriatal pathway. Exp Brain Res 93: 17–27.
Ingham CA, Hood SH, Taggart P and Arbuthnott GW (1998) Plasticity of synapses in the rat neostriatum after unilateral lesion of the nigrostriatal dopaminergic pathway. J Neurosci 18: 4732–4743.
Ito R, Dalley JW, Robbins TW and Everitt BJ (2002) Dopamine release in the dorsal striatum during cocaine-seeking behavior under the control of a drug-associated cue. J Neurosci 22: 6247–6253.
Iwakura Y, Nawa H, Sora I and Chao MV (2008) Dopamine D1 receptor-induced signaling through TrkB receptors in striatal neurons. J Biol Chem 283: 15799–15806.
Iwata S, Hewlett GH and Gnegy ME (1997) Amphetamine increases the phosphorylation of neuromodulin and synapsin I in rat striatal synaptosomes. Synapse 26: 281–291.
Jacobs EH, Smit AB, de Vries TJ and Schoffelmeer AN (2003) Neuroadaptive effects of active versus passive drug administration in addiction research. Trends Pharmacol Sci 24: 566–573.
Kebabian JW (1997) A phosphorylation cascade in the basal ganglia of the mammalian brain: Regulation by the D-1 dopamine receptor. A mathematical model of known biochemical reactions. J Neural Transm Suppl 49: 145–153.
Kelley AE, Delfs JM and Chu B (1990) Neurotoxicity induced by the D-1 agonist SKF 38393 following microinjection into rat brain. Brain Res 532: 342–346.
Kita T, Wagner GC and Nakashima T (2003) Current research on methamphetamine-induced neurotoxicity: Animal models of monoamine disruption. J Pharmacol Sci 92: 178–195.
Koizumi H, Higginbotham H, Poon T, Tanaka T, Brinkman BC and Gleeson JG (2006) Doublecortin maintains bipolar shape and nuclear translocation during migration in the adult forebrain. Nat Neurosci 9: 779–786.
Kolta MG, Shreve P, De Souza V and Uretsky NJ (1985) Time course of the development of the enhanced behavioral and biochemical responses to amphetamine after pretreatment with amphetamine. Neuropharmacology 24: 823–829.
Kruidering M, Schouten T, Evan GI and Vreugdenhil E (2001) Caspase-mediated cleavage of the Ca2+/calmodulin-dependent protein kinase-like kinase facilitates neuronal apoptosis. J Biol Chem 276: 38417–38425.
Kugawa F, Ueno A and Aoki M (2000) Apoptosis of NG108-15 cells induced by buprenorphine hydrochloride occurs via the caspase-3 pathway. Biol Pharm Bull 23: 930–935.
Licata SC and Pierce RC (2003) The roles of calcium/calmodulin-dependent and Ras/mitogen-activated protein kinases in the development of psychostimulant-induced behavioral sensitization. J Neurochem 85: 14–22.

Lowenstein EJ, Daly RJ, Batzer AG, Li W, Margolis B, Lammers R, Ullrich A, Skolnik EY, Bar-Sagi D and Schlessinger J (1992) The SH2 and SH3 domain-containing protein GRB2 links receptor tyrosine kinases to ras signaling. Cell 70: 431–442.

Mao J, Sung B, Ji RR and Lim G (2002) Neuronal apoptosis associated with morphine tolerance: Evidence for an opioid-induced neurotoxic mechanism. J Neurosci 22: 7650–7661.

Matuoka K, Shibata M, Yamakawa A and Takenawa T (1992) Cloning of ASH, a ubiquitous protein composed of one Src homology region (SH) 2 and two SH3 domains, from human and rat cDNA libraries. Proc Natl Acad Sci USA 89: 9015–9019.

McNeill TH, Brown SA, Rafols JA and Shoulson I (1988) Atrophy of medium spiny I striatal dendrites in advanced Parkinson's disease. Brain Res 455: 148–152.

Meredith GE and Steiner H (2006) Amphetamine increases tyrosine kinase-B receptor expression in the dorsal striatum. Neuroreport 17: 75–78.

Meredith GE, Ypma P and Zahm DS (1995) Effects of dopamine depletion on the morphology of medium spiny neurons in the shell and core of the rat nucleus accumbens. J Neurosci 15: 3808–3820.

Meshul CK and Tan SE (1994) Haloperidol-induced morphological alterations are associated with changes in calcium/calmodulin kinase II activity and glutamate immunoreactivity. Synapse 18: 205–217.

Mitchell IJ, Cooper AJ and Griffiths MR (1999) The selective vulnerability of striatopallidal neurons. Prog Neurobiol 59: 691–719.

Nitsch C and Riesenberg R (1995) Synaptic reorganisation in the rat striatum after dopaminergic deafferentation: An ultrastructural study using glutamate decarboxylase immunocytochemistry. Synapse 19: 247–263.

Numan S and Seroogy KB (1997) Increased expression of trkB mRNA in rat caudate – putamen following 6-OHDA lesions of the nigrostriatal pathway. Eur J Neurosci 9: 489–495.

Numan S, Lane-Ladd SB, Zhang L, Lundgren KH, Russell DS, Seroogy KB and Nestler EJ (1998) Differential regulation of neurotrophin and trk receptor mRNAs in catecholaminergic nuclei during chronic opiate treatment and withdrawal. J Neurosci 18: 10700–10708.

Packard MG and Knowlton BJ (2002) Learning and memory functions of the basal ganglia. Annu Rev Neurosci 25: 563–593.

Paulson PE and Robinson TE (1995) Amphetamine-induced time-dependent sensitization of dopamine neurotransmission in the dorsal and ventral striatum: A microdialysis study in behaving rats. Synapse 19: 56–65.

Robinson TE and Kolb B (1999) Alterations in the morphology of dendrites and dendritic spines in the nucleus accumbens and prefrontal cortex following repeated treatment with amphetamine or cocaine. Eur J Neurosci 11: 1598–1604.

Robinson TE and Kolb B (2004) Structural plasticity associated with exposure to drugs of abuse. Neuropharmacology 47(Suppl 1): 33–46.

Robinson TE, Gorny G, Savage VR and Kolb B (2002) Widespread but regionally specific effects of experimenter versus self-administered morphine on dendritic spines in the nucleus accumbens, hippocampus, and neocortex of adult rats. Synapse 46: 271–279.

Schenk GJ, Engels B, Zhang YP, Fitzsimons CP, Schouten T, Kruidering M, de Kloet ER and Vreugdenhil E (2007) A potential role for calcium/calmodulin-dependent protein kinase-related peptide in neuronal apoptosis: In vivo and in vitro evidence. Eur J Neurosci 26: 3411–3420.

Schoffelmeer AN, Hansen HA, Stoof JC and Mulder AH (1985) Inhibition of dopamine-stimulated cyclic AMP efflux from rat neostriatal slices by activation of mu- and delta-opioid receptors: A permissive role for D-2 dopamine receptors. Eur J Pharmacol 118: 363–366.

Schultz W and Dickinson A (2000) Neuronal coding of prediction errors. Annu Rev Neurosci 23: 473–500.

Shinkai T, Zhang L, Mathias SA and Roth GS (1997) Dopamine induces apoptosis in cultured rat striatal neurons; possible mechanism of D2-dopamine receptor neuron loss during aging. J Neurosci Res 47: 393–399.

Shu T, Tseng HC, Sapir T, Stern P, Zhou Y, Sanada K, Fischer A, Coquelle FM, Reiner O and Tsai LH (2006) Doublecortin-like kinase controls neurogenesis by regulating mitotic spindles and M phase progression. Neuron 49: 25–39.

Takeuchi Y, Fukunaga K and Miyamoto E (2002) Activation of nuclear Ca(2+)/calmodulin-dependent protein kinase II and brain-derived neurotrophic factor gene expression by stimulation of dopamine D2 receptor in transfected NG108-15 cells. J Neurochem 82: 316–328.

Vanderschuren LJ and Kalivas PW (2000) Alterations in dopaminergic and glutamatergic transmission in the induction and expression of behavioral sensitization: A critical review of preclinical studies. Psychopharmacology (Berl) 151: 99–120.

Vanderschuren LJ, Tjon GH, Nestby P, Mulder AH, Schoffelmeer AN and de Vries TJ (1997) Morphine-induced long-term sensitization to the locomotor effects of morphine and amphetamine depends on the temporal pattern of the pretreatment regimen. Psychopharmacology (Berl) 131: 115–122.

Vanderschuren LJ, Schoffelmeer AN, Mulder AH and de Vries TJ (1999a) Dopaminergic mechanisms mediating the long-term expression of locomotor sensitization following pre-exposure to morphine or amphetamine. Psychopharmacology (Berl) 143: 244–253.

Vanderschuren LJ, Schoffelmeer AN, Mulder AH and de Vries TJ (1999b) Lack of cross-sensitization of the locomotor effects of morphine in amphetamine-treated rats. Neuropsychopharmacology 21: 550–559.

Vanderschuren LJ, Di CP and Everitt BJ (2005) Involvement of the dorsal striatum in cue-controlled cocaine seeking. J Neurosci 25: 8665–8670.

Volkow ND, Wang GJ, Telang F, Fowler JS, Logan J, Childress AR, Jayne M, Ma Y and Wong C (2006) Cocaine cues and dopamine in dorsal striatum: Mechanism of craving in cocaine addiction. J Neurosci 26: 6583–6588.

von Campe G, Spencer DD and de Lanerolle NC (1997) Morphology of dentate granule cells in the human epileptogenic hippocampus. Hippocampus 7: 472–488.

Vreugdenhil E, Datson N, Engels B, de JJ, van KS, Schaaf M and de Kloet ER (1999) Kainate-elicited seizures induce mRNA encoding a CaMK-related peptide: A putative modulator of kinase activity in rat hippocampus. J Neurobiol 39: 41–50.

Wong DF, Kuwabara H, Schretlen DJ, Bonson KR, Zhou Y, Nandi A, Brasic JR, Kimes AS, Maris MA, Kumar A, Contoreggi C, Links J, Ernst M, Rousset O, Zukin S, Grace AA, Lee JS, Rohde C, Jasinski DR, Gjedde A and London ED (2006) Increased occupancy of dopamine receptors in human striatum during cue-elicited cocaine craving. Neuropsychopharmacology 31: 2716–2727.

Yoshikawa K, Williams C and Sabol SL (1984) Rat brain preproenkephalin mRNA. cDNA cloning, primary structure, and distribution in the central nervous system. J Biol Chem 259: 14301–14308.

Part V
Basal Ganglia Disorders: Human Studies

Paradox of the Basal Ganglia Model: The Antidyskinetic Effect of Surgical Lesions in Movement Disorders

Jose A. Obeso, Fernando Alonso-Frech, Maria Cruz Rodriguez-Oroz, Lazaro Alvarez, Raul Macias, Gerardo Lopez, and Jorge Guridi

Abstract The revitalization of pallidotomy for Parkinson's disease (PD) provided a new opportunity to examine the effects of focal lesion of the basal ganglia and the adequacy of current pathophysiological concepts. The accumulated data, both from early surgical experiences as well as contemporary pallidotomy, established a dual, antiparkinsonian and antidyskinetic effect of a similarly placed lesion of the globus pallidum medialis (GPm). Modern studies have shown that pallidotomy induces significant improvement of movement parameters, restores thalamocortical activity, and eliminates Levodopa-induced dyskinesias without causing any major deficit of movement control. In patients with hemichorea-ballism or dystonia, pallidotomy also induce marked amelioration of the dyskinesias. These observations pose two major paradoxes for the pathophysiological model of the basal ganglia (BG) (1) Pallidotomy improves parkinsonism and eliminates dyskinesias and (2) pallidotomy does not produce overt motor or behavioral deficits.

In this article, we review the data regarding the antidyskinetic effects of pallidotomy in PD and related movement disorders and discuss a possible solution to the first paradox of the classic basal ganglia model.

1 Introduction

Lesions of the basal ganglia aiming to ameliorate Parkinson's disease (PD) were started in the 1940s before the introduction of stereotactic surgery (Meyers 1942). The advances in the understanding and definition of the pathophysiology of the basal ganglia (BG) which took place during the late 1980s and 1990s (Alexander and Crutcher 1990, Albin et al. 1989; Crossman 1987; DeLong 1990) led to a revitalization of surgery for PD. Pallidotomy is thought to alleviate parkinsonism by removing excessive

J.A. Obeso (✉), F. Alonso-Frech, M.C. Rodriguez-Oroz, L. Alvarez, R. Macias, G. Lopez, and J. Guridi
Neurologia-Neurociencias, Clinica Universitaria, Universidad de Navarra, Avenida de Pio XII 36, 31008 Pamplona, Spain
e-mail: jobeso@unav.es

H.J. Groenewegen et al. (eds.), *The Basal Ganglia IX*, Advances in Behavioral Biology 58, DOI 10.1007/978-1-4419-0340-2_39,

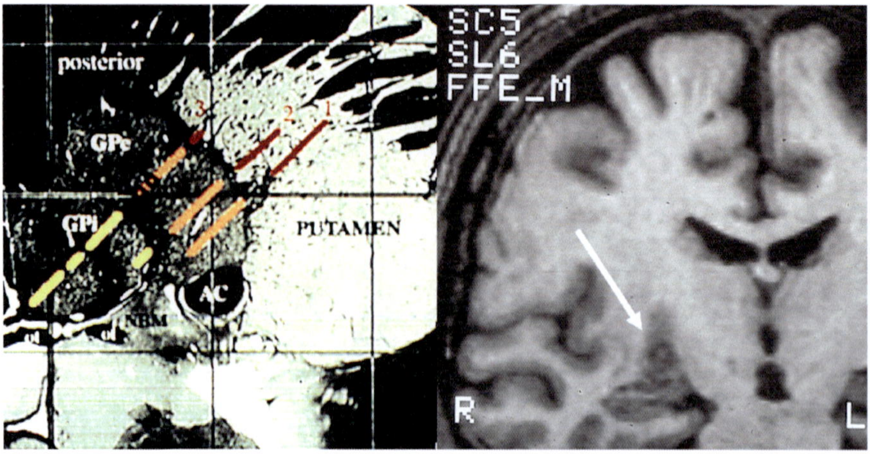

Fig. 1 *Left*: Lateral view of the human atlas of Schaltenbrand to illustrate typical microrecording trajectories during surgery. The electrodes goes through the putamen, globus pars externa (GPe), and globus pallidus pars interna (GPi). At the end of the third (in rostrocaudal or right to left order) trajectory activity from the visual tract evoked by flash stimulation could also be recorded and microstimulation led to activation of fibers in the internal capsule. All this information allowed producing a fairly centered lesion in the posteroventral GPi (*Right, white arrow*)

inhibitory activity from GPm to the nucleus ventralis oralis anterior and posterior (Voa/Vop) of the thalamus (Bakay et al. 1992; Lozano et al. 1995; Obeso et al. 1997; Vitek et al. 1998). Modern pallidotomy is usually performed by accurate definition of the nucleus by electrophysiological assessment, leading to a fairly local and well-placed lesion (Fig. 1). This provides an attractive scenario to assess the effects of a focal BG lesion on motor performance. Here we discuss the problem of the antidyskinetic effect of pallidotomy and the classic pathophysiological model of the BG.

2 Pallidotomy in Parkinson's Disease: Summary of Clinical Effects

The therapeutic efficacy of unilateral pallidotomy against the cardinal features of PD is now well documented (Laitinen et al. 1992; De Bie et al. 1999; Merello et al. 1999a; Vitek et al. 2003). In parallel, several functional studies have demonstrated an improvement of the limb contralateral to the lesion in a vast number of neurophysiological tests in patients submitted to pallidotomy. These include measurements of movement time, reaction time, peak frequency for repetitive movements (Dogali et al. 1995; Ondo et al. 1998; Baron et al. 2000) while also multijoint movements such as the "squeeze and flex" task and reaching to grasp an object are improved after pallidotomy (Obeso et al. 1996; Pfann et al. 1998; Limousin et al. 1999; Kimber et al. 1999). A few nonkinematic parameters such as the muscular stretch reflex (Narabayashi et al. 1997) and the effects of transcranial magnetic stimulation (Strafella et al. 1997) have also been counted as improved by pallidotomy. Moreover, Positron Emision Tomography (PET) measuring regional cerebral flow with H_2O–C_{15} and glucose consumption (FDG) revealed increased activation of motor cortical areas related with movement initiation and execution, and normalization of corticostriatal activity (Ceballos-Baumann et al. 1994; Grafton et al. 1995; Samuel et al. 1997; Carbon and Eidelberg 2002; Carbon et al. 2003).

3 Pallidotomy and Dyskinesias

In patients with PD, Levodopa-induced dyskinesias are very frequent and are present in nearly 100% of patients after 10 years of treatment. Pallidotomy has a profound antidyskinetic effect against "peak dose," diphasic dyskinesias, and "off dystonia" on the side contralateral to the lesion (Alkhani and Lozano 2001; Vitek et al. 2003). Indeed, the antidyskinetic effect is the most constant and permanent response to pallidotomy. Pallidotomy is also very effective against hemiballism in humans (Modesti and Van Buren 1979) as predicted from Carpenter et al.'s observations in monkeys (Carpenter et al. 1950). Contemporary pallidotomy for hemiballism of vascular origin targets the same zone of the GPm, i.e., the posterolateral motor region, than what is effective against Levodopa-induced dyskinesias in PD's patients (Suarez et al. 1997; Vitek et al. 1999). In the Centro Internacional de Restauracion Neurologica (CIREN), we encountered severe and persistent hemichorea-ballism in 8 out of 89 PD patients treated with unilateral lesion of the subthalamic nucleus (Alvarez et al. 2009). We were forced to perform pallidotomy in 8 patients out of a total of 89 PD patients included in this subthalamotomy protocol. The dyskinesias stopped immediately after pallidotomy without any adverse events or new neurological manifestations.

Pallidotomy may also improves dystonic spasms and postures, particularly in patients with genetically determined generalized Dystonia (DYT1) (Krack and Vercueil 2001). Interestingly, in such cases the lesion, which has to be bilateral to obtain significant improvement, is well tolerated and associated with persistent improvement (Merello et al. 2004). In agreement with this, our experience in CIREN is very favorable in young patients with primary generalized dystonia treated with bilateral pallidotomy.

4 The First Paradox of the Basal Ganglia Model: How Does Pallidotomy Eliminate Dyskinesias?

This paradox stems from the BG model conception that hyper- or hypoactivity of the GPm is associated with the parkinsonian and dyskinesias states, respectively, in a semilinear manner (Fig. 2). Dyskinesias are physiologically characterized by a low neuronal firing rate of GPm neurons as shown in parkinsonian monkeys (Bedard et al. 1999; Andringa et al. 1999; Papa et al. 1999) and in PD patients treated with

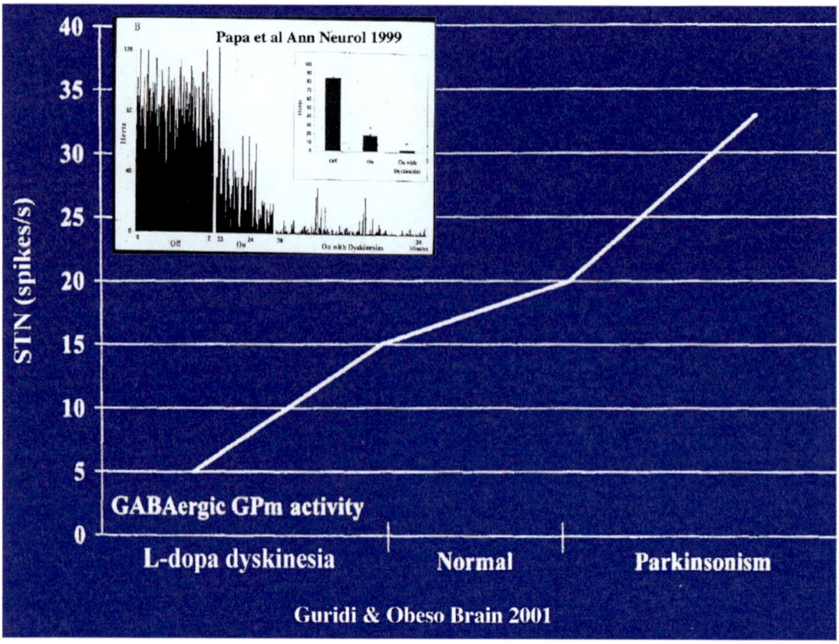

Fig. 2 Summary of the classic, firing rate-based pathophysiological model of the basal ganglia. Increased neuronal activity in the subthalamic nucleus (STN) and globus pallidus pars interna (GPi) predicts the motor state. Increased activity leads to parkinsonism and hypoactivity is associated with dyskinesias (reproduced from Guridi and Obeso 2001). The basis of this model was supported by several experimental findings. For example (*Inserted*), neuronal recording activity from the GPi in a parkinsonian monkey shows a high firing rate in the parkinsonian state and a drastic reduction after ingestion of Levodopa. Note that the neuron almost stops firing when the monkey was dyskinetic (reproduced with permission from Papa et al. 1999)

Levodopa or apomorphine (Vitek et al. 1999; Hutchison et al. 1997; Lozano et al. 2000). Thus, GPm firing rate shifted from an elevated and tonic pattern during the "off" medication state to a much lower rate and asynchronous activity during the "on" medication state. Pallidotomy should bring GPm efferent activity near to zero and, as predicted by the model, lead to worsening rather than abolition of dyskinesias.

We believe there are two recent sets of physiological observations that allow understanding and, at least conceptually, resolving the *First Paradox.*

First, changes in GPm firing rate associated with dyskinesias are not unidirectional (i.e., not all neurons dropped their discharge rate). Indeed, dyskinesias induced by dopamine agonists in monkeys and PD patients are associated with a dual (increased/decreased and vice versa) change in firing rate in the GPm, as well as burst-like discharges, prolonged pauses, and increased synchronization (Filion and Tremblay 1991; Merello et al. 1999b; Vitek and Giroux 2000).

Second, it is now well recognized that in addition to the rate of neuronal discharges in the subthalamic nucleus (STN) or GPm, the motor state in PD correlates with changes in the degree of neuronal synchronization (Brown et al. 2001; Priori et al. 2004; Israel and Bergman 2008). Recording local field potentials (Fig. 3) through macroelectrodes implanted in the STN or GPm for deep brain stimulation has shown that in patients with Levodopa-induced dyskinesias, as well as in patients with torsion dystonia there is a predominant activity in the 4–10 Hz band (Silberstein et al. 2003; Foffani et al. 2005; Alonso et al. 2006). Similar findings have been described in the substantia nigra pars reticulata in 6-OHDA-lesion rats treated with Levodopa (Meissner et al. 2006).

Thus, the dyskinetic state is characterized not only by a predominant drop in neuronal firing rate but also by an abnormal synchronization of neuronal activity in a given frequency domain. Brown and Eusebio (2008) have suggested that such oscillatory activity may represent the specific noisy signal for the occurrence of dyskinesias.

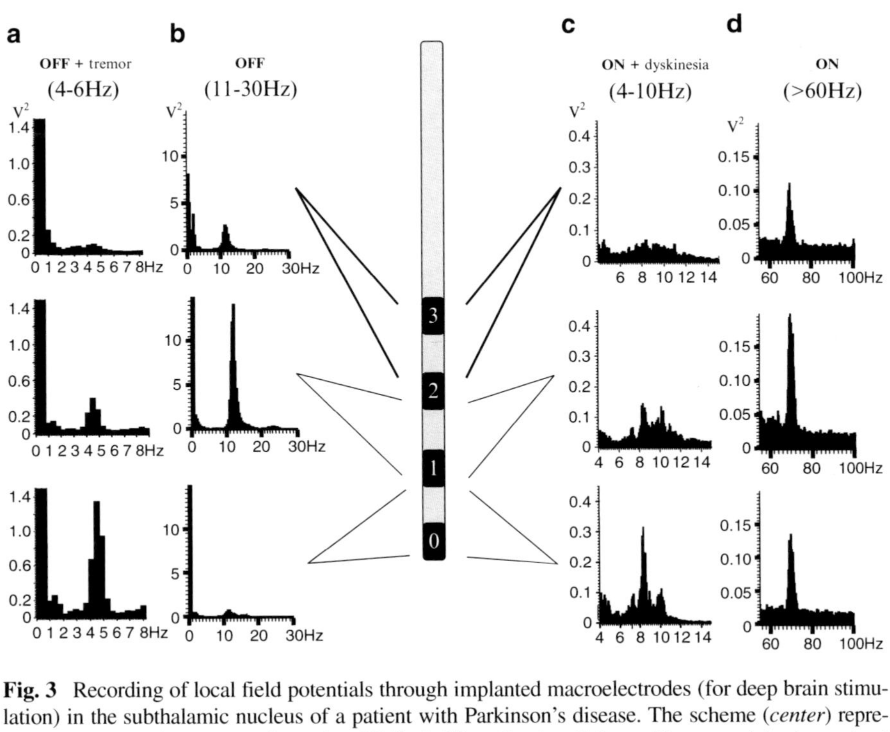

Fig. 3 Recording of local field potentials through implanted macroelectrodes (for deep brain stimulation) in the subthalamic nucleus of a patient with Parkinson's disease. The scheme (*center*) represents the three active contact electrodes. While "off" medication (*left*), oscillatory activity is maximal in the 4–6 Hz (tremor-related) and beta band (11–30 Hz). After taking Levodopa (*right*), the beta band decreases and a large increment is seen in the gamma (>60 Hz) and slow (4–10 Hz) bands. The latter was only present while patient exhibited dyskinesias (modified from Alonso et al. 2006)

5 Conclusions

Pallidal surgery may well work by interrupting the generation of abnormal firing patterns and oscillations associated with dyskinesias. This notion is in keeping with a previous, less-elaborated suggestion that pallidotomy eliminates "noisy" signals from reaching and interfering with motor areas (Marsden and Obeso 1994; Lozano et al. 2000). Considering that LID and hemichorea-ballism actually consist in the abnormal release of otherwise normal fragments of movement, it is hard to understand how elimination of such prodyskinetic signals is not associated with any motor deficit. This is indeed another well-recognized paradox of the basal ganglia. How the motor system can operate more efficiently without BG output (the *Second Paradox*)? However, due to space limits, this issue will be discussed in another forum.

References

Albin RL, Young AB and Penney JB (1989) The functional anatomy of basal ganglia disorders. Trends Neurosci 12: 366–375.

Alexander GE and Crutcher MD (1990) Functional architecture of the basal ganglia circuits: Neural substrates of parallel processing. Trends Neurosci 13: 266–271.

Alkhani A and Lozano AM (2001) Pallidotomy for Parkinson's disease: A review of contemporary literature. J Neurosurg 94: 43–49.

Alonso-Frech F, Zamarbide I, Alegre M, Rodríguez-Oroz MC, Guridi J, Manrique M, Valencia M, Arieda J and Obeso JA (2006) Slow oscillatory activity and levodopa-induced dyskinesias in Parkinson's disease. Brain 129: 1748–1757.

Alvarez L, Macias R, Pavon N, López G, Rodríguez-Oroz MC, Rodríguez R, Alvarez M, Pedroso I, Teijeiro J, Fernández R, Casabona E, García I, Guridi J, Juncos J, DeLong MR and Obeso JA (2009) Feb 9 Therapeutic efficacy of unilateral subthalamotomy in Parkinson's disease: Results in 89 patients followed for up to 36 months. J Neurol Neurosurg Psychiatry.

Andringa G, Vermeulen RJ, Drukarch B, Renier WO, Stoof JC and Cools AR (1999) The validity of the pretreated, unilaterally MPTP-treated monkeys as a model of Parkinson's disease: A detailed behavioural analysis of the therapeutic and undesirable effects of the D2 agonist quinpirole and the D1 agonist SKF 81297. Behav Pharmacol 10: 163–173.

Bakay RAE, DeLong MR and Vitek JL (1992) Posteroventral pallidotomy for Parkinson's disease. J Neurosurg 77: 487–488.

Baron MS, Vitek JL, Bakay RA, Green J, McDonald WM, Cole SA and DeLong MR (2000) Treatment of advanced Parkinson's disease by unilateral posterior GPi pallidotomy: 4-year results of a pilot study. Mov Disord 15: 230–237.

Bedard PJ, Blanchet PJ, Levesque D, Soghomonian JJ, Grondin R, Morissette M, Goulet M, Calon F, Falardeau P, Gomez-Mancilla B, Doucet JP, Robertson GS and DiPaolo T (1999) Pathophysiology of L-dopa-induced dyskinesias. Mov Disord 14(Suppl 1): 4–8.

Brown P and Eusebio A (2008) Paradoxes of functional neurosurgery: Clues from basal ganglia recordings. Mov Disord 23: 12–20.

Brown P, Oliviero A, Mazzone P, Insola A, Tonali P and Di Lázaro V (2001) Dopamine dependency of oscillations between subthalamic nucleus and pallidum in Parkinson's disease. J Neurosci 21: 1033–1038.

Carbon M and Eidelberg D (2002) Modulation of regional brain function by deep brain stimulation: Studies with positron emission tomography. Curr Opin Neurol 4: 451–455.

Carbon M, Edwards C and Eidelberg D (2003) Functional brain imaging in Parkinson's disease. Adv Neurol 91: 175–181.

Carpenter MB, Whittier JR and Mettler FA (1950) Analysis of choreoid hyperkinesia in the rhesus monkey. J Comp Neurol 92: 293–322.

Ceballos-Baumann AO, Obeso JA, Vitek JL, DeLong MR, Bakay RAE, Linazasoro G and Brooks DJ (1994) Restoration of the thalamo cortical activity after posteroventral pallidotomy in Parkinson's disease. Lancet 344: 814.

Crossman AR (1987) Primate models of dyskinesia: The experimental approach to the study of basal ganglia-related involuntary movement disorders. Neuroscience 21: 1–40.

De Bie RMA, de Haan RJ, Nijssen PCG, Rutgers WF, Beute GN, Bosch A, Haaxma R, Schmand B, Schuurman R, Staal M and Speelman J (1999) Unilateral pallidotomy in Parkinson's disease: A randomised, single blind, multicentre trial. Lancet 354: 1665–1669.

DeLong MR (1990) Primate models of movement disorders of basal ganglia origin. Trends Neurosci 13: 281–285.

Dogali M. Fazzini E, Kolodny D, Eidelberg D, Sterio O, Denivinsky and Beric A (1995) Stereotactic ventral pallidotomy for Parkinson's disease. Neurology 45: 753–761.

Filion M and Tremblay L (1991) Effects of dopamine agonist on the spontaneous activity with MPTP-induced parkinsonism. Brain Res 547: 152–161.

Foffani G, Ardolino G, Meda B, Egidi M, Rampini P, Caputo E, Baselli G and Priori A (2005) Altered subthalamo-pallidal synchronisation in parkinsonian dyskinesias. J Neurol Neurosurg Psychiatry 76: 426–428.

Grafton ST, Waters C, Sutton L, Lew MF and Couldwell W (1995) Pallidotomy increases activity of motor association cortex in Parkinson's disease. A positron emission tomographic study. Ann Neurol 37: 776–783.

Guridi J and Obeso JA (2001) The subthalamic nucleus, hemiballismus and Parkinson's disease: Reappraisal of a neurosurgical dogma. Brain 124: 5–19.

Hutchison WD, Levy R, Dostrovsky JO, Lozano AM and Lang AE (1997) Effect of apomorphine on the globus pallidus neurons in the parkinsonian patients. Ann Neurol 42: 767–775.

Israel Z and Bergman H (2008) Pathophysiology of the basal ganglia and movement disorders: From animal models to human clinical applications. Neurosci Biobehav Rev 32: 367–377.

Kimber TE, Tsa CS, Semmler J, Brophy BP and Thompson PD (1999) Voluntary movement after pallidotomy in severe Parkinson's disease. Brain 122: 895–906.

Krack P and Vercueil L (2001) Review of the functional surgical treatment of dystonia. Eur Neurol 8: 389–399.

Laitinen LV, Bergerhein T and Hariz MI (1992) Leksell's posteroventral pallidotomy in the treatment of Parkinson's disease. J Neurosurg 76: 53–61.

Limousin P, Brown RG, Jahanshahi M, Asselman P, Quinn NP, Thomas D, Obeso JA and Rothwell JC (1999) The effects of posteroventral pallidotomy on the preparation and execution of voluntary hand and arm movements in Parkinson's disease. Brain 122: 315–327.

Lozano AM, Lang AE, Galvez-Jimenez N, Miyasaki J, Duff J, Hutchison WD and Dostrovski J (1995) Effect of GPi pallidotomy on motor function in Parkinson's disease. Lancet 346: 1383–1387.

Lozano AM, Lang AE, Levy R, Hutchison W and Dostrovsky J (2000) Neuronal recordings in patients with Parkinson's disease with dyskinesias induced by apomorphine. Ann Neurol 47: S141–S146.

Marsden CD and Obeso JA (1994) The functions of the basal ganglia and the paradox of stereotaxic surgery in Parkinson's disease. Brain 117: 877–897.

Meissner W, Ravenscroft P, Reese R, Harnack D, Morgenstern R, Kupsch A, Klitgaard H, Bioulac B, Gross CE, Bezard E and Boraud T (2006) Increased slow oscillatory activity in substantia nigra pars reticulata triggers abnormal involuntary movements in the 6-OHDA-lesioned rat in the presence of excessive extracellular striatal dopamine. Neurobiol Dis 22: 586–598.

Merello M, Nouzeilles MI, Cammarota A, Betti O and Leiguarda R (1999a) Comparison of 1-year follow-up evaluations of patients with indication for pallidotomy who did not undergo surgery versus patients with Parkinson's disease who did undergo pallidotomy: A case control study. Neurosurgery 44: 461–468.

Merello M, Balej J, Delfino M, Cammarota A, Betti O and Leiguarda R (1999b) Apomorphine induces changes in GPi spontaneous outflow in patients with Parkinson's disease. Mov Disord 14: 45–49.

Merello M, Cerquetti D, Cammarota A, Tenca E, Artes C, Antico J and Leiguarda R (2004) Neuronal globus pallidus activity in patients with generalised dystonia. Mov Disord 19: 548–554.

Meyers R (1942) The modification of alternating tremors, rigidity and festination by surgery of the basal ganglia. Res Nerv Ment Dis Proc 21: 602–665.

Modesti LM and Van Buren JM (1979) Hemiballismus complicating stereotactic thalamotomy. Appl Neurophysiol 42: 267–283.

Narabayashi H (1997) Pallidotomy revisited. Analysis of posteroventral pallidotomy. Stereotact Funct Neurosurg 69: 54–61.

Obeso JA, Linazasoro G and Rothwell JC (1996) Assessing the effects of pallidotomy in Parkinson's disease. Lancet 347: 1490.

Obeso JA, Guridi J and DeLong MR (1997) Surgery for Parkinson's disease. J Neurol Neurosurg Psychiatry 62: 2–8.

Ondo WG, Desalloms M, Jankovic J and Grossman RG (1998) Pallidotomy for generalized dystonia. Mov Disord 13: 693–698.

Papa SM, Desimone R, Fiorani M and Oldfield EH (1999) Internal globus pallidus discharge is nearly suppressed during levodopa-induced dyskinesias. Ann Neurol 46: 732–738.

Pfann KD, Penn RD, Shannon KM and Corcos DM (1998) Pallidotomy and bradykinesia. Implications for basal ganglia function. Neurology 51: 796–803.

Priori A, Foffani G, Pesenti A, Tamma F, Bianchi AM, Pellegrini M, Locatelli M, Moxon KA and Villani RM (2004) Rhythm-specific pharmacological modulation of subthalamic activity in Parkinson's disease. Exp Neurol 189: 369–379.

Samuel M, Ceballos-Baumann AU, Turjanski N, Boecker H, Gorospe A, Linazasoro G, Holmes AP, DeLong MR, Vitek JL, Thomas DGT, Quinn NP, Obeso JA and Brooks DJ (1997) Pallidotomy in Parkinson's disease increases supplementary motor area and prefrontal activation during performance of volitional movements: An H2 O15 PET study. Brain 120: 1301–1313.

Silberstein P, Kühn AA, Kupsch A, Trottenberg T, Krauss JK, Wöhrle JC, Mazzone P, Insola A, Di Lazzaro V, Oliviero A, Aziz T and Brown P (2003) Patterning of globus pallidus local field potentials differs between Parkinson's disease and dystonia. Brain 126: 2597–2608.

Strafella A, Ashby P, Lozano A and Lang AE (1997) Pallidotomy increases cortical inhibition in Parkinson's disease. Can J Neurol Sci 24: 133–136.

Suarez JI, Matman LV, Reich SG, Dougherty PM, Hallett M and Lenz FA (1997) Pallidotomy for hemiballismus: Efficacy and characteristics of neuronal activity. Ann Neurol 42: 807–811.

Vitek JL, Bakay RAE, Hashimoto T, Kaneoke Y, Mewes K, Zhang JY, Rye D, Starr P, Baron M, Turner R and DeLong MR (1998) Microelectrode-guided pallidotomy: Technical approach and application for medically intractable Parkinson's disease. J Neurosurg 88: 1027–1043.

Vitek JL, Chockkan V, Zhang J-Y, Kaneoke Y, Evatt M, DeLong MR, Triche S, Mewes K, Hashimoto T and Bakay RAE (1999) Neuronal activity in the basal ganglia in patients with generalized dystonia and hemiballismus. Ann Neurol 46: 22–35.

Vitek JL, and Giroux M (2000) Physiology of hypokinetic and hyperkinetic movement disorders: model for dyskinesia. Ann Neurol 47: S131–140.

Vitek JL, Bakay RAE, Freeman A, Evatt M, Green J, McDonald W, Haber M, Barnhart H, Walay N, Triche S, Mewes K, Chockkan V, Zhang JY and DeLong MR (2003) Randomized trial of pallidotomy versus medical therapy for Parkinson's disease. Ann Neurol 53: 558–5569.

The Dynamic Relationship Between Voluntary and Involuntary Motor Behaviours in Patients with Basal Ganglia Disorders

Christian Duval, Alison Fenney, and Mandar S. Jog

Abstract The present chapter deals with technical issues and relevant questions regarding the quantification of whole-body involuntary movements (WBIM) such as Huntington's disease (HD) chorea and levodopa-induced dyskinesia (LID) in Parkinson's disease (PD). We also present results from an experiment where patients with HD chorea and age- and gender-matched control subjects had to perform a rapid alternating movement (RAM) task, which provides a measure of bradykinesia while WBIM was quantified. The RAM task was performed by subjects using their dominant and non-dominant hands independently (single-hand condition), in addition to both hands simultaneously (bimanual condition). Results showed that WBIM was higher in the HD group in all conditions, confirming the presence of chorea. A significant increase in the amplitude of chorea during RAM was also observed in the HD group in all conditions. In the control group, a significant increase in motor overflow during RAM was only observed in the bimanual condition. The range of pronation–supination cycles was systematically greater for the HD group than for control subjects in the single-hand and bimanual conditions. RAM velocity was greater in the HD group in the single-hand condition, but equal in the bimanual condition. Our results support the notion that bradykinesia is not a systematic feature of adult-onset HD. Using previous data obtained in PD patients having LID, we discuss the implications of the present findings for basal ganglia pathophysiology involved in the generation of WBIM.

C. Duval (✉) and A. Fenney
Département de Kinanthropologie, Université du Québec à Montréal, C.P. 8888, Succursale Centre-Ville, Montréal, QC, Canada, H3C 3P8
e-mail: duval.christian@uqam.ca

M.S. Jog
Clinical Neurological Sciences, London Health Science Center, University of Western Ontario, LHSC - University Hospital, 339 Windermere Road, PO Box 5339, UWO, London, Ontario, Canada, N6A 5A5

H.J. Groenewegen et al. (eds.), *The Basal Ganglia IX*, Advances in Behavioral Biology 58,
DOI 10.1007/978-1-4419-0340-2_40,

1 Introduction

Involuntary motor behaviours such as tremor, chorea or dyskinesias represent a major burden for patients. Besides the discomfort that undesired muscle contractions may cause, patients have to cope with these unwanted neural outputs when attempting to perform voluntary movements. The origin of pathological tremors, their specific characteristics, and their impact on voluntary movements have been the subject of numerous studies. Sadly, whole-body involuntary movements (WBIM) such as levodopa-induced dyskinesias (LID) in Parkinson's disease (PD) or chorea in Huntington's disease (HD) have not had nearly the same level of attention, despite their seemingly obvious debilitating effect on patients. We have only begun to understand their movement pattern characteristics (Gour et al. 2007; Fenney et al. 2008b), and more studies are needed to characterize their influence on motor performance. The usefulness of examining the dynamic relationship between involuntary and voluntary motor behaviours is twofold. First, it allows us to assess the impact of these undesired movements on the daily lives of patients, and second, it may potentially provide kinematic evidence to further improve current basal ganglia models. Here, we discuss the current methodological problems in dealing with the study of the concomitant presence of hyper- and hypokinetic behaviours in patients with movement disorders, and we present our methodological techniques to address these problems. We then introduce issues about the dynamic relationship between involuntary and voluntary motor behaviours that is specific to patients with PD having LID and patients with HD having chorea. Finally, we present further evidence showing that bradykinesia is not a systematic symptom of HD.

1.1 The Problem of Recording Whole-Body Involuntary Motor Behaviours

Several clinical scales were developed to quantify dyskinesia or chorea, the body segments that are involved, and/or the type of involuntary movements present in the patient (Fahn et al. 1987; Goetz et al. 1994; Langston et al. 1992; Reilmann et al. 2001; Katzenschlager et al. 2007). While these types of assessment are useful to monitor the clinical state of patients (Hoff et al. 1999), they provide limited information on the characteristics and patterns of involuntary movements, their relationship with other symptoms, and specifically their impact on voluntary movements. Previously used quantitative methods include electromyography (Yanagisawa and Nezu 1987), accelerometry (Hoff et al. 2001), Doppler radar (Resek et al. 1981), and rotation-sensitive movement monitors (Burkhard et al. 1999). These methodologies provided more accurate measurements, but did not offer a means to quantify WBIM simultaneously with motor performance, a crucial step in determining the dynamic relationship between the desired and undesired motor outputs.

The use of a camera-based system is also impractical since the unpredictable nature of WBIM would render the proper placement of cameras difficult, especially in experiments where patients are to be seated. In our laboratory, we have opted for a magnetic tracker system, which circumvents the line of sight issues associated with camera-based systems, while providing the necessary accuracy. This system has also the advantage of using only fifteen sensors, each providing six degrees of freedom (displacement in *x*, *y*, and *z*, in addition to pitch, roll, and yaw movements), yielding a full 3D rendering of WBIM. Using this device, we have recorded the LID of patients with PD, as well as the chorea of patients with HD.

1.2 The Problem of Recording 'Core' Bradykinesia

It has been suggested that bradykinesia can be observed in patients with PD having LID and in patients with HD having chorea (Thompson et al. 1988). This poses an interesting challenge for current basal ganglia models as none of them can fully explain the concomitant presence of both hyper- and hypokinetic behaviours in the same patient. To further complicate the situation, no one has yet determined conclusively whether the bradykinesia observed in these patients is the result of cognitive deficiencies, basal ganglia-generated motor output reduction, and/or simply the result of biomechanical interference of the involuntary descending motor output associated with WBIM upon the voluntary movement itself. Bradykinesia is a term used to include any abnormal slowness of movement during motor action. The problem here is that bradykinesia may have different origins (Berardelli et al. 2001). In patients having involuntary movements, a tradeoff between speed and accuracy may occur as they have to integrate their unwanted motor acts into the motor plan in order to predict and implement potential corrections during the movement. This amounts to dealing with the noise created by the involuntary movements. The latter phenomenon could be described as a signal-to-noise ratio (SNR) issue, where the signal is the voluntary movement and the noise represents the involuntary motor command generating WBIM. When the SNR is low, the impact of the involuntary movement could induce delays in the initiation of the movement. In this case, the patient may have to either wait for the proper phase of movement to act or fight against the involuntary movement to perform the intended movement. This phenomenon has been demonstrated in tremor when the timing of muscle contraction was under the influence of the underlying tremor (Wierzbicka et al. 1993). When the SNR is high however (i.e., the voluntary movement's amplitude or velocity surpasses that of the involuntary movement), the patient may be able to circumvent the undesired limb motion. This is analogous to the disappearance of parkinsonian resting tremor during the early phases of movement. Recent research has demonstrated that PD resting tremor may remain present during slow movements (Duval et al. 2004), but that high-velocity or large-amplitude movements such as RAM render it almost undetectable (Duval et al. 2006b).

Bradyphrenia is another possible source of bradykinesia (Cooper et al. 1994), which induces a delay in motor response from the affected patients. Other factors such as rigidity, muscle weakness, and tremor have also been implicated (Berardelli et al. 2001). Finally, there is the bradykinesia resulting from a lack of proper muscle activation, or disruption of activating/inhibiting functions of motor programs, which is the consequence of basal ganglia dysfunction. To capture the latter type of bradykinesia, the task should have a high enough SNR so as to reduce the biomechanical effect and be simple enough so as to eliminate any cognitive influence. It must also be internally generated, which is representative of basal ganglia function (Cunnington et al. 2000).

Rapid alternating movement (RAM) tasks such as the one used by our group to assess bradykinesia have been validated by ourselves (Duval et al. 2001, 2006a, b; Ghassemi et al. 2006) and others (Fimbel et al. 2005; Beuter et al. 1999a, b; Okada and Okada 1983) to measure slowness of movement in different populations and represent the ability of subjects to generate high-velocity motion and to maintain it for several seconds. This type of task involves a natural movement with a rapid learning curve; it is internally generated and has relatively large amplitude in displacement and velocity, which reduces greatly the likelihood of biomechanical effects. Finally, it is simple enough to minimize greatly the influence of cognitive problems known to occur in both diseases studied. With these characteristics, we believe that this task represents a good measure of the bradykinesia that is the result of neural disturbance within the basal ganglia–thalamo–cortical loop.

1.3 The Dynamic Relationship Between Involuntary and Voluntary Motor Behaviours in Patients with PD Having LID and in Patients with HD Having Chorea

Recently, we assessed RAM performance of patients with PD during fast pronation–supination movement at the wrist for 7 s, preceded and followed by 20-s rest periods (Ghassemi et al. 2006). Results showed that in patients with PD having mild to moderate peak-dose LID (mean disease duration of 15 years), bradykinesia indeed occurred simultaneously with WBIM. When examining the velocity and amplitude of pronation-supination cycles, we observed that both performance characteristics were worse than in controls, but not worse than that in patients with PD who had not yet experienced LID in their life time (mean disease duration of 5 years). The fact that the amplitude of LID was unrelated to the amount of bradykinesia indicated that the LID itself had a limited influence on the motor performance of these patients. In patients with HD having chorea, we found that performance characteristics such as the amplitude and velocity of RAM pronation–supination cycles were on an average superior than that of their age- and gender-matched controls (Fenney et al. 2008a). This indicates that bradykinesia is not a 'systematic' feature of HD, and that other factors, such as low SNR, cognitive impairments, or simply the transition from the choreic form of the disease towards the akinetic form, may have played a role in the bradykinesia detected in other studies. We also intended to characterize the RAM performance of both hands independently, and when both hands were performing RAM simultaneously. Here, we present these results.

2 Methods

2.1 *Participants*

Thirty participants, 15 patients with HD (12 women, age: 56 ± 12) having chorea and 15 age/gender matched control subjects (12 women, age: 55 ± 13), were tested. Patients were recruited through the London Health Sciences Movement Disorders Clinic. Control subjects were recruited from the general population. Each participant provided informed consent. Exclusion and inclusion criteria, as well as clinical details of patients, are described in detail elsewhere (Fenney et al. 2008a). In the present study, results from only 13 patients and their matched controls are presented since the data from two patients related to the bimanual condition had to be rejected for technical reasons.

2.2 *Whole-Body Involuntary Movements Quantification*

To capture whole-body chorea in three dimensions, a six-degrees-of-freedom electromagnetic measurement system, the *MotionMonitor*™ magnetic motion tracker, was employed (Innovative Sports Training, Chicago, IL). In brief, subjects were outfitted with a custom shirt, Velcro bands, gloves, and shoe covers onto which 15 sensors (receivers) that provided time series signals of both position (x, y, z) and orientation (pitch, yaw, roll) were attached. They were placed adjacent to the joint axes, such as the posterior surface of the head, the first thoracic vertebrae, sacral bone, superior spine of the scapulae, lateral surface of the forearms and upper arms, back of the hands, as well as the lateral aspects of the calves and dorsal surface of the feet. Following a calibration protocol where each body joint was identified with a marker, and based on the height, weight of the subject, and integrated anthropometric tables, an automated algorithm assigned a particular sensor to the center of mass of the limb on which it was positioned. The mean position for x, y, and z was subtracted from the actual position of x, y, and z, yielding a displacement time series around a neutral position for each of the epochs. We then squared the root mean square (RMS) values for each of the x, y, and z time series and calculated the mean of these three squared RMS values. Finally, we computed the root of this mean, thus yielding three-dimensional vectorial amplitude for each sensor. WBIM amplitude was determined by computing the sum of all these mean vectorial amplitudes over all sensors, except those allocated to the performing limb (i.e., hand, forearm, and upper arm).

2.3 Bradykinesia Quantification Using a Rapid Alternating Movements Task

Participants were instructed to maintain a 'rest' position for 20 s, i.e., arms bent at their side, holding one foam ball in each hand, and without performing voluntary movements. Then, the subjects were instructed to perform pronation–supination of the dominant hand as fast and as complete a rotation as possible, for a period of 10 s. The subjects were asked to perform RAM with their dominant hand first. Three trials were recorded in this condition with periods of rest between trials. Then, they were asked to repeat this task but with their non-dominant hand. These conditions represented the single-hand condition. Finally, the subjects were asked to perform RAM with both hands simultaneously, in an in-phase fashion (bimanual condition). For each RAM signal recorded, an automated algorithm identified each peak and trough of the pronation–supination cycles. Two characteristics were calculated: range and velocity. *Range* is the mean angular displacement over a complete pronation–supination cycle in degrees. *Velocity* represents the maximal instantaneous velocity from peak to peak and is provided in degrees per second. Only the first 7 s of the pronation-supination cycles were analysed for RAM performance as fatigue may confound results (Duval et al. 2001, 2006a, b; Ghassemi et al. 2006).

2.4 Statistics

For WBIM, in order to compare *group* (patients vs. controls), *condition* (rest vs. active), and *group x condition* interactions, we used a two-way analysis of variance (ANOVA), with repeated measures for *condition.* Post hoc analysis (Newman-Keuls) was used to indicate which group comparisons yielded significance. *T* tests were used to determine statistical significance between groups for each RAM performance characteristic. The threshold of significance was set at $p < 0.05$ for all aforementioned measures.

3 Results

3.1 Whole-Body Involuntary Movements

Figure 1 illustrates the changes of WBIM amplitude during the rest and active conditions in single-hand (dominant and non-dominant) as well as both-hands RAM trials. ANOVA on WBIM during dominant hand RAM performance revealed a significant effect of *group* ($F = 21.4$, $p < 0.05$), *condition* ($F = 31.9$, $p < 0.05$) and significant *group* × *condition* interaction ($F = 19.3$, $p < 0.05$). Post hoc analysis indicated that the HD group had higher levels of WBIM in both the rest and active conditions, confirming the presence of chorea. Also there was a significant increase

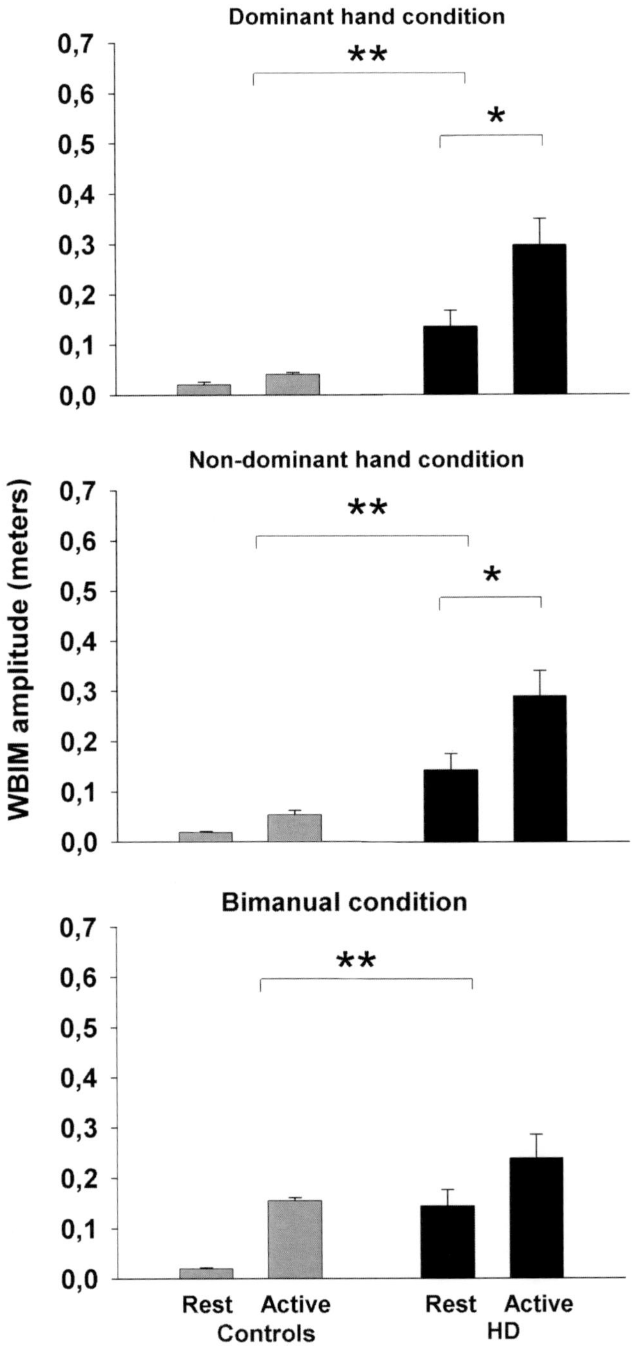

Fig. 1 Comparison of WBIM amplitude changes from rest to active conditions in controls and HD patients when the dominant hand (*top*), non-dominant hand (*middle*), and both hands are performing RAM simultaneously (*bottom*). * indicates significant differences within group. ** indicates significant differences between groups, $p < 0.05$

of WBIM from the rest to active condition in the HD group ($p < 0.05$), but not in the control group.

During non-dominant hand RAM (Fig. 1, middle), ANOVA revealed a significant effect of *group* ($F = 21.7$, $p < 0.05$), *condition* ($F = 25.5$, $p < 0.05$), and *group* × *condition* interaction ($F = 9.6$, $p < 0.05$). Once again, post hoc analysis indicated that the HD group had higher levels of WBIM in both the rest and active conditions. Also there was a significant increase of WBIM from the rest to active condition in the HD group ($p < 0.05$), but not in the control group.

Finally, bimanual RAM (Fig. 1, bottom) ANOVA revealed a significant effect of *group* ($F = 8.7$, $p < 0.05$) and *condition* ($F = 37.6$, $p < 0.05$), but no significant *group* × *condition* interaction ($F = 1.2$, $p > 0.05$). Post hoc analysis indicated that the HD group had higher WBIM amplitude, and that there was a significant increase of WBIM from the rest to active condition in the HD group and in the control group.

3.2 Rapid Alternating Movement

Figure 2 illustrates the RAM performance for *range* (top row) and *velocity* (bottom row). HD subjects had significantly greater *range* than control subjects for the dominant hand ($t = 4.663$, $p < 0.05$) and non-dominant hand ($t = 231.0$, $p < 0.05$) in the single-hand condition. Also, each hand showed greater *range* than controls

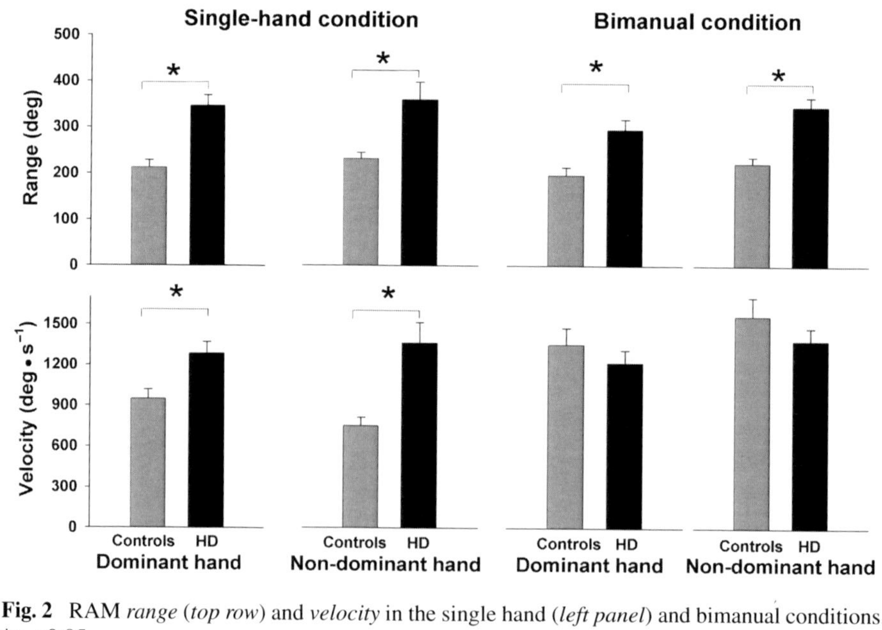

Fig. 2 RAM *range* (*top row*) and *velocity* in the single hand (*left panel*) and bimanual conditions. *$p < 0.05$

with RAM in the bilateral condition (dominant: $t = 3.467$, non-dominant; $t = 5.056$, $p < 0.05$). HD subjects had significantly greater *velocity* than control subjects for the dominant hand ($t = 3.053$, $p < 0.05$) and non-dominant hand ($t = 3.790$, $p < 0.05$) in the single-hand condition. No significant differences were found for either hands between patients and controls in the bimanual condition.

4 Discussion

In this study, we were able to show that the velocity and amplitude of RAM in HD patients were equal or better than that of control subjects, regardless of which hand was used, or whether both hands performed RAM simultaneously. These results support previous findings suggesting that bradykinesia is not a systematic symptom of adult-onset HD (Fenney et al. 2008a).

4.1 The Bradykinesia Issue in HD

Studies suggesting that bradykinesia is a feature of HD are numerous (Hefter et al. 1987; Thompson et al. 1988; Agostino et al. 1992; Phillips et al. 1996; Berardelli et al. 1999; Curra et al. 2000; Garcia-Ruiz et al. 2002; Verbessem et al. 2002; van Vugt et al. 2003, 2004). Our results seem to contradict these findings. However, it is important to note that the SNR discussed previously differed greatly between studies, as well as the level of complexity of the tasks. The previous studies indeed recorded bradykinesia in patients with HD, but it may have come from different sources, such as the aforementioned biomechanical effect, speed-accuracy tradeoff, bradyphrenia, etc. The studies also tested patients presenting with different clinical manifestations, some showing little or no chorea. This fact may indicate a different stage of disease between patients. This difference is especially relevant knowing that aggravation of bradykinesia has been associated with a decrease in chorea, which in turn is related to a worsening of reaction time (vanVugt et al. 2004). In our study, where we tested patients with measurable chorea, no sign of bradykinesia was found. Accordingly, detecting bradykinesia in patients with HD is most probably task dependent, as well as disease-stage dependent.

4.2 Implications of the Present Results for Basal Ganglia Pathophysiology

The fact that involuntary movements such as LID and chorea increase during voluntary movement (Ghassemi et al. 2006; Lemieux et al. 2007; Fenney et al. 2008a; present results) is a sign that both types of motor commands can coexist in the motor stream. In addition, there appears to be a direct relationship between the velocity of the intended movement and the increase in WBIM associated with both LID and chorea, as the RAM task induced a higher increase in WBIM than a slower manual tracking task in both diseases (Ghassemi et al. 2006; Lemieux et al. 2007; Fenney et al. 2008a). One important inference we can make from the present results is that the concomitant presence of involuntary and voluntary motor outputs, which amounts to competing neural commands, does not always generate slowness of movement. The lack of bradykinesia in the HD group, despite quantifiable chorea, attests to that fact.

The aetiology of HD and PD is undeniably different. However, what is not clear is whether a difference exists in the dynamic relationship between involuntary movements (i.e., HD chorea and PD dyskinesia) and voluntary movements in these diseases. The fact that bradykinesia is detected during the RAM task in PD patients with dyskinesia (Ghassemi et al. 2006), but is not observed in patients with HD having chorea (Fenney et al. 2008a; present results) suggests a different interaction between mechanisms generating the release of involuntary and voluntary motor commands. The answer may lie in the underlying pathophysiological alterations within the basal ganglia associated with PD and HD. Indeed, altered sensitivity of GABAergic medium spiny neurons that express predominantly D1 receptors associated with the direct pathway is believed to be mostly responsible for PD dyskinesia (Aubert et al. 2005; Carta et al. 2005; Cenci 2007), while progressive destruction of D2 expressing GABAergic medium spiny striatal neurons associated with the indirect pathway would be responsible for HD chorea (Reiner et al. 1988). This 'not so subtle' difference may explain why bradykinesia is concomitantly present with PD dyskinesia while it is not present with HD chorea. The focused attention model (Mink 1996) suggests that the role of the basal ganglia would be to facilitate the release of proper motor pattern generators (MPGs) while blocking undesirable MPGs. This would be achieved via the select ON pathway [the direct pathway from Delong's model (1990)], which would help in the release of proper motor outputs, while the hyperdirect pathway (cortico–subthalamic–globus pallidus internus) would block undesirable motor outputs and help in the scaling of movement. A control function pathway (striatal–globus pallidus externus–subthalamic pathway) would further help in preventing the release of undesired motor actions. Since striatal indirect pathway neurons are the first casualties of HD, it is quite conceivable that chorea is the consequence of an otherwise blocked motor program generator, as suggested by Mink (2003). The ability to generate RAM in HD patients with chorea probably stems from an intact direct pathway. Then, it is possible that eventual death of direct pathway neurons that is known to occur in late HD (c.f. Joel 2001) may be associated with the appearance of the rigidity and akinetic features in this disease. In PD however, the concomitant presence of bradykinesia may simply be the result of differences in activation of the direct and indirect pathways. In this case, it is conceivable that the indirect pathway would retain some of its characteristics seen in hypokinetic PD. This remains to be investigated. In any case, our results support the idea that a different dynamic relationship between involuntary and voluntary motor acts exists between HD and PD, despite some resemblance in the clinical manifestation of chorea and dyskinesias.

4.3 Clinical Implications of the Present Results

The goal of clinicians, in the treatment approach to both diseases, is to maintain optimal motor performance while avoiding hypo- or hyperkinetic side effects. For instance, the treatment of HD chorea may induce parkinsonian hypokinetic symptoms, in addition to substantial behavioural side effects (Bruneau et al. 2002; Bonelli et al. 2004). Conversely, treatment of PD symptoms may eventually generate hyperkinetic side effects such as dyskinesias. This issue is illustrated in Fig. 3.

The relevance of this study revolves around the clinical cost/benefit analysis of avoiding hyperkinetic symptoms. The present results suggest that any bradykinesia detected after dopamine antagonist administration in HD may be the consequence of drug effect, not simply an aggravation of an underlying slowness of movement that has begun to develop simultaneously with the presence of chorea. In PD, it becomes important to understand the impact of LID in order to better appreciate the clinical cost of switching to dopamine agonists, which are known to be less effec-

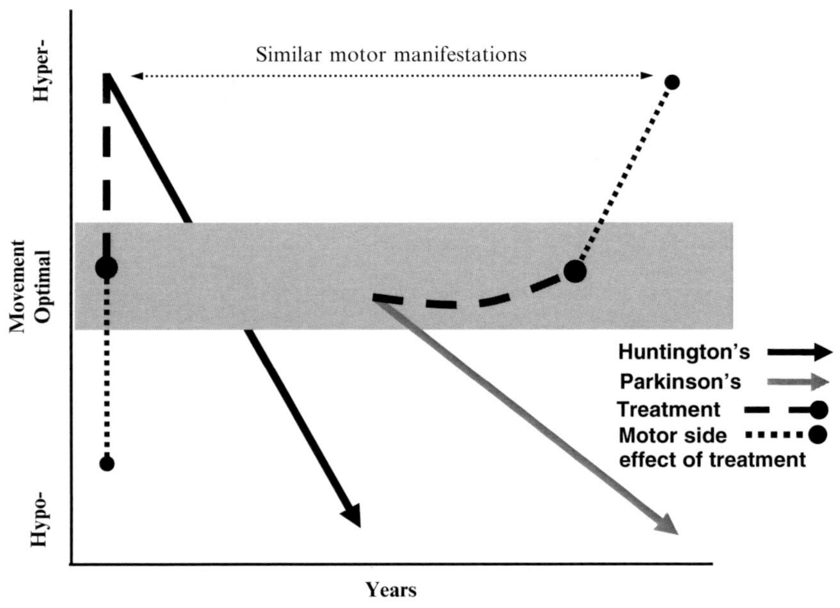

Fig. 3 Progression of motor symptoms of HD and PD. Note that both diseases tend to progress towards hypokinesia, but HD may begin with hyperkinesia. In HD, a major motor side effect of treating chorea is the possible appearance of parkinsonian symptoms that includes bradykinesia. For PD, long-term treatment with exogenous dopaminergic agents will lead to hyperkinetic motor symptoms such as LID. The challenge of treatment for both diseases is to maintain optimal motor function, without overshoot to either hypo- (in HD) or hyperkinetic (in PD) symptoms. Here, we are interested in establishing the dynamic relationship between involuntary and voluntary motor acts when both diseases present with similar clinical manifestations, i.e., WBIM (*horizontal dashed line*)

tive in treating some motor symptoms, and have a higher propensity to the induction of psychological side effects. Past results show that patients with PD do have bradykinesia along with LID (Ghassemi et al. 2006), but that the LID has little direct impact on the performance during a manual tracking task (Lemieux et al. 2007). At what time LID starts to impede voluntary movements and what strategies patients utilize to circumvent LID are under investigation by our group.

5 Conclusions

The present results support the notion that bradykinesia is not a systematic feature of adult-onset HD. Taken together with results obtained in patients with PD having LID, the present findings also support the idea that HD and PD possess distinct dynamic relationships between their respective involuntary movements and voluntary motor acts.

Acknowledgements The authors would like to thank the subjects who graciously agreed to participate in these studies. This research was partially funded by Parkinson Society Canada, the Natural Sciences and Engineering Research Council of Canada, and the Canada Foundation for Innovation.

References

Agostino R, Berardelli A, Formica A, Accornero N and Manfredi M (1992) Sequential arm movements in patients with Parkinson's disease, Huntington's disease and dystonia. Brain 115: 1481–1495.

Aubert I, Guigoni C, Hakansson K, Li Q, Dovero S, Barthe N, Bioulac BH, Gross CE, Fisone G, Bloch B and Bezard E (2005) Increased D1 dopamine receptor signaling in levodopa-induced dyskinesia. Ann Neurol 57: 17–26.

Berardelli A, Noth J, Thompson PD, Bollen ELM, Curra A, Deuschl G, van Dijk G, Töpper R, Schwarz M and Roos RAC (1999) Review: pathophysiology of chorea and bradykinesia in Huntington's disease. Mov Disord 14: 398–403.

Berardelli A, Rothwell JC, Thompson PD and Hallett M (2001) Pathophysiology of bradykinesia in Parkinson's disease. Brain 124: 2131–2146.

Beuter A, Edwards R, deGeoffroy A, Mergler D and Hundnell K (1999a) Quantification of neuromotor function for detection of the effects of manganese. Neurotoxicology 20: 355–366.

Beuter A, de Geoffroy A and Edwards R (1999b) Analysis of rapid alternating movements in Cree subjects exposed to methylmercury and in subjects with neurological deficits. Environ Res 80: 64–79.

Bonelli RM, Wenning GK and Kapfhammer HP (2004) Huntington's disease: present treatments and future therapeutic modalities. Int Clin Psychopharmacol 19: 51–62.

Bruneau MA, Lespérance P and Chouinard S (2002) Catastrophic reactions induced by tetrabenazine. Can J Psychiatry 47: 683.

Burkhard PR, Shale H, Langston JW and Tetrud JW (1999) Quantification of dyskinesia in Parkinson's disease: validation of a novel instrumental method. Mov Disord 14: 754–762.

Carta AR, Tronci E, Pinna A and Morelli M (2005) Different responsiveness of striatonigral and striatopallidal neurons to L-DOPA after a subchronic intermittent L-DOPA treatment. Eur J Neurosci 21: 1196–1204.

Cenci A (2007) Dopamine dysregulation of movement control in L-dopa-induced dyskinesia. Trends Neurosci 30: 236–243.

Cooper JA, Sagar HJ, Tidswell P and Jordan N (1994) Slowed central processing in simple and go/no-go reaction time tasks in Parkinson's disease. Brain 117: 517–529.

Cunnington R, Windischberger C, Deecke L and Moser E (2000) The preparation and execution of self-initiated and externally-triggered movement: a study of event-related fMRI. Neuroimage 15: 373–385.

Curra A, Agostino R, Galizia P, Fittipaldi F, Manfredi M and Berardelli A (2000) Sub-movement cueing and motor sequence execution in patients with Huntington's disease. Clin Neurophysiol 111: 1184–1190.

Delong MR (1990) Primate models of movement disorders of basal ganglia origin. Trends Neurosci 13: 281–285.

Duval C, Panisset M and Sadikot AF (2001) The relationship between physiological tremor and the performance of rapid alternating movements in healthy elderly subjects. Exp Brain Res 139: 412–418.

Duval C, Sadikot AF and Panisset M (2004) The detection of tremor during slow alternating movements performed by patients with early Parkinson's disease. Exp Brain Res 154: 395–398.

Duval C, Sadikot AF and Panisset M (2006a) Bradykinesia in patients with essential tremor. Brain Res 1115: 213–216.

Duval C, Panisset M, Strafella AP and Sadikot AF (2006b) The impact of ventrolateral thalamotomy on tremor and voluntary motor behavior of patient with Parkinson's disease. Exp Brain Res 170: 160–171.

Fahn S, Elton, members of the UPDRS development committee (1987) Unified Parkinson's disease rating scale In: Fahn S, Marsden M, Goldstein M and Calne DB, eds. Recent Developments in Parkinson's Disease, Vol. 2. New York: MacMillan, pp. 153–163.

Fenney A, Jog MS and Duval C (2008a) Bradykinesia is not a "systematic" feature of adult-onset Huntington's disease; implications for basal ganglia pathophysiology. Brain Res 1193: 67–75.

Fenney A, Jog MS and Duval C (2008b) Short-term variability in amplitude and motor topography of whole-body involuntary movements in Parkinson's disease dyskinesias and in Huntington's chorea. Clin Neurol Neurosurg 110: 160–167.

Fimbel EJ, Domingo PP, Lamoureux D and Beuter A (2005) Automatic detection of movement disorders using recordings of rapid alternating movements J Neurosci Methods 146: 183–190.

Garcia-Ruiz PJ, Hernandex J, Cantarero S, Bartolome M, Bernardos VS and de Yebenes JG (2002) Bradykinesia in Huntington's disease: a prospective follow up study. J Neurol 249: 437–440.

Ghassemi M, Lemieux S, Jog M, Edwards R and Duval C (2006) Bradykinesia in patients with Parkinson's disease having levodopa-induced dyskinesias. Brain Res Bull 69: 512–518.

Goetz CG, Stebbins GT, Shale HM, Lang AE, Chernik DA, Chmura TA, Ahlskog JE and Dorflinger EE (1994) Utility of an objective dyskinesia rating scale for Parkinson's disease: inter- and intrarater reliability assessment. Mov Disord 9: 390–394.

Gour J, Edwards R, Lemieux S, Ghassemi M, Jog MS and Duval C (2007) Movement patterns of levodopa-induced dyskinesias in patients with Parkinson's disease. Brain Res Bull 74: 66–74.

Hefter H, Homberg V, Lange HW and Freund HJ (1987) Impairment of rapid movement in Huntington's disease. Brain 110: 585–612.

Hoff JI, van Hilten BJ and Roos RA (1999) A review of the assessment of dyskinesias. Mov Disord 14: 737–743.

Hoff JI, van den Plas AA, Wagemans EA and van Hilten JJ (2001) Accelerometric assessment of levodopa-induced dyskinesias in Parkinson's disease. Mov Disord 16: 58–61.

Joel D (2001) Open interconnected model of basal ganglia-thalamocortical circuitry and its relevance to the clinical syndrome of Huntington's disease. Mov Disord 16: 407–423.

Katzenschlager R, Schrag A, Evans A, Manson A, Carroll CB, Ottaviani D, Lees AJ and Hobart J (2007) Quantifying the impact of dyskinesias in PD: the PDYS-26: a patient-based outcome measure. Neurology 69: 555–563.

Langston JW, Widner H, Goetz CG, Brooks D, Fahn S, Freeman T and Watts R (1992) Core assessment program for intracerebral transplantations (CAPIT). Mov Disord 7: 2–13.

Lemieux S, Ghassemi M, Jog MS, Edwards R and Duval C (2007) The influence of levodopa-induced dyskinesias on manual tracking in patients with Parkinson's disease. Exp Brain Res 176: 465–475.

Mink JW (1996) The basal ganglia: focused selection and inhibition of competing motor programs. Prog Neurobiol 50: 381–425.

Mink JW (2003) The basal ganglia and voluntary movements. Impaired inhibition of competing motor patterns. Arch Neurol 60: 1365–1368.

Okada M and Okada M (1983) A method for quantification of alternate pronation and supination of forearms. Comput Biomed Res 16: 59–78.

Phillips JG, Bradshaw JL, Chui E, Teasdale N, Iansek R and Bradshaw JA (1996) Bradykinesia and movement precision in Huntington's disease. Neuropsychologia 34: 1241–1245.

Reilmann R, Kirsten F, Quinn L, Henningsen H, Marder K and Gordon AM (2001) Objective assessment of progression of Huntington's disease: a three year follow-up study. Neurology 57: 920–924.

Reiner A, Albin RL, Anderson KD, D'Amato CJ, Penney JB and Young AB (1988) Differential loss of striatal projection neurons in Huntington's disease. Proc Natl Acad Sci USA 85: 5733–5737.

Resek G, Haines J and Sainsbury P (1981) An ultrasound technique for the measurement of tardive dyskinesia. Br J Psychiatry 138: 474–478.

Thompson PD, Berardelli A, Rothwell JC, Day BL, Dick JPR, Benecke R and Marsden CD (1988) The coexistence of bradykinesia and chorea in Huntington's disease and its implications for theories of basal ganglia control of movement. Brain 111: 223–244.

van Vugt JPP, Stijl M, Roos RAC and van Dijk JG (2003) Impaired antagonist inhibition may contribute to akinesia and bradykinesia in Huntington's disease. Clin Neurophysiol 114: 295–305.

van Vugt JPP, Piet KKE, Vink LJ, Siesling S, Zwinderman AH, Middelkoop HAM and Roos RAC (2004) Objective assessment of motor slowness in Huntington's disease: clinical correlates and 2nd year follow up. Mov Disord 19: 285–297.

Verbessem P, Eijnde BO, Swinnen SP, Vangheluwe S, Hespel P and Dom R (2002) Unimanual and bimanual voluntary movement in Huntington's disease. Exp Brain Res 147: 529–537.

Wierzbicka MM, Staude G, Wolf W and Dengler R (1993) Relationship between tremor and the onset of rapid voluntary contraction in Parkinson's disease. J Neurol Neurosurg Psychiatry 56: 782–787.

Yanagisawa N and Nezu A (1987) Pathophysiology of involuntary movements in Parkinson's disease. Eur Neurol 26 Suppl 1: 30–40.

Reduced and Modified Neuronal Activity in the Subthalamic Nucleus of Parkinson's Disease Patients with Prior Pallidotomy

Adam Zaidel, Hagai Bergman, and Zvi Israel

Abstract Parkinson's disease (PD) patients with prior pallidotomy [radiofrequency lesions in the internal segment of the globus pallidus (GPi)], whose symptoms have deteriorated, may be candidates for subthalamic nucleus (STN) deep brain stimulation (STN-DBS). In this study we analyzed the microelectrode recordings (MER, used intra-operatively for target verification) from 7 patients with prior pallidotomy (6 unilaterally and 1 bilaterally) and MERs from 12 matched PD patients without prior pallidotomy who underwent bilateral STN-DBS. The MERs were divided into three groups for comparison: (a) ipsilateral and (b) contralateral to prior pallidotomy and (c) from patients with no prior pallidotomy. For each MER trajectory, average, variance and mean successive difference [(MSD), a measure of irregularity] of the root mean square (RMS) of the STN-MER were calculated. The RMS in trajectories ipsilateral to pallidotomy demonstrated significant reduction of the mean average and MSD of STN activity when compared with trajectories from patients without prior pallidotomy, while RMS parameters contralateral to pallidotomy tended to be in between the two. The average power spectral density of 10–20 Hz oscillatory activity in the somatosensory STN was notably lower ipsilateral to pallidotomy, compared with contralateral or without prior pallidotomy. These findings highlight the critical role of direct projections from the basal ganglia to brainstem structures and suggest a possible GPi-STN reciprocal positive-feedback mechanism.

1 Introduction

We recently published a manuscript showing that prior pallidotomy reduces and modifies neuronal activity in the subthalamic nucleus (STN) of Parkinson's disease (PD) patients (Zaidel et al. 2008). This chapter recaps and extends that

A. Zaidel (✉), H. Bergman, and Z. Israel
Department of Physiology and the Interdisciplinary Center for Neural Computation,
The Hebrew University-Hadassah Medical School, Jerusalem, Israel
e-mail: adam@alice.nc.huji.ac.il

H.J. Groenewegen et al. (eds.), *The Basal Ganglia IX*, Advances in Behavioral Biology 58, DOI 10.1007/978-1-4419-0340-2_41, © Springer Science+Business Media, LLC 2009

study by including two additional patients and presenting a more advanced spectral analysis.

Based on the classical studies of Albin et al. (1989) and DeLong (1990), current models of the basal ganglia (BG) propose a closed-loop BG-thalamo-cortical network that facilitates movement via feed-forward direct and indirect striato-pallidal pathways. These models propose that movement is controlled by modulating inhibitory projections from the internal segment of the globus pallidus (GPi) and substantia nigra, pars reticulata (SNr) to the thalamus and explain the pathophysiology of PD by an over-activity of the GPi/SNr. Accordingly, excessive inhibition of the thalamo-cortical network results in the hypokinetic motor deficits typical of PD. The STN, which lies upstream to the GPi/SNr and is noted to have an abnormal increase in activity in parkinsonian monkeys (Wichmann et al. 1994), drives the GPi/SNr by excitation via the indirect pathway. The models are consistent with many experimental findings – for example, inactivation of the GPi (Lozano et al. 1995) or STN (Bergman et al. 1990) alleviates PD symptoms – but are incongruous with others. For example, thalamotomy does not lead to PD-like symptoms, and in addition to alleviating hypokinetic disorders, GPi and STN inactivation alleviate hyperkinetic disorders as well (Obeso et al. 2000).

Surgical treatment of PD includes posteroventral pallidotomy (Svennilson et al. 1960; Laitinen et al. 1992) and deep brain stimulation (DBS) of the GPi and the STN. STN-DBS has proven to be a beneficial and accepted treatment of advanced PD (Limousin et al. 1995; Starr 2002; Machado et al. 2006) and is also effective in patients who have had a prior pallidotomy, but whose symptoms have deteriorated. Kleiner-Fisman et al. (2004) found STN-DBS postpallidotomy to be no less efficacious than STN-DBS in patients without prior pallidotomy, whereas Ondo et al. (2006) reported less motor improvement in patients with prior pallidotomy.

Microelectrode recording (MER) is used to verify localization of the STN physiologically in order to implant a stimulating macroelectrode for STN-DBS (Israel and Burchiel 2004; Gross et al. 2006). The effects of pallidotomy on STN activity require further elucidation due to the apparent discrepancies between previous studies. For unilateral pallidotomy, Mogilner et al. (2002) found that STN single cell activity demonstrated a lower mean firing frequency on the side ipsilateral to prior pallidotomy compared with the contralateral side. In contrast, electrophysiological studies of the STN conducted by Kleiner-Fisman et al. (2004) revealed no significant differences.

We tested whether STN activity ipsilateral to prior pallidotomy is different in comparison to STN activity contralateral to prior pallidotomy in the same patients and to STN activity in patients who had no pallidotomy. We analyzed the raw multi-unit MERs in order to avoid possible errors and biases induced by spike detection and sorting (Lewicki 1998; Joshua et al. 2007). Insight into the effects of prior pallidotomy on STN activity may improve our understanding of the pathophysiology of PD, the mechanisms of its surgical treatments, and the functioning of the BG in general.

2 Methods

2.1 Patients

Nineteen PD patients had bilateral STN-DBS surgery, resulting in two trajectories of recording per patient (one in each hemisphere). Patients with and without prior pallidotomy had an average disease duration of 23 and 10 years respectively without any significant difference in Unified Parkinson's Disease Rating Scale (UPDRS) scores or levodopa equivalent (LDE) medications. The trajectories of recordings in the STN ipsilateral and contralateral to prior pallidotomy (in patients with prior pallidotomy) and those from patients with no prior pallidotomy were classified as ipsilateral, contralateral, and no-pallidotomy trajectories, respectively. In total, 35 trajectories were analyzed from the 19 patients. In 16 patients two trajectories were analyzed per patient. For each of the three remaining patients only one trajectory was used (one ipsilateral and one no-pallidotomy trajectory were missing due to technical difficulties and another no-pallidotomy trajectory was excluded due to STN length shorter than 2 mm. All other trajectories had an STN length > 3 mm). Six patients had received unilateral prior pallidotomies, one had received a bilateral prior pallidotomy, and 12 had received no prior pallidotomy. This resulted in 7 ipsilateral trajectories (due to a missing trajectory – as explained earlier), 6 contralateral trajectories, and 22 no-pallidotomy trajectories (due to an excluded and a missing trajectory – as explained earlier). Pallidotomies were verified on the preoperative magnetic resonance imaging (MRI) scans. The position of the implanted STN-DBS electrodes and the MER trajectory were verified on postoperative computerized tomography (CT) scans fused with the preoperative MRI scans (Framelink 4, Medtronic, Minneapolis, MN).

All patients met accepted selection criteria for STN-DBS and signed informed consent for surgery with MER. All patients were awake during surgery and no sedative was used. The patient's level of awareness was continuously assessed clinically, and when drowsy the patient was stimulated by conversation with a member of the surgical team and aroused. Data were obtained off dopaminergic medications (>12 h since last medication) and during periods of rest, but with no objective measure of tremor or the other PD signs. One of the post-pallidotomy patients (Patient E from Fig. 3) had many tremor episodes. We repeated the RMS analysis of this patient after removing the sections with clear oscillations, but this procedure did not affect the population RMS parameters analyzed in this study.

In this study a single trajectory using one or two microelectrodes was made starting from 10 mm above the estimated (center of STN) target. When two microelectrodes were used, the second was advanced in parallel (2 mm anterior) to the central (aimed at the calculated target) electrode track. Further trajectories were only made if the results of microrecording or macrostimulation were suboptimal in the first. In 27 of the 35 trajectories analyzed for this study (77%), either only one microelectrode was used or there was no STN recorded on the other microelectrode. For the eight remaining trajectories the MER with the greater STN recording span was chosen for analysis (i.e., a single trajectory per hemisphere). In all but one (no-pallidotomy) trajectory, the greater STN span and the recording analyzed were from the central (aimed at the calculated target) electrode track.

This retrospective study was authorized and approved by the Institutional Review Board of Hadassah University Hospital (Ref. code: 2334) in accordance with The Declaration of Helsinki.

2.2 Data Acquisition and Analysis

Data acquisition was conducted with the MicroGuide system (AlphaOmega Engineering, Nazareth, Israel). Neurophysiological activity was recorded via polyamide-coated tungsten microelectrodes (Frederick Haer; impedance = 0.48 ± 0.17 MΩ; measured at 1 KHz, at the beginning of each trajectory). The recorded signal was amplified by 10,000 and band-passed (250–6,000 Hz) using a hardware four-pole Butterworth filter. The signal was sampled at 12–48 KHz, by a 12-bit A/D converter using a ±5 V input range (i.e., ~0.25 μV amplitude resolution). The electrodes were advanced in small discrete steps, from 10 mm above towards the estimated center of the STN. Step size (ranging 500 μm down to 50 μm in our recordings) was controlled by the neurophysiologist in order to achieve optimal unit recording and to identify the upper and lower borders of the STN. Shorter steps were typically used when the electrode was close to the presumed location of the STN. Multiunit segments were recorded (for 5–60 s) following a 2-s signal stabilization period after electrode movement cessation. However, in many cases the human operating-room conditions have resulted in nonstable recording (e.g., due to further movement of brain tissue in relation to the electrode tip or due to neuronal injury). The data segments were, therefore, analyzed offline for stability (Gourevitch and Eggermont 2007) by visual inspection of one of the authors (AZ). When instability was observed, the longest stable section was selected from the recording, discarding the rest. All recordings chosen for further analysis were longer than 1 s (duration mean ± standard deviation: 10.15 ± 9.45 s, $n = 4{,}121$). All the statistics presented in this article use mean ± standard deviation notation.

2.3 The Root Mean Square: An STN Raw Activity Measure

The root mean square (RMS) estimate of the raw multiunit activity recorded by the microelectrode at each electrode depth was used as a measure for evaluating the STN activity. The RMS estimate (about the mean) is defined as follows (1):

$$\mathrm{RMS}(\vec{\mathrm{X}}) = \sqrt{\frac{\sum_{i=1}^{n}(X_i - \mu)^2}{n-1}} \tag{1}$$

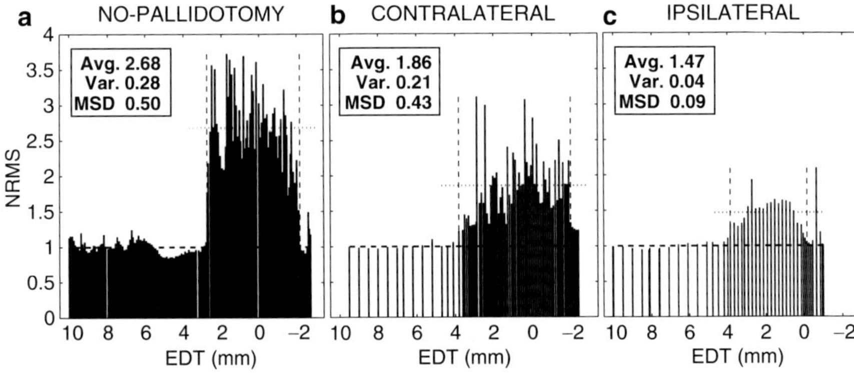

Fig. 1 The Normalized RMS (NRMS) plot of an STN trajectory: Each *solid line* represents the NRMS calculated at discrete steps of estimated distance to target (EDT) in the surgical trajectory. The *vertical dashed lines* indicate the STN borders. The *horizontal dashed and horizontal dotted lines* indicate the normalized baseline and the mean STN-NRMS, respectively. (**a**) An example of a trajectory with no prior pallidotomy, (**b**) a trajectory contralateral to pallidotomy, (**c**) the trajectory ipsilateral to pallidotomy of the same patient in (**b**)

where $\vec{X}$ is the vector of the sampled analog signal with mean μ, X_i is each sample, and n is the number of samples. Since our signal is band-pass filtered, the DC component (μ) is negligible and might exist only because of minor differences between the amplification/filtering system and the sampling/acquisition system. The RMS about the mean (standard deviation) is, therefore, an unbiased and optimal estimator of the RMS of the neuronal activity (Moran et al. 2006). RMS values are susceptible to electrode properties and other external factors (e.g., amplifier gain). Hence, normalization of the RMS is required in order to allow intertrajectory comparison. The RMS for each trajectory was normalized by the average activity from 10 mm estimated distance to target (EDT), till entry into the STN, creating a normalized RMS (NRMS; Fig. 1a–c). Once the electrode enters the STN, the NRMS increases dramatically. This is used as a marker for the point of entry into the STN (Moran et al. 2006). The point of exit was deducted in a likewise, but reverse manner – a decrease in the NRMS and return to the normalized baseline. The NRMS between the point of entry and the point of exit of the STN comprise the STN recordings and were used for RMS comparisons between the trajectories.

2.4 The Mean Successive Difference: A Measure of Irregularity

In comparison to what seemed a smoother NRMS in the STN of ipsilateral trajectories (Fig. 1c), the STN-NRMSs of the no-pallidotomy (and to a lesser degree the contralateral) trajectories (Fig. 1a, b) were more irregular, containing large jumps between consecutive steps in a trajectory. To quantify this lack of smoothness, we used the mean successive difference (MSD) as described by Cranz and Becker (1921) and von Neumann et al. (1941). The MSD is defined as the average absolute difference of the NRMS between two consecutive steps in the STN of a trajectory (2):

$$\mathrm{MSD} = \frac{\sum_{i=Q+1}^{P} \left| \mathrm{NRMS}i - \mathrm{NRMS}_{i-1} \right|}{P-Q} \tag{2}$$

where P and Q are the entry and exit point indexes of the STN, respectively (the EDT descends with trajectory progression).

2.5 Statistical Analysis

In all statistical mean comparisons a one-way analysis of variance (ANOVA) was used across the three trajectory groups (ipsilateral, contralateral, and no-pallidotomy) with statistical significance declared at $p = 0.05$. When ANOVA detected significance, a post hoc multiple comparison test using the Bonferroni correction was performed. The mean of the STN-NRMS average, variance, and MSD were compared as well as average length of the STN. Significance of step size of electrode depth in the STN of the different groups was negated by statistical analysis.

2.6 Spectral Analysis

For spectral analysis, the raw analog signal was rectified by the "absolute" operator and the mean subtracted. This procedure is required in order to expose the frequency band of interest (below 60 Hz) since the original analog data are filtered at 250–6,000 Hz (see section on data acquisition and analysis). The average power spectral density (PSD) was calculated from the top two-thirds of each STN trajectory span and thereby broadly represents the somatosensory (dorsal) area of the STN (Weinberger et al. 2006). Welch's method was used with a 1-s Hamming window and zero padding resulting in a spectral resolution of 1/3 Hz. For each recording, the PSD was normalized by the baseline activity between 55 and 95 Hz. The average PSD for ipsilateral, contralateral, and no-pallidotomy trajectories was calculated.

2.7 Software

Data analysis was carried out on custom software using MATLAB V7.1 (Mathworks, Natick, MA). The software used in this chapter can be found online (http://basalganglia.huji.ac.il/links.htm).

3 Results

Statistically significant differences were found between the ipsilateral and no-pallidotomy STN trajectories when comparing the averages of the NRMS (Fig. 2a) and MSD (Fig. 2c). Although the STN-RMS variance showed similar changes (Fig. 2b), the results were not statistically significant. The contralateral trajectories' mean average, variance, and MSD were in between the ipsilateral and no-pallidotomy trajectory values without statistical significance in either direction. The lack of significance in contralateral trajectory comparisons is attributed both to the small differences (as opposed to ipsilateral vs. no-pallidotomy) and to the small sample size. There was no significant difference in mean STN length between the three trajectories.

3.1 Average Normalized RMS

For each patient we calculated the average value of the STN-NRMS. The mean of the NRMS averages was significantly larger in the no-pallidotomy trajectory recordings (2.26 ± 0.45) than in the ipsilateral recordings: 1.77 ± 0.24; $P < 0.05$. The contralateral mean of the NRMS average was in between the two (1.92 ± 0.47) with no statistical significance in either direction. Figure 2a depicts the mean NRMS average for the different trajectories.

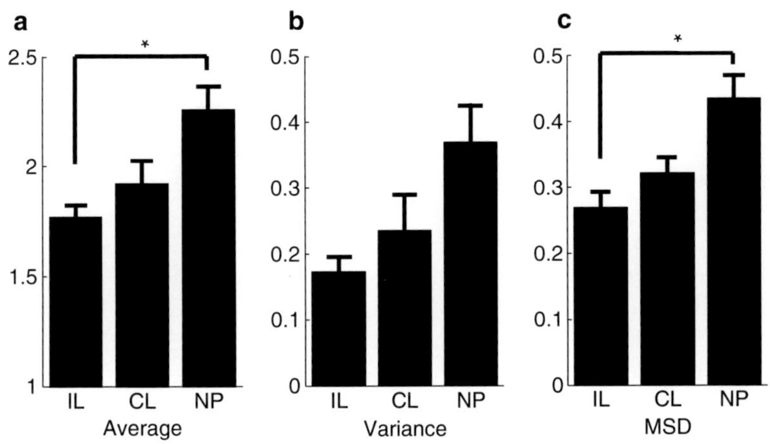

Fig. 2 Intertrajectory comparison: A bar graph comparison of average STN normalized RMS (NRMS) (**a**) average (**b**) variance, and (**c**) MSD across trajectories. Error bars indicate the standard error of the mean (SEM). Differences in average STN-NRMS and MSD between trajectories ipsilateral (IL) to pallidotomy and those with no-pallidotomy (NP) were significant (*$P < 0.05$). Contralateral (CL) trajectory values were systematically in between the values obtained for IL and NP trajectories

3.2 Variance of Normalized RMS

There were no statistically significant differences seen in the variance of the NRMS, but the trend across trajectories was the same as that seen in the average NRMS (Fig. 2b), i.e., the mean of the NRMS variance tends to be larger in the no-pallidotomy trajectory recordings (0.37 ± 0.26) than in the ipsilateral recordings: (0.17 ± 0.09). The contralateral mean NRMS variance was in between the two (0.24 ± 0.24). Figure 2b depicts the mean NRMS variance for the different trajectories.

3.3 Irregularity (MSD) of Normalized RMS

The irregularity of the NRMS was estimated by the MSD (see Sect. 2.4). The mean MSD was significantly larger in the no-pallidotomy trajectory recordings (0.43 ± 0.16) than in the ipsilateral recordings: 0.27 ± 0.10; $P < 0.05$. The contralateral mean MSD (0.32 ± 0.11) was in between the two with no significant difference in either direction. Figure 2c depicts the mean MSD of the NRMS for the different trajectories. The fact that MSD showed significant differences ($P < 0.05$) despite the nonsignificant differences in variance ($P = 0.3$) emphasizes that STN activity in no-pallidotomy trajectories is characterized by significant irregularity rather than by a larger span of RMS values.

3.4 Ipsilateral vs. Contralateral STN Activity in Cases with Unilateral Pallidotomy

The patients with unilateral pallidotomy provide an opportunity to study ipsilateral vs. contralateral trajectories in the same subject. This is particularly attractive due to considerations of control and symmetry. For five out of the six patients with unilateral pallidotomy this comparison could be made (for one patient the ipsilateral trajectory was missing due to technical problems). A bar graph comparison of the average normalized STN activity (NRMS) indicates that the average NRMS on the contralateral side was increased in four out of the five patients when compared with the ipsilateral side (Fig. 3). Although differences in the mean average NRMS were not statistically significant (probably due to small sample size), a trend of reduced activity ipsilateral to pallidotomy was evident. Interestingly, it was noted that Patient E seemed to have considerably lower (tremor dominant) UPDRS scores and levodopa-equivalent medications than the other patients. This clinical observation may explain the difference between patient E and the other patients and should be tested in future studies of these patients.

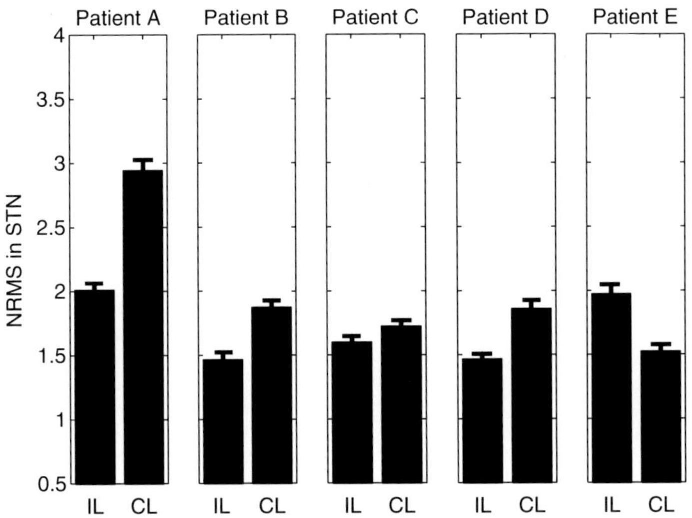

Fig. 3 Unilateral pallidotomy patients: Individual bar graph comparisons of STN normalized RMS (NRMS) for five patients with unilateral pallidotomy. Error bars indicate the standard error of the mean (SEM). For four out of the five patients the mean NRMS in STN was lower ipsilateral (IL) to pallidotomy when compared with the contralateral (CL) side

3.5 Spectral Analysis

The PSD across the STN of a no-pallidotomy patient is presented in Figure 4a. The oscillatory activity in the sensorimotor area of the STN (top two-thirds of Fig. 4a) can be clearly discerned. Periodic oscillations are evident in the average PSD (calculated for the top two-thirds of each STN) of no-pallidotomy trajectories at 10 and 15 Hz, contralateral trajectories at 5 and 14 Hz, and ipsilateral trajectories at 13 Hz (Fig. 4b). The average PSD of beta-range oscillatory activity is notably lowest for the ipsilateral trajectories, highest for the no-pallidotomy trajectories, and in between for the contralateral trajectories.

4 Discussion

This study has three main findings: The STN ipsilateral to pallidotomy has (1) altered RMS characteristics (increased regularity of successive measures), (2) reduced activity (lower RMS average), and (3) reduced beta range oscillatory activity in the dorsal somatosensory area. In extracellular recording, spike size is a function of the size and geometrical structure of the cells, the density of the cells' excitable channels, the cell environment (other cells and extracellular medium), as well as the electrode properties and the distance between the electrode and the cells (Gold et al. 2006). We have assumed that the many factors influencing the spike

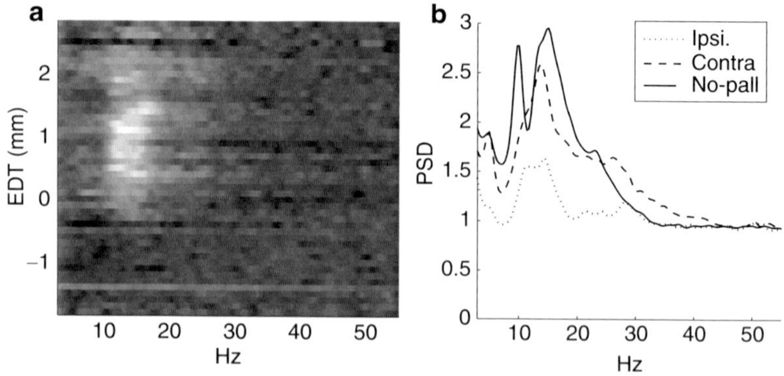

Fig. 4 Power spectral density (PSD): (**a**) A typical example of the PSD across the STN from a patient without prior pallidotomy. Power is coded by *gray level* (*white* is for high power). Expected distance to target (EDT) as in Fig. 1, with only the STN shown here. The somatosensory STN (top two-thirds) has increased beta-range oscillatory activity. (**b**) The average PSD of beta-range oscillatory activity in the top two-thirds (somatosensory) STN is notably lowest for the ipsilateral trajectories, highest for the no-pallidotomy trajectories, and in between for the contralateral trajectories

waveform average to a common value (central limit theory) and therefore our multiunit RMS measure represents the mean firing rate, cell density, or most probably both parameters. We consequently refer to the RMS as a general measure of population activity.

Our first finding of reduced MSD in the ipsilateral trajectories indicates that the STN-RMS ipsilateral to pallidotomy is less volatile and more constant. In contrast, a high irregularity in the STN population activity such as that observed in the no-pallidotomy trajectories (high MSD) could suggest high STN synchronization. Population synchrony in the STN could induce diverse states of both higher and lower RMS activity causing a large MSD, whereas fully asynchronous activity would average to a more homogenous RMS with low MSD. This is in line with the reduced oscillatory activity observed in the PSD of ipsilateral trajectories (Fig. 4b). Our observation of lower MSD and PSD of oscillatory activity ipsilateral to pallidotomy could suggest that pallidotomy reduces STN synchrony. STN synchrony has been documented as a core feature in the pathophysiology of PD (Wichmann et al. 1994; Levy et al. 2000; Brown et al. 2001; Priori et al. 2004; Fogelson et al. 2005). To verify this, STN synchrony with and without pallidotomy would need to be analyzed and compared. However, the present limitations on the duration and quality of recording in human patients do not allow this type of study on our data.

The findings of altered STN-RMS characteristics and reduced STN activity ipsilateral to pallidotomy could be the results of two separate phenomena or outcomes of the same underlying dynamics. Coupling between oscillatory activity and increased firing rate has been descried in the GPi (Bergman et al. 1994) and in the STN (Levy et al. 2000). In addition, neural simulations also indicated that average neuronal population activity and synchrony are tightly coupled (Chawla et al. 1999), and de la Rocha et al. (2007) showed that correlation between neural spike trains increases with firing rate. Hence, if the reduced MSD is a result of a reduction of synchrony as was proposed earlier, it could be suggested that the two findings come in conjunction and share a common pathway and mechanism. We suggest two possible explanations for reduced STN activity and synchronization ipsilateral to pallidotomy:

1. A major input to the GPi is the STN. Pallidotomy could cause retrograde degeneration of this massive STN to GPi projection resulting in the death of a large portion of the STN glutaminergic cell population. This could reduce overall STN activity and hence multiunit RMS measurements. Mogilner et al. (2002) found reduced STN cell density ipsilateral to pallidotomy when compared with the contralateral STN; however, this was not statistically significant. Hence, further (animal) studies are required to validate the STN retrograde degeneration hypothesis.
2. It is possible that pallidotomy reduces STN activity by interrupting a GPi-STN reciprocal positive-feedback loop, for example, via a (polysynaptic) excitatory GPi-STN pathway. This explanation would require modification of the classical direct and indirect pathway models of the BG cortical network, since within the context of these models the GPi is considered to be downstream from the STN. Via the entire BG-thalamo-cortical closed loop, reduced GPi activity would facilitate thalamo-cortical network activity. If we were to assume that the principal (projection) neurons of the cortex are the sole targets of thalamic axons, then this would lead to a net increase of STN activity and not a decrease as was found. This is based on excitatory glutaminergic projections from the cortex to the striatum, and γ-amino butyric acid (GABA)ergic projections from the striatum to the external segment of the globus pallidus (GPe) and from the GPe to the STN (the classically described indirect pathway). The effects of the hyperdirect cortical-STN pathway would lead to the same result. If on the other hand, the thalamo-cortical projections were to activate a great number of GABAergic interneurons, then increased thalamo-cortical activity could effectively reduce the activity of principal neurons of the cortex as described recently by Paz et al. (2007). This would result in decreased activity in the STN as was found in this study. Since the STN closes this loop by projecting back to the GPi, a possible (long loop) reciprocal positive-feedback mechanism can be suggested.

   An alternative (short loop) GPi-STN reciprocal positive-feedback mechanism can be suggested based on GPi to STN projections (Parent and Parent 2004) and significant rebound properties of STN neurons. GPi to STN projections were usually negated by previous anatomical studies of the basal ganglia; however, future and more sensitive axon-tracing studies might shed light on the possible existence of such GPi to STN connections. Postinhibitory rebound firing has been described in the dorsolateral thalamus of the songbird (Luo and Perkel 1999), and burst firing in STN neurons may be driven by rebound depolarization by pallidal fibers (Hallworth and Bevan 2005). Pallidotomy can possibly reduce STN burst activity by breaking a long or short GPi-STN reciprocal positive feedback loop (or both).

The reduced motor improvement of STN-DBS postpallidotomy found by Ondo et al. (2006) in conjunction with our results and those of Mogilner et al. (2002) of reduced STN activity ipsilateral to pallidotomy could suggest that clinical improvement of STN-DBS is associated with amelioration of STN overactivity or synchrony. It could be explained accordingly that STN-DBS will be more effective on a "more pathological" STN. This would also explain why the postpallidotomy patients in the Kleiner-Fisman et al. (2004) study, which showed no reduction in STN activity ipsilateral to pallidotomy, also demonstrated comparable motor improvement and efficaciousness of STN-DBS. It would seem plausible to suggest that pallidotomy reduces STN activity in a once off (set-back) manner, but does not break the continual increase associated with disease progression. Hence, pallidotomy has a palliative effect, and several years later, further treatment (such as STN-DBS) may be required.

The fact that STN-DBS is efficacious postpallidotomy (Kleiner-Fisman et al. 2004; Ondo et al. 2006) in conjunction with the classical BG models would lead to the notion that at least part of the effects of STN-DBS are mediated via the SNr. Nevertheless, prior GPi lesion (pallidotomy) should cause reduced effectiveness of STN-DBS. The fact that Kleiner-Fisman et al. (2004) found no difference in motor improvement and Ondo et al. (2006) found no difference in dyskinesia reduction with/without pallidotomy suggests that DBS antidromic activation of the cortex (Li et al. 2007) or direct projections from the STN to brainstem structures could play a more critical role than previously thought. This is in agreement with the many recent reports that suggest a significant role for the upper brainstem, and the pedunculopontine nucleus (PPN) in particular, in BG physiology and PD pathophysiology (Kojima et al. 1997; Aziz et al. 1998; Pahapill and Lozano 2000; Nandi et al. 2002).

Increased STN and GPi/SNr activity is a hallmark of PD (Wichmann and DeLong 1996) and it is well known that STN inactivation reduces GPi activity (Wichmann et al. 1994). This, together with the potential explanation of our results that pallidotomy reduces STN activity by interrupting a GPi-STN positive-feedback loop, could shed light on our understanding of overactivity and synchrony within the BG. Excessive positive feedback could aggravate GPi-STN overactivity and cause increased GPi-STN synchronization such as that observed by Brown et al. (2001). This is because volatility and excessive oscillatory activity are innate in a positive feedback loop (and would be characterized by high MSD). Breaking this loop by pallidotomy could account for the reduced activity, lower STN irregularity, and reduced beta-range oscillations seen ipsilateral to pallidotomy in this study.

5 Conclusions

We suggest that pallidotomy sets PD symptoms back by partially breaking a GPi-STN positive feedback loop. This is in agreement with our results of reduced STN-RMS and oscillatory activity and reduced MSD postpallidotomy. In this case, part of the therapeutic effect of pallidotomy could come from reduced activity and synchrony in the STN. We propose that with disease progression, STN overactivity and synchrony continue to deteriorate, and that bilateral STN-DBS further aids in alleviating PD symptoms by directly targeting the STN. STN-DBS is efficacious even postpallidotomy either via back-propagation to the cortex or STN direct projections to brainstem structures such as the PPN and not exclusively through the GPi/SNr.

Acknowledgments We thank Anan Moran for help in data acquisition, and Ya'acov Ritov and Israel Nelken for help with the statistical analysis. This research was supported in part by the "Fighting Against Parkinson" Foundation of the Hebrew University Netherlands Association (HUNA).

References

Albin RL, Young AB and Penney JB (1989) The functional anatomy of basal ganglia disorders. Trends Neurosci 12: 366–375.

Aziz TZ, Davies L, Stein J and France S (1998) The role of descending basal ganglia connections to the brain stem in parkinsonian akinesia. Br J Neurosurg 12: 245–249.

Bergman H, Wichmann T and DeLong MR (1990) Reversal of experimental parkinsonism by lesions of the subthalamic nucleus. Science 249: 1436–1438.

Bergman H, Wichmann T, Karmon B and DeLong MR (1994) The primate subthalamic nucleus. II. Neuronal activity in the MPTP model of parkinsonism. J Neurophysiol 72: 507–520.

Brown P, Oliviero A, Mazzone P, Insola A, Tonali P and Di Lazzaro V (2001) Dopamine dependency of oscillations between subthalamic nucleus and pallidum in Parkinson's disease. J Neurosci 21: 1033–1038.

Chawla D, Lumer ED and Friston KJ (1999) The relationship between synchronization among neuronal populations and their mean activity levels. Neural comput 11: 1389–1411.

Cranz C and Becker K (1921) Exterior Ballistics (translated from 2nd German edition). London: H.M.'s Stationery Office.

de la Rocha J, Doiron B, Shea-Brown E, Josic K and Reyes A (2007) Correlation between neural spike trains increases with firing rate. Nature 448: 802–806.

DeLong MR (1990) Primate models of movement disorders of basal ganglia origin. Trends Neurosci 13: 281–285.

Fogelson N, Pogosyan A, Kuhn AA, Kupsch A, van Bruggen G, Speelman H, Tijssen M, Quartarone A, Insola A, Mazzone P, Di L, Limousin P and Brown P (2005) Reciprocal interactions between oscillatory activities of different frequencies in the subthalamic region of patients with Parkinson's disease. Eur J Neurosci 22: 257–266.

Gold C, Henze DA, Koch C and Buzsaki G (2006) On the origin of the extracellular action potential waveform: a modeling study. J Neurophysiol 95: 3113–3128.

Gourevitch B and Eggermont JJ (2007) A simple indicator of nonstationarity of firing rate in spike trains. J Neurosci Methods 163: 181–187.

Gross RE, Krack P, Rodriguez-Oroz MC, Rezai AR and Benabid AL (2006) Electrophysiological mapping for the implantation of deep brain stimulators for Parkinson's disease and tremor. Mov Disord 21 Suppl 14: S259–S283.

Hallworth NE and Bevan MD (2005) Globus pallidus neurons dynamically regulate the activity pattern of subthalamic nucleus neurons through the frequency-dependent activation of postsynaptic GABAA and GABAB receptors. J Neurosci 25: 6304–6315.

Israel Z and Burchiel K (2004) Microelectrode Recording in Movement Disorder Surgery. Stuttgart: Thieme.

Joshua M, Elias S, Levine O and Bergman H (2007) Quantifying the isolation quality of extracellularly recorded action potentials. J Neurosci Methods 163: 267–282.

Kleiner-Fisman G, Fisman DN, Zamir O, Dostrovsky JO, Sime E, Saint-Cyr JA, Lozano AM and Lang AE (2004) Subthalamic nucleus deep brain stimulation for parkinson's disease after successful pallidotomy: clinical and electrophysiological observations. Mov Disord 19: 1209–1214.

Kojima J, Yamaji Y, Matsumura M, Nambu A, Inase M, Tokuno H, Takada M and Imai H (1997) Excitotoxic lesions of the pedunculopontine tegmental nucleus produce contralateral hemiparkinsonism in the monkey. Neurosci Lett 226: 111–114.

Laitinen LV, Bergenheim AT and Hariz MI (1992) Leksell's posteroventral pallidotomy in the treatment of Parkinson's disease. J Neurosurg 76: 53–61.

Levy R, Hutchison WD, Lozano AM and Dostrovsky JO (2000) High-frequency synchronization of neuronal activity in the subthalamic nucleus of parkinsonian patients with limb tremor. J Neurosci 20: 7766–7775.

Lewicki MS (1998) A review of methods for spike sorting: the detection and classification of neural action potentials. Network 9: R53–R78.

Li S, Arbuthnott GW, Jutras MJ, Goldberg JA and Jaeger D (2007) Resonant antidromic cortical circuit activation as a consequence of high-frequency subthalamic deep-brain stimulation. J Neurophysiol 98: 3525–3537.

Limousin P, Pollak P, Benazzouz A, Hoffmann D, Broussolle E, Perret JE and Benabid AL (1995) Bilateral subthalamic nucleus stimulation for severe Parkinson's disease. Mov Disord 10: 672–674.

Lozano AM, Lang AE, Galvez Jimenez N, Miyasaki J, Duff J, Hutchinson WD and Dostrovsky JO (1995) Effect of GPi pallidotomy on motor function in Parkinson's disease. Lancet 346: 1383–1387.

Luo M and Perkel DJ (1999) A GABAergic, strongly inhibitory projection to a thalamic nucleus in the zebra finch song system. J Neurosci 19: 6700–6711.

Machado A, Rezai AR, Kopell BH, Gross RE, Sharan AD and Benabid AL (2006) Deep brain stimulation for Parkinson's disease: surgical technique and perioperative management. Mov Disord 21: S247–S258.

Mogilner AY, Sterio D, Rezai AR, Zonenshayn M, Kelly PJ and Beric A (2002) Subthalamic nucleus stimulation in patients with a prior pallidotomy. J Neurosurg 96: 660–665.

Moran A, Bar-Gad I, Bergman H and Israel Z (2006) Real-time refinement of subthalamic nucleus targeting using Bayesian decision-making on the root mean square measure. Mov Disord 21: 1425–1431.

Nandi D, Stein JF and Aziz TZ (2002) Exploration of the role of the upper brainstem in motor control. Stereotact Funct Neurosurg 78: 158–167.

Obeso JA, Rodriguez-Oroz MC, Rodriguez M, DeLong MR and Olanow CW (2000) Pathophysiology of levodopa-induced dyskinesias in Parkinson's disease: problems with the current model. Ann Neurol 47: S22–S32.

Ondo WG, Silay Y, Almaguer M and Jankovic J (2006) Subthalamic deep brain stimulation in patients with a previous pallidotomy. Mov Disord 21: 1252–1254.

Pahapill PA and Lozano AM (2000) The pedunculopontine nucleus and Parkinson's disease. Brain 123: 1767–1783.

Parent M and Parent A (2004) The pallidofugal motor fiber system in primates. Parkinsonism Relat Disord 10: 203–211.

Paz JT, Chavez M, Saillet S, Deniau JM and Charpier S (2007) Activity of ventral medial thalamic neurons during absence seizures and modulation of cortical paroxysms by the nigrothalamic pathway. J Neurosci 27: 929–941.

Priori A, Foffani G, Pesenti A, Tamma F, Bianchi AM, Pellegrini M, Locatelli M, Moxon KA and Villani RM (2004) Rhythm-specific pharmacological modulation of subthalamic activity in Parkinson's disease. Exp Neurol 189: 369–379.

Starr PA (2002) Placement of deep brain stimulators into the subthalamic nucleus or globus pallidus internus: technical approach. Stereotact Funct Neurosurg 79: 118–145.

Svennilson E, Torvik A, Lowe R and Leksell L (1960) Treatment of Parkinsonism by stereotactic thermolesions in the pallidal region. A clinical evaluation of 81 cases. Acta Psychiat Neurol Scand 35: 358–377.

von Neumann J, Kent RH, Bellinson HR and Hart BI (1941) The mean square successive difference. Ann Math Stat 12: 153–162.

Weinberger M, Mahant N, Hutchison WD, Lozano AM, Moro E, Hodaie M, Lang AE and Dostrovsky JO (2006) Beta oscillatory activity in the subthalamic nucleus and its relation to dopaminergic response in Parkinson's disease. J Neurophysiol 96: 3248–3256.

Wichmann T and DeLong MR (1996) Functional and pathophysiological models of the basal ganglia. Curr Opin Neurobiol 6: 751–758.

Wichmann T, Bergman H and DeLong MR (1994) The primate subthalamic nucleus. III. Changes in motor behavior and neuronal activity in the internal pallidum induced by subthalamic inactivation in the MPTP model of parkinsonism. J Neurophysiol 72: 521–530.

Zaidel A, Moran A, Marjan G, Bergman H and Israel Z (2008) Prior pallidotomy reduces and modifies neuronal activity in the subthalamic nucleus of Parkinson's disease patients. Eur J Neurosci 27: 483–491.

Inhibition of Neuronal Firing in the Human Substantia Nigra Pars Reticulata in Response to High-Frequency Microstimulation Aids Localization of the Subthalamic Nucleus

Myriam Lafreniere-Roula, William D. Hutchison, Mojgan Hodaie, Andres M. Lozano, and Jonathan O. Dostrovsky

Abstract Deep brain stimulation (DBS) has been playing an increasing role in the treatment of various movement disorders with the most common being Parkinson's disease (PD). Currently, the preferred target for treating PD motor symptoms is the subthalamic nucleus (STN). Microelectrode recordings are frequently used to aid in the determination of the optimal target for implanting the DBS electrode. Changes in cellular activity are typically used to detect the borders of the STN during microelectrode recordings. Although the dorsal border of the STN is usually clear, its ventral border with the substantia nigra pars reticulata (SNr) is sometimes more difficult to identify. Our previous studies of the effects of microstimulation in the internal globus pallidus (GPi), which is functionally similar to SNr, and STN revealed that firing in GPi but not STN neurons is readily inhibited by low current stimulation through an adjacent microelectrode. The aim of the current study was to examine and compare the aftereffects of local high-frequency microstimulation through the recording electrode on the firing of STN and SNr neurons to determine whether this might be a useful technique for differentiating STN from SNr. Neurons in the SNr and STN were identified as well isolated high-amplitude spikes and were stimulated extracellulary through the recording microelectrode with 0.5-s trains of high-frequency (200 Hz) and low current (<5 μA). In the majority (89%) of SNr neurons, this type of stimulation led to a period of inhibition lasting several hundreds of milliseconds following the end of the train. A much smaller proportion of STN neurons (9%) was similarly affected by this type of stimulation. These findings indicate that almost all SNr neurons but few STN neurons display prolonged inhibition following very low current microstimulation through the recording electrode. This characteristic provides a useful additional finding that can be used to identify the border between STN and SNr.

M. Lafreniere-Roula, W.D. Hutchison, M. Hodaie, A.M. Lozano, and J.O. Dostrovsky (✉)
Department of Physiology, University of Toronto, Medical Sciences Building Room 3302, 1 King's College Circle, Toronto, ON, Canada M5S 1A8
e-mail: j.dostrovsky@utoronto.ca

H.J. Groenewegen et al. (eds.), *The Basal Ganglia IX*, Advances in Behavioral Biology 58,
DOI 10.1007/978-1-4419-0340-2_42,

1 Introduction

1.1 Parkinson's Disease and the Basal Ganglia

Parkinson's disease (PD) is characterized by a combination of tremor, bradykinesia, and akinesia, which are thought to arise as a consequence of the degeneration of dopaminergic neurons in the substantia nigra pars compacta (Jellinger 1987). The mainstay of therapy is dopamine replacement via the administration of the dopamine precursor levodopa or dopaminergic agonists. Following long-term dopaminergic therapy, however, patients frequently develop uncontrollable movements (dyskinesias) that can become quite severe.

High-frequency stimulation of the subthalamic nucleus (STN) termed as STN-HFS has been increasingly used to relieve PD symptoms in cases resistant to pharmacotherapy or where medication-induced dyskinesias are very severe [reviewed in Benabid (2003)]. Although STN-HFS deep brain stimulation (DBS) is efficacious, the mechanisms by which it works are not yet clear. Since the clinical effects of DBS are similar to those produced by a lesion of the same region, it was initially thought that high-frequency DBS inhibited neuronal activity [e.g., see review by Dostrovsky et al. (2002)]. In vitro studies and modeling have suggested, however, that HFS can also excite local neuron populations. Thus, the effect of HFS may be to not only inhibit pathological activity but also to replace it with a regular firing pattern or a combination of these effects [discussed in Filali et al. (2004), Garcia et al. (2005), and Liu et al. (2008)].

In addition to the variety of possible effects of DBS on local cellular activity (see Garcia et al. 2005; Liu et al. 2008), the difficulty in understanding the effects of STN-HFS is also due to the complexity of the basal ganglia circuitry. The classical model to describe activity within basal ganglia circuits is centered around two pathways, the direct and indirect pathways, from the striatum to the internal segment of the globus pallidus (GPi) and the substantia nigra pars reticulata (SNr). The GPi and SNr are the two output nuclei through which the basal ganglia circuitry exerts its influence on other brain structures. The SNr and GPi share many similar features in terms of their afferents as well as the physiological characteristics of their constituent neurons. Importantly, both nuclei mainly comprise GABAergic neurons and receive major GABAergic inputs from the putamen and the external segment of the globus pallidus (GPe) [reviewed in Smith et al. (1998)].

1.2 Effects of High-Frequency Stimulation

To better understand the mechanisms underlying the effects of high-frequency DBS in STN and GPi (the usual target for DBS in dystonia patients), our group has studied the effects of microstimulation on the activity of neurons in these nuclei in patients undergoing stereotactic surgery for DBS electrode implantation. In these studies, single stimuli and short trains of stimuli were delivered through an electrode placed about 600 μm away from the recording electrode. In the first study in GPi (Dostrovsky et al. 2000) we found that the firing of almost all of the neurons was inhibited for about 15–25 ms after each pulse near threshold and that the onset of the inhibition was usually very short (shorter than the stimulation artifact caused by each current pulse). Intensities of 25 μA or lower were frequently effective in eliciting the inhibition. Increasing the intensity of stimulation resulted in inhibition lasting up to about 100 ms in some cells. Increasing the frequency of stimulation gave rise to a more pronounced suppression of firing which outlasted the period of stimulation but the aftereffects of high-frequency trains (100–300 Hz) were not studied. We have proposed that the short duration inhibitory effects induced by single pulses are due to stimulation-induced release of gamma-aminobutyric acid (GABA) from the terminals of neurons in the putamen and the GPe. In contrast, single low-intensity stimuli were generally ineffective in inhibiting STN neurons (Filali et al. 2004). Stimulation at high frequencies (100–300 Hz) and intensities was found to produce inhibition following the stimulus train in 42% of the 60 cells tested. Furthermore, in 44% of the cases where HFS produced inhibition there was an early inhibition followed by rebound excitation and a further inhibitory period, suggesting that the inhibitions observed are due to hyperpolarization, although the mechanism is unclear (see Garcia et al. 2005).

In the present study, we investigated and compared the aftereffects of HFS in the STN and the SNr. We hypothesized that SNr-HFS would give rise to more robust inhibition and at much lower stimulation thresholds than STN-HFS and that this feature could be used to help determine the boundary between these two nuclei during surgery.

2 Methods

2.1 Surgery and Recordings

The data presented in this study were obtained from 28 awake patients undergoing surgery for uni- or bilateral implantation of DBS electrodes in the STN to reduce their parkinsonian motor symptoms. All procedures were approved by the University Health Network Ethical Review Board at the University of Toronto and patients gave written informed consent. The patients were withdrawn from dopaminergic medication for 12 h preceding the surgery. Surgical procedures have been described in detail elsewhere (Lozano et al. 1996). Briefly, a stereotactic frame was fixed under local anesthesia and a burr hole made in the skull on each side to be implanted. A preliminary target was determined using imaging data and a guide tube was inserted, aimed at this preliminary target. A pair of gold and platinum-plated tungsten microelectrodes (25 μm exposed tip) was driven through the thalamus, STN, and SNr, with each electrode being controlled by an independent hydraulic microdrive (Levy et al. 2007). Lateral separation of the electrodes was approximately 600 μm. Data were recorded using Guideline 3,000 system amplifiers (Axon Instruments) and captured digitally using Spike2 software (Cambridge Electronic Devices).

The dorsal border of the STN was identified intraoperatively as the depth at which the baseline noise increased, indicating higher cellular density and firing rates (Hutchison et al. 1998). The ventral border of STN was assumed to be the site of the last irregularly firing neuron preceding a gap of usually at least 1 mm before encountering the high frequency (about 70 Hz in SNr compared with 25–45 Hz in STN), more regularly firing neurons characteristic of SNr (Hutchison et al. 1998). However, in some cases this sequence was not clear and there was some uncertainty with respect to the ventral border of STN and dorsal border of SNr (see Sect. 4). The location of cells examined was plotted on Schaltenbrand and Wahren atlas maps (Schaltenbrand and Wahren 1977) scaled to the anatomical landmarks (anterior commissure and posterior commissure) specific to each patient.

Only large amplitude, well-isolated spikes within each nucleus were selected for microstimulation tests. Stimulation was performed first with a 0.5-s train at 200 or 300 Hz. Initial current intensity was usually 2 or 3 μA. Stimulation at higher intensities was tried if there was no inhibition with the original trial but currents did not exceed 10 μA. In general, stimulation in STN and SNr at intensities higher than 10 μA appeared to damage the neuron or at least lead to long-term (many seconds/minutes) cessation of firing and was avoided.

2.2 Analysis

Recordings were high-pass filtered (200 Hz) prior to analysis with Spike2 software (Cambridge Electronic Devices, Cambridge, UK). The response to stimulation was examined on neurons recorded by the stimulating electrode itself. Because of the required switch between the recording and stimulating modes of the amplifier, there is a period of approximately 200 ms (177 or 207 ms depending on which of the two amplifiers was used) following the end of the stimulus train where recording is not possible. This period is included in the measurement of inhibition duration, which was measured as the time between the last current pulse in the train (assessed by looking at the stimulation artifact on a second microelectrode) and the first action potential recorded on the stimulating electrode after the return of the amplifier to the recording mode. This measurement assumes that the inhibition observed after the return of the amplifier to the recording mode was also present while the amplifier was still in stimulating mode although no current was delivered. This is a reasonable assumption given that it has been shown, using stimulation from a second, nearby electrode (and thus where recordings are available throughout the stimulation period) that inhibition is present from the termination of the train and sustained for several hundreds of ms [for example, see Fig. 1a in Filali et al. (2004)]. When multiunit activity was present, duration of inhibition was measured for the largest unit in the recording.

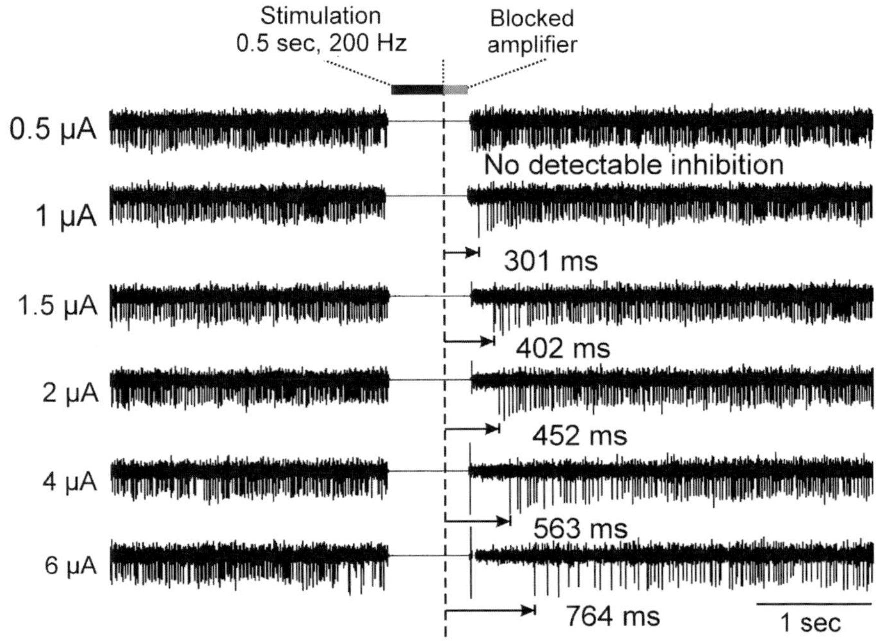

Fig. 1 Example of microstimulation-induced inhibition of an SNr neuron. Traces show raw extracellular recordings of a single SNr neuron stimulated at a variety of currents while keeping the train frequency at 200 Hz and the train duration at 0.5 s. The time at which the train was delivered is shown by the *black box* on top. The *gray box* indicates the period, following the cessation of the train, during which recording is not possible since stimulation was performed through the recording electrode and the amplifier has a delay of about 200 ms in switching from the stimulating mode to the recording mode. Firing was inhibited with stimuli of 1 µA and larger. Inhibition lasted from about 300 to over 700 ms depending on the amount of current that was applied

3 Results

We analyzed the effects of microstimulation on the firing of 53 neurons assumed to be in the SNr and 64 neurons in the STN of 28 PD patients.

Low-intensity, high-frequency stimulation through the recording electrode resulted in a prolonged inhibition of the firing of most of the neurons recorded in the SNr. Figure 1 shows an example of the effects of stimulation with 0.5-s trains at 200 Hz and current intensity varying from 0.5 (top trace) to 6 µA (bottom trace). Following this period, firing was inhibited for several hundreds of milliseconds when stimuli equal to or larger than 1 µA were delivered. For this cell, the total duration of inhibition, measured from the last current pulse to the first action potential lasted from approximately 300 to over 700 ms depending on the stimulation current intensity.

The long-duration inhibition of SNr neuronal activity following termination of the train was observed only when high-frequency stimulation was employed. Figure 2 shows an example of the effects of stimulation with 0.5-s trains of 4-μA pulses with the frequency varying from 200 (top trace) down to 20 Hz (second to last trace) and back to 200 Hz (bottom trace). As in Fig. 1, recording capability resumes after a period of approximately 200 ms following the cessation of the train. Following this period, firing was inhibited for several hundreds of milliseconds when the train frequency was 50 Hz or higher. There was no evidence for prolonged inhibition of firing following stimulation with trains of 20 or 30 Hz. The duration of inhibition was usually related positively to the frequency of stimulation with higher frequency trains giving rise to longer inhibition.

Overall, 89% (47 of 53) of neurons identified to be in the SNr on the basis of classical criteria were inhibited following high-frequency trains of low-current (<5 μA) stimuli. The remaining SNr neurons did not show changes in their firing frequency or pattern following the train (tested up to 10 μA) compared with the time immediately before stimulation. The duration of inhibition was variable across neurons even when stimulated at the same parameters. For example, the mean duration

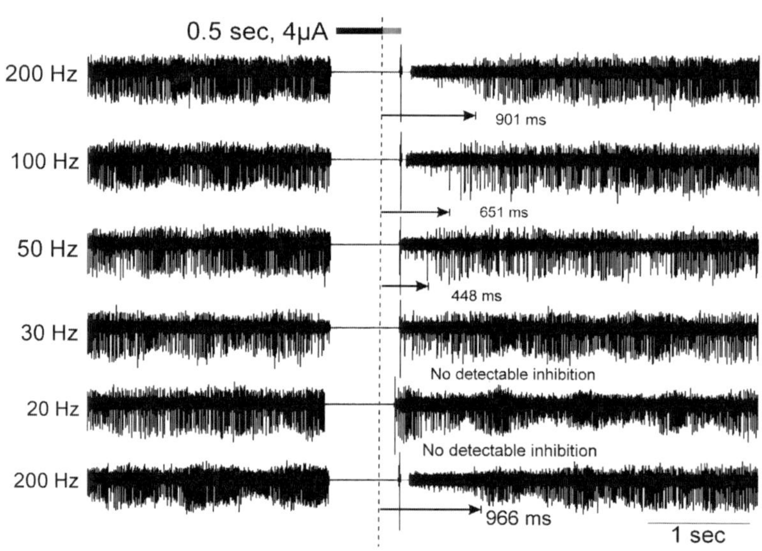

Fig. 2 SNr neurons are inhibited only by high-frequency stimulation. Raw recordings of an SNr neuron stimulated with 0.5-s trains at 4 μA and frequencies varying between 200 and 20 Hz. Stimulation at or above 50 Hz gave rise to a period of silence of a few hundred milliseconds. Stimulation below 50 Hz, however, did not have an effect on the firing. The duration of inhibition increased with the frequency of stimulation. This effect was dependent on the frequency of stimulation and not on the fact that consecutive stimulations were performed since a repeat stimulation at 200 Hz at the end of the sequence gave rise to inhibition similar to what was observed when it was first delivered at the beginning of the sequence

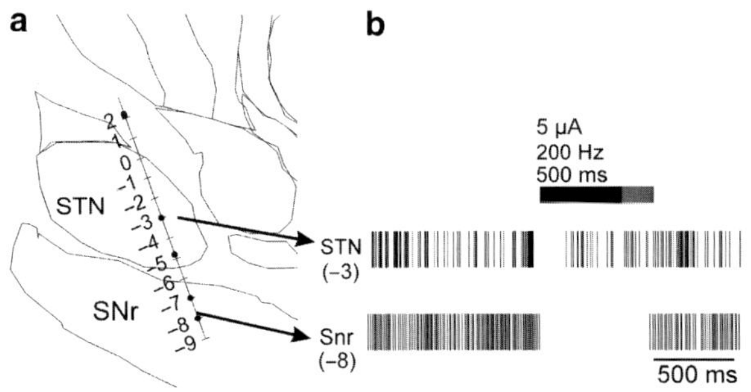

Fig. 3 Microelectrode trajectory and sample recordings in STN and SNr. (**a**) Parasagittal view of the anatomical reconstruction based on the Schaltenbrand–Wahren atlas showing the location of the STN and SNr, the trajectory of the electrode in this particular patient (depth relative to preoperative target indicated in mm), and the locations of recorded neurons (*solid dots*). (**b**) Raster plot of firing from two neurons located along the track shown in (**a**) and stimulated with a train at 5 μA. The neuron located in the STN (depth of −3) was not inhibited by the 5-μA train while the neuron located in the SNr (depth of −8) was clearly inhibited for almost 500 ms before resuming its activity

of inhibition in 23 SNr neurons stimulated with 0.5-s trains at 5 μA was 572 ms (range 217–1,254 ms). In contrast to SNr, when similar stimuli were applied to 64 STN neurons only 9% (6 of 64 neurons) showed evidence of a poststimulation inhibitory period. The mean duration of inhibition in six STN neurons that were inhibited at 5 μA was 346 ms (range 226–527 ms).

Figure 3 shows a sample track through the STN and SNr and the location of the recorded neurons based on the anatomical reconstruction. Only the neuron located most ventrally along the recording track, whose firing was typical of an SNr neuron, was inhibited by microstimulation (5 μA, 200 Hz, 0.5-s train).

4 Discussion

4.1 SNr Inhibition

The majority of the SNr neurons tested were inhibited by short high-frequency stimulus trains delivered from the recording electrode. The duration of the posttrain inhibition was several hundreds of milliseconds but varied across cells. Inhibition tended to be longer when stronger currents were applied (e.g., Fig. 1) and when train frequency was higher (e.g., Fig. 2). Poststimulation neuronal firing was observed to be inhibited only at frequencies of 50 Hz or higher. To our knowledge this is the first report showing that high-frequency electrical stimulation through the recording microelectrode can induce prolonged inhibition of SNr neurons and can be used as a useful test of neuron location during surgery for DBS.

Stimulation in the GPi of patients with various movement disorders with short trains of high-frequency, low-amplitude currents also results in inhibition lasting several hundreds of milliseconds (unpublished observations). In both the GPi and the SNr, the long-lasting inhibition of firing may be due to activation of GABAergic afferents from the striatum and GPe by the stimulation train and release of GABA. In a previous study (Dostrovsky et al. 2000) we have reported that single pulse microstimulation in the GPi produces a short-lasting inhibition in neurons recorded several hundreds of microns away by another microelectrode, and we have also observed this in SNr (unpublished observations). The long-duration inhibition outlasting high-frequency stimulus trains may be due to temporal summation of the presumed inhibitory postsynaptic potentials induced by release of GABA following each stimulus pulse and possibly also due to substantial accumulation of the released GABA in the synaptic clefts and spread outside the synaptic junction to act at extrajunctional $GABA_A$ and $GABA_B$ receptors, although other mechanisms such as potassium accumulation extracellularly, prolonged inactivation of voltage-gated channels, and/or increased K^+ conductance, for example, may be responsible (however, one would expect that the latter mechanisms would also be prominent in STN where stimulation does not generally give rise to this type of inhibition). It is likely that at low frequencies there is a short inhibitory period following the termination of the train but this cannot be observed with the technique used in the present study since the amplifier remains switched off for about 200 ms following the stimulation. It is not surprising that similar effects are seen in both SNr and GPi since these two nuclei are anatomically and functionally similar (Albin et al. 1989; Alexander and Crutcher 1990).

4.2 STN Inhibition

We found that, in comparison to SNr neurons, only a minority (9%) of STN neurons were inhibited by low-current high-frequency trains. This is consistent with previous findings that the threshold for inhibition of STN neurons by single pulses from a nearby microelectrode or macroelectrode is usually high (75–100 μA) (Filali et al. 2004). However, on the basis of these previous results, we had expected to find even fewer inhibited STN neurons than we did. The differences in stimulation parameters between both studies may account for this. In the previous study, neurons were stimulated with a microelectrode located approximately 600 μm away from the recording electrode while in the present study all stimuli were delivered through the recording electrode, and thus the current density would be much higher since the stimulus is applied much closer to the recorded neuron and thus effective in exciting presynaptic terminals at low intensities. In addition, the assessment of whether neurons were in the STN or the SNr was done intraoperatively based on classical criteria of background neuronal activity, firing rate of neurons, and presence of gaps between recordings of cellular activity.

It is possible that some of the neurons assigned to the STN were in fact in SNr or vice versa (see later).

4.3 Determining STN Borders

The use of microelectrode recordings for determining STN borders is based on the observation of the firing characteristics of the neurons encountered along the electrode trajectory as well as their density. As mentioned in Sect. 2, entry into the STN is usually accompanied by an increase in background neuronal activity followed by the appearance of multiunit large spike activity. There is usually a gap in cellular activity between the ventral part of the STN and the dorsal part of the SNr. Furthermore, STN neurons can usually be distinguished from SNr neurons by their firing rates (higher in SNr) and patterns (more regular in SNr) (Hutchison et al. 1998), and this can be used to determine exit from the STN and entry into the SNr. However, it is our impression that in the ventral STN the neuronal firing rates are generally slower and firing pattern less typical than those observed in dorsal STN and furthermore that the density of neurons drops so that there can be fairly large gaps between recordings of successive neurons. It is also our experience that some neurons in SNr fire at much lower rates than typical and the cell density is low. Thus, classification of neurons in this region as being in STN or SNr can sometimes be difficult leading to uncertainty as to where the ventral border of the STN truly is.

The finding reported in this paper indicates that inhibition of firing by short trains of low-amplitude, high-frequency stimulation through the recording electrode can be used as a quick and simple intraoperative assessment of whether the SNr has been reached during microelectrode recordings for implantation of DBS electrodes. This provides an additional characteristic that aids in determining the borders of STN and SNr. To be able to use this technique one needs to use an amplifier that allows high-frequency stimulation at very low currents (less than 5 μA) and that switches back automatically to recording mode in less than 200 ms (the shorter this period the better).

Several other methods for determining the borders of STN have recently been reported. These methods are based on intraoperative online calculation of either multiple cell density (MSD) (Kano et al. 2006), the normalized root mean square (NRMS) of the microelectrode recording signal (Moran et al. 2006), or the power of the high-frequency neuronal background (Novak et al. 2007). Unlike auditory assessment of background activity, which requires experience, computational methods are performed automatically and are independent of operator skill. Moreover, these methods generally do not rely on online spike detection and sorting, which is harder to automate. On the other hand, these methods typically require longer continuous recordings than are necessary for simple auditory evaluation of firing rate and pattern, and long recordings are not always possible in the context of surgery. In addition, these methods are based on an assessment of background activity that can be variable across PD patients due to the severity of the disease.

5 Conclusion

In conclusion, inhibition of firing by short trains of low-amplitude, high-frequency stimulation through the recording electrode can be used as a quick and simple intraoperative assessment of whether the SNr has been reached during microelectrode recordings for implantation of DBS electrodes. This provides an additional characteristic that aids in determining the ventral border of the STN. The mechanism underlying the inhibition of firing may be the activation of GABAergic striatal and pallidal afferents by the stimulation.

References

Albin RL, Young AB and Penney JB (1989) The functional anatomy of basal ganglia disorders. Trends Neurosci 12: 366–375.

Alexander GE and Crutcher MD (1990) Functional architecture of basal ganglia circuits: neural substrates of parallel processing. Trends Neurosci 13: 266–271.

Benabid AL (2003) Deep brain stimulation for Parkinson's disease. Curr Opin Neurobiol 13: 696–706.

Dostrovsky JO, Hutchison WD and Lozano AM (2002) The globus pallidus, deep brain stimulation, and Parkinson's disease. Neuroscientist 8: 284–290.

Dostrovsky JO, Levy R, Wu JP, Hutchison WD, Tasker RR and Lozano AM (2000) Microstimulation-induced inhibition of neuronal firing in human globus pallidus. J Neurophysiol 84: 570–574.

Filali M, Hutchison WD, Palter VN, Lozano AM and Dostrovsky JO (2004) Stimulation-induced inhibition of neuronal firing in human subthalamic nucleus. Exp Brain Res 156: 274–281.

Garcia L, D'Alessandro G, Bioulac B and Hammond C (2005) High-frequency stimulation in Parkinson's disease: more or less? Trends Neurosci 28: 209–216.

Hutchison WD, Allan RJ, Opitz H, Levy R, Dostrovsky JO, Lang AE and Lozano AM (1998) Neurophysiological identification of the subthalamic nucleus in surgery for Parkinson's disease. Ann Neurol 44: 622–628.

Jellinger K (1987) The pathology of parkinsonism. In: Marsden CD and Fahn S (eds) Movement Disorders 2. Oxford: Butterworth-Heinemann, pp 124–165.

Kano T, Katayama Y, Kobayashi K, Kasai M, Oshima H, Fukaya C and Yamamoto T (2006) Detection of boundaries of subthalamic nucleus by multiple-cell spike density analysis in deep brain stimulation for Parkinson's disease. Acta Neurochir Suppl 99: 33–35.

Levy R, Lozano AM, Hutchison WD and Dostrovsky JO (2007) Dual microelectrode technique for deep brain stereotactic surgery in humans. Neurosurg 60: 277–283.

Liu Y, Postupna N, Falkenberg J and Anderson ME (2008) High frequency deep brain stimulation: What are the therapeutic mechanisms. Neurosci Biobehav Rev 32: 343–351.

Lozano A, Hutchison W, Kiss Z, Tasker R, Davis K and Dostrovsky J (1996) Methods for microelectrode-guided posteroventral pallidotomy. J Neurosurg 84: 194–202.

Moran A, Bar-Gad I, Bergman H and Israel Z (2006) Real-time refinement of subthalamic nucleus targeting using Bayesian decision-making on the root mean square measure. Mov Disord 21: 1425–1431.

Novak P, Daniluk S, Ellias SA and Nazzaro JM (2007) Detection of the subthalamic nucleus in microelectrographic recordings in Parkinson disease using the high-frequency (> 500 hz) neuronal background. Technical note. J Neurosurg 106: 175–179.

Schaltenbrand G and Wahren W (1977) Atlas for Stereotaxy of the Human Brain. Stuttgart, Germany: Thieme.

Smith Y, Bevan MD, Shink E and Bolam JP (1998) Microcircuitry of the direct and indirect pathways of the basal ganglia. Neuroscience 86: 353–387.

Activity of Thalamic Ventralis Oralis Neurons in Rigid-Type Parkinson's Disease

Chihiro Ohye, Sumito Sato, and Tohru Shibazaki

Abstract The neural mechanisms of rigidity and dystonia in relation to the neural activity of the thalamic ventralis oralis (Vo) nucleus, where the pallidal (GP) projection is thought to be, are one of our research interests. At the previous IBAGS meetings we have reported that the activity of the Vo area differed between rigid and tremor-dominant type of Parkinson's disease (PD). At that time we analyzed global background activity and concluded that it was lower in rigid-type PD than in tremor-dominant type PD.

In the present study, activity of small spikes in the background activity recorded during the course of stereotactic selective thalamotomy was measured by an analyzer (PowerLab). It was revealed that the spontaneous activity of Vo in rigid-type PD was about 30% less than that in tremor-dominant type PD, with a statistically significant difference at $p < 0.0087$.

1 Introduction

At the previous IBAGS meetings, we have presented data indicating that the basic activity of the thalamic ventralis oralis (Vo) nucleus measured during the course of stereotactic surgery for Parkinson's disease (PD) was less in rigid-type PD than in tremor-dominant type PD (Ohye 1991, 1993; Ohye et al. 1991, 1994, 2002). At that time, the neuronal noise was measured globally and the integrated sum of the electrical activities was compared between groups (Fang et al. 2006). In the present study, we performed further quantitative analyses of the neuronal activity of the Vo nucleus in PD patients with dominant tremor or rigidity, respectively.

C. Ohye (✉), S. Sato and T. Shibazaki
Hidaka Hospital, Functional and Gamma Knife Surgery Center, 886 Nakao-machi, Takasaki, Gunma, Japan
e-mail: hidaka.sps@tiara.ocn.ne.jp

H.J. Groenewegen et al. (eds.), *The Basal Ganglia IX*, Advances in Behavioral Biology 58, DOI 10.1007/ 978-1-4419-0340-2_43, © Springer Science+Business Media, LLC 2009

2 Subjects and Methods

Six cases of rigid-type PD, six cases of tremor-dominant type PD, one case suffering from writer's cramp, one case with dystonia, and one case of essential tremor were included in this study (Table 1). All patients received a stereotactic thalamotomy: Vim-Vo thalamotomy in the rigid group and Vim thalamotomy in the tremor-dominant group. Operative procedures were always the same (Ohye 1997a). We used microrecording-aided selective thalamotomy. The relevant processes to this neuronal activity analysis are data acquisition (microrecording) and data analysis as described later.

Orientation of the recording electrode is the most important process in our selective thalamotomy. We use the lateral part of the Vim nucleus, 2 mm medial to the thalamo-internal capsular border, 5 mm anterior to PC, at the level of the intercommissural (IC) line as zero point (tentative target). Standard angle to approach is about 45° referring to the IC line in lateral view and about 10° or less in frontal view. In this way, the electrode passes from the cortical area by way of the thalamic reticular nucleus, the dorsal nucleus (Vo), and the Vim nucleus to the ventral caudal (Vc) nucleus theoretically. So we can obtain the activity of Vo and Vim in the same case (Fig. 1).

According to our operating principles, a recording electrode is set at 3 mm posterior to the center (core) electrode, the direction of which is oriented toward the real zero point for operative planning but not used for recording process. This is because we now know that the posterior track is more useful to find better neurophysiological information to finally decide upon the optimal target (see Ohye 1997a, 1997c).

Table 1 Patients list

	Patients list				
	Age	Gender	Diagnosis	Date Op	Op
Rigid type					
MO	58	M	Writer's cramp	2003/10/15	L-Vim-Vo
HI	53	M	PD	2003/10/22	L-Vim-Vo
MW	51	F	PD	2004/2/18	L-Vim-Vo
MS	40	F	PD	2004/2/25	L-Vim-Vo
ST	34	M	PD	2005/8/3	R-Vim-Vo
KU	64	M	PD	2005/10/19	R-Vim-Vo
KS	39	F	PD	2005/11/16	R-Vim-Vo
NW	26	M	Dystonia	2007/3/7	L-Vim-Vo
Tremor type					
RN	59	F	PD	2007/1/7	L-Vim
HT	72	M	PD	2007/2/21	L-Vim
JO	48	M	PD	2007/3/28	R-Vim
HT	71	M	PD	2007/5/9	R-Vim
IT	63	M	PD	2007/5/30	L-Vim
SY	61	F	PD	2007/8/8	R-Vim
KS	60	M	Ess Tr.	2005/5/11	L-Vim

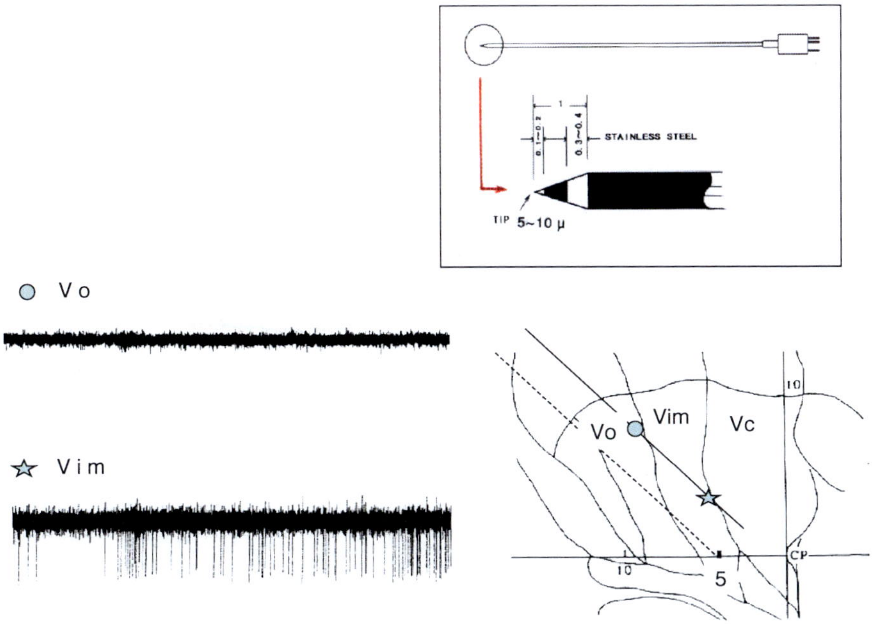

Fig. 1 Semischematic of typical spontaneous activity of thalamic dorsal Vo and ventral Vim areas, respectively, showing different background activity patterns. On the right side, the approximate trajectory of the recording electrodes and the recording points of each type of spontaneous activity within the thalamus are plotted on the sagittal plane of the thalamus. In the upper right corner, the inset shows a detail of the recording electrode

By means of the posterior electrode, we can distinguish the dorsal thalamus (maybe Vo, usually from 12 to 7–8 mm above zero), the ventral thalamus (mostly Vim from 6 to 0 mm), and sometimes the Vc nuclei. Discrimination between these nuclei is usually not difficult because of the differences in background activity and in the characteristics of the neural response to natural stimulation (Ohye 1997b). Thus, in typical cases of PD, we can determine the position of the electrode tip with a high degree of certainty. In general, characteristics of Vo activity are smaller spikes, which do not respond to natural stimulation of the contralateral body part, in particular passive movement, although there sometimes is a response to active movement (Jasper and Bertrand 1966). By contrast, in the Vim zone, recordings exhibit very active high-amplitude spike discharges, which show rhythmic bursts in conjunction with the tremor oscillation of the contralateral limb and/or the kinesthetic response to passive movement of the contralateral limbs. For these different activities and responses in Vo and Vim, there is morphological basis (Hirai et al. 1989).

The recording electrode was of a bipolar concentric type (steel–steel), with outer diameter being 0.3 mm, the tip being 0.1–0.2 mm sharp with an electrical resistance less than 100 KΩ, and interpolar distance about 0.5 mm (Fig. 1). The electrode was introduced with the aid of a micromanipulator that enables selecting a single spike from the background of multiple spikes. Recorded electrical activity was led to an oscilloscope in parallel with the necessary surface EMGs. Subsequently, thalamic activity and selected EMG were fed to a data analyzer (PowerLab). The recorded data were reproduced later and analyzed by means of PowerLab, which has several facilities to make spike histograms (discriminator, amplitude histogram, rate meter). Other programs included cycle variable and period program, which showed the number of spikes discriminated by amplitude or measured the interspike interval consecutively during a certain selected time period. In this paper, we describe the results obtained by the spike histogram program. The cycle variable and period program were used for reference purposes only.

In each case operated under local anesthesia, thalamic activity recorded from the dorsal thalamic area, presumably from the Vo nucleus and from the further ventral part (Vim nucleus), was compared mainly with respect to the number of spikes counted in 1 s (frequency). Since Vim activity had already been studied in detail before (Hirai et al. 1989; Ohye 1990, 1997b; Ohye et al. 1989), it was not difficult to identify based upon both spontaneous activity and evoked activity in response to passive movement (kinesthetic neurons in Vim, but not in Vo) (Jasper and Bertrand 1966; Ohye 1997b).

3 Results

A typical example of a spike frequency analysis is shown in Fig. 2. First, a spike histogram is used to visualize the whole spike distribution for a certain spike width during the selected period (5 s) (Fig. 2, left). In addition, an amplitude histogram is made (Fig. 2, right). Then, the number of spikes in each second is shown by histogram for the period of 5 s. In Fig. 3, the upper part (raw record) corresponds to the

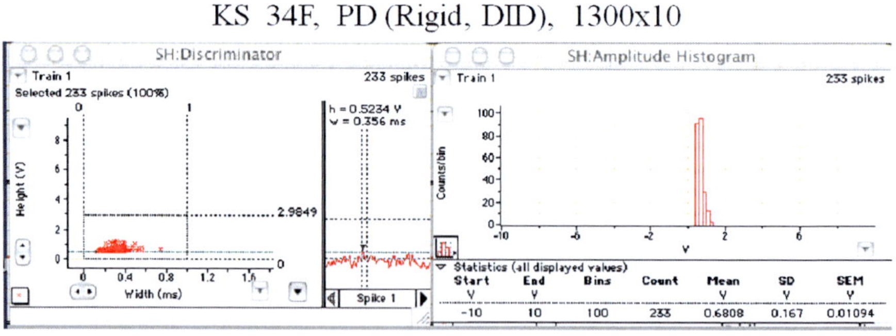

Fig. 2 Example of the basic data of thalamic Vo activity in rigid-type PD analyzed by means of the PowerLab program. The distribution of the collected spikes (*left*) and the amplitude histogram (*right*) of the Vo spikes are shown

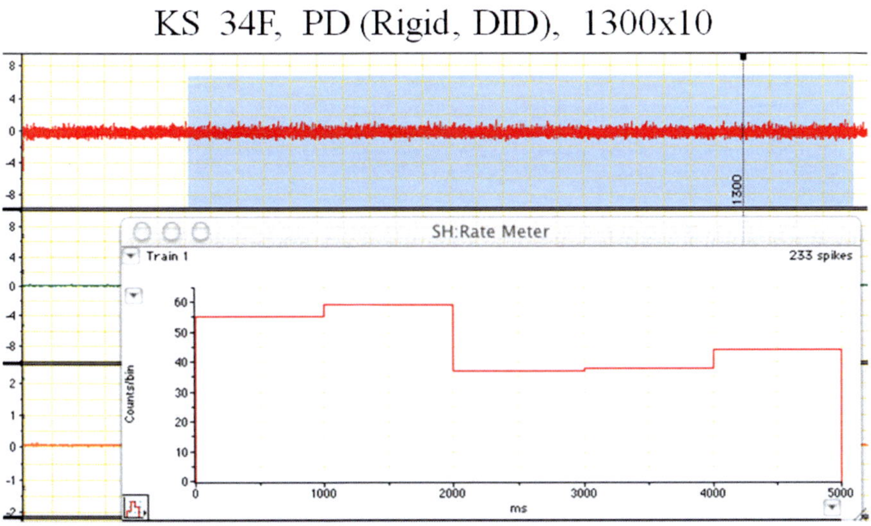

Fig. 3 Activity of Vo neurons recorded from the same patient as in Fig. 2, but now analyzed by the rate meter to count spike numbers. Randomly selected raw data for a period of 5 s are shown in the *upper part* of the figure. The *lower part* illustrates the histogram of spike numbers (derived from the upper raw data), per second

lower part histogram of the spike numbers per second during the same selected period of 5 s. In this way, we examined each case. In this way, in the Vo area, at 21 points in tremor-type PD and at 26 points in rigid-type PD, stable recording points were selected for further treatment.

Comparing the Vo activity in rigid-type PD (Fig. 3) and tremor-dominant type PD (as Fig. 4), the former is less active than the latter. Also, in the more ventral part of Vim, it is likely that the activity in Vim in rigid-type PD (Fig. 5) is less than that in tremor-type PD (Fig. 6). In Fig. 7, all calculated spike frequency values are plotted against each of the recording points. The mean value of Vo activity in rigid-type PD was 47.8 ± 21.1 and that in tremor-type PD 67.9 ± 28.6. The activity of Vo in rigid-type PD is about 70% of that in tremor-type PD. This difference was statistically significant ($p < 0.0087$ by t test).

4 Discussion

Qualitative measurement of the activity of small spikes in Vo revealed that it was less active in rigid-type PD than that in tremor-dominant type PD. This finding confirms the results of a previous study comparing the integrated background activity of the Vo area between two different clinical types of PD (rigid vs. tremor type) claiming that the activity Vo is less in the former than in the latter (Ohye et al. 2002).

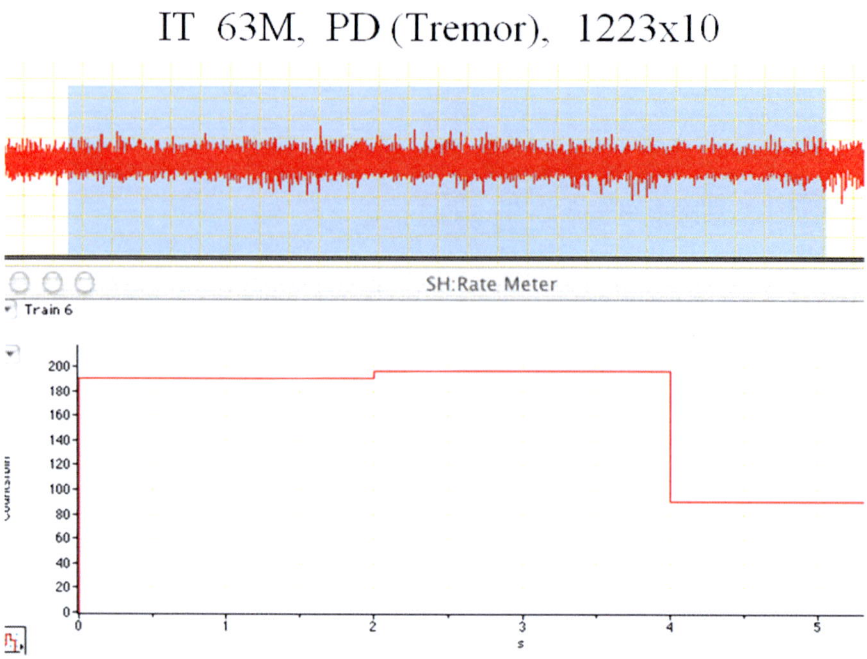

Fig. 4 Analysis of the number of spikes in the Vo area in a tremor-dominant case of PD. Data analysis and mode of illustration are the same as in Fig. 3. Note that the basal activity is higher than in rigid-type PD (see Fig. 3)

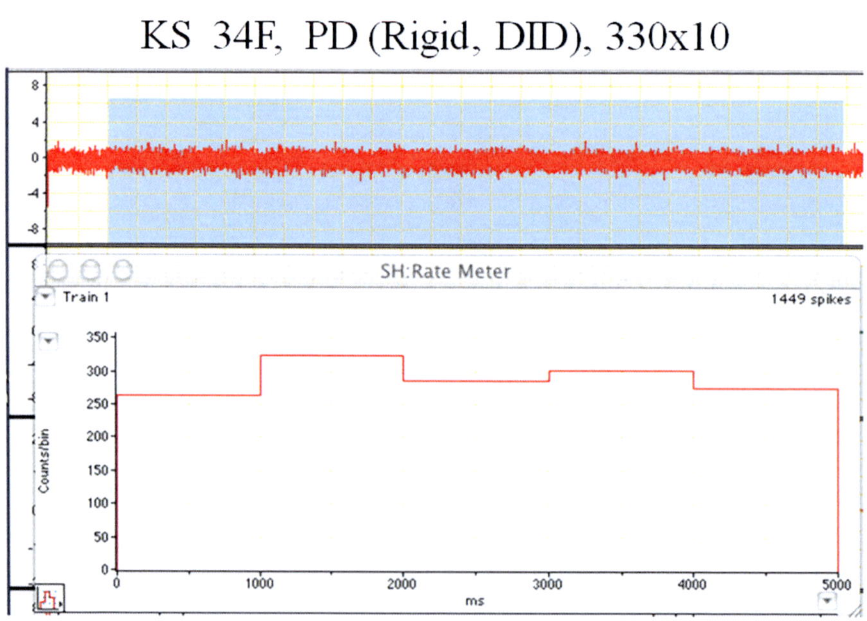

Fig. 5 Activity of Vim neurons recorded from the same patient as in Figs. 2 and 3 but now from the ventral area of Vim. The number of spikes was counted in the same way as in Fig. 3 and illustrated in a similar manner

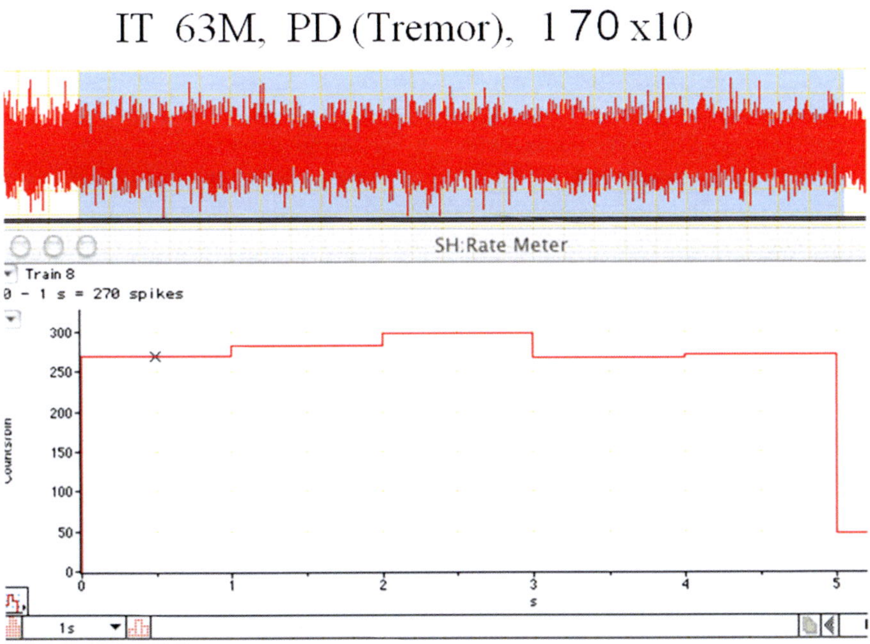

Fig. 6 Analysis of the number of spikes in the Vim area in the same tremor-dominant case of PD as illustrated in Fig. 4. Data analysis and mode of illustration are the same as in Fig. 3. In this case, Vim spikes were more intense than in the case of rigid-type PD, but also higher than in the Vo region in the same case (see Fig. 4)

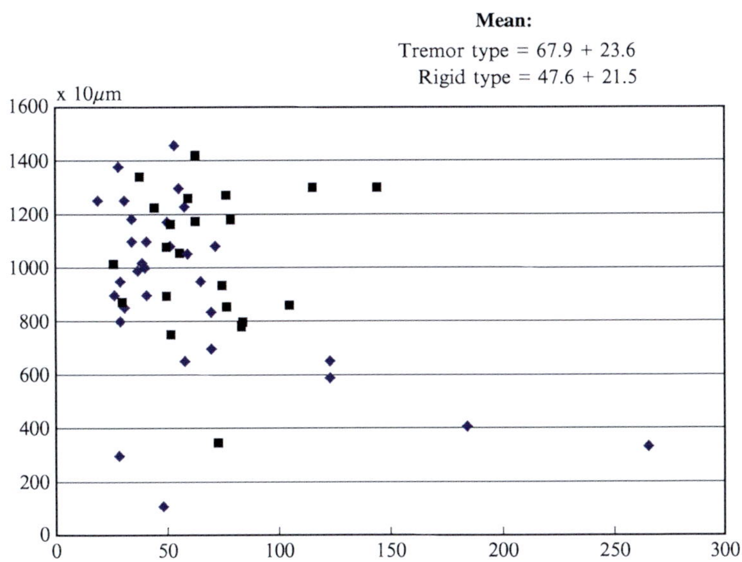

Fig. 7 Summary diagram for the present study. Spike frequencies obtained from eight cases of rigid-type PD and seven cases of tremor-dominant type PD were plotted in this diagram. The abscissa is the number of spikes per second (frequency) and the ordinate is the recording point: zero being set at the level of the intercommissural line (see Fig. 1). *Squares* indicate the tremor-dominant PD case and diamonds mark the rigid-type PD cases

The results of the present study were obtained mainly by a small spike analysis system. With this device, we can calculate (count) the number of small spikes of over 20 μV that occurred in the background activity during a given period of recording. The number of data sampling points in the presumed Vo zone was random from one point to four to five points per person, depending on the condition of each patient.

As explained in our previous study, our approach to the thalamic Vim nucleus is through the dorsally located Vo. The spontaneous activity of these structures changes depending on, at least partly, the cellular size and dimension (Hirai et al. 1989); the dorsal area, Vo, is less active than the ventral area, Vim. This difference is a very reliable landmark to distinguish these structures. Therefore, the identification of the recorded point was almost guaranteed.

In the present study, the activity of the small spikes in the background activity recorded by so-called semimicroelectrodes was considered. The role of such small spike activity is not yet clarified. In fact, the small spikes in background activity have been almost neglected. But as shown in this study, activity of small spikes seems to be correlated with the clinical state of PD, low levels of Vo activity in rigid cases, and high levels of Vo activity in tremor-dominant cases. At present, we cannot explain the consequence of this kind of difference in activity levels in the motor thalamic nucleus Vo, which is connected to the cortical motor areas (Ohye 1990; Hazrati and Parent 1992; Fang et al. 2006; Hamani et al. 2006). This will require further analysis. Interestingly, results of recent studies on neural noise suggest that neural noise may have a significant role in information transmission or signal processing (White et al. 2000; Manwani 2001; Miller and Troyer 2002; Stacey and Durand 2002).

References

Fang PC, Stepniewska I and Kaas JH (2006) The thalamic connections of motor, premotor, and prefrontal areas of cortex in a prosimian primate (*Otolemur garnetti*). Neuroscience 143: 987–1020.

Hamani C, Dostrovsky JO and Lozano AM (2006) The motor thalamus in Neurosurgery. Neurosurg 58: 146–158.

Hazrati L-N and Parent A (1992) The striatopallidal projection displays a high degree of anatomical specificity in the primate. Brain Res 592: 213–227.

Hirai T, Ohye C, Nagaseki Y and Matsumura M (1989) Cytometric analysis of the thalamic ventralis intermedius nucleus in humans. J Neurophysiol 61: 478–487.

Jasper HH and Bertrand G (1966) Thalamic units involved in somatic sensation and voluntary movements in man. In: Purpura D and Yahr MD (eds) The Thalamus. Columbia University Press, New York, pp 365–390.

Manwani A (2001) Detecting and estimating signals over noisy and unreliable synapses information. Theoretic analysis. Neural Comp 13: 1–33.

Miller KD and Troyer TW (2002) Neural noise can explain extensive, power law nonlinearities in neural response functions. J Neurophysiol 87: 653–659.

Ohye C (1990) The thalamus. In: Paxinos G (ed) The Human Nervous System. Academic Press, San Diego, pp 439–468.

Ohye C (1991) Positron emission tomographic study in Parkinson's disease: Rigid vs tremor type. In: Nagatsu T, Narabayashi H and Yoshida M (eds) Parkinson's Disease. From Clinical Aspects to Molecular Basis. Springer, New York, pp 179–186.

Ohye C (1993) Dynamic aspects of striatothalamic connection studied in cases with movement disorder. In: Narabayashi H, Nagatsu T, Yanagisawa N and Mizuno Y (eds) Parkinson's Disease. From Basic Research to Treatment. Raven Press, New York, pp 78–87.

Ohye C (1997a) Thalamotomy for Parkinson's disease and other types of tremor. In: Gildenberg PL and Tasker RR (eds) Textbook of Stereotactic and Functional Neurosurgery. McGraw-Hill, New York, pp 1167–1178.

Ohye C (1997b) Functional organization of the human thalamus. In: Steriade M, Jones EG and McCormick DA (eds) Thalamus, Vol.2. Amsterdam, Elsevier, pp 517–542.

Ohye C (1997c) Neural noise recording in functional neurosurgery. In: Gildenberg PL and Tasker RR (eds) Textbook of Stereotactic and Functional Neurosurgery. McGraw Hill, New York, pp 941–947.

Ohye C, Shibazaki T, Hirai T, Wada H, Hirato M and Kawashima Y(1989) Further physiological observations on the ventralis intermedius neurons in the human thalamus. J Neurophysiol 61: 488–500.

Ohye C, Shibazaki T, Hirato M, Kawashima Y, Matsumura M and Shibasaki T (1991) Neural activity of the basal ganglia in Parkinson's disease studied by depth recording and PET scan. In: Bernardi G (ed) The Basal Ganglia III. Plenum, New York, pp 385–393.

Ohye C, Hirato M, Kawashima Y, Hayase N and Takahashi A (1994) Neuronal activity of the human basal ganglia in parkinsonism compared to other motor disorders. In: Percheron G, McKenzie J and Feger J (eds) The Basal Ganglia IV. Plenum, New York pp 386–393.

Ohye C, Shibazaki T and Andou Y (2002) Pallidothalamic relation in Parkinson's disease – Microrecording study. In: Nicholson LFB and Faull RLM (eds) The Basal Ganglia VII. Academic/Plenum, New York, pp 611–616.

Stacey WE and Durand DM (2002) Noise and coupling affect signal detection and bursting in a simulated physiological neural network. J Neurophysiol 88: 2589–2611.

White JA, Rubinstein JT and Kay AR (2000) Channel noise in neurons. Trends Neurosci 23: 131–137.

Motor and Non-motor Effects of PPN-DBS in PD Patients: Insights from Intra-operative Electrophysiology

Alessandro Stefani, Salvatore Galati, Mariangela Pierantozzi, Antonella Peppe, Livia Brusa, Vincenzo Moschella, Francesco Marzetti, and Paolo Stanzione

Abstract Three decades of basic research have focused on the multiple functions sub-served by the pedunculopontine nucleus (PPN) in mammals. Yet, far from understood is the impact that lesioning PPN or modulating PPN-fugal pathways have on motor, limbic and/or associative domains. Recently, we have pioneered the low-frequency deep brain stimulation (DBS) of pontine tegmental areas in severely parkinsonian patients, aiming at providing new insights in the knowledge of this puzzling region. Here we show that, under PPN-DBS, significant amelioration of axial and hypokinetic signs occurs (although to a lesser extent than following STN-DBS in the same patients), together with a normalization of the spinal H reflex. Furthermore, PPN-DBS improves REM sleep behaviour disorder and attentive and cognitive executive performances.

As a first step to understand the limited motor response to PPN-DBS, systematic intra-operative recordings in STN were performed during PPN-DBS at 25 Hz. Almost each STN cell showed significant and long-lasting changes of the mean firing frequency during PPN stimulation. However, PPN-ON caused two conflicting effects: a dramatic decrease of the ongoing firing in bursting STN neurons and a large excitatory effect in irregular and tonic neurons. If dampening of STN bursting units seems to be in accord with the PPN therapeutic role, the simultaneous excitatory influence in non-bursting cells might counteract the efficacy on motor signs.

As a further step to understand the mechanisms underlying non-motor benefits, FDG-PET imaging was routinely performed in different conditions (PPN-OFF, PPN-ON). These investigations might clarify whether PPN-ON influences the subcortico-cortical pathways responsible for learning processes and goal-directed behaviours. Preliminary data confirm the possibility that PPN-DBS affects multiple ascending pathways involving intralaminar thalamic nuclei and the ventral tegmental area.

A. Stefani (✉), S. Galati, M. Pierantozzi, A. Peppe, L. Brusa, V. Moschella, F. Marzetti, and P. Stanzione
Clinica Neurologica, Department of Neuroscience, University Tor Vergata, Rome, Italy
e-mail: Stefani@uniroma2.it

H.J. Groenewegen et al. (eds.), *The Basal Ganglia IX*, Advances in Behavioral Biology 58,
DOI 10.1007/978-1-4419-0340-2_44, © Springer Science+Business Media, LLC 2009

These findings confirm a complex interplay between PPN, basal ganglia and cortical regions. However, the size of the inserted electrode, extension of the electrical field and inter-individual anatomical differences of the surgical targeting do not allow us to draw any definite conclusions.

1 Introduction

The human pedunculopontine nucleus (PPN) is located in the ponto-mesencephalic region delimited in its lateral boundaries by the lemniscus medialis and in its medial margins by the superior cerebellar peduncle (Olszewski and Baxter 1982; Muthusamy et al. 2007). In rats, PPN extends from the posterior pole of the substantia nigra (SN) back to the lateral tip of the superior cerebellar peduncle (Jackson and Crossman 1983; Garcia-Rill 1991). Histochemical evidence suggests that PPN has two main sub-regions: the pars compacta (PPNc) and the pars dissipata (PPNd) (Mesulam et al. 1989). These compartments are characterized by the predominance of different cell populations (Takakusaki et al. 1996; Bevan and Bolam 1995), the most clearly identifiable being the acetylcholine (ACh) neurons in the Ch5 area (Mesulam et al. 1989). Ch5 neurons send, among others, sensory information to the thalamus and the SN, pars compacta (SNc). On the other hand, the non-cholinergic compartment (characterized by the presence of GABAergic and, mostly, glutamatergic neurons) receives corticostriatal inputs and projects to structures involved in motor control, such as the internal pallidum (GPi), the SN pars reticulata (SNr) and the subthalamic nucleus (STN) as part of the corticostriatal loop system (Calzavara et al. 2007).

Experimental data in animals showed a clear role of PPN in shaping motor performance (Garcia-Rill et al. 1987; Kelland and Asdourian 1989; Takakusaki et al. 2003). Interestingly, a moderate degree of PPN neuronal degeneration occurs in human extrapyramidal disorders such as idiopathic Parkinson's disease (PD) (Zweig et al. 1989; Pahapill and Lozano 2000). However, PPN is by no means simply identified as a locomotor structure, and a wide body of evidence shows that it is critically involved in the analysis of reinforcement, learning and attention (Florio et al. 1999; Winn 2005). In addition, diffuse ascending cholinergic projections from the brainstem affect rapid-eye-movement sleep (Rye 1997). Further, experimental evidence suggests that the PPN is also involved in anti-nociception and startle reactions (Reese et al. 1995).

In PD as well as in PD disease models (6-OHDA and MPTP), a classical scientific cornerstone consists of attributing a key pathogenetic role to overactive GPi and SNr inhibitory output to the thalamic relay nuclei in producing hypokinesia. Similar observations may apply to GPi and SNr projections towards the PPN: a hypoactive (and apoptotic) PPN is unable to send its physiological excitatory output to the SNc, contributing to basal ganglia dysbalance (Breit et al. 2001). This hypothesis was tested in MPTP-lesioned non-human primates, where low-frequency stimulation of PPN area was effective in alleviating akinesia (Nandi et al. 2002).

Likewise, we have recently demonstrated that the PPN may indeed be targeted with reliable safety in humans as well, and that PPN activation is well associated with STN (Mazzone et al. 2005; Stefani et al. 2007). However, PPN is not established yet as an alternative clinical target to STN or GPi, since the impact of PPN stimulation alone on global motor scores remains modest and shows a small decline of efficacy upon chronic stimulation (Stefani et al. 2007).

The present chapter describes the most clear-cut motor and non-motor effects that low-frequency (25 Hz) PPN stimulation promotes in PD patients.

2 Methods

The six PD patients selected for simultaneous bilateral implantation of STN and PPN met the UK PDS Brain Bank diagnostic criteria for idiopathic PD. They all presented severe axial signs (scarcely responsive to dopaminergic agents), although a disabling freezing was present in only three. Written, informed consent was obtained from the patients, and the Local Ethics Committee approved the protocols.

2.1 Neurosurgery

The surgical procedure is described elsewhere (Stefani et al. 1997; Peppe et al. 2004; Mazzone et al. 2005). Briefly, the electrodes (Medtronic 3389) were implanted in PPN and STN in each hemisphere through a double-arch system (Peppe et al. 2004). Target coordinates and trajectories were determined as follows: for STN – the mid-point of the anterior commissure–posterior commissure (AC–PC) line, 11–12 mm lateral to the midline of the third ventricle, 4 mm below AC–PC (the angle in the sagittal plane was 80°–85° and was 75°–80° in the coronal plane to obtain an extra-ventricular and extra-capsular trajectory), and for PPN, a simple indication of a fixed angle range in the sagittal plane sounds improper, given the high inter-individual variability. The key landmark to minimize surgical risks is the floor of the fourth ventricle (parallel to the brainstem axis). Therefore, in each patient, the trajectory was performed parallel to the floor of the fourth ventricle. However, for better coordinates and the clear description of PPN (and its caudal representation, the nucleus of the tegmenti pedunculopontine, PPTg), consider Fig. 1 and Mazzone et al. 2008.

2.2 Patient Evaluation

The careful clinical evaluation of this new combination of targets required a complex series of observations for each patient [see Fig. 3 in Stefani et al. (2007)]. Briefly, comparative evaluations allowed us to establish as optimal the following

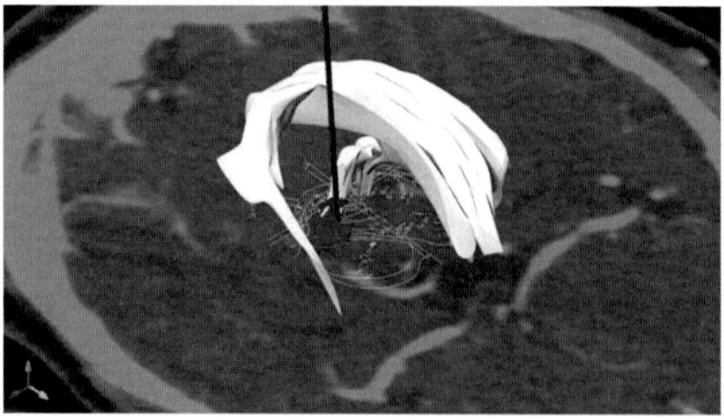

Fig. 1 Three-dimensional reconstruction of the PPN, pars disseminata (*blue*) and pars compacta (*yellow*) with representation of the lead trajectory (*dark red*). Note the lamina quadrigemina (*white*) and the third ventricle (*light brown*). Background: axial slide of angio-CT scan with overlying 2D slides from Schaltenbrand–Wahren brain atlas (Tc 0, Tc –3)

standard stimulus parameters: for PPN (bipolar contacts 0–1/4–5) = 60-μs pulse width, 25 Hz, 1.5–2 V; for STN (monopolar contact 1 or 2 and 5 or 6) = 90-μs pulse width, 185 Hz, 1.5–2.4 V. OFF-therapy evaluations were performed after an overnight therapy suspension (CAPIT protocol; Langston et al. 1992) and started between 2 and 3 months post-surgery. The specific stimulation condition (STN or PPN or both) was randomly activated and kept blinded to the single neurologist in charge of the clinical evaluations. DBS (PPN or STN or combined activation) was maintained for about 24 h in order to avoid additive or artefactual response.

Each clinical evaluation included the motor section of the Unified Parkinson's Disease Rating Scale (UPDRS-III). To evaluate gait and posture, we focused on the relevant UPDRS-III items (items 27–30) plus, more recently, a three-dimensional approach using an optoelectronic system (SMART system, BTS Padova, Italy) to measure the three coordinates of retro-reflective markers. Six video cameras were placed along an 8-m walkway; the working volume ($2 \times 3 \times 6$ m^3) is calibrated by sweeping it with the three-marker wand provided, that is, by successively moving the wand up and down several times parallel to each axis. After three-dimensional (3D) calibration, the spatial accuracy of the system is less than 0.5 mm. For correct positioning of markers a common 'Davis' protocol was used. Twenty-three spherical markers (10 mm diameter) were attached to the subject's body with double-sided tape, according to the marker configuration of the *Davis* model, except for the calves; thigh markers were attached approximately 7–10 cm away from the skin on iron sticks.

Cognitive evaluations were performed by a neuropsychologist [blind to the DBS condition (PPN-ON/STN-OFF vs. PPN-OFF/STN-OFF)] in the morning, after overnight drug therapy suspension (CAPIT). Re-tests were performed by using parallel test forms to avoid learning-related phenomena. The order of presentation of the parallel forms was appropriately counterbalanced. Cognitive functions were assessed by using the following tests: the California Verbal Learning test and Digit Span test for memory, the Trail Making test (TMT) for attention and executive functions, FAS for verbal fluency, the object naming test for naming and the classical Rey figure for visuo-spatial ability.

In four out of six double-implanted PD patients (mean age 62.8 ± 2.2; disease duration 11.8 ± 3.5 years) sleep was extensively studied. Three sleep scales were administered: Parkinson's Disease Sleep Scale (PDSS), Pittsburgh Sleep Quality Index (PSQI) and Epworth Sleepiness Scale (ESS). All patients underwent the following modalities: STN-ON + PPN-ON, STN-ON + PPN-OFF, STN-ON + PPN-cyclic (nighttime-ON). Each condition was maintained for 2 weeks, and patients compiled PDSS, ESS and PSQI at the end of each step. Stimulation parameters remained constant as well as L-DOPA dosage.

Polysomnography (PSG) recordings were carried out in a single case (not included in the group of six patients described earlier) using a dynamic 32-channel system polygraph with standard montage before surgery (Romigi et al. 2006). Post-surgery, to minimize DBS-induced artefacts during PSG, STN-DBS was delivered bipolar without a significant worsening in comparison to monopolar stimulation, and the following montage was utilized: F3-C4, F4-C3 and O1-O2, according to the Iranzo protocol (Iranzo et al. 2002). PSG sessions consisted of two consecutive 24-h PSGs carried out using the first night as an adaptation night in each condition (de novo, pre-surgery, STN-ON, PPN-ON). The scorers were blind to the order of the post-surgery recordings (excerpts in Table 2).

2.3 *Peri-operative Recordings*

This chapter deals mainly with STN extracellular activity recorded in PD patients (three of those undergoing stereotactic surgery), in particular before and during PPN-DBS (right after PPN implantation). Extracellular units were amplified (ISO-DAM8; World Precision Instruments, Hertfordshire, UK), sampled (50 KHz) on-line with our artefact suppression method (Galati et al. 2006), stored into a computer connected to a CED 1401 interface and analyzed off-line using the Spike 2 Analysis Program (Cambridge Electronic Design, Cambridge, UK). PPN-DBS, through a Medtronic external device Model 3625, consisted of 60-μs width and 2–3-V pulses, delivered bipolar (contacts 0–1) at 10–25 Hz. STN spike features are described in detail elsewhere (Galati et al. 2008).

In 48 STN neurons, the spiking activity was evaluated 3–10 min before, during (10 min) and after PPN-DBS (3–10 min, Fig. 2). Differences in the spontaneous discharge pattern of STN neurons were determined by comparing the inter-spike interval histograms (ISIH). ISIH analysis was performed using a 1-ms bin width, mean interval, coefficients of variation (CV = standard deviation of the mean interval/mean interval) and asymmetry index (mode/mean interval). To investigate whether there was any clustering of spikes in spontaneous discharge, auto-correlograms (AutoCrls) were constructed for each unit before and during PPN-DBS; exemplary

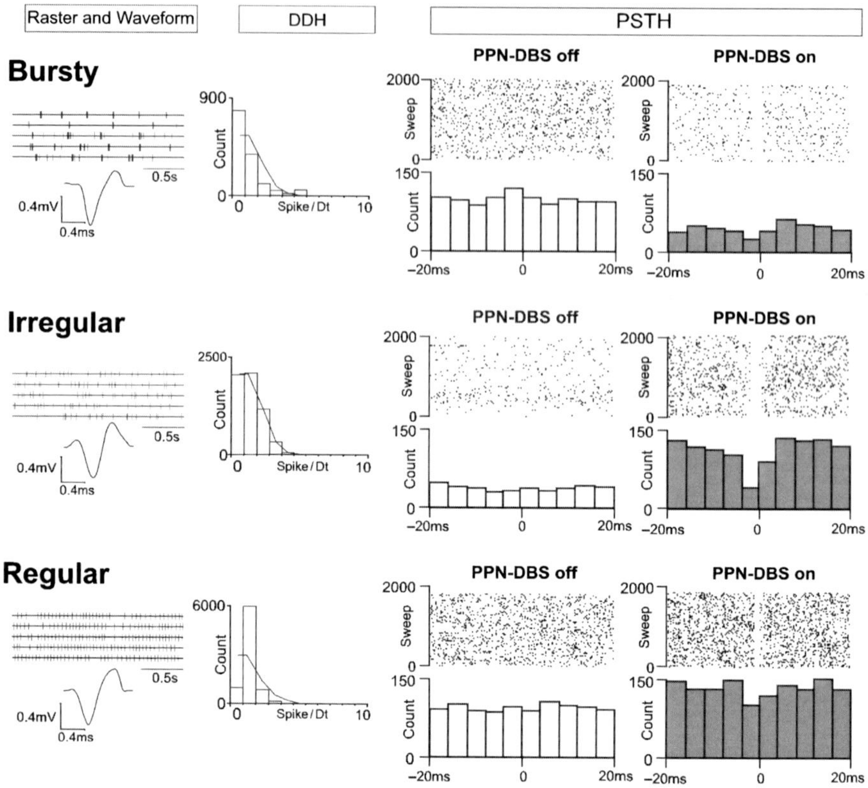

Fig. 2 On the left, electrophysiological identification characterization of three main modality patterns of STN unit firing discharges: bursty, irregular and regular. On the right, examples of rate meter recordings showing the clear-cut effects promoted by the condition PPN-ON. Twenty-five hertz PPN-DBS decreased the STN firing activity when in bursty units (*upper line*), whilst increasing firing rate in regular and irregular neurons units (Galati et al. 2008)

post-stimulus histograms were examined during either 10 or 25-Hz PPN stimulation [for details and statistical analysis, see Galati et al. (2008)].

3 Results

3.1 Acute and Long-Lasting Motor Effect

The main aim of PPN implantation was to provide specific benefits on gait (mostly in patients whose freezing was scarcely responsive to the best drug combinations). So far, the results in this small cohort lead to doubtful conclusions, since the responses obtained are highly variable. On the one hand, all patients reported a significant amelioration of gait sub-items when PPN and STN were switched ON

Table 1 Early and long-lasting effects of PPN-DBS (25 Hz) and PPN plus STN-DBS (185 Hz) on gait sub-items (summation of items 27–30 from UPDRS-III)

Patient	OFF	PPN-DBS (1 week)	PPN-DBS (12 months)	STN and PPN-DBS (1 week)	STN and PPN-DBS (12 months)
1	12	6	9	5	8
2	13	5	8	4	6
3	10	3	6	2	6
4	13	5	9	4	9
5	13	5	8	6	8
6	10	5	6	4	5
Mean	11.8	4.8	7.7	4.2	7
SD	1.4	0.9	1.3	1.3	1.5

A prolonged efficacy (>40%) was observed only in patients 2, 3 and 6

together (Stefani et al. 2007), whilst the impact of PPN stimulation alone was slight to modest in four and rather impressive only in the other three patients. When gait analysis was performed with a three-dimensional blinded approach, we could confirm the effect of PPN and STN in recovering the kinetic and kinematic alterations observed in PD patients during OFF-therapy-OFF-DBS, expressed as percentage differences with healthy subjects (Peppe et al., unpublished observations). This reduction was particularly important when both stimuli were simultaneously switched ON, as an additional action of PPN on STN activity, which is even more pronounced when PPN is added to STN and L-DOPA therapy.

Table 1 summarizes gait subscores, comparing 'acute' (1 week) and chronic (12 months) motor effects induced by PPN-ON and STN-ON (185 Hz). Noticeably, these 'declining' gait responses occurred despite the homogeneous (6 out of 6!) PPN-mediated change of Hoffman reflex threshold [up to normalization, see Pierantozzi et al. (2008)] and a great improvement of their quality of life (Stefani et al. 2007).

3.2 Cognitive Effects

PPN-ON significantly improved cognitive functions in the following domains: verbal long-term memory assessed with California Verbal Learning test and delayed recall and executive functions as revealed by TMT and FAS (data not shown). A clear sensation of 'well-being' was also reported by all patients as PPN stimulation was turned on. These preliminary observations are being correlated with changes of cerebral metabolic activity, as studied and quantified by FDG-PET (Brusa et al., 2008).

3.3 Effects on Sleep

In the PPN-cyclic condition, the group of four PD patients reported a clear improvement of nocturnal motor restlessness (PDSS item 4: PPN-cyclic 7.5 ± 3.3, PPN-ON

Table 2 Polysomnographic parameters under STN-DBS vs. PPN-DBS

Sleep parameters	DBS-OFF	STN-ON[a]	PPN-ON[a]
Sleep efficiency (%)	74.5	88.9	90
Stage 1 (%)	26	15.9	13.4
Stage 2 (%)	38	57	50.9
Slow wave stages (%)	6.8	8.9	11.2
REM%	4	5.3	12.9
Awakenings (*n*)	53	19	17
REM (*n*)	2	3	4

STN-DBS parameters = 2.8 V, 185 Hz, 90 μs, bipolar STIM (contact 2+, 1−); PPN-DBS parameters: 2 V, 25 Hz, 60 μs, bipolar STIM (contact 1+, 0−)
[a]Data represent mean of two nights (the second one of each PSG protocol)

5.3 ± 2, PPN-OFF 5.5 ± 3; PDSS item 5: PPN-cyclic 9 ± 1, PPN-ON 5 ± 3, PPN-OFF 6.9 ± 4), psychosis (PDSS items 6: PPN-cyclic 8.3 ± 3, PPN-ON 7.3 ± 1.8, PPN-OFF 7.7 ± 1.8; PDSS item 7: PPN-cyclic 10, PPN-ON 8 ± 1.4, PPN-OFF 8.4 ± 1.7) and daytime sleepiness (PDSS item 15: PPN-cyclic 8 ± 3, PPN-ON 5.9 ± 1, PPN-OFF 6.3 ± 2.2) (data not shown). ESS scores equally indicated reduced daytime sleepiness in PPN-cyclic (PPN-cyclic 4 ± 2.8, PPN-ON 5.5 ± 2.9, PPN-OFF 8.9 ± 2.8).

In another recently implanted PD patient (UM, 49 years), we had the opportunity to monitor PSG alterations before and after surgery and determined whether the effects of low-frequency PPN-DBS differed from those detected under 185-Hz STN-DBS (Table 2). PSG parameters, in the pre-surgery phase, revealed a clear sleep disruption (number awakenings, increased stage 1 sleep, loss of REM sleep, Table 2). After surgery, PPN-ON (25 Hz) and STN-DBS, to a lesser extent, promoted a better stability and continuity of nocturnal sleep, expressed by the increase from 74 to 90% of sleep efficiency, a mild reduction of stage 1, a 30% increase of stage 2 and a 70% decrease of awakenings (Romigi et al. 2008). Appealingly, under PPN-DBS these changes were associated with a relevant increase in REM sleep (up to 13%) (Table 2). By contrast, STN-DBS (185 Hz), albeit providing a better control of nocturnal akinesia, failed to affect the occurrence and duration of REM epochs.

3.4 *Effects of PPN-DBS on STN Neurons*

On the basis of the firing pattern analysis, STN single units (*n* = 48) were divided into the following subgroups: (1) Burst-like firing activity (45%, characterized by the appearance of burst-like clustering of action potentials separated by periods of absence of discharge or by low-frequency activity, Fig. 2), (2) irregular firing activity (30%, characterized by a flat, i.e., random, distribution of the ISIs, which sometimes showed mild positive skewing and an auto-correlogram with less than two peaks,

Fig. 2) and (3) regular/tonic firing activity (25%, displaying a symmetric peak in the ISIH and an auto-correlogram with at least three identifiable peaks, Fig. 2).

Out of the neurons displaying a clustered, train-like discharge activity, 20 cells (90%) showed a statistically significant decrease in firing rate during PPN-DBS (Fig. 2). This response had a mean latency of 35.73 ± 15.02 s (under steady state 2 V STIM) and a mean magnitude of 62% ± 17. No variation in waveform or firing pattern was observed during PPN-DBS (not shown). The firing rate partially returned to pre-stimulus level after 3-min OFF-STIM. Post-stimulus histogram (PSTH) and raster plots did not reveal any short-term inhibition around each PPN stimulus (data not shown).

In contrast, the activity of all but one (13/14) of the irregularly discharging units exhibited a statistical significant increase in spike frequency from 21% (±8.2) to 34% (±14) with a latency of 10 ± 7.26 s (Fig. 2). No changes in morphology of waveforms, ISIHs and auto-correlograms were evident during PPN-DBS. The firing rate did not recover fully to pre-stimulus levels after the end of PPN-DBS. PSTH and raster plots did not show any increase of firing probability within 40 ms before or after each PPN stimulus (data not shown).

Similarly, 25-Hz PPN-DBS caused a statistically significant increase in the mean firing rate of most tonic regular units (10/12, 83.33%). This response had a mean latency of 25.3 ± 7.8 s and a mean magnitude of 20% (±14%). The waveform, ISIHs and auto-correlograms did not change during PPN-DBS. However, no long-term slow oscillations were observed during PPN-DBS (Fig. 2).

In each unit 'subtype' we also tested ten PPN-STIM ($n = 2$ for any given patterning) and never detected significant changes of firing rate and PSTH (data not shown).

4 Discussion

The data so far collected indicate that PPN-DBS has a profound impact on different motor and non-motor functions. On the one hand, it was shown that PPN-ON may ameliorate hypokinetic signs (about 30%), but the results were modest when compared with STN-mediated benefits (Stefani et al. 2007). In addition, the improvement in gait control was fluctuating and not fully convincing. Actually, four out of the six original patients took advantage of a cyclic PPN activation (overnight) together with a continuous STN stimulation. On the other hand, we found significant changes in sleep macrostructure as well as an interesting beneficial impact on some, mainly executive, cognitive functions.

4.1 Motor Effects

As far as the slightly disappointing clinical motor effects are concerned, the peri-operative recordings may offer some clues. First, the ability of PPN 25-Hz stimulation to affect the on-going firing rate in almost all STN neurons is an unequivocal proof of the strong interplay between the two stations. Whether PPN-STN connections are direct or indirect through multi-synaptic mechanisms (i.e., through GPi) cannot be fully resolved by electrophysiology performed on humans in the surgery room. However, the evidence of a PPN-mediated change of firing frequency in 90% of recorded STN neurons without altering PSTH or shifting the firing pattern would seem to indicate a PPN-driven *direct* mechanism that simultaneously affects different STN neurons. Alternatively, the described changes of STN firing discharge may require complex alterations, involving other basal ganglia stations such as GPi or 'extra-basal ganglia' areas such as the intralaminar thalamus.

The second relevant finding is the lack of PSTH changes in STN cells during 25 and 10-Hz PPN stimulation; thus, it is plausible that firing rate changes are not related to changes in fast excitatory neurotransmission (at least in our experimental conditions of stimulus intensity) but, rather, to neuromodulatory transmitter(s) not strictly responsible for fast transmission. The cholinergic nature of the PPN-STN pathway has been supported by morphological studies (Lavoie and Parent 1994), also suggesting the possibility of co-expression of ACh and glutamate. Indeed, a direct axo-dendritic synapse, as occurs between PPN terminals and dendrites of STN neurons, might modulate the on-going excitability by releasing ACh without any short latency effect at this stimulation intensity. The direct excitation described on this pathway could be also ascribed to the release of glutamate but the latter usually requires different stimulus intensities.

Third, the delivery of PPN-ON at 25 Hz excited irregular and tonic cells and inhibited neurons dominated by train-like pattern. An increased *burstiness* is acknowledged as a disease hallmark correlated with PD pathophysiology (Bergman et al. 1994); the PPN-mediated reduction of firing discharge in bursty cells is in line with this theory. Hence, the dramatic inhibition of STN bursts may underlie, at least in part, the efficacy of PPN-DBS. On the other hand, the simultaneous presence of a strong PPN-mediated excitation on about 50% STN neurons may counterbalance the beneficial decline in *burstiness*.

4.2 Non-motor Effects

Regardless of the questionable impact on the motor parkinsonian signs, PPN-DBS demonstrated a striking influence on nocturnal sleep structure and cognitive performance.

In several previous manuscripts, no beneficial increase in REM phase was detected under STN-DBS, even in the presence of longer periods of uninterrupted sleep and reduction of sleep fragmentation. STN-mediated benefits were attributed to a better nocturnal mobility. By contrast, our data revealed a selective recovery of REM sleep induced by low-frequency PPN-DBS.

The most parsimonious explanation of PPN-mediated effects on sleep structure is that low-frequency PPN-DBS modifies the functional activity of its subpopulation of cholinergic neurons, activating muscarinic receptors and rebalancing REM-sleep physiology through the projection to the thalamus (Rye 1997). However, the electrode catheter size and the spread of the electrical field, compared with the limited extension of brainstem structures, do not exclude the possible involvement of the locus coeruleus (whilst the lack of PPN-DBS-mediated effects on the slow wave phase seems to point out a critical role of serotonin). In any case, our result is the first in vivo non-lesional confirmation of the ancient theory by Moruzzi's group on 'midpoint' brain stem structure involvement in sleep (Batini et al. 1958).

In addition, the improvement of executive functions, just following PPN-ON, together with subjective 'good feelings' highlights interesting other possibilities. First, PPN sends an abundant contingent of fibres towards the intra-laminar thalamic nuclei, whose involvement in limbic/motor interfaced circuitries is well documented. Second, the reported beneficial effect we observed by means of neuropsychological evaluations could be referred to the spread of the electrical effect to nearby nuclei and in particular to the locus coeruleus, known to be involved in arousal and attention due to its diffuse noradrenergic projections directed to the cortex. Third, as emphasized by the group of Winn (Anderson et al. 2006), this nucleus could modulate projections to the prefrontal cortex, ventral striatum and accumbens through its ascending output directed toward dopaminergic neurons of the ventral tegmental area and SNc, to the intralaminar nuclei of thalamus and to STN and GP. This interpretation takes for granted several assumptions, such as the presumption that we are modulating *only or mainly* the PPN itself. This, of course, is not the case, because the active part of the catheter (>7 mm) overcomes the rostrocaudal extension of PPN (about 4 mm in humans). A well-designed long-lasting study on effects influenced by each electrical contact (in monopolar modality) could provide more specific results or interpretations. That said, further understanding of the PPN fine functional anatomy could be achieved with single unit or intracellular recordings from PPN sub-portions in mammalian disease models (Florio et al. 2007).

5 Conclusions

5.1 Is PPN-DBS (at 10–25 Hz in Our Protocol) Re-activating Impaired Pathways?

The evidence by which PPN-DBS helps to restore a more physiological sleep structure and to improve attentive and executive functions suggests (albeit heuristically) that the stimulation, instead of simply abolishing a gross over-excitability, imposes new activity patterns in otherwise impaired/silent fibre pathways. Our findings highlight once more that DBS should not be considered as a simple 'resetting' or jamming device (Stefani et al. 2005). Moreover, the present findings suggest that PPN should not be viewed as a mere station for processing descending output to the spinal cord

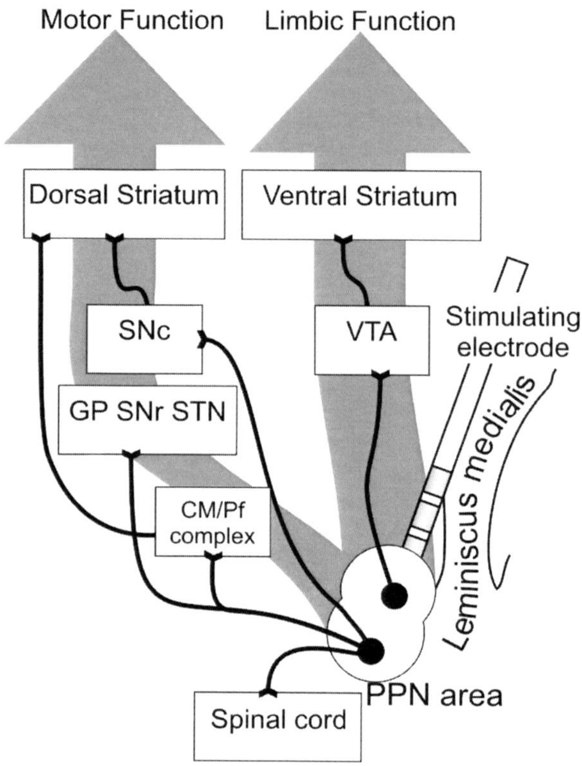

Fig. 3 Simplified scheme of PPN-fugal projections as supported by clinical findings

(Pierantozzi et al. 2008). Instead, it appears to exert also a strong influence onto ascending pathways, as hypothesized by Mena-Segovia et al. (2004), who assume the PPN and the BG structures to be a functional integrated system. Furthermore, the ongoing imaging studies (which seem to correlate cognitive benefits with increased metabolic activity in prefrontal areas as well as ventral striatum) strongly support Winn's views, indicating PPN as an active player in 'limbic-reward' circuitries (Winn 2005) (Fig. 3).

5.2 *Is PPN-DBS Mostly Affecting Non-motor and Not Strictly Dopamine-Centred Functions?*

Strong evidence supports the critical involvement of non-motor non-dopaminergic brain-stem areas in the natural history of extra-pyramidal disorders (Braak et al. 2003). Our studies demonstrate that PPN-DBS promotes a general improvement of sleep efficiency and cognitive performance, hard to attribute to a mere change of endogenous dopamine release and availability. For example, REM sleep alterations are probably related to degenerative processes of non-dopaminergic circuitries and indeed largely unaffected by dopamine-centred therapy or standard stereotactic neurosurgery. Hopefully, new research lines in rodents (i.e., unequivocal identification of PPN neuronal subtypes and their complex relationship with surrounding structures) will elucidate these aspects.

References

Anderson HL, Latimer MP and Winn P (2006) Intravenous self-administration of nicotine is altered by lesions of the posterior, but not anterior, pedunculopontine tegmental nucleus. Eur J Neurosci 23: 2169–2175.

Batini C, Moruzzi G, Palestini M, Rossi GF and Zancheti A (1958) Presistent patterns of wakefulness in the pretrigeminal midpontine preparation. Science 128: 30–32.

Bergman H, Wichmann T, Karmon B and DeLong MR (1994) The primate subthalamic nucleus. II. Neuronal activity in the MPTP model of parkinsonism. J Neurophysiol 72: 507–520.

Bevan MD and Bolam JP (1995) Cholinergic, GABAergic, and glutamate-enriched inputs from the mesopontine tegmentum to the subthalamic nucleus in the rat. J Neurosci 15: 7105–7120.

Braak H, Del Tredici K, Rüb U, de Vos RA, Jansen Steur EN and Braak E (2003) Staging of brain pathology related to sporadic Parkinson's disease. Neurobiol Aging 24: 197–211.

Breit S, Bouali-Benazzouz R, Benabid AL and Benazzouz A (2001) Unilateral lesion of the nigrostriatal pathway induces an increase of neuronal activity of the pedunculopontine nucleus, which is reversed by the lesion of the subthalamic nucleus in the rat. Eur J Neurosci 14: 1833–1842.

Brusa L, Morchella V, Stanzione P, Ceravolo R, Pierantozzi M, Galati S, Stefani A (2008) PPN-DBS improves cognitive performance in Parkinsonian patients. Eur J Neurol (submitted for publication).

Calzavara R, Mailly P and Haber SN (2007) Relationship between the corticostriatal terminals from areas 9 and 46, and those from area 8A, dorsal and rostral premotor cortex and area 24c: an anatomical substrate for cognition to action. Eur J Neurosci 26: 2005–2024.

Florio T, Capozzo A, Puglielli E, Pupillo R, Pizzuti G and Scarnati E (1999) The function of the pedunculopontine nucleus in the preparation and execution of an externally-cued bar pressing task in the rat. Behav Brain Res 104: 95–104.

Florio T, Scarnati E, Confalone G, Minchella D, Galati S, Stanzione P, Stefani A and Mazzone P (2007) High-frequency stimulation of the subthalamic nucleus modulates the activity of pedunculopontine neurons through direct activation of excitatory fibres as well as through indirect activation of inhibitory pallidal fibres in the rat. Eur J Neurosci 25: 1174–1186.

Galati S, Mazzone P, Fedele E, Pisani A, Peppe A, Pierantozzi M, Brusa L, Tropepi D, Moschella V, Raiteri M, Stanzione P, Bernardi G and Stefani A (2006) Biochemical and electrophysiological changes of substantia nigra pars reticulata driven by subthalamic stimulation in patients with Parkinson's disease. Eur J Neurosci 23: 2923–2928.

Galati S, Scarnati E, Mazzone P, Stanzione P and Stefani A (2008) PPN-DBS promotes excitation and inhibition in human PD STN. Neuroreport 19: 661–666.

Garcia-Rill E (1991) The pedunculopontine nucleus. Prog Neurobiol 36: 363–389.

Garcia-Rill E, Houser CR, Skinner RD, Smith W and Woodward DJ (1987) Locomotion-inducing sites in the vicinity of the pedunculopontine nucleus. Brain Res Bull 18: 731–738.

Iranzo A, Vallderiola F, Santamaria J, Tolosa E and Rumià J (2002) Sleep symptoms and polysomnographic architecture in advanced Parkinson's disease after chronic bilateral subthalamic stimulation. J Neurol Neurosurg Psychiatr 72: 661–664.

Jackson A and Crossman AR (1983) Nucleus tegmenti pedunculopontinus: efferent connections with special reference to the basal ganglia, studied in the rat by anterograde and retrograde transport of horseradish peroxidase. Neuroscience 10: 725–765.
Kelland MD and Asdourian D (1989) Pedunculopontine tegmental nucleus-induced inhibition of muscle activity in the rat. Behav Brain Res 34: 213–234.
Langston JW, Widner H, Goetz CG, Brooks D, Fahn S, Freeman T and Watts R (1992) Mov Disord 7: 2–13.
Lavoie B and Parent A (1994) Pedunculopontine nucleus in the squirrel monkey: distribution of cholinergic and monoaminergic neurons in the mesopontine tegmentum with evidence for the presence of glutamate in cholinergic neurons. J Comp Neurol 344: 190–209.
Mazzone P, Lozano A, Stanzione P, Galati S, Peppe A, Scarnati E and Stefani A (2005) Implantation of human pedunculopontine nucleus: a safe and clinically relevant target in Parkinson's disease. Neuroreport 16: 1879–1883.
Mazzone P, Sposato S, Insola A, Di Lazzaro V and Scarnati E (2008). Stereotactic surgery of nucleus tegmenti pedunculopontini. Br J Neurosurg 22: S33–S40.
Mena-Segovia J, Bolam JP and Magill PJ (2004) Pedunculopontine nucleus and basal ganglia: distant relatives or part of the same family? Trends Neurosci 27: 585–588.
Mesulam MM, Geula C, Bothwell MA and Hersh LB (1989) Human reticular formation: cholinergic neurons of the pedunculopontine and laterodorsal tegmental nuclei and some cytochemical comparisons to forebrain cholinergic neurons. J Comp Neurol 283: 611–633.
Muthusamy KA, Aravamuthan BR, Kringelbach ML, Jenkinson N, Voets NL, Johansen-Berg H, Stein JF and Aziz TZ (2007) Connectivity of the human pedunculopontine nucleus region and diffusion tensor imaging in surgical targeting. J Neurosurg 107: 814–820.
Nandi N, Aziz TZ, Giladi N, Winter J and Stein JF (2002) Reversal of akinesia in experimental parkinsonism by GABA antagonist microinjections in the pedunculopontine nucleus. Brain 125: 2418–2430.
Olszewski JD and Baxter DW (1982) Cytoarchitecture of the Human Brain Stem. Karger, Basel.
Pahapill PA and Lozano AM (2000) The pedunculopontine nucleus and Parkinson's disease. Brain 123: 1767–1783.
Peppe A, Pierantozzi M, Bassi A, Altibrandi MG, Brusa L, Stefani A, Stanzione P and Mazzone P (2004) Stimulation of the subthalamic nucleus compared with the globus pallidus internus in patients with Parkinson disease. J Neurosurg 101: 195–200.
Pierantozzi M, Calmieri MG, Galati S, Stanzione P, Peppe A, Troppi D, Brusa L, Pisani A, Moschella V, Marciani MG, Mazzone P and Stefani A (2008) Pedunculopontine nucleus deep brain stimulation changes spinal cord excitability in Parkinson's disease patients. J Neural Transm 115: 731–735.
Reese NB, Garcia-Rill E and Skinner RD (1995) The pedunculopontine nucleus – auditory input, arousal and pathophysiology. Prog Neurobiol 47: 105–133.
Romigi A, Stanzione P, Marciani MG, Izzi F, Placidi F, Cervellino A, Giacomini P, Brusa L, Grossi K and Pierantozzi M (2006) Effect of cabergoline added to levodopa treatment on sleep–wake cycle in idiopathic Parkinson's disease: an open label 24-hour polysomnographic study. J Neural Transm 113: 1909–1913.
Romigi A, Placidi F, Peppe A, Pierantozzi M, Izzi F, Brusa L, Galati S, Moschella V, Marciani MG, Mazzone P, Stanzione P and Stefani A (2008) Pedunculopontine nucleus stimulation influences REM sleep in Parkinson's disease. Eur J Neurol 15: e64–e65.
Rye DB (1997) Contributions of the pedunculopontine region to normal and altered REM sleep. Sleep 20: 757–788.
Stefani A, Stanzione P, Bassi A, Mazzone P, Vangelista T and Bernardi G (1997) Effects of increasing doses of apomorphine during stereotaxic neurosurgery in Parkinson's disease: clinical score and internal globus pallidus activity. J Neural Transm 104: 895–904.
Stefani A, Fedele E, Galati S, Pepicelli O, Frasca S, Pierantozzi M, Peppe A, Brusa L, Orlacchio A, Hainsworth AH, Gattoni G, Stanzione P, Bernardi G, Raiteri M and Mazzone P (2005) Subthalamic stimulation activates internal pallidus: evidence from cGMP microdialysis in PD patients. Ann Neurol 57: 448–452.

Stefani A, Lozano AM, Peppe A, Stanzione P, Galati S, Tropepi D, Pierantozzi M, Brusa L, Scarnati E and Mazzone P (2007) Bilateral deep brain stimulation of the pedunculopontine and subthalamic nuclei in severe Parkinson's disease. Brain 130: 1596–1607.

Takakusaki T, Shiroyama T, Yamamoto T and Kitai ST (1996) Cholinergic and noncholinergic tegmental pedunculopontine projection neurons in rats revealed by intracellular labelling. J Comp Neurol 371: 345–361.

Takakusaki K, Habaguchi T, Ohtinata-Sugimoto J, Saitoh K and Sakamoto T (2003) Basal ganglia efferents to the brainstem centers controlling postural muscle tone and locomotion: a new concept for understanding motor disorders in basal ganglia dysfunction. Neuroscience 119: 293–308.

Winn P (2005) How best to consider the structure and function of the pedunculopontine tegmental nucleus: evidence from animal studies. J Neurol Sci 248: 234–250.

Zweig RM, Jankel WR, Hedreen JC, Mayeux R and Price DL (1989) The pedunculopontine nucleus in Parkinson's disease. Ann Neurol 26: 41–46.

Observation of Involuntary Movements Through Clinical Effects of Surgical Treatments

Fusako Yokochi, Makoto Taniguchi, Toru Terao, Ryoichi Okiyama, and Hiroshi Takahashi

Abstract Involuntary movements are of numerous types. It is difficult to treat involuntary movements by medication, and the mechanisms underlying involuntary movements are unclear. Stereotactic surgery has been effective for the treatment of involuntary movements. Outcomes of surgical treatments are excellent indicators for understanding the clinical differentiation or pathophysiological mechanism of involuntary movements. The clinical observations of involuntary movements following stereotactic surgery are described in this chapter.

1 Introduction

Involuntary movements are of numerous types and it is often difficult to distinguish between the different types of involuntary movements by clinical examination. Furthermore, it can be difficult to treat them by medication. Consequently, involuntary movements have also been treated by stereotactic surgery similarly to parkinsonian tremor. Deep brain stimulation (DBS) has been introduced in stereotactic surgery, and the range of indications in treating involuntary movements has become wider. However, the brain mechanisms underlying involuntary movements are still unclear. The target of stereotactic surgery is in some cases determined from the results of previous experiments and experience. Here, we consider the factors involved in treating involuntary movements by stereotactic surgery based upon our previous experience of its clinical effects.

F. Yokochi (✉), M. Taniguchi, T. Terao, R. Okiyama, and H. Takahashi
Department of Neurology, Tokyo Metropolitan Neurological Hospital,
2-1-16 Musashidai, Fuchu, Tokyo 183-0042, Japan
e-mail: fyokochi-tmnh@umin.ac.jp

H.J. Groenewegen et al. (eds.), *The Basal Ganglia IX*, Advances in Behavioral Biology 58,
DOI 10.1007/978-1-4419-0340-2_45,

2 Methods

2.1 Subjects

Twenty-seven patients presenting with involuntary movements, excluding those with Parkinson's disease, were treated by stereotactic surgery. The patients with involuntary movements included 15 with essential tremor, 1 with multiple sclerosis, 6 with hereditary generalized dystonia, 2 with focal dystonia, and 3 suffering from neuroacanthocytosis.

2.2 Surgery

Surgery was performed using a Leksell stereotactic frame; tentative targets were determined by MRI, and neural activities were recorded by microrecording during the operation. The operation was performed under general anesthesia for patients with generalized dystonia and neuroacanthocytosis and under local anesthesia for the other patients.

2.3 Stereotactic Targets

The targets of stereotactic surgery were determined on the basis of the type of involuntary movement and the body part affected by the involuntary movements. The target for relief of tremor was the thalamus; the target in cases of dyskinesia and choreoballism was the internal pallidum. In patients with dystonia, the body part affected by dystonia should be considered to make an appropriate selection of the target for surgical treatment. The surgical target in the case of generalized dystonia is the internal pallidum, whereas it is the thalamus in the case of focal dystonia of the extremities.

2.4 EMG Analysis

Involuntary movements were analyzed using surface EMG carried out before and after surgery. Before the operation surface EMG showed the type and distribution of involuntary movements, whereas after surgery the clinical outcome can be confirmed by the surface EMG recordings.

3 Results

3.1 *Tremor*

3.1.1 Essential Tremor

Fifteen patients with essential tremor were subjected to stereotactic surgery. In 12 patients with an isolated limb tremor, a unilateral operation of the most affected side, usually the right hand, was performed. The type of surgical procedure in these cases was a thalamotomy in six patients, a pallidotomy in one patient, and nucleus ventralis internedius (Vim) DBS in five patients. In three patients with a combination of limb tremor and cervical tremor, bilateral operations were performed. In two of these patients, bilateral Vim DBS was performed. In the third patient (33 years) with both a limb tremor and cervical tremor, thalamotomy was performed on one side and pallidotomy on the other side, because bilateral thalamotomy can have adverse effects such as dysphonia. Left Vim thalamotomy was performed first, followed by a right pallidotomy. The left thalamotomy diminished the right cervical tremor. Following the right-sided pallidotomy, the left cervical tremor diminished when the patient was in a relaxed position such as when in a supine or relaxed sitting position, but reappeared when the patient straightened his back.

3.1.2 Multiple Sclerosis

Severe postural, action, and intentional tremors were observed in a 34-year-old male patient with multiple sclerosis. Tremor was more prominent in the right limbs than in the left limbs. MRI scanning revealed multiple sclerosis lesions near the right red nucleus. Tremor was observed to be more severe during intended limb action such as holding a glass of water than when only maintaining the posture such as holding the arm, and his condition was diagnosed as Holmes tremor. During surgery, tremor-related neural activities were recorded in the left Vim nucleus. Vim DBS was effective in diminishing the tremor. The alleviation of postural tremor and intentional tremor depends on stimulation intensity. Figure 1 shows the light tracks of his right shoulder, elbow, and wrist during movement in the finger-to-nose test, improving upon DBS.

3.2 *Dystonia*

3.2.1 Generalized Dystonia

The six patients with primary dystonia include five patients with DYT1 and one with hereditary non-DYT1 dystonia. Abnormal movements associated with generalized dystonia can include other involuntary movements. By definition, postural dystonia

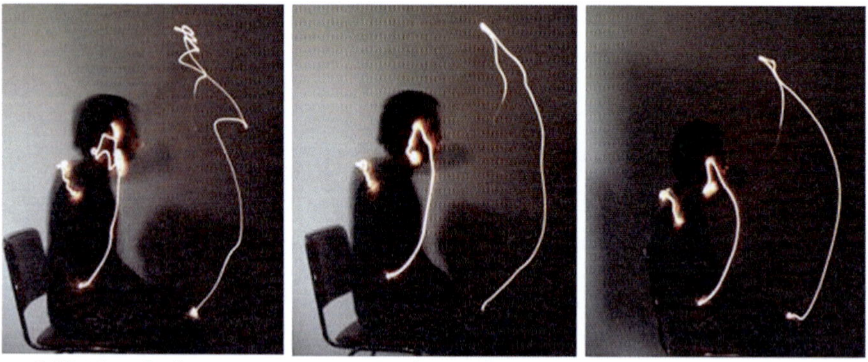

Fig. 1 The images illustrate the light tracks of the right shoulder, elbow, and wrist during movement in the finger-to-nose test in a patient with multiple sclerosis in three conditions. The OFF condition denotes no stimulation. The other two conditions involve stimulation with either 3.5 or 4.6 V with a pulse width of 90 μs and frequency of 160 Hz. Alleviation of postural and intentional tremor depends on the intensity of the stimulation

is observed during posture, and action dystonia is observed during action. The dystonic movement is phasic and observed throughout the body; it is similar to dyskinesia. Other involuntary movements associated with generalized dystonia are tremor or myoclonus. All six patients with generalized dystonia were treated with bilateral pallidal DBS. Spontaneous dystonic movements diminished first, followed later by a reduction of postural dystonia. Finally, action dystonia decreased, although action dystonia of the hand during writing was usually sustained to some extent.

3.2.2 Focal Dystonia

Two patients with focal limb dystonia were treated by unilateral thalamic DBS. One patient with Machado–Joseph disease had painful dystonia of the left upper limb and another patient showed action dystonia of the right upper limb during repetitive movements (task-specific dystonia). In both patients, DBS electrodes were positioned in the nucleus ventralis oralis posterior (Vop) and Vim areas of the thalamus. Bipolar stimulation of the Vop and Vim areas improved focal dystonia in both patients.

3.3 Choreoballistic Involuntary Movements

Three patients with neuroacanthocytosis were treated with bilateral pallidotomy or bilateral pallidal DBS. Before surgery, all patients showed severe involuntary movements including flexion extension of the trunk, ballistic involuntary movements of the extremities, dyskinesias of the tongue, and autophagia. The patients were unable to sit, stand, or eat independently. In two of the patients, bilateral pallidotomies were performed in two separate surgical sessions. In one patient, bilateral pallidal DBS was performed in a single session under general anesthesia. All types of involuntary movement observed in the three patients with neuroacanthocytosis were abolished after the operation (Yokochi and Burbaud 2008).

4 Discussion

Tremors, whether resting, postural, action, or intentional tremors, are known to improve with thalamotomy or Vim DBS (Kumar et al. 2003). However, postural tremor does not improve with pallidotomy. It is recognized in many published studies and from our clinical experience that subthalamic nucleus (STN) DBS improves parkinsonian tremor. There are also some reports showing that STN DBS can lead to an improvement of essential tremor (Plaha et al. 2004). Yokochi et al. (2005) studied the effects of STN DBS on tremor and speculated that it does not directly improve parkinsonian tremor. Parkinsonian tremor does not immediately stop after starting STN DBS, but first shows oscillation of tremor amplitude and then gradually diminishes, sometimes over a few months. Apparently, the effects of Vim DBS and STN DBS on tremor are different, and it is postulated that Vim DBS inhibits the current rising tremor, whereas STN DBS modulates the tremor circuit. Pallidotomy or pallidal DBS can improve parkinsonian tremor under levodopa medication (Yokochi, unpublished observations). In conclusion, the Vim nucleus is a key nucleus in the tremor circuit for connecting the output of the cerebellum to the cortex, and the thalamo-cortical pathway is involved in all modalities of tremor.

In the present study, generalized dystonia showed improvement with pallidal DBS. Each of the various types of involuntary movements associated with generalized dystonia, including spontaneous dystonic movements, postural dystonia, and action dystonia, had a different time course of improvement after starting pallidal DBS. Dystonic movement diminished first, followed by postural dystonia. Finally, action dystonia decreased, although action dystonia of the hand during writing was to some extent sustained. In other words, action dystonia during hand movements is not fully suppressed by pallidal DBS. Figure 2 shows a hypothetical diagram of the connections in the brain related to the three different types of dystonia (from Segawa et al. 2002). The output of the internal pallidum includes ascending and descending pathways. Postural dystonia may be related to the descending pathways from the internal pallidum. Dystonic movements may be related to ascending pathways, such as the pallido-thalamo-cortical pathways, and action dystonia could appear through the thalamo-cortico-spinal pathways. Focal limb dystonia associated with Machado–Joseph disease and task-specific dystonia of the upper limb showed improvement with Vop-Vim DBS. Focal dystonia such as writer's cramp can be improved by the surgical treatment of the Vop-Vim nucleus (Fukaya et al. 2007). The output from the internal pallidum converges in the nucleus ventralis

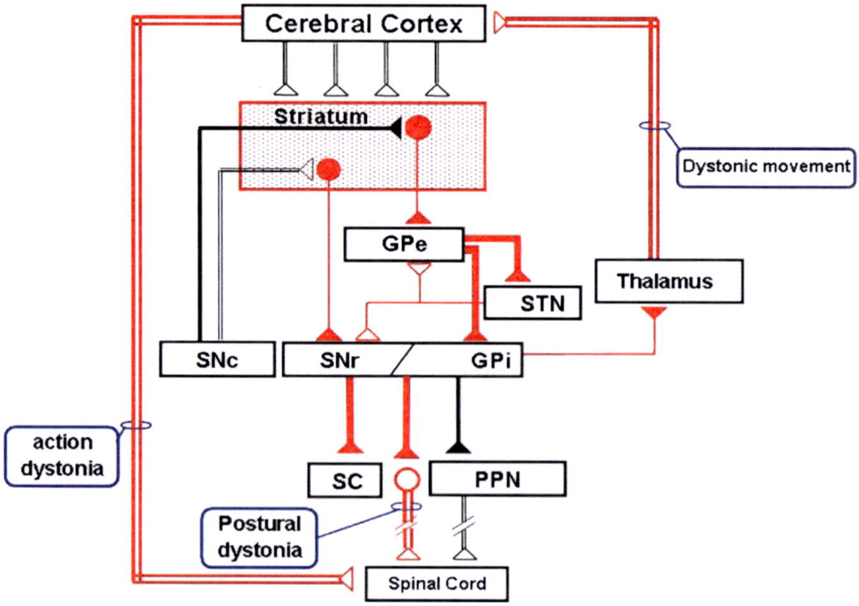

Fig. 2 The schematic was proposed by Segawa and Nomura to explain the mechanism of generalized dystonia observed in DYT 1. Personal communication (2005); modified from Segawa et al. 2002

lateralis (VL), and focal dystonia is assumed to be related to the pallido-thalamo-cortical pathways.

Choreoballistic movements of the face, tongue, trunk, and limbs showed improvement with pallidotomy or pallidal DBS. It has been reported that ballism can be improved by the surgical treatment of either the internal pallidum or the thalamic VL nucleus (Krauss and Mundinger 1996). Choreoballistic involuntary movements are assumed to be related to the pallido-thalamo-cortical pathways.

The type of involuntary movement and the distribution of involuntary movements over the body should be taken into account when selecting a target for stereotactic neurosurgical treatment. The Vim nucleus is a target in the case of tremor of all modalities, including parkinsonian tremor, essential tremor, or cerebellar tremor. The internal pallidum is the target of choice in the surgical treatment of dystonia and choreoballistic movements. However, in the case of dystonia the distribution of involuntary movement over the body is important as well. The two motor systems, namely the lateral motor system and the medial motor system, could be differentially involved in the pathophysiology of dystonia (Kuypers 1982). Dystonia of the extremities may be regulated by the lateral motor system, and the target of choice is the thalamus (VL-Vim nucleus). Dystonia of the face and/or trunk may involve the medial motor system, and the optimal target is the internal pallidum.

5 Conclusions

The factors involved in the choice of the target of stereotactic neurosurgery for involuntary movements are the type of involuntary movement and the body part(s) affected by the involuntary movements. In the case of tremor or choreoballistic movements, the most important factor is the type of involuntary movement, whereas in the case of dystonia, it is a combination of the body part(s) affected by the dystonia and the type of dystonia.

References

Fukaya C, Katayama Y, Kano T, Nagaoka T, Kobayashi K, Oshima H and Yamamoto T (2007) Thalamic deep brain stimulation for writer's cramp. J Neurosurg 107: 977–982.

Krauss JK and Mundinger F (1996) Functional stereotactic surgery for hemiballism. J Neurosurg 85: 278–286.

Kumar R, Lozano AM, Sime E and Lang AE (2003) Long-term follow-up of thalamic deep brain stimulation for essential and parkinsonian tremor. Neurology 61: 1601–1604.

Kuypers HG (1982) A new look at the organization of the motor system. Prog Brain Res 57: 381–403.

Plaha P, Patel NK and Gill SS (2004) Stimulation of the subthalamic region for essential tremor. J Neurosurg 101: 48–54.

Segawa M, Hoshino K, Hachimori K, Nishiyama N and Nomura Y (2002) A single gene for dystonia involves both or either of the two striatal pathways. In: Nicholson LFB and Faull RLM (eds) The Basal Ganglia VII, Kluwer/Plenum, New York. pp 155–163.

Yokochi F and Burbaud P (2008) Neurosurgery for neuroacanthocytosis. In: Walker RH (ed) Neuroacanthocytosis syndromes II. Springer, Berlin.

Yokochi F, Okiyama R, Taniguchi M, Takahashi H and Hamada I (2005) Effect of deep brain stimulation on tremor. In: Bolam JP, Ingham CA, Magill PJ (eds) The Basal Ganglia VIII. Springer, Berlin. pp 407–413.

Index

D

E

F

G

R